Proceedings in Technology Transfer

The book series "Proceedings in Technology Transfer" provides a valuable platform for researchers, practitioners, and policymakers to share their experiences and insights related to technology transfer. Our aim is to facilitate knowledge exchange between academic researchers and industry professionals, accelerating the introduction of new technologies and driving economic growth.

Our series publishes books and conference proceedings that focus on original research findings, best practice recommendations, emerging trends, and technology evaluations. The series covers all topics which explore the transfer of technology in various industry sectors that require innovations and economic growth.

Join us in advancing the technology transfer and its impact on innovation. Publish your proceedings with us and offer insights and expertise to professionals in your field.

Dzung Hoang Tien · Vijender Kumar Solanki ·
Jamaluddin Mahmud · Thi Dieu Linh Nguyen
Editors

Proceedings of the 1st International Conference on Sustainability and Emerging Technologies for Smart Manufacturing

SETSM 2024, 27–28 April, Hanoi, Vietnam

Springer

Editors
Dzung Hoang Tien
School of Mechanical and Automotive Engineering
Hanoi University of Industry
Bac Tu Liem, Hanoi, Vietnam

Jamaluddin Mahmud
School of Mechanical Engineering
Universiti Teknologi MARA
Shah Alam, Malaysia

Vijender Kumar Solanki
Department of Computer Science and Engineering
Stanley College of Engineering and Technology For Women
Hyderabad, TG, India

Thi Dieu Linh Nguyen
Department of Science and Technology
Hanoi University of Industry
Bac Tu Liem, Hanoi, Vietnam

ISSN 2948-2321 ISSN 2948-233X (electronic)
Proceedings in Technology Transfer
ISBN 978-981-97-7082-3 ISBN 978-981-97-7083-0 (eBook)
https://doi.org/10.1007/978-981-97-7083-0

This Springer imprint is published by the registered company Springer Nature Singapore Pte Ltd.
The registered company address is: 152 Beach Road, #21-01/04 Gateway East, Singapore 189721, Singapore

Contents

About the Editor

Vijender Kumar Solanki, PhD, is an Associate Professor in the Computer Science and Engineering Department at Stanley College of Engineering and Technology for Women (Autonomous), Hyderabad, TG, India. He has over 17 years of academic experience in network security, IoT, Big Data, Smart City, and IT. Before his current role, he was associated with Apeejay Institute of Technology, Greater Noida, UP; KSR College of Engineering, Tamil Nadu, India; Institute of Technology & Science, Ghaziabad, UP, India; and CMR Institute of Technology, Hyderabad, India. He is a Senior Member of IEEE.

He has attended an orientation program at UGC-Academic Staff College, University of Kerala, Thiruvananthapuram, Kerala, and a Refresher course at the Indian Institute of Information Technology, Allahabad, UP, India.

He has authored or co-authored more than 120 research articles published in various journals, books, and conference proceedings in soft computing. He has also edited or co-edited 35 books and conference proceedings.

He received a PhD in Computer Science and Engineering from Anna University, Chennai, India, in 2017, an ME, MCA from Maharshi Dayanand University, Rohtak, Haryana, India, in 2007 and 2004, and a bachelor's degree in science from JLN Government College, Faridabad Haryana, India, in 2001.

He is the Book Series Editor of Internet of Everything (IoE): Security and Privacy Paradigm, CRC Press, Taylor & Francis Group, USA; Artificial Intelligence (AI): Elementary to Advanced Practices Series, CRC Press, Taylor & Francis Group, USA; IT, Management & Operations Research Practices, CRC Press, Taylor & Francis Group, USA; Bio-Medical Engineering: Techniques and Applications with Apple Academic Press, USA; and Computational Intelligence and Management Science Paradigm, (Focus Series) CRC Press, Taylor & Francis Group, USA.

He is Editor-in-Chief of International Journal of Machine Learning and Networked Collaborative Engineering (IJMLNCE) ISSN 2581-3242; International Journal of Hyperconnectivity and the Internet of Things (IJHIoT), ISSN 2473-4365, IGI-Global, USA, Co-Editor Ingenieria Solidaria Journal ISSN (2357-6014). He can be contacted at spesinfo@yahoo.com or vijendersolanki@ieee.org.

Design of Adaptive Fuzzy-PID for Adaptive Cruise Control of Electric Vehicle Using DC Motor: Theory and Experiment

Van Du Phan, Van Quyet Phan, Thai Son Dang, Ngoc Hoang Trinh, Phuc Ngoc Nguyen, Ngoc Minh Luong, Ba Uy Nguyen, Quoc Cuong Phan, Ha Phan Bui, Phi Cuong Anh Nguyen, Van Nguyen Phan, and Dinh Thanh Dang

Abstract Electric vehicles (EVs) are becoming increasingly popular, however, achieving smooth and efficient speed control remains a significant challenge. This paper introduces an innovative approach by proposing an AFUPID (adaptive fuzzy-PID) technique for regulating the speed of EVs utilizing a DC motor. The paper begins by presenting the hardware architecture and modeling of the DC motor within an EV. Subsequently, it critically analyzes the limitations of conventional PID and fuzzy logic control methods. The proposed methodology is an AFUPID controller, which intelligently adjusts PID coefficients based on fuzzy inference mechanisms. Simulation results verify the effectiveness of the AFUPID controller, showcasing a smoother response and reduced errors when compared to standard PID and fuzzy controllers. Furthermore, experimental results validate the exceptional performance of the proposed method in real-world implementation. In essence, this study provides a potential solution for robust adaptive speed control in EVs through the integration of fuzzy inference and PID control, addressing the challenges of achieving optimal speed regulation in EVs.

Keywords Electric vehicle · DC motor · PID · Adaptive cruise control · PID-fuzzy logic controller

---

V. D. Phan (✉) · V. Q. Phan · T. S. Dang · N. H. Trinh · P. N. Nguyen · N. M. Luong · B. U. Nguyen · Q. C. Phan · H. P. Bui · P. C. A. Nguyen · V. N. Phan · D. T. Dang
Vinh University, Vinh City, Nghe An 43100, Vietnam
e-mail: dupv@vinhuni.edu.vn

D. H. Tien et al. (eds.), *Proceedings of the 1st International Conference on Sustainability and Emerging Technologies for Smart Manufacturing*, Proceedings in Technology Transfer, https://doi.org/10.1007/978-981-97-7083-0_1

1 Introduction

Electric vehicles have emerged as a promising solution to mitigate global carbon emissions from the transportation sector [1, 2]. However, a critical aspect of electric vehicles (EVs) lies in ensuring adaptive speed control for the traction motor under diverse road conditions. Hence, the importance of efficient cruise control systems has a crucial role in enhancing energy efficiency for electric vehicles. This paper focuses on the modeling and adaptive speed control system in EVs.

Two primary kinds of motors are regularly used in EVs: DC motors and induction motors. However, induction motors exhibit strong nonlinearity and difficulties in speed adjustment. Due to their straightforward structure, flexible control capability, high starting torque, and broad speed range, DC motors are deemed more suitable for electric vehicle propulsion systems [3–5]. This paper specifically examines the electric vehicle propulsion system based on a DC motor.

To meet the speed control requirements for electric vehicles, numerous scholars have proposed various methods, including Proportional–Integral–Derivative (PID) controller, backstepping, fuzzy logic, neural networks, and more. Among these, PID controllers and fuzzy logic controllers are recognized as the two primary methods for motor speed control [6–9]. However, conventional methods often suffer from fixed control gains, limiting the output performance of the system. Therefore, this study introduces an AFUPID controller designed to achieve optimal cruise control performance. The AFUPID controller continuously adjusts PID coefficients based on fuzzy inference mechanisms.

The paper's three main contributions are: First, addressing the modeling and hardware architecture of the EV speed control system. Second, proposing an AFUPID controller tailored for electric vehicle systems. Third, presenting results that demonstrate the superiority of the proposed controller, validated through detailed simulations and experiments.

2 System Implementation

2.1 Hardware Architecture

The hardware design processes aim to maximize flexibility in the surrounding environment for the electric vehicle (EV) model testing, as depicted in Fig. 1. In this paper, the EV model is designed for experimentation in the most stable environment to avoid unwanted external influences.

Fig. 1 Hardware platform (1—Arduino Uno, 2—driver L298H, 3—lithium battery, 4—DC motor integrated encoder, 5—PC)

The fundamental components of the EV model testing include one Arduino board corresponding to a central controller, one L298 H-bridge circuit corresponding to a power converter, a 12 V power supply derived from three lithium polymer (LiPo) cell batteries, each LiPo cell with a voltage capacity of 3.7 V. It is noted that a JGB37-520 gearbox motor equipped with an encoder to feedback speed signal, ensuring the stable and precise control of the motor as well as the wheel of EV. Lastly, a USB connecting cable is employed for communication between the personal computer (PC) and Arduino to control the system. All hardware connections are made through specialized connectors to enhance system stability and flexibility. The block diagram of the EV model testing is presented in Fig. 2.

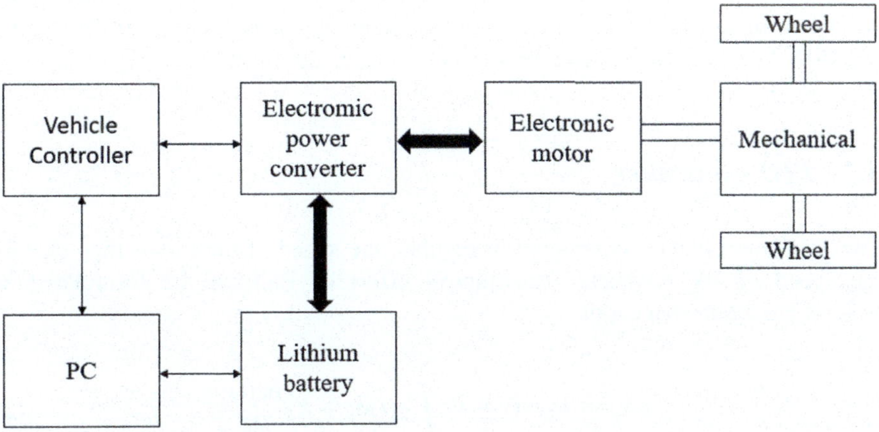

Fig. 2 Block diagram of EV model testing

Table 1 Specifications of the DC motor

Parameters	Value	Parameters	Value
Voltage constant (k_t, K_e)	0.02 Nm/A	Inductance (L)	0.5 H
Inertia motor (J)	0.01 kg m^2/s^2	Factor of friction (b)	0.1
Resistance (R)	2 Ω		

2.2 *Modeling of the DC Motor*

In this section, the mathematical model of the DC motor is represented as follows [5]:

$$\begin{aligned} \frac{d^2\theta}{dt^2} &= \frac{1}{J}\left(k_t i - b\frac{d\theta}{dt}\right) \\ \frac{di}{dt} &= \frac{1}{L}\left(-Ri + v - K_e\frac{d\theta}{dt}\right) \end{aligned} \tag{1}$$

where J, b, k_t, and K_e are inertia torque, the coefficient of friction of mechanical components, positive gains, respectively. R, L, v, and i are the resistance of the winding wire, the coefficient of self-inductance, the voltage applied and the current flowing to the motor coil, respectively. θ is the position of the rotational axis.

From the constructed DC electric motor model above, the specifications of the motor model are presented in Table 1.

3 Control Design

In this section, a AFUPID controller is formulated to enhance control quality which will be compared with two other controllers, namely, fuzzy logic controller (FLC) and PID. The overview speed adjust system diagram is illustrated in Fig. 3.

3.1 *PID Controller*

The PID controller is constructed to regulate the speed of the motor then change the speed of EV's wheels. The control effort of PID can be formulated as follows:enlargethispage40pt

$$u(t) = K_p e(t) + K_i \int_0^t e(t)dt + K_d \frac{de(t)}{dt} \tag{2}$$

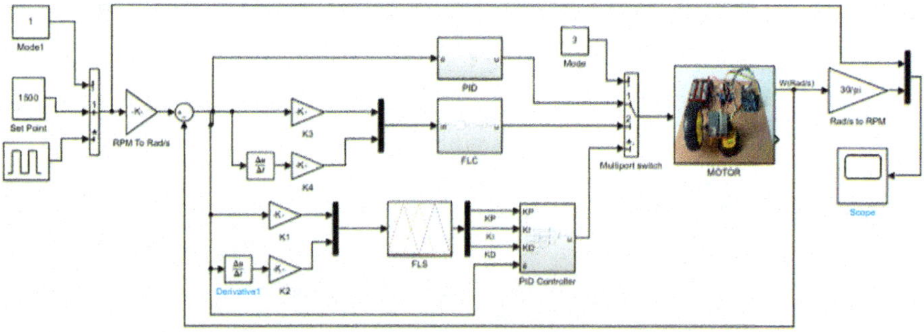

Fig. 3 Block diagram of AFUPID control system for DC motor control

where $e(t)$ is the tracking error. K_p, K_i, and K_d is the designed gain.

3.2 *Fuzzy Logic Controller*

The FLC is designed to regulate the speed of the motor that utilizes the Mamdani approach. It is constructed with one output (u) and two inputs (e, de). The membership functions (MF) for the inputs consist of 5 triangle MFs with the standard input range [− 1; 1], known as, Positive Big (PB), Positive Small (PS), Zero (ZE), Negative Big (NB), and Negative Small (NS). The outputs consist of 7 triangle MFS, known as, Positive Large (PL), Positive Medium (PM), Positive Small (PS), Zero (ZE), Negative Large (NL), Negative Medium (NM), and Negative Small (NS). The fuzzy MFs for the inputs and outputs are described in Fig. 4.

In this study, the control rule table for the FLC is presented in Table 2.

3.3 *AFUPID Controller*

The focal point of this study is to construct a AFUPID controller, a combination of the two controllers introduced above. The controller will be designed with 2 inputs and 3 outputs, where two inputs (e, de), and the three outputs correspond to the parameters of Δk_p, Δk_i, Δk_d to be fed into the PID controller in Eq. (2). The MFs for the inputs consist of Positive Big (PB), Positive Small (PS), Zero (ZE), Negative Big (NB), and Negative Small (NS). The outputs consist of 7 MFs: Very Large (VL), Medium Large (MLa), Large (L), Medium (M), Medium Low (ML), Low (L), and Very Low (VL). The fuzzy MFs for the inputs and outputs are illustrated in Figs. 5 and 6.

Then the following equations to calculate the control gains K_p, K_i, and K_d in Eq. (2) is expressed by:

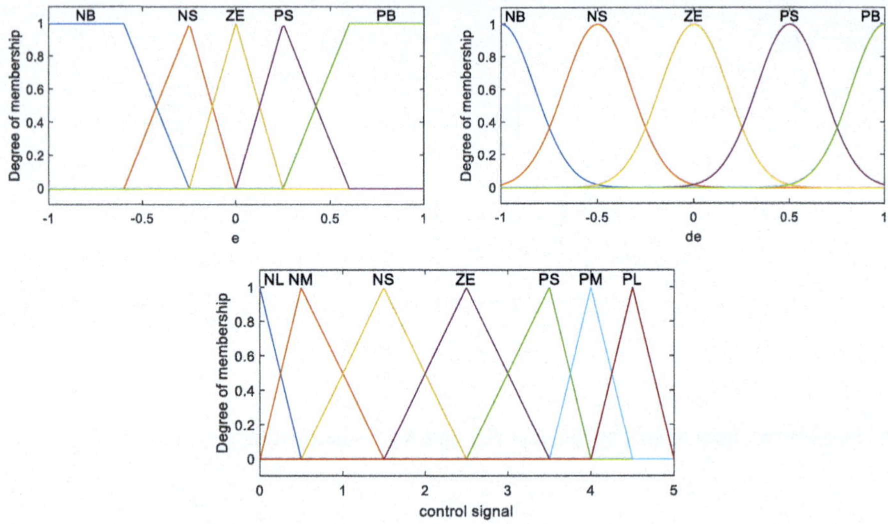

Fig. 4 MFs of two inputs (*e* and *de*) and output (control signal) of the FLC

Table 2 Control rules for FLC

u		*e*				
		NB	NS	ZE	PS	PB
de	NB	NL	NL	NM	NS	ZE
	NS	NL	NM	NS	ZE	PS
	ZE	NM	NS	ZE	PS	PM
	PS	NS	ZE	PS	PM	PL
	PB	ZE	PS	PM	PL	PL

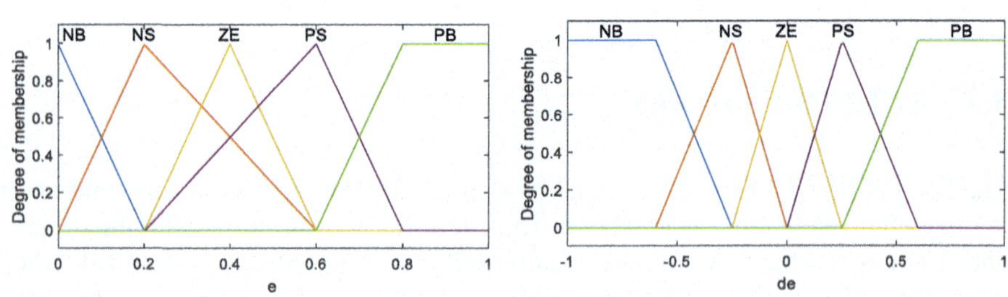

Fig. 5 MF of the inputs (*e* and *de*) of the AFUPID controller

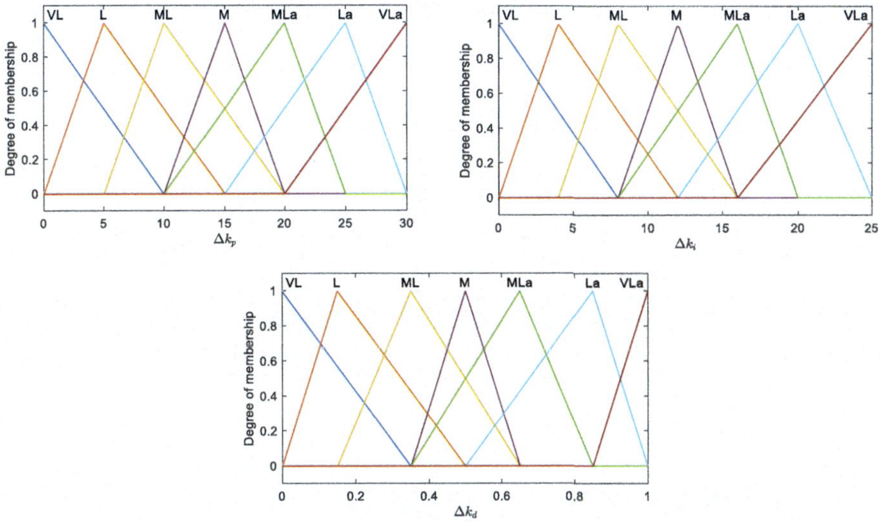

Fig. 6 MFs of the output (Δk_{p}, Δk_{i}, Δk_{d}) of the AFUPID controller

$$\Delta k_j = \frac{K_j - K_{j\,\text{min}}}{K_{j\,\text{max}} - K_{j\,\text{min}}}, \quad j = p, i, \text{ and } d. \tag{3}$$

The ranges of K_j ($j = p$, i, and d) are $[K_{j\,\text{min}}, K_{j\,\text{max}}]$. The control rule table for the AFUPID controller is presented in Tables 3, 4, and 5.

Table 3 Rules applied in the PID controller for K_{p}

e/*de*	NB	NS	ZE	PS	PB
NB	VLa	VLa	VLa	VLa	VLa
NS	MLa	MLa	MLa	La	VLa
ZE	VL	VL	L	ML	ML
PS	MLa	MLa	MLa	La	VLa
PB	VLa	VLa	VLa	VLa	VLa

Table 4 Rules applied in the PID controller for K_{i}

e/*de*	NB	NS	ZE	PS	PB
NB	M	M	M	M	M
NS	ML	ML	ML	ML	ML
ZE	L	L	PVS	L	L
PS	ML	ML	ML	ML	ML
PB	M	M	M	M	M

Table 5 Rules applied in the PID controller for K_d

e/*de*	NB	NS	ZE	PS	PB
NB	VL	ML	M	La	VLa
NS	ML	MLa	La	VLa	VLa
ZE	M	La	La	VLa	VLa
PS	MLa	VLa	VLa	VLa	VLa
PB	VLa	VLa	VLa	VLa	VLa

In this study, the "centroid" approach is utilized for defuzzification stage to achieve K_p, K_i, and K_d, that are fed to the PID controller and using Eqs. (2) and (3) to regulate the DC motor of the EV.

4 Main Result

4.1 Simulation Result

To validate the effectiveness of the suggested method, the simulation and experiment studies were conducted in the MATLAB environment. In the modern EV, the function of adaptive cruise control has appeared. The driver can select several options of this function such as tracking preset speed and tracking the previous EV on the road. Hence, in this study, the motor speed driven for EV was set according to two cases. Case study 1 presents the output response following the constant speed which utilizes a constant block in MATLAB Simulink and set to 450 [rpm]. Case study 2 employed a square wave block with a maximum value of 500 [rpm] and a pulse width of 50% that illustrates the tracking the previous EV. The response plots for the three controllers (PID, fuzzy controller, and proposed controller) are displayed in Figs. 7 and 8.

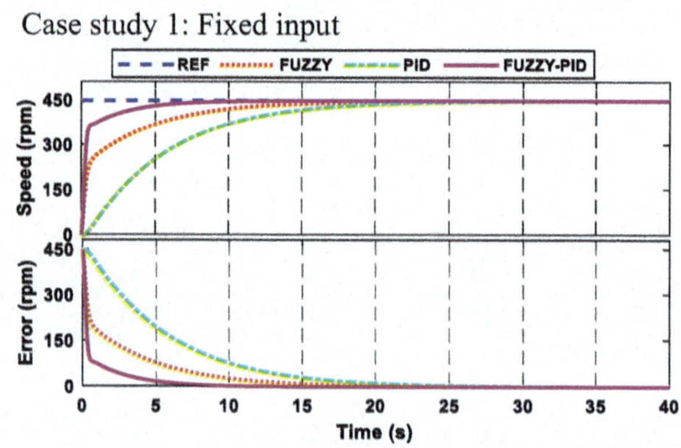

Fig. 7 Simulation results of the fixed input signal

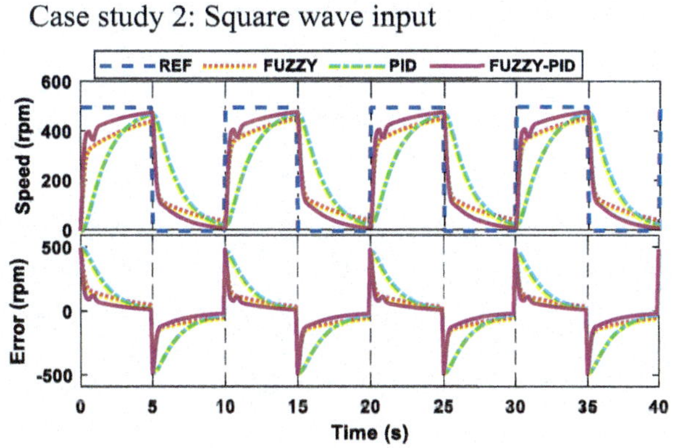

Fig. 8 Simulation results of the square wave input signal

From Fig. 7, it can be observed that the PID has the worst performance as compared to other controllers because of the fixed control gains. The self-tuning control gain, achieved by combining PID and fuzzy logic techniques, optimizes the performance of the proposed controller. From Fig. 8, we can see that the output speed responses of the PID and fuzzy become poor when the reference signal has significant change. Meanwhile the suggested controller still brings the better tracking performance as compared to relevant controller.

4.2 *Experiments Results*

To evaluate the efficiencies of the proposed algorithm, the experiment trials are studied. During the experimental, the desired speed of two case studies are the same with in the simulation section. Hence, the response and tracking errors graph of the experimental phase are displayed in Figs. 9 and 10.

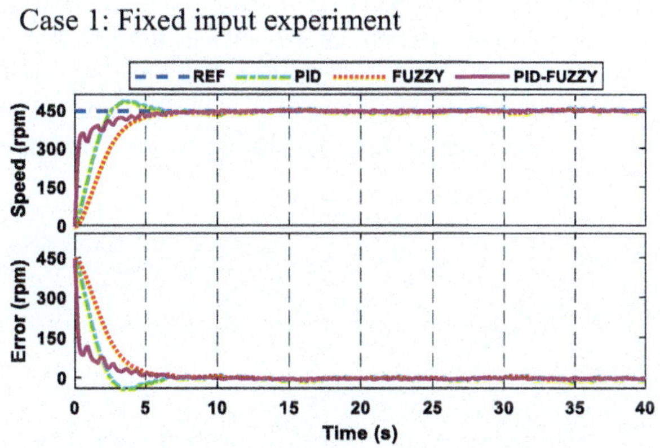

Fig. 9 Experiment results of the fixed input signal

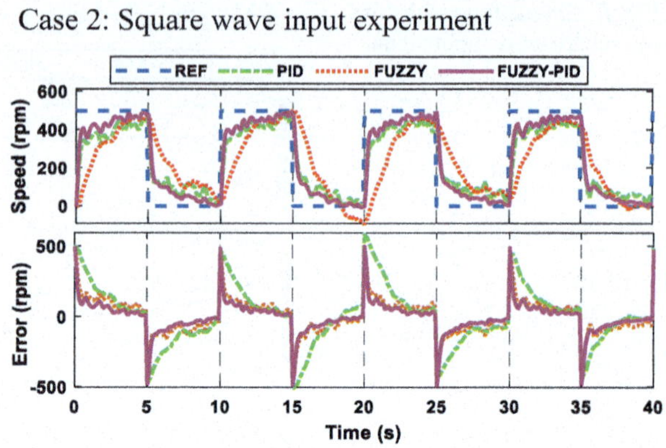

Fig. 10 Experiment results of the square wave input signal

From the results of Figs. 9 and 10, it is evident that big transient, overshoot, and rise time are achieved by the other two methods. Meanwhile, the quality of the suggested method achieves the best performance. The suggested algorithm guarantees that the tracking error has not overshoot and satisfactory performance is obtained. It provides the output speed nearest to the desired signal. The reason is that it combines the advantage of two separate technique, namely the PID and the fuzzy logic technique. Consequently, the proposed controller demonstrates accuracy, paving the way for new advancements in the EV applications.

5 Conclusion

In this article, a description of the hardware and software system of an EV model has been presented to implement adaptive cruise speed control through various control approach. In the suggested methodology, the gains of the PID are adjusted utilizing fuzzy logic methods. The study demonstrates that the AFUPID achieves both the accuracy of a traditional PID controller and the flexibility of a fuzzy logic controller. Simulation and experimental results exhibited that the suggested control strategy obtains good control performance for the DC motor applied in EV with accurately tracking speed, a fast response time and no overshooting. The steady-state error is decreased to zero, indicating better efficiency compared to other presented controllers.

References

1. Jang, Y.T., Jovanovic, M.M.: A contactless electrical energy transmission system for portable-telephone battery chargers. IEEE Trans. Ind. Electron. **50**, 520–527 (2013)
2. Bishop, J.D.K., Molden, N., Boies, A.M.: Real-world environmental impacts from modern passenger vehicles operating in urban settings. Int. J. Transp. Dev. Integr. **1**(2), 203–211 (2017)
3. Meyer, P., Germano, P., Markovic, M., Perriard, Y.: Design of a contactless energy transfer system for desktop peripherals. IEEE Trans. Ind. Appl. **47**, 1643–1651 (2011)
4. Anjali, A.R.: Control of three phase BLDC motor using fuzzy logic controller. Int. J. Eng. Res. Technol. (IJERT) **2**(7) (2013)
5. Kandiban, R., Arulmozhiyal, R.: Design of AFUPID controller for speed control of BLDC motor. Int. J. Soft Comput. Eng. **2**(1) (2012)
6. Ansari, U., Alam, S., Jafri, S.M.N.: Modelling and control of three phase BLDC motor using PID with genetic algorithm. In: UK Sim 13th International Conference on Modelling and Simulation, pp. 189–194 (2011)
7. King, P.J., Mamdami, E.H.: The application of fuzzy control system to industrial process. Automatica 235–242 (1977)
8. Abd-Alkarim, S.F.: Application of fuzzy logic in servo motor. Al-Khwarizmi Eng. J. **3**(2), 8–16 (2007)
9. Yesilbag, E., Ergene, L.T.: Field oriented control of permanent magnet synchronous motors used in washers. In: 2014 16th International Power Electronics and Motion Control Conference and Exposition, pp. 1–6 (2014)

Novel Thermoelectric Generator Design for Recovering Waste Heat

Duc Tran Duy, Duc Pham Xuan, Tan Nguyen Tien, Dien Vu Minh, Quang Khong Vu, and Vinh Nguyen Duy

Abstract Thermoelectric generator (TEG) devices have attracted worldwide scientists for several decades. However, the application of TEG worldwide has been limited due to its low power density and complex system layout. The purpose of this research is to improve the TEG power mounted in proton exchange membrane fuel cell (PEMFC) vehicles. Hence, the cases of channel structure of TEG are modeled and simulated to compare among cases. In these cases, the experimental bench model is set up to demonstrate precisely with numerical simulation results. Furthermore, this experimental model utilized an input heat source of 100 °C, assuming whether the release heat of the fuel cell devices or cooling systems of the engine. All results of simulation and experiments are compared to evaluate precisely.

Keywords Thermal electric generator · Maximum power point tracker · Heat exchanger solvent · Heat and transfer · Boiling point · Temperature difference

1 Introduction

Energy loss, pollutant, and strict-peremptory statutes significantly affect the transportation sector. In the current diesel engine, exhaust energy loss is 41%, and engine cooling loss is 20% if the total energy of the engine created by an internal combustion

D. T. Duy · D. P. Xuan · V. N. Duy (✉)
Faculty of Vehicle and Energy Engineering, Phenikaa University, Hanoi, Vietnam
e-mail: Vinh.nguyenduy@phenikaa-uni.edu.vn

T. N. Tien
Department of Student Affairs, Hanoi University of Industry, Hanoi, Vietnam

D. V. Minh
Center for Automotive Technology and Driving Training, Hanoi University of Industry, Hanoi, Vietnam

Q. K. Vu
Department of Vehicle and Energy Conversion Engineering, School of Mechanical Engineering, Hanoi University of Science and Technology, Hanoi, Vietnam

D. H. Tien et al. (eds.), *Proceedings of the 1st International Conference on Sustainability and Emerging Technologies for Smart Manufacturing*, Proceedings in Technology Transfer, https://doi.org/10.1007/978-981-97-7083-0_2

cycle is 100% [1]. The two mentioned losses are entirely thermal. TEG, a green technology, has recently been utilized in vehicles to regenerate heat loss from exhaust waste heat resources or engine cooling systems (up to 100 °C of heat loss from engine cooling systems) [2]. Furthermore, the generated temperature of the fuel cell's coolant system, up to over 100 °C, causes energy waste and a dramatic converter of the membrane, which decreases the performance of the PEM fuel cell [3]. This waste heat is utilized to input the heat source of TEG to create electricity. Also, TEG technology is not noisy because of the absence of movable mechanisms and extra cost reduction in maintenance [2]. Thus, many studies of TEG were applied in vehicles [4–8]. However, TEG still has many challenges, such as quick-thermal loss and permissible-limited temperature of TEG systems to business popularly [2, 9]. Hence, this project improves the performance of TEG and applies them in vehicles to trade more widely. In the distant future, the results of this project will be utilized to build experimental models for various TEGs and vehicle-mounted TEGs.

2 Fundamentals of Thermal Electric Generator

Numerous temperature-dependent energy conversion phenomena occur in the thermoelectric generator, including Thomson heat flow, Seebeck voltage, Joule heating, and the Peltier effect [10–17]. The system absorbs a predetermined quantity of thermal energy from the heating plate and establishing a temperature gradient between the two operation sides. The temperature disparity generates an electrical potential between the n and p connections [10]. A thermopile consists of many thermocouples linked in series electrically and parallel thermally. A thermocouple is constructed by connecting two dissimilar materials with opposite Seebeck coefficients at their ends. As a result, an electric voltage can be calculated as Eq. (1) [10, 11].

$$V_{\text{out}} = N * \alpha * \Delta T \tag{1}$$

where ΔT and N are, respectively, the temperature gradient and number of thermocouple (TC) modules, and α is the Seebeck coefficients. The total internal resistance of TCs linked in series with one another is proportional to their number N. Therefore, although a large number of TCs will increase the voltage that the TEGs supply, the effect on internal resistance is the opposite.

The three distinct TEG design approaches are determined by the orientation of the thermocouples on the substrate concerning the direction of heat flow: (i) an arrangement of lateral TCs and lateral heat flow; (ii) an arrangement of vertical TCs and vertical heat flow; and (iii) an arrangement of vertical TCs and lateral heat flow [11]. In the first TEG configuration, lateral TCs are utilized to convert a lateral heat flux. This design, also called planar TEG, involves the deposition, printing, or patterning of thermocouples on the surface of the substrate. One of the main advantages of this approach is the ability to regulate thermocouple arm parameters. Additionally, it is compatible with thin-film deposition, which facilitates the manufacturing of longer and thinner thermocouples compared to alternative methods.

Furthermore, the implementation of elongated TC limbs in this design enhances the resistance in comparison with alternative TEG designs. This causes a temperature gradient to build along the TC arms, which causes an output voltage to rise. TCs are positioned vertically between the heat source and sink in the second TEG configuration. As a result, heat moves along the substrates and thermoelement arms vertically. This type of TEG is the most widely used due to its advantage in simple design and high output voltage. As the heat moves vertically, TCs positioned laterally on the substrate create the final TEG pattern, known as mixed [12]. The vertical heat transfer was triggered by incorporating micro-cavities beneath the thermocouple arms into the substrate. A polyimide polymer-based flexible foil lift-off method or a comparable metal-oxide semiconductor standard technology might be used to implement this approach in silicon.

3 Experimental Methods

3.1 Methods

Research methods use experimental methods to prove the simulation results. The simulation process is carried out step by step as follows:

- The model is on CAD according to the TEMs from the manufacturer, as seen in Fig. 1. TEM dimensions are 40 × 40 × 4 (mm). Parameters of real models are carefully measured and drawn using a commercial design software.
- The meshing process is performed after the model has been previously modeled. The grid size will be adjusted to accommodate the size of the simulation model.

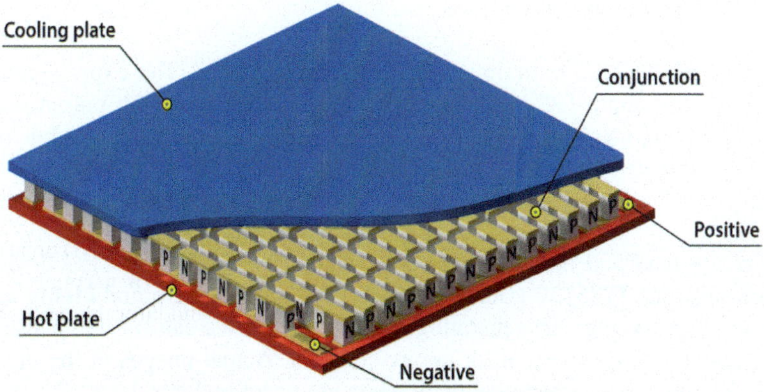

Fig. 1 TEM module

- Set physical properties and boundary conditions for the simulation model.
- Perform simulations and export simulation results. Simulation results include the effects of thermal variation and different channel profiles.
- Validation of simulation results based on the experimental results.

The course of the experiment was carried out as follows:

- Building an experiment model according to a simulation model.
- Installing of electronic control boards.
- Performing experiments and acquire experimental results.
- Comparing the simulation and experiment results; thus relevant analyses and evaluations can be performed.

3.2 Models

3.2.1 Simulation Model

To evaluate effects of thermal variation in TEGs, one TEM module is utilized during the simulation, see Fig. 1. The TEM includes many P-type and N-type modules linked together by multi-conjunction as the copper link. The end of TEM is two plates, a hot plate and a cooling plate. Those are non-electrical materials and good conductors of heat.

Two models are used to assess the effect of channel profiles, including Model #1 and Model #2; see Fig. 2. Both models contain nine TEM plates and are placed on top of the cooling block. The cooling block of Model #1 uses tube profiles. Otherwise, pin profiles are designed to Model #2. Air is drawn into the channel led by the fan, and the end of the TEM hot plates is heated as the heat source. This thermal value is 363.15 K.

3.2.2 Experimental Model

The model of the TEG experiment is illustrated in Fig. 3. The experimental suite consists of a program DC electronic load, a computer receiving output data, and TEGs. The computer reads data and controls the temperature of the hot plate of TEGs via a circuit board, see Fig. 4. The outcome of TEG power is included in the DC electronic load program. All results are stored in the computer for evaluation and analysis.

To create load and measure the power characteristics of the TEG device, this research used the IT8513C+ load generator. This is a modern and highly accurate power load measuring and generating device. The device can set the measurement mode to keep voltage, current, or power constant. Actual images of the measuring equipment and specific technical parameters are given in Table 1.

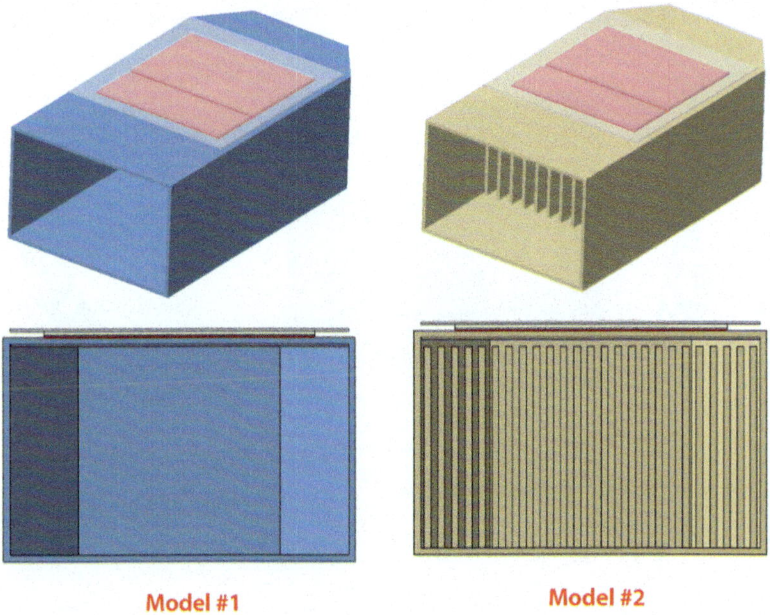

Fig. 2 Models utilized to assess channel profiles

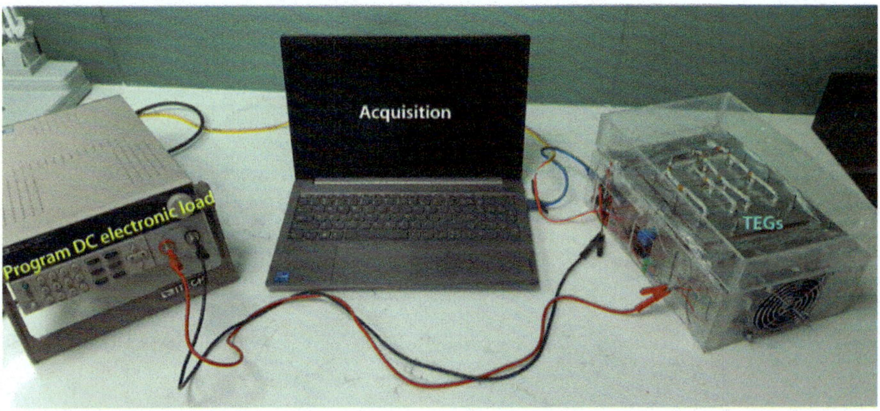

Fig. 3 Experimental model

This study used NTC 10K 3950 and Type K temperature sensors to measure temperature during the experiment. Accordingly, these temperature sensors were placed at the inlet and outlet fluid lines on the surfaces of the heat exchanger. Characteristics of the NTC 10K 3950 sensor and Type K temperature sensor are given in Table 2.

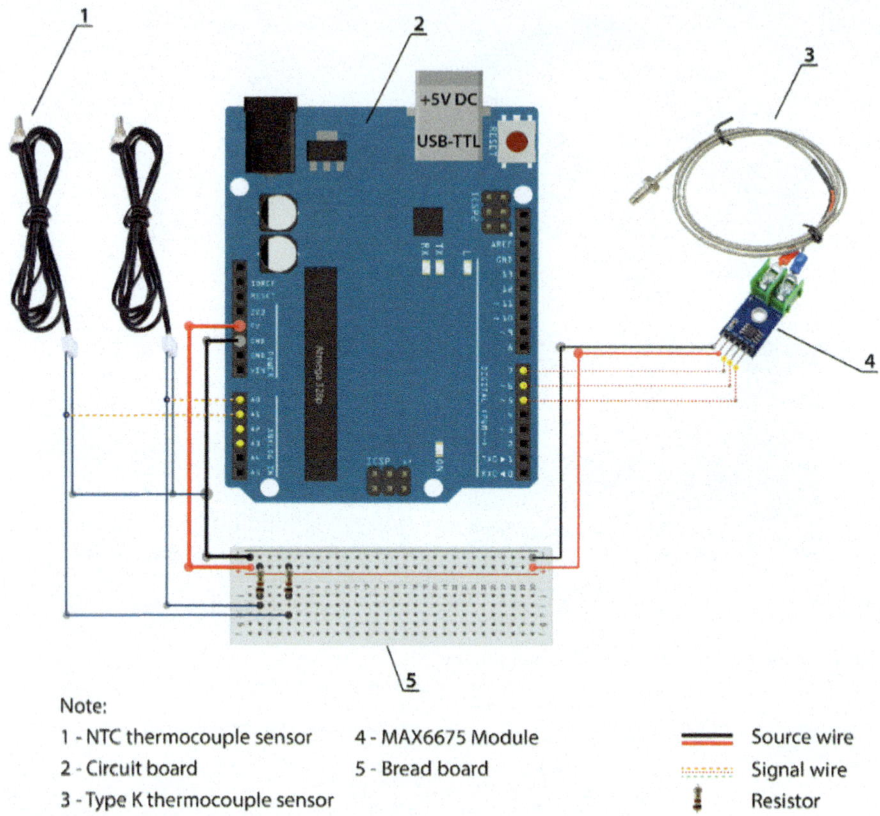

Fig. 4 Interface of electronic circuit

Table 1 Technical parameters of the IT8513C+ load module

Parameter	Value
Operating temperature	0–40 °C
Working humidity	20–80%
Working attitude	Up to 2000 m above sea level
Power supply	110 V/220 V
Frequency	Up to 40 kHz
Error	0.1 mV, 0.1 mA
Maximum power	600 W

Figure 4 shows the electronic circuit receiving temperature-sensing signals. This dataset will be sent to the computer via the TTL protocol on a USB.

Table 2 Characteristics of temperature sensors

Parameters	NTC 10K 3950	Type K
Working temperature range (°C)	– 40 to 200	0–1024
Error	0.5%	0.25 °C
Power supply (V)	3–5.5	3–5.5
Limitation	Use for the insulating head	± 5 μA
Signal cable type	2 cables	2 cables

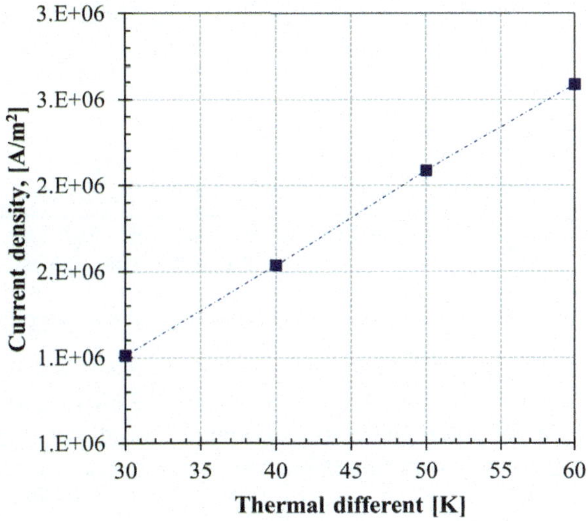

Fig. 5 Simulation results of thermal effects

4 Results and Discussion

4.1 *Evaluation of Thermal Effects*

Simulation results of temperature variability are shown in Fig. 5. This result shows that the higher the temperature difference between TEG's hot and cold sides, the more electricity produced will increase. The increased electricity production is due to the Seebeck effect on temperature variations of TEMs. The results are consistent with the experimental simulations on two models, as seen in Fig. 6.

4.2 *Evaluation of Channel Profiles*

Figure 6 indicates that the power and voltage of Model 2 are higher than Model 1 at an amperage value of 0.1 A. The voltage-2 and power-2 curves characterize Model

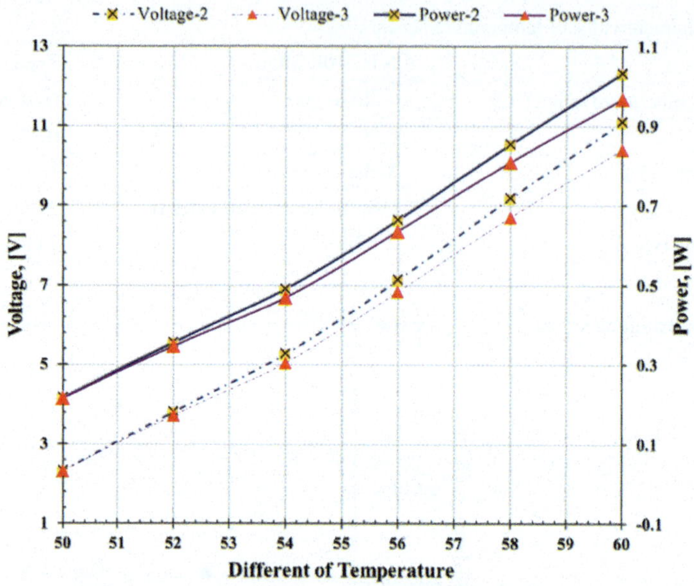

Fig. 6 Effect of thermal differences of two models in experiments

#2. The remaining voltage-3 and power-3 curves characterize Model #1. As can be seen, the power and voltage of model 2 are slightly higher than that of model 1. This is consistent with simulation because the average temperatures at the two sides of the TEMs of Model #1 and Model #2 are 363.13 K, 360.23 K and 363.13 K, 358.59 K, respectively. As a result, the temperature difference between the two sides of Model #2 is greater than that of Model #1.

Figure 7 shows a model that simulates airflow in the cooling block. The maximum velocity of Model #1 is 2.90 m/s. Meanwhile, in Model #2, this value is 3.49 m/s. Otherwise, pins utilized in Model #2 create multiple eddies throughout the pins. This improves the efficiency of convection. Then, the thermal difference between the ends of the hot and cooling plates of TEMs increases. As the results, see Fig. 6, the power of TEGs of Model #2 is higher than Model #1.

4.3 Characteristics of Model #2

Characteristics of Model #2 in experiments are denoted in Fig. 8. In Fig. 8, Voltage-1–4 and Power-1–4 curves correspond to thermal values of the hot plate as 90–84 °C. At 90 °C, the voltage and the power of TEGs are at the top of other temperatures. In addition, when the amperage value is increased, the voltage will decrease, and the power will increase to a certain value, which will subsequently decrease. The maximum voltage and power of TEGs is approximately 12 V and 1.4 W, respectively.

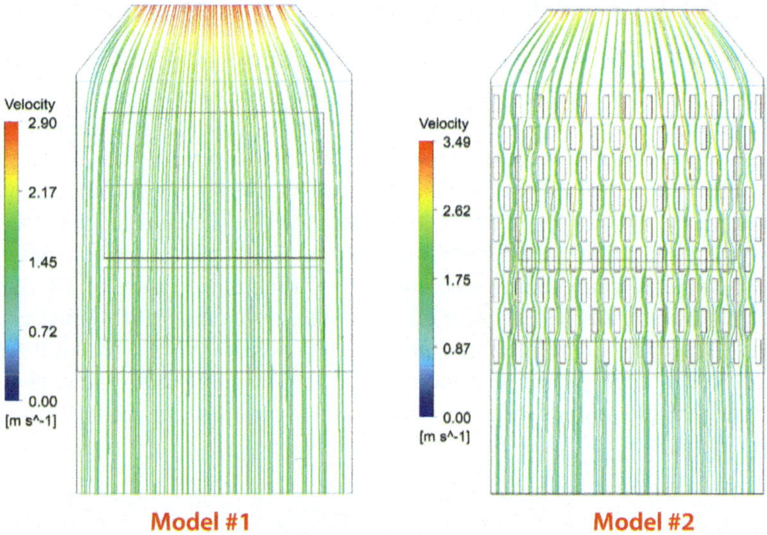

Fig. 7 Streamline simulation of Models #1 and #2

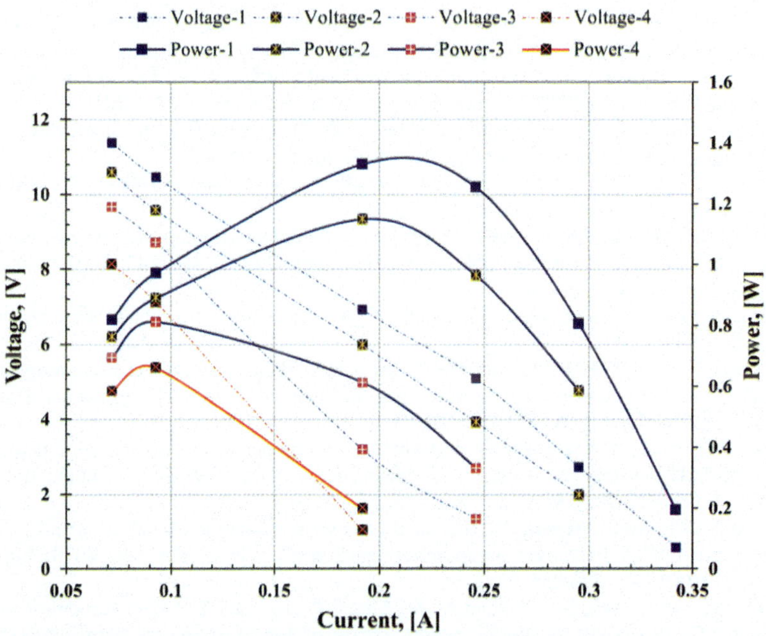

Fig. 8 Characteristics curves of Model #2

5 Conclusions

The thermoelectric generator is a potential product for utilizing waste heat to produce electrical energy. Experimental TEGs are fabricated to evaluate device characteristics based on the wasting heat of automobiles. The results indicate that the higher the thermal difference, the more the amperage increases. Otherwise, the pins set in Model #2 increase the efficiency of convective heat transfer, leading to increased TEG capacity. The top power of 9 TEMs connected in Model #2 is approximately 1.4 W at 90 °C hot plate temperature. Because the working temperature of TEMs is still low for recovering waste heat from engine cooling systems or PEM fuel cells, the following research direction of the project is to select TEMs with high working temperatures to apply waste heat recovery on the exhaust pipe of vehicles. At the same time, suitable profiles should be researched to optimize convective heat transfer on equipment.

References

1. MC2D (Organization): IEEE Industry Applications Society, IEEE Power Electronics Society, IEEE Power & Energy Society, IEEE Magnetics Society, and Institute of Electrical and Electronics Engineers. In: 2015 Tenth International Conference on Ecological Vehicles and Renewable Energies (EVER), Grimaldi Forum, Monaco, 31 Mar–2 Apr 2015
2. Jouhara, H., et al.: Thermoelectric generator (TEG) technologies and applications. Int. J. Thermofluids **9** (2021). https://doi.org/10.1016/j.ijft.2021.100063
3. Zhang, J., et al.: High-temperature PEM fuel cells. J. Power Sources **160**(2), 872–891 (2006). https://doi.org/10.1016/j.jpowsour.2006.05.034
4. Espinosa, N., Lazard, M., Aixala, L., Scherrer, H.: Modeling a thermoelectric generator applied to diesel automotive heat recovery. J. Electron. Mater. **39**(9), 1446–1455 (2010). https://doi.org/10.1007/s11664-010-1305-2
5. Hussain, Q.E., Brigham, D.R., Maranville, C.W.: Thermoelectric exhaust heat recovery for hybrid vehicles (2009)
6. Martins, J., Goncalves, L.M., Antunes, J., Brito, F.P.: Thermoelectric exhaust energy recovery with temperature control through heat pipes. In: SAE 2011 World Congress and Exhibition (2011). https://doi.org/10.4271/2011-01-0315
7. Crane, D., et al.: TEG on-vehicle performance and model validation and what it means for further TEG development. J. Electron. Mater. **42**(7), 1582–1591 (2013). https://doi.org/10.1007/s11664-012-2327-8
8. Sugiartha, N., Sastra Negara, P.: Technical feasibility evaluation on the use of a Peltier thermoelectric module to recover automobile exhaust heat. J. Phys. Conf. Ser. (2018). https://doi.org/10.1088/1742-6596/953/1/012090
9. Chen, M., Andreasen, S.J., Rosendahl, L., Kær, S.K., Condra, T.: System modeling and validation of a thermoelectric fluidic power source: proton exchange membrane fuel cell and thermoelectric generator (PEMFC-TEG). J. Electron. Mater. **39**(9), 1593–1600 (2010). https://doi.org/10.1007/s11664-010-1270-9
10. Sabry, M., Lashin, A., Al Turkestani, M.: Experimental and simulation investigations of CPV/TEG hybrid system. J. King Saud Univ. Sci. **33**(2), 101321 (2021). https://doi.org/10.1016/j.jksus.2020.101321

11. Bensafi, M., Ameur, H., Kaid, N., Hoseinzadeh, S., Memon, S., Sohani, A.: Experimental study of electric power generation with concentrated solar thermoelectric generator. Electronics **11**(12), 1867 (2022)
12. Jaziri, N., Gutzeit, N., Bartsch, H., Boughamoura, A., Müller, J., Tounsi, F.: LTCC-based Y-type thermoelectric generator with an improved heat flow guide for automotive waste heat recovery. Sustain. Energy Fuels **6**(9), 2330–2342 (2022). https://doi.org/10.1039/D2SE00048B
13. Heikes, R.R., Ure, R.W., Jr.: Thermoelectricity: Science and Engineering. Interscience Publishers, New York (1961)
14. Drebushchak, V.A.: The Peltier effect (2008)
15. Demirel, Y.: Nonequilibrium Thermodynamics: Transport and Rate Processes in Physical, Chemical and Biological Systems, 3th edn. Elsevier B.V., UK (2014)
16. Jacovelli, P.B., Norris, R.C., Canada, C.E., Zinke, O.H.: The thermocouple revisited: the Thomson effect. J. Non-Equilib. Thermodyn. **44**(4), 333–353 (2019). https://doi.org/10.1515/jnet-2018-0062
17. Chen, J., Yan, Z., Wu, L.: The influence of Thomson effect on the maximum power output and maximum efficiency of a thermoelectric generator. J. Appl. Phys. **79**(11), 8823–8828 (1996). https://doi.org/10.1063/1.362507

Modulating Optoelectronics Characteristics of Janus SiSSe Monolayer by Mechanical Strain

Tran The Quang, Nguyen Hoang Linh, Bui Thi Hoa, Dinh The Hung, Nguyen Minh Son, To Toan Thang, and Do Van Truong

Abstract Utilizing the *ab inito* calculation, we focus on exploring the mechanical behaviors and optoelectronics characteristics of the Janus SiSSe monolayer. First, a dynamic stability test was performed, and the results showed the system was completely stable. The obtained results show that the Janus SiSSe monolayer has small elastic constants and a good load-bearing capacity. Through hybrid functional calculations, the energy band structures of the Janus SiSSe have been explored and show the properties of the semiconductor material. Under mechanical strains, energy band structures and the bandgap energy have significantly changed, with a maximum increase of ~ 32% and a maximum decrease of ~ 53% compared to the natural state. Besides, we also have investigated the absorption spectrum under mechanical strains. Vertical strain leads to a maximum improvement of ~ 3.5 times compared to the natural state. The findings provide helpful information and are practical for the application of the Janus SiSSe in nanoelectronics and optical devices.

Keywords Mechanical strain · Energy band structure · Density function theory · Optoelectronics characteristics · Janus SiSSe monolayer

T. T. Quang (✉) · B. T. Hoa
Faculty of Technology and Engineering, Thai Binh University, Thai Binh 410000, Vietnam
e-mail: tranthequang12@gmail.com

N. H. Linh · N. M. Son · T. T. Thang · D. V. Truong
Hanoi University of Science and Technology, Hanoi 100000, Vietnam

D. T. Hung
Phenikaa University, Hanoi 100000, Vietnam

D. H. Tien et al. (eds.), *Proceedings of the 1st International Conference on Sustainability and Emerging Technologies for Smart Manufacturing*, Proceedings in Technology Transfer, https://doi.org/10.1007/978-981-97-7083-0_3

1 Introduction

The release of the first two-dimensional material (2D)—graphene—has created a revolution in materials science worldwide [1]. With distinctive features in mechanical, optical absorption, thermoelectric, and electronic properties, monolayer materials offer numerous advantages for applications in robotics, optics, nanoelectronics, actuators, and various other fields [2–4]. Consequently, there has been extensive research and development in the realm of 2D materials. Presently, this field continues to expand, with numerous novel structures being generated. Some famous 2D materials include chalcogenide-based materials, MXenes, oxides of transition metals, α–black phosphorene, layered hydroxide salts (LDHs), and Janus structures, etc. [2, 3, 5, 6].

Janus structures include many atomic strong bonds, wherein metal atoms consistently reside in the center of the structures and the outermost surface is surrounded by nonmetal atoms [7, 8]. This family has been interesting in materials science due to many advantages for practical applications such as the process of converting heat into electricity [9], photocatalytic water splitting [10, 11], and optoelectronics devices [12, 13]. Some typical Janus structures include MoSSe, WSTe, SnSO, HfSSe, and SnSSe, their promise is excellent proof of the potential of applying 2D materials in real life [12, 14–16].

The Janus SiSSe monolayer is composed of Si atoms positioned between S and Se atoms, forming a unique structural arrangement. It combines elements from two fundamental structures, SiS_2 and $SiSe_2$ (both denoted as SiX_2), which have been previously investigated and have demonstrated potential for advanced optical and electronic applications. Linh et al. have reported a nonlinear response of the band gap and absorption coefficients to mechanical strains in these structures [3]. Moreover, SiX_2 materials are promising candidates in the thermoelectric sector, facilitating the conversion of waste heat into electricity in high-efficiency generators [6]. However, research on the Janus SiSSe monolayer remains limited, with only basic parameters explored thus far [17]. Additionally, existing calculation methods may not provide sufficient accuracy to determine the material's full applicability.

In this study, we focus on exploring the mechanical characteristics and the in-plane strain-induced in the optoelectronics characteristics of the Janus SiSSe monolayer. To achieve exceptionally precise outcomes, we employed the Heyd–Scuseria–Ernzerhof (HSE) exchange-potential in conjunction with hybrid functional assessments to investigate optoelectronics characteristics. This report thoroughly examines and analyzes the effects of in-plane strains on the optoelectronics properties of the Janus SiSSe monolayer using density functional theory (DFT).

2 Methodology

The Schrödinger equation describes the fundamental interactions within the system and finds application in calculations in solid-state physics. To solve this equation, DFT is extensively employed to handle the multitude of variables inherent in the problem. Through self-consistent calculations of the quantum problem, an electron density function is obtained, providing the information regarding energy, stress, energy structure, and optical coefficients. All calculations in this study regarding mechanical behaviors and optoelectronics characteristics of the Janus SiSSe monolayer were performed by the Quantum Espresso package [18]. To describe the interactions between electrons in the system, this study utilized Perdew–Burke–Ernzerhof functional (PBE) within the framework of the generalized gradient approximation (GGA) [19]. We selected an energy threshold cutoff of 60Ry and 720Ry for the electron density to ensure sufficient interaction space in the optimized conditions with stress of 5×10^{-2} GPa and force lower than 10^{-6} Ry/a.u. at 0 K. The ($15 \times 15 \times 1$) **k**-point grids were selected following the Monkhorst–Pack method for the calculation process. Additionally, to avoid the interaction between layers, a vacuum was set to 30 Å. To calculate the optoelectronics properties of the Janus SiSSe monolayer, we used a hybrid functional with HSE exchange–correlation and Wannier90 interpolation to achieve the highest accuracy, align closely with the values of the experiment [20]. A **k**-mesh for the polarization plane is chosen to be $40 \times 40 \times 1$ to assurance convergence for calculating optical coefficients.

3 Results and Discussions

3.1 Optimize and Stabilize Structures

The Janus SiSSe monolayer is structurally optimized under standard conditions, resulting in the lattice constant of $a = b = 3.40$ Å and the layer thickness of $d = 1.46$ Å are presented in detail in Table 1. These results have similarities with previous investigations [17]. To demonstrate the stability of the Janus SiSSe monolayer, we have performed two testing methods. Firstly, a phonon vibration spectrum is calculated within the reciprocal space. The phonon spectrum of the Janus SiSSe monolayer possess nine oscillation branches with distinct oscillation frequencies. Therein, it has three acoustic branches at low-frequency ranges and six optical ones at frequency ranges. The maximum frequency of the Janus SiSSe monolayer is 290 cm^{-1}. In addition, the vibration results also demonstrate the entirely dynamic stability of the structure without any negative frequency vibrations in the Brillouin zone (Fig. 1).

Table 1 Lattice constants—a and b (Å), thickness—d (Å), elastic constants—C (N/m)

a and b	d	C_{aa}	C_{bb}	C_{ab}	$C_{\gamma\gamma}$
3.40	1.46	105.05	105.05	35.23	34.33

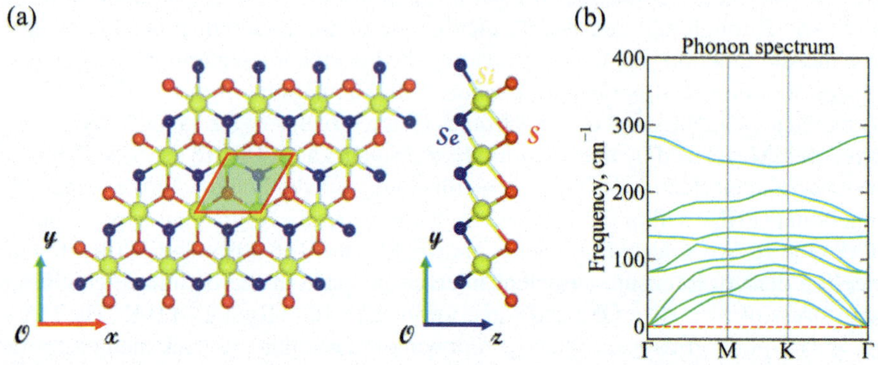

Fig. 1 **a** The top and side views after the optimization process and **b** the phonon spectrum with high-symmetry points in the Brillouin area of the Janus SiSSe monolayer

The second method we use to prove the stability of the structure is elastic stability by calculating the elastic constants. Based on the energy and mechanical strain relationship in the elastic domain, we have found the elastic constants C by **Thermo–pw.x** code [21] as given in Table 1. The obtained results show that the elastic coefficients of the Janus SiSSe monolayer are quite small and satisfy Born's criteria ($C_{aa} > |C_{ab}| > 0$; $C_{\gamma\gamma} > 0$) [22]. Compared with some previous materials such as HfSSe, ZrSSe, SiS_2, and $SnTe_2$, elastic constants of the Janus SiSSe monolayer are quite small [2, 6, 16, 23]. These results demonstrate that the Janus SiSSe monolayer has good elasticity and flexible mechanical properties.

3.2 *Mechanical Behaviors*

To illuminate the lattice isotropy in the elastic domain for the Janus SiSSe monolayer, we present the relationship between elastic modulus—Y and Poisson's ratio—P following the tensile vector as shown in following equations [2]:

$$Y(\theta) = \frac{C_{aa}C_{bb} - C_{ab}^2}{C_{aa}\sin(\theta)^4 + C_{bb}\cos(\theta)^4 + \left(\frac{C_{aa}C_{bb}-C_{ab}^2}{C_{\gamma\gamma}} - 2C_{ab}\right)\sin(\theta)^2\cos(\theta)^2} \tag{1}$$

$$P(\theta) = -\frac{\left(C_{aa} + C_{bb} - \frac{C_{aa}C_{bb} - C_{ab}^2}{C_{\gamma\gamma}}\right)\sin(\theta)^2\cos(\theta)^2 - C_{ab}\left(\sin(\theta)^4 + \cos(\theta)^4\right)}{C_{aa}\sin(\theta)^4 + C_{bb}\cos(\theta)^4 + \left(\frac{C_{aa}C_{bb} - C_{ab}^2}{C_{\gamma\gamma}} - 2C_{ab}\right)\sin(\theta)^2\cos(\theta)^2} \tag{2}$$

where C is elastic constants of the Janus SiSSe structure and θ is the angle created by the investigated vector with respect to the x-axis.

Figure 2 shows that the crystal lattice of the Janus SiSSe monolayer is completely symmetry in the elastic domain with polar contours being perfect circles. We also show some mechanical parameters regarding shear modulus, and layer modulus of the Janus SiSSe monolayer in Table 2 to indicate the flexible mechanical properties of this structure.

Moreover, we have calculated the stress and strain relationship to explore the ideal strength and mechanical behaviors of the Janus SiSSe monolayer as shown in Fig. 3. We note that only basic biaxial strain is performed on the material, since this is a typical strain, it is easily created in reality and calculated based on the formula $\varepsilon = (L - L_0)/L_0$ with L and L_0 being the parameters of the crystal lattice at equilibrium and strained states. Based on obtained results, the Janus SiSSe structure has shown quite good mechanical characteristics with a quite large fracture strain, up to $\varepsilon_{\text{bia}} = 0.16$. In the early stages, the material's stress gradually increases and shows linearity at small strained levels. When the strained state reaches $\varepsilon_{\text{bia}} = 0.16$, the stress is the highest. After reaching the maximum stress state, the material's stress tends to decrease almost linearly. Besides, the energy of the system is also presented through the red line in Fig. 3. It is easy to see that the energy of the coefficient increases

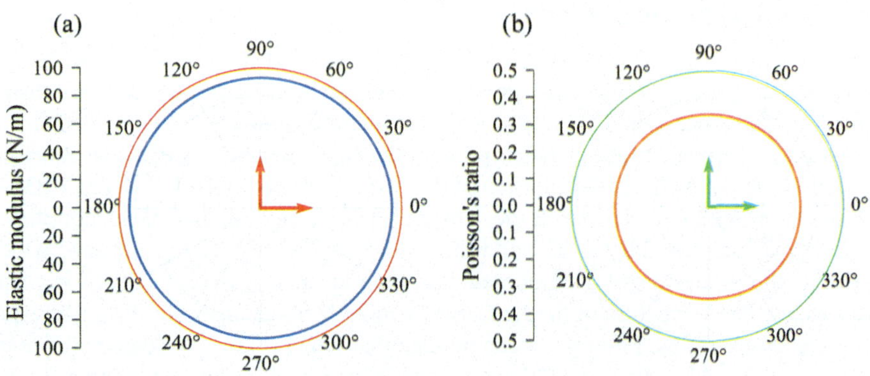

Fig. 2 Polar contours for **a** elastic modulus and **b** Poisson's ratio of the Janus SiSSe monolayer

Table 2 Elastic modulus—Y (N/m), Poisson's ratio—P, shear modulus—G (N/m), layer modulus—γ_{2D} (N/m), ultimate tensile strength—UTS (N/m), fracture stress—ε^f

Y	P	G	γ_{2D}	UTS_{bia}	$\varepsilon^f_{\text{bia}}$
93.73	0.33	34.33	70.14	2.65	0.16

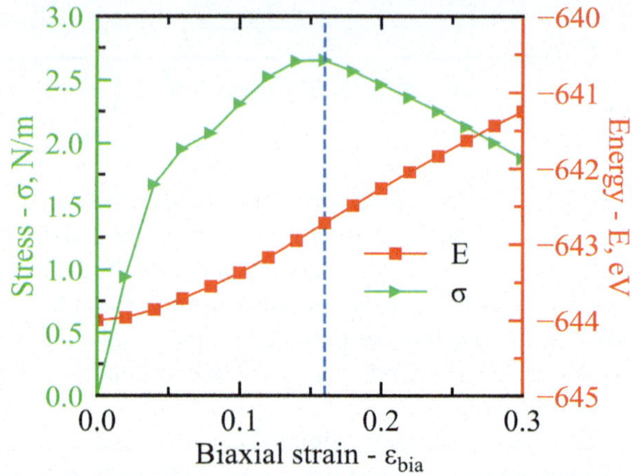

Fig. 3 The relationship between the biaxial strain with the stress (green line) and the energy system for the Janus SiSSe monolayer (red line)

almost linearly with the value of the tensile strain. The relationship between energy and strain gradually increases, proving that the material can withstand high tensile strain, without degenerate states appearing. Compared to some previously studied monolayer structures such as $SiSe_2$ ($\varepsilon^f = 0.14$) [2], and ReS_2 ($\varepsilon^f = 0.14$) [24], the Janus SiSSe monolayer has quite the high fracture strain, which can be widely used in practical devices.

3.3 Modulating Optoelectronic Characteristics by Mechanical Strain

We explore the electronic structures of the Janus SiSSe structure at equilibrium (Fig. 4). The lowest energy point of the conduction region (CP) is located in the *M*–*K* path, while the highest energy point of the valence region (VP) is distributed at Γ point. The inconsistent distribution of the two extreme points—CBM and VBM—shows the potential to modify the electronic properties of the Janus SiSSe monolayer. Additionally, the width of the band gap achieves 1.32 eV at equilibrium, which exhibits the properties of a semiconductor material. Under the impacts of in-plane strains, the band structure of the Janus SiSSe monolayer changes significantly, especially at large strains. Figure 4, it can be seen that CBM is always located at *K* point to $\varepsilon_{bia} = 0.10$. After that, the CP has moved from *K* to Γ point, meanwhile the VP has moved its position from Γ point to Γ–*M* path immediately under strains. Therefore, the energy structures are very sensitive to the external impact, easily modified and improved one's interaction with the ambient environment. Collated electronic properties of some previous Janus structures such as SnSO, HfSSe, SnSSe [12, 14, 16], we detect that the Janus SiSSe material is too suitable for producing electronic devices.

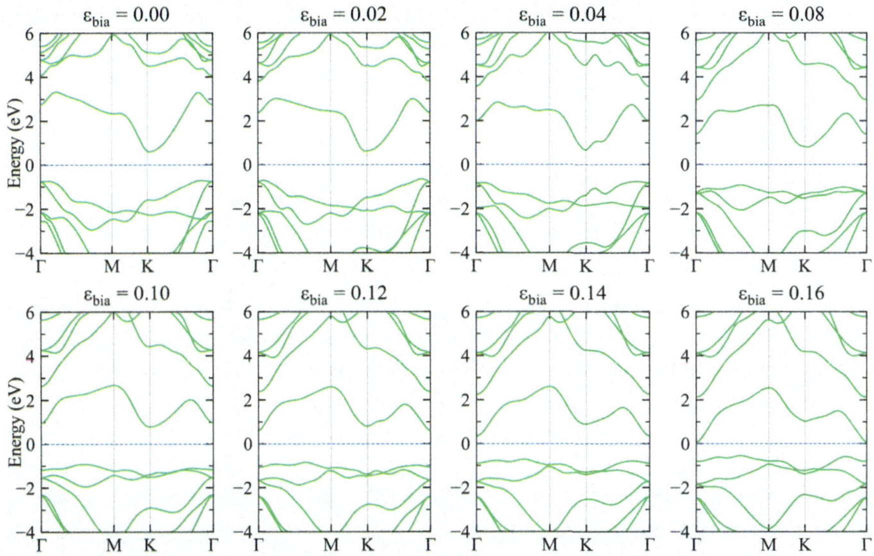

Fig. 4 Calculated energy band structures were investigated by HSE functional at a number of values. The Fermi energy is blue

The scatted function between bandgap energy and biaxial strains is present in Fig. 5. Our obtained results indicate that the bandgap energy changes flexibly under mechanical strains. At the small strains, the bandgap energy of this monolayer tends to increase modestly to a 1.74 eV (maximum increase ~ 32% compared to equilibrium). After reaching the highest one, the band gap has a rapid decrease until achieving the ideal value of 0.61 eV (decrease ~ 53% compared to equilibrium). The flexible adjustment of electronic properties is fascinating and useful information for the improvement of materials for electronic applications.

Fig. 5 The bandgap energy-biaxial strains curve of the Janus SiSSe monolayer

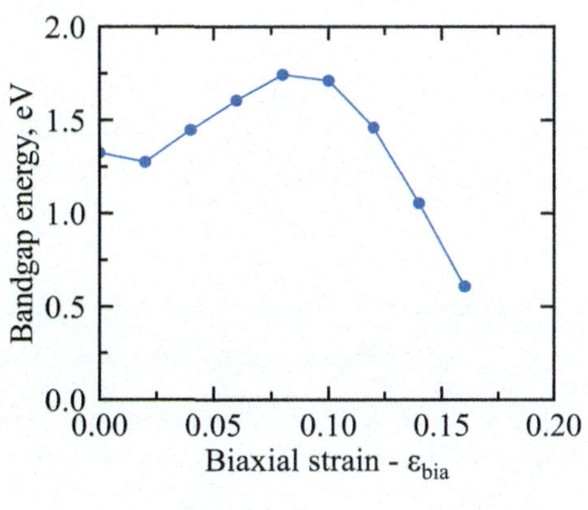

The optical characteristics of the Janus SiSSe monolayer are explored through the absorption spectrum. The light absorption spectrum discloses the ability applications of the materials in optical devices and is a crucial key to applying the light absorption ability of materials for optical devices and is calculated based on the equation [2, 11]:

$$\alpha = \frac{\sqrt{2}}{c}\omega\left[\sqrt{\varepsilon_{\mathrm{r}}(\omega)^2 + \varepsilon_{\mathrm{i}}(\omega)^2} - \varepsilon_{\mathrm{r}}(\omega)\right]^{\frac{1}{2}} \tag{3}$$

where ε_{r} and ε_{i} are real and imaginary parts of the dielectric function that create the optical characteristics. ω is the angular frequency. α is the absorption coefficient.

Figure 6 shows the light absorption spectrum according to the absorption states in the parallel plane $\alpha_1^{//}(\omega) = \alpha_2^{//}(\omega)$ and the perpendicular plane $\alpha_3^{\perp}(\omega)$. In all strained cases, the material's absorption spectrum on-plane has a greater value than the perpendicular plane. For the parallel plane, biaxial strains do not improve the light absorption coefficient with photons of low energy (IR). However, in the VS region and part of the ultraviolet light (UV), biaxial strain optimistically contributes to improving the absorption ability of the Janus SiSSe structure. In the VS region, the photon energy absorptivity $\alpha_1^{//}(\omega) = \alpha_2^{//}(\omega)$ is improved ~ $1.5 \div 3.5$ times compared to equilibrium. Similarly, in the ultraviolet light region, the photon energy absorptivity $\alpha_1^{//}(\omega) = \alpha_2^{//}(\omega)$ is enhanced significantly ~ 1.6 times compared to equilibrium. For the perpendicular plane, the photon energy absorptivity $\alpha_3^{\perp}(\omega)$ is also upgraded, focusing only on the VS region and the UV region. The photon energy absorptivity $\alpha_3^{\perp}(\omega)$ increases up to ~ 1.7 times in the VS region and ~ 1.5 times in the UV region. Thus, based on changes in the nanostructure, light absorption spectrums of the Janus structure have undergone positive changes, increasing the ability to absorb photons. This is one of the extremely worthy impacts serving the manufacture of advanced optical devices.

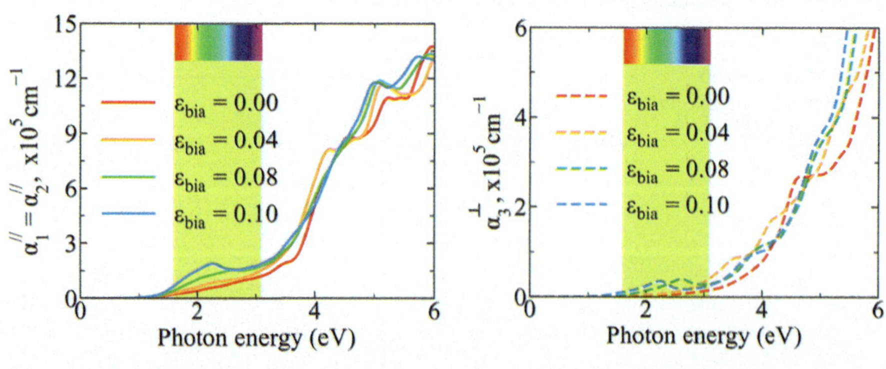

Fig. 6 Light absorption spectrums of the Janus SiSSe monolayer were calculated by HSE functional. The visible light in the range of 1.61–3.1 eV (VS) that fill by yellow

4 Conclusions

Through the results of simulating calculations of fundamental interactions between particles of an atomic system using the DFT calculation, the mechanical behaviors and optoelectronics characteristics of the Janus SiSSe monolayer have been clarified. Possessing small elastic constants, the Janus SiSSe monolayer has demonstrated the ability to withstand high strength with an ideal strain of $\varepsilon = 0.16$. On account of the in-plane strains, the optoelectronics characteristics of the Janus SiSSe monolayer have positive changes such as changing the energy band structure, and the absorption ability with the photon energy. The band gap energy of the Janus SiSSe monolayer has the flexible, nonlinear function of the in-plane strain. The electronic properties can be enhanced up to ~ 32% and declined up to ~ 53% at any strain. Especially, under the impact of the in-plane strains, the absorption ability of the Janus SiSSe structure is improved significantly up to 1.7 times in the VS region. The interesting phenomenon is important and useful information for experimental fabrication studies of electronic and optical devices.

References

1. Novoselov, K.S., Geim, A.K., et al.: Electric field effect in atomically thin carbon films. Science **306**, 666–669 (2004)
2. Linh, N.H., Quang, T.T., et al.: Prediction of mechanical, electronic and optical properties of monolayer 1T Si-dichalcogenides via first-principles theory. Mater. Today Commun. **36** (2023)
3. Linh, N.H., Son, N.M., et al.: First-principles investigation on the electromechanical properties of monolayer 1H Pb-dichalcogenides. Korean J. Mater. Res. **33**, 189–194 (2023)
4. Liu, Q., Cui, S., Bian, R., Pan, E., Cao, G., Li, W., Liu, F.: The integration of two-dimensional materials and ferroelectrics for device applications. ACS Nano **3**, 1778–1819 (2024)
5. Wang, Q., O'Hare, D.: Recent advances in the synthesis and application of layered double hydroxide (LDH) nanosheets. Chem. Rev. **112**, 4124–4155 (2012)
6. Wang, Y., Gao, Z., et al.: Ultralow lattice thermal conductivity and electronic properties of monolayer 1T phase semimetal $SiTe_2$ and $SnTe_2$. Phys. E **108**, 53–59 (2019)
7. Li, S., Jiang, X., Wang, Z., Song, S., Cai, Z., Leung, K.C.-F., Zeng, T.: Free-standing Janus nanosheets: ultrafast self-assembly and versatile biphase-application. Adv. Funct. Mater. (2024)
8. Vu, T.V., Hoi, B.D., et al.: Induced out-of-plane piezoelectricity and giant Rashba spin splitting in Janus $WSiZ_3H$ (Z = N, P, As) monolayers toward next-generation electronic devices. J. Appl. Phys. **135** (2024)
9. Guo, S.D., Guo, X.S., et al.: Predicted Janus SnSSe monolayer: a comprehensive first-principles study. Phys. Chem. Chem. Phys. **21**, 24620–24628 (2019)
10. Riis-Jensen, A.C., Deilmann, T., et al.: Classifying the electronic and optical properties of Janus monolayers. ACS Nano **13**, 13354–13364 (2019)
11. Linh, N.H., Quang, T.T., et al.: First-principles study of electronic and optical properties and photocatalytic performance of MS monolayer under strain. JST Eng. Technol. Sustain. Dev. **33**, 44–51 (2023)
12. Nguyen, H.T.T., Tuan, V.V., et al.: Electronic and optical properties of a Janus SnSSe monolayer: effects of strain and electric field. Phys. Chem. Chem. Phys. **22**, 11637–11643 (2020)
13. Linh, N.H., Quang, T.T., et al.: Theoretical computational of electronic and transport properties and optical conductivity of monolayer NiS_2 under mechanical strain. JST Eng. Technol. Sustain. Dev. **33**, 34–41 (2023)

14. Barhoumi, M., Lazaar, K., et al.: A DFT study of Janus structure of S and Se in HfSSe layered as a promising candidate for electronic devices. J. Mol. Graph. Model. **96**, 107511 (2020)
15. Evarestov, R.A., Kovalenko, A.V., et al.: First-principles study on stability, structural and electronic properties of monolayers and nanotubes based on pure Mo(W)S(Se)$_2$ and mixed (Janus) Mo(W)SSe dichalcogenides. Phys. E Low-Dimens. Syst. Nanostruct. **115** (2020)
16. Vu, T.V., Phuc, H.V., et al.: A theoretical study on elastic, electronic, transport, optical and thermoelectric properties of Janus SnSO monolayer. J. Phys. D Appl. Phys. **54** (2021)
17. Tanisha, T.T., Hiramony, N.T., et al.: First-principles study on the structural, electronic and optical properties of Janus SiSSe monolayer. Presented at the 2022 12th International Conference on Electrical and Computer Engineering (ICECE) (2022)
18. Giannozzi, P., Baroni, S., et al.: QUANTUM ESPRESSO: a modular and open-source software project for quantum simulations of materials. J. Phys. Condens. Matter **21**, 395502 (2009)
19. Becke, A.D.: Density-functional exchange-energy approximation with correct asymptotic behavior. Phys. Rev. A Gen. Phys. **38**, 3098–3100 (1988)
20. Mostofi, A.A., Yates, J.R., et al.: An updated version of wannier90: a tool for obtaining maximally-localised Wannier functions. Comput. Phys. Commun. **185**, 2309–2310 (2014)
21. Dal Corso, A.: Elastic constants of beryllium: a first-principles investigation. J. Phys. Condens. Matter **28**, 075401 (2016)
22. Mouhat, F., Coudert, F.-X.: Necessary and sufficient elastic stability conditions in various crystal systems. Phys. Rev. B **90** (2014)
23. Zhao, Q., Guo, Y., et al.: Elastic, electronic, and dielectric properties of bulk and monolayer ZrS_2, $ZrSe_2$, HfS_2, $HfSe_2$ from van der Waals density-functional theory. Phys. Status Solidi (b) **254** (2017)
24. Sun, H., Agrawal, P., et al.: A first-principles study of the relationship between modulus and ideal strength of single-layer, transition metal dichalcogenides. Mater. Adv. **2**, 6631–6640 (2021)

A Solution for Converting Gasoline Engines to Use Ammonia with a Combination Power System

Van Dung Nguyen, Andrey Y. Dunin, Thin Quynh Nguyen, Dinh Thi Luong, and Huy Truong Thai

Abstract Due to the lower volumetric energy density and the lower heating value of an ammonia-air mixture, ammonia-fueled engine to suffer from a power deficit in comparison with gasoline engines. Therefore, this study was carried out of the use of a Combination Power System (CPS) to increase the power output of Spark Ignition (SI) engines using ammonia. In this case, it is necessary to select a rational combination of the required power of the SI engine and the electrical machines. Two-cylinders 2CH 7.6/7 and four-cylinders 4CH 7.6/7 engines were tested with ammonia and gasoline fuels. During the test the mixtures of ammonia-air and gasoline-air stoichiometric ($\lambda = 1$), the compression ratio (CR = 9.9) is kept unchanged. The results show that, the use of a CPS makes it to reduce the swept volume and the fuel consumption of the SI engine in a cycle drive during medium and high load operating conditions, while the average effective engine efficiency increases. Although the fuel consumption of the SI engine in a cycle drive is reduced by 18%, and the average effective efficiency is increased by 29%, the CPS provides required power.

Keywords Combination Power System (CPS) · Ammonia fuel · Liquefied ammonia · Spark ignition engine

V. D. Nguyen (✉) · A. Y. Dunin
Department of Heat Engineering and Automotive Engines, Moscow Automobile and Road Construction State Technical University (MADI), Moscow, Russia
e-mail: nguyenvandung1996kchy@gmail.com

T. Q. Nguyen
Faculty of Mechanical Engineering, University of Transport and Communications, Hanoi, Vietnam

D. T. Luong
Institute of Vehicle and Energy Engineering, Military Technical Academy, Hanoi, Vietnam

H. T. Thai
Logistic Commanding Faculty, Air Defense-Air Force Academy, Hanoi, Vietnam

D. H. Tien et al. (eds.), *Proceedings of the 1st International Conference on Sustainability and Emerging Technologies for Smart Manufacturing*, Proceedings in Technology Transfer, https://doi.org/10.1007/978-981-97-7083-0_4

1 Introduction

Climate changes due to greenhouse gas emissions have become increasingly severe. Most cars run on fossil fuels that makes cars a major contributor of greenhouse gases, especially carbon dioxide. Therefore, reducing the environmental impact of transportation by converting internal combustion engines to alternative fuels is an important goal for automakers [1].

Alternative fuels are environmental-friendly in use, and they are also renewable in nature. Among several alternative fuels as a gasoline replacement in spark ignition engines, with a focus on ammonia, which has recently received a lot of interest from researchers [1–3].

Ammonia is common in nature. It is lighter than air, and easily liquefied due to the strong hydrogen bonding between molecules. Gaseous ammonia turns to a colorless liquid, which boils at – 33.1 °C (– 27.58 °F) and freezes to colorless crystals at – 77.7 °C (– 107.86 °F). Low ammonia viscosity promotes better spraying and mixing with air compared to gasoline [4].

Ammonia can be used directly as a fuel in mono-fuel or bi-fuel modes. Moreover, ammonia with hydrogen from ammonia decomposition can be used (Fig. 1).

The thermodynamic properties of ammonia and gasoline are compared as given in Table 1. It can be seen that the low heating value of gasoline is more than 2 times and the volumetric energy density of gasoline is more than 1008 times higher than that of ammonia. Therefore, when the engine converts to using ammonia, it is necessary to increase the fuel amount in the cycle and change the throttle angle in the stoichiometric condition that maintains the engine power as the original engine.

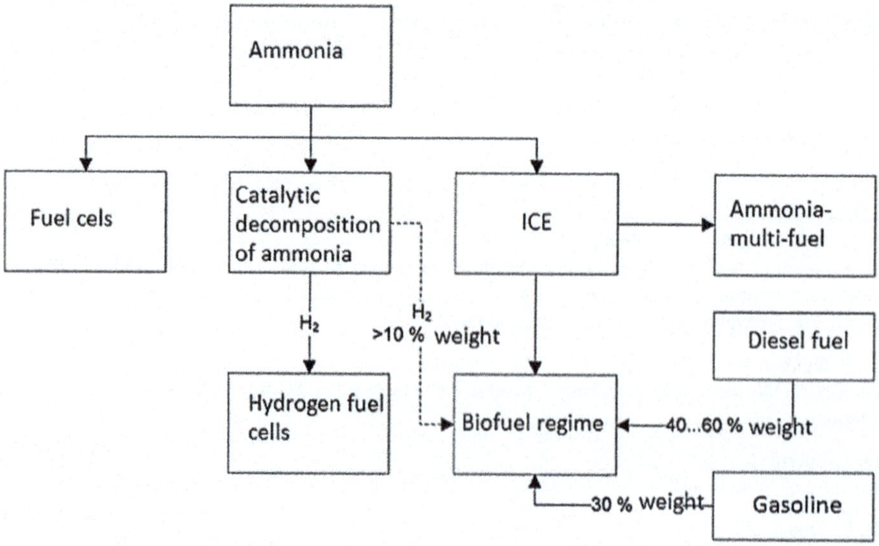

Fig. 1 Possibilities of using ammonia in vehicles

Table 1 Thermodynamic properties of different types of fuels [4–8]

Property	Gasoline	Ammonia
Lower heating value (MJ/kg)	44.5	18.8
Volumetric energy density at 0.1 MPa and 298 K (kg/cub m)	736	0.73
Specific isobaric heat capacity (kJ/(kg K))	2.22	2.19
Auto-ignition temperature (K)	503	930
Octane rating (RON)	90–98	130
Mass carbon content (%)	85.5	–
Mass hydrogen content (%)	14.5	17.7
Mass nitrogen content (%)	–	82.3

In addition, ammonia has an auto-ignition temperature nearly twice as high and especially the octane rating of ammonia is up to 130. Those are the advantages of using ammonia for spark ignition engines.

2 Object of Study and Research Model

To resolve the power deficit when converting the SI engine to ammonia, we propose to apply the CPS to increase the power output of ammonia internal combustion engines.

It is necessary to select a rational combination of the required power of the SI engine and electrical machines (generator and electric motor), battery capacity, and their weight affect the movement of the car (Fig. 2).

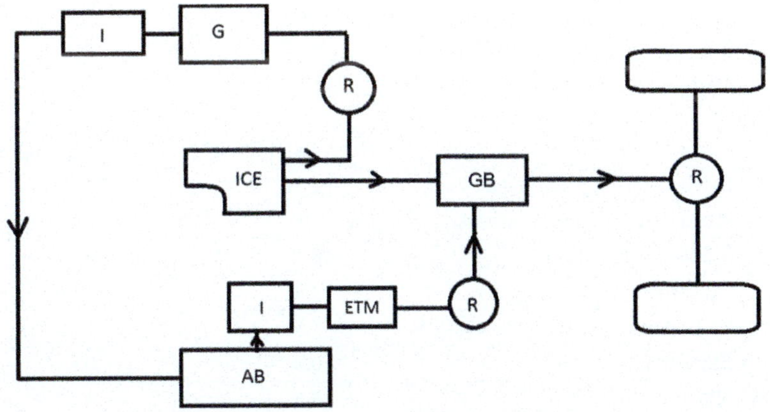

Fig. 2 Proposed scheme of a combined power installation: ICE—internal combustion engine; GB—gearbox; R—reductor; ETM—electric traction motor; I—invertor; G—generator; AB—accumulator battery

Table 2 gives variants of combination of CPS and the parameters of their power plants.

Table 2 Variants of combination of CPS and the parameters of their power plants [9–11]

Variants of combination of CPS			
Variant	Internal combustion engine	Electric traction motor	Generator
1	4CH 7.6/7	4AHK160S4У3	4AK160S8У3
2	2CH 7.6/7	4AHK200L4У3	4AK200L8У3
Internal combustion engine (ICE)			
1	Model	4CH 7.6/7	2CH 7.6/7
2	Displacement (cm^3)	1289	649
3	Diameter (mm)	76	76
4	Stroke (mm)	71	71
5	Compression ratio	9.9	9.9
6	Power (Hp)	64.6	30
7	Torque (Nm)	94.8	44
8	Weight (kg)	127	63.5
Electric traction motor (ETM)			
1	Model	4AHK160S4У3	4AHK200L4У3
2	Rated stator supply voltage (V)	200	200
3	Rated stator supply voltage frequency (Hz)	50	50
4	Number of pole pairs	4	4
5	Synchronous speed (min^{-1})	1500	1500
6	Rated power on the rotor shaft (kW)	14	45
7	Efficiency	0.865	0.9
8	Slip	5	3.5
9	Weight (kg)	140	315
Generator (G)			
1	Model	4AK160S8У3	4AK200L8У3
2	Rated stator supply voltage (V)	220	220
3	Rated stator supply voltage frequency (Hz)	50	50
4	Number of pole pairs	8	8
5	Synchronous speed (min^{-1})	750	750
6	Rated power on the rotor shaft (kW)	5.5	18.5
7	Efficiency	0.8	0.86
8	Slip	6.5	3.5
9	Weight (kg)	170	320
Weight of CPS (kg)		437	698.5

Fig. 3 New European Driving Cycle (NEDC)

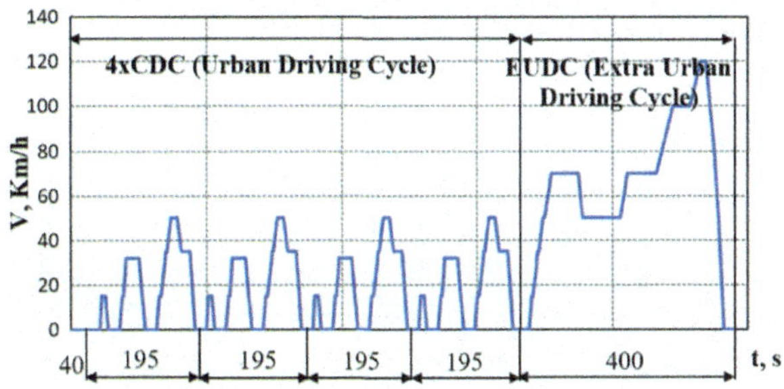

To evaluate engine emissions that have been improved using CPS, the study conducted an analysis of exhaust gas composition according to the New European Driving Cycle (NEDC). The NEDC is shown in Fig. 3.

It can be seen from Table 2 that to ensure the operation of SI engine in the vehicle moving on the selected driving cycle (Fig. 3) including charging of the battery by the generator, the use of a 2CH 7.6/7 engine compared to a 4CH 7.6/7 increases the mass of the CPS (from 437 to 698.5 kg) and the whole average vehicle weight. It is necessary to increase the electric motor and generator powers while keeping the battery capacity constant.

3 Research Results

A decrease the number of cylinders and an increase in the mass of CPS determines the performance of the SI engine with high power utilization factor *K* (Fig. 4). The power utilization factor *K* determines the ratio of the instantaneous effective power to its value at the maximum open throttle, while the engine speeds and the air/fuel ratio are unchanged.

The *K* values differ little from each other (Fig. 4) when converting SI engine from gasoline to ammonia. At the same time, due to an increase in the octane number from 92 to 130, the proportion of detonating fuel dx decreases (Figs. 5 and 6). During the simulation, the ignition timing remained unchanged in all operating modes and amounted to 22° crank angle before TDC. Detonation recognition in the mathematical model of the SI operating cycle was carried out using the Daud and Eyzat formula [11].

The high dx values when the 4CH 7.6/7 engine is running on gasoline (Fig. 5) are explained by the increase in load (Fig. 4) in the driving modes of the vehicle, the weight of which has increased taking into account the use of the CPS (the mass of the standard SI engine is 127 kg, the mass of the CPS with the SI engine is 437 kg). When using SI engine with a smaller number of cylinders (2CH 7.6/7) and the associated increase in the mass of the CPS to 698.5 kg, the proportion of detonating fuel increases in a greater number of driving cycle modes (Figs. 5 and 6).

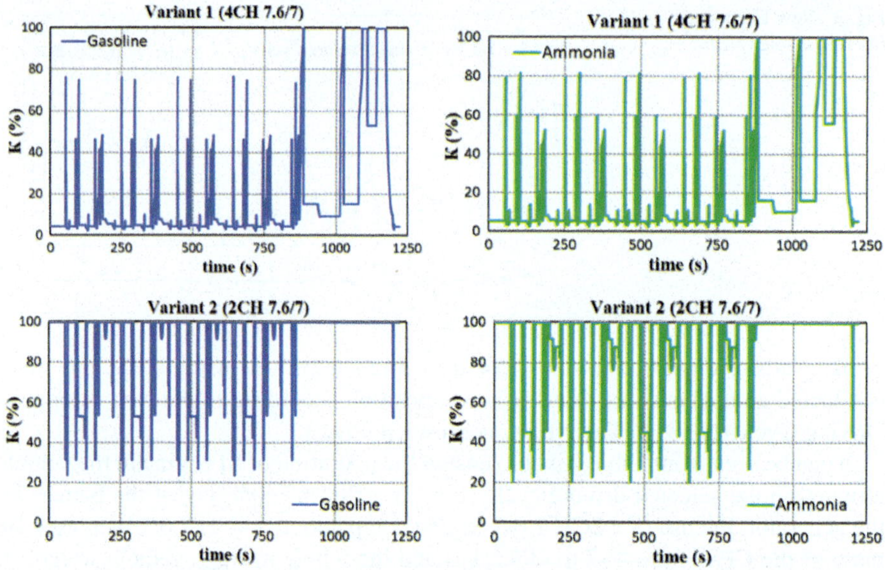

Fig. 4 Power utilization factor *K* (%) of SI engines 4CH 7.6/7 and 2CH 7.6/7 when running on gasoline and ammonia

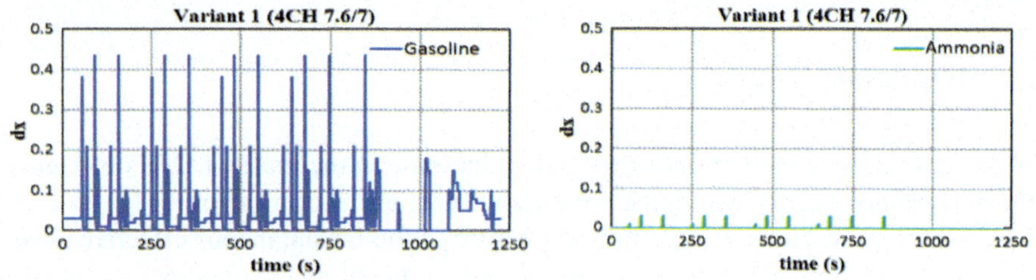

Fig. 5 Knock probability determination in the SI engine 4CH 7.6/7 when running on gasoline and ammonia

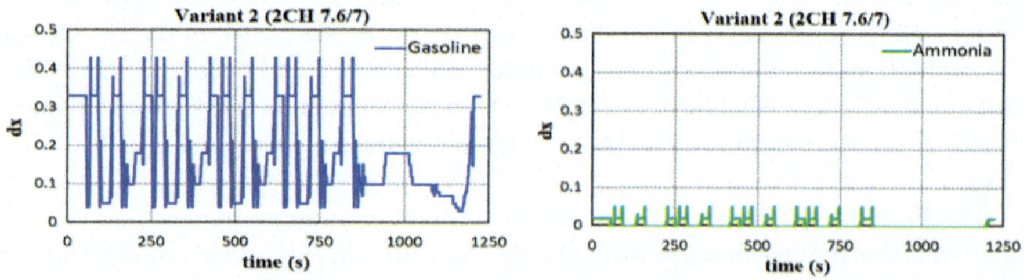

Fig. 6 Knock probability determination in the SI engine 2CH 7.6/7 when running on gasoline and ammonia

In the case of ammonia, this increase has less impact on engine operation due to the low dx values.

As shown in Fig. 7, the deficit torque in the 2CH 7.6/7 engine is compensated by the torque of the ETM. If the ETM is frequently operated, it will increase the operating time ratio of the ETM, which increases the K factor.

Because the gas exchange losses reduce, the effective efficiency of the 2CH 7.6/7 engine as a part of the CPS with higher K value compared to 4CH 7.6/7 engine increases. As a result, the effective efficiency of the 2CH 7.6/7 engine is higher than that of the 4CH 7.6/7 engine in the driving cycle modes (Fig. 8).

An increase in effective efficiency and a decrease in the cylinder number on SI engines when working as a part of CPS ensure a decrease in fuel consumption of the SI engine in the driving cycle. The fuel consumption value of the 2CH 7.6/7 engine is 18% less than the 4CH 7.6/7 (Fig. 9). Due to the volumetric energy density and the lower heating value of the ammonia are lower than those of gasoline, the fuel

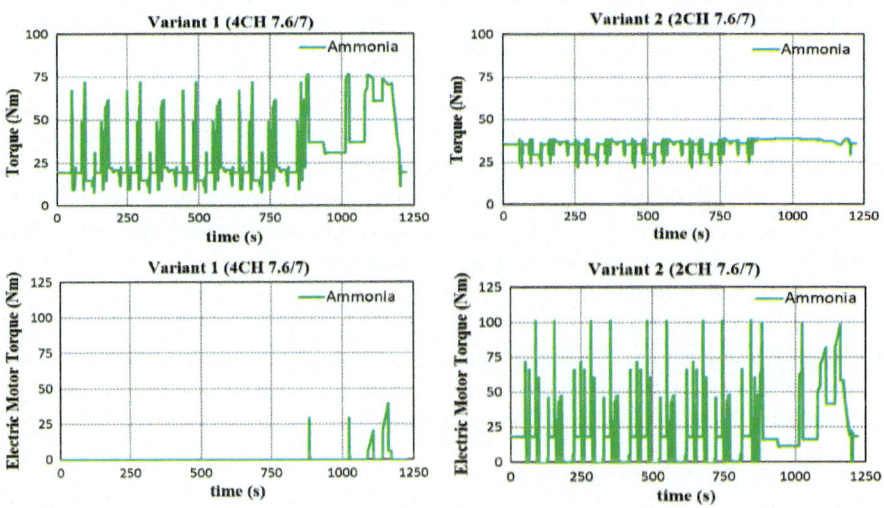

Fig. 7 Engine torque and ETM torque when operating on ammonia

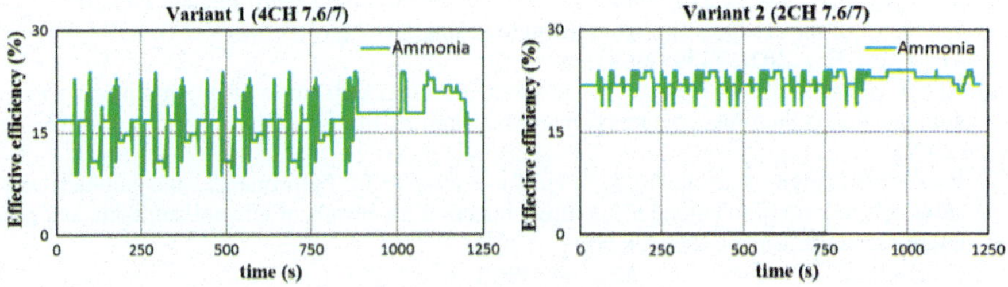

Fig. 8 Effective engine efficiency when operating on ammonia

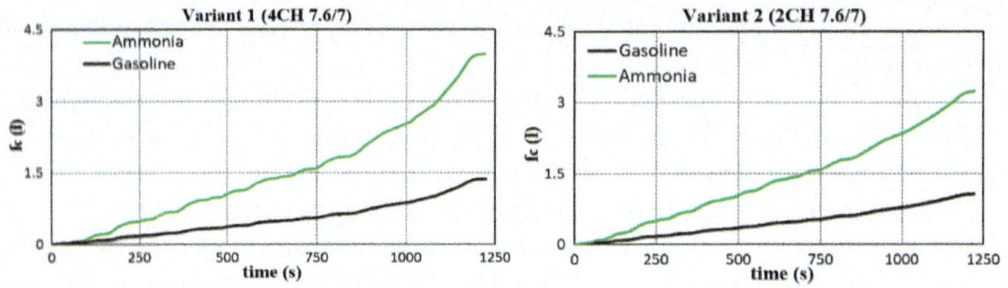

Fig. 9 Fuel consumption of SI engine during the NEDC

consumption will increase when the engine operates in modes that use ammonia (Fig. 9).

4 Conclusion

Ammonia is a fuel environmental-friendly and it is also renewable in nature. Converting internal combustion engines using ammonia fuels is a way to reduce greenhouse gas, and hence the environmental impact of transportation.

Due to the increase in octane rating from 92 (RON of gasoline) to 130 (RON of ammonia), ammonia has higher anti-knock properties. In addition, its storage conditions are not as demanding in terms of safety as hydrogen.

The CPS makes it reduce the swept volume, and the fuel consumption of the SI engine in a cycle drive during medium and high load operating conditions while the average effective engine efficiency increases. Although the fuel consumption of the SI engine in a cycle drive is reduced by 18%, and the average effective efficiency is increased by 29%, the CPS provides required power.

References

1. Thin Quynh, N., Duc, L.H., Dunin, A.Y., Shatrov, M.G.: Engine performance, emissions characteristics of the compression ignition engine using HHO gas (Brown's gas). IOP Conf. Ser. Mater. Sci. Eng. **2021**, 1159 (2021)
2. Kurien, C., Mittal, M.: Review on the production and utilization of green ammonia as an alternate fuel in dual-fuel compression ignition engines. Energy Convers. Manag. **251**, 114990 (2022)
3. Pelé, R., Brequigny, P., Bellettre, J., Mounaïm-Rousselle, C.: Performances and pollutant emissions of spark ignition engine using direct injection for blends of ethanol/ammonia and pure ammonia. Int. J. Engine Res. (2023)
4. Meulenbelt, J.: Ammonia fuel. Medicine **40**(2), 94–95 (2012)
5. Legasov, V.A.: Hydrogen energy. Nature **3**, 3–17 (1977)
6. Sarafraz, M.M., Christo, M.M.: Sustainable three-stage chemical looping ammonia production (3CLAP) process. Energy Convers. Manag. **229**, 113735 (2021)

7. Feng, Y., et al.: Low-temperature auto-ignition characteristics of NH_3/diesel binary fuel: ignition delay time measurement and kinetic analysis. Fuel **281**, 118761 (2020)
8. Gridin, V.M.: Calculation of parameters and characteristics of asynchronous motors. In: [Electronic Resource]: Guidelines for Completing Homework in the Course "Electrical Engineering and Electronics", pp. 4–8 (2011)
9. Engine VAZ-2108, VAZ-21080. https://www.motorist.ru/. Accessed 15 Dec 2023
10. Engine VAZ-1111, VAZ-11113. https://www.motorist.ru/. Accessed 15 Dec 2023
11. Daud, A.M., Eyzat, P.: Four—octane—number method for predicting the anti knock behavior of fuels and engines. SAE Trans. (1978)

Analyzing Frontal Car Collisions Through Simulation Utilizing HyperWorks Software

Kien Trung Nguyen, Manh Duc Nguyen, Hieu Duc Le, Hieu Minh Pham, Linh Ngoc Pham, and Tuan Minh Pham

Abstract The car body frame can be considered as the primary protective shield for passenger safety during frontal collisions. This study focuses on the deformation and failure of the front car body frame when colliding with a rigid barrier. The accident scenario is simulated using explicit dynamic Finite Element software. The geometric design of the car body frame is created using HyperWork software and imported into the Finite Element software within HyperCrash, where a rigid block is designed as an absolute rigid barrier. The car body is simulated using two materials, elasto-plastic modeled with the Johnson–Cook Law 2. The study evaluates a collision at a speed of 56 km/h with a rigid barrier, examining the displacement, velocity, stress, deformation, and energy of the Kia Pride. The results show that at the point of impact, the velocity decreases gradually immediately after the collision. The reaction force of the rigid barrier at the impact area reaches 1.2E+07 kN, causing deformation of the car's front end. The internal energy of the car increases from 0 to approximately 1.75E+08, while the kinetic energy decreases from 2.25+08 to approximately 0. The crush behavior of the car body frame, energy dissipation, and vehicle deceleration from the crash simulation was observed.

Keywords HyperCrash simulation software · Collision analysis · CAE

K. T. Nguyen (✉) · M. D. Nguyen · H. D. Le · H. M. Pham
Faculty of Transportation Engineering, School of Mechanical and Automotive Engineering, Hanoi University of Industry, No. 298 Cau Dien Street, Bac Tu Liem District, Hanoi, Vietnam
e-mail: kiennt@haui.edu.vn

L. N. Pham
Faculty of Engineering Foundation, School of Mechanical and Automotive Engineering, Hanoi University of Industry, No. 298 Cau Dien Street, Bac Tu Liem District, Hanoi, Vietnam

T. M. Pham
Department of Vehicle Engineering, National Formosa University, 64, Wenhua Road, Huwei Township 63208, Yunlin County, Taiwan, R.O.C.

D. H. Tien et al. (eds.), *Proceedings of the 1st International Conference on Sustainability and Emerging Technologies for Smart Manufacturing*, Proceedings in Technology Transfer, https://doi.org/10.1007/978-981-97-7083-0_5

1 Introduction

Car accidents are occurring every day. Most drivers believe they can avoid such troublesome situations. However, it is crucial to consider the statistics, with 11,626 deaths and 15,292 injuries reported in 2023 [1].

These figures emphasize the need to enhance automobile safety during accidents. Car bumpers, positioned at the front and rear of vehicles, act as protective shields for the chassis, designed to withstand impacts without compromising safety. They safeguard critical components and safety equipment, minimizing physical damage during collisions [2].

During collisions with rigid objects, car bumpers may not effectively absorb impact alone, typically made of steel, aluminum, rubber, or plastic and mounted at both the front and rear of passenger vehicles. In low-speed crashes, they absorb shock to minimize vehicle damage, some using energy absorbers or brackets, while others incorporate foam cushioning. While designed to prevent or minimize physical damage during low-speed collisions, they are not intended as structural components to enhance crashworthiness or occupant protection [3].

In the field of automotive collision simulation, extensive research has been conducted worldwide, including studies and scientific papers. In Vietnam, Tín [4] utilized HyperWorks software for durability analysis and optimization of the Thaco City B60 bus frame design. Quang et al. [5] employed GrabCad software to simulate car collisions, revealing increased deformation zones and reduced passenger survival space at higher speeds, leading to complete vehicle shell damage.

Lopes et al. [6] analyzed collision errors by employing the ECE R29 standard and test monitoring techniques, such as accelerometers and Digital Image Correlation (DIC), to validate numerical models created with VPS/PAMCRASH® FEM software. Klyavin's research [7] simulated car collisions with lighting poles to assess head injury criteria (HIC) using LS-Dyna finite element analysis. This involved developing CAD and 3D FEM models of various lighting poles to conduct nonlinear simulations under different scenarios, allowing for the estimation of safety measures for potential accident scenarios. In Kuwait, Abdel-Nasser and colleagues [8] investigated the impact of vehicles colliding with lighting poles through finite element modeling simulations, focusing on factors such as acceleration, contact forces, and energy absorption in the vehicle's front region. Their study highlighted the effectiveness of using new materials in lighting pole construction to improve passenger safety by reducing vehicle speed and absorbing higher impact energy.

Based on research findings, traditional software like SolidWorks and ANSYS are commonly used for direct impact simulations, offering cost and time benefits. However, the potential of newer software such as HyperWorks in crash simulation remains underexplored. Recent studies have mainly focused on simulating collisions within vehicle frames, leaving uncertainties about its applicability to other aspects like mechanics and materials. Therefore, the authors decided to utilize HyperCrash software to focus on building models, simulating collisions, and evaluating results on a Kia Pride model, emphasizing the importance of understanding collision impacts on factors such as displacement, velocity, energy, and vehicle deformation to enhance transportation vehicle safety and performance.

2 Methodology

2.1 Research Methods

Finite Element Method (FEM)

The finite element method and numerical computation techniques are advanced methods used to solve physical and engineering problems, commonly employed in computer-based computations [9]. While initially proposed in the 1940s, the finite element method faced limited usage due to computational constraints. However, with the significant progress in computer technology, it has now become widely utilized across engineering disciplines, playing a vital role in durability and materials analysis and research [10].

In modern finite element analysis, problem-solving generally consists of two stages: pre-processing and post-processing. Pre-processing includes tasks like mesh generation and setting up boundary conditions for the model, while post-processing involves solution, analysis, and result manipulation.

Explicit Method-Collision Problem

The explicit method is a numerical technique utilized in computational analysis to approximate solutions of differential equations over time [11]. It is specifically employed for dynamic problems with short durations, like instantaneous collisions or high-speed events such as car crashes, free falls, and complex multiple connections involving destruction. The equilibrium equation for instantaneous dynamics is expressed as:

$$[M]\left\{\frac{\mathrm{d}^2u}{\mathrm{d}t^2}\right\} + [C]\left\{\frac{\mathrm{d}u}{\mathrm{d}t}\right\} + [K]\{u\} = \{F_{\mathrm{ext}}(t)\}$$

where:

M Mass matrix;
C Damping matrix;
K Stiffness matrix.

Basic quantities such as displacement, velocity, and acceleration serve as foundational metrics, derived from initial conditions over time. From these, key parameters like element stresses, plastic deformation, and interaction forces, alongside kinetic and potential energies, can be deduced. HyperWorks, a finite element platform, employs the explicit method for collision simulations conducted across various dimensions (1D/2D/3D).

Fig. 1 Kia Pride vehicle

Table 1 Dimensional specifications of the Kia Pride automobile

No.	Technical parameter	Value	Unit
1	Overall length [L]	3565	mm
2	Overall width [B]	1605	mm
3	Overall height [H]	1460	mm
4	Wheelbase length [$L0$]	2295	mm

2.2 *Research Objectives*

Through the research process, the authors have chosen the Kia Pride model with the following specifications (Fig. 1; Table 1).

Step 1: Model Pre-processing

To prepare the Kia Pride model for collision simulation in HyperWorks software, simplification steps are necessary [12–14]. These involve removing insignificant details using the Defeature tool, addressing common CAD errors when importing the 3D model. Following error correction, the entire vehicle model undergoes meshing, with 2D meshing applied to thin-sheet components and 3D meshing to parts with similar dimensions. Mesh quality is assessed using the element quality tool, with any errors corrected to ensure suitability for simulation (Fig. 2).

After meshing, comparing the mesh parameters with the standard quality mesh table in Fig. 3 reveals that with the applied mesh parameters, the problem is likely to converge easily.

Step 2: Implementing Boundary Conditions Using HyperCrash and Solving the Problem

From the meshed model in the HyperWorks software, it is necessary to perform a conversion of the model to the Radioss solver to export the model file with a .rad extension. This file can then be imported into HyperCrash to apply boundary conditions and solve the problem (Fig. 4).

After successfully exporting the Kia Pride model file in .rad format, proceed to import that model file into HyperCrash.

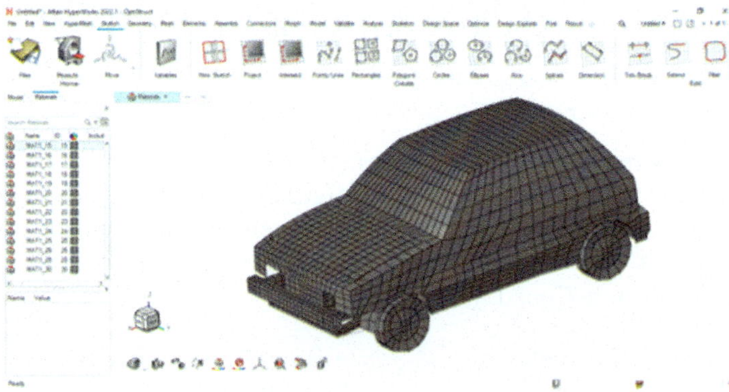

Fig. 2 Kia Pride model has been meshed in the HyperWorks software

Fig. 3 Mesh quality standard

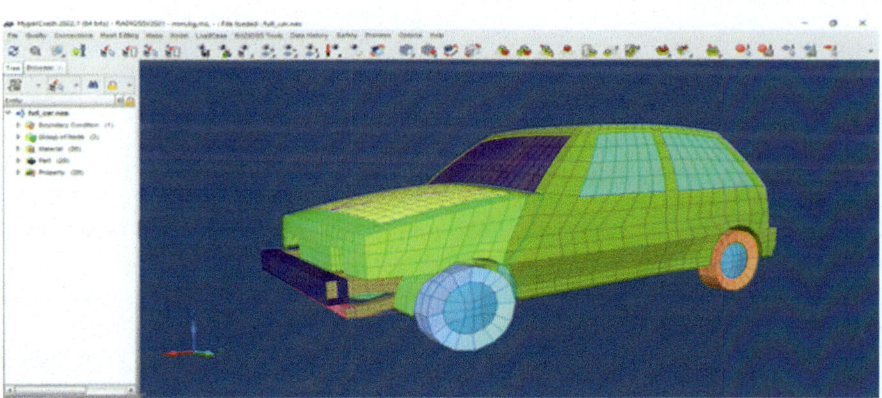

Fig. 4 The model has been imported into HyperCrash

Within the HyperCrash interface, apply boundary conditions by:

- Assigning materials such as glass for the windshield, steel for the vehicle's shell, and rubber for the tires.

The car body primarily utilizes high-strength steel (HSS), with strengths ranging from 210 to 550 MPa, falling under the categories of hard steel and high-strength steel. After reviewing practical applications of steel codes, the suitable material for the car body was selected and defined, belonging to the elasto-plastic group and following the Johnson–Cook Law 2 behavior.

• Elastic modulus: 210,000 MPa • Plastic strain at yielding: 425 MPa • Elastic strain to yield: 200 MPa	• Hardening parameter: 450 MPa • Poisson's ratio: 0.3

- Creation of a rigid wall column for collision with the vehicle, establishing a rigid road surface beneath the vehicle's tires.
- Establishing a contact interface throughout the entire Kia Pride model.
- Assigning an initial velocity of 56 km/h.
- Creating output cards to export simulation results accordingly.
- Executing the simulation task using the Altair compute console solver support system.

Step 3: Result Output and Evaluation in Automotive Engineering Simulation

The computed simulation results are presented in various forms, including images and charts.

3 Results and Discussion

- *Displacement of the Collision Model*

The maximum displacement is 707.650 mm at the node with ID 3004, while the minimum displacement is 0 mm at the node with ID 4761.

In regions of motion and high velocity, such as the wheel areas, it is evident that displacement is maximized (Fig. 5).

In the direct collision area with the rigid wall, despite experiencing the highest deformation, the displacement is zero due to the wall's stationary status (considered as rigid), resulting in zero velocity at that point (as depicted in the figure).

- *Von Mises Stress Distribution*

In this region, the material assigned is steel with the following parameters (Table 2).

Maximum Von Mises Stress = 424.989 MPa at element ID 4127.

Fig. 5 Displacement of the collision model

Table 2 Material parameters of the rigid wall

Value	Elastic modulus	Poisson's ratio	Density	Maximum stress
Unit	2.1e+5 (MPa)	0.3	7.9e−9 (kg/m^3)	425 (MPa)

From the visual representation of the simulation results, it is evident that the collision area experiencing the highest Von Mises stress is the front part of the vehicle colliding with the rigid wall (Fig. 6).

The collision's maximum stress, at 424.989 MPa, falls below the material's maximum stress (425 MPa), indicating that the material remains within its load-bearing limit. According to the Von Mises theory, plastic deformation occurs only

Fig. 6 Von Mises stress distribution

Fig. 7 Visualization of plastic deformation

when the stress surpasses the maximum load threshold, leading to irreversible deformation. Hence, since the Von Mises stress does not exceed the limit, there is no plastic deformation in any part of the vehicle due to the collision.

- *Plastic Deformation*

Based on the visual representation, the maximum plastic deformation point is observed to be 2.184 mm.

In the simulated scenario, the front of the car exceeds the material's plastic deformation limit upon collision. This indicates that the material at the collision point, designated as the steel of the car, is destroyed and unable to revert to its original shape after the impact (Fig. 7).

- *Velocity Distribution*

The visual display indicates that the highest velocity upon collision is 33,366.566 mm/s (approximately 120 km/h) at the impact point of the car's front. Immediately after the collision, the velocity at this point gradually decreases (Fig. 8).

- *Chart Illustrating the Collision Model's Energy*

During the collision, the car's internal energy peaks at around 1.75E+08, while its kinetic energy drops from 2.25E+08 to nearly 0, indicating a conversion to deformation energy and heat. Despite these changes, the total energy of the vehicle remains nearly constant at approximately 2.25E+08. When the total energy of the system remains constant (TE), it indicates energy conservation during the collision, signifying that the total energy before and after the collision remains unchanged. This highlights the collision model's effectiveness in describing real-life collision events (Fig. 9).

- *The Results Depicted in This Energy Chart Align with the Theoretical Expectations*

Fig. 8 Velocity distribution across the entire vehicle

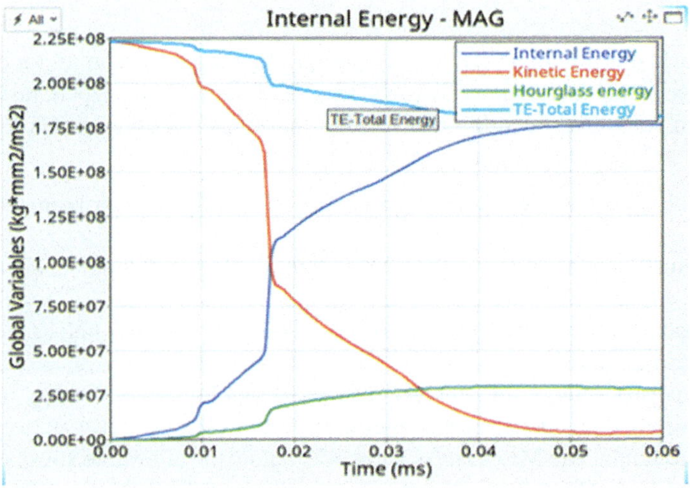

Fig. 9 Chart illustrating the collision model's energy

The graph illustrates that approximately 0.017–0.018 s after the collision, the reaction force from the rigid wall, or the force acting on the front of the vehicle at the point of impact, peaks at 1.2E+07 kN, leading to deformation of the vehicle's front. Subsequently, the reaction force gradually diminishes to nearly 0 kN post-collision, indicating the cessation of interaction between the rigid wall and the vehicle. Initially, there is a rapid momentum transfer upon collision, resulting in a sharp rise in the reaction force. As the collision progresses and the vehicle absorbs energy, the reaction force decreases until reaching equilibrium, signifying the conclusion of the interaction between the vehicle and the rigid wall (Fig. 10).

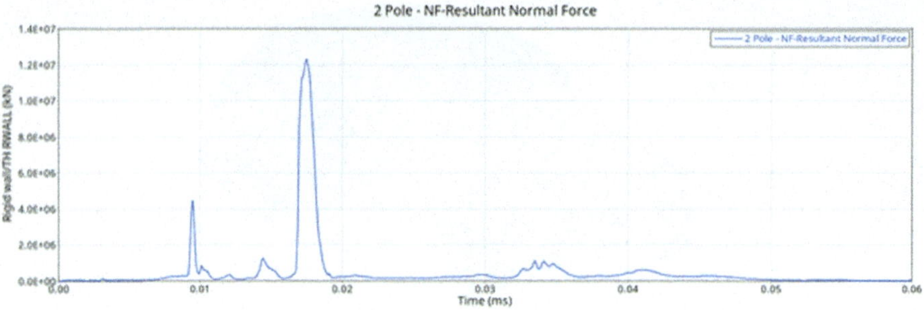

Fig. 10 Reaction force chart at the rigid wall

4 Conclusion

In summary, this research establishes a strong theoretical foundation for examining automotive frame factors in collision scenarios. Using Altair HyperCrash collision simulation software, the study investigated the impact between a Kia Pride and a rigid wall at 56 km/h, yielding significant insights:

- Post-collision, the velocity at the impact point decreased to approximately 120 km/h.
- Maximum displacement was 707.650 mm, with a minimum displacement of 0 mm.
- The peak reaction force from the rigid wall reached 1.2E+07 kN, causing frontal area deformation.
- The internal energy of the car increased while kinetic energy decreased, maintaining total energy nearly constant.
- The maximum Von Mises stress value was below the allowable stress of the steel material, indicating its suitability for vehicle assembly and providing direction for exploring alternative materials for automotive body construction.

References

1. BCA: Tình hình tai nạn giao thông, cháy, nổ và xử lý hành chính về trật tự, an toàn giao thông năm 2023 (2024). Accessed 01/02/2024. https://bocongan.gov.vn/tin-tuc/tinh-hinh-tai-nan-giao-thong-chay-no-va-xu-ly-hanh-chinh-ve-trat-tu-an-toan-giao-thong-nam-2023-t37769.html
2. Keni, L.G., Singh, V., Singh, N., Thyagi, A., Kalburgi, S., Chethan, K.N.: Conceptual design and analysis of a car bumper using finite element method. Cogent Eng. **8**(1), 1976480 (2021)
3. Joshi, N., Tiwari, R., Seshagiri Rao, G.V.R.: Design and analysis of front bumper for light passenger vehicles. Int. J. Rapid Res. Eng. Technol. Appl. Sci. **2**(4) (2016). ISSN (Online): 2455-4723
4. Tín, L.C.: Ung dung phan mem hyperworks toi uu hoa khung xe bus Thaco City B60, DH BK-DH DN (2019)

5. Quang, N.T., Anh, N.T., Thành, P.V.: Ungdung GrabCad mo phong va cham o to, HAUI. HN (2020)
6. Lopes, R., Ramos, N.V., Cunha, R., Maia, R., Rodrigues, R., Parente, M.P.L., Moreira, P.M.: Coach crashworthiness and failure analysis during a frontal impact. Eng. Fail. Anal. **151**, 107369 (2023)
7. Klyavin, O., Michailov, A., Borovkov, A.: Finite element modeling of the crash-tests for energy absorbing lighting columns. In: ENOC-2008, Saint Petersburg, Russia, June 2008
8. Abdel-Nasser, Y., Alrajhi, J., Alardhi, M., Alkhulaifi, K.: Frontal crash simulation of vehicles against lighting columns in Kuwait using FEM. Int. J. Traffic Transp. Eng. **2**(5), 101–105 (2013)
9. Thinh, T.I., Khoa, N.N.: Phuong phap phan tu huu han, NXBGDVN, chap. 4, pp. 31–39 (2007)
10. Duc, N.D., Mai, D.N.: Suc ben vat lieu, NXB KHKT, HN, chap. 2–3, pp. 18–43 (2011)
11. Altair Engineering: Explicit Analysis Radioss Ebook. Academic Program (2018)
12. Altair: Radioss Help. Altair HyperWorks, 1 (2022)
13. Altair: HyperMesh Help. Altair HyperWorks, 1 (2022)
14. Altair: HyperCrash Help. Altair HyperWorks, 2 (2022)

A Novel Approach of Mining High Utility Itemsets

Nong Thi Hoa

Abstract Mining High Utility Itemsets finds itemsets from transaction databases that are usually sold together and their profit overcomes a given threshold. Some popular items for everyday life appear in many transactions. Therefore, itemsets containing all popular items usually are High Utility Itemset (HUI). These HUIs is known for a long time. Therefore, algorithms of mining HUIs do not need to mine them. Previous studies still mine all HUIs. As a result, they consume both CPU time and memory. In this paper, we present a novel effective algorithm of mining HUIs overcomes this disadvantage. It is called **npHUIM**. We use compact lists to store items and itemsets. First, our algorithm finds popular items, unpopular items, unpromising items, and unpromising 2-itemsets after scanning the database. Next, it scans the database to mine HUIs. It only generates itemsets that contain zero or one popular item and not contain unpromising items or unpromising 2-itemsets. Then, it adds/updates these itemsets to lists. Then, npHUIM presents HUIs based on testing utility of generated itemsets. We use pruning strategies based on utility of transactions to remove both unpromising items and unpromising itemsets. Experiments are conducted on benchmark datasets. Our algorithm is compared with an equivalent algorithm. Experimental results show that the proposed algorithm is better than the compared algorithm. Moreover, npHUIM significantly reduces number of HUIs.

Keywords High utility itemset · Mining HUI · Knowledge discovery · Business intelligence · Data mining

N. T. Hoa (✉)
East Asia University of Technology, Hanoi, Vietnam
e-mail: hoantl@eaut.edu.vn

D. H. Tien et al. (eds.), *Proceedings of the 1st International Conference on Sustainability and Emerging Technologies for Smart Manufacturing*, Proceedings in Technology Transfer, https://doi.org/10.1007/978-981-97-7083-0_6

1 Introduction

Mining data from transaction databases in shopping provides useful knowledge for business. Mining HUI provides itemsets that are bought together and earn high profits. From that knowledge, shops recommend related items for customers. There are too many HUIs in databases. Therefore, decreasing number of presentation of HUIs is essential topics of mining HUIs.

Newest papers can be divided on two categories including decreasing number of representation of HUI, and the improving both storing itemsets and pruning the search space. In the first category, mining closed HUIs [1–4], maximal HUIs [4, 5], minimal HUIs [6] have developed to drop the number of representation of HUIs. In the second one, new structures help to generate larger itemsets more simple and prune the search space better [7–10]. Moreover, models and algorithms in artificial intelligence help to improve the searching of HUIs [11–13]. Transaction databases contain many popular items for everyday life. As a result, itemsets only containing popular items are HUIs. Knowledge about these HUIs is known. Therefore, it is not essential to generate itemsets only containing popular items as well as mine HUIs from these itemsets. Newest studies still mine these HUIs. In this paper, we present a novel algorithm of mining HUIs that avoids generating these itemsets. Our algorithm uses the list structure to store itemsets. Each list only stores k-itemsets containing k items. npHUIM consists of the following steps. First, we scans the database to find popular items, unpromising items, and unpromising itemsets. Next, we generate itemsets containing zero or one popular item from each transaction. Then, formulate their utility and update to utility lists. After scanning all transactions, the proposed algorithm exports HUIs based on the utility of itemsets. Experiments have been conducted on benchmark datasets including *mushroom, chess, connect*. npHUIM is compared with FHM+. Experimental results show that our algorithm is better than FHM+. Moreover, number of HUIs of npHUIM also sharply decreases.

The rest of this paper is organized as follows. In Sect. 2, we present our approach for mining HUIs. Section 3 shows and explains experiment results. Finally, conclusions are written in Sect. 4.

2 Our Approach

We provide some definitions that use in our algorithm.

High utility itemset: Given a threshold of utility (*minutil*), an itemset X is high utility itemset if utility of X is greater or equal to *minutil*.

Popular item: Give a minimal frequent threshold (*minsupp*), an item is a popular item if its frequency overcomes *minsupp*. Frequency of an itemset X is noted *supp*(X).

Transaction weight utility (twu): Give an itemset X, *twu*(X) is the sum of utilities of transactions containing X.

Transaction weighted downward closure property: If $twu(X) < minutil$, all supersets $Y \supset X$ are not high utility itemsets.

Our algorithm only generates itemsets containing zero or one popular items. Meaning, we avoid generating itemsets containing all popular items. Promising itemsets still contain one popular item because new items can be popular or an old popular item and some other items are sold together.

We use compact lists to store promising itemsets. Its structure includes sets of items, and utility of itemset. Each list only contains k-itemsets where $k = 2, \ldots, n$ which n is largest number of items in a transaction.

Our algorithm consists of following steps. First, we list popular items, unpopular items, unpromising items, and unpromising 2-itemsets. Next, our approach generates promising itemsets and update set of items and utility of these itemsets to lists. Promising itemsets contain zero or one popular item, and do not include an unpromising items or an unpromising 2-itemset. Then, it presents HUIs from promising itemsets if their utility overcomes *minutil*.

2.1 Classifying Items and Listing Unpromising 2-Itemsets

Our approach uses two thresholds including *minutil* (the minimal utility of a HUI) and *minsupp* (the minimal frequency of a popular item). We scan the database to obtain information of each item I and 2-itemsets including $twu(I)$, $supp(I)$, and $twu(X)$ where X is a 2-itemset. Classifying items, 2-itemsets is done by following rules.

- An item I is a popular item if $supp(I) \geq minsupp$. Otherwise, I is unpopular.
- An item I' is promising if $twu(I') \geq minutil$. Otherwise, I' is unpromising.
- A 2-itemset X is promising if $twu(X) \geq minutil$. Otherwise, X is unpromising.

We use transaction weighted downward closure property to write two last rules. Unpromising items are store in an array (1-L) and unpromising 2-itemsets are store in a list (2-L). They are used to generate new itemsets. Structure of 2-L consist of two items. PI is an array to store popular items of the database and UI is an array to store unpopular items of the database items.

2.2 Mining HUIs

Our algorithm rescans the database to mine HUIs. For each transaction T, we run following steps:

- Build PI(T) and UI(T) where PI(T) is a set of items that are in PI and T, and UI(T) is a set of items that are in UI and T. UI(T) also remove unpromising items in 1-L.

- Generate k-itemsets containing 0 or 1 item in PI(T) in which k is in $[2, r + 1]$ where r is the number of items in UI(T). An itemset Y is not generated from a transaction if (i) $Y \supseteq Z$ where $Z \in$ 2-L, or (ii) exist $I \in Y$ and $I \in$ 1-L, or (iii) each $I \in Y$ and $I \in$ PI(T).
- Add/update promising k-itemsets to lists.

After scanning the database, test itemset's utility to present HUIs.
We generate k-itemsets from PI(T) and UI(T) based on two below rules.

- k-itemsets are generated from items in UI(T).
- $(k - 1)$-itemsets are generated from items in UI(T). Then, combine a $(k - 1)$-itemsets with an item in PI to become k-itemsets.

After generating an itemset X, we compute its utility and update the utility list that contains X. If X is new, add X to the utility list. Otherwise, update utility of X.

2.3 *Discussion*

Our algorithm decreases number of promising itemsets as well as number of HUIs. Its advantages includes

- Avoid generating itemsets containing all popular items.
- Apply the pruning strategy based on utility of transactions.
- Store k-itemsets in a list to search generated itemsets more quickly.

Therefore, npHUIM decreases both CPU time and memory consumption. Experiments are conducted to prove the effectiveness of it.

3 Experiments

We conduct experiments on three benchmark datasets including *mushroom, chess, connect* from SPMF datasets [14]. Our algorithm is compared to FHM that is the best algorithm for mining HUIs [7]. FHM+ limits the number of items in each itemset during mining HUIs [10]. Number of items in each itemset of two algorithms is set to from 2 to 5 to decrease time for computing. Two algorithms are code in Java. FHM+ is downloaded from SPMF algorithms [14]. Results are obtained from a computer with Core i5×2 Processor, 8 GB RAM and Window Pro 10.

3.1 *Results of* Mushroom

Each 1000 transactions of *mushroom* are used for subtests. We set *minutil* = 100 for two algorithms. *minsupp* = 150 is the threshold to decide one item be popular or unpopular. Each result of a subtest is the mean of three tests. The CPU time is shown in Figs. 1 and 2 presents memory consumption.

Data from Fig. 1 show that time of our algorithm depends on semantics of data, while time of FHM+ is stable. For transactions from 1 to 1000, time of npHUI is **20** times lower than FHM+. Similarly, npHUIM is **7.4** times, **19.9** times, **9.8** times, **1.5** times lower at four subtests (1001–2000, 3001–4000, 5001–6000, and 4001–5000). For transactions (2001–3000 and 7001–8000), npHUIM is slightly higher than FHM+ (1.02 times and 1.08 times). Time of npHUIM on all subtests is **19,291 ms**, while FHM+ is **37,659 ms**. Meaning, the time of our algorithm is **1.95 times lower** than FHM+.

Figure 2 shows memory consumption of npHUIM depends on semantics of data, while one of FHM+ increases steadily. Our algorithm uses **0.49** MB for three subtests (1–1000, 3001–4000, 5001–6000), **3.94** MB for two subtests (2001–3000, 4001–5000), and **7.39–7.44** MB for three subtests (1001–2000, 6001–7000, 7001–8000). Memory consumptions of FHM+ increase from **160.8** to **188.4** MB from the first subtest to the last one. Memory of all subtests of npHUIM is **31.6** and **1398.1 MB** is memory of FHM+. Meaning, memory of the proposed algorithm is **44.3 times lower**

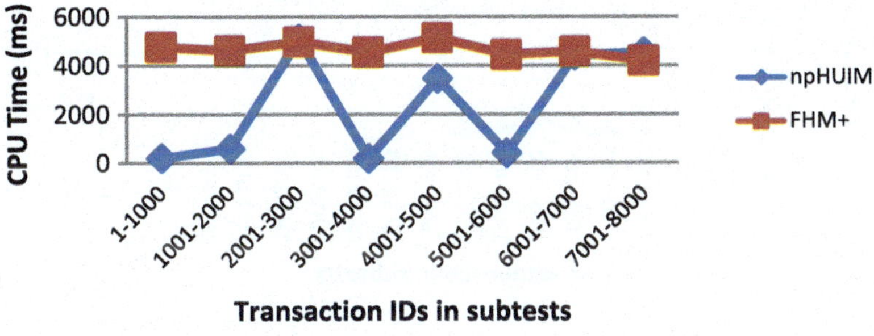

Fig. 1 CPU time of *mushroom*

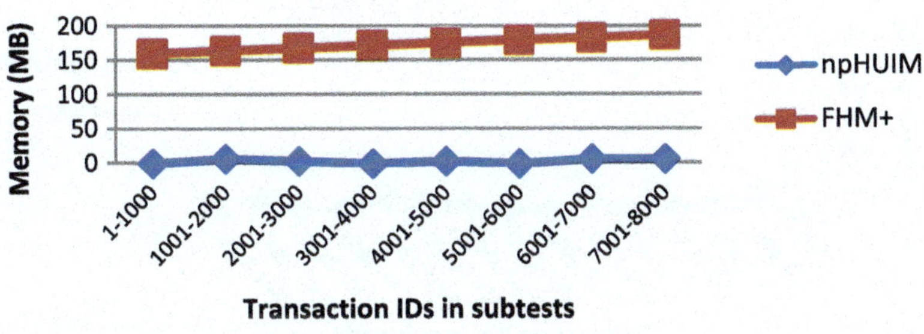

Fig. 2 Memory consumption of *mushroom*

than FHM+. Number of HUIs of pHUI is **256.8** times **lower** than FHM+. Meaning, our algorithm sharply drops the number of HUIs. Data from Figs. 1 and 2 shows that both time and memory of npHUIM are better than FHM+.

3.2 *Results of* Chess

Transactions of *chess* are divided into small datasets containing 200 transactions. *minutil* = 100 is set for two algorithms and *minsupp* = 95 is the threshold of frequency of items. Figures 3 and 4 present CPU time and memory consumption.

Figure 3 shows that time of npHUIM depends on semantics of data, while time of FHM+ is stable at 8000 ms. For transactions from 1 to 1600 (eight subtests), time of our algorithm is **10.3** times, **2.0** times, **9.1** times, **34.1** times, **12.7** times, **6.1** times, **15.5** times, **18.7** times lower than FHM+. Moreover, the time of npHUIM is **1.3** times, **1.6** times lower than FHM+ for two subtests (transactions in 2001–2200, 2601–2800). Time of FHM+ is **1.1** times lower than npHUIM in four subtests (1601–1800, 1801–2000, 2201–2400, 2401–2600). Time of our algorithm on all subtests is **66,908 ms**, while FHM+ is **125,365 ms**. Meaning, the time of npHUIM is **1.87 times lower** than FHM+.

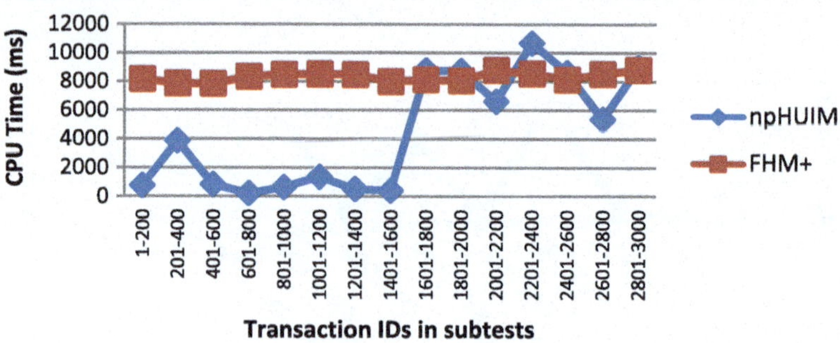

Fig. 3 CPU time of *chess*

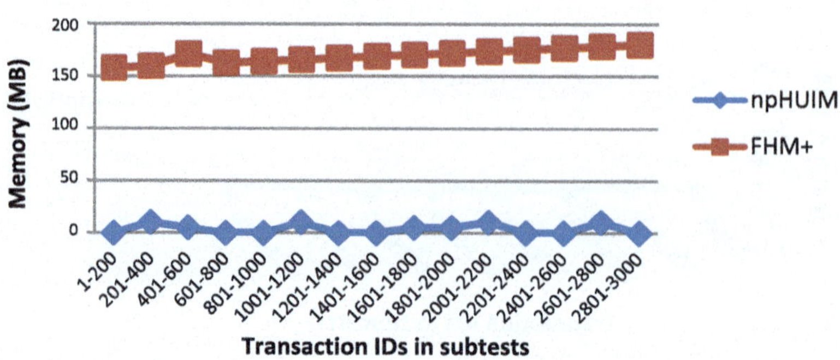

Fig. 4 Memory of *chess*

Data of Fig. 4 show memory of the proposed algorithm depends on semantics of data, while one of FHM+ increases steadily. npHUIM uses **0.488 MB** for eight subtests (1–200, 601–800, 801–1000, 1201–1400, 1401–1600, 2201–2400, 2401–2600, 2801–3000), **5.69–5.71 MB** for three subtests (401–600, 1601–1800, 1801–2000), and **10.91–10.94 MB** for four subtests (201–400, 1001–1200, 2001–2200, 2601–2800). Memory of FHM+ increases from **158.6** to **181.3 MB** from the first subtest to the last one. Memory of all subtests of npHUIM is **64.7** and **2559.2 MB** is memory of FHM+. Meaning, memory of npHUIM is **39.5 times lower** than FHM+. Number of HUIs of npHUIM is **2090.6 times lower** than FHM+. Data from Figs. 3 and 4 shows that both time and memory of our algorithm are lower than FHM+. Moreover, npHUIM significantly decreases the number of HUIs.

3.3 Results of Connect

We extract 3000 last transactions of *connect* for mining HUIs. Similarly, these transactions are divided into small datasets containing 100 transactions. We set *minutil* = 100 for two algorithms and *minsupp* = 50 is the threshold of frequency of items. The CPU time and memory consumption are shown in Figs. 5 and 6.

Similarly, Fig. 5 shows that the time of our algorithm depends on semantics of data, while time of FHM+ is stable. In 29 subtests, the time of npHUIM is lower than FHM+ except one subtest. For transactions from 1901 to 2000, time of npHUIM is 1.4 times higher than FHM+. Time of npHUIM on all subtests is **113,460.5 ms**, while FHM+ is **249,394.5 ms**. Meaning, the time of our algorithm is **2.19 times lower** than FHM+. Similarly, data of Fig. 6 show memory of npHUI depends on semantics of data, while one of FHM+ increases steadily. Memory consumptions of FHM+ increase from **183.2** to **214.6 MB** from the first subtest to the last one. Memory of all subtests of npHUIM is **149.9** and **5923.8 MB** is memory of FHM+. Meaning, memory of pHUI is **39.5 times lower** than FHM+. Number of HUIs of pHUI **is 4630.7 times lower** than FHM+. Figures 5 and 6 show that both time and memory of our algorithm is better than FHM+.

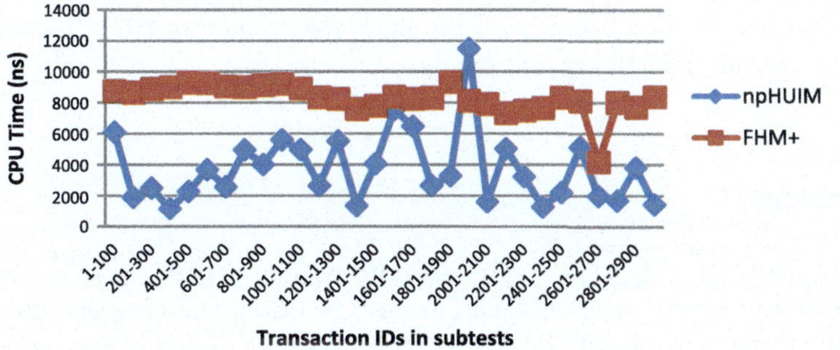

Fig. 5 CPU time of *connect*

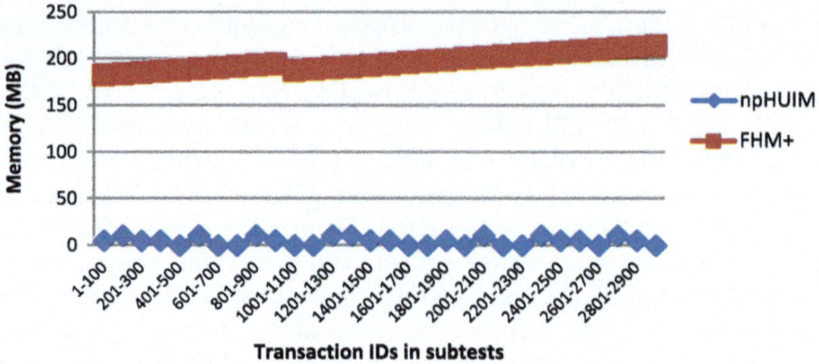

Fig. 6 Memory of *connect*

Data from Figs. 1, 2, 3, 4, 5 and 6 show that our algorithm is better than FHM+. Time of npHUIM is approximately **2 times lower** than FHM+. Memory of npHUIM is approximately from **39.5** times to 44.3 times **lower**.

4 Conclusions

In this paper, we propose a novel effective algorithm for mining HUIs to avoid generating itemsets containing all popular items. The list structure is used to store information. Three lists store popular items, unpopular items, and unpromising items. Many lists store promising itemsets that each list only contains k-itemsets to reduce time for searching generated itemsets. Our algorithm consists of the following steps. First, scan the database to list popular items, unpopular items, unpromising items, and unpromising 2-itemsets. Next, explore each transaction to generate promising itemsets and update their information in lists. Then, test the utility of itemsets to present HUIs. Experiments are conducted on benchmark datasets including *mushroom, chess, connect*. Our algorithm is compared to FHM+. Results of experiments show that both time and memory of the proposed algorithm is lower than FHM+. Moreover, memory consumption and the number of HUIs of npHUIM sharply decrease.

We will investigate to design a new algorithm to mining HUIs appearing in a period on datasets containing time labels.

References

1. Fournier-Viger, P., Zida, S., Lin, J.C.W., Wu, C.W., Tseng, V.S.: EFIM closed: fast and memory efficient discovery of closed high-utility itemsets. In: Machine Learning and Data Mining in Pattern Recognition, pp. 199–213. Springer, Cham (2016)

2. Wu, C.W., Fournier-Viger, P., Gu, J.Y., Tseng, V.S.: Mining closed + high utility itemsets without candidate generation. In: Proceedings of Conference on Technologies and Applications of Artificial Intelligence, pp. 187–194 (2015)
3. Dam, T.L., Ramampiaro, H., Nørvåg, K., Duong, Q.H.: Towards efficiently mining closed high utility itemsets from incremental databases. Knowl.-Based Syst. **165**, 13–29 (2019)
4. Tseng, V.S., Wu, C.W., Fournier-Viger, P., Yu, P.S.: Efficient algorithms for mining the concise and lossless representation of high utility itemsets. IEEE Trans. Knowl. Data Eng. **27**(3), 726–739 (2015)
5. Wu, C.W., Fournier-Viger, P., Gu, J.Y., Tseng, V.S.: Mining compact high utility itemsets without candidate generation. In: High-Utility Pattern Mining: Theory, Algorithms and Applications, pp. 279–302. Springer, Cham (2019)
6. Fournier-Viger, P., Lin, J.C.W., Wu, C.W., Tseng, V.S., Faghihi, U.: Mining minimal high-utility itemsets. In: Database and Expert Systems Applications, pp. 88–101. Springer, Cham (2016)
7. Fournier-Viger, P., Wu, C.W., Zida, S., Tseng, V.S.: FHM: faster high-utility itemset mining using estimated utility co-occurrence pruning. In: Foundations of Intelligent Systems, pp. 83–92. Springer, Cham (2014)
8. Duong, Q.H., Fournier-Viger, P., Ramampiaro, H., Norvag, K., Dam, T.L.: Efficient high utility itemset mining using buffered utility-lists. Appl. Intell. **48**(7), 1859–1877 (2018)
9. Liu, Y., Wang, L., Feng, L., Jin, B.: Mining high utility itemsets based on pattern growth without candidate generation. Mathematics **35**(9) (2021)
10. Fournier-Viger, P., Lin, J.C.W., Duong, Q.H., Dam, T.L.: FHM+: faster high-utility itemset mining using length upper-bound reduction. In: Trends in Applied Knowledge-Based Systems and Data Science. Lecture Notes in Computer Science, vol. 9799, pp. 115–127. Springer (2016)
11. Nawaz, M.S., Fournier-Viger, P., Yun, U., Wu, Y., Song, W.: Mining high utility itemsets with hill climbing and simulated annealing. ACM Trans. Manag. Inf. Syst. **13**(1), 1–22 (2021)
12. Song, W., Li, J., Huang, C.: Artificial fish swarm algorithm for mining high utility itemsets. In: Advances in Swarm Intelligence, pp. 407–419. Springer, Cham (2021)
13. Song, W., Li, J.: Discovering high utility itemsets using set-based particle swarm optimization. In: Advanced Data Mining and Applications, pp. 38–53. Springer, Cham (2020)
14. Fournier-Viger, P., Gomariz, A., Soltani, A., Lam, H., Gueniche, T.: SPMF: opensource data mining platform (2014). http://www.philippe-fournier-viger.com/spmf

Optimizing Power Consumption and Moisture Content of Avocado Slices Using a Heat Pump Dryer

Thi Minh Hue Pham, Khac Trung Chu, Ngoc Thai Huynh, Ut Hien Thi Nguyen, Ngoc Chien Vu, Ho Nguyen, and Quoc Manh Nguyen

Abstract In this work, the power consumption and ensure the moisture content of dried avocado slices were minimized by response surface method based on drying technology of heat pump. Experiments were conducted to determine the ability consume electrical energy when reaching the moisture content of dried avocado slices. The obtained data was also used to construct an objective function describing the influence of temperature and drying speed of the drying agent on the drying process. The results of the analysis of variance also confirmed the influence of drying conditions on the power consumption and the moisture content of dried avocado slices. Optimal results were obtained the power consumption and the moisture content of dried avocado slices was 4.8 kW and 6.2%, respectively, when drying with temperature and drying speed of 46.03 °C and 1.15 m/s, respectively.

Keywords Experimental drying of avocado slices · Power consumption · The moisture content of dried avocado slices · Heat pump drying technology · Surface response method · Analysis of variance

T. M. H. Pham · K. T. Chu
Hanoi University of Industry, Hanoi, Vietnam

N. T. Huynh · U. H. T. Nguyen (✉)
Ho Chi Minh City University of Industry and Trade, Ho Chi Minh City, Vietnam
e-mail: hienntu@huit.edu.vn

N. C. Vu
Nha Trang University, Nha Trang, Vietnam

H. Nguyen
Nguyen Tat Thanh University, Ho Chi Minh City, Vietnam

Q. M. Nguyen
Hung Yen University of Technology and Education, Khoai Chau, Hung Yen, Vietnam

D. H. Tien et al. (eds.), *Proceedings of the 1st International Conference on Sustainability and Emerging Technologies for Smart Manufacturing*, Proceedings in Technology Transfer, https://doi.org/10.1007/978-981-97-7083-0_7

1 Introduction

Avocado possess high economic and nutritional value. Fresh avocados are mainly used for smoothies or fruit jams to provide nutrients for the body and dryed avocados are mainly used in the field of food and cosmetics. Dryed avocados are easily preserved, have a long shelf life and are transported to many places in the country. Dried avocados are often exported to China and Thailand. Recent studies on dried avocado slices have focused on factors such as drying speed, color, fat content, moisture level, and other characteristics. These aspects vary due to different drying processes and the thickness of the material. The study on vacuum-dryed avocado [1] examined and compared samples dried at 60 °C in an oven to those dried under a vacuum. As a result, samples were badly shriveled and discolored. Vacuum-dried avocado has low moisture content. Compared with the oven-dried samples, the vacuum-dried lyophilized powder has a higher fat content. Through the above study, dried avocados are dried at 60 °C, the color is darkened and the fat content is reduced, the hot drying method is not suitable. Vacuum drying has good color, high fat, low moisture, but this method requires high technology and consumes a lot of money. Husen et al. [2] investigated the enhancement of polyphenol content and antioxidant capacity in superheated steam-dried avocado. The results of this drying process are compared with the sublimation drying process. Although freeze-drying is a well-known method to produce high-quality food products, the results show that all antioxidant activities and polyphenol content are enhanced when steam-dried overheated avocados. However, color and some nutrients in this study were not mentioned by Temu [3] studied the effect of temperature and avocado slice thickness on the drying speed of avocado and the amount of oil in the dried avocado at 50, 60, and 70 °C, as well as the avocado slice thickness of 2 and 5 mm. The drying rate increases with temperature but decreases with the slice thickness. Based on the result, the size of the avocado slice (2 mm), the faster the drying process. However, color and some nutrients are not mentioned. The quality of dried avocado; depends mainly on the drying method. The outstanding advantage of the heat pump drying method is that it limits the discoloration of the product. Chua et al. [4] reported that when drying with a heat pump, the author could adjust the drying time and temperature to limit the discoloration of the product due to the effect of non-enzymatic coloring reactions. Through experimental studies, the intensity of discoloration was reduced by 87% when drying potatoes, 75% when drying guava, and 67% when drying bananas compared with drying with hot air. Due to the reduction in drying time, the ascorbic acid content was increased by 20% when drying guava with a heat pump compared to traditional drying with hot air at the same drying temperature. Ingvald Strommen and Kramer [5] reported that a heat pump drying system could reduce energy by about 60–80% compared to another traditional drying system operating at the same temperature range. Chou et al. [6], a heat pump drying system, significantly saves energy, especially when utilizing different waste heat sources. The heat pump dryer [7] was utilized to dry herbs; the experiment indicated that the energy consumption and drying time were deficient. The dryer productivity depended on the surface load, drying air velocity, and drying air temperature. The heat pump dryer and an infrared-assisted heat pump dryer were combined [8] to determine dryer efficacy, estimate dryer energy and efficiency, and examine the drying kinetics of grated carrots. Apple, banana, grape, jujube, kiwifruit, and pineapple were dried using solar, infrared, vacuum, and ultrasound-intensified heat pump dryers [9]; the heat pump drier produced high-quality dried fruits, and the drying conditions could be adjusted.

The above studies show that drying sliced avocados with a heat pump dryer is a direction that is very consistent with current reality. They suggest one of the methods to reduce energy costs for the drying process and improve product quality. However, the drying process at different modes affects the drying time, energy consumption, labor, nutritional composition, color, and product taste. In this work, we determined the optimal drying conditions of sliced avocados to achieve the minimum energy consumption and product moisture while maintaining the color, taste, and nutritional composition.

2 Material and Method

2.1 Material and Equipment for Experiment

The Booth avocados were utilized in this investigation originated from the Americas, grown in Vietnam, choose the ripe ones and then slice them with a thickness of 2 mm. Experimental equipment is a dryer made using heat pump drying technology with the principle of operation: Fan (10) sucks air and moisture of the drying material in the drying chamber and blows into the indoor unit (8), current humid air through the indoor unit will condense into water, and the water is taken out through the duct. The air flow continues to enter the outdoor unit (9) and through the resistor (11) to heat up to the required temperature. The air stream with high temperature and low will enter the drying chamber to receive moisture from the drying material thanks to the fan pushing to the indoor unit to separate the moisture out. Thus, the drying agent is 100% recirculated and the moisture in the drying material is condensed at the indoor unit and taken out through the duct, which will make the avocados gradually reduce moisture and dry (Fig. 1).

The heat pump dryer is designed with dimensions of 600 × 800 × 1350 mm, a total of 6 drying trays, each tray has dimensions of 500 × 700 × 20 mm. The machine uses 1 blower with capacity of 0.5HP, 1 compressor with capacity: 2HP, indoor unit capacity: 1.5HP, main condenser capacity: 1HP, auxiliary condenser capacity: 1.5HP, capacity auxiliary resistor: 2 kW. The machine uses voltage 220 V/50 Hz, drying agent temperature is adjusted from 35 to 50 °C, drying agent speed: 0.3–2.0 m/s.

Temperature controller code Eliweli-China brand, Korean FOX-1H humidity, measuring range 20–90% RH, resolution of ± 1%, input sensor HS-220, voltage used 230 V AC, 50/60 Hz to measure humidity drying agent.

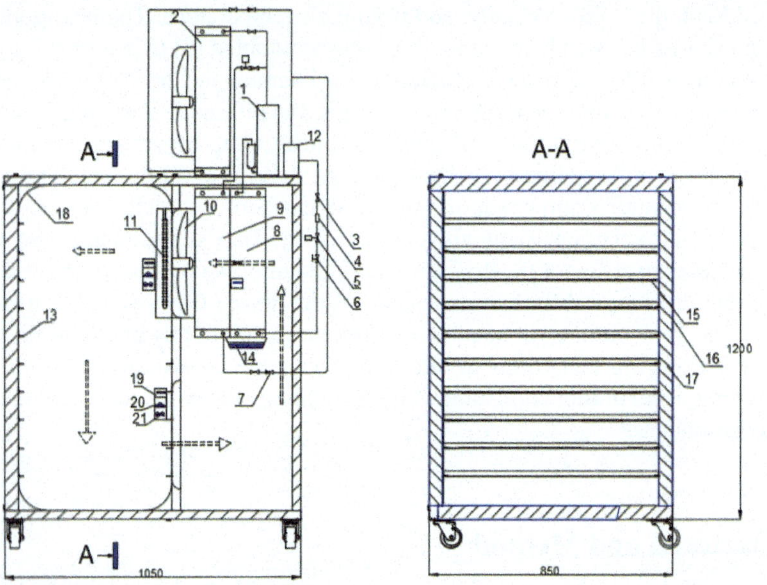

1. Heating machine
2. Auxiliary condenser
3. Block valve
4. Filters,
5.Electromagnetic valve
6. Throttling valve
7. One-way valve
8. Cooler,
9. Heated main
10. Axial fan
11. Resistor
12. High-pressure storage tank
13. Drying tray rack
14. Condensate trough
15. Drying material storage tray
16. Drying chamber wall
17. Insulation
18. Path direction
19. Temperature sensor
20. Humidity sensor
21. Wind speed sensor

Fig. 1 Layout diagram of heat pump dryer equipment

The device for measuring the drying speed of the UKAS brand (Taiwan) has a measuring range of from 0.01 to 50 m/s, a resolution of 0.01 m/s to measure the drying speed. Taiwan's HZ TP-A-5000 g electronic balance has a resolution of $\pm$ 1 g, measuring limit of 5000 g to weigh the product after drying. Electronic caliper from Japan's Mitutoyo for measuring drying material thickness, measuring range from 0 to 150 mm, error $\pm$ 0.01 mm. Power consumption is determined by an electronic Watt meter with a scale from 10 to 99,999 kWh, with an error of $\pm$ 0.01 kWh to determine the power consumption for each drying batch (Fig. 2).

2.2 *Analysis of Variance and Analysis of Regression*

This method was performed by Minitab software. The obtained results were listed in tables and equations showing that the relationship between output parameters depending on the design variables. The output parameter is the electrical energy

Fig. 2 Experimental dryer

consumed and the moisture content is significantly affected by the drying temperature and drying speed. The results were presented in Sect. 3.

3 Results and Discussion

3.1 Setup Experiment

The drying temperature (*T*) and velocity (*V*) were two design variables in this investigation. 13 experimental cases were design as an orthogonal array by Minitab. The design variables and their level of the design variables were listed in Table 1. In this table, presented *T* variable with five level 38.964, 40, 42.5, 38.964, and 45 °C, *V* variable has five level as following 1.14645, 1.254, 1.5, 1.75, and 1.85355 m/s.

The results of 13 experiments were presented in Table 2. In this table indicated that when changed *T* and *V*, consumed electrical power (*W*) and moisture content of product (%) (*E*) were also changed. That problem proved that the design variables have affected significantly on two outputs.

Table 1 Design variables and their levels

Design variables	Symbol	Unit	Level 1	Level 2	Level 3	Level 4	Level 5
Temperature	*T*	°C	38.9645	40	42.5	45	46.0355
Velocity	*V*	m/s	1.14645	1.254	1.5	1.75	1.85355

Table 2 Orthogonal array and results of experiment

Experiment order	Design variables		Output	
	T (°C)	V (m/s)	Consumed electrical power (kWh)	Moisture content(%)
1	42.5	1.5	9.30	10.50
2	40	1.25	8.13	14.70
3	42.5	1.5	8.76	9.50
4	40	1.75	9.12	8.10
5	45	1.25	5.57	6.10
6	42.5	1.14645	6.78	6.50
7	40	1.25	8.56	14.60
8	46.0355	1.5	6.78	6.50
9	45	1.75	9.97	11.11
10	42.5	1.5	9.27	10.05
11	42.5	1.5	9.30	10.14
12	42.5	1.14645	8.13	14.70
13	38.9645	1.5	8.76	9.50

3.2 *Analysis of Variance*

The effects of two design variables on the consumed electrical power were also confirmed by analysis of variance as presented in Table 3. In this table, the values of F-value are greater than 2 and the values of P-value are less than 0.05. This problem pointed out that two design variables were important and have significantly affected on the consumed electrical power. Therefore, in investigation about drying of sliced avocados using a heat pump dryer can't ignore two design variables. The results of analysis of variance were also obtained R-square of 90.37%, R-square (adj) of 85.56%, R-square (pred) of 62.57% as listed in Table 4.

Besides, the effects of two design variables on the moisture content were also confirmed by analysis of variance as presented in Table 5. In this table, the values of F-value are greater than 2 and the values of P-value are less than 0.05. This problem pointed out that two design variables were important and have significantly affected on the moisture content. Therefore, in investigation about drying of sliced avocados using a heat pump dryer can't ignore two design variables. The results of analysis of variance for moisture content were also obtained R-square of 94.7%, R-square (adj) of 90.92%, R-square (pred) of 67.37% as listed in Table 6.

Table 3 Results of analysis of variance *W* versus *T*, *V*

Source	DF	Adj SS	Adj MS	*F*-value	*P*-value
Model	4	17.4892	4.37230	18.77	0.000
Linear	2	7.9786	3.98930	17.13	0.001
T	1	1.3446	1.34462	5.77	0.043
V	1	6.6340	6.63399	28.48	0.001
Square	1	6.6036	6.60358	28.35	0.001
T * *T*	1	6.6036	6.60358	28.35	0.001
2-way interaction	1	2.9070	2.90703	12.48	0.008
T * *V*	1	2.9070	2.90703	12.48	0.008
Error	8	1.8633	0.23291		
Lack-of-fit	4	1.5469	0.38672	4.89	0.077
Pure error	4	0.3164	0.07910		
Total	12	19.3525			

Table 4 Model summary of variance *W* versus *T*,*V*

S	*R*-sq	*R*-sq (adj)	*R*-sq (pred)
0.482608	90.37%	85.56%	62.57%

Table 5 Results analysis of variance *E* versus *T*, *V*

Source	DF	Adj SS	Adj MS	*F*-value	*P*-value
Model	5	76.9575	15.3915	25.02	0.000
Linear	2	28.5726	14.2863	23.22	0.001
T	1	20.9431	20.9431	34.05	0.001
V	1	7.6295	7.6295	12.40	0.010
Square	2	14.6869	7.3434	11.94	0.006
T * *T*	1	4.5459	4.5459	7.390	0.030
V * *V*	1	8.2897	8.2897	13.48	0.008
2-way interaction	1	33.6980	33.6980	54.78	0.000
T * *V*	1	33.6980	33.6980	54.78	0.000
Error	7	4.3060	0.6151		
Lack-of-fit	3	3.5665	1.1888	6.43	0.052
Pure error	4	0.7395	0.1849		
Total	12	81.2634			

Table 6 Model summary of variance *E* versus *T*,*V*

S	*R*-sq	*R*-sq (adj)	*R*-sq (pred)
0.784308	94.70%	90.92%	67.37%

3.3 Results Regression Analysis

The results of analysis of regression were also confirmed that two design variables have affected on the consumed electrical power and were described by Eq. (1). In this equation, the consumed electrical power depends on the drying temperature and drying velocity. According to Eq. (1), when the drying temperature increased, the consumed electrical power increased. That problem coincided with the investigation [1, 2, 7–9].

$$W = -181.5 + 10.93T - 543V - 0.1546T * T + 1.364T * V \quad (1)$$

The Pareto chart identified the influence of the design variable on the consumed electrical power as presented in Fig. 3. According to this figure, the drying velocity has significantly on the consumed electrical power, next to the value of the square of the temperature, the interaction of temperature and velocity and finally is temperature.

The results of analysis of regression were also confirmed that two design variables have affected on the moisture content and were described by Eq. (2). In this equation, the moisture content depends on the drying temperature and drying velocity. According to Eq. (2), when the drying velocity increased, the moisture content increased. That problem coincided with the investigation [1, 2, 7–9].

$$E = 145.1 + 3.38T - 253.7V - 0.1293T * T + 17.47V * V + 4.644T * V \quad (2)$$

The Pareto chart identified the influence of the design variable on the moisture content as presented in Fig. 4. According to this figure, the drying velocity has significantly on the consumed electrical power, next to the value of the square of the temperature, the interaction of temperature and velocity and finally is temperature.

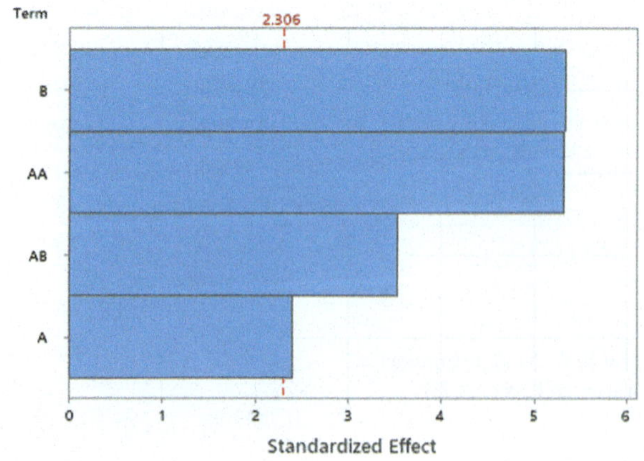

Fig. 3 Pareto chart of the standardized effects of on the consumed electrical power

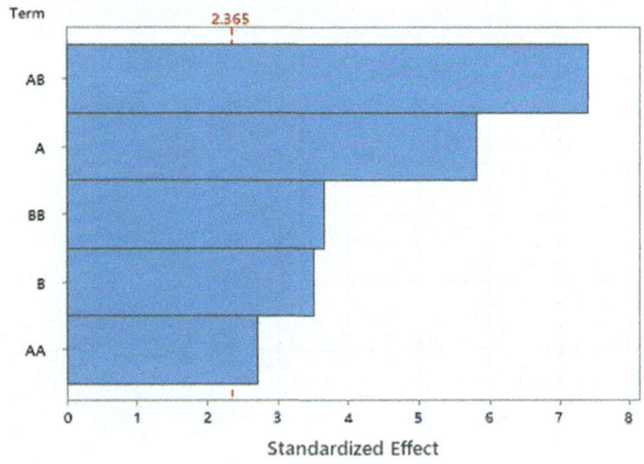

Fig. 4 Pareto chart of the standardized effects of on the moisture content

Table 7 Parameters

Response	Goal	Lower	Target	Upper	Weight	Importance
E	Minimum		6.10	14.70	1	1
W	Minimum		5.57	9.97	1	1

Table 8 Solution

Solution	*T*	*V*	*E* fit	*W* fit	Composite desirability
1	46.0355	1.14645	3.92239	3.71065	1

3.4 Optimal Results

In order to reduce the cost of drying avocado by heat pumps, it is necessary to reduce the influence of the two variable on the consumed electrical power and moisture content. This problem implies that the consumed electrical power and moisture content are "the smaller is the better". The target of *W* and *E* required interval from 5.57 to 9.97 and 6.1 to 14.7, respectively, by analysis in Minitab software as presented in Table 7. The optimal results of *W* and *E* obtained 3.71065 and 3.92239 at the values of temperature and velocity are 46.0355 °C and 1.14645 m/s, respectively, as presented in Table 8.

3.5 Confirm Optimal Results

After the optimal case was selected with the optimal combination parameters of the design variables design at the optimal level, the experiment was performed with the design variables at the optimal level. The results obtained the optimal drying

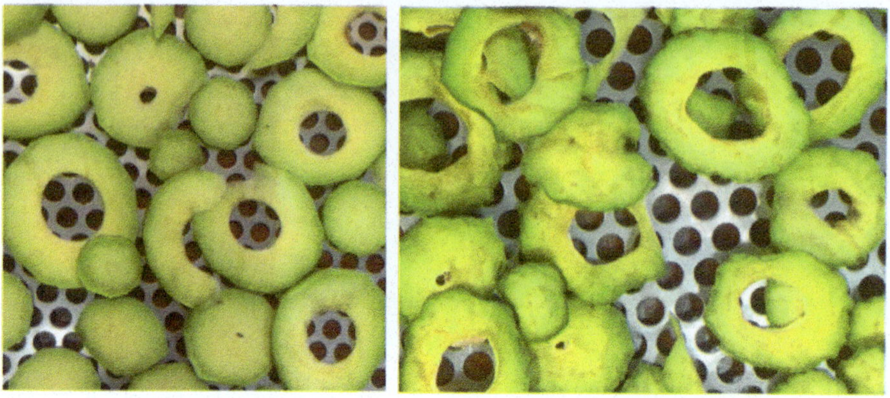

a) Before drying b) After drying

Fig. 5 Avocado slices before and after drying at 46 °C and velocity of 1.15 m/s

temperature obtained of 46.03 °C (adjustable temperature ranges from 45 to 47 °C) and the optimal drying speed of 1.146 m/s (adjustable speed ranges from 1.1 to 1.2 m/s). Result of consumption electricity power and moisture content was 4.8 kWh with ± 20% and 6.2 with ± 30%, respectively. Characteristic aroma, beautiful bright color, natural, vitamin C content: 35.7 mg/100 g (Fig. 5).

4 Conclusions

In this article, we shown that optimal parameter for drying sliced avocado by using a heat pump dryer is drying temperature of 46.03 °C, drying speed of 1.146 m/s, the power consumption is 4.8 kWh, and the moisture content is 6.2%. The products were estimated with characteristics aroma, bright and natural color, vitamin C content: 35.7 mg/100 g. The error between the optimal results of the regression analysis method of power consumption and moisture content is 6.2%.

References

1. Mujaffar, S., Dipnarine, T.-A.: The production of a dried avocado (*Persea americana*) powder, pp. 44–54 (2020)
2. Husen, R., Andou, Y., Ismail, A., Shirai, Y., Hassan, M.A.: Enhanced polyphenol content and antioxidant capacity in the edible portion of avocado dried with superheated-steam. Int. J. Adv. Res. Sci. Eng. **2**, 241–248 (2014)
3. Temu, A.: Effect of temperature and slice size on avocado pulp drying rate and oil yield. Tanz. J. Eng. Technol. **34**, 116–124 (2013)

4. Chua, K.J., Mujumdar, A.S., Chou, S.K., Hawlader, M.N.A., Ho, J.C.: Convective drying of banana, guava and potato pieces: effect of cyclical variations of air temperature on drying kinetics and color change. Dry. Technol. **18**, 907–936 (2000)
5. Ingvald Strommen, I., Kramer, K.: New applications of heat pumps in drying processes. Dry. Technol. **12**, 889–901 (1994)
6. Chua, K.J., Chou, S.K., Yang, W.M.: Advances in heat pump systems: a review. Appl. Energy **87**, 3611–3624 (2010)
7. Fatouh, M., Metwally, M.N., Helali, A.B., Shedid, M.H.: Herbs drying using a heat pump dryer. Energy Convers. Manag. **47**, 2629–2643 (2006)
8. Aktaş, M., Khanlari, A., Amini, A., Şevik, S.: Performance analysis of heat pump and infrared–heat pump drying of grated carrot using energy-exergy methodology. Energy Convers. Manag. **132**, 327–338 (2017)
9. Salehi, F.: Recent applications of heat pump dryer for drying of fruit crops: a review. Int. J. Fruit Sci. **21**, 546–555 (2021)

A Study Using Advanced Simulation Techniques and Statistical Analysis for Enhanced Precision of Cylinder Cup in Sheet Metal Deep Drawing Processes

Tran Dinh Van, Nguyen Anh Khoa, Nguyen Duc Toan, and Hoang Minh Vu

Abstract Utilizing ABAQUS simulations and finite element analysis, this research aims to refine sheet metal deep drawing by reducing costs and improving product quality. The study incorporates the TAGUCHI method for iteration and defect prediction, while ANOVA evaluates the effect of process parameters on outcomes. Emphasis is placed on the mechanical behavior of SUS304 stainless steel in simulations to optimize tool design and process parameters. The Forming Limit Diagram and statistical analysis via Taguchi and ANOVA highlight the importance of parameters like die fillet radius in preventing defects. Ultimately, the research achieves a systematic enhancement of the deep drawing process, resulting in high-precision, low-error manufacturing.

Keywords FEM · TAGUCHI · ANOVA · SUS304 stainless steel

1 Introduction

Sheet metal forming is a crucial manufacturing process utilized across automotive, aerospace, construction, and appliance industries, involving bending, deep drawing, and shearing techniques to shape metal sheets into various geometries using specialized tools like punches and dies [1]. In deep drawing, controlling factors such as cost, time-to-market, and quality are essential for competitiveness, with the initial design phase being a significant cost determinant. The adoption of Computer-Aided Engineering (CAE) and Finite Element Analysis (FEA) during early design stages

T. D. Van · N. A. Khoa · N. D. Toan (✉)
School of Mechanical Engineering, Ha Noi University of Science and Technology, No 1 Dai Co Viet, 100000 Hanoi, Vietnam
e-mail: toan.nguyenduc@hust.edu.vn

T. D. Van · H. M. Vu
East Asia University of Technology, Trinh Van Bo, Xuan Phuong, Nam Tu Liem, 129630 Hanoi, Vietnam

D. H. Tien et al. (eds.), *Proceedings of the 1st International Conference on Sustainability and Emerging Technologies for Smart Manufacturing*, Proceedings in Technology Transfer, https://doi.org/10.1007/978-981-97-7083-0_8

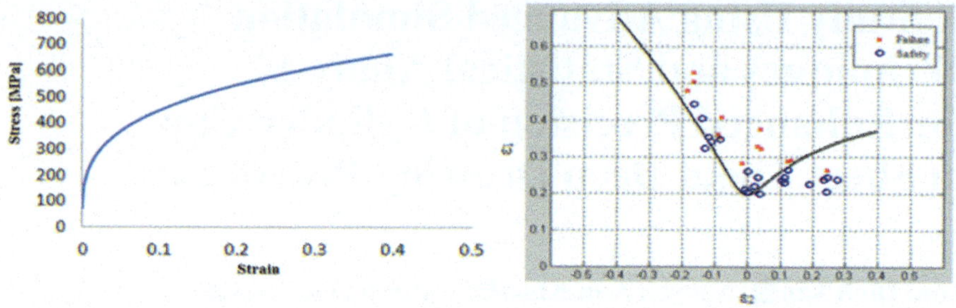

Fig. 1 Stress-strain curve and Forming Limit Curve (FLC) diagrams for SUS304 material

is vital for "getting it right the first time," allowing for the optimization of the manufacturing process and defect detection [2, 3]. CAE leverages software like LS-Dyna and AutoForm for virtual simulations [4, 5], with studies by Zhu et al. [6] and Lee et al. [7] demonstrating the use of CAE in addressing forming issues and predicting deformation. This paper aims to predict and minimize failure in deep drawing cylindrical cups through a combination of FEA and the Taguchi algorithm, highlighting the Forming Limit Curve (FLC) as a key predictive tool for assessing the influence of technological and geometrical parameters on product quality [8, 9].

2 Simulation of the Sheet Metal Deep Drawing Process

This section focuses on the simulation of the sheet metal deep drawing process, specifically targeting the stamping process employed to shape sheet metal components into a cylindrical cup form, utilizing SUS304 stainless steel material. The simulation aims to facilitate an in-depth analysis of the mechanical characteristics exhibited by SUS304 during the stamping operation. Stringent precision in controlling technical parameters is essential to ensure the final product conforms to specified shape and dimensions while preserving material integrity. The stress–strain curve of SUS304 material (Fig. 1) is utilized during simulation, illustrating the relationship between stress and strain under load conditions.

2.1 Stress–Strain Analysis

Figure 1 showcases the stress–strain curve diagram for SUS304 material, a fundamental tool in understanding the material's response to load conditions during the deep drawing process.

2.2 *Forming Limit Diagram (FLD) and Deformation Analysis*

In the evaluation of potential issues, such as wrinkling, tearing, and additional deformations within the sheet metal deep drawing process, a pivotal assessment involves comparing simulation results against the Forming Limit Diagram (FLD) (Fig. 1). This graphical representation delineates the deformation limits that sheet metal can sustain without succumbing to tearing. In the simulation framework, sheet deformation is approached as a plane strain problem, with both major ($\varepsilon 1$) and minor ($\varepsilon 2$) deformations meticulously considered.

The Forming Limit Curve (FLC) for SUS304 material is established through the implementation of the standard dome test. This critical step involves conducting precise measurements of both major and minor strains at locations characterized by both cracked and uncracked conditions [10]. Subsequently, these meticulously acquired output figures are meticulously charted on a bi-axial graph. The demarcation between the safe points and the cracking points on this graph precisely represents the Forming Limit Diagram (FLD) for the material under investigation.

2.3 *ABAQUS Simulation Setup*

This research leverages the robust capabilities of ABAQUS software [11] to simulate the stamping procedure for a round-bottomed component crafted from SUS304 stainless steel sheet. The detailed model of the die, punch, and blank holder, is visually presented in Fig. 2. This simulation configuration adheres to the principle of a fixed forming die, while the blank holder and punch are granted vertical freedom of movement. These include the 1 mm blank thickness and the coefficient of friction between contact surfaces, meticulously set at 0.12. The blank thickness directly impacts the overall formability of the material, while the coefficient of friction plays a pivotal role in governing the material flow dynamics during the intricate stages of the forming operation.

These analyses are conducted based on the specified input parameters and guided by the Forming Limit Curve (FLC) specific to SUS304 material. The outcomes of this simulation endeavor hold the key to optimizing both tool design and process parameters. This optimization, in turn, translates into an enhanced quality of the formed part and a fortified defense against defects that may compromise the integrity of the final product. Armed with the insights gained from the ABAQUS simulation, the subsequent steps involve a meticulous optimization of tool design and process parameters. This optimization endeavor is driven by a strategic aim to not only improve the overall quality of the formed part but also to proactively prevent potential defects. The ABAQUS simulation acts as a virtual testing ground, offering valuable

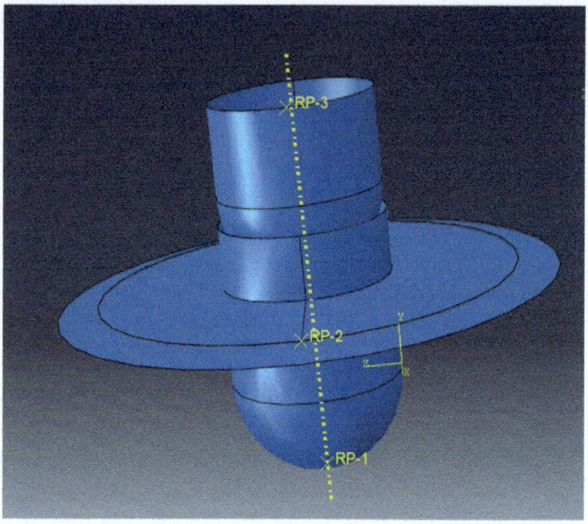

Fig. 2 3D model of cylinder cup in the ABAQUS environment

data-driven guidance for refining the intricate details of the stamping process. By iteratively adjusting parameters based on simulation outcomes, the research aims to achieve a robust and optimized stamping operation for the production of high-quality sheet metal components.

2.4 *Initial Sheet Metal Dimensions and Formability*

The choice of a blank size assumes paramount significance in influencing the formability and structural integrity of the final product during deep drawing operations. Informed by an analytical approach, the selected blank size with a predetermined radius of $r = 60$ mm. The Forming Limit Diagram (FLD) emerges as a pivotal predictive tool harnessed to simulate and anticipate critical phenomena inherent to the stamping process. Figure 4 serves as a visual representation of the FLD in action, offering insights into the consequences of forming conditions on the risk of failures, such as wrinkling and tearing. Figure 3 not only illustrates the predictive power of the FLD but also accentuates the repercussions of inaccurate selection of forming conditions. Parameters such as restraining force, die radius, the gap between the blank holder and the die, and geometric intricacies of the product wield substantial influence. Failures, indicated by an FLDCRT value surpassing 1, unveil the potential risk of product tearing, underscoring the critical role of precision in forming condition selection. Accurate determination of initial sheet metal dimensions, coupled with insights from the FLD, provides a strategic vantage point for navigating the complexities of the stamping process. This informed decision-making process lays the foundation for a robust and optimized deep drawing operation, where each parameter is meticulously tailored to foster enhanced formability and minimize the likelihood of defects.

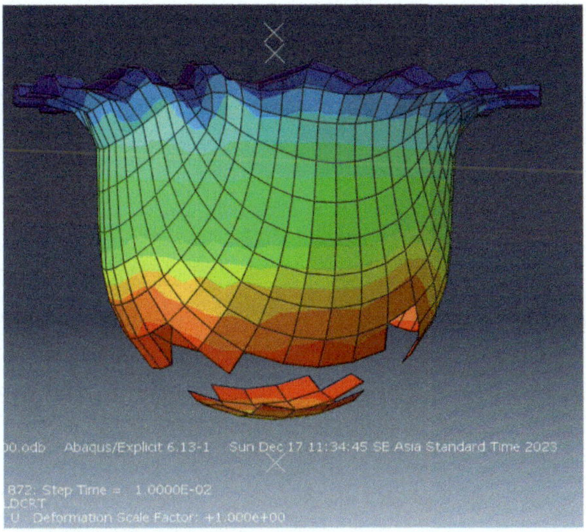

Fig. 3 Detail of tearing observed during simulation

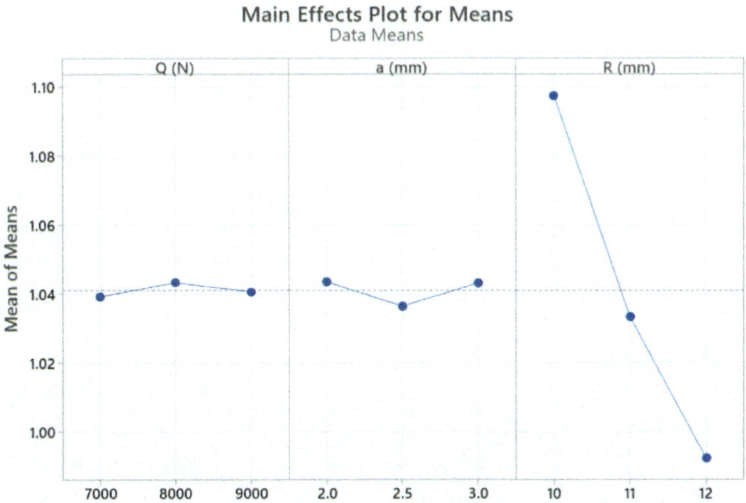

Fig. 4 The impact of parameters *Q*, *a*, *R* on the deep drawing process

2.5 FEM Simulation and Taguchi Method

Critical parameters, such as blank holder force (Q), clearance between the blank holder and the die (a), and die radius (R), are systematically varied in FEM simulation (Table 1). The Taguchi orthogonal array algorithm reduces the number of experiments, with Table 2 displaying the outcomes. The analysis of variance (ANOVA) (Fig. 4 and Table 3) evaluates the impact of these parameters on the Forming Limit Diagram for Critical Tearing (FLDCRT) value.

In the course of the simulation procedure, each parameter underwent selection at three distinct levels. Subsequently, an L9 orthogonal array was employed to formulate the experimental statistical methodology. A minimum of 9 experiments becomes

Table 1 Parameters and their level in FEM simulation

Parameters	Level		
	1	2	3
Clamping force—Q (N)	7000	8000	9000
Clearance between the blank holder and the die (mm)	2.5	2.5	3.0
Radius of the die (mm)	10	11	12

Table 2 Taguchi L9 orthogonal array table

No.	Parameter			FLDCRT	
	Q (N)	a (mm)	R (mm)	Value	S/N
1	7000	2	10	1.097	– 0.8041326
2	7000	2.5	11	1.024	– 0.2059991
3	7000	3	12	0.997	0.03045394
4	8000	2	10	1.042	– 0.3573544
5	8000	2.5	11	0.989	0.09519596
6	8000	3	12	1.099	– 0.8199538
7	9000	2	10	0.992	0.07326964
8	9000	2.5	11	1.096	– 0.7962111
9	9000	3	12	1.034	– 0.2904108

Table 3 ANOVA calculation

Parameter	Average value η			SS	Distribution
	1	2	3		
Q (N)	– 0.3266*	– 0.3607	– 0.3378	0.00182	0.002
a (mm)	– 0.3627	– 0.3023*	– 0.3600	0.00698	0.0006
R (mm)	– 0.8068	– 0.2846	0.0662*	1.15807	0.992

*Optimum level

imperative to comprehensively scrutinize the impacts stemming from variations in parameters during simulation. The integration of the Taguchi orthogonal array algorithm serves as a strategic approach to streamline the experimental design process. By harnessing this algorithm, the number of essential experimental configurations is significantly reduced, transitioning from the initial 27 (calculated as 3^3 to account for three parameters at three levels each) to a more manageable set of 9 experiments. This reduction not only optimizes resource utilization but also maintains the integrity of the experimental exploration.

2.6 Analysis of Variance (ANOVA) Examination

The Taguchi experimental design model was employed to scrutinize the influence of key parameters—blank holder force (Q), clearance between the blank holder and the punch (a), and the die radius (R)—on the Forming Limit Diagram Critical (FLDCRT) value. The analytical findings, illustrated in Fig. 5, reveal that restraining force and clearance, ranging from 7000 to 8000 N and 2.0 to 3.0 mm, respectively, exhibit negligible variations in FLDCRT, consistently remaining above 1 and only experiencing marginal fluctuations between 1.03 and 1.05. In contrast, the fillet radius of the die (R) exerts a conspicuous impact. The progressive increase of R from 12 to 14 mm consistently reduces FLDCRT, indicating an inverse relationship between R and FLDCRT. To pinpoint the optimal parameter set, the Analysis of Variance (ANOVA), detailed in Table 3, was employed to elucidate the relationships between parameters and the observed FLDCRT values. The method assesses the significance of each parameter on the outcome and its influence on tearing deformation. The ANOVA results, computed using a summation formula (SS) (Eq. 1), are indicative of the degree of impact of each parameter on FLDCRT.

$$\mathrm{SS} = 3\left(m_{j1} - m\right)^2 + 3\left(m_{j2} - m\right)^2 + \left(m_{j3} - m\right)^2 \tag{1}$$

where: $m_{ij} = \frac{1}{3}\sum_{n}^{3}(n_i)_j$; $m = \frac{1}{9}\sum_{n=1}^{9} n_i$.

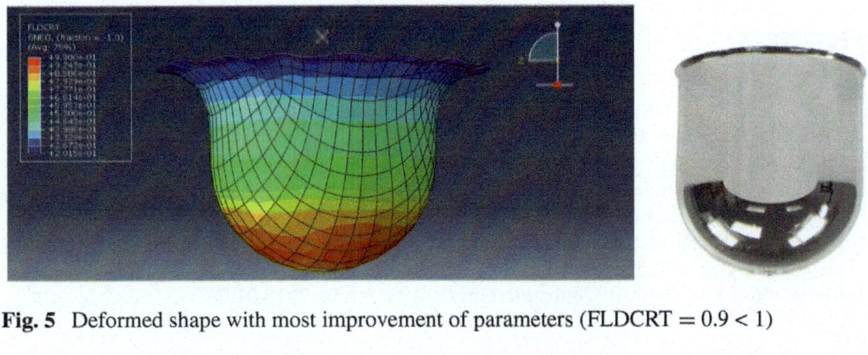

Fig. 5 Deformed shape with most improvement of parameters (FLDCRT = 0.9 < 1)

The ANOVA outcomes emphasize the imperative role of precise control over the die radius (*R*) to avert product tearing. The parameter set featuring a restraining force of 7000 N, a clearance between the blank holder and the die of 2.5 mm, and a die fillet radius of 12 mm emerges as the most optimal, producing a superior product without tearing incidents (Fig. 5). The analysis underscores that blank holder force (*Q*) and clearance between the blank holder and the punch (*a*) contribute minimally to FLDCRT variation (0.2% and 0.6%, respectively), emphasizing their limited impact on the deep drawing process. Conversely, the die radius (*R*) substantially dominates the variance distribution (99.6%), underscoring its critical role in shaping the outcome of the deep drawing process. This comprehensive variance analysis underscores the significance of meticulous parameter control, particularly the die fillet radius, to mitigate production incidents and uphold the final product's quality.

3 Conclusion

The amalgamation of ANOVA analysis, the Taguchi experimental model, and finite element simulation unveils substantial opportunities for refining the metal sheet deep drawing process by meticulously adjusting process parameters and geometric characteristics. Statistical scrutiny distinctly elucidates the influence of blank holder force (*Q*), clearance between the blank holder and the die (*a*), and notably, the die radius (*R*) on the outcomes of the forming process. The efficacy of finite element simulation emerges as a robust predictive tool, enabling precise anticipation of phenomena in the production process and, consequently, economizing time and costs associated with physical testing. The heightened reliability of simulation outcomes establishes a robust basis for discerning optimal process conditions, with a paramount goal of enhancing product quality and augmenting production efficiency. The integration of statistical methodologies and simulation showcases a systematic advancement in the forming process, culminating in the fabrication of components characterized by elevated precision and a diminished error rate throughout the manufacturing continuum.

References

1. Groover, M.P.: Part II engineering materials. In: Fundamentals of Modern Manufacturing: Materials, Processes, and Systems, pp. 98–132 (2010)
2. Fu, M.W., Yong, M.S., Tong, K.K., Muramatsu, T.: A methodology for evaluation of metal forming system design and performance via CAE simulation. Int. J. Prod. Res. **44**(6), 1075–1092 (2006). https://doi.org/10.1080/00207540500337643
3. Kwon, H.J., Kwon, H.K.: Computer aided engineering (CAE) simulation for the design optimization of gate system on high pressure die casting (HPDC) process. Robot. Comput. Integr. Manuf. **55**, 147–153 (2019). https://doi.org/10.1016/j.rcim.2018.01.003

4. Hattalli, V.L., Srivatsa, S.R.: Sheet metal forming processes—recent technological advances. Mater. Today Proc. **5**(1), 2564–2574 (2018). https://doi.org/10.1016/j.matpr.2017.11.040
5. Şen, N., Baykal, Y.: Development of car wishbone using sheet metal tearing process via the theory of inventive problem-solving (TRIZ) method. J. Braz. Soc. Mech. Sci. Eng. **41**(10) (2019). https://doi.org/10.1007/s40430-019-1884-7
6. Zhu, Q., Huang, S., Wang, D., Li, J., Hua, F., Yang, P.: Numerical simulation and experimental study of warm hydro-forming of magnesium alloy sheet. J. Manuf. Process. **80**, 43–53 (2022). https://doi.org/10.1016/j.jmapro.2022.05.054
7. Lee, H.R., Lee, M.G., Park, N.: Effect of evolutionary anisotropic hardening on the prediction of deformation and forming load in incremental sheet forming simulation. Thin-Walled Struct. **193**, 111231 (2023). https://doi.org/10.1016/j.tws.2023.111231
8. Duc-Toan, N., Young-Suk, K.: Combined numerical and experimental study to predict the forming limit curve of boron steel sheets at elevated temperatures. Proc. Inst. Mech. Eng. Part B J. Eng. Manuf. **234**(1–2), 189–203 (2020). https://doi.org/10.1177/0954405419843783
9. Nguyen, D.T., Kim, Y.S., Jung, D.W.: Finite element method study to predict spring-back in roll-bending of pre-coated material and select bending parameters. Int. J. Precis. Eng. Manuf. **13**(8), 1425–1432 (2012). https://doi.org/10.1007/s12541-012-0187-z
10. Hecker, S.S.: Simple technique for determining forming limit curves. Sheet Metal Ind. **52**, 671–676 (1975)
11. Hibbit, D., Karlsson, B., Sorensen, P.: ABAQUS user's manual, version 6.13.1. Providence, RI: AMS (2015)

Modeling and Simulation of Automotive Dynamics with the Effect of Lateral Forces

Nguyen Xuan Tuan and Bui Van Hai

Abstract In the process of moving, automobiles are influenced by many factors that destabilize the trajectory of motion, including mining factors and traffic environmental conditions. When analyzing the stability of the trajectory of movement of a car should be considered Automotive is a dynamical system with input effects (such as steering angle, horizontal wind, …) and stability of output parameters. The paper builds a horizontal dynamics model with the action of lateral forces on automobile and applies Matlab/Simulink software to simulate and evaluate results. The study of the factors affecting the lateral forces on the dynamics of automobile motion helps to stabilize the trajectory of the car to design and bring automatic control systems to improve the stability and direction control of the car to meet the current trend of driving automation.

Keywords Orbital motion · Lateral automotive dynamics model · Matlab/Simulink · Dynamics simulation model

1 Introduction

When analyzing the stability of an automobile's trajectory, we consider the automobile as a dynamical system with input effects (such as steering angle, transverse wind force, velocity, friction, etc.) and the stability of output parameters [1]. The state of motion in different conditions related to the driving speed, the horizontal force of the wind can make the car destabilize the direction of movement such as: deviating from the lane, overturning the car. In order to investigate these states of motion, it is necessary to study the dynamics of horizontal movement of vehicles [2]. Studies related to horizontal automobile dynamics include: stability control research [3, 4]; Calculate the transverse velocity and transverse force of the tire [5]; One-track transverse dynamics for automobiles [6]. The factors affecting the dynamics of the car

N. X. Tuan (✉) · B. V. Hai
Hanoi University of Industry, Hanoi, Vietnam
e-mail: tuannx@haui.edu.vn

D. H. Tien et al. (eds.), *Proceedings of the 1st International Conference on Sustainability and Emerging Technologies for Smart Manufacturing*, Proceedings in Technology Transfer, https://doi.org/10.1007/978-981-97-7083-0_9

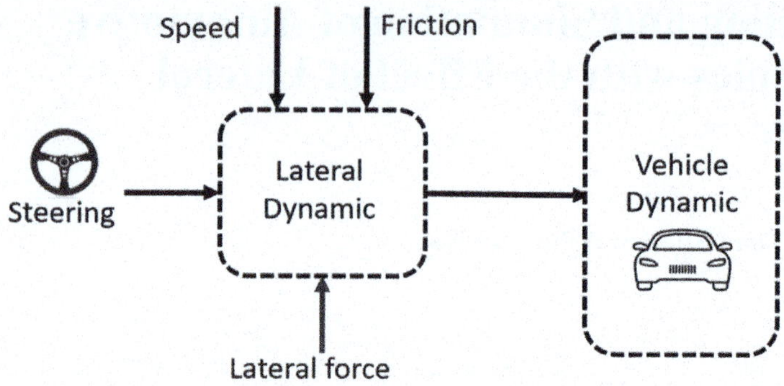

Fig. 1 The influence of external factors on automotive dynamics

horizontally can be mentioned as: lateral forces (horizontal wind force, road surface tilt, car movement speed, friction forces in the system) when the car moves [7, 8]. The content of the paper refers to the influence of the lateral force component on the transverse dynamics of the automobile in two positions of force action: Lateral forces placed in front of G (center of gravity) and placed behind G of the automobile. The influence of external factors on horizontal dynamics in particular and automotive dynamics in general (see Fig. 1).

2 Dynamics Models and Simulation Models

2.1 Dynamics Model

Study of dynamics models in the horizontal plane of automobile (road plane) with the assumption that the wheels have an elasticity coefficient [6, 9] (see Fig. 2).

Applying Newton's Law 2 and Dalambe's principle, writing the equation of differential motion of automobile. When measuring the system of differential equations written in terms of x_v, y_v, and body rotation angle β.

$$\ddot{x}_v = \dot{y}_v\dot{\beta} + \frac{1}{m_v}\left[(F_{x2} + F_{x1})\cos\delta_f - \left(F_{y4} + F_{y3}\right)\sin\delta_f + F_{x4} + F_{x3}\right] \quad (1)$$

$$\ddot{y}_v = -\dot{x}_v\dot{\beta} + \frac{1}{m_v}\left[(F_{x2} + F_{x1})\sin\delta_f + \left(F_{y4} + F_{y3}\right)\cos\delta_f + F_{y4} + F_{y3} + F_L\right] \quad (2)$$

$$\ddot{\beta} = \frac{1}{I_z}\begin{bmatrix} L_f(F_{x2} + F_{x1})\sin\delta_f + L_f\left(F_{y2} + F_{y1}\right)\cos\delta_f - L_f\left(F_{y4} + F_{y3}\right) \\ -\frac{S_b}{2}(F_{x2} - F_{x1})\cos\delta_f + \frac{S_b}{2}\left(F_{y2} - F_{y1}\right)\sin\delta_f - \frac{S_b}{2}(F_{x4} - F_{x3}) + F_L \cdot L_w \end{bmatrix} \quad (3)$$

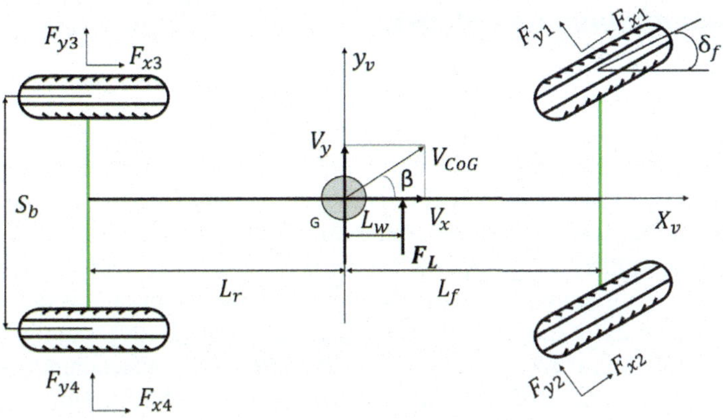

Fig. 2 Transverse dynamics models of automobile with lateral forces F_L

We have: m_v, I_z: Mass and moment inertia of the automobile; δ_f: front-wheel rotation angle; F_{x1}, F_{y1}, F_{x3}, F_{y3}: force applied in *x*, *y* directions on the left wheels; F_{x2}, F_{y2}, F_{x4}, F_{y4}: the force acting in the *x*, *y* directions on the right wheels; F_L: lateral force; L_f, L_r: distance from *G* to the front and rear-wheels; L_w: the distance of the force setpoint to *G*; β: the angle of rotation; y_v, x_v: displacement of the vehicle in the direction of *x*, *y*.

With the assumption that V_x = const, the differential equation is rewritten as follows:

$$\ddot{y}_v = -\frac{1}{m_v v_x}(C_f + C_r)\dot{y}_v - \frac{1}{m_v v_x}(L_f C_f + L_r C_r)\dot{\beta} - v_x\dot{\beta} + \frac{C_f}{m_v}\delta_f \quad (4)$$

$$\ddot{\beta} = -\frac{1}{I_z v_x}(L_f^2 C_f + L_r^2 C_r)\dot{\beta} - \frac{1}{I_z v_x}(L_f C_f - L_r C_r)\dot{y}_v + \frac{L_f C_f}{m_v}\delta_f \quad (5)$$

The transverse force acting on elastic tires is calculated (6):

$$F_y = C_f\alpha_f = C_f\left(\delta_f - \frac{\dot{y}_v + L_f\dot{\beta}}{v_x}\right) \quad (6)$$

where:

V_x vehicle velocity;
C_f, C_r wheel stiffness.

Dynamics Eqs. (4) and (5) are used to simulate the effects of F_L lateral forces on automotive dynamics.

2.2 Build a Simulation Model

The simulation model is built with the following assumptions: Automobile in linear motion, constant velocity of motion $V_x = 60$ km/h; The simulation was performed for two scenarios: the point where the force is placed on the front relative to the center of gravity of the car (CASE 1), and the point where the force is placed on the rear relative to the center of gravity of the car (CASE 2).

The model input is car speed, lateral force, and steering angle equal to 0. The model output is horizontal acceleration, lateral displacement, and rotation angle at the center of gravity of the vehicle. Block diagram of the model (see Fig. 3).

Simulate the lateral force acting in two different force placement positions on the dynamics of the car [10]. Simulation model with Matlab/Simulink software (see Fig. 4).

Parameters referenced and used to simulate Table 1 [9].

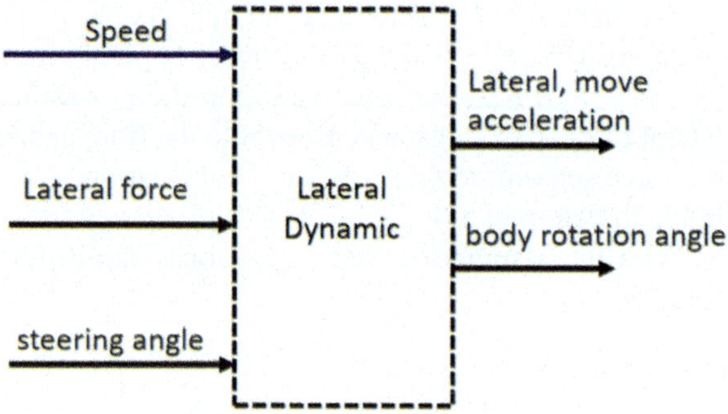

Fig. 3 Diagram of automobile dynamics relations in the horizontal direction

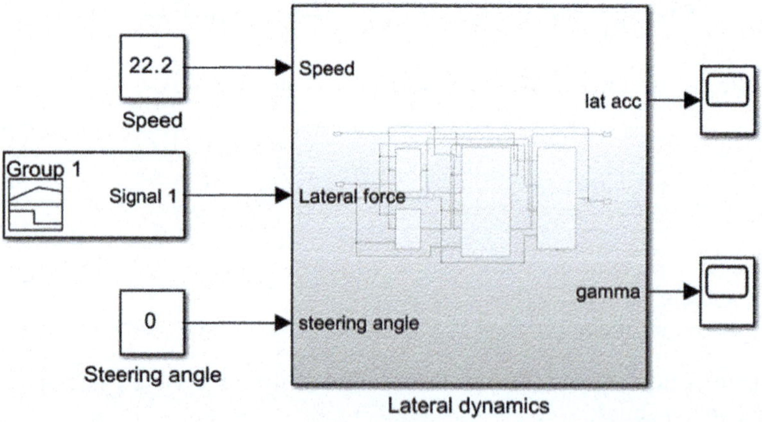

Fig. 4 Model simulates automotive lateral dynamics

Table 1 Simulation parameters

Parameter	Symbol	Unit	Value
Moment of inertia	I_z	kg m^2	2500
Mass	m_V	kg	3970
Stiffness of the front-wheels	C_f	N/rad	100.000
Stiffness of the rear-wheels	C_r	N/rad	110.000
Distance from G to the front-wheels	L_f	m	1.24
Distance from G to the rear-wheels	L_r	m	1.30

3 Results and Discussion

Survey for two cases: The lateral force F_L placed before G (Case 1) and after G (Case 2) when the car is moving in a straight line.

The input signal for the simulation is the lateral force applied to the car in the period from 4 ÷ 5 s where the force value is 3000 N (see Fig. 5).

The output signal includes: The horizontal acceleration acting on the car in Case 1 reaches the maximum value of 1.3 m/s^2. Case 2 reaches the maximum value of 0.8 m/s^2 (see Fig. 6). The biggest deviation of the two cases is 0.5 m/s^2 (see Fig. 7). Automobiles deviated from the original lane of movement in CASE 1 and CASE 2 reached maximum values of 0.19 and 0.1 m (see Fig. 8). The largest deviation of the two cases is 0.09 m (see Fig. 9). The car is rotated at an angle, changing the direction of movement, then rotated to a linear position in Case 1, reaching a maximum value of 0.038 rad; Case 2 reaches the maximum value – 0.012 rad (see Fig. 10). The biggest deviation of the two cases is 0.05 rad (see Fig. 11).

The maximum output values are summed (Table 2).

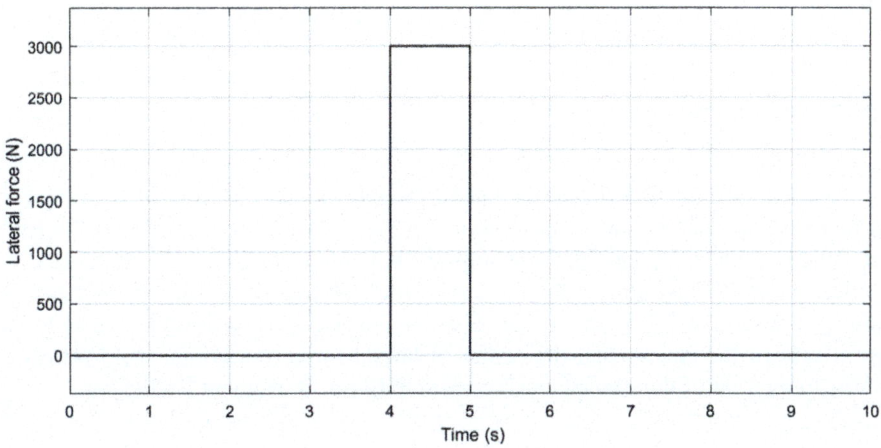

Fig. 5 Simulated input signal (F_L)

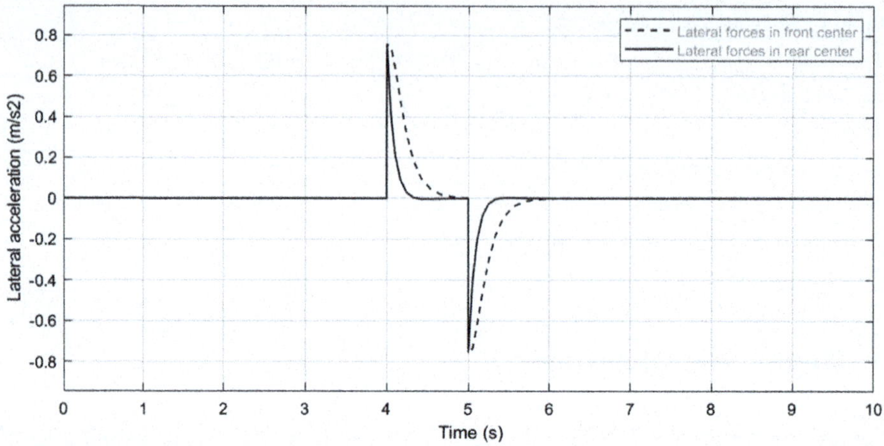

Fig. 6 Lateral acceleration

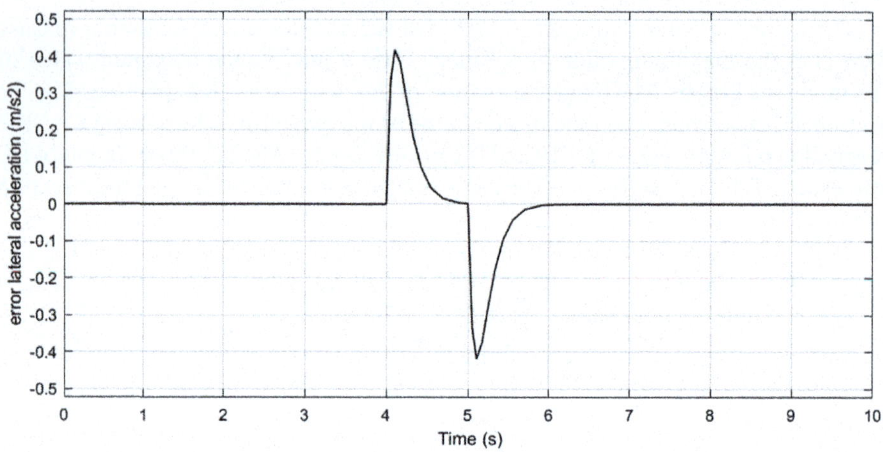

Fig. 7 Error lateral acceleration

The results show that: when the lateral force setpoint is in front of *G* of the car, it increases the deflection of movement thereby reducing the stability of the car's movement. When the lateral force setpoint is behind *G* of the car, it reduces the deflection of movement, thereby increasing the stability of the car's movement. To improve motion stability for the car, *G* of the car needs to be designed close to the front axle of the car, so that the center of lateral force (horizontal wind pressure) shifts to the rear of the car, increasing the stability of movement for the car.

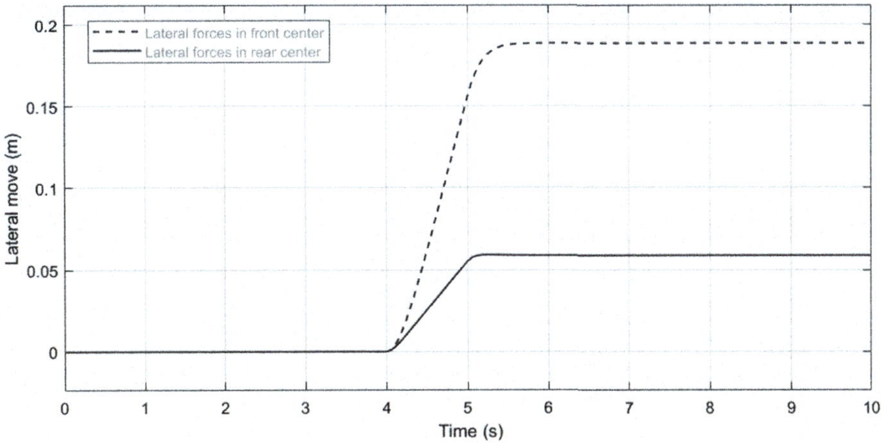

Fig. 8 Lateral move

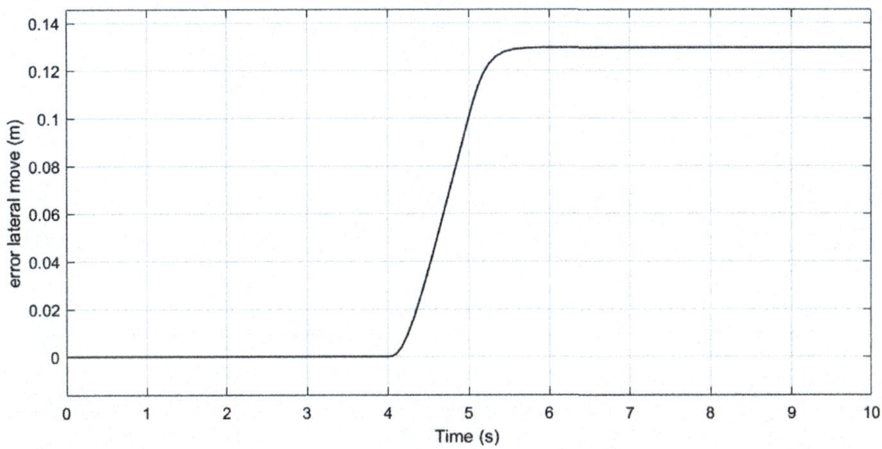

Fig. 9 Error lateral move

4 Conclusions

To experiment on real automobile requires a lot of factors, such as automobile, experimental equipment, test tracks, and drivers, very expensive and time-consuming. Therefore, it is necessary to build a computer simulation model, which allows simulation in a variety of conditions as well as reduces cost and execution time. The study changed the transverse force setpoint to the car's transverse dynamics model in two cases: the transverse force setpoint in front and behind the automobile's center of gravity. The simulation results show that the degree of influence of lateral forces when changing the force placement position is very different. This shows that the

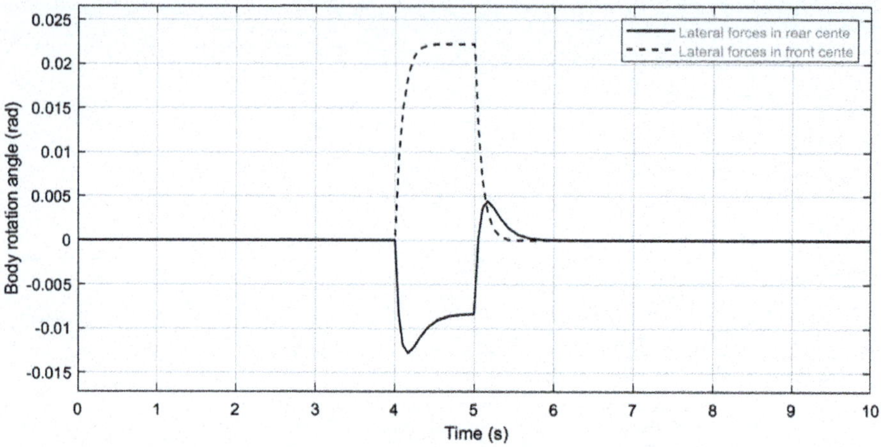

Fig. 10 Body rotation angle

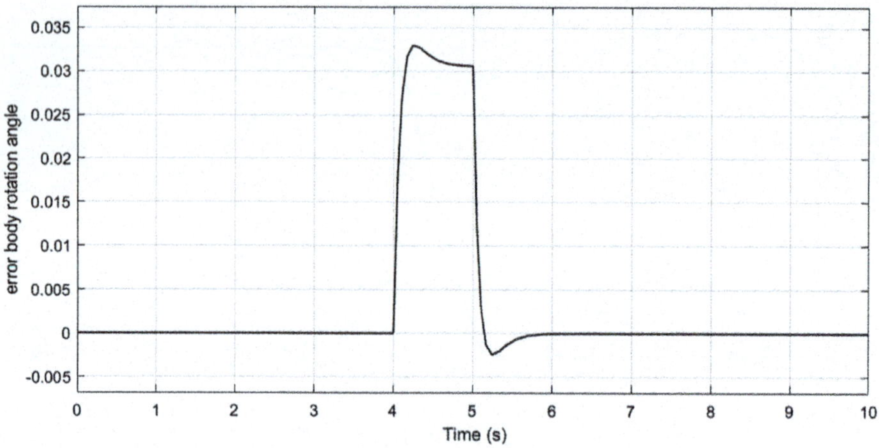

Fig. 11 Error body rotation angle

Table 2 Aggregate results

Parameter	Case 1	Case 2	Error
Lateral acceleration (m/s^2)	1.3	0.8	0.5
Lateral move (m)	0.19	0.1	0.09
Body rotation angle (rad)	0.038	– 0.012	0.05

design and layout of systems to change *G* position of the vehicle is important, and this study has shown that putting the *G*-spot forward is necessary.

References

1. Plöchl, M., Edelmann, J.: Driver models in automobile dynamics application. Veh. Syst. Dyn. ISSN 0042-3114 Print/ISSN 1744-5159 Online (2007). https://doi.org/10.1080/00423110701432482
2. Abdullah, M.A., Jamil, J.F., Mohan, A.E.: Vehicle Dynamics Modeling & Simulation. Centre for Advanced Research on Energy (CARe), Faculty of Mechanical Engineering, Universiti Teknikal Malaysia Melaka, Malacca. ISBN 978-967-0257-78-5 (2016)
3. Xiao, F., Hu, J., Jia, M., Zhu, P., Deng, C.: A novel estimation scheme of tyre–road friction characteristics based on parameter constraints on varied-μ roads. Measurement **194**, 111077 (2022)
4. Kumar, M., Hildebrandt, A.-C., Strauss, P., Kraus, S., Stiller, C., Zimmermann, A.: An optimal lateral trajectory stabilization of vehicle using differential dynamic programming. In: 2020 IEEE Intelligent Vehicles Symposium (IV), Las Vegas, NV, USA, pp. 623–630 (2020). https://doi.org/10.1109/IV47402.2020.9304582
5. Ifqir, S., Ichalal, D., Ait-Oufroukh, N., Mammar, S.: Vehicle lateral velocity and lateral tire-road forces estimation based on switched interval observers. In: The 2020 American Control Conference (ACC 2020), Denver, CO, July 2020, pp. 3023–3028 (2020)
6. Rosa, F.D., Mastinu, G.: Analysis of the lateral dynamics of a vehicle and driver model running straight ahead. Nonlinear Dyn. **92**, 97–106 (2018). https://doi.org/10.1007/s11071-017-3478-1
7. Guo, J., Chu, L., Zhang, X., Zhou, F.: The integrated control of semiactive suspension and electronic stability program based on fuzzy logic method. Appl. Mech. Mater. **29**(32), 2059–2064 (2010)
8. Rill, G.: Vehicle Dynamics. Lecture Notes, Fachhochsule Regenburg, University Applied Sciences (2005)
9. Mehdizadeh, E., Kabganian, M., Kazemi, R.: A new force feedback for steer-by-wire vehicles via virtual vehicle concept. In: IEEE Conference on Decision and Control and European Control Conference (CDC-ECC), Orlando, FL, USA, 12–15 Dec 2011
10. Miloradovic, D., et al.: Simulation of vehicle's lateral dynamics using nonlinear model with real inputs. IOP Conf. Ser. Mater. Sci. Eng. **659**, 012060 (2019). https://doi.org/10.1088/1757-899X/659/1/012060

Simulate Airbag Folding Using Ansys LS-DYNA Software

Hieu Pham Minh, Viet Tran Duc, Tien Tang Van, Huu Hoang Van, Quynh Luu Ba, Phuong Do Thai, Hoang Chu Vuong, and Khanh Pham Quoc

Abstract Safety is of utmost importance when it comes to automobile manufacturing, especially when it comes to protecting passengers in the event of a collision. To this end, the passive safety system plays a critical role in minimizing injuries. The airbag system is a key component of this system, and manufacturers are constantly striving to improve its effectiveness. This particular study examines the pressure, volume, and reaction time of two airbag folding methods—thin fold and spiral fold. The results indicate that the spiral fold method offers faster volume stability, while the pressure and temperature difference is only marginally affected (about 0.6%). The goal of this research is to gather input and feedback to determine the most optimal folding method, with the spiral fold method being the primary focus.

Keywords Airbag · Thin fold · Spiral fold · Ansys LS-DYNA

1 Introduction

The automotive sector has made impressive strides in enhancing passenger safety by leveraging scientific, technological, and engineering breakthroughs. Airbags have emerged as a critical element in mitigating the likelihood of passenger harm during a collision. Numerical simulation has become a crucial instrument in developing and designing airbags, guaranteeing their optimal performance and dependability. The article discusses a simulation study on airbag folding using LS-DYNA software's 3D nonlinear finite element method. The study compares the deployment process and airbag parameters with actual explosion results [1]. Additionally, the article examines

---

H. P. Minh · V. T. Duc (✉) · T. T. Van · H. H. Van · Q. L. Ba · P. D. Thai · H. C. Vuong
Faculty of Transportation Engineering, School of Mechanical and Automotive Engineering, Hanoi University of Industry, No. 298 Cau Dien Street, Bac Tu Liem District, Hanoi, Vietnam
e-mail: Viettd1809@gmail.com

K. P. Quoc
Department of Mechanical and Computer-Aided Engineering, National Formosa University, No. 64 Wenhua Rd, Huwei Township, Yunlin County, Taiwan 632

D. H. Tien et al. (eds.), *Proceedings of the 1st International Conference on Sustainability and Emerging Technologies for Smart Manufacturing*, Proceedings in Technology Transfer, https://doi.org/10.1007/978-981-97-7083-0_10

the use of airbags to reduce human injuries resulting from accidents. The deployment of airbag systems reduces death and injury rates in traffic accidents. Furthermore, the authors simulated the operation of airbags using LS-DYNA software and presented the results of the study.

Previously, there were research articles on this issue such as the article "Airbag simulation with LS-DYNA Past-present-future" which presents the results and discusses the calculation methods of airbag hybrid jetting, airbag ALE, and airbag particles. From there, evaluate the algorithms that allow the simulation. The interaction between discrete air flows and the airbag structure [2]. The corpuscular (C) method has proven to be a very promising new development path for airbag deployment simulations in LS-DYNA. In particular, the application of the method is rather simple. Switching from an *AIRBAG_HYBRID input deck toward *AIRBAG_PARTICLE is very easy and by no means as complicated as setting up an ALE input deck. The present examples show that the accuracy and agreement with the experimental results of the new method are very good. Next is the article "Airbag: A Safety Restraint System of an Automobile" which studies airbags to reduce human injuries by reducing the force exerted by the steering wheel, windows, dashboards, etc. The deployment of airbag systems reduces death and injury rates in traffic accidents [3]. The continuous decline in the number of fatalities and injuries in traffic incidents can be attributed to the evolution of the airbag system. The subsequent development of seat belts has become an undeniable norm. Presently, the integration of the 3-point automatic seat belt, seat belt tensioner, and airbag forms a meticulously designed passenger protection system. Proper implementation of these safety restraint systems, in accordance with regulations and due diligence, has the potential to significantly reduce both the fatality rate and the occurrence of serious injuries during road accidents.

In this paper, we focus on simulating the "Thin fold" and "Spiral fold" of airbags in LS-DYNA. We consider important factors such as airbag geometry, materials, loads, and boundary conditions to ensure simulation accuracy and reliability. The simulation results are compared with each other to come up with a more effective method. This research work hopes to provide important information about airbag simulation in LS-DYNA and contribute to improving the performance and safety of airbags in the automotive industry.

2 Theoretical Basis

2.1 The Operating Principle of the Airbag System

Car airbags work like rocket boosters to ensure driver safety during collisions. A sensor system sends signals to the airbag controller, which uses a high-pressure pump to fill the airbag. The master system has three stages that occur within 0.04 s, from the moment of collision to when the airbag deploys.

The Airbag Control Unit (ACU) manages the sensors that detect collisions and sends a signal to deploy the airbags. It analyzes the data to adjust safety features and activates the airbag inflator when a collision exceeds a particular level.

During a car crash, a detonator ignites gas particles which inflate the airbag to cushion the impact. The detonator produces a 1–3 A electric current in under 2 ms to initiate this process.

Compressed air causes rapid airbag inflation at 300 km/h. Air escapes through small holes, deflating the bag in just 100 ms, often unnoticed by vehicle occupants.

2.2 *Airbag Fabric*

The LS-DYNA theoretical tutorial describes the fabric material model used in airbag analysis. Fabrics exhibit nonlinear properties and are not capable of withstanding tension and bending. Initially, the fabric has a low level of tension, which gradually increases until the material becomes taut.

This behavior is modeled by expressing the fiber modulus as a function of fiber strain [1] (Fig. 1).

When analyzing the stress of materials with small deformations, the Cauchy Stress Tensor is used. The tension state is then updated incrementally in the material coordinates [4]

$$\sigma_{\mathrm{L}}^{n+1} = \sigma_{\mathrm{L}}^{n} + \Delta\sigma_{\mathrm{L}}^{n+\frac{1}{2}} \tag{1}$$

In which,

$$\Delta\sigma_{\mathrm{L}}^{n+\frac{1}{2}} = \begin{bmatrix} \Delta\sigma_{11} \\ \Delta\sigma_{22} \\ \Delta\sigma_{12} \\ \Delta\sigma_{23} \\ \Delta\sigma_{31} \end{bmatrix} \begin{bmatrix} Q_{11} & Q_{12} & 0 & 0 & 0 \\ Q_{12} & Q_{22} & 0 & 0 & 0 \\ 0 & 0 & Q_{44} & 0 & 0 \\ 0 & 0 & 0 & Q_{55} & 0 \\ 0 & 0 & 0 & 0 & Q_{66} \end{bmatrix} \begin{bmatrix} d_{11} \\ d_{22} \\ d_{12} \\ d_{23} \\ d_{31} \end{bmatrix}_{\mathrm{L}} \Delta t \tag{2}$$

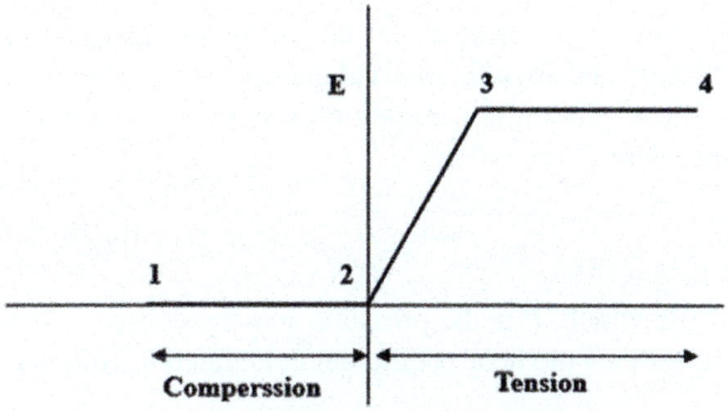

Fig. 1 Idealized Young's modulus as a function of fiber strain

Q_{ij} is determined:

$$Q_{11} = \frac{E_{11}}{1 - v_{12}v_{21}},\ Q_{22} = \frac{E_{22}}{1 - v_{12}v_{21}},\ Q_{12} = \frac{v_{12}E_{11}}{1 - v_{12}v_{21}},$$
$$Q_{44} = G_{12},\ Q_{55} = G_{23},\ Q_{66} = G_{31}$$

Apply symmetry

$$v_{21} = v_{12}\frac{E_{22}}{E_{11}} \tag{3}$$

In which,

v_{ij} is the Poisson coefficient for the perpendicular strain in the j direction for the material subjected to tension in the i direction
E_{ij} is Young's modulus in direction i
G_{ij} is the shear modulus.

2.3 Airbag Interior

To find the volume of an airbag, use the control block method and establish a control volume enclosed by a control surface. The airbag fabric serves as the control surface and the internal air space is the control volume. Calculate the control volume using Green's theorem by measuring the airbag's components' location, alignment, and dimensions [1].

$$\iiint \emptyset\frac{\partial\psi}{\partial x}\mathrm{d}x\mathrm{d}y\mathrm{d}z = -\iiint \psi\frac{\partial\emptyset}{\partial x}\mathrm{d}x\mathrm{d}y\mathrm{d}z + \oint \emptyset\psi n_x \mathrm{d}\Gamma \tag{4}$$

2.4 Pressure and Internal Energy

The airbag volume is determined by an equation of state that relates pressure to gas density and specific internal energy.

The gamma law equation of state, as described in [3], is used to determine the pressure in the airbag.

$$P = (\gamma - 1)\cdot p \cdot e \tag{5}$$

In which, P is the pressure, p is the density, e is the internal energy of the gas, γ represents the heat coefficient calculated as follows:

$$\gamma = \frac{C_p}{C_v} \tag{6}$$

The equation for internal energy that corresponds to the two states is determined in the following manner:

$$e_2 = e_1[v_2/v_1]^{(1-\gamma)} \tag{7}$$

With v_1 and v_2 being the volume of the two states. The flow rate over time is determined by the law of conservation of mass

$$\frac{dM}{dt} = \frac{dM_i}{dt} - \frac{dM_o}{dt} \tag{8}$$

In which,

M_i is the volume of incoming air
M_o is the mass of gas released.

3 Simulation Airbag

3.1 Model Setup

The authors created a 3D model of the airbag and established the folding steps of thin fold and Spiral fold. Use LS-DYNA software to create surfaces, gas blocks, and mesh the model.

3.2 Setup Materials and Boundary Conditions

The authors selected and simulated a driver airbag with a diameter of 660 mm with 2 vents. Besides, there are also parameters that characterize the material properties of the airbag fabric (Table 1).

Table 1 Material properties of airbags

Parameter	Symbol	Unit	Value
Mass density	ρ_{fabric}	–	7.86e−7
Young's modulus	E	–	0.3
Thickness	t	mm	0.4
Poisson's ratio	υ	–	0.2

Table 2 Thermodynamic properties of nitrogen and its surroundings

	Parameter	Symbol	Unit	Value
Nitrogen	Specific heat capacity at constant volume	C_v	–	196.168
	Specific heat capacity at constant pressure	C_p	–	276.6
	Temperature	T	°K	800
Air	Density	P_{air}	g/cc	1.22
	Atmospheric pressure	P_e	Pa	1.013E+05

Next are the parameters of the characteristics of the gas and the environment surrounding the airbag such as temperature, specific heat capacity at volume, specific heat capacity at pressure, gas density, and ambient pressure are presented in Table 2.

4 Results

4.1 Pressure

Airbags are crucial safety features in vehicles that prevent passengers from hitting hard surfaces during a collision. This study assesses the airbag system's properties using the ideal gas equation and evaluates its pros and cons in two scenarios. Figures 4 and 5 illustrate the airbag's pressure parameters.

In Fig. 2, the airbag system pressure with spiral fold fluctuates during initial air intake. Max pressure is 0.162E−3 (GPa) at 9 s and decreases to 0.102E−3 (GPa) at 12 s. After 20 s, the pressure stabilizes between 0.116E−3 (GPa) and 0.125E−3 (GPa). The pressure is concentrated along the airbag's edge when fully deployed.

Figure 5 shows that the airbag system with thin fold style has pressure fluctuations initially, reaching a maximum of 0.16E−3 GPa at 9.49 ms and stabilizing between 0.12E−3 and 0.127E−3 GPa after 20 s. Figure 3 shows that the airbag pressure is concentrated at the edge when fully open (Figs. 4 and 5).

Fig. 2 Spiral fold airbag

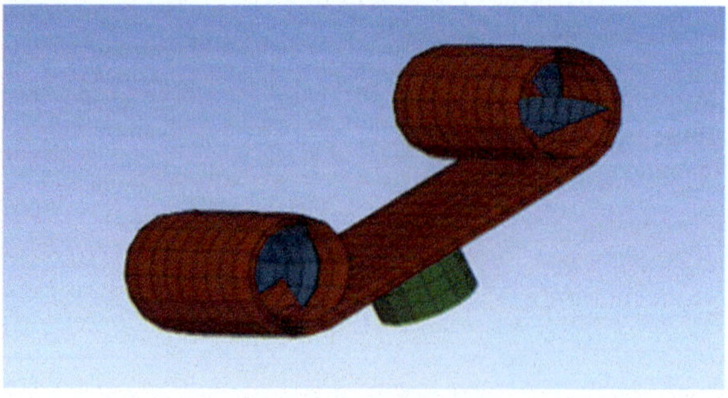

Fig. 3 Thin fold airbag

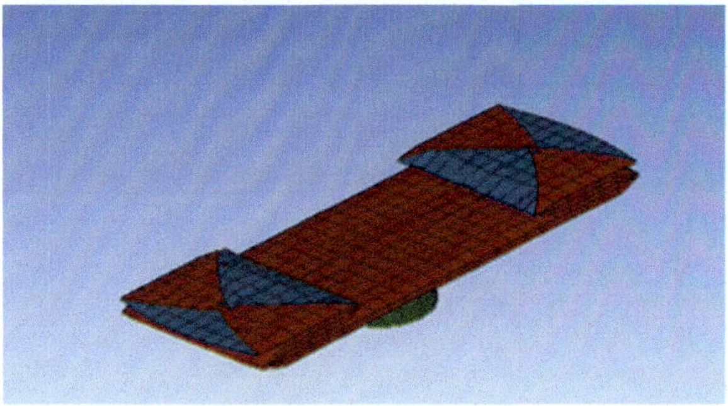

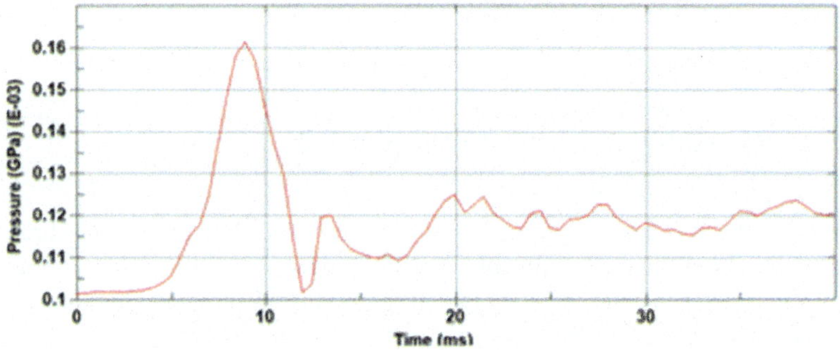

Fig. 4 Pressure chart with spiral folding

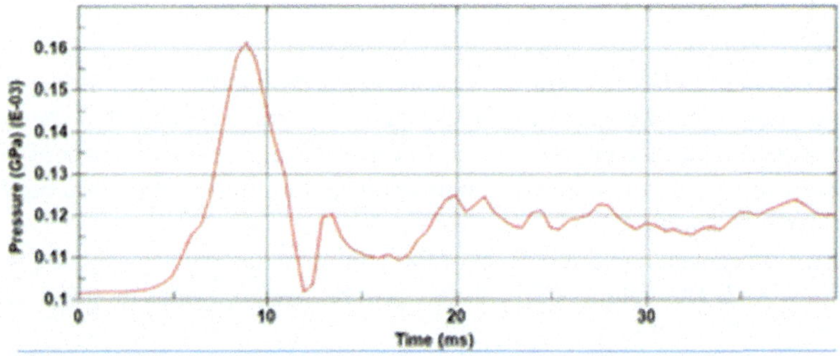

Fig. 5 Pressure chart with thin folding

In Fig. 6, we observe distinct pressure variations between two folding techniques. The airbag system's pressure parameters display fluctuations in the initial 20 s with both methods before gradually stabilizing.

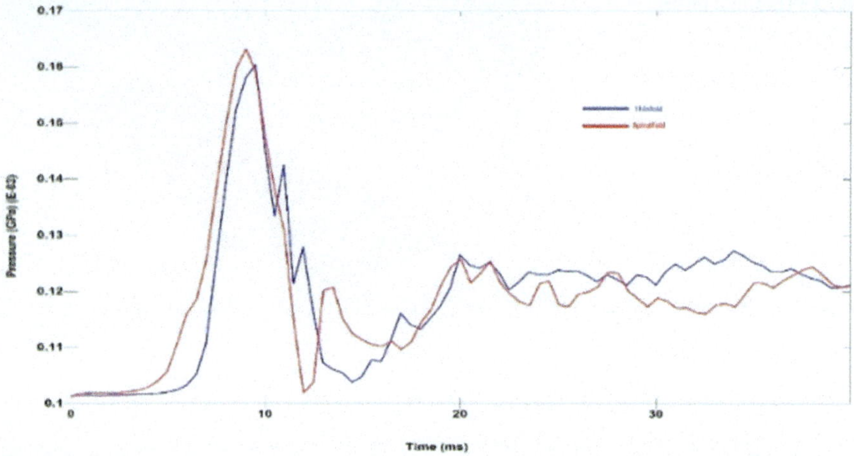

Fig. 6 Pressure comparison chart between 2 folding methods

However, the spiral fold approach shows a more sudden drop in pressure parameters after reaching the peak value at 10 s, in contrast to the thin fold method. Additionally, the pressure achieved when the airbag is fully deployed using the spiral fold method is somewhat lower than that of the thin fold approach.

4.2 *Volume*

The data depicted in Fig. 7 outlines the volume measurements of an airbag that has been folded utilizing the spiral fold technique. The graph reveals that the airbag's volume experiences a steady increase during the initial phase and peaks at the 21 s mark. After this point, the volume remains constant until the simulation concludes, ranging from 51.2 to 54.2 L.

Figure 8 presents a lucid depiction of the volume parameters associated with the thin fold method. The volume of the airbag progressively elevates in the initial stage

Fig. 7 Volume chart with spiral folding

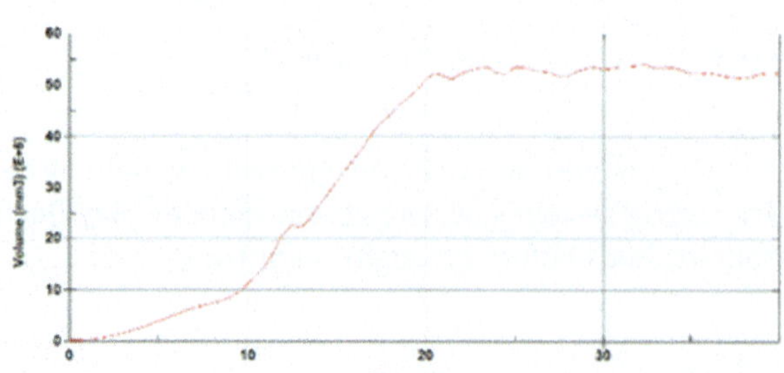

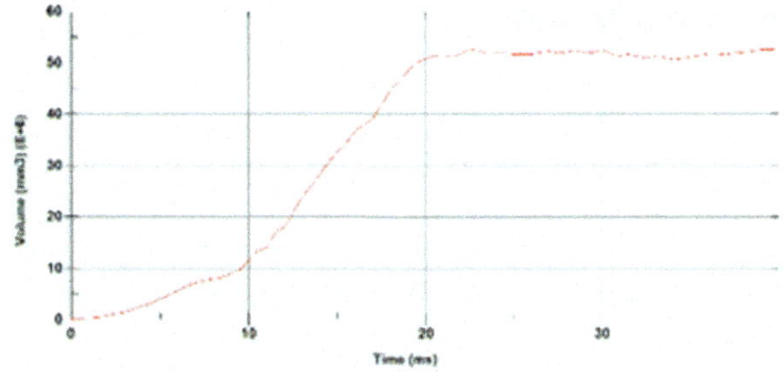

Fig. 8 Chart with thin folding

of the procedure. At the 20 s mark, the volume parameter of the airbag attains a steady threshold. Subsequently, throughout the simulation, the volumetric parameter of the airbag system remains confined to a range of 50.7–52.7 L.

4.3 Temperature

Figure 9 compares temperature parameters of spiral fold and thin fold methods during airbag opening. Although they're similar initially, they differ after the fifth second. The temperature of the thin fold method fluctuates more from the 5th to the 20th second, but stabilizes around 810–820 K after 20 ms due to isothermal process. The temperature of the spiral fold method stabilizes around 795–815 K. The solver takes time to process the problem, causing significant temperature changes in the initial stage.

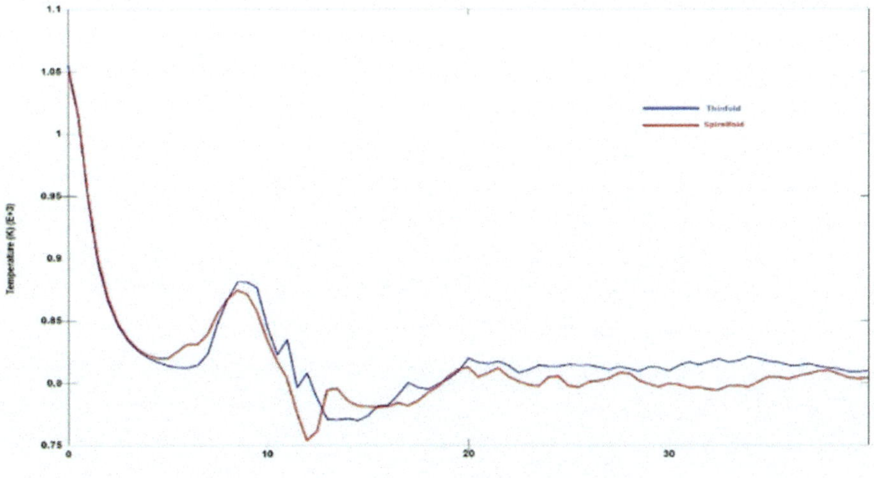

Fig. 9 Temperature comparison chart between 2 folding methods

5 Conclusion

The simulation of two airbag folding methods, thin fold and spiral fold, was conducted using Ansys LS-DYNA software and theoretical principles. The authors analyzed various parameters, including pressure, displacement, velocity, and internal energy, to compare the performance of these two methods. Based on the simulation results, the authors have identified the strengths and weaknesses of each method:

- The spiral fold method produces higher and more consistent pressure than the thin fold method when the airbag is activated. This method injects air faster, resulting in higher displacement, velocity, and acceleration than the spiral fold method. However, spiral fold provides greater displacement. The spiral fold method dominates in terms of kinetic and internal energy values.
- During the peak deployment of the airbag, typically occurring after 30 ms, the thin fold method generally has a larger surface area than the spiral fold method.

References

1. LS-DYNA3D (Non-linear Dynamic Analysis of Structures in 3 Dimensions) Theoretical
2. Manual, tr. 581–591. Livermore Software Technology Corporation (2006)
3. Hirth, A., Haufe, A., Olovsson, L.: Airbag simulation with LS-DYNA past-present-future (2007)
4. Shaikh, T.N., Chaudhari, S., Rasania, H.: Air bag: a safety restraint system of an automobile

Non-singular Terminal Sliding Mode Control of the Steering Wheels of 4WD4WS Mobile Robot

Kien Nam Dang, Vu Nguyen, and Huy Quoc Vu

Abstract Most of the path following research for 4WD4WS mobile robots is conducted with two control loops, the kinematic loop and the dynamic loop. This article proposes a method using Non-Singular Terminal Sliding Mode Control (NSTSM) to improve the efficiency of dynamic control loop. The proposed control method is not only proved to be stable under disturbances but also has finite time convergence which was calculated explicitly and visually shown in simulation.

Keywords Mobile robot · Path following · Four-wheel-drive/four-wheel-steer · SMC · NTSM · Dynamic control

1 Introduction

4WD4WS mobile robots with redundant drive system (4 drive motors and 4 steering motors) are able to navigate challenging terrain and complex environments while still maintaining high agility and stability. Therefore, they are being used more widely.

Path following is a common problem in mobile robot control and many researches have been done for many years. Most of the cases, authors built two control loops: kinematic and dynamic.

With the path following problem for 4WD4WS, many authors create dynamic control system of 4WD4WS mobile robot [1–5] but they only consider the dynamic of the robot's body in detail, not the real nature of a DC motor for each wheel.

K. N. Dang · H. Q. Vu
Control, Automation in Production and Improvement of Technology Institute (CAPITI), Academy of Military Science and Technology (AMST), Hanoi, Vietnam
e-mail: vuquochuy@capiti.info

V. Nguyen (✉)
Faculty of Electrical Engineering, Hanoi University of Industry, Hanoi, Vietnam
e-mail: vutudonghoa@yahoo.com.vn

D. H. Tien et al. (eds.), *Proceedings of the 1st International Conference on Sustainability and Emerging Technologies for Smart Manufacturing*, Proceedings in Technology Transfer, https://doi.org/10.1007/978-981-97-7083-0_11

Research in [1] proposed two control loops in details, where front and rear steering wheels are controlled independently but left steering wheels and right steering wheels have the same angle which imposing side slip phenomenon at the part of the tire in contact with the road surface. Research in [2] proposed a conventional Sliding Mode Control (SMC) for dynamic control loop, in which steering wheels will exponentially converge to desired values. Although the system is stable, the convergence time is undetermined, while 4WD4WS is a redundant drive system the response speed of each motor is required very fast and each of them need to be well tolerated with disturbances.

Therefore, the authors propose to use NSTSM control, which has finite convergence time, in the dynamic control loop of each steering wheel to enhance efficiency of the path following problem for the 4WD4WS robot, while the kinematic control loop will use a geometry control method as in [6, 7]. When the kinematic control loop finds the desired steering angles so that the lateral error of the system is minimal, the dynamic control loop of each steering wheel will follow those angles and give the actual steering angle back to the kinematic loop.

2 Methodology

2.1 Two Control Loops in the Path Following Problem of 4WD4WS Mobile Robot

As shown in Fig. 1, the kinematic controller creates the desired steering angles which can be done by using a geometric control method like in [1, 2] or an improved method that also uses exponential functions in virtual target guidance algorithm to make the system's convergence time to be finite. A 4WD4WS mobile robot has 4 wheels, each of those can be driven by two motors, one for the traction purposes, one for steering purposes. In this article, robot body is considered to have a constant velocity so the traction control is not the author's objective, but the steering control. The steering motor of each wheel is considered as a DC motor and has the same characteristics. Then the dynamic steering controller is going to built based on the DC motor nature of each steering wheel.

2.2 Dynamic Model of a Steering Wheel

As a DC motor, the steering wheel of a 4WD4WS mobile robot will have this dynamic system of equations [8]:

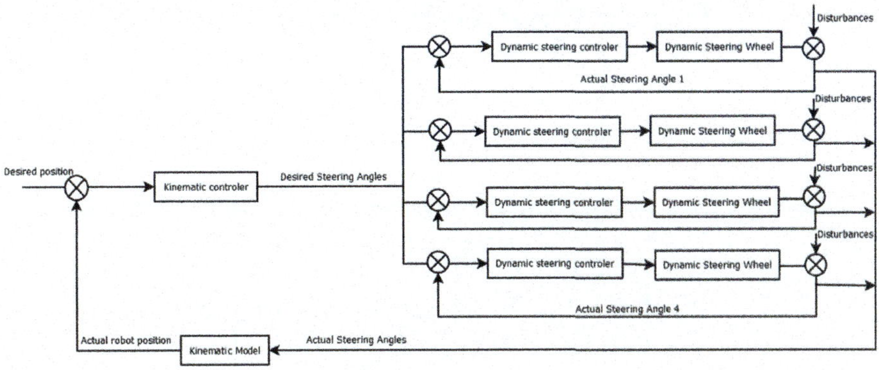

Fig. 1 Two control loops in path following problem of 4WD4WS mobile robot

$$\dot{x} = Ax + Bu = \begin{bmatrix} 0 & 1 & 0 \\ 0 & -\frac{b}{J} & \frac{k_m}{J} \\ 0 & -\frac{k_e}{L} & -\frac{R}{L} \end{bmatrix} \begin{bmatrix} x_1 \\ x_2 \\ x_3 \end{bmatrix} + \begin{bmatrix} 0 \\ 0 \\ \frac{1}{L} \end{bmatrix} u \tag{1}$$

where x_1 is the motor shaft rotation angle, x_2 is the angular velocity of the motor shaft, x_3 is the current in the armature coil. The input u is the voltage of the armature winding. The parameters R and L are the resistance and electromagnetic induction of the armature coil, k_e is the speed constant, b is the viscous friction coefficient, J is the moment of inertia and k_m is the generating torque constant.

System (1) needs to be transformed to a normal form to apply sliding mode control method, then as in [8] the new system has the form:

$$\dot{x}_f = \begin{bmatrix} 0 & 1 & 0 \\ 0 & 0 & 1 \\ 0 & -\frac{k_m k_e + bR}{JL} & -\frac{JR+bL}{JL} \end{bmatrix} \begin{bmatrix} x_{1,f} \\ x_{2,f} \\ x_{3,f} \end{bmatrix} + \begin{bmatrix} 0 \\ 0 \\ \frac{k_m}{JL} \end{bmatrix} u \tag{2}$$

$$\text{Set}: \begin{cases} a_2 = \frac{k_m k_e + bR}{JL} \\ a_3 = \frac{JR+bL}{JL} \\ b_3 = \frac{k_m}{JL} \end{cases} \tag{3}$$

In the dynamic loop of path following problem of a mobile robot, each DC motor of steering wheels must follow a desired steering angle then system (2) must change to a deviation form.

$$\text{Set}: \; e_1 = x_{1f} - x_{1d} \tag{4}$$

Then from (2) also get:

$$\begin{cases} e_2 = x_{2,f} - x_{2d} = \dot{x}_{1,f} - \dot{x}_{1d} = \dot{e}_1 \\ e_3 = x_{3,f} - x_{3d} = \dot{x}_{2,f} - \dot{x}_{2d} = \dot{e}_2 \\ \dot{e}_3 = -a_2e_2 - a_3e_3 + b_3u + a_2x_{2d} + a_3x_{3d} - \dot{x}_{3d} \end{cases} \tag{5}$$

$$\text{Set}: \; D = a_2x_{2d} + a_3x_{3d} - \dot{x}_{3d} \tag{6}$$

One can get the deviation form of the steering wheel dynamic equations:

$$\dot{e} = \begin{bmatrix} 0 & 1 & 0 \\ 0 & 0 & 1 \\ 0 & -a_2 & -a_3 \end{bmatrix} \begin{bmatrix} e_1 \\ e_2 \\ e_3 \end{bmatrix} + \begin{bmatrix} 0 \\ 0 \\ b_3 \end{bmatrix} u + \begin{bmatrix} 0 \\ 0 \\ D \end{bmatrix} \tag{7}$$

2.3 NSTSM Control of Steering Wheels

To apply NSTSM control for system (7), choose sliding variable as follows:

$$\sigma = c_1e_1 + c_2e_2 \tag{8}$$

where $c_1, c_2 > 0$ and were chosen based on a linearized form of an original TSM control, then c_1, c_2 is changing according to the value of e_1. Assume that the original TSM form of σ_{org} is come from:

$$\sigma_{org} = c_e e_1^{\frac{p}{q}} + e_2 \tag{9}$$

where $c_e > 0$ and unchanged; p *and* q are odd integers.

Assume that actual deviation of steering angle is perturbed by some bounded error:

$$e_{1a} = e_1 + d_e \tag{10}$$

where

$$|d_e| \le M, M > 0 \tag{11}$$

Actual sliding variable:

$$\sigma_a = \sigma + c_1d_e \tag{12}$$

Then, sliding variable's derivatives:

$$\dot{\sigma} = c_1\dot{e}_1 + c_2\dot{e}_2 = c_1e_2 + c_2e_3 \tag{13}$$

$$\ddot{\sigma} = c_1\dot{e}_2 + c_2\dot{e}_3 = c_1e_3 + c_2(-a_2e_2 - a_3e_3 + b_3u + D) \tag{14}$$

Use the 2nd order sliding mode with a non-singular terminal sliding variable as follow:

$$S = \sigma_a^{p/q} + c\dot{\sigma}, \quad c > 0 \tag{15}$$

Then, its derivative:

$$\dot{S} = \frac{p}{q}\sigma_a^{(p-q)/q}\dot{\sigma} + c\ddot{\sigma} \tag{16}$$

There are two parts of control input, equivalent part or continuous part will keep the sliding variable stay on the sliding surface and the discontinuous part that drag it to the sliding surface, or:

$$u = u_{eq} + u_d \tag{17}$$

Then apply u_{eq} to (14) and (16) one can get the equation:

$$\dot{S} = 0 \tag{18}$$

From what one can achieve u_{eq}:

$$u_{eq} = \frac{-1}{cc_2b_3}\left(\frac{p}{q}\sigma^{(p-q)/q}\dot{\sigma} + c[c_1e_3 + c_2(-a_2e_2 - a_3e_3 + D)]\right) \tag{19}$$

When apply u to (16) then the equivalent part will actually neutralize most of the terms of $\dot{S}$:

$$\dot{S} = cc_2b_3u_d + \frac{p}{q}\sigma^{(p-q)/q}\dot{\sigma} - \frac{p}{q}\sigma_a^{(p-q)/q}\dot{\sigma} \tag{20}$$

By using NSTSM with p, q such that $p = 2 * q - 1$ with very large values (20) can be approximated as:

$$\dot{S} = cc_2b_3u_d + \frac{p}{q}\sigma\dot{\sigma} - \frac{p}{q}\sigma_a\dot{\sigma} = cc_2b_3u_d - c_1\frac{p}{q}d_e\dot{\sigma} \tag{21}$$

From (13), by nature of a DC motor, the value of $\dot{\sigma}$ can be bounded, assume that:

$$|\dot{\sigma}| \leq N \tag{22}$$

Then, by choosing Lyapunov function $V = \frac{1}{2}S^2$, one can obtain:

$$\dot{V} = S\dot{S} = cc_2 b_3 u_d S - c_1 \frac{p}{q} d_e \dot{\sigma} S \tag{23}$$

u_d can be chosen to make $\dot{V} \leq 0$:

$$u_d = -K\text{sign}(S) \tag{24}$$

where

$$K = \frac{c_1 \frac{p}{q} MN}{cc_2 b_3} + \eta, \quad \eta > 0 \tag{25}$$

Then:

$$\dot{V} = -c_1 \frac{p}{q}(MN|S| - d_e \dot{\sigma} S) - \eta cc_2 b_3 |S| \leq -\eta cc_2 b_3 |S| \leq 0 \tag{26}$$

Then with the control input as in (17), (19) and (24), the system is stable according to Lyapunov.

To calculate convergence time of the system, let the sliding variable be zero:

$$S = \sigma^{p/q} + c\dot{\sigma} = 0 \tag{27}$$

Taking one term to the other side:

$$\sigma^{p/q} = -c\dot{\sigma} = -c\frac{\mathrm{d}\sigma}{\mathrm{d}t} \tag{28}$$

Transform (28) to a more convenient form:

$$-c\sigma^{-p/q}\mathrm{d}\sigma = \mathrm{d}t \tag{29}$$

Integrate both sides over t and σ:

$$-c\int_{\sigma_{(t_0)}}^{0} \sigma^{-p/q}\mathrm{d}\sigma = \int_{t_0}^{t_0+t_r} \mathrm{d}t \tag{30}$$

In which, t_r is the time to bring σ from the initial value $\sigma_{(t_0)}$ at the time t_0 where sliding variable $S = 0$ to zero value.

Then convergence time of σ can be determined:

$$t_r = c\frac{\sigma_{(t_0)}^{1-\frac{p}{q}}}{1 - \frac{p}{q}} \tag{31}$$

Similarly to (31), t_0, convergence time of S, can be calculated as:

$$t_0 = \frac{|S_0|}{cc_2 b_3 K} \tag{32}$$

From (9), the final convergence time t_e of e_1, e_2 is calculated similarly to (31):

$$t_e = \frac{q}{c_e(q-p)} e_{1(t_r)}^{\frac{q-p}{q}} \tag{33}$$

Then, total convergence time of the system is calculated as:

$$t_{total} = t_0 + t_r + t_e \tag{34}$$

3 Simulation and Discussion

The article chooses to simulate two cases: when the steering motor is used independently and when four steering motors are used in 4WD4WS mobile robot to follow a circular trajectory on Matlab Simulink software with the same initial conditions for two control methods: traditional SMC and NSTSM control. The parameters that were used in this article can be seen at literature [8].

In addition, the article also demonstrates the noise resistance of the control method by adding different noise to each steering wheel, where each noise was added by using a bounded value (1.8°) with random frequency during the time of simulation. Simulation results of first case are shown from Figs. 2, 3, 4 and 5 with desired angle is a sine function. Noise, that was added to the system (second case), has the form as shown from Figs. 6, 7, 8 and 9.

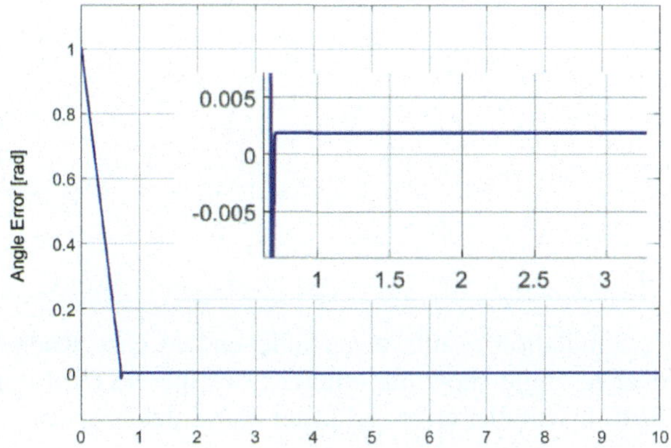

Fig. 2 Error of traditional SMC

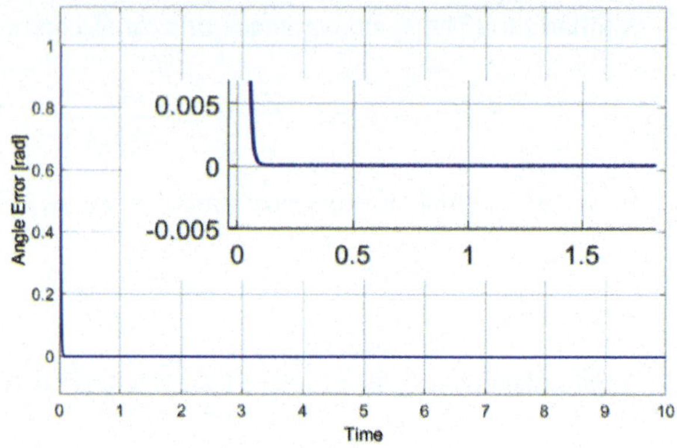

Fig. 3 Error of NSTSM control

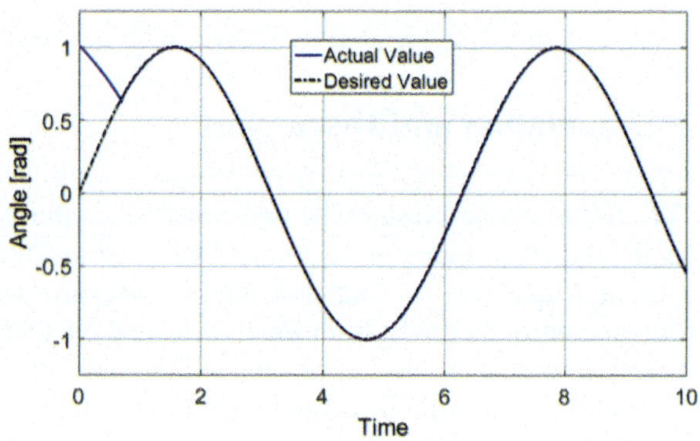

Fig. 4 Actual and desired angles of traditional SMC

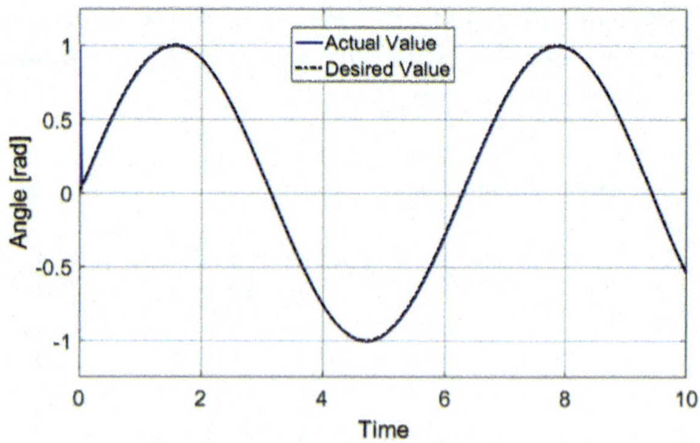

Fig. 5 Actual and desired angles of NSTSM control

Simulation results of an independent case when both controllers track a sine desired angle show the efficiency of the NSTSM control method over traditional SMC by smaller error and faster convergence speed.

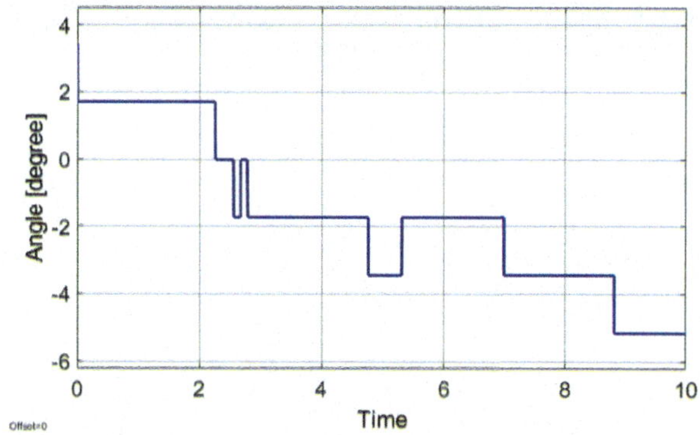

Fig. 6 Noise of the first steering wheel

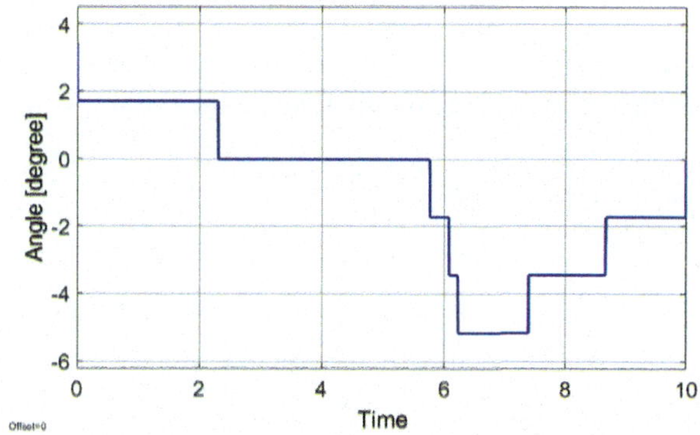

Fig. 7 Noise of the second steering wheel

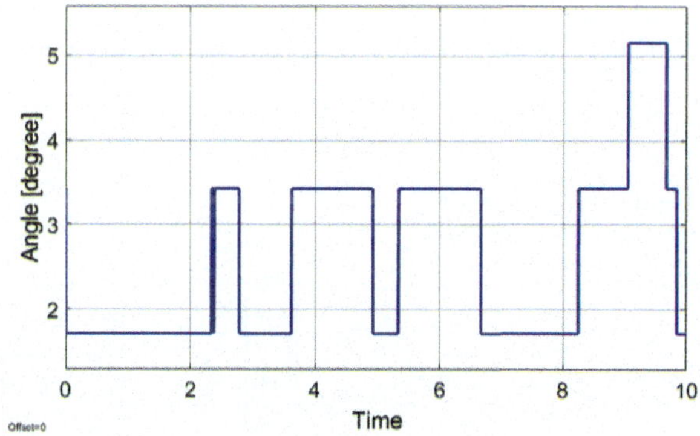

Fig. 8 Noise of the third steering wheel

Simulation results of second case when apply NSTSM control and traditional to four steering wheels of a 4WD4WS mobile robot case are shown from Figs. 10, 11, 12, 13, 14, 15, 16, 17, 18 and 19. The results also show the differences between the two methods, when traditional SMC needs more time to get to a stable error value and this error value is also much larger than the NSTSM control method. Looking

Fig. 9 Noise of the forth steering wheel

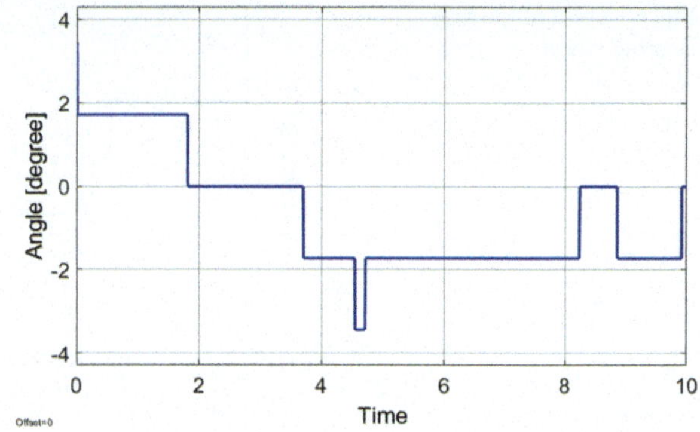

into details, each steering angle error from traditional SMC keeps a value with the same direction with the noise that drags it out of the stable value. Actually, SMC control is not able to totally tolerate the noise.

Fig. 10 Error of the first steering wheel (SMC)

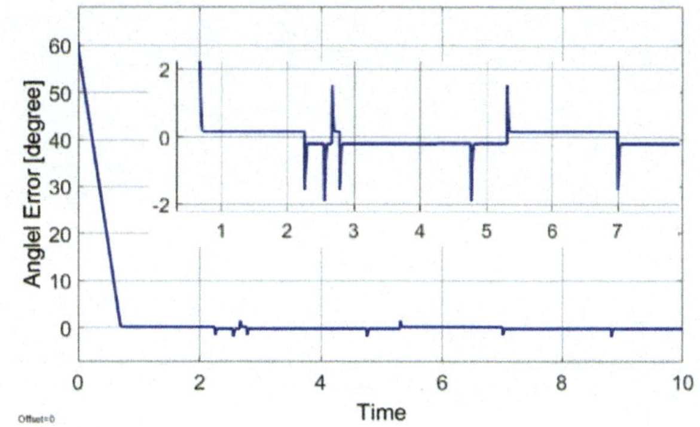

Fig. 11 Error of the first steering wheel (NSTSM)

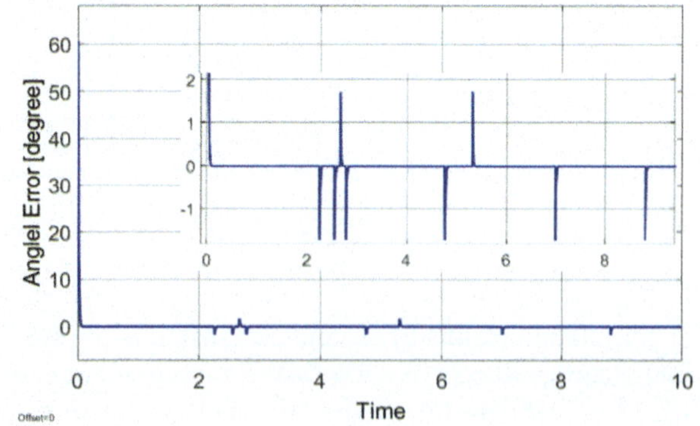

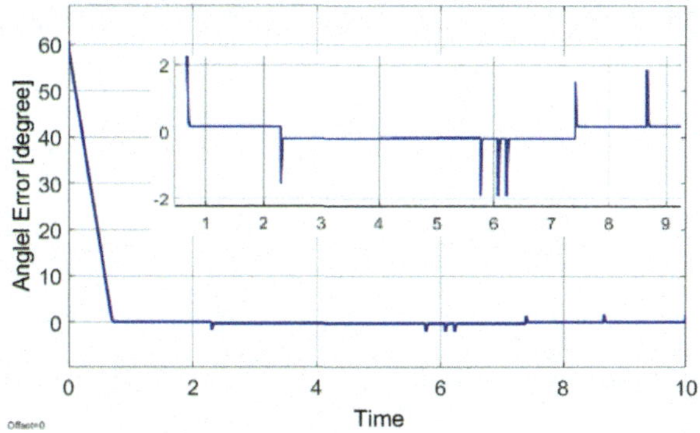

Fig. 12 Error of the second steering wheel (SMC)

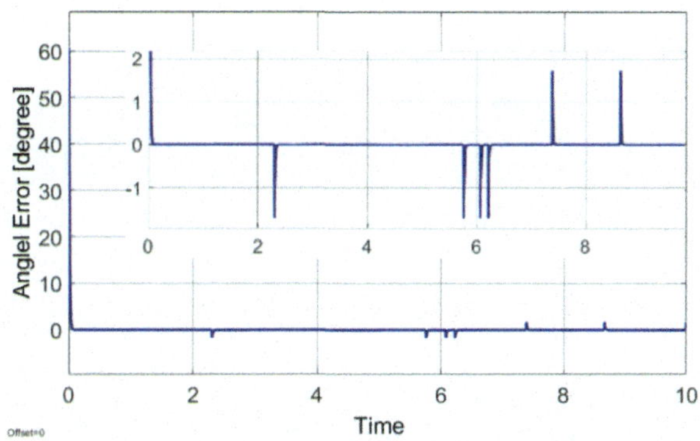

Fig. 13 Error of the second steering wheel (NSTSM)

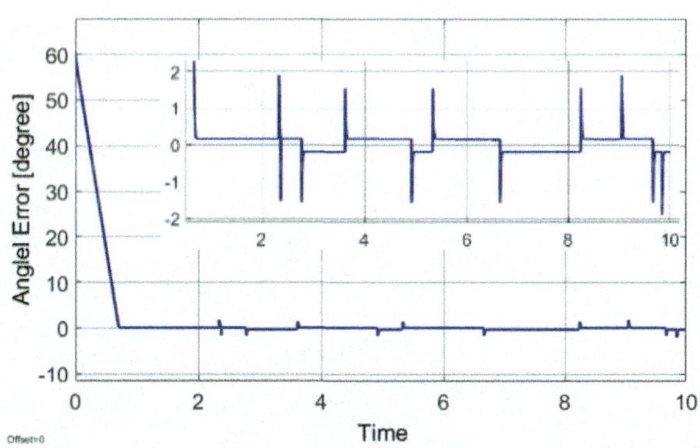

Fig. 14 Error of the third steering wheel (SMC)

The total lateral error that is shown in Fig. 18 and Fig. 19 also provides more proofs of the efficiency when applying NSTSM control method. Lateral error of NSTSM control method is close to – 0.02m or 2 cm when traditional SMC has lateral error about -0.045m.

Fig. 15 Error of the third steering wheel (NSTSM)

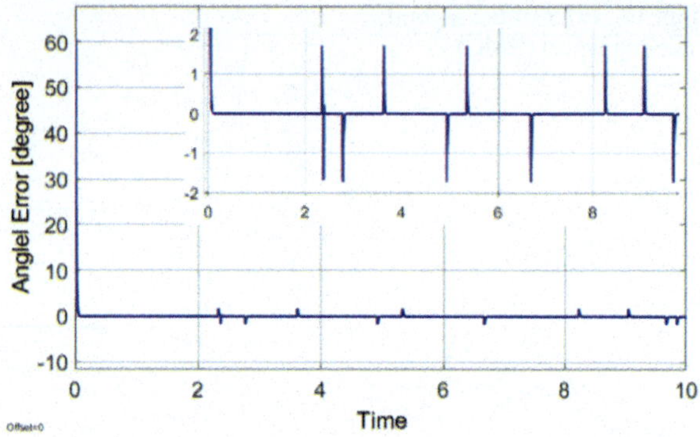

Fig. 16 Error of the forth steering wheel (SMC)

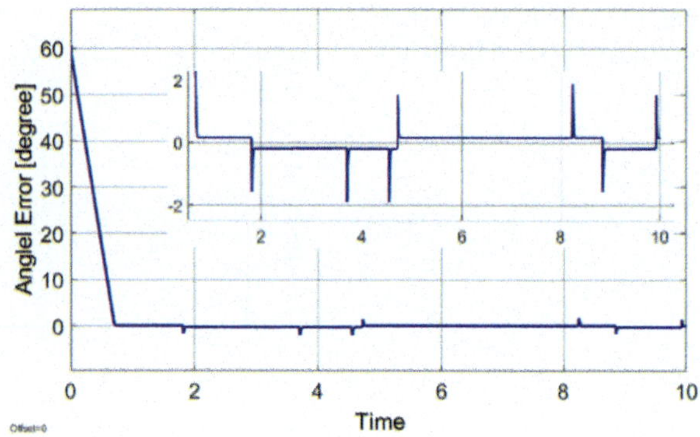

Fig. 17 Error of the forth steering wheel (NSTSM)

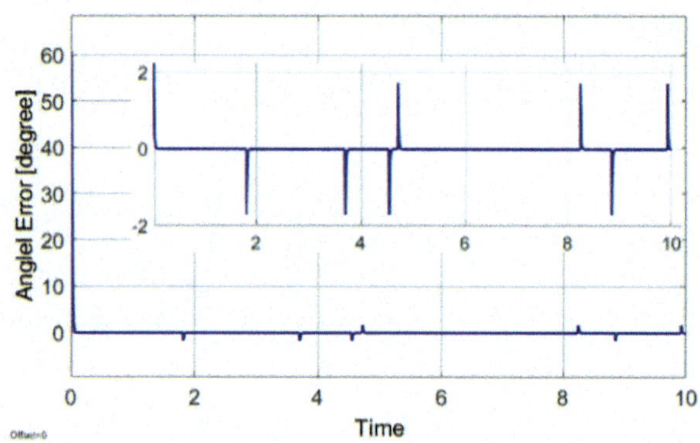

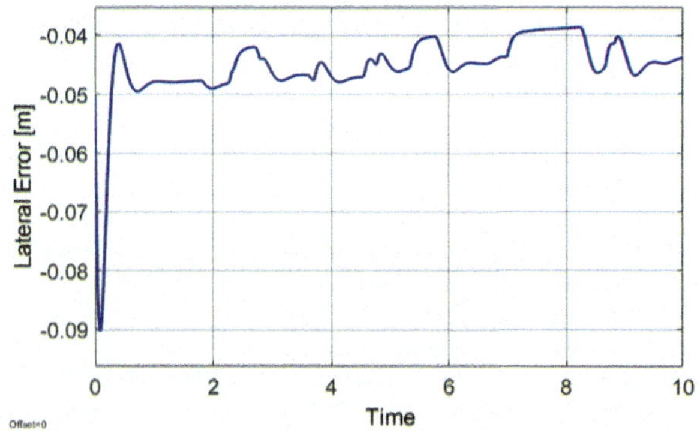

Fig. 18 Lateral error of 4WD4WS mobile robot (SMC)

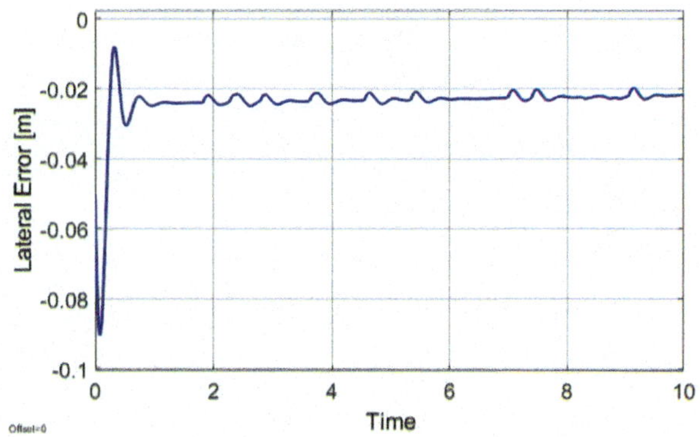

Fig. 19 Lateral error of 4WD4WS mobile robot (NTSM control)

4 Conclusion

The article has presented a method of using NSTSM control for the steering wheels of 4WD4WS mobile robot, whereby the quality of trajectory tracking is improved compared to traditional SMC, especially the convergence time of the system is proved to be finite and explicitly calculated, while ensuring the robustness of the system against disturbances. With that result, this method can be applied to other wheeled mobile robots which also have steering wheels, ensuring high efficiency. Research results are rigorously mathematically proven and visualized by simulation.

References

1. Dai, P., Katupitiya, J.: Force control for path following of a 4WS4WD vehicle by the integration of PSO and SMC. Vehicle Syst. Dyn. 56(11):1682–1716. https://doi.org/10.1080/00423114.2018.1435888
2. Ming-Han, L., Tzuu-Hseng, S.L.: Kinematics, dynamics and control design of 4WIS4WID mobile robots. J. Eng. **2015**, 6–16 (2015)

3. Wang, R., Hu, C., Wang, Z., Yan, F., Chen, N.: Integrated optimal dynamics control of 4WD4WS electric ground vehicle with tire-road frictional coefficient estimation. Mech. Syst. Signal Process. **60–61**, 727–741 (2015). ISSN 0888-3270. https://doi.org/10.1016/j.ymssp.2014.12.026
4. Jeong, Y., Yim, S.: Model predictive control-based integrated path tracking and velocity control for autonomous vehicle with four-wheel independent steering and driving. Electronics **10**, 2812 (2021). https://doi.org/10.3390/electronics10222812
5. Zhang, Z., Yang, C., Zhang, W., Xu, Y., Peng, Y., Chi, M.: Motion control of a 4WS4WD path-following vehicle: dynamics-based steering and driving models. Shock Vibr. **2021**, Article ID 8861159, 13. https://doi.org/10.1155/2021/8861159
6. Dang, N.K., Nguyen, V.: Dynamic virtual target guidance algorithm for path following control of a 4WD4WS mobile robot. Int. J. Multidiscip. Res. Growth Eval. **4**(4), 896–902. E-ISSN 2582-7138. https://doi.org/10.54660/.IJMRGE.2023.4.4.896-902
7. Dang, N.K., Nguyen, V.: Path following control for four-wheel-drive/four-wheel-steer mobile robot (4WD4WS) by front and rear independent driving method. J. Milit. Sci. Technol. 21–28. https://doi.org/10.54939/1859-1043.j.mst.FEE.2023.21-28
8. Huspeka, J.: Second Order sliding mode control of the DC motor. In: Fikar, M., Kvasnica, M. (eds.) Proceedings of the 17th International Conference on Process Control '09, Štrbské Pleso, Slovakia, pp. 134–139 (2009)

Testing of Thermal Management System Intended for Automotive Traction Electric Drives by Means of X-in-the-Loop Technology Featuring VPN-Connected Remote Facility

Ilya Kulikov, Kirill Karpukhin, and Denis Endachev

Abstract The paper presents the concept, design, and verification of a cyber-physical facility intended for testing of the thermal management system (TMS) designed for an electric vehicle with individually driven wheels. The basic functionality of the test system is implemented using Component-in-the-Loop (CiL) technology with the TMS embodied as a physical entity that interacts with virtual models of the electric drive components by means of devices emulating thermal and hydraulic behaviors of those components. The operating environment of the TMS is physically imitated by a model-driven air fans and a climatic chamber. Expanding the system's functionality is achieved by means of X-in-the-Loop technology that enables remote test facilities, housing different components of the vehicle and its powertrain, to connect and interact in real time through computer networks. Using that technique, the CiL system is connected to a remote laboratory that houses a setup intended to test the individual-wheel-drive controller in Software-in-the-Loop (SiL) mode employing virtual models of the electric vehicle, its powertrain, and the driving environment. Interaction between the CiL and the SiL facilities allows testing the TMS in operating modes occurring when the vehicle runs through a driving cycle in specified ambient conditions. The test results presented in the paper prove the developed X-in-the-Loop architecture to be operational and adequately replicating behavior of the studied system.

Keywords Electric vehicle · Thermal management system · Individual-wheel traction electric drive · Component-in-the-loop · X-in-the-loop · Cyber-physical testing

---

I. Kulikov · K. Karpukhin (✉) · D. Endachev
NAMI Russian State Scientific Research Center, Moscow, Russia
e-mail: K.Karpukhin@nami.ru

D. H. Tien et al. (eds.), *Proceedings of the 1st International Conference on Sustainability and Emerging Technologies for Smart Manufacturing*, Proceedings in Technology Transfer, https://doi.org/10.1007/978-981-97-7083-0_12

1 Introduction

A cyber-physical testing technology called Component-in-the-Loop (CiL) has been in use in research and development (R&D) of automotive thermal management systems (TMS) for over a decade now. The known descriptions of implementing this technology and using it in R&D encompass such applications as TMS for engines [1, 2], electric and hybrid powertrains [3, 4], as well as climatic systems for passenger vehicles [5] and refrigeration systems for cargo vehicles [6]. In particular, the works [1, 2] describe a CiL system intended to test an engine whose TMS is augmented with a heat regeneration system. The virtual part of the system includes a vehicle represented by a model of its passenger compartment and the driving environment modeled in terms of its thermal properties. Another example is presented in the articles [3, 4], where CiL technology is used for designing a TMS of an electric vehicle. In the described CiL system, the TMS interacts with physical thermal emulators representing the driveline components (traction electric drive and traction battery), as well as the passenger compartment and the driving environment. The emulators are interfaced with the software models of the corresponding entities. The functionality of the system allows studying thermal regulation of the whole vehicle including its powertrain and the habitable space. The paper [5] gives a description of a work focused on detailed elaboration of a thermal management system for the vehicle's inner space. The physical part of the CiL facility consists of an air conditioning unit and its accessories installed within a climatic chamber and equipped with measuring hardware. Thermal conditioning of the operating environment is implemented by a device playing a role of an interface between the tested hardware and the virtual part of the CiL system that simulates the thermal behavior of the vehicle's inner space and the ambient conditions. Yet another challenging task effectively addressed by CiL methods is testing refrigeration units intended for heavy-duty vehicles. Due to large dimensions of the thermally regulated space, it is a complicated and costly approach to emulate ambient conditions by placing a refrigerated trailer within a climatic chamber. The CiL solution to this problem presented in [6] is to simulate both the trailer and the environment as virtual entities while having the refrigerating unit housed within a relatively small chamber that controls the parameters of the input air in accordance with the output of the virtual model.

While CiL technology has its undoubted advantages and brings R&D to a new technical level, its conventional implementation is confined to a single laboratory where a CiL system resides. On the other hand, some R&D projects have geographically scattered nature with the test facilities for different components of the developed system located in remote regions and countries. To enable all the participants to test their components as parts of a single system interacting with one another, one can resort to connecting them by means of a computer network whose transmitting capacity is sufficient to use the exchanged data in real time operation of the involved facilities. One such effort, called XiL for EV project, has a number of automotive research and producing organizations all across Europe cooperating to implement advanced technologies within the chassis of a production electric vehicle

[7]. XiL or X-in-the-Loop has been introduced by the works [8, 9] as a concept for cyber-physical systems whose components, both hardware and software, reside at remote facilities connected by networks and interacting in real time in such a way that their combination reproduce the entire developed system (in particular, an electric vehicle), its user(s), and the operating environment. Further elaboration of this concept, including synthesis of network architectures, is described in the articles supporting XiL for EV project [10, 11], while the work [12] and its predecessors propose a method to virtually transfer thermal interactions between remote parts of XiL systems.

The part of the project assigned to NAMI Research Center was to design a TMS for the target electric vehicle and to implement it within a CiL facility that should be further incorporated into the XiL environment to interact with other participants of the project [13]. The sequel of the paper describes the concepts and the designs of both the TMS and the CiL system, as well as their main parts, physical and virtual, and the interface with an external XiL facility. This is followed by a description of testing the TMS by means of the CiL system connected to a remote laboratory via the VPN-based XiL network. The final section outlines the main results and conclusions of the work.

2 System Concept, Design, and Implementation

The main outcome of the project is a passenger electric vehicle (EV) equipped with four in-wheel traction electric drives implementing such advanced functions as individual-wheel traction and regenerative braking, and torque vectoring [11, 14]. Four "inverter – in-wheel motor" pairs operating independently require an appropriate thermal management system that maintains their temperatures within recommended ranges across all functional modes of the powertrain [15]. Several possible TMS layouts were analyzed by means of numerical experiments that simulated the EV, its powertrain, and the TMS operating in different driving cycles and temperature conditions [16, 17]. The competing layouts were compared with one another using criteria of energy efficiency, defined by power consumption of the TMS' components, and design feasibility defined by the number of employed components and complexity of their arrangement. The analysis led to the conclusion that the most favorable combination of these properties is offered by the layout featuring four parallel electrohydraulic pumps each feeding an inverter and a motor whose cooling jackets are connected in series. A schematic of this layout, which was approved for implementation within the CiL setup, is shown in Fig. 1. The numbers 1–4 represent the pumps, 5–8 denote the inverters, and 9–12 indicate the motors. Additionally, standard TMS elements are shown including the radiator (13) with its fan (14) and the expansion tank (15). Operating states of the TMS (i.e., cold/hot) are switched by a three-way valve (16). The TMS components are interconnected by the hydraulic pipelines depicted in blue.

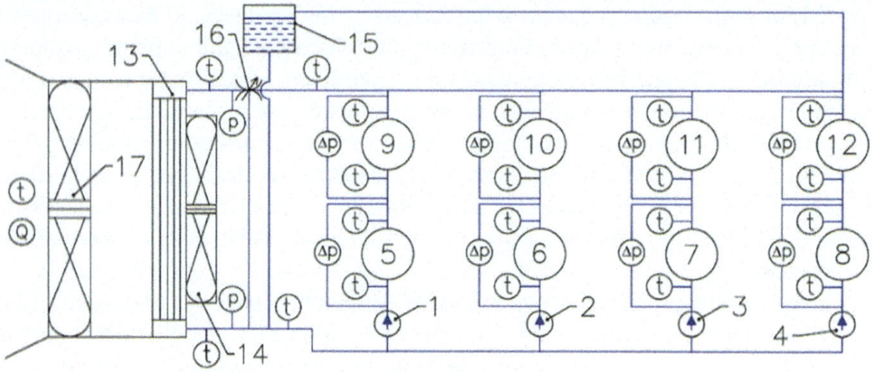

Fig. 1 Schematic of the TMS embedded into the test bench

The main concept underlying the developed CiL system is embodying the TMS and its operating environment as physical entities while simulating the EV, the powertrain, and the driving conditions by virtual models. The hardware part and the virtual part are interfaced in real time using both physical and data interactions. To this effect, the TMS has been embedded into a test bench where the cooling jackets of the inverters and the motors are emulated by devices that regulate the temperature and the pressure of the passing coolant in accordance with behavior of the corresponding driveline components simulated by the virtual models. Furthermore, the incoming air stream is emulated by the array of fans (position 17 in Fig. 1) that are also controlled by a virtual model in accordance with the simulated driving conditions. The operational environment of the TMS is replicated by placing the test bench inside a climatic chamber and conditioning it during a specified time to stabilize the required temperature regime. To provide feedback information to both the virtual part of the CiL system and the TMS controller, multiple sensors are installed in representative points of the TMS and the test bench (see Fig. 1) to measure temperatures (t), absolute (p) and differential (Δp) pressures of the coolant, and the speed of the airflow (Q).

Figure 2 shows control and data-exchange loops of the CiL test facility by an example of an electric drive component simulated in both the physical and virtual domains (the similar principle applies to all of the involved components). The diagram also shows a remote facility interacting with the CiL system via Internet using a virtual private network (VPN) established as a part of the XiL architecture. That external facility, located in Technical University of Ilmenau (which is ca. 1855 km away from the CiL laboratory at NAMI), houses a setup for software-in-the-loop (SiL) tests of the powertrain controller that implements individual-wheel torque distribution functions. SiL tests are conducted in a virtual environment that simulates the EV, its powertrain, and driving conditions. The TMS is not part of this simulator; instead, it is connected to the powertrain model as a remote object whose role is played by the CiL system at NAMI. Bilateral data exchange is established between the two facilities. The SiL setup transmits operating variables of the electric drive components resulted from the simulated driving schedule and ambient conditions. These variables include the rpms and torques of the electric drives. The data are received by the CiL facility and used as the inputs to the virtual models to calculate the thermal power generated within the electric drive components. From the CiL hardware, the model receives values of the measured variables characterizing the operating regime of the physical emulator, specifically, the coolant temperatures at the inlet ($T_{f,\text{in}}^{\text{meas}}$ in Fig. 2) and outlet ($T_{f,\text{out}}^{\text{meas}}$) of the emulator and the differential pressure of the fluid (Δp). The latter is used for indirect evaluation of the coolant flow rate Q_f using an experimentally obtained characteristic $Q_f = f(\Delta p)$. Based on the input data, the model calculates the output temperature of the component ($T_{f,\text{out}}^{\text{mod}}$), which is then compared with the measured one ($T_{f,\text{out}}^{\text{meas}}$). The resulting difference is relayed to the regulator that compensates it by generating a pulse-width modulated (PWM) signal, which is fed into the emulator to control the fluid heaters contained within. This process aligns the actual output temperature of the coolant with the reference value calculated by the model, making the physical emulator behave like an actual electric drive component in terms of thermal dynamics.

The main element of a virtual component that calculates temperature dynamics is a mathematical model consisting of two lumped masses: the fluid passing through the cooling jacket, and the component's body enveloped by air. It is assumed that convective heat transfer predominantly governs the thermal interaction between these masses. In the model, it should be considered that the actual heat flux originates within a specific volume (e.g., in motor's stator windings) that is separated from the fluid by walls and other elements whose combined thickness determines the heat transfer dynamics. However, models with lumped masses are not able to reflect this mechanism, as they simulate heat sources affecting the whole entities rather than their

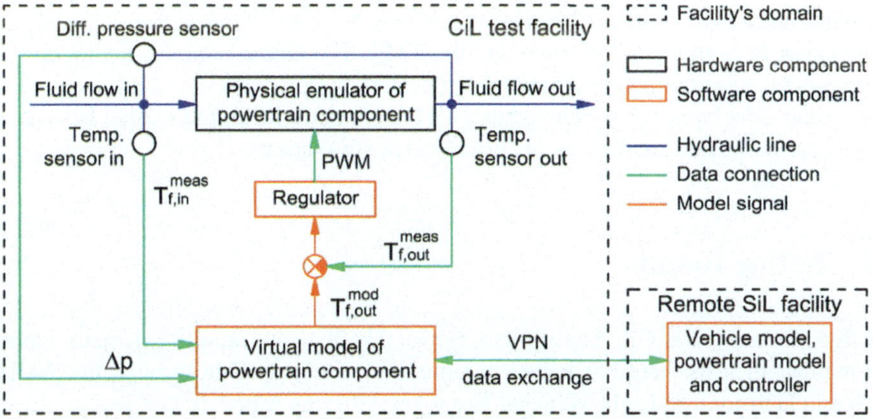

Fig. 2 Interactions between the software and hardware components of the CiL-TMS system connected and synchronized with a remote SiL facility via VPN-based XiL network

parts. To circumvent this drawback and imitate the distance from the heat source to the coolant flow, the thermal power can be shared between the lumped masses of the fluid and the component's body using a weighed proportion. The closer the heat source is to the coolant, the larger the corresponding weighting factor, and vice versa. With these considerations, the equations governing a two-mass thermal system are as follows:

$$\begin{cases} \frac{d\tilde{T}_f}{dt} = \frac{c_f \cdot \dot{m}_f (T_{f,in} - T_{f,out}) + k_f \cdot P_{heat} - \alpha_{f-b} \cdot A_{f-b} (\tilde{T}_f - \tilde{T}_b)}{c_f \cdot m_f} \\ \frac{d\tilde{T}_b}{dt} = \frac{\alpha_{f-b} \cdot A_{f-b} (\tilde{T}_f - \tilde{T}_b) + k_b \cdot P_{heat} - \alpha_{b-a} \cdot A_{b-a} (\tilde{T}_b - \tilde{T}_a)}{c_b \cdot m_b} \end{cases} \quad (1)$$

where the indices f, b, and a denote fluid, body, and air respectively, $\tilde{T}$ is the temperature of a lumped mass; $T_{f,\text{in}}$ and $T_{f,\text{out}}$ are the fluid temperatures at the inlet and outlet of the cooling jacket; $c_{\dots}$ and $m_{\dots}$ are the specific thermal capacity and the mass of a lumped entity; $\alpha_{\dots-\dots}$ is the coefficient of convective heat transfer between two adjacent lumped masses; $A_{\dots-\dots}$ is the area of the surface conducting the convective heat transfer; P_{heat} is the heat power generated within the powertrain component; k_b and k_f are the power weighting factors assigned to the component's body and the coolant, respectively. These factors should satisfy the condition $k_f + k_b = 1$.

The coefficients of heat transfer used in the model were calculated from empirical formulae taking into account material properties and geometry of the electric drive components, as well as the fluid temperature and flow regime [13, 16].

Figure 3 shows the laboratory at NAMI hosting the CiL facility. Section 1 of the figure demonstrates the test bench that incorporates the TMS and its controller, the emulators of the electric drive components, the fans imitating the incoming airflow, and the power electronics used for controlling the fluid heaters. Section 2 depicts the climatic chamber that houses the test bench during experiments, which require a specific temperature regime to be maintained accurately. To the right one can see the control equipment of the chamber including the display informing of the current conditions within. Section 3 shows the means for controlling and monitoring the operation of both the CiL system and the TMS. The laptop contains the models of the control systems and the virtual part of the CiL system working simultaneously in real time. The laptop also establishes a VPN connection with the remote laboratory and communicates the data during cooperative simulations.

3 Testing Results

After preparing the CiL facility and connecting it to its remote SiL counterpart, a number of tests were conducted to simulate the EV running through the WLTC (World Harmonized Light Vehicles Test Cycle) driving schedule that was approved as the basic test procedure of the project. Figure 4 demonstrates the results obtained in one of the test runs. The ambient air temperature was maintained at + 24 °C. Both physical (measured) and virtual (calculated) variables are shown in the graphs

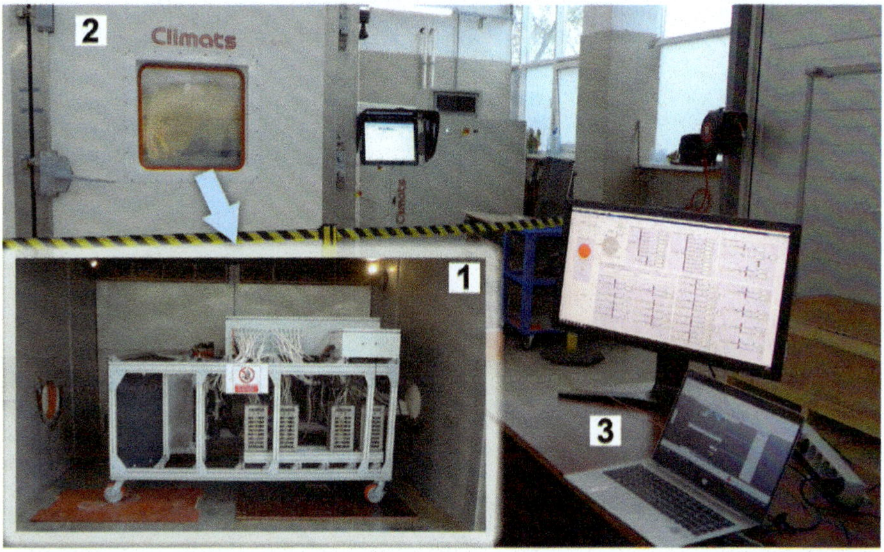

Fig. 3 CiL-TMS test facility

including (from top to bottom) the vehicle velocity (Veh. vel.) calculated by the SiL model, the velocity of the air passing through the radiator calculated by the CiL model and measured by the sensor installed in the test bench. The wheel torques at the front and rear wheels (Wheel trq) were calculated with the SiL model, driven by the torque distribution controller. The heat powers generated by the electric drive components associated with the front and rear wheels were calculated by the CiL models. The wheel torques and the heat powers visually coincide for the front and rear wheels, as the torque distribution was close to symmetrical. The two bottom plots show the measured and calculated temperatures (temp.) of the emulated electric drive components exemplified by the front right (FR) electric motor and the front right inverter.

From the graphs of the calculated and measured output temperatures, it can be concluded that the thermal emulators provided an acceptable tracking quality. For the motor, this is substantiated by the absolute error not exceeding 1.1 °C and the mean relative error of 0.77%. For the inverter, these errors amounted to 0.38 °C and 0.39%, respectively, attributed to a significantly lower dissipated power than that of the motor and, consequently, lower thermal dynamics required of the fluid heaters enabling them to operate with higher accuracy. The air speed tracking quality is characterized by the mean relative error of 9.2%, which can be considered acceptable.

The data exchange between the CiL and the SiL facilities was maintained with a ping time of 100–110 ms meaning that 50–55 ms were required for a data packet to reach its destination. Although such delay is excessive for rapid processes as wheel slip control, it proved adequate for relatively slow thermal dynamics of driveline components.

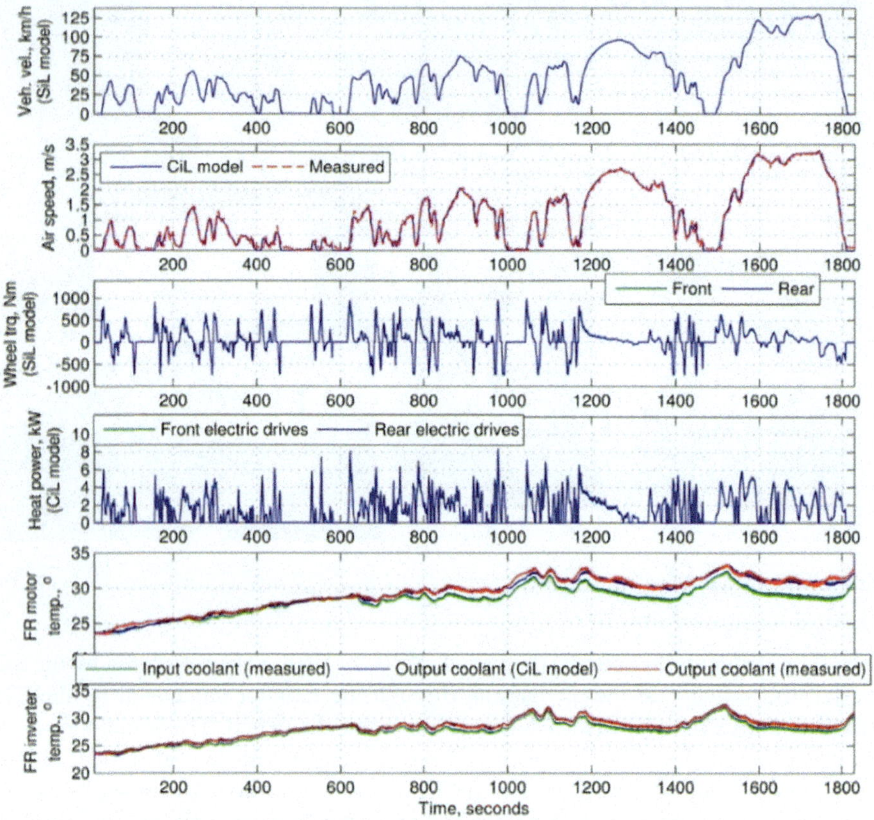

Fig. 4 Results of testing the TMS in WLTC driving cycle using the CiL setup and the remote SiL facility connected and synchronized via VPN

4 Conclusions

The thermal management system designed for the target electric vehicle features four independently regulated electrohydraulic pumps, each supplying an "inverter-motor" pair with the cooling jackets connected in series. The TMS was embedded into a test bench serving as the physical part of the CiL system. The virtual part simulates the electric drive components in terms of their thermal dynamics. The interface between the virtual and the physical parts of the system constitutes devices emulating temperature and hydraulic behaviors of the electric drive components in accordance with the variables calculated by the virtual models. Ambient conditions of the TMS are emulated by means of model-driven fans exerting the incoming airflow and a climatic chamber providing a stable temperature regime.

The CiL control architecture incorporates several closed loops to regulate the fluid heaters. The feedbacks are provided by temperature and pressure sensors, whose signals are fed into the virtual models to calculate the thermal dynamics of the electric drive components. The differences between the calculated and the measured fluid temperatures are compensated by the PWM-regulators, which control the fluid heaters making the temperature dynamics of the emulators replicate the actual behavior of the driveline components.

The functionality of the CiL system has been expanded by means of X-in-the-Loop technology that allows connecting remote testing facilities into one synchronized system using VPN-based computer networks. In this manner, the CiL setup has been connected to a remote laboratory housing a virtual environment for SiL testing of wheel-torque distribution functions. The SiL model of the EV, driving under specified conditions, provides information on the powertrain operating regimes necessary for the CiL models to calculate the heat dissipation of the powertrain components and enable simulation of the thermal processes driving the operation of the CiL facility.

An experimental study was conducted, with the CiL and the SiL facilities connected and synchronized via VPN, to verify characteristics of the XiL architecture. The results of simulating the EV operation in WLTC driving cycle demonstrated the system's operability and functionality. Based on the obtained errors, the tracking quality of the model-driven emulating devices was deemed satisfactory.

References

1. Baltzer, S., Lichius, T., Gissing, J., Jeck, P., Eckstein, L., Küfen, J.: Hardware-in-the-loop (HIL) simulation with Modelica—a design tool for thermal management systems. In: Proceedings of the 10th International Modelica Conference, Lund, Sweden, pp. 401–408, 10–12 Mar 2014
2. Gross-Weege, C., Lichius, T., Baltzer, S., Abel, D.: Control Design for a thermal hardware-in-the-loop test bench for automobile thermal management systems. IFAC-PapersOnLine **48**(15), 441–447 (2015)
3. Leighton, D.: Combined fluid loop thermal management for electric drive vehicle range improvement. SAE Int. J. Passeng. Cars Mech. Syst. **8**(2), 711–720 (2015)
4. Rugh, J.P.: Integrated vehicle thermal management—combining fluid loops in electric drive vehicles. Technical presentation NREL/PR-5400–58161, Vehicle Technologies Program Annual Merit Review (2013)
5. Eisele, M.R., Hwang ,Y., Radermacher, R.: Small-scale dynamic test facility for automotive thermal management systems. In: Vehicle Thermal Management Systems Conference and Exhibition (VTMS10), pp.523–535. Woodhead Publishing (2011)
6. Otten, R., Li, B., Alleyne, A.: Hardware-in-the-loop load emulation for air-conditioning and refrigeration systems. In: International Refrigeration and Air Conditioning Conference, Purdue, USA, July 12–15 2010, Paper 1101 (2010)
7. Schreiber, V., Ivanov, V., Augsburg, K., Noack, M., Shyrokau, B., Sandu, C., Els, P.S.: Shared and distributed X-in-the-loop tests for automotive systems: feasibility study. IEEE Access **6**, 4017–4026 (2018)
8. Albers, A., You, Y., Klingler, S., Behrendt, M., Zhang, T., Song, K.: A new validation concept for globally distributed multidisciplinary product development. In: Proceedings of the 20th International Conference on Industrial Engineering and Engineering Management, Bangkok, Thailand, pp. 231–242, 10–13 Dec 2013. Springer, Berlin (2013)

9. Tibba, G., Malz, C., Stoermer, C., Nagarajan, N., Zhang, L., Chakraborty, S.: Testing automotive embedded systems under X-in-the-loop setups. In: Proceedings of the IEEE/ACM International Conference on Computer-Aided Design (ICCAD '16), Austin, TX, USA, pp. 1–8, Nov 7–10 Nov 2016
10. Schreiber, V., Büchner, F., Lehne, C., Ivanov, V.: X-in-the-loop-Ansatz zur Entwicklung von Elektrofahrzeugen. MTZ Extra **26**, 18–23 (2021)
11. Beliautsou, V., Alfonso, J., Giltay, J., Büchner, F., Shyrokau, B., Castellanos, J. A., Ivanov, V.: Validation of integrated EV Chassis controller using a geographically distributed X-in-the-loop network. In: 2022 IEEE Vehicle Power and Propulsion Conference (VPPC), pp. 1–7 (2022)
12. Leitenberger, F., Matthiesen, S.: Thermal testing and system reliability: transferring thermal interactions by heat conduction through a peltier-based thermal coupling system. Energies **17**, 1127 (2024)
13. Kulikov, I., Karpukhin, K., Kurmaev, R.: X-in-the-loop testing of a thermal management system intended for an electric vehicle with in-wheel motors. Energies **13**, 6452 (2020)
14. Zavatsky, A.M., Keller, A.V., Shadrin, S.S., Makarova, D.A., Furletov, Y.M.: Development of an electric all-wheel-drive simulation model used to test torque distribution algorithms. Energies **16**(20), 7144 (2023)
15. Karpukhin, K.E., Kurmaev, R., Terenchenko, A., Struchkov, V., Tsimbaluk, M.: Aspects of construction of combined thermo-statics system for electric vehicle. ARPN J. Eng. Appl. Sci. **11**(23), 13674–13680 (2016)
16. Kulikov, I., Karpukhin, K.: Studying energy efficiency of thermal management systems designed for electric vehicles with in-wheel motors. Int. J. Emerg. Trends Eng. Res. **8**(6), 2654–2662 (2020)
17. Terenchenko, A., Karpukhin, K., Kurmaev, R.: Features of operation of electromobile transport in the conditions of Russia. In: 28th International Electric Vehicle Symposium and Exhibition 2015, EVS 2015 28, e-Motional Technology for Humans (2015)

Application of CVCC Model in Researching Mixture Formation and Fuel Combustion Process

Truong Nguyen Phi, Chien Nguyen Huy, Toan Nguyen Van, Dien Vu Minh, and Tuan Nguyen Van

Abstract Constant volume combustion chambers (CVCC) are prominent in fuel combustion research, especially in monitoring mixture formation and combustion processes. With the ability to maintain constant space and control the internal environment, the combustion chamber provides a precise platform for observing essential factors such as chemical reactions and pressure while the fuel is agitated. The results show that important information on fuel performance and mixture stability is collected, providing quality data for developing and validating mathematical models. This research focuses on simulating the mixture formation process inside a constant volume combustion chamber, an essential step toward conventional engine development. In this way, research can improve the performance and reduce emissions of conventional engines, promoting a deeper understanding of the fuel combustion process and assisting in technology development. Greener and sustainable for the auto industry.

Keywords Constant volume combustion chambers · Fuel · Mixture formation and combustion processes

1 Overview

Diesel engines have become an important energy source globally, with higher efficiency than gasoline engines. Diesel engines are considered a solution to reduce climate change from vehicles [1, 2]. Due to this high efficiency, engine manufacturers are focusing on further developing solutions to reduce environmental pollution from

---

T. N. Phi · C. N. Huy · T. N. Van · D. V. Minh
School of Mechanical and Automotive Engineering, HaNoi University of Industry, No. 298, Cau Dien Street, Bac Tu Liem, Hanoi 100000, Vietnam

T. N. Van (✉)
School of Transportation Engineering, University of Transport Technology, No. 54 Trieu Khuc Street, ThanhXuan District, Hanoi 100000, Vietnam
e-mail: nguyenvantuan@utt.edu.vn

D. H. Tien et al. (eds.), *Proceedings of the 1st International Conference on Sustainability and Emerging Technologies for Smart Manufacturing*, Proceedings in Technology Transfer, https://doi.org/10.1007/978-981-97-7083-0_13

diesel engines [3]. Within the diesel engine domain, precise fuel/air mixture formation is paramount for regulating combustion dynamics, thus impacting environmental sustainability and operational effectiveness [4, 5]. Constant volume chambers represent invaluable assets for fuel research [6]. Solutions for forced-vortex combustion chambers or dual-vortex combinations have been introduced through design and optimization to create optimal fuel/air mixtures and minimize environmental pollutants from diesel engines. Research results on optimal vortex formation in this combustion chamber have shown feasible results in reducing its emissions [7]. To better understand the fuel/air mixture formation mechanism in a forced-vortex combustion system, the diffusion flame was visualized in a constant volume vessel. The main objective of this study was to evaluate the effects of temperature and pressure on the injection process characteristics, along with the fuel–air mixture properties [8]. Later, the research was expanded by using a conventional Bosch pipe system to inject diesel fuel into the combustion chamber. It aims to probe into the formation of biodiesel combustion mixtures employing a constant volume chamber alongside optical visualization. A comprehensive analysis of spray characteristics, particularly those affecting mixture formation such as spray evaporation and hindrance, will be undertaken, highlighting key factors such as Fast compressor, Ignition delay, nozzle, Spray system, Air movement, and Surrounding environment [9].

Different designs of piston crowns were subjected to numerical simulations to assess combustion and exhaust gas emissions. These simulations incorporated a combustion model capable of accommodating the spontaneous combustion of heterogeneous mixtures, along with a detailed phenomenological soot model. Experimental measurements were conducted using the two-color method in a constant volume vessel under high temperature and pressure conditions to analyze the soot formation characteristics linked to various piston tip shapes [10]. Based on an understanding of the impact of the molecular structure of biofuels on soot formation and particle morphology, the impact of alcohol, ether, esters, and biodiesel on injection and combustion processes in a constant volume chamber in recent years is analyzed. Through the comprehensive evaluation of the effects, alcohol shows the best injection, atomization, and combustion performance, while biodiesel shows the worst performance [11].

Simulating the mixture formation and fuel combustion process in a constant volume combustion chamber plays an important role. This helps build an accurate and reliable database to support empirical research on this process in CVCC. Simulation helps understand important factors such as chemical reactions and pressure and provides a rigorous basis for adjusting results during experiments. Besides, it also ensures that the mixture formation and combustion processes in CVCC are reliably evaluated through prior simulation. This is important in optimizing experimental conditions and ensuring consistency between experimental results and simulation models.

2 Theoretical Basis of Simulation in CVCC

2.1 CVCC Pressure

Controlling the gas composition before the combustion process plays an important role. The principle of Dalton's Law is applied to measure partial pressure. To make this process more accurate, the compressibility of the gas mixture is determined based on the partial pressure of each component in the gas mixture, as described in the formula [12].

2.2 Low Flammability of Fuel

After the initial combustion stage, the excess of O_2 plays an essential role in the spontaneous combustion of injected diesel fuel. To ensure adequate O_2 supply, low concentrations of the pre-flammable gas mixture (especially C_2H_2) can be calculated from the pure mixture using the Lower Flammability Limit (LFL) as described in the formula. This method of calculating the flammability limit was based on Le Chatelier's formula [13].

$$\mathrm{LFL}(\%) = \frac{C_p N_2 \Delta T_{\mathrm{ad}}}{-\Delta h_c} \tag{1}$$

In there:

C_p, N_2: Specific heat of nitrogen; ΔT_{ad}: Temperature of the adiabatic flame

Δh_c: Enthalpy.

2.3 Temperature of an Adiabatic Flame

In the adiabatic combustion process, the temperature of the combustion products was determined as the temperature of the adiabatic flame. This is the maximum temperature value that can be achieved when there is an equilibrium between products and reactants, according to the given formula [14].

$$\sum_{\mathrm{Prod}} N_i\left(h_f + \Delta h\right)_i = \sum_{\mathrm{React}} N_i\left(h_f + \Delta h\right)_i \tag{2}$$

h_{f}: Is the enthalpy change and Δh is the other changes in enthalpy.

Table 1 Properties of simulated fuel

Density at 15 °C	8379
Kinematic viscosity at 40 °C (mm^2/s)	2.5
Calorific value (MJ/kg)	42.83
Oxygen content (%)	0
Cetane number value	51.3
Surface tension at 20 °C, (N/m)	0.023
Air to fuel ratio (A/F)	15.0
Latent heat of vaporization (kJ/kg)	250
Boiling point (°C)	180–360

3 CVCC Combustion Chamber Model and Boundary Conditions

3.1 CVCC Combustion Chamber Model

The numerical simulation model is built based on the geometric design parameters of the combustion chamber with constant volume. The parameters of the fuel charging and combustion process are also set to be consistent with the parameters of the actual model. Experience. Therefore, the numerical simulation results can be used to compare, evaluate, and analyze the experiment results. The tested CVCC combustion chamber has a combustion chamber diameter of 80 mm and a combustion chamber length of 90 mm. The intake and exhaust valves' diameter is 5/5 mm.

Based on the geometric parameters in Table 1, the computational mesh model is built using the 3D-CAD tool integrated into Star-CCM+. According to the provisions of the In-Cylinder CAD model, it includes three main components: Cylinder—is the volume where the combustion reaction of fuel occurs in the combustion chamber; IntakeValve and ExhaustValve—are valves that feed fuel and discharge the product mixture after combustion. The spatial domain where the combustion reaction (Cylinder) occurs, and the intake and exhaust valves in the simulation model are illustrated in Fig. 1.

3.2 Fuel Used in Research

The fuel used in the simulation is diesel oil with properties as shown in Table 1

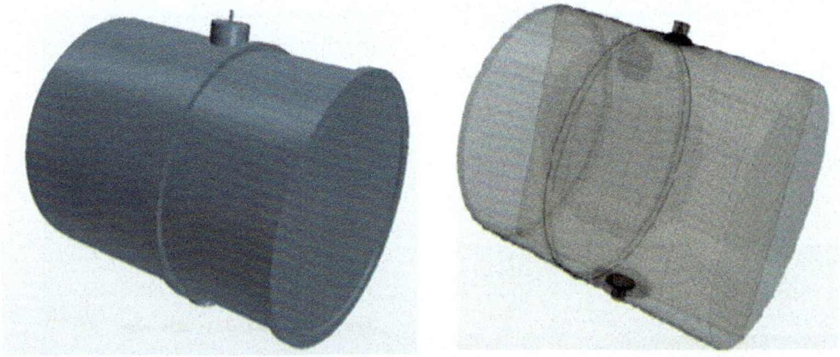

Fig. 1 CVCC simulation model

3.3 Boundary Conditions

The combustion process in an engine is a process that evolves. In the current simulation model, the PISO (Pressure-Implicit with Splitting of Operators) algorithm is used to solve the main equations sequentially. Accordingly, the velocity component equations are solved in turn before solving the continuity, energy, and combustion reaction equations. The combustion process of the fuel is simulated using the ECFM-3Z model, which allows the simulation of complex turbulent mixtures, flame propagation, diffuse combustion, and pollutant emissions typical for Modern mechanisms of internal combustion engines.

Simulate the effect of temperature on fuel's mixture formation and combustion process, including fuel injection time, injection end time, and fuel evaporation time. Temperature varies from 300 to 1200 K with divisions: 300, 450, 600, 900, and 1200 K. Amount of fuel injected into the combustion chamber: 0.039 (gram).

4 Results and Discussion

4.1 Evaluate the Reliability of the Model

Figure 2 shows the experimental and simulated pressure graphs. The results show that the graphs between simulation and experiment have similar trends. The peak pressure value in simulation and experiment differs by 3%. This proves that the model is secure when simulating the combustion process of fuel in CVCC.

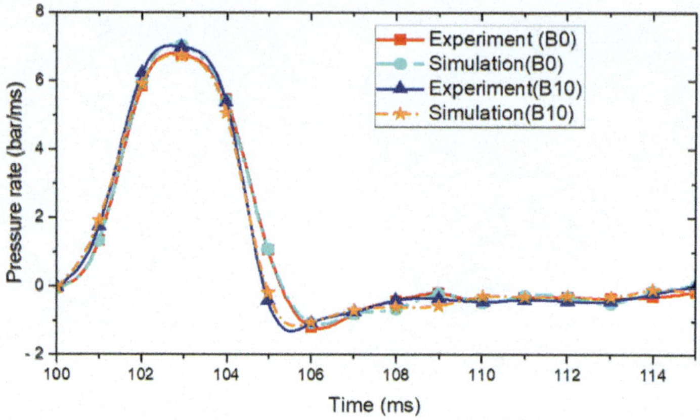

Fig. 2 Experimental and simulated pressure graph

4.2 *Evaporation Process of Fuel in CVCC*

The pressure evolution in the volumetric combustion chamber remains constant over the period 1e-07 (s). Examining the pressure dynamics within the combustion chamber reveals an initial phase where the pressure remains constant. However, upon introducing fuel to blend with the chamber gas, a subsequent decrease in pressure occurs, as indicated by the blue trend. This decline results from the fuel absorbing energy from the combustion chamber. Following this, the pressure gradually rebounds and continues to increase. This phenomenon stems from the energy absorption by the fuel, leading to its evaporation and subsequent mixing with the gas already present in the combustion chamber. Following this, the mixture ignites, leading to a rapid rise in pressure within the combustion chamber, as indicated by the red coloration. The visualization depicts an initial ignition along the walls of the combustion chamber, followed by ignition propagation throughout the remaining air mixture. This phenomenon is driven by the elevated temperature of the combustion chamber walls relative to other areas, which facilitates the initiation of ignition centers within the air mixture in the chamber.

On the other hand, the pressure in the combustion chamber is not uniform in all positions. This result has excellent significance when conducting experiments. Ensuring the accuracy of experimental findings necessitates measuring pressure at various locations within the combustion chamber, as depicted in Figs. 3 and 4.

Figures 5 and 6 clearly describe the heating process in the CVCC, from fuel injection to its conversion to vapor and combustion, with fluctuations in temperature. Initially, when fuel is injected, the temperature inside the CVCC decreases. Then there is a sudden rise, the temperature graph increases rapidly and plateaus, then gradually decreases with decreasing slope. The fuel is then dispersed and vaporized, combining with the gas in the CVCC. The image also shows temperature unevenness in the combustion chamber, with areas near the walls of the combustion chamber

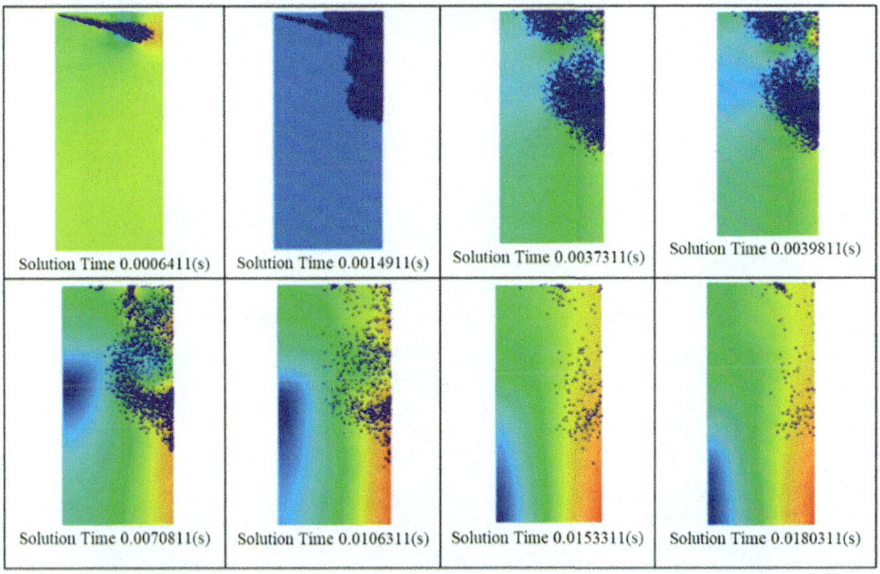

Fig. 3 Image of pressure evolution in CVCC

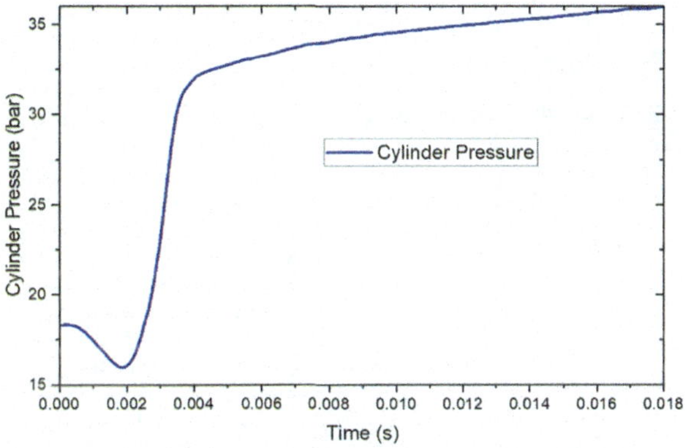

Fig. 4 Graph showing pressure in CVCC. The process of increasing Pressure

having higher temperatures than other locations. The cause of this phenomenon is due to the combustion chamber wall having a higher temperature than the space inside the combustion chamber and due to the heat transfer process from the combustion chamber wall to the inside of the CVCC.

Combustion process in internal combustion engines. According to research [15], the rate of pressure and temperature increases quickly and then decreases rapidly due to the change in volume of the combustion chamber and heat transfer to the outside of the cylinder wall. The combustion process in CVCC increases temperature and pressure quickly. It decreases slowly because the chamber volume is kept the same,

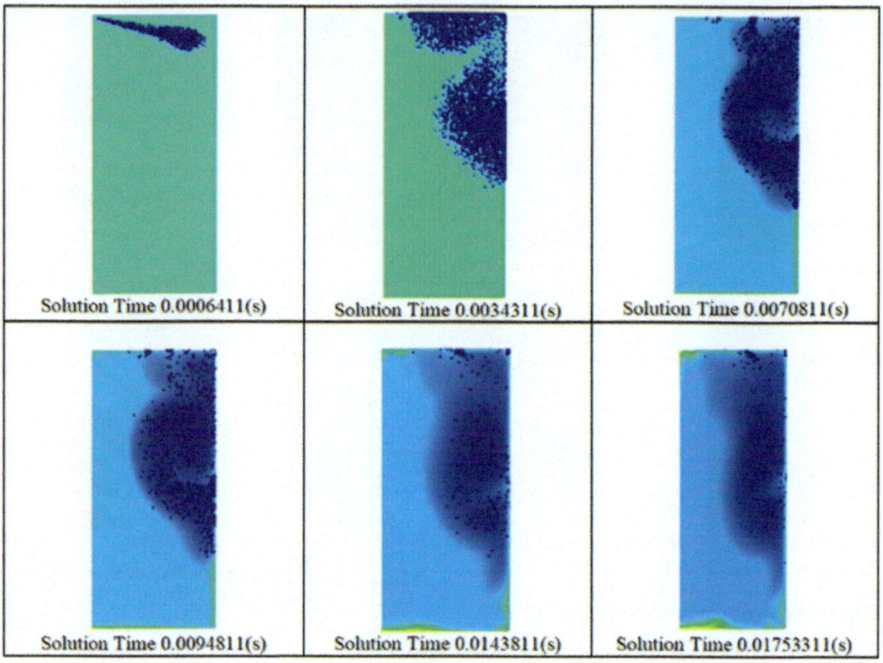

Fig. 5 Image of temperature evolution in CVCC

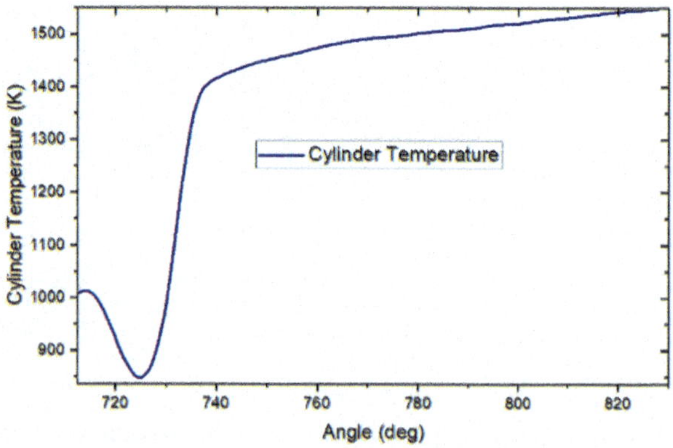

Fig. 6 Graph of temperature evolution in CVCC

and the decrease in pressure and temperature is due to energy loss due to heat transfer to the walls of the combustion chamber.

5 Conclusion

The constant volume combustion chamber model demonstrates its unique benefits in predicting experimental results on traditional engines. It describes the advantages of studying various fuels and analyzing the influence of different factors. The model provides the ability to adjust initial factors, such as pressure, temperature, air/fuel ratio, mixing, and ignition time, to reflect the performance of the study under real-world conditions… This is a great advantage compared to performing experimental experiments. Therefore, using the CVCC model allows for assessing the quality of the mixture and predicting the temperature and pressure of the mixture in the combustion chamber over time. This result provides a basis for calculations and improvements on traditional engines to increase economic and technical efficiency while reducing toxic emissions into the environment.

References

1. Şener, R., Zafer Gül, M.: Optimization of the combustion chamber geometry and injection parameters on a light-duty diesel engine for emission minimization using multi-objective genetic algorithm. Fuel **304**(15), 121379 (2021)
2. De Risi, A., Donateo, T., Laforgia, D.: An Innovative methodology to improve the design and the performance of direct injection diesel engines. Int. J. Engine Res. **5**(5), 425–441 (2004)
3. Yu, L., Jun, L., Chao, J.: Fuel spray and combustion characteristics of butanol blends in a constant volume combustion chamber. Energy Convers. Manage. **105**, 1059–1069 (2015)
4. Agarwal, A.K., Som, S., Shukla, P.C., Goyal, H., Douglas, L.: In-nozzle flow and spray characteristics for mineral diesel, Karanja, and Jatropha biodiesels. Appl. Energy **156**(15), 138–148 (2015)
5. Li, X., Wei, Y., Su, L., Liu, F.: Mixing and combustion mechanisms within lateral swirl combustion system (LSCS) in a DI diesel engine. Appl. Therm. Eng. **123**, 7–18 (2017)
6. Olsen, D.B., Lisowski, J.M.: Prechamber NOx formation in low BMEP 2-stroke cycle natural gas engines. Appl. Thermal Eng. **29**(4), 687–694 (2019)
7. Lee, S.W., Lee, H.S., Park, Y.J., Cho, Y.S.: Combustion and emission characteristics of HCNG in a constant volume chamber. J. Mech. Sci. Technol. **25**(2), 489–494 (2011)
8. Su, L., Li, X., He, X., Liu, F.: Experimental research on the diffusion flame formation and combustion performance of forced swirl combustion system for DI diesel engines. Energy Convers. Manage. **106**, 826–834 (2015)
9. Jaat, M., Khalid, A., Sapit, A., Basharie, S., M., Andsaler, A., R., Ramsy, H.: Effects of temperature and ambient pressure on spray characteristics of biodiesel combustion. Appl. Mech. Mater. 773–774 (2015)
10. Jung, J.Y., Park, S., Bae, C.: Combustion characteristics of gasoline and n-butane under lean stratified mixture conditions in a spray-guided direct injection spark ignition engine. Fuel **187**, 146–158 (2017)
11. Kim, S., Fukuda, D., Shimo, D., Kataoka, M., Nishida, K.: Simultaneous improvement of exhaust emissions and fuel consumption by optimization of combustion chamber shape of a diesel engine. Int. J. Engine Res. **18**(5–6) (2017)
12. Shi, Z., Lee, C., F., Wu, H., Li, H., Wu, Y., Zhang, L., Liu, F.: Effect of nozzle diameter on macroscopic spray behavior of heavy-duty diesel engine under cold-start conditions. Atomization Sprays **29**(8), 741–762 (2019)
13. Mott, R.L., Untener, J.A.: Applied Strength of Materials. CRC Press, Boca Raton (2021)

14. Mashuga, C., Crowl, D.: Derivation of Le Chatelier's mixing rule for flammable limits. Process Safety Progress **19**(2), 112–117
15. Nghia, N.T., Phong, T.D., Khoa, N.X.: Simulation study on the effect of biodiesel ratio derived from waste cooking oil on performance and emissions of a single cylinder diesel engine. Eng. Technol. Sustain. Develop. **32**(3), 042–050 (2022)

Free Vibration Analysis on Inner/Outer Ring-Stiffened Combined Shells Made by Functionally Graded Material

Manh Cuong Nguyen, Quoc Hung Nguyen, and Tuan Hai Nguyen

Abstract This paper is devoted to investigating the dynamic mechanical behavior of composite shells composed of modern Functionally Graded Materials (FGMs) using the Continuous Element Method (CEM). The study focuses on analyzing submarine-like composite shell configurations, comprising cylindrical shells, conical shells, and inner/outer ring stiffeners constructed from FG materials. The natural frequencies of these structures are evaluated considering the diverse mechanical properties of FG materials, including coefficients a, b, c, and various exponent values (p). The findings demonstrate the superiority of CEM in terms of computational efficiency and accuracy, even at high frequencies.

Keywords Mechanics of materials · FGM · Continuous element method · Dynamic stiffness matrix · Combined shell · Ring stiffener · Shell vibration

1 Introduction

The combined shell structure including conical shells, cylindrical shell, and inner/outer ring stiffeners is one of the popular structures in various engineering and technology fields: submarine, aerospace, and construction. One significant advancement in this field is the utilization of modern Functional-graded Materials (FGM) to construct diverse hierarchical shell structures. FGMs exhibit continuously varying properties in space, enabling substantial customization of material structures with the potential to enhance their performance and functionality in demanding applications.

M. C. Nguyen (✉) · Q. H. Nguyen · T. H. Nguyen
School of Mechanical Engineering, Hanoi University of Science and Technology, 1 Dai Co Viet Street, Hai Ba Trung District, Hanoi, Vietnam
e-mail: cuong.nguyenmanh1@hust.edu.vn

D. H. Tien et al. (eds.), *Proceedings of the 1st International Conference on Sustainability and Emerging Technologies for Smart Manufacturing*, Proceedings in Technology Transfer, https://doi.org/10.1007/978-981-97-7083-0_14

Up to now, there have been various methods for analyzing the dynamics of different types of shells made by isotropic and laminate composite material such as: analytical method [1–6], Meshless method [7], Finite Element approach [8]. The vibration analysis of combined shells has been referenced in previous works [5, 6]. However, these studies did not consider structures equipped with both inner and outer ring stiffeners and fabricated from FGM.

This research endeavors to explore the dynamic mechanical behavior of combined conical-cylindrical-conical shells composed of FGM, incorporating inner and outer ring stiffeners. The investigation employs the CEM to analyze the structural dynamics. This method saves the computational time and data storage volume and assures the exactness for all frequency zones. It has been presented to overcome difficulties of other approaches, especially Finite Element Method when dealing with medium and high frequencies. Many CEM studies have been performed for isotropic, composite, and FGM shell structures [9–12] but the Continuous Element model for combined shell with inner and outer ring stiffeners has never been presented before.

Here, a new continuous element model will be constructed for determining the natural frequencies of the studied FGM structures. In addition, the effect of various material parameters is also under investigation. The understanding of all these mechanics factors will provide crucial insights for the practical design and application of FGM materials.

2 FG Material Model

The considered FGM here is a continuous material composed of a mixture which can be a combination of two materials from various types of substances. The FG material has a continuous variation in the ratio of metal and ceramic along the thickness of the material. The Young modulus $E(z)$, Poisson's ratio $\mu(z)$, and mass density $\rho(z)$ change continuously as below:

$$\begin{aligned} E(z) &= (E_c - E_m)V_c + E_m \\ \mu(z) &= (\mu_c - \mu_m)V_c + \mu_m \\ \rho(z) &= (\rho_c - \rho_m)V_c + \rho_m \end{aligned} \tag{1}$$

In this context, the subscript "c" refers to ceramic components, "m" denotes metallic components, and V_C signifies the volume fraction distribution of ceramic material, which adheres to the distributions with four-parameter power-law denoted as $\text{FGM}_{I(p,a,b,c)}$ [4]:

$$\text{FGM}_{I(p,a,b,c)} : V_c = \left[1 - a\left(\frac{1}{2} + \frac{z}{h}\right) + b\left(\frac{1}{2} + \frac{z}{h}\right)^c\right]^p \tag{2}$$

The change in the volume fraction ratio between metal and ceramic along the shell thickness depends on the values of parameters *p, a, b, c*. A particular instance of Eq. (2) arises when $p = 0$ or $p = \infty$, resulting in the shell material becoming isotropic and homogeneous. This implies that the material layer will respectively become entirely ceramic with $E(z) = E_C$, $V_C = 1$ or entirely metallic with $E(z) = E_m$, $V_m = 1$. In this study, two cases are examined for analysis: one where $a = 1$, $b = 0$, $c = 2$, and another where $a = 1$, $b = 1$, $c = 2$.

3 Theory of FGM Conical Shell

3.1 Kinematic Relations

Consider a FGM cone in (x, θ, z) coordinate system described in Fig. 1. Let x represent the axis aligned with the generatrix, θ denote the coordinate corresponding to the rotational angle, and z signify the orthogonal axis perpendicular to the cone's surface. The radius R_x of the cone at a point located at a distance x from the origin is determined through the following computation:

$$R_x = R_1 + x \sin \alpha \tag{3}$$

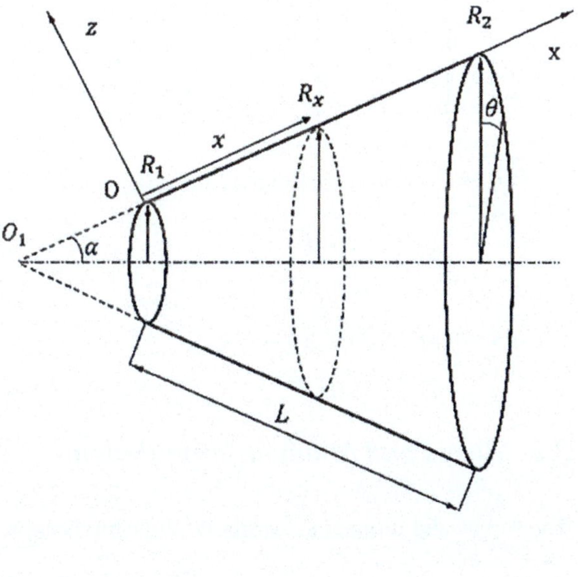

Fig. 1 A general conical shell model

Using the Mindlin-Reissner theory, the displacements at any point on the middle FGM shell surface are expressed by [1]:

$$\begin{aligned} u(x,\theta,z,t) &= u_0(x,\theta,z,t) + z\phi_x(x,\theta,z,t) \\ v(x,\theta,z,t) &= v_0(x,\theta,z,t) + z\phi_\theta(x,\theta,z,t) \\ w(x,\theta,z,t) &= w_0(x,\theta,z,t) \end{aligned} \tag{4}$$

In (4), u, v, w are the displacement along x, z, and θ directions. Angles ϕ_x and ϕ_θ represent the rotation and angular displacements around the θ and x axes, respectively and t is the variable of time. The strain–displacement relationships of the shell are:

$$\begin{aligned} &\varepsilon_x = \frac{\partial u_0}{\partial x}; \quad k_x = \frac{\partial \phi_x}{\partial x} \\ &\varepsilon_\theta = \frac{1}{R}\left(u_0 \sin\alpha + \frac{\partial v_0}{\partial \theta} + w_0 \cos\alpha\right); \quad k_{x\theta} = \frac{\partial \phi_\theta}{\partial x} - \frac{\partial \sin\alpha}{R}\phi_\theta \\ &\varepsilon_{x\theta} = \frac{\partial v_0}{\partial x} + \frac{1}{R}\frac{\partial u_0}{\partial \theta} - \frac{\partial \sin\alpha}{R}v_0; \quad k_\theta = \frac{1}{R}\left(\phi_x \sin\alpha + \frac{\partial \phi_\theta}{\partial \theta}\right) \\ &\gamma_{xz} = \frac{\partial w_0}{\partial x} + \phi_x; \quad \gamma_{\theta z} = \frac{-\cos\alpha}{R}v_0 + \frac{1}{R}\frac{\partial w_0}{\partial \theta} + \phi_\theta \end{aligned} \tag{5}$$

The stress–strain relationship in a conical shell is written as:

$$\begin{Bmatrix} \sigma_x \\ \sigma_\theta \\ \tau_{x\theta} \\ \tau_{xz} \\ \tau_{\theta z} \end{Bmatrix} = \begin{bmatrix} Q_{11}(z) & Q_{12}(z) & 0 & 0 & 0 \\ Q_{12}(z) & Q_{11}(z) & 0 & 0 & 0 \\ 0 & 0 & Q_{66}(z) & 0 & 0 \\ 0 & 0 & 0 & Q_{66}(z) & 0 \\ 0 & 0 & 0 & 0 & Q_{66}(z) \end{bmatrix} \begin{Bmatrix} \varepsilon_x \\ \varepsilon_\theta \\ \gamma_{x\theta} \\ \gamma_{xz} \\ \gamma_{\theta z} \end{Bmatrix} \tag{6}$$

with $Q_{ij}(z)$ being the coefficient of elasticity defined as follows:

$$Q_{11}(z) = \frac{\boldsymbol{E}}{1-\boldsymbol{\mu}^2}, \quad Q_{12}(z) = \frac{\boldsymbol{\mu E}}{1-\boldsymbol{\mu}^2}, \quad Q_{66}(z) = \frac{\boldsymbol{\mu E}}{2[1+\boldsymbol{\mu}]} \tag{7}$$

where $\boldsymbol{E}$ and $\boldsymbol{\mu}$ are functions of z.

3.2 *Stress and Moment Formulation*

The forces and moments—strain relationship is expressed in matrix form:

$$\begin{Bmatrix} N_x \\ N_\theta \\ N_{x\theta} \\ M_x \\ M_\theta \\ M_{x\theta} \\ Q_x \\ Q_\theta \end{Bmatrix} = \begin{bmatrix} A_{11} & A_{12} & 0 & B_{12} & B_{12} & 0 & 0 & 0 \\ A_{12} & A_{11} & 0 & B_{21} & B_{12} & 0 & 0 & 0 \\ 0 & 0 & A_{66} & 0 & 0 & B_{66} & 0 & 0 \\ B_{11} & B_{12} & 0 & D_{11} & D_{12} & 0 & 0 & 0 \\ B_{21} & B_{11} & 0 & D_{21} & D_{11} & 0 & 0 & 0 \\ 0 & 0 & B_{66} & 0 & 0 & D_{66} & 0 & 0 \\ 0 & 0 & 0 & 0 & 0 & 0 & kF_{44} & 0 \\ 0 & 0 & 0 & 0 & 0 & 0 & 0 & kF_{55} \end{bmatrix} \begin{Bmatrix} \varepsilon_x \\ \varepsilon_\theta \\ \varepsilon_{x\theta} \\ k_x \\ k_\theta \\ k_{x\theta} \\ \gamma_{xz} \\ \gamma_{\theta z} \end{Bmatrix} \tag{8}$$

The analysis employs a shear correction factor, $k = 5/6$, to adjust shear effects, which $k = 5/6$ is the shear correction factors. N_x, N_θ, and $N_{x\theta}$ represent in-plane force resultants, while M_x, M_θ, and $M_{x\theta}$ denote moment resultants. Q_x and Q_θ indicate resultant transverse shear forces. The FGM under investigation is considered to be linearly elastic. Here, the tensile stiffness A_{ij}, bending stiffness D_{ij}, and tensile-bending coupling stiffness B_{ij} are expressed as:

$$A_{ij} = \int_{-\frac{h}{2}}^{\frac{h}{2}} Q_{ij}(z)\mathrm{d}z, \quad i,j = 1,2,6 \quad B_{ij} = \int_{-\frac{h}{2}}^{\frac{h}{2}} zQ_{ij}(z)\mathrm{d}z, \quad i,j = 1,2,6$$

$$D_{ij} = \int_{-\frac{h}{2}}^{\frac{h}{2}} z^2 Q_{ij}(z)\mathrm{d}z, \quad i,j = 1,2,6 \quad F_{ij} = \int_{-\frac{h}{2}}^{\frac{h}{2}} Q_{ij}(z)\mathrm{d}z, \quad i,j = 4,5 \tag{9}$$

3.3 Equilibrium Equations

The equilibrium equations for a cone according to the First-order Shear Deformation Theory (FSDT) are given by:

$$\begin{aligned}
&\frac{\partial N_x}{\partial x} + \frac{\sin\alpha}{R}(N_x - N_\theta) + \frac{1}{R}\frac{\partial N_{x\theta}}{\partial \theta} = I_0\ddot{u}_0 + I_1\ddot{\phi}_x \\
&\frac{\partial N_{x\theta}}{\partial x} + \frac{2\sin\alpha}{R}N_{x\theta} + \frac{1}{R}\frac{\partial N_{x\theta}}{\partial \theta} + \frac{\cos\alpha}{R}Q_\theta = I_0\ddot{v}_0 + I_1\ddot{\phi}_\theta \\
&\frac{\partial Q_x}{\partial x} + \frac{1}{R}\frac{\partial Q_\theta}{\partial \theta} + \frac{\sin\alpha}{R}Q_\theta - \frac{\cos\alpha}{R}N_\theta = I_0\ddot{w}_0 \\
&\frac{\partial M_x}{\partial x} + \frac{\sin\alpha}{R}(M_x - M_\theta) + \frac{1}{R}\frac{\partial M_{x\theta}}{\partial \theta} - Q_x = I_1\ddot{u}_0 + I_2\ddot{\phi}_x \\
&\frac{\partial M_{x\theta}}{\partial x} + \frac{2\sin\alpha}{R}M_{x\theta} + \frac{1}{R}\frac{\partial M_\theta}{\partial \theta} - Q_x = I_1\ddot{v}_0 + I_2\ddot{\phi}_\theta
\end{aligned} \tag{10}$$

where $I_i = \sum_{k=1}^{N} \int_{z_k}^{z_{k+1}} \rho^{(k)} z^i \mathrm{d}z$, $i = 0 - 2$; $\rho^{(k)}$ are the density of mass of the shell material and I_0, I_1, I_2 are the mass inertias.

4 Continuous Element for FGM Conical Shell

4.1 State Vectors

The development of displacements, forces, and moment resultants using Levy's series are applied to determine the vibration of FGM conical shell as follows:

$$\begin{aligned}&\{\boldsymbol{u}_0, \boldsymbol{w}_0, \boldsymbol{\phi}_x, \boldsymbol{N}_x, \boldsymbol{M}_x, \boldsymbol{Q}_x\}\\&= \sum_{m=1}^{\infty} \{u_m(x), w_m(x), \phi_{xm}(x), N_{xm}(x), M_{xm}(x), Q_{xm}(x)\} \cos(m\theta) e^{i\omega t}\\&\{\boldsymbol{v}_0, \boldsymbol{\phi}_\theta, \boldsymbol{N}_\theta, \boldsymbol{N}_{x\theta}, \boldsymbol{M}_{x\theta}, \boldsymbol{Q}_\theta\}\\&= \sum_{m=1}^{\infty} \{v_m(x), \phi_{\theta m}(x), N_{\theta m}(x), N_{x\theta m}(x), M_{x\theta m}(x), Q_{\theta m}(x)\} \sin(m\theta) e^{i\omega t}\end{aligned} \quad (11)$$

The state vector for the symmetric circumferential modes is chosen as shown below:

$$y_{sm} = \{u_m, v_m, w_m, \phi_{xm}, \phi_{\theta m}, N_{xm}, N_{x\theta m}, Q_{xm}, M_{xm}, M_{x\theta m}\}_S^T \quad (12)$$

Substituting Eq. (11) in Eqs. (8) and (10) results in obtaining 13 equations in term of x and ω, after eliminating $\cos(m\theta)$, $\sin(m\theta)$, $e^{i\omega t}$:

$$\frac{\mathrm{d}y_{sm}}{\mathrm{d}x} = \boldsymbol{A}_{sm} \boldsymbol{y}_{sm} \quad \text{in which } \boldsymbol{A}_{sm} \text{ is a } 10 \times 10 \text{ matrix.} \quad (13)$$

4.2 The Dynamic Stiffness Matrix Construction

Following Eq. (13) the dynamic transfer matrix $\boldsymbol{T}_{sm}$ is calculated by:

$$\boldsymbol{T}_{sm} = e^{\int A_{sm} dx} = \begin{bmatrix} \boldsymbol{T}_{11} & \boldsymbol{T}_{11} \\ \boldsymbol{T}_{21} & \boldsymbol{T}_{22} \end{bmatrix} \quad (14)$$

Ultimately, the dynamic stiffness matrix $(\omega)K(\omega)$ is assessed in the following manner:

$$\boldsymbol{K}(\omega) = \begin{bmatrix} -\boldsymbol{T}_{12}^{-1}\boldsymbol{T}_{11} & \boldsymbol{T}_{12}^{-1} \\ \boldsymbol{T}_{21} - \boldsymbol{T}_{22}\boldsymbol{T}_{12}^{-1}\boldsymbol{T}_{11} & \boldsymbol{T}_{22}\boldsymbol{T}_{12}^{-1} \end{bmatrix} \quad (15)$$

It is noted that the above-mentioned DSM $\boldsymbol{K}(\omega)$ can be employed for a single FGM cylinder ($\alpha = 0$), cone ($\alpha \neq 0$), inner ring-stiffener ($\alpha = -90°$), or outer ring-stiffener ($\alpha = 90°$). The stiffness matrix of the complex FGM shell is constructed by an assembly of the stiffness matrices of each single element (cylindrical shell, conical shell, ring stiffener). The assembly of stiffness matrices must satisfy the continuity conditions between the two surfaces of the two elements, which follows [11]. Then natural frequencies of the considered complex FGM shell will be determined using harmonic responses as detailed in [9].

5 Computational Results and Analysis

5.1 Comparative Analysis Results

For validating the accuracy of our research, results in [6] using analytical solutions will be compared with those obtained using the presented formulations. The structural characteristics under consideration entail a clamped–clamped cylindrical shell constructed from a FGM comprising two constituents: zirconia and aluminum. The parameters defining the cylindrical shell and material properties are: $L/R = 2$, $h/R = 0.1$, $R = 1$ m, $E_m = 70$ GPa, $\mu_m = 0.3$, $\rho_m = 2707$ kg/m^3, $E_c = 168$ GPa, $\mu_c = 0.3$, $\rho_c = 5700$ kg/m^3. Additionally, the volume fraction parameters are specified as $a = 1$, $b = 0.5$, $c = 2$, and $p = 5$. The first seven frequencies of the studied cylindrical shell are shown in Table 1. The errors are evaluated as follows: Error = |Present − Ref| × 100/Ref.

It is seen in Table 1 that small discrepancies between two approaches are obtained which confirms the accuracy of the CE model. Thus, the developed formulation can be used for studying vibration of FGM shell structures.

Table 1 Free frequencies (Hz) of cylindrical shells with $FGM_{I(a=1,b=0.5,c=2,p=5)}$

Method	n						
	1	2	3	4	5	6	7
Ref. [6]	526.75	380.22	369.25	479.25	670.59	915.48	1201.03
Present	561.4	398.4	370.5	468.2	651.5	890.5	1171.9
Error (%)	6.57	4.78	1.25	0.34	2.84	2.72	2.42

5.2 CEM for FGM Conical-Cylindrical-Conical Shell with Ring Stiffeners

The developed model will now be employed to predict the vibration characteristics of two complex shell structures: one with inner ring stiffeners and another with both inner and outer ring-stiffeners. Subsequently, natural frequencies will be computed and analyzed to underscore the advantages of CEM.

5.2.1 Vibration of FGM Conical-Cylindrical-Conical Shell with Inner Ring Stiffeners

Here, a combined conical-cylindrical-conical shell reinforced by two inner ring stiffeners as illustrated in Fig. 2 is in consideration using Clamped-Free boundary condition. The geometric dimensions are presented in Table 2. The used FGM is Mat I consists of Silicon (Si) and Nickel (Ni) with properties provided in Table 3:

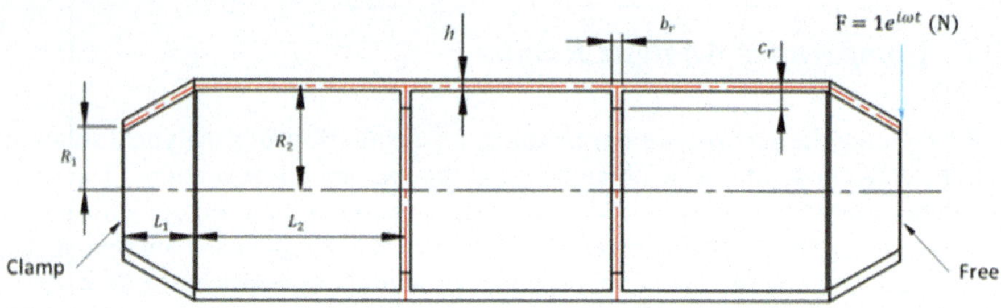

Fig. 2 Investigated shell structure

Table 2 Dimensions of the investigated shell structure

Characteristics	Geometrical properties
Cone radius R_1 (m)	0.09
Cylinder radius R_2 (m)	0.14
Thickness h (m)	0.015
Ring height b_r (m)	0.015
Ring width c_r (m)	0.02
Cylinder shell length L_2 (m)	0.27
Conical shell length L_1 (m)	0.1
Total length (m)	1.01
Angle α (°)	26.57

Table 3 FG material properties (Mat I)

	Material	
	Si3N4	Nickel
E (GPa)	322.27	205.098
μ	0.24	0.31
ρ (kg/m^3)	2370	8900

The analysis results for the first eight natural frequencies of the model utilizing Material I are depicted in Table 4 and in Fig. 3. It is evident from both the tabulated data and the graphical representation that increasing the exponent p for Material I leads to a substantial reduction in the natural frequencies. This is due to the fact that the Aluminum component has significantly lower stiffness and density (approximately 2 times lower) compared to the ceramic component, Zirconia.

Table 4 Free frequencies of the studied model using Mat I ($a = 1, b = 0, c = 1$):

Function type	Mode (n, m)	Power-law distribution paramter p				
		0.05	0.2	1	5	20
FGM_I	1,1	320	276	201	159	146
	2,1	1305	1120	821	660	605
	2,2	1925	1656	1215	966	887
	1,2	1973	1697	1240	973	900
	0,1	2314	1991	1456	1142	1054
	2,3	2991	2574	1886	1491	1372
	3,1	3191	2738	2009	1617	1482
	3,2	4034	3470	2549	2035	1864

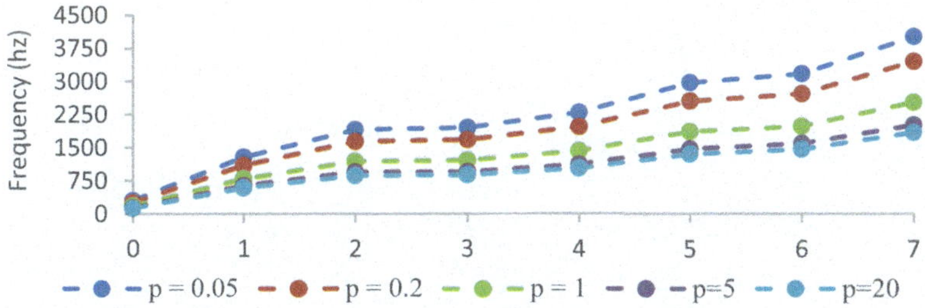

Fig. 3 The impact of the exponent p value (Material I—$a = 1, b = 0, c = 1$) on the free frequencies of the Functionally Graded Material combined stiffened shells

5.2.2 Vibration Analysis of Functionally Graded Material Conical-Cylindrical-Conical Shell Structure with Inner and Outer Ring Stiffeners

In the subsequent section, our methodology is applied to analyze a more complex shell configuration, denoted as Model 2. This structure closely resembles the one described previously, with the exception that an inner ring stiffener (located closer to the clamped end) is substituted with an outer ring stiffener (refer to Fig. 4). The parameters characterizing this combined shell are provided in Tables 2 and 3, and the structure is subjected to a Clamped-Free boundary condition. The analytical results pertaining to the initial eight natural frequencies of Model 2, utilizing Material I with the volume fraction distribution parameters V_c $(a = 1, b = 1, c = 2)$, are presented in Table 5 and illustrated in Fig. 5. In the case of $a = 1, b = 1, c = 2$, the distribution curve of the material does not vary significantly within the range $0.05 \leq p \leq 0.2$, in contrast to the range of p from 5 to 20.

Table 5 and Fig. 5 have shown that the natural frequencies of Model 2 have slightly decreased compared to the original model, with the most significant reduction observed in modes 3 and 2. In general, changing an inner ring of first model to an outer ring has resulted in a reduction in the natural frequencies compared to the first model.

For both studied complex structures, FE models necessitate fine meshes to achieve high precision. In contrast, CE models employ only 7 elements (3 cylindrical, 2 conical, and 2 ring elements) for the same structures, offering significant storage data volume savings. Consequently, the application of CE aids in overcoming numerical computation challenges in medium and high-frequency ranges, similar to those faced in FEM. This advantage is particularly pronounced in complex Functionally Graded Material structures, where a substantial number of meshing elements would be necessary.

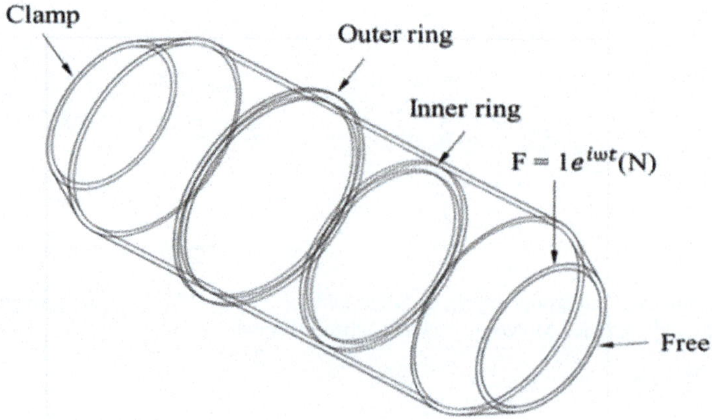

Fig. 4 Combined shell Model 2 with outer and inner rings

Table 5 Free frequencies of Model 2 calculated by CEM using Mat I ($a = 1, b = 1, c = 2$):

Function type	Mode (*n*, *m*)	Power-law distribution parameter *p*				
		0.05	0.2	1	5	20
FGM_I	1,1	332	319	271	189	149
	2,1	1392	1339	1144	813	643
	2,2	1904	1831	1693	1100	870
	1,2	2080	1998	1693	1178	933
	0,1	2443	2346	1988	1383	1096
	2,3	3152	3030	2575	1805	1429
	3,1	3375	3248	2775	1976	1563
	3,2	3950	3801	3245	2303	1821

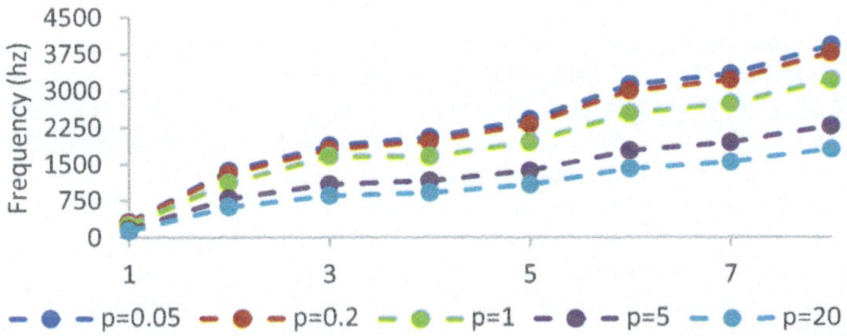

Fig. 5 Effect of different values of *pp* on the natural frequencies of Model 2, utilizing Material I with the parameters $a = 1, b = 1, c = 2$

6 Conclusion

This study explores the free vibration behavior of combined conical-cylindrical-conical shells reinforced by inner and outer ring stiffeners, constructed from Functionally Graded Material, using the developed Continuous Element model. The obtained natural frequencies are validated against literature values. The effects of FG material parameters are thoroughly investigated. Various test cases confirm that the Continuous Element approach, requiring minimal meshing elements, reduces model complexity and data storage volume. Particularly, the Continuous Element model excels in accurately modeling complex FGM shells, especially in high-frequency domains where other methods encounter challenges. Moreover, our model has the potential to be extended to analyze complex Functionally Graded Material or porous shell structures interacting with elastic foundations, fluids, or sound.

Acknowledgements This research is funded by Hanoi University of Science and Technology (HUST) under project number T2022-PC-034.

References

1. Reddy, J.: Theory and analysis of elastic plates and shells, 2nd edn. CRC Press, New York (2007)
2. Tornabene, F.: Free vibration analysis of functionally graded conical, cylindrical shell and annular plate structures with a four-parameter power-law distribution. Comput. Methods Appl. Mech. Engrg **198**, 2911–2935 (2009)
3. Viola, E., Tornabene, F.: Free vibrations of three parameter functionally graded parabolic panels of revolution. Mech. Res. Commun. **36**, 587–594 (2009)
4. Tornabene, F., Viola, E.: Free vibration analysis of functionally graded panels and shells of revolution, Meccanica **44**, 255–281 (2009)
5. Shu, Z., Jin, G., Shi, S., Ye, T., Jia, X.: A unified solution for vibration analysis of functionally graded cylindrical, conical shells and annular plates with general boundary conditions. Int. J. Mech. Sci. **80**, 62–80 (2014)
6. Peng, Q., Wang, Q., Chen, Z., Zhong, R., Liu, T., Qin, B.: Dynamic stiffness formulation for free vibration analysis of rotating cross-ply laminated combined elliptical-cylindrical-conical shell. Ocean Eng. **269**(1), 113486 (2023)
7. Zhao, X., Liew, K.M.: Free vibration analysis of functionally graded conical shell panels by meshless method. Compos. Struct. **93**, 649–664 (2011)
8. Nam, L.B., Cuong, N.M., Thinh, T.I., Dat, T.T., Trung, V.D.: Continous element formulation for composite ring-stiffened cylindrical shells. Vietnam J. Sci. Technol. **56**(4), 515 (2018)
9. Thinh, T.I., Nguyen, M.C.: Dynamic stiffness matrix of continous element for vibration cross-ply laminated composite cylindrical shells. Comp. Struct. **98**, 93–103 (2013)
10. Tounsi, D., Casimir, J.B., Abid, S., Tawfiq, I., Haddar, M.: Dynamic stiffness formulation and response analysis of stiffened shells. Comput. Struct. **132**, 75–83 (2014)
11. Nam, L.T.B., Cuong, N.M., Thinh, T.I.: Continuos element formulation for vibration of thick composite annular plates and rings. In: The National Conference on Mechanic, Hanoi (2014)
12. Anh, N.D., Vinh, L.Q., Cuong, N.M., Hien, V.Q.: Dynamic analysis of FGM conical shells surrounded by Pasternak elastic foundations. In: The 3rd International Confrence on Transport Infrastructure & Sustainable Development (TISDIC 2019). Construction Publishing House (2019)

Topology Optimization for Resin 3D Printed Products to Prevent Weak Connections

Trong Hieu Bui, Hoang Phi Dang, and Huu Nghi Huynh

Abstract Currently, resin 3D printing technology is being applied in many different important industrial fields with great potential. However, the process of designing and manufacturing products using this technology still has some issues that need to be researched and perfected, such as the phenomenon of weak bonds in the product structure. This phenomenon will cause damage to the manufactured part or reduce mechanical properties due to adhesion and the ability of the material layers to bond together. The article proposes a method using the Topology optimization module of Ansys software to simulate and eliminate weak bonds through stress constraints to prevent damage during the manufacturing process. The simulation process shows that, on the MBB (Messerschmitt-Bolkow-Blohm) beam model, a mass limit of 50% will give a model with a reasonable structure. At the same time, the stress constraint at a value of 2 MPa will eliminate weak connection locations in the structure of the test model. The results of the research help improve manufacturing capabilities and improve the durability of resin 3D printed products.

Keywords 3D printer resin · MBB · Weak bonds

T. H. Bui
Division of Machine Design, Faculty of Mechanical Engineering, Ho Chi Minh City University of Technology (HCMUT), 268 Ly Thuong Kiet Street, District 10, Ho Chi Minh City, Vietnam

T. H. Bui · H. P. Dang · H. N. Huynh (✉)
Vietnam National University Ho Chi Minh City, Linh Trung Ward, Thu Duc City, Vietnam
e-mail: hhnghi@hcmut.edu.vn

H. P. Dang · H. N. Huynh
Division of Manufacturing Engineering, Faculty of Mechanical Engineering, Ho Chi Minh City University of Technology (HCMUT), 268 Ly Thuong Kiet Street, District 10, Ho Chi Minh City, Vietnam

D. H. Tien et al. (eds.), *Proceedings of the 1st International Conference on Sustainability and Emerging Technologies for Smart Manufacturing*, Proceedings in Technology Transfer, https://doi.org/10.1007/978-981-97-7083-0_15

1 Introduction

Liquid plastic 3D printing technology includes a group of processes that use laser or UV energy to solidify liquid plastic materials to form products such as SSLA (Stereolithography), DLP (Digital Light Projector), and LCD (Liquid Crystal Display). This technology has been and is being applied in many different important fields, including dentistry and jewelry. In addition, they are also applied to products with complex structures and light weights in the aerospace and automobile industries [1]. With great potential and benefits from the ability to accurately manufacture complex parts [2], 3D Resin printing technology is continuing to develop and promises to bring many applications in the many fields of the manufacturing industry. One of the important issues when manufacturing with 3D Resin printing technology is that the product is susceptible to small damages, broken at weak bonds locations, directly affecting the quality. This phenomenon often occurs due to adhesion forces during the manufacturing process that affect the bond between material layers to the print bed and between material layers together on the same product. Therefore, many scientists are focusing on research to find a solution to this problem. Wang and colleagues [3] proposed a method to optimize the product structure to support the 3D resin printing process using DLP technology. The authors considered adhesion forces to prevent stress concentrations and weak bonds, with the goal of improving the mechanical properties of the product. The study uses an interpolated finite element model that can change the design of the test specimen to simulate the adhesion force and develop an algorithm to optimize the stress-constrained topology. The simulation process was validated by fabricating test samples according to the results obtained, which showed a clear improvement. Li and Shen [4] proposed a heterogeneous lattice structure optimization method that takes into account the spatial stress state of the products. Solid structures are stressed in a static state, and lattice units of different stiffness are combined based on local stress levels. Refined topology optimization results in improved mechanical performance compared to homogeneous lattice designs. The simulation results were validated by manufacturing samples using SLA technology and the bending strength test results using the three-point method demonstrated the feasibility of the solution. In addition, there are also some scientists researching structure optimization in the field of metal 3D printing, typically Xu and Liu [5] have proposed a method to optimize the topology, bonding to limit errors caused by residual stresses during product manufacturing to avoid product damage phenomena such as cracking, delamination, and warping using finite element analysis based on deformations. At the same time, the authors also built a topology optimization algorithm with residual stress constraints. The results showed that the above method is effective in controlling residual stress and deformation. Some researchers also focus on structural optimization methods, for example, Zhu et al. [6] proposed a topology optimization method based on effective control of the mechanical capacity. They fabricated prototypes using SLA technology and tested their mechanical performance. The finite element method was also used to analyze the structural responses. The results showed that there was agreement between simulation and experiment. This result helps reduce time for the product design phase. Research by Ajanthan et al. [7] proposed a structural optimization method with the goal of reducing product weight. The aim of optimization is to achieve the best design with respect to a different set of criteria or constraints. In this study, the authors simulated structural optimization using the finite element method on Ansys software with a number of different test specimens according to IS456:2000 standards. The results showed that the sample meets the necessary requirements and the weight reduction did not affect the product's functionality. Xue et al. [8] proposed a shape structure optimization method to minimize the amount of material. This method combined the SIMP (Solid Isotropic Material with Penalization) method and the GA (Genetic Algorithm) method called the SIMP-GA method. As a result, the material distribution after being optimized using the SIMP-GA method and the objective function value obtained were both better than when performed using the SIMP and BESO methods. The authors concluded that SIMP-GA can easily obtain light-weight structures with higher performance. Meanwhile, Liu et al. [9] proposed a topology optimization method involving stress constraints for the FDM method. The P-norm function with stability transformation was used to construct the global stress constraint by iterative corrections to eliminate the gap between the maximum local stress and the P-norm stress. Experiments performed on MBB beams, which have been topologically optimized, show that the deposition direction and construction direction have an influence on the structural strength. A survey of the literature showed that scientists have used many different approaches and methods to achieve the goal of optimizing structures to reduce weight and minimize errors that occur during the manufacturing process products using 3D printing technology. The results showed that structural optimization can help improve the quality and strength of resin 3D printed products.

2 Methodologies and Data

The goal of the research team is to propose a method to optimize the shape structure of products to be manufactured using Resin 3D printing technology by constraining stress to reduce weight and eliminate weak bonds in product texture. The model used for the simulation process is the MBB (Messerschmitt-Bolkow-Blohm) beam model. This is a model often used by scientists when performing topology optimization researches. The simulation process is performed on the Topology optimization module of Ansys software with two conditions: with and without stress constraints. The simulation results are verified by fabricating test specimen on the open source Voxelab Polaris 3D printer and re-evaluating on the Perfactory P4K90 industrial 3D printer from Envisiontec. The parameters used in the simulation process include two types: fixed parameters and variable parameters selected based on the survey of published works and practical experience of the research team. The experimental values of each parameter are presented in Table 1. The structural optimization

Table 1 Parameters set for the simulation process

Parameters	Symbol	Unit	First attempt	Second attempt	Third attempt	Fourth attempt
Stress constraints	–	Mpa	10	5	3	2
Mass constraints	–	–		50%		
Model dimension	X	mm	60			
	Y	mm	360			
	Z	mm	2			
Load	F	N	1.5			
Mesh size	–	mm	1			
Number of mesh elements	–	–	43,200			

simulation process is shown in the flow chart Fig. 1. The steps are as follows: The sample after design will be divided mesh with a mesh size of 1 mm and the number of elements is 43,200. Then, the model is set with loads and boundary conditions (Fig. 2). Once the above conditions are satisfied, the simulation process continues to perform finite element analysis using the Static Structural Module and topology optimization using the Topology optimization Module on Ansys software with mass and stress constraint conditions. The parameters used in this step are presented in Table 1. The results obtained will be saved and evaluated.

3 Results and Discussion

3.1 Simulation Results of Structural Optimization

In case of unconstrained stress. The simulation results presented in Fig. 3a show that the model after optimization with a mass constraint of 50% has a clear structure and no fracture phenomenon. Therefore, a mass constraint level of 50% is reasonable to proceed to the next step of stress and displacement analysis. The results of stress and displacement analysis show that the Von-Mises stress value and the maximum displacement value of the model after optimization and the model without optimization are not significantly different Table 2. Therefore, the structure of the model after optimization is considered safe and material removal will not greatly affect the structure of the model. At the same time, through the above results, the research team also preliminarily predicted that areas with weak bonds are thin, easily broken struts such as the location zoned in Fig. 3a. Therefore, this result (sample I) is used to create a validation sample to compare the model when there is a stress constraint (sample 2).

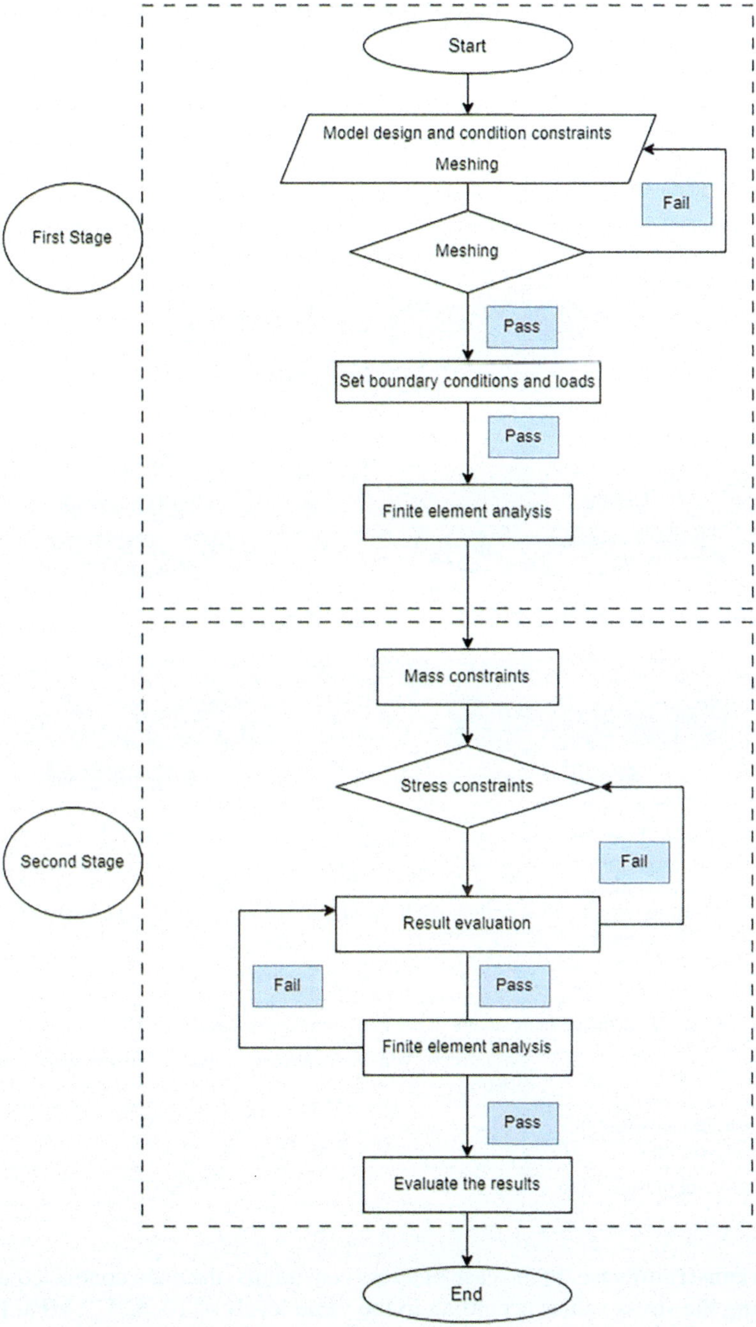

Fig. 1 Simulation process flow chart

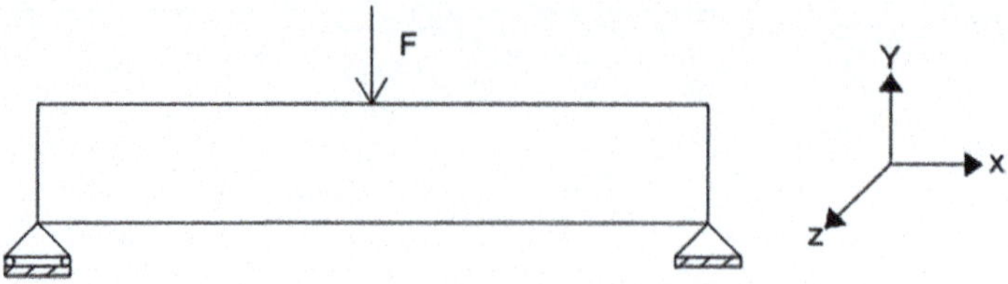

Fig. 2 MBB beam model

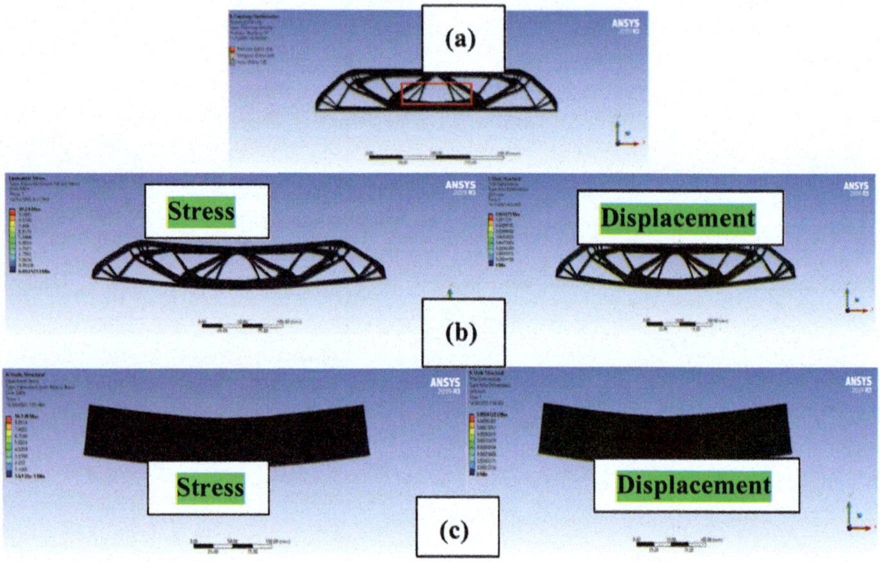

Fig. 3 Results of model optimization with 50% mass constraint (**a**), results of stress and displacement analysis of the model after optimization (**b**) and stress and displacement analysis results of the unoptimized model (**c**)

Table 2 Stress and displacement values in the case of unconstrained stress

Case	Maximum Von-Mises stress (MPa)	Largest displacement (mm)
Optimized model with 50% volume constraint	10.24	0.00128
Model without optimization	10.138	0.00091

Stress constraint case. In the case of stress constraints, the team conducted tests by changing the stress values according to the value levels of 10, 5, 3, 2 MPa. Results Fig. 4 shows that the model structure at the value level of 10, 5 MPa does not change significantly in shape and weak bonds still exist (for example, the locations are zoned in Fig. 4a, b). When the stress value was at 3Mpa, the weak bonds were significantly lost. Finally, with a stress value of 2 MPa, the structure of the model has clearly changed, weak bond areas appearing on the model without stress constraints seem to have been eliminated and replaced by connections more certain (for example, the zoning position in Fig. 4d). This demonstrates that when stress constraints are present, weak bonds become stronger. Therefore, the research team chose a model constrained to stress at 2 MPa to conduct stress and displacement analysis. The analysis results in Fig. 5 show that the maximum stress and displacement values of the model after optimization and the model without optimization do not have a large difference (Table 3). Therefore, the structure of the model after optimization is considered safe and material removal will not greatly affect the structure of the model. The results in this case (sample II) are used to create a control sample and compare with (sample I).

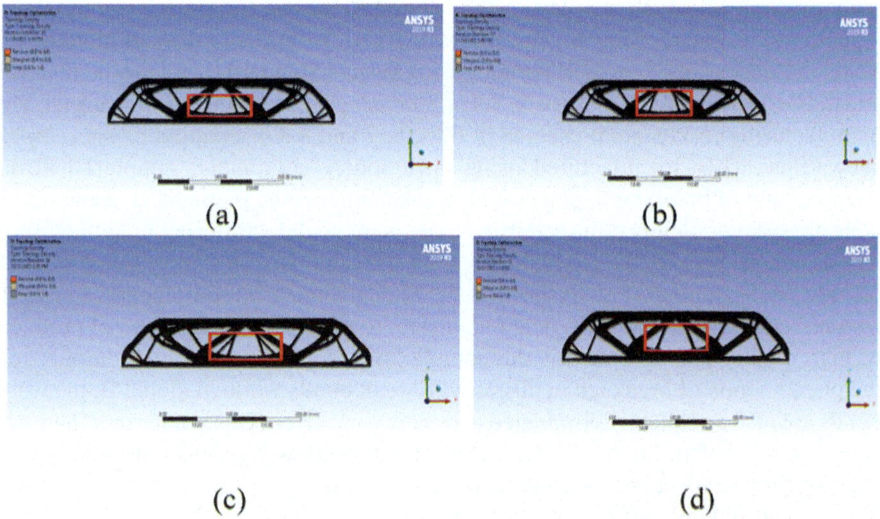

Fig. 4 Optimization results with maximum Von-Mises stress constraints at the following values: **a** 10 MPa, **b** 5 MPa, **c** 3MPa, and **d** 2 MPa

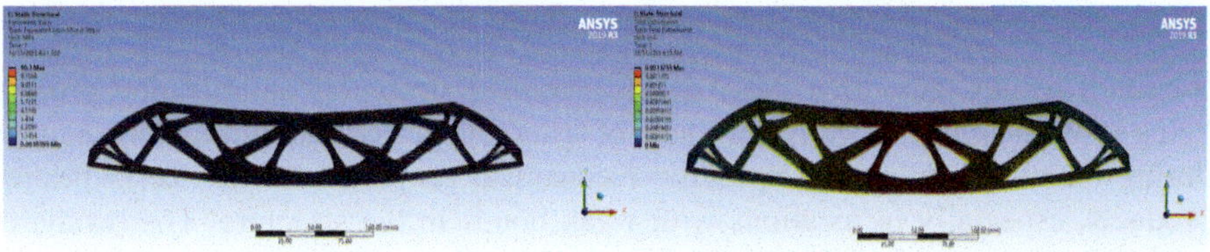

Fig. 5 Stress and displacement analysis results of the maximum Von-Mises stress constraint optimization model of 2 MPa

Table 3 Stress and displacement values in the case when stress is constrained

Case	Maximum Von-Mises stress (MPa)	Largest displacement (mm)
Optimized model with a stress limit of 2 MPa	10.3	0.00132
Model without optimization	10.138	0.00091

3.2 *Experimental Results*

Experimental samples with the largest envelope size of 60 × 10 × 0.5 mm were made in two cases: with 50% mass constraint, no stress constraint (sample I—Figs. 6 and 8) and with mass constraint capacity of 50%, has the maximum stress constraint with a value of 2 MPa (sample II—Figs. 7 and 8) on two different types of equipment to test, evaluate, compare and contrast. These two types of devices include: the Voxelab Polaris open source 3D printer using Esun's Standard Resin plastic material and the Perfactory P4K90 industrial 3D printer using E-Shell 200 Series material. The manufacturing results are presented in Figs. 6, 7 and 8. Observation of the samples shows that, when manufactured using open source equipment, sample I had the phenomenon of breaking in places with weak connections. In model II, weak bonds were almost completely eliminated and there were no defects during manufacturing. This is consistent with the simulation results. The comparison sample manufactured on industrial machines (Fig. 8) shows that, although industrial equipment is standardized in materials and technological parameters to allow the manufacture of products of high quality and complexity, the complexity is higher. In model I, there is still the phenomenon of breakage in places with weak bonds while in model II, the weak bonds are also completely eliminated and there is no error. This proves that the weak connection problem on the structure does not depend on the device used, and both experimental results are consistent with the structural optimization simulation results on Ansys-Topology optimization software.

4 Conclusions

Research results show that the stress-constrained structural optimization process helps solve the multi-objective problem of reducing weight and minimizing damage during manufacturing and improving mechanical properties of 3D Resin printed products, especially in locations with weak bonds in the structure. The results of the MBB beam simulation process on Ansys-Topology optimization software with the mass constraint condition of 50% (sample I) and stress at a value of 2 MPa (sample II) show the appearance of weak connection areas. In the structure in the 50% mass constrained model and those weak connection areas have been replaced by stronger connections in the 2 MPa stress-constrained model. Experimental results

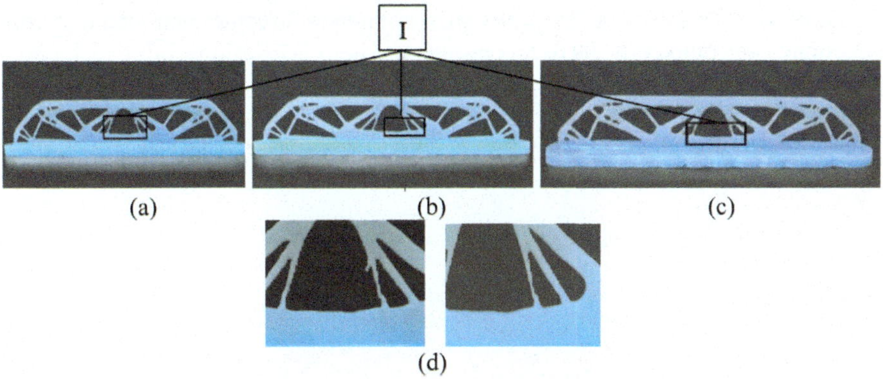

Fig. 6 Optimized print sample with 50% weight constraint, 1st print (**a**), 2nd print (**b**), 3rd print (**c**), Extract from I (**d**)

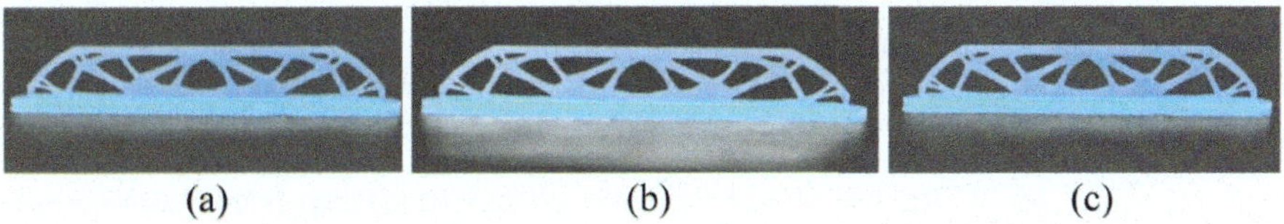

Fig. 7 Optimized printed sample with mass constraint and maximum stress constraint of 2 MPa, 1st printing (**a**), 2nd printing (**b**), 3rd printing (**c**)

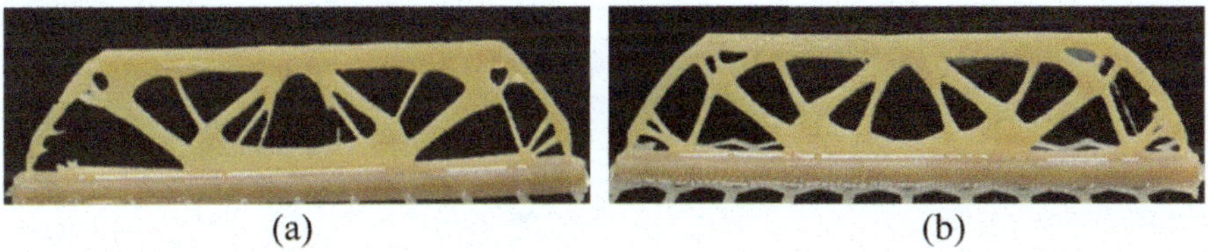

Fig. 8 Samples fabricated using an industrial printer: **a** sample with 50% mass constraint and **b** sample with additional stress constraint of 2 MPa

with two samples manufactured on open source 3D printing equipment and industrial equipment are consistent with the results of the simulation process. Accordingly, the fracture phenomenon in the model without stress constraints (sample I) has been eliminated in the model with stress constraints (sample II), which shows that the method of structural optimization by Stress constraints both achieve the goal of weight reduction and improve the problem of weak bonds in the structure of Resin 3D printed products. The authors also found that the method proposed in this study can be used to optimize the structure for other types of resin 3D printed products besides the MBB beam model. In addition, to use 3D Resin printing technology to manufacture products with complex structures in actual production, it is necessary to consider the interactive effects of other factors such as support structure and manufacturing direction… to completely solve this problem to help 3D Resin printing technology become more widely applied in many different fields.

Acknowledgements We acknowledge Ho Chi Minh City University of Technology (HCMUT), VNU-HCM for supporting this study.

References

1. Lakkala, P., Munnangi, S.R., Bandari, S., Repka, M. Additive manufacturing technologies with emphasis on stereolithography 3D printing in pharmaceutical and medical applications: a review. Int. J. Pharmaceutics X **5** (2023)
2. Mohanavel, F.: The roles and applications of additive manufacturing in the aerospace and automobile sector. Mater Today Proc. **47**, 405–409 (2021)
3. Wang, F., Liu, S., Li, T.: Topology optimization for digital light projector additive manufacturing addressing the in-situ structural strength issue. Polymers **15**, 3573 (2023)
4. Li, F., Shen, S.: Solid stress-distribution-oriented design and topology optimization of 3D-printed heterogeneous lattice structures with light weight and high specific rigidity. Polymers **14**, 2807 (2022)
5. Xu, F., Liu, S., Ma, T.: Residual stress constrained self-support topology optimization for metal additive manufacturing. Comput. Methods Appl. Mech. Eng. (2022)
6. Zhu, F., Yang, S., Zhang, T.: Backbone cup —a structure design competition based on topology optimization and 3D printing. Int. J. Simul. Multisci. Des. Optim. (2016)
7. Abhijith F.: Topology optimization of deep beam using ansys. IJSSR **9**(3), 177–186 (2020)
8. Xue, F., Yu, S., Zhang, T.: A novel method for structural lightweight design with topology optimization. Energies (2021)
9. Ansys: https://www.ansys.com/. Last accessed 15 Dec 2023

A Numerical Study on the Thermo-electrohydrodynamic Performance of ECF Micro-pumps

The Khanh Lai, Khanh Duong Tran, and Ich Long Ngo

Abstract This study presents a numerical method on the dynamic behaviors of the flow in an electro-conjugate fluid (ECF) micro-pump under the effect of temperature and electric double layer (EDL). The finite element method (FEM) was used to examine the thermo-electrohydrodynamic performance of the ECF micro-pump with the applied voltage ranging from 2 to 6 kV. The numerical results indicated that the temperature increases linearly with the applied voltage but decreases with increasing fluid Reynolds number (Re_F). Additionally, characteristic curves of ECF micro-pumps are defined under the influence of temperature. The effect of EDL in the ECF micro-pump is also considered, and the increase in average velocity is approximately 14.1% with the presence of the EDL. A new correlation of temperature with Re_F and applied voltage is first proposed. These results play a critical role in understanding the dynamic behavior of the ECF flow and controlling it in mechanical engineering, biomedical engineering, and especially in microelectronics cooling.

Keywords Electro-conjugate fluid · Finite element method · Electrothermal flow

1 Introduction

A micro-pump using electro-conjugate fluid (ECF) is a non-micro-pump. ECF is known as a functional and dielectric fluid [1]. A jetting flow of ECF is generated between electrodes when subjected to a high DC voltage. This means that changing the voltage or current can easily control the ECF. Many applications of ECF micro-pump such as an inchworm robot [2], a novel flexible finger [3], a droplet transportation system for micro total analysis system (μTAS) [4], and a droplet generator powered on-chip [5]. The selection of ECF micro-pump structures, electrodes, and working media can be found in the study review by Peng et al. [6]. In general, the

T. K. Lai · K. D. Tran · I. L. Ngo (✉)
School of Mechanical Engineering, Hanoi University of Science and Technology, No. 01, Dai Co Viet, Hai Ba Trung, Hanoi, Vietnam
e-mail: long.ngoich@hust.edu.vn

D. H. Tien et al. (eds.), *Proceedings of the 1st International Conference on Sustainability and Emerging Technologies for Smart Manufacturing*, Proceedings in Technology Transfer, https://doi.org/10.1007/978-981-97-7083-0_16

applicability depends on altering the shape of the electrode pairs [7]. This signifies that manufacturing plays a crucial role. However, developing ECF micro-pumps remains a significant challenge due to technological limitations.

A CFD approach has attracted more attention. It can be used to investigate the intricate behavior of a fluid flow. For example, the effect of the electric double layer in ECF flow was first considered using a numerical model [7]. This method was also used to analyze the behavior of ECF flow with square electrodes. The numerical and experimental results agree with observing the ECF flow fields. Then, the finite element method (FEM) in COMSOL Multiphysics was used to investigate the dynamic phenomena of fluid flow in an ECF micro-pump. It was found that the Onsager effect greatly affected the appearance of the ECF jet. However, to the best of our knowledge, no research has considered the performance of ECF micro-pumps under electrothermal effects.

This paper aims to investigate the dynamic behavior of ECF fluid flow in ECF micro-pump with electrothermal effect. The FEM was used to solve this problem. FEM is a numerical method for solving problems that have a continuous or spatially continuous structure, such as those involving mechanics, heat transfer, electromagnetism, and problems with liquids and gases. It can handle complex problems that are difficult or impossible to solve analytically. It improves the modeling process and allows designers to save time and money. The influence of Joule heating on velocity, pressure, concentration, and efficiency. Correlations have been developed to estimate the temperature. Additionally, many guidelines have been proposed to effectively predict this ECF micro-pump's optimal performance.

2 Methodology

The numerical model used for the simulation problem is shown in Fig. 1 for a typical pair of electrodes. This figure shows the teardrop-shaped emitter and narrow-gap collector. Table 1 gives the boundary condition (BC) used for this problem. Thanks to the outstanding development of technology, the Microelectromechanical Systems (MEMS) method is widely applied to design ECF micro-pumps. MEMS technology allows the fabrication of miniature devices with dimensions ranging from micrometers to millimeters. Specifically, μm-sized ECF micro-pumps have been successfully manufactured by Kim et al. [8] and Mao et al. [5].

The behaviors of ECF flow in ECF micro-pumps can be simulated by solving continuity, momentum, electrostatics, transport of species, and energy. These governing equations are shown in Eqs. (1)–(5).

$$\nabla^* \cdot u^* = 0 \tag{1}$$

$$\rho\left(u^* \cdot \nabla^* u^*\right) = \nabla^* \cdot \left\{-p^* I + \mu\left[\nabla^* u^* + \left(\nabla^* u^*\right)^{\mathrm{T}}\right]\right\} + F\rho_{\mathrm{v}}^* E^* \tag{2}$$

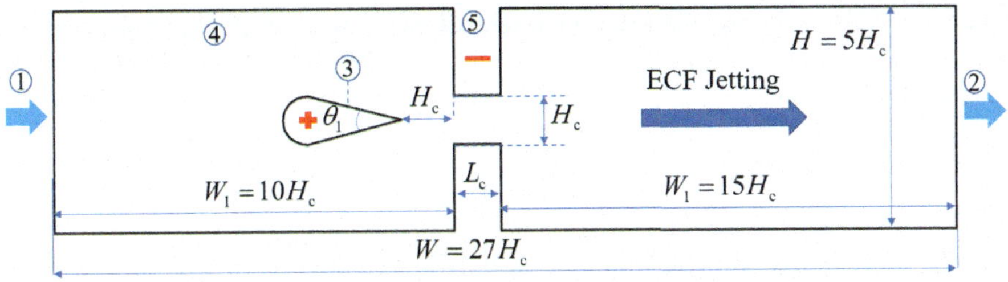

Fig. 1 Schematic of numerical model and boundary conditions (BC)

Table 1 BC for ECF micro-pump with NACA electrodes

No. of BC	Laminar flow	Transport of species	Electrostatic	Heat transfer in fluid
①	$u = U_{in}$	$\partial c/\partial x = 0$	$\partial \phi/\partial x = 0$	$T = 1$
②	$p_{out} = 0$	$\partial c/\partial x = 0$	$\partial \phi/\partial x = 0$	$\partial T/\partial x = 0$
③	$u_s = 0$	$c_E = \lambda \mathrm{Nu_E}(E_S - E_{thresP})\mathrm{sign}(\phi_P)$	$\phi = \phi_0$	$q_0 = \mathrm{Nu_f}(1 - T)$
④	$u_s = 0$	$\partial c/\partial y = 0$	$\partial \phi/\partial y = 0$	$q_0 = \mathrm{Nu_f}(1 - T)$
⑤	$u_s = 0$	$c_N = 0$	$\phi = 0$	$q_0 = \mathrm{Nu_f}(1 - T)$

$$\nabla^* \cdot (\varepsilon E^*) = F\rho_v^* \tag{3}$$

$$\nabla^* \cdot (u^* c_i^*) = \nabla^* \cdot \left\{ D_i \nabla^* c_i^* + \frac{F}{RT^*} D_i z_i c_i^* \nabla^* \phi^* + \frac{\sigma}{F} \nabla^* \phi^* \right\} \tag{4}$$

$$\rho c_p (u^* \cdot \nabla^* T^*) = \nabla^* \cdot \{k_F \nabla^* T^*\} + \sigma (E^* \cdot E^*) \tag{5}$$

where ρ, μ, u, p, σ, c_p, k_F, ε_0, and ε_r are fluid density, dynamic viscosity, velocity, pressure, conductivity, specific heat, thermal conductivity, permittivity constant, and relative permittivity, respectively. The last term on the right-hand side of Eq. (3) is an electrostatic force acting on the fluid flow under the electric field strength $E = -\nabla\phi$, where ϕ denotes electric potential. Space charge density can be generally expressed as, $\rho_v = F \cdot \sum z_i c_i$. D_i, z_i, u_{mi}, and F in Eq. (2) represent diffusion coefficient, charge number, mobility, and Faraday's constant, respectively. These governing equations are solved under the non-dimensional forms with the dimensionless variables shown in Eq. (6).

$$x = \frac{x^*}{H_c};\quad y = \frac{y^*}{H_c};\quad u = \frac{u^*}{u_{in}};\quad p = \frac{p^*}{\rho U_{in}^2};\quad c = \frac{c^*}{c_0};\quad \phi = \frac{\phi^*}{\phi_0};$$
$$T = \frac{T^*}{T_{in}};\quad \phi_0 = V_e^*;\quad c_0 = \frac{k_P^* E_{thresP}^*}{F} \tag{6}$$

where H_c, U_{in}, c_0, ϕ_0 are defined as characteristic length (typically ranging from 100 to 200 μm), velocity, concentration of charges, and applied voltage, respectively. As a result, the non-dimensional forms governing Eqs are shown in Eqs. (7)–(11),

$$\nabla \cdot u = 0 \tag{7}$$

$$(u \cdot \nabla u) = \nabla \cdot \left\{ -pI + \frac{1}{\mathrm{Re_F}} \left[\nabla u + (\nabla u)^{\mathrm{T}} \right] \right\} + \mathrm{Mc}\rho_{\mathrm{v}} E \tag{8}$$

$$-\nabla \cdot \left(\frac{1}{\mathrm{De}^2} \nabla \phi \right) = \rho_{\mathrm{v}} \tag{9}$$

$$\nabla \cdot (uc_{\mathrm{i}}) + \nabla \cdot \left\{ -\frac{1}{\mathrm{Pe}} \nabla c_{\mathrm{i}} - \frac{\phi_{\mathrm{c}}}{\mathrm{Pe}} \frac{1}{T} z_{\mathrm{i}} c_{\mathrm{i}} \nabla \phi \right\} = \frac{1}{\mathrm{Re_E}} \nabla \cdot \left\{ \frac{1}{\mathrm{De}^2} \nabla \phi \right\} = -\frac{\rho_{\mathrm{v}}}{\mathrm{Re_E}} \tag{10}$$

$$(u \cdot \nabla T) = \nabla \cdot \left\{ \frac{1}{\mathrm{Pe_H}} \nabla T \right\} + \frac{Q_{\mathrm{T}}}{\mathrm{Pe_H}} (E \cdot E) \tag{11}$$

These important parameters are shown in Eq. (12). $\mathrm{Re_F}$, $\mathrm{Re_E}$, Mc, Pe, and PeH, which are a fluid Reynolds number, electric Reynolds number, a dimensionless number for mobility ratio, the Péclet number for fluid, and the Péclet number in heat transfer. Additionally, Q_{T} is the Joule heating parameter.

$$\mathrm{Re_F} = \frac{\rho u_{\mathrm{in}} H_{\mathrm{c}}}{\mu}; \quad \mathrm{Re_E} = \frac{\varepsilon u_{\mathrm{in}}}{\sigma H_{\mathrm{c}}}; \quad \mathrm{Mc} = \frac{F c_0 \phi_0}{\rho u_{\mathrm{in}}^2}; \quad \mathrm{Pe} = \frac{u_0 H_{\mathrm{c}}}{D_{\mathrm{i}}};$$
$$\mathrm{Pe_H} = \frac{H_{\mathrm{c}} \rho c_{\mathrm{p}} U_{\mathrm{in}}}{k_{\mathrm{F}}}; \quad Q_{\mathrm{T}} = \frac{\sigma \phi_0^2}{T_{\mathrm{in}} k_{\mathrm{F}}} \tag{12}$$

In Eq. (13), De and λ_{D} are the Debye number and Debye length.

$$\lambda_{\mathrm{D}}^2 = \frac{\varepsilon \phi_0}{F c_0}; \quad \mathrm{De} = \frac{H_{\mathrm{c}}}{\lambda_{\mathrm{D}}}; \quad \varepsilon = \varepsilon_0 \varepsilon_{\mathrm{r}}; \quad \phi_{\mathrm{c}} = \frac{F \phi_0}{R T_{\mathrm{in}}}; \quad \mathrm{Nu_E} = \frac{k_{\mathrm{P}}^* \phi_0}{F c_0 H_{\mathrm{c}}}; \quad \kappa = \frac{k_{\mathrm{N}}}{k_{\mathrm{P}}} \tag{13}$$

The particular BCs capture the charge distribution injected from an electrode, as shown in Eq. (14).

$$c_{\mathrm{E}} = \lambda \mathrm{Nu_E} (E_{\mathrm{S}} - E_{\mathrm{thresP}}) \mathrm{sign}(\phi_0); \; c_{\mathrm{N}} = \lambda \kappa \mathrm{Nu_E} (E_{\mathrm{S}} - E_{\mathrm{thresN}}) \mathrm{sign}(\phi_0) \tag{14}$$

where $\kappa = k_{\mathrm{n}}/k_{\mathrm{p}}$, k_{p}, and k_{n} are proportional constants for injected cation (positive) and anion (negative) density. E_{s} and E_{thresp}, E_{thresn} correspond to the electrostatic and threshold field strength, respectively. Additionally, λ in Eq. (14) is a correction factor considering measurement errors and discrepancies between experiment and numerical simulation results. $\mathrm{Nu_E}$ parameters have been compiled in Eq. (13).

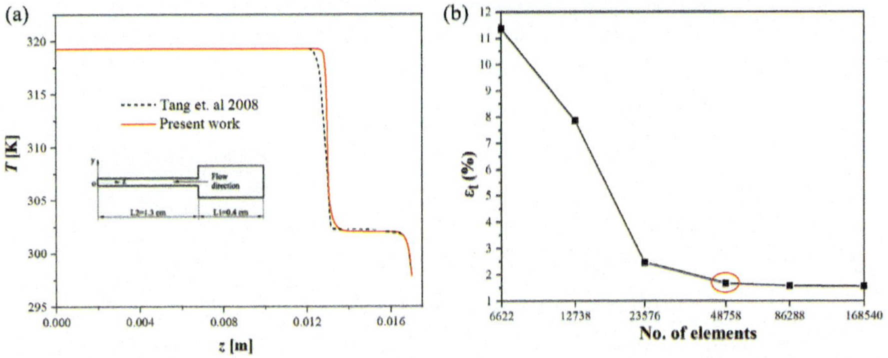

Fig. 2 **a** Joule heating-induced temperature along the centerline of the system [9] and **b** grid convergence test ($Re_F = 2.6 \times 10^{-2}$, $Re_E = 2.8 \times 10^{-3}$, Pe = 3.27, Mc = 9812, $Pe_H = 1.5$)

The numerical mathematics outlined previously has been proven reliable by comparing the present results with those from Tang et al. [9]. The simulation problem has been solved using COMSOL Multiphysics [10]. The joule heating-induced temperature along the system's centerline (Fig. 2a). This result indicated that the numerical model is in excellent agreement with the ones obtained by Tang et al. [9] under the same operating conditions. The maximum deviation is only 1.5% at $z =$ 0.0125 [m]. Therefore, the numerical model has been evaluated extensively to obtain a premise for further calculations in this study.

For the grid convergence test, the pressure difference between the inlet and outlet was calculated for six mesh initial models #1, #2, #3, #4, #5, and #6 with the number of elements of 6622, 12,738, 23,576, 48,758, 86,288, and 168,540, respectively. The relative error in pressure difference (ε_t) was calculated. The #4 mesh model has 48,758 elements, and the relative error is only about 1.67% compared to the finer model, even with approximately double the number of elements. Therefore, the mesh model of M#4 was used for further computations.

3 Results and Discussions

3.1 Distribution of Temperature in ECF Micro-pump

The relationship between the temperature at emitter T and the applied voltage ϕ_0 of the ECF micro-pump at various Re_F numbers from 2.6×10^{-4} to 2.6×10^{-2} is shown in Fig. 3. Consequently, the temperature increases as the applied voltage increases. For Re_F, it is the opposite; the temperature decreases as the Re_F number increases.

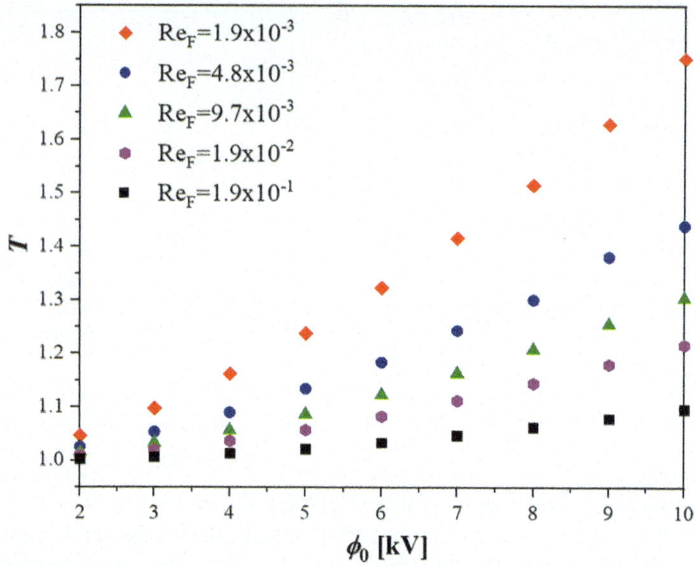

Fig. 3 Temperature at the emitter as a function of the applied voltage for various Re_F

Figure 4 shows a visualization of the conclusions mentioned previously. In this figure, the temperature in a blue rectangle on the surface of the electrodes is most significant. Because this is the region where the current discharged from the emitter to the collector is the strongest. Figure 4 also indicates that the temperature at the electrode surfaces approximates the temperature at the outlet.

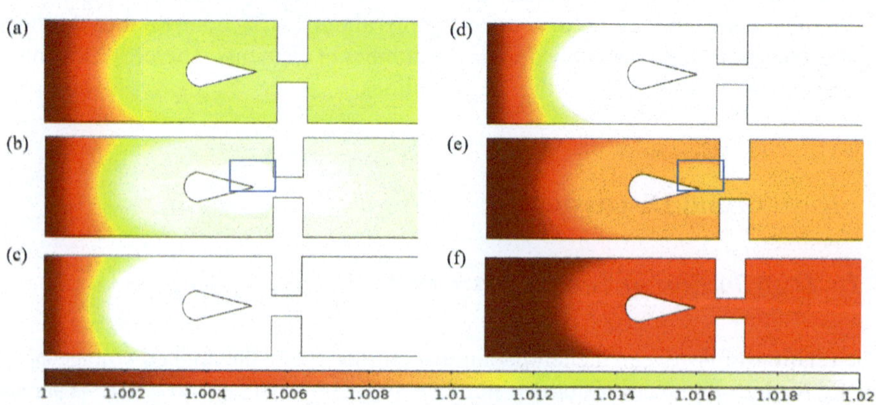

Fig. 4 Temperature distributions in ECF micro-pump with **a** $\phi_0 = 3$ kV; **b** $\phi_0 = 3.5$ kV; **c** $\phi_0 = 4$ kV at $Re_F = 2.6 \times 10^{-3}$ and **d** $Re_F = 2.6 \times 10^{-3}$; **e** $Re_F = 6.7 \times 10^{-3}$; **f** $Re_F = 2.0 \times 10^{-2}$ at $\phi_0 = 4$ kV

Moreover, a general correlation was developed to determine the average temperature efficiently, as shown in Eq. (15). The minimum determination coefficient R^2 is around 0.91. From the physical viewpoint, it can be found from Eq. (15) that (1) T decreases as $\mathrm{Re_F}$ increases; (2) T also increases with increasing ϕ_0, and T equals zero when $\phi_0 = 0$. This means there is no external source for driving the ECF micro-pump; thus, the micro-pump stops.

$$T = 0.993 + 0.00193\frac{\phi_0}{\mathrm{Re_F^{0.208}}} \tag{15}$$

3.2 *Dynamic Behavior of ECF Micro-pump Under Temperature Effect*

The kinetic and dynamic behaviors of ECF micro-pumps are of most interest and main focus. Indeed, Δp, Q, N_p, and η_p are quantities evaluated and used when operating the micro-pump. Therefore, characteristic curves for various applied voltages will be examined and quantitatively assessed, as shown in Fig. 5. These characteristic curves generally have the same trend as those of vane or wing pumps but with a lower range of values. Specifically, Fig. 5a shows that the differential pressure-flow rate (Δp-Q) characteristic curve lifts when ϕ_0 increases from 2 to 4 kV. This means that the operating range of the ECF micro-pump is extended, or power is increased ($N_\mathrm{p} = \Delta p \times Q$). Figure 5b shows that, in general, the power tends to increase as the applied voltage increases; it reaches its maximum and then gradually declines when Q continues to increase. The maximum power values are connected by red points, as shown in Fig. 5b. When the applied voltage increases, these points move to the right-hand side. Indeed, the maximum power values at the operating conditions of $\phi_0 = 2$, 3, and 4 kV are $Q = 60$, 75, and 175 ml/h, respectively.

From the results outlined above, an optimal point exists in a particular operating condition of the ECF micro-pump. Therefore, this advantage can be used to maximize the performance of the ECF micro-pump depending on specific applications. Finally, these characteristic curves show that the ECF micro-pump can be easily controlled from the external power source.

3.3 *Effect of Electric Double Layer*

The electrical double layer (EDL) is a phenomenon that occurs at the interface between a solid surface and a liquid, typically an electrolyte solution. To model the phenomenon of EDL in the ECF micro-pump, the BC of electromotive velocity is defined in COMSOL Multiphysics [10]. As a result, a non-dimensional form of

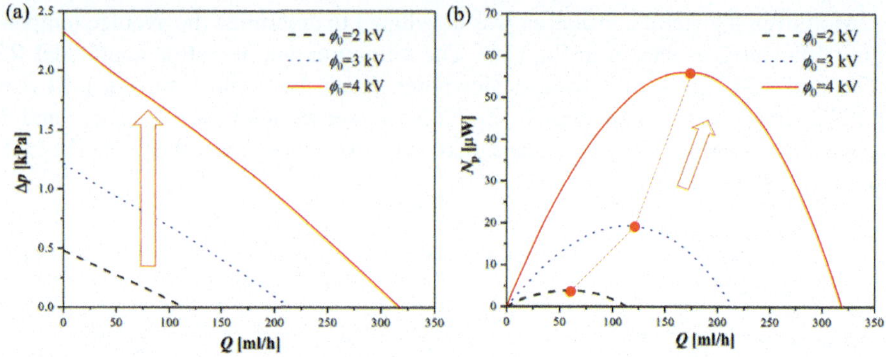

Fig. 5 **a** Δp-Q and **b** N-Q characteristic curves for various values of applied voltages of ECF micro-pump ($\mathrm{Re_F} = 0$–9.37, $\mathrm{Re_E} = 0$–0.99, $\mathrm{Mc} = 0.10$–∞, $\mathrm{Pe} = 0$–1147, $\mathrm{Pe_H} = 0$–525.16)

electroosmotic mobility is derived as $\mu_{\mathrm{slip}} = \mu_{\mathrm{eo}}\phi_0/U_{\mathrm{in}}H_{\mathrm{c}}$. Because of the influence of EDL, the velocity of the micropump tends to increase near the boundary layer, as shown in the red ellipse of Fig. 6.

Figure 7 shows the variation of average velocity measured in the entire computational domain with the applied voltage ϕ_0 at $\mathrm{Re_F} = 2.6 \times 10^{-2}$. The difference value gradually increases, with an average difference of about 14.1%. Due to the influence of EDL, the interaction between positive and negative charges along the micro-pump wall increases. Therefore, the average velocity value increases compared to without

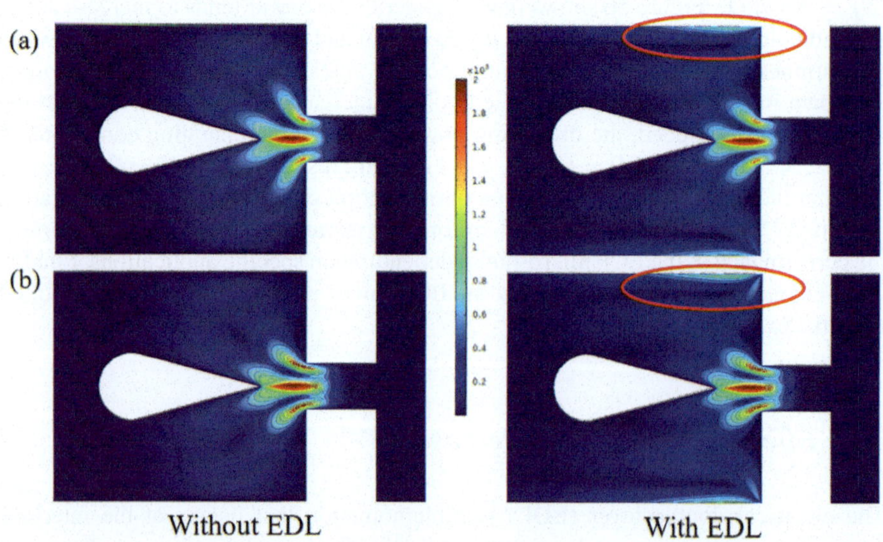

Fig. 6 Velocity distribution in ECF micro-pump **a** $\phi_0 = 4$ kV and **b** $\phi_0 = 5$ kV (The left side without EDL, the right side with EDL) at $\mathrm{Re_F} = 2.6 \times 10^{-2}$, $\mathrm{Re_E} = 2.8 \times 10^{-3}$, $\mathrm{Pe} = 3.27$, $\mathrm{Pe_H} = 1.5$, and $\mu_{\mathrm{slip}} = 2520$–$3150$

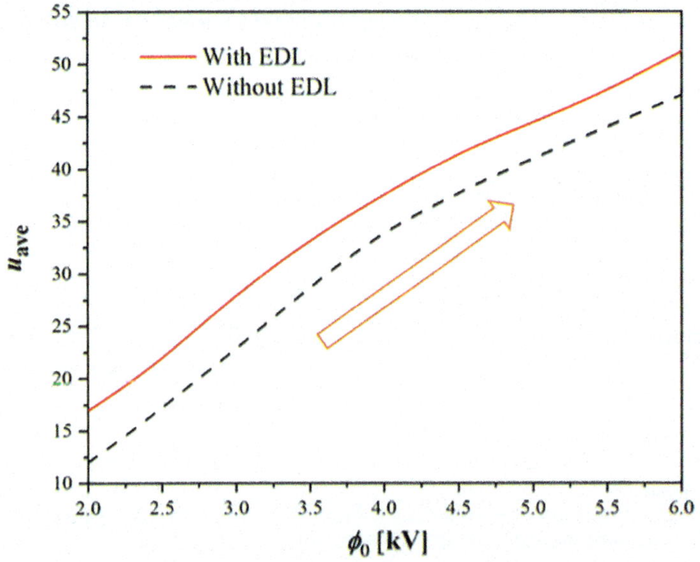

Fig. 7 Average velocity as a function of applied voltage at $Re_F = 2.6 \times 10^{-2}$

EDL. It was found from Fig. 7 that the average velocity increases when the applied voltage increases. In addition, the influence of EDL on the average velocity in the ECF micro-pump will influence Δp, T, Q, N_p, and many other factors in the ECF micro-pump. This will motivate further studies. From there, there is an overview of ECF micro-pumps.

4 Conclusion

The study of the effect of temperature in the ECF micro-pump was carried out using the FEM. The dynamic behaviors of the ECF micro-pump were investigated. Consequently, the majority results of this study are as follows:

(1) A numerical model is first proposed for simulating the thermo-electrodynamic flow in ECF micro-pumps.
(2) Characteristic curves of ECF micro-pumps are found under the influence of temperature.
(3) The temperature increases linearly with the increase in applied voltage but decreases with increasing Re_F. Additionally, generalized correlation was first proposed to predict the temperature value in the ECF micro-pump quickly and effectively (Eq. 15).

These results have important implications for studying the behavior of ECF flow under the influence of temperature and EDL and for precise control of micro-pumps for applications in fields such as mechanical engineering, engineering, biomedical, and especially microelectronics cooling.

Acknowledgements This research is funded by Vietnam National Foundation for Science and Technology Development (NAFOSTED) under grant number 107.03-2021.37.

References

1. Raghavan, R.V., et al.: Electrokinetic actuation of low conductivity dielectric liquids. Sens. Actuators B Chem. **140**, 287–294 (2009)
2. Ueno, S., et al.: Micro inchworm robot using electro-conjugate fluid. Sens. Actuators A **216**, 36–42 (2014)
3. Mori, K., et al.: Control of a novel flexible finger using an electro-conjugate fluid with built-in angle sensor. Sens. Actuators A **183**, 75–83 (2012)
4. Abe, Y., et al.: Droplet μTAS using electro-conjugate fluid—feedback position control of multiple droplets in flow channel matrix. Sens. Actuators A **198**, 1–7 (2013)
5. Mao, Z., Yoshida, K., Kim, J.W.: Developing O/O (oil-in-oil) droplet generators on a chip by using ECF (electro-conjugate fluid) micropumps. Sens. Actuators B Chem. **296,** 126669 (2019)
6. Peng, Y., et al.: A review on electrohydrodynamic (EHD) pump. Micromachines **14**, 321 (2023)
7. Iijima, Y., et al.: Numerical simulation of electro-conjugate fluid flow considering electric double layer. Mech. Eng. J. **2**, 15–00341 (2015)
8. Kim, J.-W., Yamada, Y., Yokota, S.: Micro ECF (electro-conjugate fluid) hydraulic power sources based on the modular design of TPSEs (triangular prism and slit electrode pairs). Int. J. Adv. Manuf. Technol. **106**, 627–639 (2019)
9. Tang, G., Yang, C.: Numerical modeling of Joule heating-induced temperature gradient focusing in microfluidic channels. Electrophoresis **29**, 1006–1012 (2008)
10. COMSOL: COMSOL multiphysics user guide series COMSOL multiphysics user guide ed. Series Editor (2019)

Electromechanical Properties of GaN Monolayer

Dinh The Hung, Tran The Quang, Nguyen Hoang Linh, and Do Van Truong

Abstract In this research, we focus on determining the electromechanical properties of the Gallium Nitride (GaN) material using density functional theory (DFT). The GaN monolayer exhibits structural stability through two tests of elastic constants and phonon dispersions. Obtained results show that the GaN structure is quite durable with the highest mechanical fracture strain up to 24%. In the elastic domain, the soft mechanical properties of the GaN monolayer are isotropic, and the polar contours of elastic parameters are perfect circles. Besides, we find out that the GaN monolayer is a semiconductor material and energy band structures have the indirect behavior. Under the impacts of mechanical strains on the lattice crystal, the GaN monolayer's bandgap energy tends to gradually decline compared to the equilibrium state. At the uniaxial strains of 24% along the *x*-axis and 18% along the *y*-axis, the bandgap energy decreases by 100% and 90%, respectively. Thereby, the characteristics of the high semiconductor ability of the GaN monolayer as well as great potential in replacing silicone semiconductors in future are shown.

Keywords GaN monolayer · 2-D material · Energy band structures · Indirect semiconductor · Density functional theory (DFT)

D. T. Hung (✉)
Faculty of Mechanical Engineering and Mechatronics, Phenikaa University, Hanoi, Vietnam
e-mail: hung.dinhthe@phenikaa-uni.edu.vn

T. T. Quang
Faculty of Engineering and Technology, Thai Binh University, Thai Binh, Vietnam

N. H. Linh · D. Van Truong
Hanoi University of Science and Technology, Hanoi, Vietnam

D. H. Tien et al. (eds.), *Proceedings of the 1st International Conference on Sustainability and Emerging Technologies for Smart Manufacturing*, Proceedings in Technology Transfer, https://doi.org/10.1007/978-981-97-7083-0_17

1 Introduction

Recently, thanks to the two-dimensional (2-D) materials' electronic properties and various surface effects, these materials have demonstrated their unique properties in the world of materials science. As a result, it shows great and widespread potential in both daily life and industrial products such as electronic devices, optical devices, chemical sensors, or micro-electromechanical systems (MEMS) [1–7]. Among the known semiconductor materials, the GaN monolayer is the most commonly used, and it can replace silicone in the near future. The GaN structure is a semiconductor material that is widely used in micro- to nanosizes due to its huge band gap, high charge carrier transmission speed, and high-temperature resistance. The GaN monolayer is often used in high-frequency electrical and optoelectronic devices, radio transmitters, mobile fast charging, energy conversion, and battery charging in hybrid and electric vehicles [8–10]. There have been several studies on the physical properties of the GaN structure [11, 12]. Meanwhile, GaN material with a two-dimensional structure has been only studied on its electronic and optical properties. Author Behzad et al. have researched the strain and sheet thickness of the GaN monolayer's optical properties. However, this research has not shown the nonlinear relationship between stress and strain to prove the flexible mechanical properties of this material in practical devices [13]. Sanders et al. also did a similar study on the impact of strain on monolayer and bilayer GaN's optical properties. In that study, the GaN monolayer only investigated the strain from -5% to $+5\%$ and also did not clearly show the relationship between strain to the structure and electronic properties of GaN [14]. Meanwhile, mechanical properties are a necessary key and the basic background to reveal the applicability of the GaN structure in practice, which has not yet been clarified. Besides, the strain on electronic properties is one of the predictions needed to enhance understanding regarding the behavior of the GaN structure in practical devices. Therefore, in this research, the DFT is applied to analyze the properties and the impacts of mechanical strains on the electronic characteristics of the GaN monolayer. The obtained results provide important information in applying mechanical strain to adjust the bandgap energy of GaN materials in particular and 2-D semiconductor materials in general.

2 Simulation

According to the DFT and by applying the Quantum Espresso (QE) software, the electromechanical properties of the GaN monolayer are explored. The atoms in the structure interact with each other through a Perdew–Burke–Ernzerhof (PBE) exchange–correlation functional with the generalized gradient approximation (GGA) [15]. The plane wave function limited with an energy cutoff of 80 Ry is chosen so that the structure satisfies the energy convergence conditions. The Monkhorst–Pack method is applied to divide the k-point in the Brillouin zone as grid $20 \times 20 \times 1$ [16].

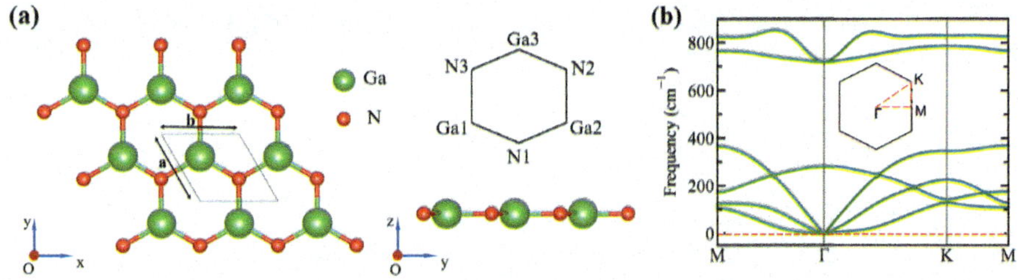

Fig. 1 Schematic of **a** structure of GaN material boundary along xOy, yOz plane, and position of atoms in the GaN monolayer and **b** phonon dispersion curves in Brillouin region

Figure 1a shows the atomic structure of the GaN structure equilibrated under ideal conditions (0 K) with force less than 10^{-6} Ry/a.u and stress less than 5×10^{-2} GPa.

Based on Hooke's law [17], the GaN monolayer's elastic modulus and Poisson's ratio are determined by formulas (1) and (2) as below:

$$E_{xx} = E_{yy} = \frac{C_{11}C_{22} - C_{12}^2}{C_{11}} \tag{1}$$

$$\upsilon_{xy} = \upsilon_{yx} = \frac{C_{12}}{C_{11}} \tag{2}$$

3 Results and Discussions

3.1 Structure Optimization and Stable Equilibrium of GaN Material

The input parameters regarding atomic positions and lattice crystal vectors of the GaN monolayer were balanced at ideal conditions (0 K). The lattice constant's findings are listed in Table 1, and they are similar to previous studies [18–21]. Geometrically, the GaN monolayer is created from a hexagonal structure, similar to the graphene or the BN structure. The Ga and N atoms are arranged alternately to form each hexagonal ring as shown in Fig. 1a. In spatial distribution, Ga and N atoms are all located on the same plane. After optimizing the structure, the GaN monolayer was tested for stability and viability in practice by two methods. The first method is to investigate the dynamic stability based on phonon lattice vibrations of the structure. Figure 1b illustrates the phonon vibration spectrum of the GaN monolayer. Through analysis of the vibration spectrum, it is shown that the structure is completely stable due to the no appearance of any negative vibrations in the Brillouin zone. The maximum vibration frequency of the structure achieved is 854.92 cm^{-1}. Besides, we used static testing standards through elastic constants. According to Born's standard, the GaN

Table 1 A a—lattice constants (Å), C_{ij}—elastic constants (N/m), E—elastic modulus (N/m), υ—Poisson's ratio of the GaN monolayer

a	$C_{11} = C_{22}$	C_{12}	C_{66}	$E_{xx} = E_{yy}$	$\upsilon_{xy} = \upsilon_{yx}$
3.21	143.85	69.07	37.35	110.64	0.48

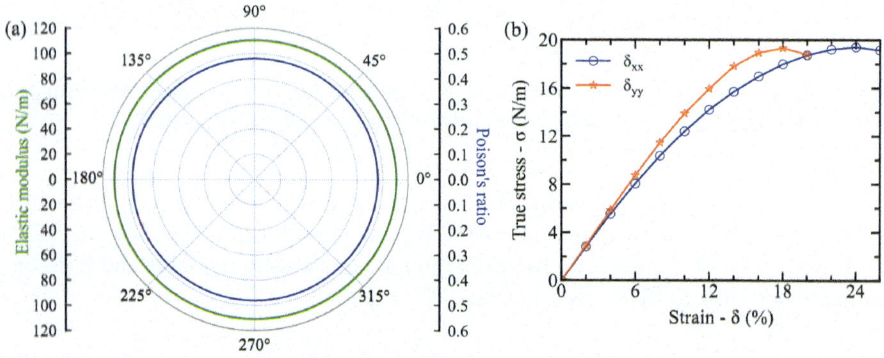

Fig. 2 **a** Polar contours of elastic modulus and Poison's ratio depending on the strain of the GaN monolayer and **b** the relationship between mechanical strains and true stress of the GaN monolayer following the uniaxial strains

monolayer is stable with elastic constants satisfying the conditions $C_{11} > |C_{12}| > 0$ and $C_{66} > 0$ [22]. Thus, through two stability criteria, phonon lattice vibration and elastic constants, it has been shown that the GaN monolayer has high stability.

After confirming the stability of the material, they proceed to determine material parameters such as elastic modulus (E) and Poisson's ratio (υ) which are shown in Table 1 and Fig. 2a. As a result, the crystal lattice structure of GaN is completely symmetric in the elastic domain with elastic modulus and Poisson's ratio dispersions having the same value at all strains. The results show the complete symmetry of the GaN monolayer as well as directly decide the elastic mechanical properties. Our results are completely appropriate to the earlier studies [20, 21, 23].

3.2 Mechanical Properties

The mechanical strain used in this study comes from the variation in the dimensions of the lattice vectors and is calculated according to the formula $\delta = (L - L_0)/L_0$, in which L_0 and L perform the lattice vectors before and after strain, respectively. Figure 2b shows the relationship between the uniaxial strains and the true stress of the GaN monolayer. Our results show that the GaN monolayer could withstand tensile strain quite well with strain up to 18% and 24% for the uniaxial strain along the y-axis and the x-axis, respectively. In the early steps, the relationship between the strains of the two directions is similar. When the strain reaches a value of approximately

6%, the behavior of the material changes and the strains tend to separate from each other. The critical stress of strains in the *y*-axis is always higher than the *x*-axis at the same values. However, the fracture strain of the *x*-axis is larger than the *y*-axis, with the fracture strain $\delta_{xx} = 24\%$ and the critical stress of $\sigma = 19.44$ N/m. Compared with some previously studied structures such as SiS_2 ($\delta^{frac} = 16\%$) [24], ReS_2 ($\delta^{frac} = 14\%$) [25], PbS_2 ($\delta^{frac} = 20\%$) [26], the damage resistance of the GaN monolayer is quite good and could withstand large destructive strain.

Figure 3 shows the distances and angles between atoms of the GaN monolayer in space. At the natural state, the distance of Ga and N atoms is $d_{\text{Ga-N}} = 1.85$ (Å), and the obtained result is similar to previous studies [20, 21, 23]. Under the changes of lattice vectors by uniaxial strains along the *x*- and *y*-axis, the distance between Ga and N changes from 0.26 to 12.51% and from 0.44 to 16.41%, respectively, compared to the initial value. And it reaches the highest at the variation $\delta_{yy} = 18\%$ corresponding to 16.41%. However, when having tensile in the *x* direction, the bond length between Ga_1 and N_1 has the smallest change when all bond length values are less than 0.9%. Besides, the angles created by Ga and N atoms under the impact of uniaxial strains along the *x*- and *y*-axis have changed from − 10.28 to + 5.14% and from − 10.54 to + 21.07% at the facture strain, respectively, compared to the original value ($\angle Ga_1N_1Ga_2 = \angle N_2Ga_1N_1 = 120°$). This change depends on the tensile strain direction (*x* or *y*), which leads to the angles $\angle Ga_1N_1Ga_2$ and $\angle N_2Ga_1N_1$ showing an increasing or decreasing trend, respectively, according to the deformation steps. Among them, $\angle N_2Ga_1N_1$ shows a rapidly increasing trend and reaches 145.29° at the strain $\delta_{xx} = 24\%$ 24%. However, this same angle when subjected to deformation in the *x* direction shows a decreasing trend reaching 108° at $\delta_{yy} = 18\%$, thereby showing the mechanical flexibility of the GaN monolayer.

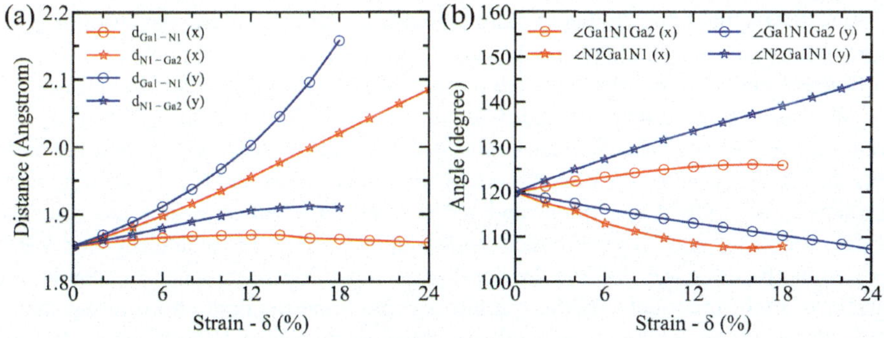

Fig. 3 Distances and angles of atoms in space

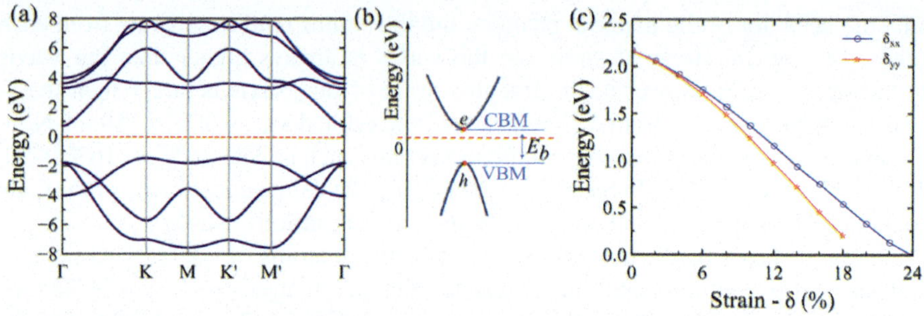

Fig. 4 **a** Band gap at $\delta = 0\%$, **b** the method to determine the bandgap energy of the GaN monolayer, and **c** the energy band gap of the GaN monolayer as the difference of strain

3.3 Electronic Properties

We investigate the electronic properties of the GaN monolayer under the changes of lattice vectors by PBE functional, and it is an important key to apply in the nanoelectronic devices. Figure 4a shows the energy band structures of the GaN monolayer at several strain values. Our results show that the GaN monolayer is a semiconductor with a band gap of 2.17 eV and has the indirect semiconductor behavior at the natural state. This result is consistent with previous calculation results [11, 18]. The bandgap is calculated by distance from the valence band's highest value to the conduction band's lowest value as in Fig. 4b. Specifically, the changes in the bandgap energy under the uniaxial strains are shown in Fig. 4c. Under these impacts, the bandgap energy is changed, and it tends to decrease about 100% and about 90% compared to equilibrium for uniaxial strains along the *x*- and *y*-axis, respectively. At initial strains, the bandgap energy almost does not change. When the strain in both directions reaches $\delta = 6.00\%$, the bandgap energy in the two directions has a certain difference and the more the strain increases, the more this difference increases.

We can see the effects of strains on the energy band structure clearly by the relationship between mechanical strains and bandgap of the GaN monolayer as in Fig. 5. For strains along the x-axis, the GaN monolayer shows indirect semiconductor properties with CBM at Γ point and VBM at *K* point. When strains increase, the GaN monolayer still shows indirect semiconductor properties with CBM and VBM positions. Besides, we also see the band gap tends to decrease when increasing the strain in both directions. At the fracture strain of the uniaxial strain along the *x*-axis, the bandgap energy of the GaN monolayer decreases up to 0 eV (about 100%). Meanwhile, the bandgap energy of the GaN monolayer only decreases by about 90% compared with the natural state. Therefore, we have determined that the GaN monolayer has anisotropic electronic properties and is sensitive to strains. This is an exciting finding to apply to nanoelectronic devices.

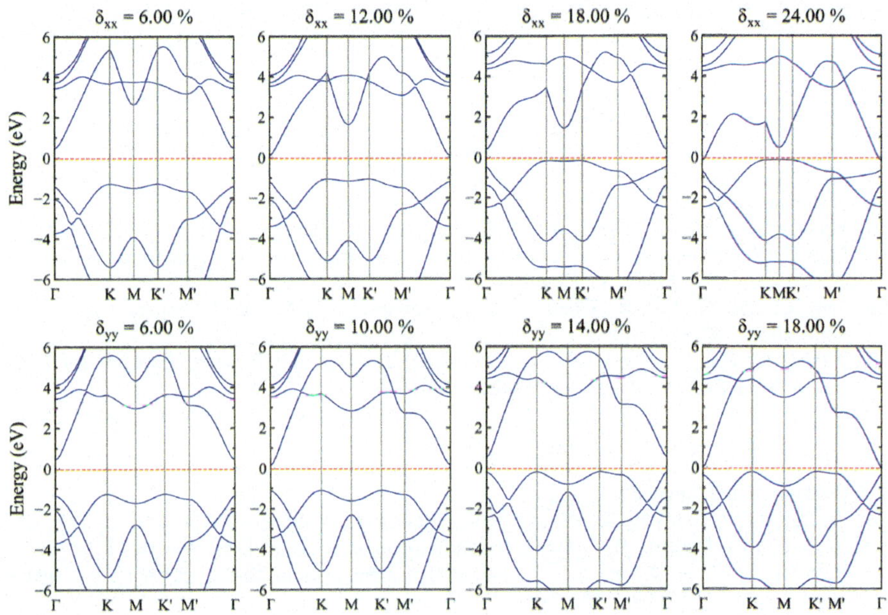

Fig. 5 Energy band structures of the GaN monolayer with several strains along the x and y directions. The Fermi energies (red dashed line) are set to zero for all strains

4 Conclusion

The GaN monolayer was investigated for its mechanical properties and the strain on the energy band structures through the density functional theory. The stability of the structure is investigated with two standards dynamic and static. The GaN structure exhibits the properties of an indirect semiconductor material with VBM (CBM) dependently distributed and a bandgap energy of 2.17 eV. Meanwhile, we find the fracture strain and the critical stress for tensile strain along the x-axis better than the y-axis and reach the critical point of stress at the strain of 24% and 18% in the x- and y-axis, respectively. Under the strain, the GaN monolayer shows that the bandgap energy decreases to 0.22 and 0.00 eV at fracture strain along the y- and x-axis, respectively. Our results also show that the electronic characteristics of the GaN monolayer have flexible and sensitive properties with mechanical strains, which is of great significance in the application of GaN monolayers to electronic devices as well as the possibility of replacing silicone semiconductors in the near future.

References

1. Cao, V.A., Kim, M., Hu, W., Lee, S., Youn, S., Chang, J., Chang, H.S., Nah, J.: Enhanced piezoelectric output performance of the SnS2/SnS heterostructure thin-film piezoelectric nanogenerator realized by atomic layer deposition. ACS Nano **15**, 10428–10436 (2021)
2. Zhang, L., Wang, N., Li, Y.: Design, synthesis, and application of some two-dimensional materials (2023)
3. Khan, K., Tareen, A.K., Aslam, M., Wang, R., Zhang, Y., Mahmood, A., Ouyang, Z., Zhang, H., Guo, Z.: Recent developments in emerging two-dimensional materials and their applications. J. Mater. Chem. C Mater. **8**, 387–440 (2020)
4. Cai, X., Luo, Y., Liu, B., Cheng, H.M.: Preparation of 2D material dispersions and their applications (2018)
5. Zhang, X., Hou, L., Ciesielski, A., Samorì, P.: 2D Materials Beyond Graphene for High-Performance Energy Storage Applications, (2016).
6. Jia, L., Wu, J., Zhang, Y., Qu, Y., Jia, B., Chen, Z., Moss, D.J.: Fabrication technologies for the on-chip integration of 2D materials (2022)
7. Zheng, W., Lee, L.Y.S.: Beyond sonication: advanced exfoliation methods for scalable production of 2D materials (2022)
8. Al Balushi, Z.Y., Wang, K., Ghosh, R.K., Vilá, R.A., Eichfeld, S.M., Caldwell, J.D., Qin, X., Lin, Y.C., Desario, P.A., Stone, G., Subramanian, S., Paul, D.F., Wallace, R.M., Datta, S., Redwing, J.M., Robinson, J.A.: Two-dimensional gallium nitride realized via graphene encapsulation. Nat. Mater. **15**, 1166–1171 (2016)
9. Sahu, T.K., Sahu, S.P., Hembram, K.P.S.S., Lee, J.K., Biju, V., Kumar, P.: Free-standing 2D gallium nitride for electronic, excitonic, spintronic, piezoelectric, thermoplastic, and 6G wireless communication applications. NPG Asia Mater. **15** (2023)
10. Jia, Y., Shi, Z., Hou, W., Zang, H., Jiang, K., Chen, Y., Zhang, S., Qi, Z., Wu, T., Sun, X., Li, D.: Elimination of the internal electrostatic field in two-dimensional GaN-based semiconductors. NPJ 2D Mater. Appl. **4** (2020)
11. Zhang, Z., Chai, C., Song, Y., Kong, L., Yang, Y.: A DFT study on physical properties of III-V compounds (AlN, GaN, AlP, and GaP) in the P3121 phase. Mater. Res. Express. **8** (2021)
12. Javaheri, S., Boochani, A., Babaeipour, M., Naderi, S.: Electronic and optical properties of GaN under pressure: DFT calculations. Int. J. Mod. Phys. B. **31** (2017)
13. Behzad, S.: Effects of strain and thickness on the electronic and optical behaviors of two-dimensional hexagonal gallium nitride. Superlattices Microstruct. **106**, 102–110 (2017)
14. Sanders, N., Bayerl, D., Shi, G., Mengle, K.A., Kioupakis, E.: Electronic and optical properties of two-dimensional GaN from first-principles. Nano Lett. **17**, 7345–7349 (2017)
15. Perdew, J.P., Burke, K., Ernzerhof, M.: Generalized gradient approximation made simple (1996)
16. Monkhorst, H.J., Pack, J.D.: Special points for Brillonin-zone integrations (1976)
17. Wei, Q., Peng, X.: Superior mechanical flexibility of phosphorene and few-layer black phosphorus. Appl. Phys. Lett. **104** (2014)
18. Zhang, Y.: Many-body calculations of excitons in two-dimensional GaN. Crystals (Basel) **13** (2023)
19. Hidayati, S., Sholihun: Strain effects on the band structures of monolayer GaN from the density functional theory. In: Materials Science Forum, pp. 144–149. Trans Tech Publications Ltd. (2022)
20. Ben, J., Liu, X., Wang, C., Zhang, Y., Shi, Z., Jia, Y., Zhang, S., Zhang, H., Yu, W., Li, D., Sun, X.: 2D III-Nitride materials: properties, growth, and applications (2021)
21. Ye, C., Peng, Q.: Mechanical stabilities and properties of graphene-like 2D III-nitrides: a review (2023)
22. Mouhat, F., Coudert, F.X.: Necessary and sufficient elastic stability conditions in various crystal systems. Phys. Rev. B Condens. Matter Mater. Phys. **90** (2014)
23. Onen, A., Kecik, D., Durgun, E., Ciraci, S.: GaN: From three- to two-dimensional single-layer crystal and its multilayer van der Waals solids. Phys. Rev. B. **93** (2016)

24. Linh, N.H., Quang, T.T., Son, N.M., Van Thanh, V., Truong, D.V.: Prediction of mechanical, electronic and optical properties of monolayer 1T Si-dichalcogenides via first-principles theory. Mater. Today Commun. **36**, 106553 (2023)
25. Sun, H., Agrawal, P., Singh, C.V.: A first-principles study of the relationship between modulus and ideal strength of single-layer, transition metal dichalcogenides. Mater. Adv. **2**, 6631–6640 (2021)
26. Linh, N.H., Son, N.M., Quang, T.T., Hoi, N.V., Thanh, V., Truong, D.V.: First-principles investigation on the electromechanical properties of monolayer 1H Pb-Dichalcogenides. Korean J. Mater. Res. **33**, 189–194 (2023)

A DNS Simulation of the Low-Reynold Flow Through an Elliptical Cylinder Using the Lattice Boltzmann Method

Van Tuyen Vu, Quang Dung Ta, Viet Dung Duong, and Ich Long Ngo

Abstract This paper presents a numerical investigation into the dynamic behaviors of fluid flow through an elliptical cylinder. The lattice Boltzmann method is implemented in the direct numerical simulation. The aspect ratio (AR) varies from 0.25 to 1.0, and the attack angle (α) ranges from 0° to 90°. The numerical results were thoroughly validated by comparing them with both the experimental and numerical results. Consequently, the behavior of vortex formation is complex and depends on both AR and α. Moreover, the vortex formation regime changes significantly; i.e., the steady wake (pattern I) disappears, the Karman wake (pattern II) shrinks, and the Karman wake with the downstream secondary wake (pattern III) extends toward the higher AR. The results indicate the numerical method's superiority in effectively capturing the vortex structure of the incompressible fluid flow in potential uses, such as heat transfer, bridge cables, offshore structures, and pipe racks.

Keywords Direct numerical simulation · Lattice Boltzmann method · Elliptical cylinder · Vortex structure

1 Introduction

In recent decades, research on the dynamic behavior of the flow through the bluff body has attracted due to its significance in a wide range of applications. The related investigations are useful for better understanding and knowledge about fluid flow through such a system. The behavior of fluid flow over an elliptical cylinder (EC) is more complex and unpredictable than the canonical configuration, a circular cylinder [1]. Raman et al. [2] indicated that the fluid flow over an isolated EC is chosen instead of circular cylinders. Both the aspect ratio (AR) and the angle of attack (α) play crucial

V. T. Vu · Q. D. Ta · I. L. Ngo (✉)
Hanoi University of Science and Technology, Hanoi, Vietnam
e-mail: long.ngoich@hust.edu.vn

V. D. Duong
Vietnam National University, University of Engineering and Technology, Hanoi, Vietnam

D. H. Tien et al. (eds.), *Proceedings of the 1st International Conference on Sustainability and Emerging Technologies for Smart Manufacturing*, Proceedings in Technology Transfer, https://doi.org/10.1007/978-981-97-7083-0_18

roles in the flow behavior around the EC. Li et al. [3] analyzed the influence of AR on flow over an EC. The numerical results indicate that an increase in AR leads to a decrease in both the mean drag coefficient and the root-mean-square of the lift coefficient for the elliptic cylinder. Moreover, various methods are used to investigate the flow over the cylindrical structures. Wang et al. [4] used a 2-D Reynolds-averaged Navier–Stokes (*RANS*) model and stated that this model is unable to capture the vortex shedding at the downstream. Nguyen et al. [5] applied direct numerical simulation (*DNS*) and large eddy simulations (*LES*) in analyzing incompressible flows over moving cylindrical structures. In recent years, the lattice Boltzmann method (*LBM*) has advanced as a feasible and optimistic numerical method for simulating fluid dynamics systems. Its effectiveness has been demonstrated in fluid flow applications, including interfacial dynamics and complex boundaries. Chen et al. [6] and Yu et al. [7] indicated the positive aspects of *LBM*. As far as we are aware, no research considers the capability of the LBM to fully capture the vortex structure of fluid flow over an EC compared with other methods, e.g., the finite volume method (*FVM*).

The objective of this study is to consider the advantages of the *LBM* in direct numerical simulation. The results obtained are compared with the previous investigation using the finite volume method.

2 Numerical Methodology

This paper used the D2Q9 model, which has nine velocity-positioned particles on a two-dimensional lattice. Each lattice point is assigned specific values to capture the particle's velocity effectively. In *LBM*, the computation of the particle distribution function $f_i(x, t)$ is based on the original discrete Boltzmann equation, as depicted in Eq. (1). Notably, the LBM uses open-source code with C++ language programming. This method has a highly parallelizable computation ability and is fully compatible with the DNS. These features become advantages compared with commercial software like RANS or LES.

$$f_i(\boldsymbol{x} + e_i\Delta t, t + \Delta t) - f_i(\boldsymbol{x}, t) = -\frac{1}{\tau}\left[f_i(\boldsymbol{x}, t) - f_i^{eq}(\boldsymbol{x}, t)\right] \tag{1}$$

Here x, Δt, and t are spatial points, time steps, and time, respectively. On the right hand, the Bhatnagar–Gross–Krook collision [8] utilizes a single relaxation time denoted by τ and $f_i^{eq}(x, t)$ denotes the equilibrium distribution function in the D2Q9 model [6], as given by Eq. (2):

$$f_{\mathrm{i}}^{eq} = \rho w_{\mathrm{i}}\left[1 + 3e_i \cdot \boldsymbol{u} + \frac{9}{2}(e_i \cdot \boldsymbol{u})^2 - \frac{3}{2}\boldsymbol{u} \cdot \boldsymbol{u}\right] \tag{2}$$

The density (ρ) and velocity (u) in *LBM* are derived from populations through their moments, as illustrated in Eq. (3):

$$\rho = \sum_i f_i, \; \rho \boldsymbol{u} = \sum_i e_i f_i \tag{3}$$

The relationship between ν and Δt can be expressed through the formula provided in Eq. (4):

$$\nu = c_s^2\left(\tau - \frac{1}{2}\right)\Delta t \tag{4}$$

In this equation, the lattice sound speed c_s can be determined by utilizing the state equation for an ideal gas to calculate the pressure field, $p = \rho c_s^2$. In *LBM*, the normalized output macroscopic variables are expressed by $p^* = (p - p_0)/\left(\rho U_\infty^2\right)$. The momentum exchange method is employed to determine the total fluid forces (F (F_x, F_y)) exerted on the surface of the 2-D rigid body within the boundary cell layer [7].

The important variables, such as drag coefficient (C_D), lift coefficient (C_L), mean drag coefficient ($\overline{C}_D$), mean drag coefficient ($\overline{C}_L$), the root-mean-square value of lift coefficient (C'_L), the Strouhal number (*St*), and the root-mean-squared streamwise (u^*_{rms}), are accordingly computed by Eq. (5).

$$C_D = \frac{F_x}{0.5\rho U_\infty^2 D_2}, \quad C_L = \frac{F_y}{0.5\rho U_\infty^2 D_2}, \quad \overline{C}_D = \frac{1}{N}\sum_1^N C_D, \quad \overline{C}_L = \frac{1}{N}\sum_1^N C_L,$$
$$C'_L = \sqrt{\frac{1}{N}\sum_1^N \left(C_L - \overline{C}_L\right)^2}, \quad St = \frac{f \cdot D_2}{U_\infty}, \quad u^*_{rms} = \frac{1}{U_\infty}\sqrt{\frac{1}{N}\sum_1^N \left(u - u_{avg}\right)^2} \tag{5}$$

Here, N denotes the number of time instances in a history dataset.

Figure 1 depicts the computational domain with 100*D* (length) and 60*D* (width) dimensions. This computational domain has a blockage ratio of less than 2%, satisfying the sufficiently required blockage ratio ($B = D/W$ with W denotes the computational domain's width) threshold of 6% [9]. Additionally, Fig. 1 presents the boundary conditions (*BC*). The Dirichlet-type *BC* is used at the inlet ($u = U_\infty$ and $v = 0$), and the Neumann-type BC was implemented at both the upstream and downstream boundaries. An outlet BC $\partial u/\partial x = \partial v/\partial x = 0$ was specified.

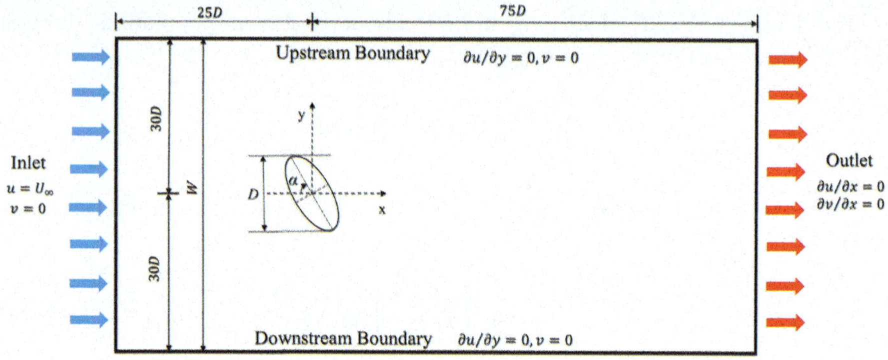

Fig. 1 Numerical model and the boundary conditions

3 Results and Discussions

3.1 Model Validation

The mesh convergence was implemented by changing the block level and the cell numbers in each block. The details of this mesh are shown in the research of Duong et al. [10]. The four mesh models from coarse model (#1) to the finest level (#4) are considered in this study. The deviation of mesh models #1, #2, and #3 compared with mesh model #4 is 1.23%, 0.96%, and 0.27% for $\overline{C}_D$, respectively, as shown in Table 1. The other hydrodynamic coefficients have a zero deviation. Based on these results, mesh model #2, which is highlighted in Table 1, is chosen for further simulation in this study.

Table 2 shows the numerical model validated for the important hydrodynamic coefficients ($\overline{C}_D$, C'_L, and *St*) through an isolated EC at Re = 150. The numerical codes are validated to those of Duong et al. [10] and Thompson et al. [11]. It was found from Table 2 that the maximum deviation is less than 3%. This model was also validated for $\overline{C}_D$, C'_L, and *St* of the fluid flow through a single circular cylinder at Re = 150. The numerical results were compared to those obtained by Qu et al. [12] and Chen et al. [13]. As a result, the maximum deviation is less than 4%. Consequently, the numerical model was well validated, and it can be used for further computations in the present study.

Table 1 Mesh convergence test, AR = 0.5, $\alpha = 90°$, Re = 150

Mesh Block level/cells	#1 3/30^2	#2 3/50^2	#3 4/40^2	#4 3/70^2
$\overline{C}_D$	1.896 (1.23%)	**1.855 (0.96%)**	0.178 (0.27%)	0.173
C'_L	0.541 (0%)	**0.541 (0%)**	0.541 (0%)	0.541
St	0.196 (0%)	**0.196 (0%)**	0.196 (0%)	0.196

Table 2 Validation for an isolated elliptical cylinder, AR = 0.5, $\alpha = 90°$, Re = 150

Authors	$\overline{C}_D$	C'_L	St
Duong et al. [10]	1.837	0.54	0.193
Thompson et al. [11]	1.78	–	0.189
Present work	1.885	0.541	0.196
Maximum Deviation (%)	1.70	0.18	2.62

3.2 The Vorticity and Root-Mean-Square Streamwise Velocity Field

Figure 2 illustrates the comparison of vorticity fields between the findings of Shi et al. [14] (left) and the present results (right). The AR = 0.25 was kept and $\alpha = 0°$ in Fig. 2a, b, $\alpha = 30°$ in Fig. 2c, d, and $\alpha = 90°$ in Fig. 2e, f. The AR = 0.50 and $\alpha = 75°$ is shown in Fig. 2g, h. According to Shi et al. [14], three distinct patterns were observed. Pattern I, characterized by a steady wake, was observed in Fig. 2a. Pattern II, which consists of a Karman wake with trailed by a steady wake, was observed in Fig. 2c, g. Lastly, pattern III, characterized by a Karman wake with trailed by a secondary wake, was observed in Fig. 2e.

A different pattern between the two results can be seen. First, only the Karman wake forms at the downstream cylinder shown in Fig. 2b, d, whereas no steady wake is found in Fig. 2a, c, according to Shi's results. Second, pattern III was found for both results, as shown in Fig. 2e, f. The change of flow regime is due to the moving of the stagnation point and stagnation angle when α and AR increase the stagnation angle, respectively. Moreover, the vortex intensity in the present result is greater than that of Shi's result. Finally, pattern II (Fig. 2g) becomes pattern III (Fig. 2h). Based on these results, the present numerical method is shown to have more advantages than that of Shi et al. in capturing the vortex shedding downstream. This result also shows the accuracy of the DNS method compared with other numerical approaches.

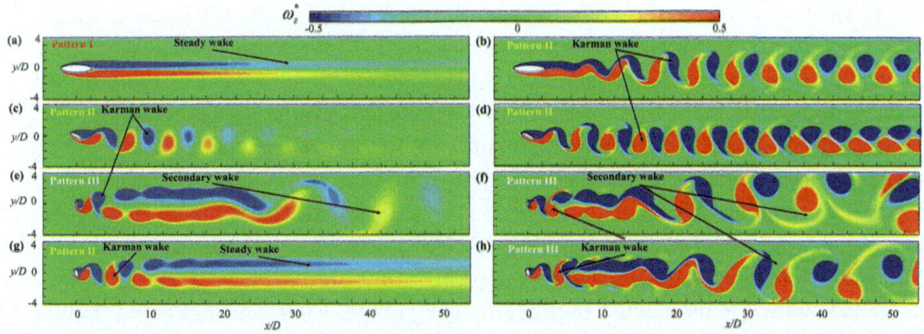

Fig. 2 The vorticity fields $\omega_z{}^* = \omega_z D/U_\infty$: **a, b** AR = 0.25, $\alpha = 0°$, **c, d** AR = 0.25, $\alpha = 30°$, **e, f** AR = 0. 25, $\alpha = 90°$ and **g, h** AR = 0.50, $\alpha = 75°$

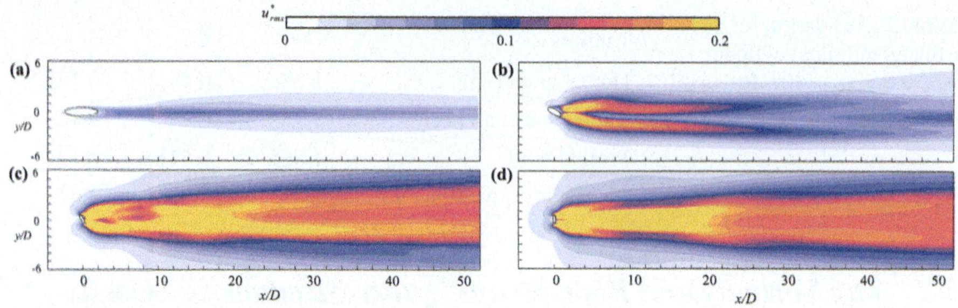

Fig. 3 The root-mean-square streamwise velocity fields at AR = 0.25 for various values of α: **a** $\alpha = 0°$, **b** $\alpha = 30°$, **c** $\alpha = 75°$ and **d** $\alpha = 90°$

This result predicts that the flow regime map can be changed, which will be presented in the next section.

Figure 3 illustrates the root-mean-square velocity fields for the various values of α (0°, 30°, 60°, and 90°) and a constant AR = 0.25. The intensity of u^*_{rms} indicated the ability to appear vortices. The fluctuation at the top and bottom cylinders is small, but it is greater downstream. This implies that the Karman wake street can be formed the downstream. The intensity behind the cylinder increases when the α increases from 0° to 90°. It can be seen the dependence on α for the change of flow field. These results can predict the trend of the hydrodynamic coefficients increase when α increases. The change of u^*_{rms} is the same with AR = 0.5 and AR = 0.75. This can be found clearly in the next sections.

3.3 The Hydrodynamic Force and the Strouhal Number

Figure 4 shows the values of $\overline{C}_{\text{D}}$, $\overline{C}_{\text{L}}$, C'_{L}, and St for the EC in terms of α. AR of 0.25, 0.50, 0.75, and 1.00 was used. Apart from AR = 1.00, $\overline{C}_{\text{D}}$ generally increases with α for all values of AR, as shown in Fig. 4a. When AR decreases, the slope of the curve increases. In particular, at about $\alpha = 30°$, the values $\overline{C}_{\text{D}}$ changed. When $\alpha < 30°$, $\overline{C}_{\text{D}}$ increases with increasing AR, and this is the opposite when $\alpha > 30°$. Figure 4b shows that $\overline{C}_{\text{L}}$ increases from 0° to maximum value α and gradually decreases to 90°. In particular, the dashed line connects the maximum values in each case. This indicated that $\overline{C}_{\text{L}}$'s maximum value tends to move on the higher α when AR increases. The deviation between the maximum value AR = 0.25 and AR = 0.75 is 78.74%. In Fig. 4c, C'_{L} is non-linear and increases with increasing α from 0° to 90°. When AR increases, C'_{L} increases at small α (< 30°) and high α (> 60°). Conversely, C'_{L} decreases with AR increasing in the remaining range. Figure 4d indicated that the St is slightly changed when α increases. The tendency with various values α increases at AR = 0.5 and 0.75. At AR = 0.25, St increases to $\alpha = 45°$ and decreases. The variation of AR from 0.25 to 0.75 increases St at all α. In particular, at AR = 1.00, $\overline{C}_{\text{D}}$, $\overline{C}_{\text{L}}$, C'_{L}, and St are independent of α.

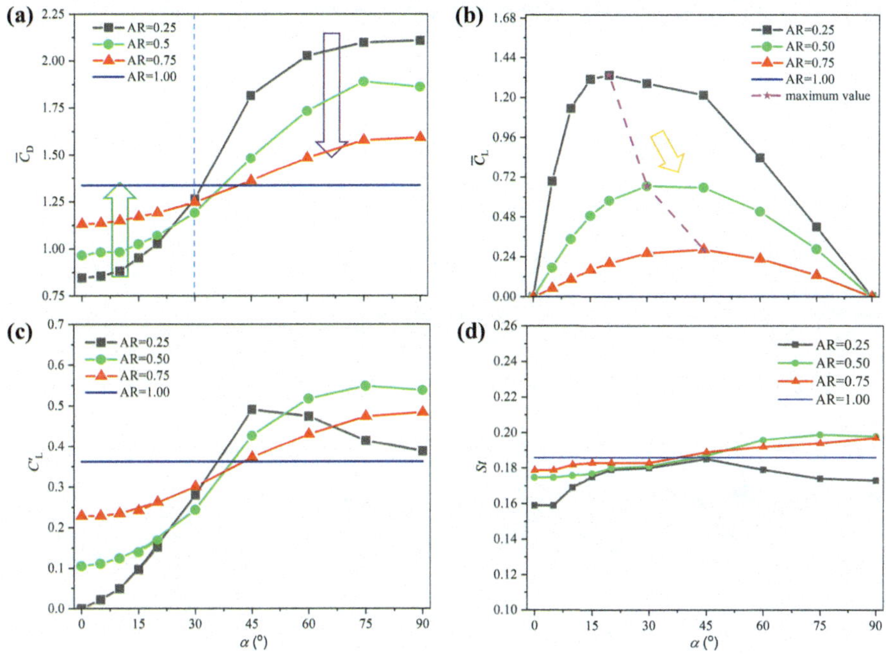

Fig. 4 The variations of $\overline{C}_D$, $\overline{C}_L$, C'_L, and *St* with α for different AR values. The arrow represents the trend of AR

The hydrodynamic coefficients and the *St* numbers in this study were also compared with those obtained by Shi et al. [14], as presented in Fig. 5. It was found that the two results have the same trend, although the discrepancy between them is available. At all values of AR, the maximum deviations for $\overline{C}_D$, $\overline{C}_L$, C'_L, and *St* are 4.14%, 13.6%, 6.24%, and 3.30%, respectively. The variation is a result of the disparity in the numerical methods utilized, causing a shift in the vortex flow patterns presented in the next paragraph.

Figure 6 shows a flow regime map of flow through an EC. In this map, α varies from 0° to 90°, whereas AR ranges from 0.25 to 1.00. The green areas represent pattern I and pattern III in Shi's results. The yellow area is the expansion of pattern III in this study. A notable disparity is evident in the phase map of Shi et al. [14] and the present results. Particularly, the pattern I found in Shi's study does not exist in the present study using the *DNS* method. Additionally, the region of Pattern II shrinks while Pattern III extends toward the higher AR. This indicates the difference between the two numerical methods. Therefore, this result needs to be validated thoroughly by further studies, particularly experimental ones in future.

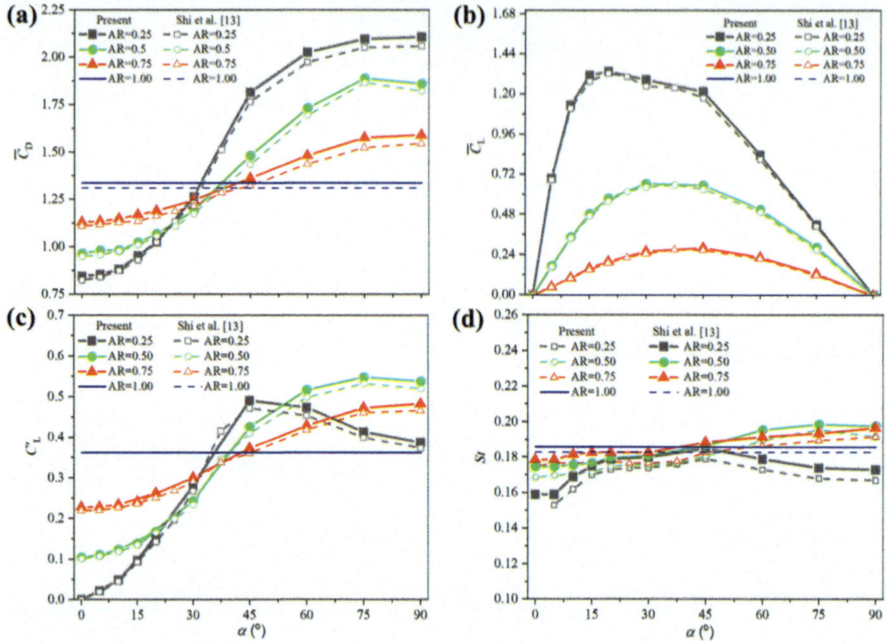

Fig. 5 Comparing with Shi's results for variations of $\overline{C}_D$, $\overline{C}_L$, C'_L, and St for various values of α

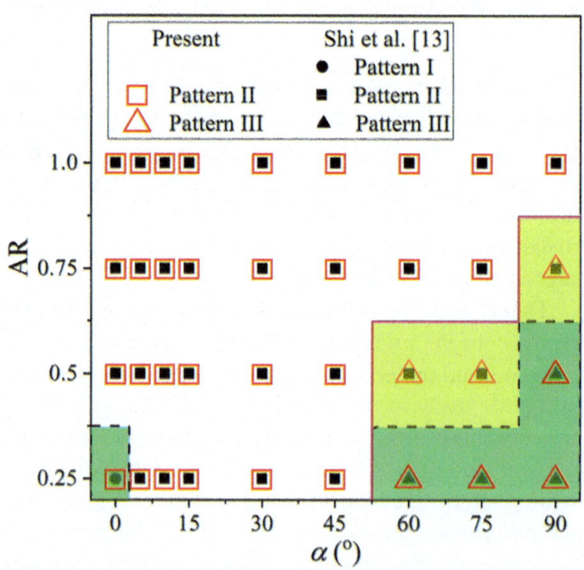

Fig. 6 Flow regime map in the parametric α. The dashed black lines and the magenta solid line represent the boundaries between the different patterns of Shi et al. [14] and the present. The green areas represent the area of pattern I and pattern III in Shi's results. The yellow area is the expansion of the area of pattern III

4 Conclusion

The numerical investigation of fluid flow through an EC was conducted using the lattice Boltzmann method in direct numerical simulation at Re = 150. The angle of attack α varies from 0° to 90°. The numerical model was thoroughly validated for both circular and ECs. Consequently, first, the vortex regime patterns differ from those of Shi's study. Particularly, the steady wake (pattern I) disappears, the Karman wake (pattern II) shrinks, and the Karman wake with the downstream secondary wake (pattern III) extends toward the higher AR. These results imply the benefits of *LBM* in capturing vortex structures compared to other methods like the FVM used by Shi et al. Second, the behavior of vortex formation is complex and depends on both AR and α. The results obtained are useful in understanding the vortex structure in the incompressible fluid flow in potential uses, such as heat transfer, bridge cables, offshore structures, and pipe racks.

Acknowledgements This research is funded by Vietnam National Foundation for Science and Technology Development (NAFOSTED) under grant number 107.03-2021.37.

References

1. Vijay, K., Srinil, N., Zhu, H., Bao, Y., Zhou, D., Han, Z.: Flow-induced transverse vibration of an elliptical cylinder with different aspect ratios. Ocean Eng. **214**, 107831 (2020)
2. Raman, S. K., Arul Prakash, K., Vengadesan, S.: Effect of axis ratio on fluid flow around an elliptic cylinder—a numerical study. J. Fluids Eng. **135**(11) (2013)
3. Li, Y., Li, B., Li, T., Zhang, Y.: Numerical study on the flow around an elliptic cylinder with different aspect ratios in the subcritical regime using large eddy simulation. Phys. Fluids **36**(1) (2024)
4. Wang, L., Alam, M.M., Zhou, Y.: Two tandem cylinders of different diameters in cross-flow: effect of an upstream cylinder on wake dynamics. J. Fluid Mech. **836**, 5–42 (2018)
5. Nguyen, V.-T., Nguyen, H.H.: Detached eddy simulations of flow induced vibrations of circular cylinders at high Reynolds numbers. J. Fluids Struct. **63**, 103–119 (2016)
6. Chen, H., Chen, S., Matthaeus, W.H.: Recovery of the Navier-Stokes equations using a lattice-gas Boltzmann method. Phys. Rev. A **45**(8), R5339 (1992)
7. Chen, Y., Cai, Q., Xia, Z., Wang, M., Chen, S.: Momentum-exchange method in lattice Boltzmann simulations of particle-fluid interactions. Phys. Rev. E **88**(1), 013303 (2013)
8. Chen, S., Doolen, G. D.: Lattice Boltzmann method for fluid flows. Anuu. Rev. Fluid Mech **30**(1), 329–364 (1998)
9. Zheng, Q., Alam, M.M.: Intrinsic features of flow past three square prisms in side-by-side arrangement. J. Fluid Mech. **826**, 996–1033 (2017)
10. Duong, V.D., Nguyen, V.D., Nguyen, V.T., Ngo, I.L.: Low-Reynolds-number wake of three tandem elliptic cylinders. Phys. Fluids **34**(4), 043605 (2022)

11. Thompson, M.C., Radi, A., Rao, A., Sheridan, J., Hourigan, K.: Low-Reynolds-number wakes of elliptical cylinders: from the circular cylinder to the normal flat plate. J. Fluid Mech. **751**, 570–600 (2014)
12. Qu, L., Norberg, C., Davidson, L., Peng, S.-H., Wang, F.: Quantitative numerical analysis of flow past a circular cylinder at Reynolds number between 50 and 200. J. Fluids Struct. **39**, 347–370 (2013)
13. Chen, W., Ji, C., Alam, M.M., Williams, J., Xu, D.: Numerical simulations of flow past three circular cylinders in equilateral-triangular arrangements. J. Fluid Mech. **891**, A14 (2020)
14. Shi, X., Alam, M., Bai, H.: Wakes of elliptical cylinders at low Reynolds number. Int. J. Heat Fluid Flow **82**, 108553 (2020)

Macroscopic Elastic Moduli of the Random Circle-Inclusion Model with Spring-Layer Imperfect Interfaces

Van-Luat Nguyen, Xuan-Khoa Hoang, and Tien-Tai Bui

Abstract The article studies some methods to calculate the effective elastic moduli of materials in 2D with spring-layer imperfect interfaces. Based on the polarisation approximation method (PA), this work develops a new formula to compute the macroscopic elastic moduli (MEM) of the randomly circular inclusion model with the spring-layer (S-L) imperfect interface model. From that, the obtained algebraic expressions are simple and effective tools for estimating the effective moduli of the random circle-inclusion model with the S-L imperfect interface model. The FFT simulation method is constructed to determine the effective moduli of this model. Besides that, the differential approximation (DA) also builds differential equations to compute the effective moduli of the random circle-inclusion model with the S-L imperfect interface. The FFT simulation results are compared with DA and PA approximation methods to show the reliability of the methods.

Keywords Macroscopic elastic moduli (MEM) · Imperfect interface · Spring-layer interface · Polarisation approximation · FFT simulation

1 Introduction

In studies on the MEM of composite materials, it is often assumed that the contact surface of the material is perfect. However, the imperfect contact surfaces have a great effect on the effective elastic moduli (or the MEM) of the compound materials. Therefore, the elastic problem of materials with imperfect surfaces is an important topic of interest to many scientists. The MEM of compound materials with imperfect interface models has been studied by many scientists as shown in the literature [1–10]. These models were initially proposed based on phenomenological arguments. Currently, the imperfect surface models have been rigorously demonstrated and clarified on mathematical and mechanical foundations. There are many research models

V.-L. Nguyen (✉) · X.-K. Hoang · T.-T. Bui
Hanoi University of Industry, Hanoi, Vietnam
e-mail: nguyenvanluat@haui.edu.vn

D. H. Tien et al. (eds.), *Proceedings of the 1st International Conference on Sustainability and Emerging Technologies for Smart Manufacturing*, Proceedings in Technology Transfer, https://doi.org/10.1007/978-981-97-7083-0_19

for the MEM of compound materials with imperfect interface models. In the S-L imperfect surface model that will be studied in this work, the displacement vector has a jump at the interface whilst the stress vector is continuous across this interface [9, 10]. A composite material model studied in this work is random circle-fibre reinforced in the matrix (2D) with the S-L imperfect interface. The proposed solution for these material models is based on exact asymptotic methods combined with imperfect surface properties to construct algebra expressions for the S-L imperfect interface model. In Sect. 2, the PA method is introduced to construct the explicit algebra expressions to determine the effective moduli of circle inclusions with the S-L imperfect interface model. From there, the macro-elastic moduli of the composite materials can be determined. In Sect. 3, the DA approach schemes are developed to derive the differential equations to determine the MEM with the S-L imperfect interface model. In Sect. 4, the FFT simulation is constructed to determine the MEM of the random transversely isotropic fibre inclusions with spring-layer imperfect interfaces, which were presented to estimate the MEM of materials by Michel [15], Bonnet [17], and Moulinec [16]. In Sect. 5, some numerical examples are given for comparison between FFT, DA, and PA methods followed by the conclusion.

2 The PA Method with S-L Imperfect Interfaces

Consider the two-component materials in 2D with random suspension inclusions in the matrix, in which the circular inclusions (volume ratio v_I) have the bulk modulus K_I, shear modulus μ_I.. The matrix phase (volume ratio v_M) has the elastic moduli K_M, μ_M. The polarisation approximation expressions for the MEM of the two-component materials have the form:

$$K^{\text{eff}} = \frac{1}{\frac{v_I}{K_I+\mu_M} + \frac{v_M}{K_M+\mu_M}} - \mu_M \tag{1}$$

$$\mu^{\text{eff}} = \frac{1}{\frac{v_I}{\mu_I+\mu_M^*} + \frac{v_M}{\mu_M+\mu_M^*}} - \mu_M^*, \quad \mu_M^* = \frac{K_M\mu_M}{K_M+2\mu_M} \tag{2}$$

After transformation, expressions (1) and (2) can be written as:

$$K^{\text{eff}} = \frac{K_I K_M + \mu_M K_I v_I + \mu_M K_M v_M}{\mu_M + K_M v_I + K_I v_M} \tag{3}$$

$$\mu^{\text{eff}} = \frac{\mu_I\mu_M + \mu^*\mu_I v_I + \mu^*\mu_M v_M}{\mu^* + \mu_M v_I + \mu_I v_M}, \quad \mu^* = \frac{K_M\mu_M}{K_M+2\mu_M} \tag{4}$$

Apply Eqs. (3) and (4) to compute the MEM of the coated circular inclusion with a thin coating. The circle inclusion with the S-L imperfect interface can be replaced by the coated circle-inclusion model (Fig. 1). Then, we use the equivalent inclusion

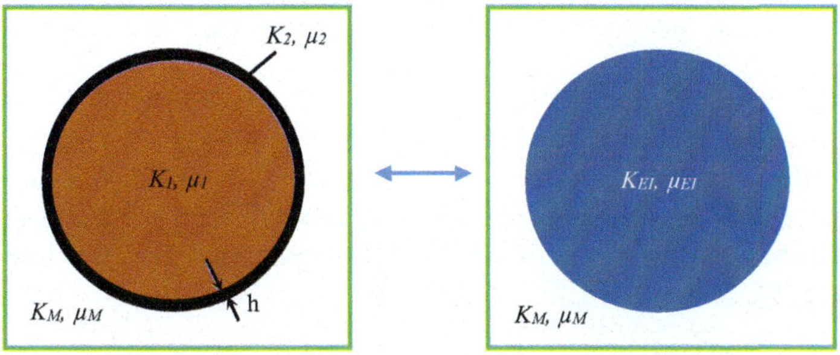

Fig. 1 A coated circle inclusion model (left) and equivalent circle inclusion (right)

method by replacing the coated circle inclusion to become the equivalent inclusion (EI). Equations (3) and (4) can be presented as:

$$K_{\mathrm{EI}} = \frac{K_1K_2 + \mu_2K_1v_1^* + \mu_2K_2v_2^*}{\mu_2 + K_2v_1^* + K_1v_2^*} \tag{5}$$

$$\mu_{\mathrm{EI}} = \frac{\mu_1\mu_2 + \mu^*\mu_1v_1^* + \mu^*\mu_2v_2^*}{\mu^* + \mu_2v_1^* + \mu_1v_2^*} \tag{6}$$

Here, $\mu^* = \frac{K_2\mu_2}{K_2+2\mu_2}$, $v_1^* = v_1(v_1+v_2)^{-1}$, $v_2^* = v_2(v_1+v_2)^{-1}$

We consider only the circular inclusion with a thin coating with the volume ratio $v_2 \ll 1$, $v_1 + v_2 = 1$, the circular inclusion is phase 1, and the circular shell (phase 2) is considered the matrix. Therefore, the circle inclusion with the S-L imperfect interface can be replaced by the coated circle inclusion with the infinitely thin coating of thickness h ($h \to 0$).

Symbol R_1 is the radius of circle inclusion 1 ($R_2 = R_1 + h$), and we have a geometric correlation:

$$\frac{v_1+v_2}{v_1} = \frac{1}{v_1} = \frac{R_2^2}{R_1^2} = \left(1+\frac{h}{R_1}\right)^2 = 1 + \frac{2h}{R_1} + O\left(\frac{h^2}{R_1^2}\right),$$
$$\frac{1}{v_1} = \frac{1}{1-v_2} = 1 + v_2 + O(v_2^2) \Rightarrow v_2 = \frac{2h}{R_1}v_1 \tag{7}$$

In composite materials with the S-L imperfect interface model, we assume that the mechanical behaviour of the shell S between two phases M and I_1 is the linear S-L interface model. The stress vector ($\boldsymbol{\sigma}\cdot\mathbf{n}$) is continuous across the contact surface between the phases whilst the displacement vector field ($\mathbf{u}$) has a jump across the interface. In this model, the displacement vector field is not continuous across the imperfect interface S:

$$\mathbf{u}^M \neq \mathbf{u}^I \tag{8}$$

The stress vector field is continuous across the imperfect interface *S*:

$$t_i^M = t_i^{I1}, \quad t_i = \sigma_{ij} n_j, \quad t_i = C_{ij}^s (u_j^M - u_j^{I1}) \tag{9}$$

Here, **n** is the unit normal vector of the interface, σ stands for the Cauchy stress tensor field, and $\mathbf{C}^s$ is a second-order tensor characterising the stiffness of *S*.

The interface stiffness tensor of the S-L imperfect interfaces has the form (Benveniste [6]; Zhu et al. [9]):

$$C_{ij}^s = \frac{1}{h} C_{ijkl}^c n_k n_n = \lambda_s n_i n_j + \mu_s (\delta_{ij} - n_i n_j), \quad (i, j, k = 1, 2) \tag{10}$$

λ_s is the normal stiffness, and μ_s is the shear stiffness of the imperfect interface.

From (10), the effective moduli of the S-L interface model are given by (Zhu [9]):

$$K_2 = h\left(\lambda_s - \frac{4}{3}\mu_s\right), \quad \mu_2 = h\mu_s, \quad v_2 = \frac{2h}{R_1}, \quad \frac{h}{R} \to 0 \tag{11}$$

Now substituting (7) and (11) into (5) and (6) (in the limit $h \to 0$, $v_1 = 1$), we get the approximate algebraic expressions K_{EI}, μ_{EI} of the circle inclusions with the S-L imperfect interface:

$$K_{\mathrm{EI}}^{\mathrm{SL}} = \frac{K_1 K_2 + \mu_2 K_1 v_1 + \mu_2 K_2 v_2}{\mu_2 + K_2 v_1 + K_1 v_2} = \frac{K_1 R_1 \left(\lambda_s - \frac{1}{3}\mu_s\right)}{R_1 \left(\lambda_s - \frac{1}{3}\mu_s\right) + 2K_1} \tag{12}$$

And

$$\mu^* = \frac{K_2 \mu_2}{K_2 + 2\mu_2} = \frac{h\left(\lambda_s - \frac{4}{3}\mu_s\right)\mu_s}{\lambda_s + \frac{2}{3}\mu_s} \tag{13}$$

$$\mu_{\mathrm{EI}}^{\mathrm{SL}} = \frac{\mu_1 \mu_2 + \mu^* \mu_1 v_1 + \mu^* \mu_2 v_2}{\mu^* + \mu_2 v_1 + \mu_1 v_2} = \frac{R_1 \left(\lambda_s - \frac{1}{3}\mu_s\right)\mu_s \mu_1}{R_1 \left(\lambda_s - \frac{1}{3}\mu_s\right)\mu_s + \mu_1 \left(\lambda_s + \frac{2}{3}\mu_s\right)} \tag{14}$$

According to (3), one obtains the elastic bulk modulus of the random circular inclusions with the S-L imperfect interface:

$$K^{\mathrm{eff}} = \frac{K_{\mathrm{EI}}^{\mathrm{SL}} K_M + \mu_M \left(K_{\mathrm{EI}}^{\mathrm{SL}} v_I + K_M v_M\right)}{\mu_M + K_M v_I + K_{\mathrm{EI}}^{\mathrm{SL}} v_M} \tag{15}$$

Similarly, using Eq. (4), the elastic shear modulus of the random circle inclusions with the S-L imperfect interfaces in the matrix has a particular expression:

$$\mu^{eff} = \frac{\mu_{EI}^{SL} \mu_M + \mu^* \left(\mu_{EI}^{SL} v_I + \mu_M v_M\right)}{\mu^* + \mu_M v_I + \mu_{EI}^{SL} v_M}, \quad \mu^* = \frac{K_M \mu_M}{K_M + 2\mu_M} \tag{16}$$

Equations (15) and (16) will be used as simplified to estimate the MEM of the random circular inclusions with the S-L imperfect interface model.

3 The DA Method with S-L Imperfect Interfaces

We start with the two-component material model with random circular inclusions in the matrix, in which the effective moduli of the circle inclusions are K_I, μ_I, and the effective moduli of phase M are K_M, μ_M. The process of constructing the differential equation for the dilute suspension materials starts with the base phase M. At each step of the process, we add infinite volumes according to the volume ratio $v_I\Delta\xi$ ($\Delta\xi \ll 1$) of random circle inclusions into the material of the previous step, which has the volume ratio $v_I\xi$ of the circle inclusions (ξ increases from 0 to 1). After each differentiation process, the new material obtained is considered as the dilute suspension model with circle-inclusion phases (I), which have the volume ratio:

$$\Delta v_I = \frac{v_I\Delta\xi}{1 + v_I\Delta\xi} \tag{17}$$

According to Eshelby formulae [18], the MEM of the dilute suspension of circular inclusions is written as:

$$K^{\text{eff}} = K_M + (K_I - K_M)v_I \cdot D_{KI}(K_I, \mu_I, K_M, \mu_M) \tag{18}$$

$$\mu^{\text{eff}} = \mu_M + (\mu_I - \mu_M)v_I \cdot D_{\mu I}(K_I, \mu_I, K_M, \mu_M), \quad v_I \ll 1 \tag{19}$$

Here, D_{KI} and $D_{\mu I}$ are the shape functions of the inclusion phase, and for the circular inclusion composite in two dimensions, we have:

$$D_{\text{KI}}(K_I, \mu_I, K_M, \mu_M) = \frac{K_M + \mu_M}{K_I + \mu_M},$$
$$D_{\mu I}(K_I, \mu_I, K_M, \mu_M) = \frac{\mu_M + \frac{K_M\mu_M}{K_M + 2\mu_M}}{\mu_I + \frac{K_M\mu_M}{K_M + 2\mu_M}} \tag{20}$$

Assume that the MEM of the material in the current step are K, μ. From Eqs. (17)–(19), we obtain the MEM of the material in the next step:

$$K + \mathrm{d}K = K + (K_I - K)\frac{v_I\Delta\xi}{1 + v_I\Delta\xi}D_{KI}(K_I,\ \mu_I,\ K,\ \mu) \tag{21}$$

$$\mu + \mathrm{d}\mu = \mu + (\mu_I - \mu)\frac{v_I\Delta\xi}{1 + v_I\Delta\xi}D_{\mu I}(K_I,\ \mu_I,\ K,\ \mu) \tag{22}$$

The volume ratio of the new composite:

$$v_I \mathrm{d}\xi = \frac{v_I t + v_I \Delta\xi}{1 + v_I \Delta\xi} - v_I \xi = \frac{v_I \Delta\xi}{1 + v_I \Delta\xi}(1 - v_I \xi) \tag{23}$$

Since Eqs. (21–23), we get the two differential equations for the effective bulk modulus K and effective shear modulus μ of the material:

$$\frac{\mathrm{d}K}{\mathrm{d}\xi} = (K_I - K)\frac{v_I}{1 - v_I \Delta\xi} \cdot D_{KI}(K_I, \mu_I, K, \mu) \tag{24}$$

$$\frac{\mathrm{d}\mu}{\mathrm{d}\xi} = (\mu_I - \mu)\frac{v_I}{1 - v_I \Delta\xi} \cdot D_{\mu I}(K_I, \mu_I, K, \mu) \tag{25}$$

$$K(0) = K_M, \ \mu(0) = \mu_M, \ 0 \le t \le 1. \tag{26}$$

We can use Eqs. (24) and (25) to estimate the MEM of the compound materials with the S-L imperfect interface. By replacing $K_{\mathrm{EI}}^{\mathrm{SL}}$, $\mu_{\mathrm{EI}}^{\mathrm{SL}}$ with K_I, μ_I, the differential approximations are the solutions $K^{\mathrm{eff}} = K(1)$, $\mu^{\mathrm{eff}} = \mu(1)$ of the two differential equations:

$$\frac{\mathrm{d}K}{\mathrm{d}\xi} = (K_{\mathrm{EI}}^{\mathrm{SL}} - K)\frac{v_I}{1 - v_{\mathrm{EI}} \Delta\xi} \cdot D_{\mathrm{KI}}(K_{\mathrm{EI}}^{\mathrm{SL}}, \mu_{\mathrm{EI}}^{\mathrm{SL}}, K, \mu) \tag{27}$$

$$\frac{\mathrm{d}\mu}{\mathrm{d}\xi} = (\mu_{EI}^{SL} - \mu)\frac{v_I}{1 - v_I \Delta\xi} \cdot D_{\mu I}(K_{\mathrm{EI}}^{\mathrm{SL}}, \mu_{\mathrm{EI}}^{\mathrm{SL}}, K, \mu) \tag{28}$$

Here,

$$K_{EI}^{SL} = \frac{K_1 R_1 \left(\lambda_s - \frac{1}{3}\mu_s\right)}{R_1 \left(\lambda_s - \frac{1}{3}\mu_s\right) + 2K_1},$$
$$\mu_{EI}^{SL} = \frac{R_1 \left(\lambda_s - \frac{1}{3}\mu_s\right)\mu_s \mu_1}{R_1 \left(\lambda_s - \frac{1}{3}\mu_s\right)\mu_s + \mu_1 \left(\lambda_s + \frac{2}{3}\mu_s\right)} \tag{29}$$

4 FFT Simulation

The FFT numerical for the MEM of composite materials uses Green's tensor and exact algebra expressions of the shape function in Fourier space. The FFT simulation algorithm to determine the MEM of the composite material has been introduced in [12–17]. In this work, the FFT simulation will be applied to compute the MEM of the random circle inclusions with the S-L imperfect interface. By substitution scheme, we consider the circular inclusions with the S-L imperfect interface model as an equivalent circle inclusion, in which the MEM of equivalent circle inclusion is given by (29).

5 Applications

We apply the FFT numerical, PA, and DA methods to compute the MEM of the random circle inclusion ($I1$) in the matrix with the S-L imperfect interface. To illustrate the application of the methods, we take $K_M = 1$, $\mu_M = 0.5$, $K_{I1} = 100$, $\mu_{I1} = 60$, and consider three examples:

A. $\lambda_s = 50,\ \mu_s = 20$
B. $\lambda_s = 30,\ \mu_s = 10$
C. $\lambda_s = 20,\ \mu_s = 6$.

In the FFT numerical algorithm, we use the Matlab program to create 60 circular inclusions in the matrix phase as shown in Fig. 2, in which no circles are overlapping. The smallest distance between the circle-inclusion phases in the unit cell is 0.01. We perform FFT numerical simulations for a two-phase material model with randomly circular inclusions in the matrix, in which the circle inclusion has elastic moduli (k_{EI}, μ_{EI}) given by expressions (12) and (14). The coordinates of the centre circle inclusion do not change, but the outer radii vary from 0.0230 to 0.0520 or the volume ratio of the circle inclusions varies from 0.1 to 0.5097. The grid 256 × 256 of the unit cell is used in the FFT calculation. The FFT result compared with PA and DA methods for the MEM of the random circle-inclusion model which are reported in Figs. 3, 4 and 5. From the results obtained, we can observe that the FFT, DA, and PA results converge almost coincidentally, and we also see the strong dependence of the surface parameters on the MEM. In case of (A) $\lambda_s = 50$, $\mu_s = 20$, the volume ratio v_I of the inclusions increases and the MEM also increases. In case of (B) $\lambda_s = 30$, $\mu_s = 10$, the volume ratio v_I of the inclusion phase increases but the MEM decreases, and then, they increase slightly. This result does not follow the rules of the composite material model with a perfect contact surface. In case (C) $\lambda_s = 20$, $\mu_s = 6$ as the volume ratio v_I of the inclusion phase increases, we see that the MEM decreases. This result is also in contrast to the composite material model with a perfect contact surface. The results obtained show the great influence of surface parameters on the MEM of the composite material in the case of imperfect contact surfaces.

6 Conclusions

The MEM of materials with the S-L imperfect surface is a complex study because it depends greatly on the properties of the imperfect surface with discontinuous conditions at the contact surface. This article has built the explicit algebra expressions for effective moduli of the random circle inclusions in 2D with spring-layer imperfect

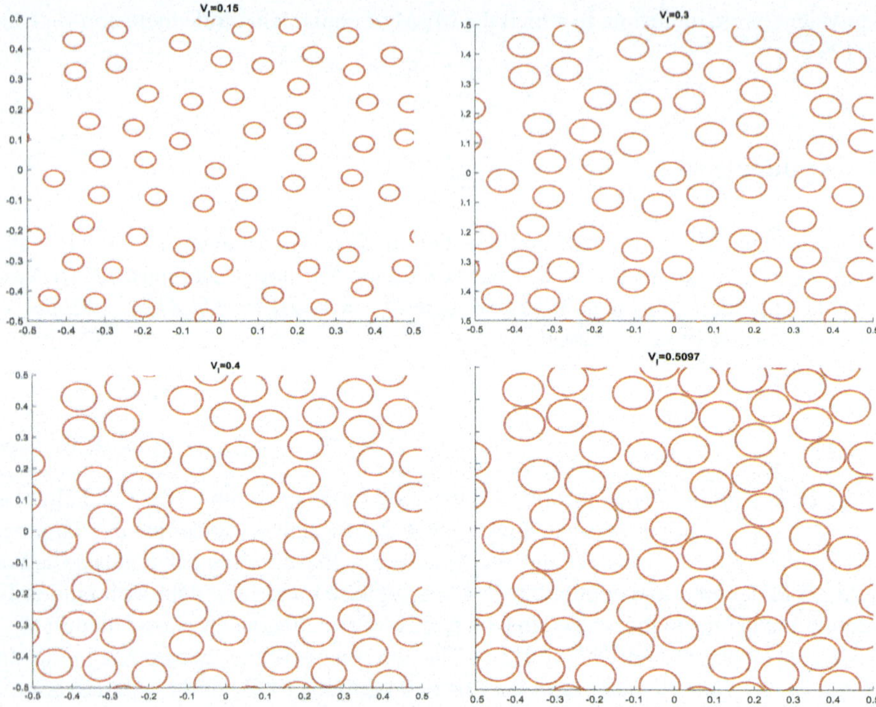

Fig. 2 The random circle-inclusions model

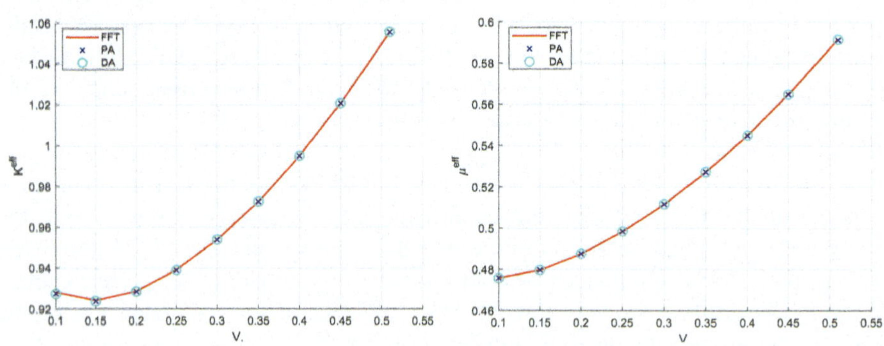

Fig. 3 The elastic bulk modulus (left) and shear modulus (right) of the case (A) with spring-layer imperfect interfaces $K_M = 1$, $\mu_M = 0.5$, $K_{I1} = 100$, $\mu_{I1} = 60$, $\lambda_s = 50$, $\mu_s = 20$

interfaces. From there, different methods have been used to calculate the MEM of composite materials with the S-L imperfect interface including FFT, DA, and PA methods. The calculation results in Figs. 3, 4 and 5 show the effectiveness of the applied methods and the influence of imperfect contact surface parameters between phases on the macro-elastic moduli of materials. The AD and PA approximations have

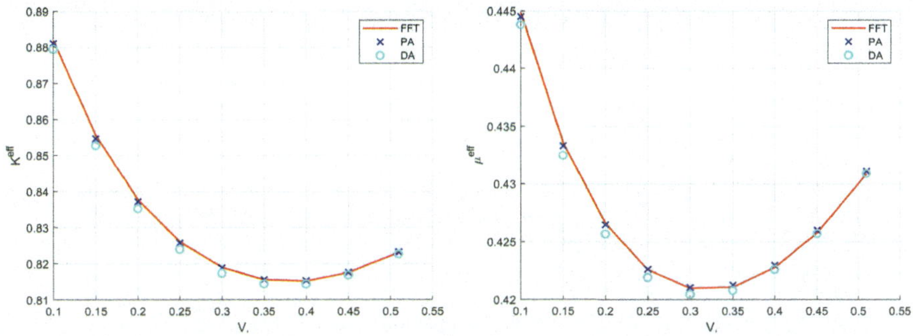

Fig. 4 The elastic bulk modulus (left) and shear modulus (right) of the case (B) with spring-layer imperfect interfaces $K_M = 1$, $\mu_M = 0.5$, $K_{I1} = 100$, $\mu_{I1} = 60$, $\lambda_s = 30$, $\mu_s = 10$

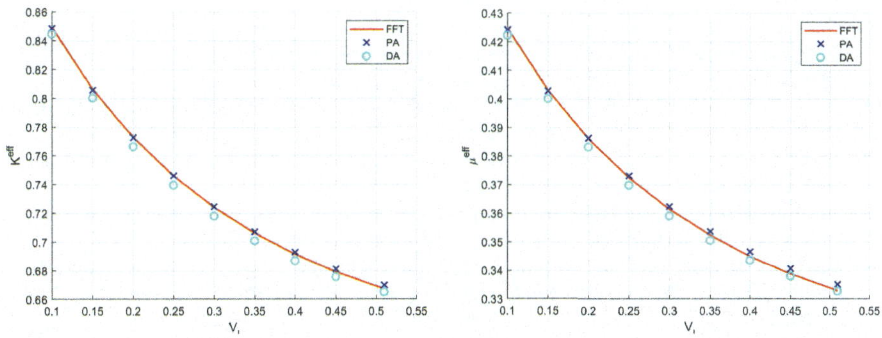

Fig. 5 The elastic bulk modulus (left) and shear modulus (right) of the case (C) with spring-layer imperfect interfaces $K_M = 1$, $\mu_M = 0.5$, $K_{I1} = 100$, $\mu_{I1} = 60$, $\lambda_s = 20$, $\mu_s = 6$

quite clear and simple algebraic expressions for application to determine effective moduli for composite materials with the S-L imperfect interfaces. Hence, it can be more useful to engineers for initial predictions of macro-elastic moduli of the materials with the S-L imperfect interface.

References

1. Hashin, Z.: The spherical inclusion with imperfect interface. J. Appl. Mech. ASME **58**, 444–449 (1991)
2. Hashin, Z.: Extremum principles for elastic heterogeneous media with imperfect interfaces and their application to bounding of effective moduli. J. Mech. Phys. Solids **40**, 767–781 (1992)
3. Duan, H.L., Yi, X., Huang, Z.P., Wang, J.: A unified scheme for prediction of effective moduli of multiphase composites with interface effects. Theor. Framework Mech. Mater. **39**, 81–93 (2007)

4. Duan, H.L., Wang, J., Huang, Z.P., Karihaloo, B.L.: Size-dependent effective elastic constants of solids containing nanoinhomogeneities with interface stress. J. Mech. Phys. Solids **53**, 1574–1596 (2005)
5. Gu, S.T., He, Q.C.: Interfacial discontinuity relations for coupled multifield phenomena and their application to the modelling of thin interphases as imperfect interfaces. J. Mech. Phys. Solids **59**, 1413–1426 (2011)
6. Benveniste, Y.: A general interface model for a three-dimensional curved thin anisotropic interphase between two anisotropic media. J. Mech. Phys. Solids **54**, 708–734 (2006)
7. Brisard, S., Dormieux, L., Kondo, D.: Hashin-Shtrikman bounds on the bulk modulus of a nanocomposite with spherical inclusions and interface effects. Comput. Mater. Sci. **48**, 589–596 (2010)
8. Chen, T., Dvorak, G.J., Yu, C.C.: Solids containing spherical nano-inclusions with interface stresses: effective properties and thermal-mechanical connections. Inter. J. Solids Struct. **44**, 941–955 (2007)
9. Zhu Q Z, Gu S T, vonnet J Y, Shao J F, He Q C. Three-dimensional numerical modelling by XFEM of spring-layer imperfect curved interfaces with applications to composite materials. *Int. J. Numer. Meth. Engng,* 00:1–6, (2000).
10. Pham, D.-C., Nguyen, T.-K., Tran, B.-V.: Macroscopic elastic moduli of spherically-symmetric-inclusion composites and the microscopic stress-strain fields. Int. J. Solids Struct. **169**, 141–165 (2019)
11. Pham, D.-C., Vu, L.-D., Nguyen, V.-L.: Bounds on the ranges of the conductive and elastic properties of randomly inhomogeneous materials. Phil. Mag. **93**, 2229–2249 (2013)
12. Nguyen, V.-L.: The effective conductivity of elliptic inclusion with lowly and highly conducting interface model. Mech. Res. Commun. **123**, 103902 (2022)
13. Van-Luat Nguyen. Estimating macroscopic conductivity of randomly elliptic inclusion model with imperfect interfaces. *Mathematics and Mechanics of Solids*, (2023).
14. Van-Luat Nguyen. Solutions for elastic moduli of three-phase composite with random distribution of coated-ellipse inclusions. *Functional Composites and Structures*, Volume 4, Number 4, (2022).
15. Michel, J., Moulinec, H., Suquet, P.: Effective properties of composite materials with periodic microstructure: a computational approach. Comput. Methods. Appl. Mech. Engrg **172**, 109–143 (1999)
16. Moulinec, H., Suquet, P.: A fast numerical method for computing the linear and nonlinear mechanical properties of composites. C. R. Acad. Sci. **1994**(318), 1417–1423 (1994)
17. Bonnet, G.: Effective properties of elastic periodic composite media with fibers. Journal of the Mechanics and Physicsof Solids **55**, 881–899 (2007)
18. Eshelby, J.D.: The determination of the elastic field of an ellipsoidal inclusion and related problems. Proc. R. Soc. Lond. A **241**, 376–396 (1957)
19. Mura, T.: Micromechanics of Defects in Solids. Martinus-Nijhoff, Dordrecht (1987)
20. Nemat-Nasser S, Hori M. Micromechanics: overall properties of heterogeneous materials. *Amsterdam-New York: Elsevier*, 786p, (1999).

Analysis, Simulation of the Welding System in Industrial Production Using Tecnomatix Plant Simulation Software

Thi Minh Hue Pham, Son Xuan Phung, Thi Lan Pham, Thi Thanh Mai Nguyen, Dongyi Lin, and Viet Anh Nguyen

Abstract The research team collected and analyzed data related to the welding process, workstations, material flow, required number of workers, and actual operation time at the factory. Subsequently, we designed and simulated the system using Tecnomatix Plant Simulation software to observe the productivity and output of the machines in the system. According to these data, businesses can implement lean production strategies, improve productivity, reduce waiting times and bottlenecks without wasting resources, and devise a well-thought-out plan for the design.

Keywords Bottlenecks · Lean · Output · Productivity · Simulate the system

1 Introduction

Nowadays, the use of assembly lines and manufacturing systems to produce products is no longer a novelty for countries around the world. Countries with developed industrial bases have invested in and equipped these processing systems with full modernity, designing them with highly accurate mechanical structures and a wide variety of flexible robots in the production lines. A study was conducted on a welding control system based on Siemens PLC software to promote rapid development in the automotive manufacturing industry in Guangzhou, China, by Minhhai Yuan, Weijun Huang, Xiteng Zhang, and Shumin Tian [1]. This system aims to enhance competitiveness, meet production process requirements, effectively reduce costs, and improve production efficiency. With the support of the Information Center for Science and Technology, KS. Le Hong Loan and Nguyen Hoai Phuong have researched "Designing an automatic welding system for loading ships" [2].

T. M. H. Pham · S. X. Phung · T. L. Pham · T. T. M. Nguyen · D. Lin · V. A. Nguyen (✉)
School of Mechanical and Automotive Engineering, Hanoi University of Industry, Hanoi City, Vietnam
e-mail: anhhy1704@gmail.com

Anxi College of Tea Science, Fujian Agriculture and Forestry University, Quanzhou 362400, China

D. H. Tien et al. (eds.), *Proceedings of the 1st International Conference on Sustainability and Emerging Technologies for Smart Manufacturing*, Proceedings in Technology Transfer, https://doi.org/10.1007/978-981-97-7083-0_20

The analysis and simulation of the welding system in industrial production using software aim to provide data that allow the company to improve the system, streamline the production process, and maximize profits. Tecnomatix Plant Simulation is a software that simulates various fields, including industrial machinery, manufacturing systems, industrial and production systems, and other specialized applications [3]. By utilizing Tecnomatix Plant Simulation for simulation, businesses can create a database to implement lean solutions effectively, avoiding unnecessary expenses, and enabling better design planning [4].

2 Method

The research is based on data collection methods, combined with real-time observation of the production situation at Factory No. 1, Thanglong Metalwares Joint Stock Company, to determine the welding process of a product. The appropriate layout is arranged according to the method of determining the quantity and standard of the product. There are four basic layout forms: product-oriented layout, group-oriented layout, process-oriented layout, and hybrid layout [5]. After selecting the layout form, the next step is to determine the workstations in the system, the types of machinery, and the corresponding quantities for each type. We apply the method of calculating the required number of workers [6]:

- Working time of a worker in one day:

$$G = 8h - \mathrm{GV_{bq}}\ \text{(hours)} \tag{1}$$

In which, $\mathrm{GV_{bq}}$ is the average number of hours that a worker rests in one working day. GVbq is based on the expected working schedule, usually ranging from 0.5 to 1 h.

- Number of working days in a month of a worker:

$$N = k - \mathrm{NV_{bq}}\ \text{(days)} \tag{2}$$

In which, $\mathrm{NV_{bq}}$ is the average number of working days that a worker rests in a year.

- Actual working time of a worker in a month:

$$T_{\mathrm{bq}} = N \times G\ \text{(hours)} \tag{3}$$

In which, G is the average number of working hours of a worker in a day.

- Required number of workers:

$$CN = \sum_{i=1}^{m} Q_i * DM_{ti} * \frac{1}{T_{bq}} \text{ (workers)} \tag{4}$$

In which, DM_{ti} is the production rate (number of products per one labor hour).

Using the simulation method with Tecnomatix Plant Simulation software, the cycle time and failure rates of machines in the system can be set up. Model simulations can then be carried out to observe the performance of the machines and the output of each production line [7]. Statistical methods can be applied to synthesize the collected data, and the data can be combined with graphical representations to gain a visual understanding and conduct an in-depth analysis of bottleneck points and their causes. Based on the analysis, solutions can be proposed to address the identified bottlenecks and improve the overall system efficiency.

3 Results and Discussion

3.1 Data Collection and Analysis

3.1.1 Welding Process

The welding system is organized into six workstations and three production lines. Line 1 is dedicated to welding the K2T products, Line 2 is for welding the K0Z products, and Line 3 is for welding the GGZ products.

- Station 1: Input material (source) with an input quantity of 3000 workpieces per shift, lasting 7 h. The raw materials will be distributed to three lines: K2T, G0Z, and GGZ, in proportions of 33%, 34%, and 33% respectively.
- Station 2: Automated machine welding station consisting of three K2T robot welding machines: K2T_1, K2T_2, and K2T_3. Each welding machine is equipped with a buffer to prevent backlogs and ensure a smooth production process.
- Station 3: Manual welding station with three hand welding machines: K2T_A_1, 2, 3.
- Station 4: Smoothing station with three K0 welding machines and three K2 welding machines. At this station, three buffers are designed before the K0 welding machines.
- Station 5: Quality inspection station with two SubR machines and SubL machine.
- Station 6: Finished product station (export).

3.1.2 Layout Design

The layout at Factory No. 1, Thanglong Metalwares Joint Stock Company, is a hybrid layout, which combines both product-oriented and process-oriented layouts (Fig. 1).

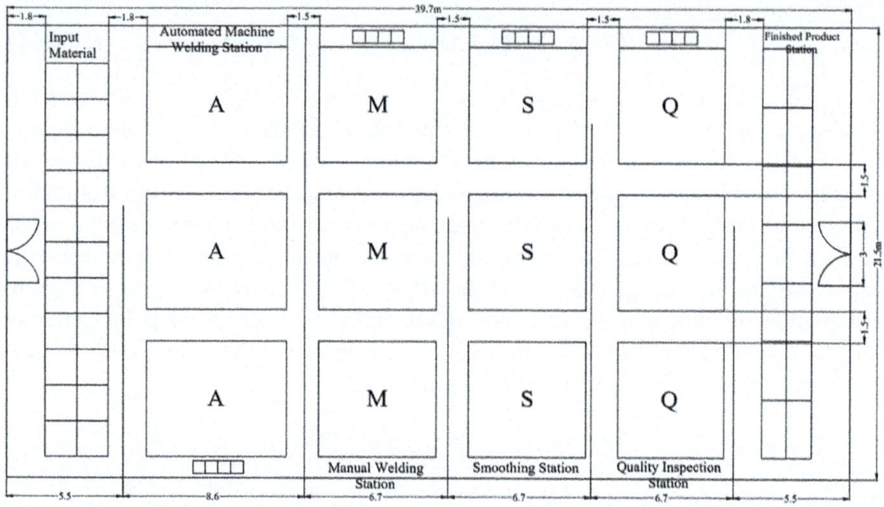

Fig. 1 Overall layout drawing of the welding system

3.1.3 Movement Lines

The overall layout design at Factory No. 1, Thanglong Metalwares Joint Stock Company, follows an I-shaped line with a straight flow from receiving to transportation and vice versa. This layout is considered highly optimized as it effectively utilizes the entire length of the workshop. It ensures that similar products are kept separated by assembly line formats, and it minimizes congestion by eliminating unnecessary back-and-forth movements. To facilitate smooth operations and efficient workflow, small movement lines between workstations and aisles are designed to be parallel or perpendicular. This design allows for the seamless movement of materials and products between different stages of the production process.

3.1.4 Required Number of Workers

At Factory No. 1, Thanglong Metal Wares Joint Stock Company, the welding system mainly processes three main car bracket components: K2T, K0Z, and GGZ, with annual outputs of 876,000; 897,800; and 835,000 products per year, respectively. The standard time for these three products is 0.267 h, 0.27 h, and 0.274 h, respectively. The company operates for 300 days per year, and on average, each worker is absent for 10 days annually. During a typical shift, workers take a one-hour break. Each worker operates in three shifts, with each shift lasting 8 h.

- The working time of a worker in one day:

$$G = 8 - \mathrm{GV}_{\mathrm{bd}} = 8 - 1 = 7 \text{ (hours)}$$

- Number of working days in a year of a worker:

$$N = 300 - NV_{bq} = 300 - 10 = 290 \text{ (days)}.$$

- Actual working time of a worker in a year:

$$T_{bq} = N \times G = 290 \times 7 = 2030 \text{ (hours)}.$$

- Required number of workers:

$$\text{CN} = \sum_{i=1}^{m} Q_i \cdot DM_i \cdot \frac{1}{T_{bq}}$$

$$= (87600 \times 0.267 + 897800 \times 0.27 + 835000 \times 0.274) \times \frac{1}{2030} = 115.7$$

116 workers are required to complete the tasks. As there are three shifts in a day, the number of workers per shift will range from 36 to 39 workers.

3.2 System Simulation

Using Tecnomatix Plant Simulation software, the research team simulated the welding system of Factory No. 1, Thanglong Metal Wares Joint Stock Company, as shown in Figs. 2 and 3 (Table 1).

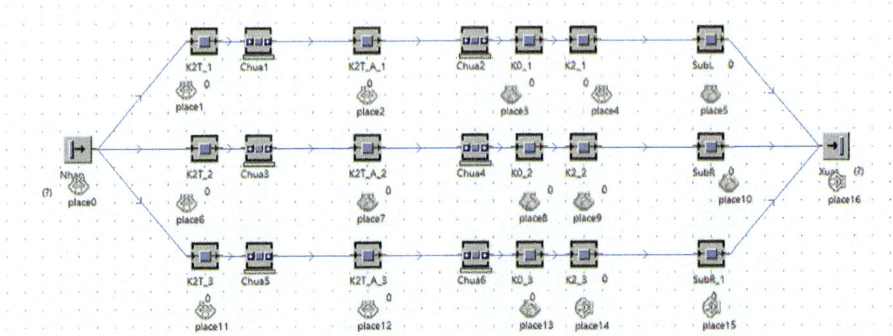

Fig. 2 2-D image of the welding system simulated using Tecnomatix Plant Simulation

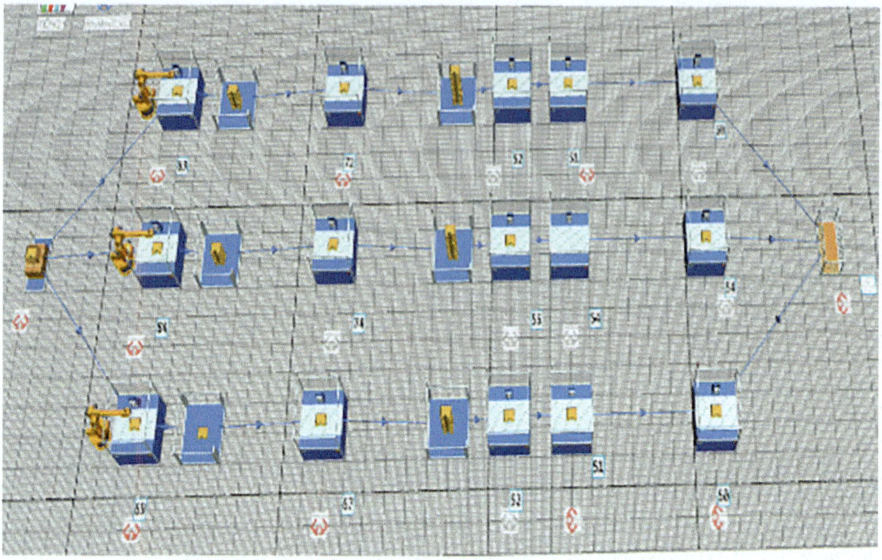

Fig. 3 3-D image of the welding system simulated using Tecnomatix Plant Simulation

Table 1 Parameters setup for machines in the welding system

Production line	Machine	Quantity	Operation Time	Setup time	Availability (%)	Cycle time
K2T	K2T_1	1	2:10	0:50	95	3:00
	K2T_A_1	1	2:25	0:20	95	2:45
	K0_1	1	2:10	0:45	90	2:55
	K2_1	1	2:45	0:45	95	3:30
	SubL	1	3:10	0:40	95	3:50
K0Z	K2T_2	1	2:15	0:45	97	3:00
	K2T_A_2	1	2:30	0:25	97	2:55
	K0_2	1	2:25	0:45	95	3:10
	K2_2	1	2:55	0:50	95	3:45
	SubR	1	3:05	0:40	95	3:45
GGZ	K2T_3	1	2:18	0:40	87	2:58
	K2T_A_3	1	2:28	0:30	90	2:58
	K0_3	1	2:30	0:40	85	3:10
	K2_3	1	3:10	0:30	95	3:40
	SubR_1	1	3:10	0:50	95	4:00

The working hours of the workshop are organized into three shifts per day, with each shift lasting 8 h: 6:00 AM–2:00 PM, 2:00 PM–10:00 PM, and 10:00 PM–6:00 AM. Each shift includes a 15-min break and a 45-min lunch break, which adds up to a total of 1 h break time for each shift. The simulation of the welding system's operation is carried out within this 7-h timeframe. The results of the simulation are presented in Figs. 4, 5, and 6.

To facilitate the analysis of the data from Figs. 4, 5, and 6, we summarize the obtained data in Table 2. In this table, we list the values for the working rate, blocked rate, waiting rate, and failure rate for each operational parameter.

The working rates of the oscillation machines in the welding system fall within the range of 62–90%. However, there is a concern regarding the relatively high blocked rate, which frequently occurs in the initial stations of the production line. Notably, at the K2T assembly line, the blocking rate of the K2T_1 and K0_1 welding machines reaches as high as 30%. Figure 7 offers a more comprehensive representation of the blocked and waiting rates of the machines within the welding system.

The analysis reveals that the smoothing machines K0 in the production lines exhibit the highest blocked rate, which is 23% higher than the average rate of 11.6%. As a result, the K0 machine at station 4 has been identified as a bottleneck in the system. The main issue in the welding system arises from significant variations in the working cycle time, leading to performance discrepancies, increased blocking time for upstream machines, and waiting time for downstream machines. This accumulation of workpieces causes production congestion, reducing labor productivity, and increasing the risk of product quality-related errors.

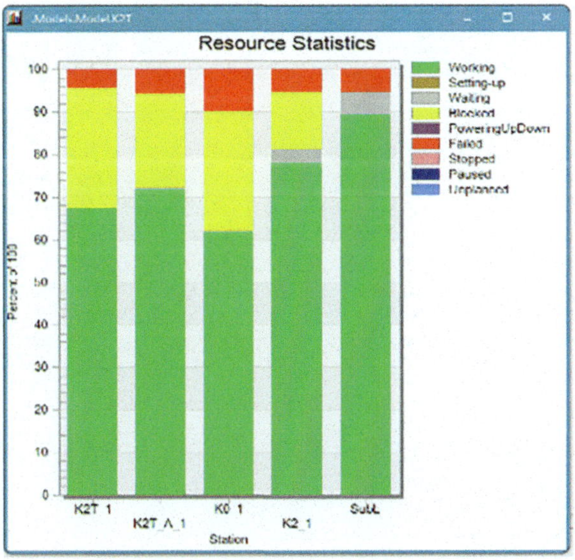

Fig. 4 The operating parameters of the K2T

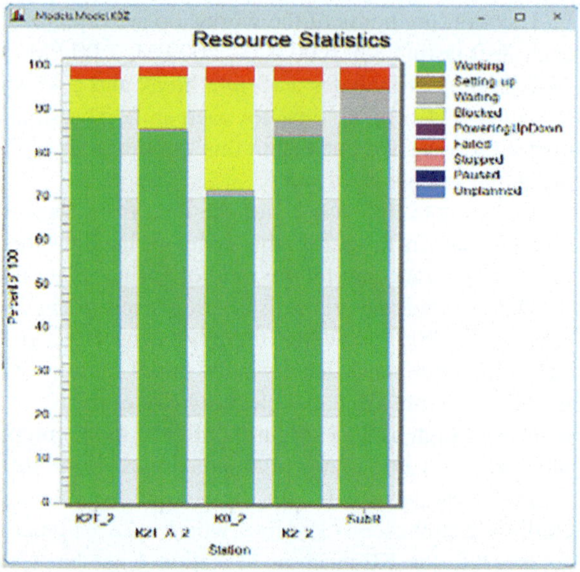

Fig. 5 The operating parameters of the K0Z

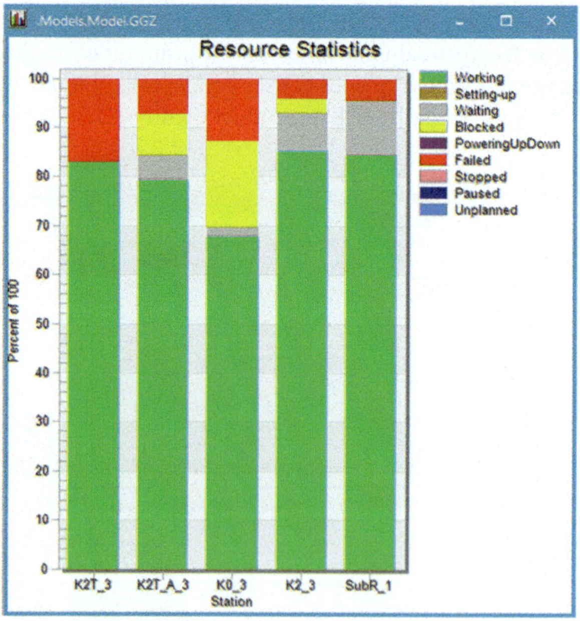

Fig. 6 The operating parameters of the GGZ

Table 2 Summary of simulation results for the welding line

Production line	Machine	Cycle time	Working rate (%)	Blocked rate (%)	Waiting rate (%)	Failure rate (%)	Output per hour	Total output per shift
K2T	K2T_1	3:00	65	30	0	5	144	1008
	K2T_A_1	2:45	70	25	0	5		
	K0_1	2:55	62	30	0	8		
	K2_1	3:30	78	14	2	6		
	SubL	3:50	90	0	4	6		
K0Z	K2T_2	3:10	88	10	0	2	147	1032
	K2T_A_2	2:55	85	12	1	2		
	K0_2	3:10	70	24	2	4		
	K2_2	3:45	86	6	3	5		
	SubR	3:45	88	0	7	5		
GGZ	K2T_3	2:58	83	0	0	17	137	960
	K2T_A_3	2:58	79	5	8	8		
	K0_3	3:10	68	16	2	14		
	K2_3	3:40	85	2	7	6		
	SubR_2	4:00	84	0	8	8		

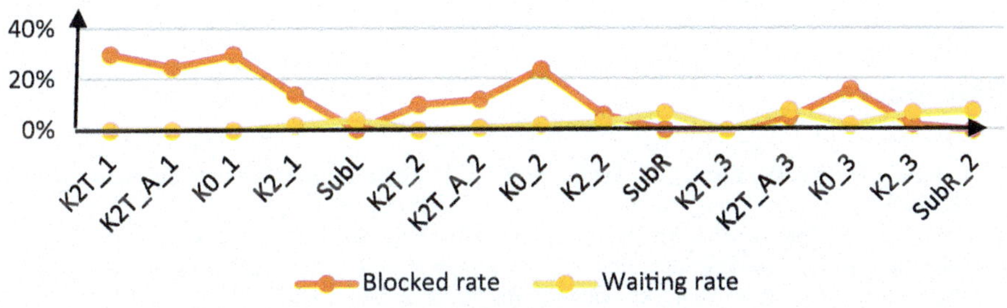

Fig. 7 Chart showing the blocked rate and waiting rate of machines in the welding system

The welding system exhibits a higher waiting time rate at the downstream stations in each production line, ranging from 2 to 8% more compared to the stations at the beginning. This waiting time is considered wasteful and requires streamlining to optimize the production process.

To enhance the efficiency of the welding system's production lines, Kim Khí Thăng Long Company should focus on shortening the cycle time, reducing transportation time, and optimizing machine performance. To minimize waiting time and prevent congestion, increasing the buffer capacity from the current 20 units to 30 units or adding more buffers is essential.

4 Conclusions

The research team collected the necessary data for the welding system including movement lines, the number of machines, workstations, required workers, and product processing times. Following the simulation, the results for each production line were as follows: For the K2T production line, the total output is 1008 products per shift, working rate is 73%, failure rate is 6%, waiting rate is 1%, and blocked rate is 20%; for the KOZ production line, total output is 1032 products per shift, working rate is 83%, failure rate is 6%, waiting rate 3%, and blocked rate is 10%; for the GGZ production line, total output is 960 products per shift, working rate is 80%, failure rate is 5%, waiting rate is 5%, and blocked rate is 11%. These simulation results serve as a valuable basis for the company to make improvements in their welding production lines. By designing a more efficient production system, the company can significantly reduce working time, enhance machine utilization, and ultimately increase revenue and profitability. This solution will facilitate smoother material flow between stations, leading to reduced waiting times and improved overall production efficiency.

References

1. Yuan, M., Huang, W., Zhang, X., Tian, S.: Design of welding control system based on Siemens PLC (2019)
2. Loan, L.H., Phuong, N.H.: Designing automatic welding system only loads the ship (2020)
3. Siderska, J.: Application of tecnomatix plant simulation for modeling production and logistics processes. In: Business, Management and Education, pp 68–73 (2016)
4. Shintre, R.A., Naik, G.R., Bhujbal, M.: Facility and plant layout using tecnomatrix plant simulation. Int. Res. J. Eng. Technol. (IRJET)
5. Lam, L.N.Q.: Industrial System Location and Layout Design. Ho Chi Minh City National University Publishing House, pp. 36–59 (2013)
6. Chung, N.V.: Labor Measurement and Job Design for Industrial Production. Ho Chi Minh City National University Publishing House, pp. 30–36 (2009)
7. Martijn, K.R.: Simulation Modelling Using Practical Examples: A Plant Simulation Tutorial. University of Twente, pp. 131–165

Development of the Concept of Digital Twin of Temperature Control System of Electric Vehicles Category L

Sergey Bazavluk, Rinat Kurmaev, Kirill Karpukhin, Alexey Terenchenko, and Givi Nadareishvili

Abstract The introduction of computer-aided design systems into the vehicle development process makes it possible to reduce the time between creating a concept and receiving the finished product. Application of a digital twin in the creation of electric vehicles allows not only to reduce design time, but also to improve the key characteristics of the product, to study the operation of the vehicle in a wide variety of driving and operating modes due to connection with its physical twin throughout the entire life cycle. This paper proposes the concept of a digital twin of the temperature control system for an electrified vehicle of category L and reviews the experience of creating a test bench for the temperature control system of an electric vehicle connected to a virtual environment. Use of data on the creation of a stand for the temperature control system of an electric vehicle will allow the development of a full-fledged experimental stand for assessing the accuracy of a digital twin of the temperature control system for an electrified vehicle of category L in extreme climatic conditions. The necessary mathematical models included in the digital twin were identified. This digital twin will be useful for various companies involved in the development and production of electrified vehicles, as well as for organizations engaged in research activities in the field of electrical engineering and automotive engineering.

Keywords Electric vehicle · Digital twin · Temperature control system · Experimental stand · X-in-the-loop

1 Introduction

The development of modern competitive vehicles is impossible without usage of computer technologies and computerized development processes. One of the first stages of vehicle development is creation of 3-D design model and mathematical models of the vehicle, which allows testing various design solutions. With further

S. Bazavluk (✉) · R. Kurmaev · K. Karpukhin · A. Terenchenko · G. Nadareishvili
NAMI Russian State Scientific Research Center, Moscow, Russian Federation
e-mail: sergey.bazavluk@nami.ru

D. H. Tien et al. (eds.), *Proceedings of the 1st International Conference on Sustainability and Emerging Technologies for Smart Manufacturing*, Proceedings in Technology Transfer, https://doi.org/10.1007/978-981-97-7083-0_21

development of the project and refinement of the vehicle design, the mathematical model is supplemented based on the test results of the real vehicle to bring behavior of the model closer to behavior of the real object.

The introduction of computer-aided design systems into the car development process allows reducing the time between the creation of a concept and the start of manufacturing the finished product. This system involves the integration of engineering graphics (CAD), engineering calculations (CAE) and automation of preparation and production management (CAM) systems into a single closed system. The next main step for the automotive industry was usage of product lifecycle management (PLM) systems—that is, systems that calculate the entire set of processes performed from the moment the society's needs for a certain product are identified to the disposal of the product after its use. However, in such systems, the manufacturer does not have full feedback from the product during its operation. Some of the design solutions used in the product, as well as control algorithms systems, are sometimes obtained empirically, which often reduces the efficiency or increases the cost of the product. After the vehicle development stage, its mathematical model is no longer used or updated. The concept of a "digital twin" implies the use of a virtual model together with its physical counterpart at the stages of testing, fine-tuning, operation and disposal.

The digital twin includes a set of mathematical models of varying levels of complexity, which are updated based on the results of full-scale experiments. This allows you to obtain the first full-scale sample of the product and predict its behavior throughout its entire life cycle. An important property of a digital twin is a constantly updating representation of the actual physical product or process. During operation of a physical product, information from its sensors, reports from users and other data are continuously transmitted to the digital twin. In response, the digital twin transmits various forecasts and estimates that improve the operation and maintenance of the real object [1].

2 Literature Review

NASA first proposed the digital twin, as a concept, in 1970 in preparation for manned flights, but only in 2003 in the works of Grieves [2], it received wider development, as the rapid development of computing and digital technologies began. According to the original concept, a digital twin must meet three basic requirements:

1. Possibility of visualization, analysis of components and their research.
2. Realism of behavior during virtual tests, showing high convergence with real tests.
3. Possibility of receiving feedback from a double.

Bochert and Rosen believe that the digital twin is an abstract concept that allows for improved design and production of products [3]. Another opinion involves the implementation of simulation methods in the model [4], that is, a description of processes as they would occur in reality and with which computational experiments can be carried out. Borovkov [5], outlining the requirements for digital twins, believes that they are built on the basis of mathematically formalized methods of the theory of elasticity, strength, reliability, etc., and the technology of digital twins must be accompanied by a methodology that includes theoretical, software and technical methods and tools for developing simulation models. The experience of development companies ANSYS, Siemens, Gamma Technologies confirms the reliability of these statements, since their computer complexes are based on simulation models. The concept of a digital twin (DT) is one of the most promising concepts that can contribute to the rapid development of the automotive industry. The application of digital data, in the context of the industry under consideration, can be divided into several areas: (1) use in the development and manufacture of vehicles; (2) study of the chemical and physical aspects of batteries used in transport; (3) use to improve traffic and infrastructure [6].

When developing modern vehicles, including vehicles of category L, usage of DT is a promising direction. For example, usage of this technology makes it possible to monitor the degradation of certain operational properties of a vehicle throughout the entire life cycle of the product. The work [7] describes the concept of DT for dynamic systems; in particular, two approaches are considered: DT based solely on physics and DT, which is a so-called gray box. The application of this concept makes it possible to improve the operational characteristics of a physical twin during its operation and recalculation of its digital copy. In work [8], the emphasis is placed on the creation and study of recalculation of the DT of a vehicle/vehicle component or software on various target devices—a built-in microcontroller or a cloud service—this issue is important and relevant in relation to non-stationary objects that have very limited power of microcontrollers in their composition.

On the other hand, the DT concept is also applicable in the assembly of vehicles and the manufacture of their components. This technology allows you to reduce the number of defective products, reduce downtime; in addition, this approach allows you to reduce the amount of time required to commission a production line or an entire plant [9, 10]. In addition, usage of DT makes it possible to evaluate the behavior of a technological system over time, as well as to evaluate deformations, stresses, and the state of the surface of a manufactured product or workpiece after using of certain technological operations [11].

One of the important issues of electrified transport is the design and manufacture of high-voltage batteries. The use of digital twins in this area can significantly improve the performance characteristics of batteries, reduce the cost of manufacturing and the entry of the finished product to the market, and the use of digital data during battery operation increases the safety of this unit [12, 13].

Thus, the digital twin is an integral part of the modern automotive industry. The main bias falls on electrified vehicles (EV), including category L. The main problem for EV is maintaining a given temperature range of high-voltage or traction battery (TB) cells under different climatic conditions, for which they use thermostating systems. Using of the digital twin concept in the development of temperature control systems will make it possible to more accurately correspond to real operating conditions and ensure reliable and efficient operation of electrified vehicles of category L in the extreme climate conditions. For example, in the Russian Federation, the amplitude of ambient temperatures can range from minus 67.8 to + 45.2 °C. And in the Socialist Republic of Vietnam [14], there is high humidity during the rainy season and high ambient temperatures in the summer up to + 50 °C.

The innovation of this research is the development of a digital twin of an electric vehicle of category L with the creation of an experimental stand for conducting virtually physical tests using a virtually physical environment. Such a stand will allow testing in remote laboratories in real time and modulate different operating conditions.

3 Prerequisites for Creating the Concept of a Digital Twin of Temperature Control System

To develop a digital twin of the temperature control system of electric vehicle category L, it is planned to use the experience of creating an experimental stand to study the functioning of the electric vehicle temperature control system connected to a virtual environment. The hardware of this stand includes directly the components of the thermostating system (radiator with fan, pumps and pipes), and the thermostatic components of the electric drive are replaced by simulator devices that create heat generation and hydraulic resistance corresponding to the components of the electric drive. Depending on the test methodology, components can be implemented as physical components located in a remote laboratory or implemented as virtual models [15–17]. The scheme of the experimental stand is shown in Fig. 1.

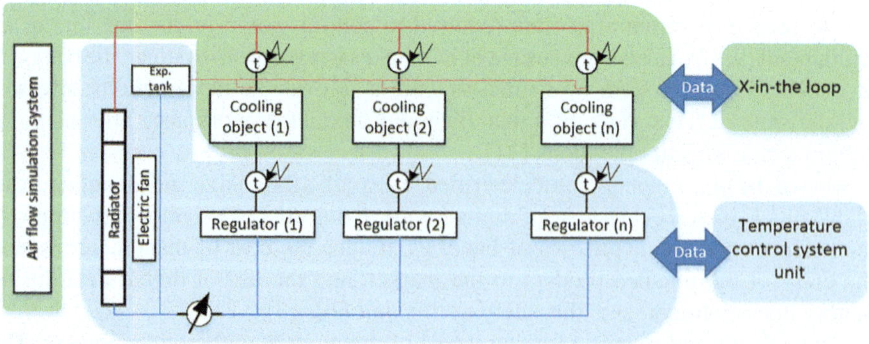

Fig. 1 Scheme of the experimental stand for studying the temperature control system for high-voltage components of the traction drive of the electric vehicle

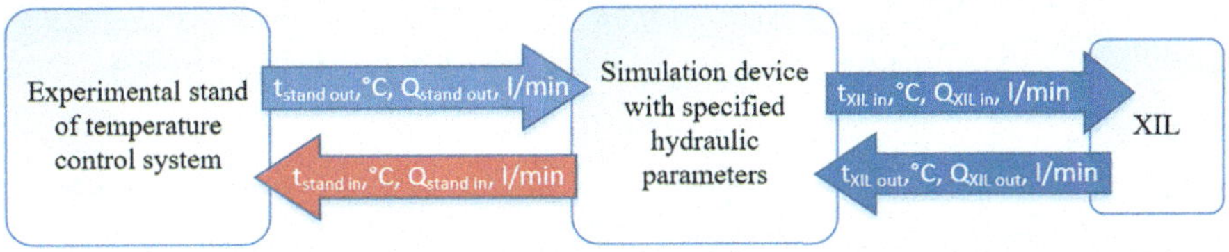

Fig. 2 Diagram of a device that simulates thermal loads and hydraulic resistance of high-voltage components of an electric vehicle's traction drive

This stand is part of the X-in-the-loop (XIL) virtual–physical environment, which allows for joint research and development of vehicle components and systems in the real time that interact with each other in various laboratories and research centers located remotely from each other.

In the process of conducting research, the stand receives real-time data from the XIL environment about the current thermal–hydraulic state at the outlet of thermostating objects, as well as information about the environment (temperature and humidity) and the speed of the electric vehicle. In turn, the stand transmits data on the temperature at the inlet to the simulator devices and the coolant flow rate to the X-in-the loop environment.

The main elements responsible for the accuracy of the results of the study of the temperature control system are devices that simulate the thermal and hydraulic characteristics of real objects. These devices have a control algorithm in software and hardware that is not connected with the other components of the stand. The circuit diagram of the simulation device is shown in Fig. 2.

To create an air flow simulating the movement of an electric vehicle, the stand has a system for simulating air flow. This system has independent control in accordance with the test cycle and recreates the movement of the electric vehicle at different speeds in real time.

The circuit of the temperature control system for high-voltage components of the traction drive of an electric vehicle (Fig. 3) was implemented on the experimental stand. Regulation of the temperature of devices simulating thermostating objects relative to specified coolant temperature ranges is carried out by changing the liquid flow and air flow.

4 Task Plan for Creating Digital Twin of the Temperature Control System of Electrified Vehicle of Category L

Based on previous experience in creating an experimental stand to study the functioning of the temperature control system of an electric vehicle in a virtual environment, we can conclude that in order to develop a digital twin of the temperature control system for an electrified vehicle of category L, it is necessary to develop the following:

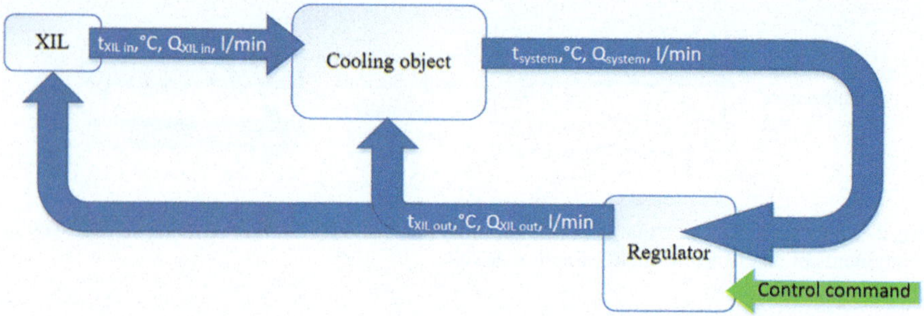

Fig. 3 The circuit of the temperature control system for high-voltage components of the traction drive of an electric vehicle

- the concept of a digital twin of the EV thermostating system of category L;
- mathematical model of electrified vehicle of category L;
- mathematical model of the EV thermostating system of category L;
- mathematical model of the operating environment of the EV digital twin of category L;
- control algorithm for the digital twin of the EV thermostating system of category L.

To assess the accuracy of the digital twin, as well as to research and create temperature control systems for electrified vehicles of category L, it is necessary to develop an experimental stand that allows virtual–physical tests. The stand must include a temperature control system and simulators of the thermal load from the battery and traction drive components (electric motor and inverter). This will allow, by regulating the operating modes of electric pumps of the temperature control system and electric radiator fans in the liquid circuits of the experimental stand, to maintain a given temperature under extreme operating conditions. Liquid and air flow regulators are controlled by PWM duty cycle values. The signal for setting the regulator is the liquid flow rate for the pump and the coolant temperature at the fan inlet. The measured liquid flow and coolant temperature values are transmitted to the controller as feedback. The difference between the required and measured values is transmitted to the controllers as a control error. It is necessary to conduct a comprehensive study of the temperature control system for electric vehicles of category L in extreme climatic conditions with a high temperature range and high ambient humidity to create a digital twin of the temperature control system in a virtual–physical environment.

Figure 4 shows diagrams of two independent cooling circuits for high-voltage components of EV category L, which will be implemented at the experimental stand. Each circuit maintains its own temperature range to ensure efficient operation of high-voltage components. Each circuit includes electric pump (1), expansion tank (2), cooling jackets for components, radiator with electric fan (4), temperature sensor, tubes and connecting hoses. In Fig. 4a, inverter (5) and electric motor (6) are also marked, and in Fig. 4b, cooling circuit of TB (3) is marked.

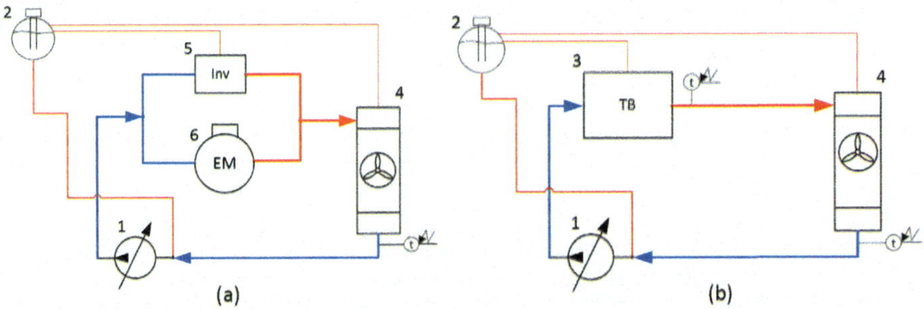

Fig. 4 **a** Cooling circuit of the electric motor and inverter, **b** cooling circuit of TB

5 Conclusion

Results of research the temperature control system for electric vehicles of category L may be in demand for further research and development in the field of electrification of transport and the creation of a sustainable and environmentally friendly transport system. Also, these results will contribute to the development of innovative technologies to improve the performance and reliability of thermal control systems and the creation of competitive and environmentally friendly vehicles. This digital twin will be useful for various companies involved in the development and production of electrified vehicles, as well as for organizations engaged in research activities in the field of electrical engineering and automotive engineering.

References

1. Denisov, A.: Digital twins as the basis for the digital transformation of the technical operation of cars within the framework of the fourth technological revolution. Tech. Regul. Transp. Constr. **3**(42), 165–168 (2020) (in Russian)
2. Grieves, M.: Origins of the Digital Twin Concept. https://www.researchgate.net/publication/307509727_Origins_of_the_Digital_Twin_Concept. Last accessed 18 Dec 2023
3. Boschert, S.: Digital twin—a simulation aspect. In: Mechatronic Futures. Springer International Publishing. https://link.springer.com/chapter/10.1007/978-3-319-32156-1_52016. Last accessed 21 Dec 2023
4. Petrov, A.: Imitation as the basis of digital twin technology. Vestn. Irkutsk State Tech. Univ. **22**(10), 56–66 (2018)
5. Borovkov, A., Burdakov, S., Klyavin, O., Melnikova, M., Mikhailov, A., Nemov, A., Palmov, V., Silina, E.: Computer Engineering: Textbook. Publishing House of the Polytechnic University, St. Petersburg (2012)
6. Deng, S., Ling, L., Zhang, C., Li, C., Zeng, T., Zhang, K., Guo, G.: A systematic review on the current research of digital twin in automotive application. Int. Things Cyber Phys. Syst. **3**, 180–191 (2023)
7. Tripura, T., Garg, S., Chakraborty, S.: Digital twin for dynamical systems. In: Rabczuk, T., Bathe, K.J. (eds.) Machine Learning in Modeling and Simulation. Computational Methods in Engineering & the Sciences 2023, vol. 1227, pp. 255–296. Springer, Cham (2023)

8. Blech, J.: Towards digital twins for the description of automotive software systems. Electr. Proc. Theoret. Comput. Sci. **312**, 20–28 (2019)
9. Soori, M., Arezoo, B., Dastres, R.: Digital twin for smart manufacturing. A review. Sustain. Manufact. Serv. Econ. **2**, 100017 (2023)
10. Tüzün, A., Hackenberg, G.: Quality assurance of digital twins: an experience report in the automotive industry. In: Quality of Information and Communications Technology. QUATIC 2023. Communications in Computer and Information Science, vol. 1871. Springer, Cham (2023)
11. Cappellini, C., Giorleo, L., Allegri, G., Attanasio, A., Ceretti, E.: A digital twin approach to automotive wheel flow forming process. In: Managing and Implementing the Digital Transformation. ISIEA 2022. Lecture Notes in Networks and Systems, vol 525. Springer, Cham (2022)
12. Utama, P., Haq, I.N., Leksono, E., Juristian, M., Alim, G., Pradipta, J.: Development of digital twin platform for electric vehicle battery system. Int. J. Sustain. Transp. Technol. **6**(1), 7–11 (2023)
13. Bilansky, J., Lacko, M., Pastor, M., Marcinek, A., Durovsky, F.: Improved digital twin of li-ion battery based on generic MATLAB model. Energies **16**, 1194 (2023)
14. Khanh, C.N., Balasooriya, S., Kavalchuk, I., Kolbasov, A., Karpukhin, K., Terenchenko, A.: Novel adaptive control method for BLCD drive of electric bike for Vietnam environment. IOP Conf. Ser. Mater. Sci. Eng. 819 (2020)
15. Kulikov, I., Karpukhin, K., Kurmaev, R., Ivanov, V.: X-in-the-loop technology for research and development of electric vehicles. Trudy NAMI **2**, 6–14 (2021) (in Russian)
16. Terenchenko, A., Karpukhin, K., Kurmaev, R.: Features of operation of electromobile transport in the conditions of Russia. In: 28th International Electric Vehicle Symposium and Exhibition 2015, EVS 2015 28, e-Motional Technology for Humans (2015)
17. Zavatsky, A., Keller, A., Shadrin, S., Makarova, D., Furletov, Y.: Development of an electric all-wheel-drive simulation model used to test torque distribution algorithms. Energies **16**(20), 7144 (2023)

Fuel Deposits and Methods for Their Control

Mikhail Zlenko, Alexey Terenchenko, and Givi Nadareishvili

Abstract Deposits in a general sense are an inevitable consequence of the processes of mixture formation and combustion in the engine cylinder, complex physical and chemical processes with hydrocarbon "participants" in this process—fuel and oil. Moreover, certain qualitative characteristics of fuels, for example, group composition, presence or absence of detergent additives, can have a significant impact on the amount of deposits. Scientific and methodological support for testing fuels for their tendency to form deposits is also important. The complex of works carried out at NAMI Russian State Scientific Research Center resulted in creation of a series of test methods for automotive fuels, which are analogs or equivalents of the generally accepted CEC and ASTM methods, but at the same time devoid of the disadvantages associated with the use of these methods. Comparative tests showed good convergence of results with the analogs. At the same time, all logistical support for testing both gasoline and diesel fuels is carried out on a domestic basis using Russian engines. In the period of general confusion in the European engine industry, uncertainty in the prospects of the automotive industry as a whole, the creation of an independent service for certification testing of fuels for the Russian market and the development of our own methods for assessing the tendency of fuels to deposits using our own engines become an important both scientific and practical tasks.

Keywords Fuel · Deposits · Crankcase gases · Crankcase ventilation system · Test methods

1 Introduction

Among the variety of requirements for motor gasoline for automobile manufacturers and ordinary consumers, the most important are those properties that have "hidden" or non-obvious features that appear after a relatively long time with a negative effect

M. Zlenko · A. Terenchenko (✉) · G. Nadareishvili
NAMI Russian State Scientific Research Center, Moscow, Russia
e-mail: terenchenko@nami.ru

D. H. Tien et al. (eds.), *Proceedings of the 1st International Conference on Sustainability and Emerging Technologies for Smart Manufacturing*, Proceedings in Technology Transfer, https://doi.org/10.1007/978-981-97-7083-0_22

on the engine and that are associated with difficulties in identifying the causes of engine failures or malfunctions [1–3]. These features include deposits that form on or inside engine components.

Deposits in a general sense are an inevitable consequence of the processes of mixture formation and combustion in the engine cylinder, complex physical and chemical processes with hydrocarbon "participants" in this process—fuel and oil. Numerous studies have proven that the main "building material" of deposits is fuel, or rather the products of its incomplete evaporation and/or combustion.

Conditions for complete evaporation of fuel in the intake system or in the engine cylinder are not always created. For example, with a warm engine, the temperature of the walls of the intake channels does not exceed 70–90 °C. In the engine cylinder, the surfaces forming the combustion chamber can also have a temperature within a very wide range—from negative (in winter) to 200–250 °C at full load; therefore, in most cases during normal operation, the main part of fuel enters the intake pipe and engine cylinder in a fine droplet state [4].

2 Types of Fuel Deposits

By their nature, deposits are varied, but within the framework of this article, they can be divided into two groups—these are conditionally "cold" deposits (Fig. 1), formed in the low-temperature ($T < 200$ °C) zone of the internal combustion engine, in particular in the intake tract, i.e., where there is no direct contact with the flame of burning fuel in the combustion chamber, and conditionally "hot" deposits (Fig. 2) that are formed on the surfaces that form the combustion chamber. In a port fuel injection (PFI) engine, fuel is injected into the intake manifold air medium through the fuel injector in the form of fine droplets. On the way to the cylinder, due to heating from the wall of the intake manifold, the temperature of the working air/fuel mixture increases and the fuel droplets lose part of their volume due to the evaporation of light fractions.

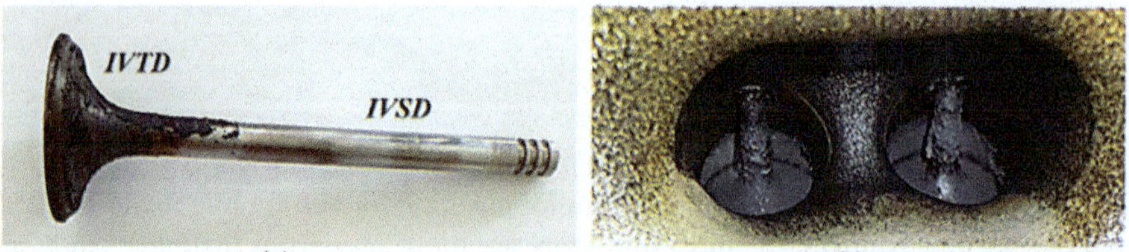

(a) (b)

Fig. 1 "Cold" deposits: **a** deposits on the valve tulip (IVTD) and stem (IVSD); **b** lacquered deposits on the walls of the inlet channel

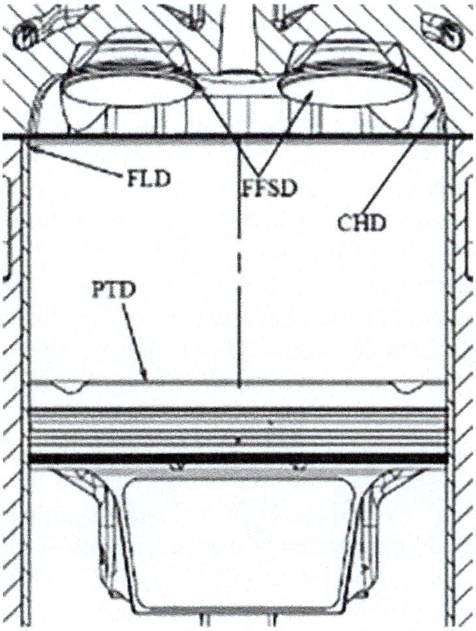

Fig. 2 "Hot" deposits in the combustion chamber

However, a relatively small portion of the fuel is subject to evaporation. The rest of the fuel will continue to remain in the liquid-droplet (aerosol) phase. Having reached the valve surface, the fuel droplets begin to evaporate—i.e., lose part of their light fractions, but only to the extent that will be determined by the temperature of the intake valve tulip. This temperature varies widely—from negative during winter operation (start-up, early warm-up) to 150–200 °C at medium load conditions, and the temperature of the valve flame face side can reach 600 °C during prolonged engine operation at full load. In this case, the temperature of the valve stem lies in the range of 100–140 °C. Thus, when the engine operates in a "normal" city traffic mode, favorable conditions are created in the intake system for the formation of deposits on the surface of the intake valve. Drops of fuel fall on the valve tulip and stem and partially evaporate. Some of the unevaporated fuel settles on the surface of the valve. Subsequent even short-term heating of the valve to a higher temperature will lead to further evaporation of higher-boiling fractions, and in the end, only heavy fuel fractions with a boiling temperature of 150–200 °C will remain on the surface of the tulip. These fractions are more viscous and more "sticky" with good adhesion to the valve material.

The second and equally important source of the deposit building material, especially for direct injection engines (DI engines), is also fuel, but coming into the intake from the crankcase ventilation system (CVS). Crankcase gases (CCG) have a complex composition and are a multicomponent mixture of gases, an important component of which is aerosol—finely dispersed hydrocarbon particles of fuel and oil origin suspended in the gas environment of the crankcase space mixed with vapors of water, air and combustion products.

Engine oil can also be a source of deposits formation. Oil in the form of a film can enter the intake port from the gap between the valve stem and the valve guide, flowing onto the valve tulip, and, interacting with fuel particles, can create its own kind of deposits on the valve tulip and valve stem. The process of deposit formation can be compared to layer-by-layer synthesis (by analogy with the operation of a 3-D printer)—the gradual layering of particles of the building material, which are the sticky, unevaporated remains of tiny drops of fuel. Thus, it is incorrect to call the deposits on the valve "soot" or "carbon" deposits. The resulting mass is basically not a product of fuel combustion, but a "heavy" hydrocarbon compound that arose as a result of a kind of fuel "boiling" or "distillation" on the surface of the intake valve. In addition to throttling the flow of the working fluid at the inlet and the associated deterioration in cylinder filling, this "stalagmite" can periodically act as a kind of sponge and absorb drops of fuel from the nozzle, thereby disrupting the control of the air–fuel mixture. The engine becomes "dull" and loses vehicle dynamics control.

"Hot" deposits include all deposits that arise as a result of direct contact of fuel with burning gases in the combustion chamber. It has been proven that the main building material of deposits is fuel, or more precisely, the products of incomplete oxidation of the fuel that gets on the walls of the combustion chamber, but does not completely evaporate, for example during the process of cold start and warm-up. The bulk of the hot deposits is carbon deposits formed in the combustion chamber, see Fig. 2. In practical work and for research purposes, the concept of "total combustion chamber deposits"—TCD (total combustion chamber deposits), is used, which is the sum of the masses of the sediments or deposits indicated in Fig. 2:

$$\mathrm{TCD} = \mathrm{CHD} + \mathrm{FFSD} + \mathrm{FLD} + \mathrm{PTD}\ (\mathrm{mg/cyl.})$$

CHD is the cylinder head deposit; FFSD is the flame face side deposits; FLD is the fire land deposits; PTD is the piston top deposits.

For the deposits in the combustion chamber, we use the term "carbon deposits" or "coke", since in this case we are dealing with products directly related to combustion in the engine cylinder. The borderline position of the fuel injectors causes the presence of two types of deposits—"hot" on the outer surface of the spray tip (EFID—external fuel injector deposits) and "cold" inside the injector (IFID—internal fuel injector deposits), see Fig. 3.

With regard to a diesel injector, it is also necessary to distinguish between external "hot" deposits (EDID—external diesel injector deposits) and internal "cold" deposits (IDID—internal diesel injector deposits), the size of which significantly depends on the design of the spray tip. Figure 4 shows design differences between the two most common types of injector nozzles—SAC type and VCO (valve covered orifice) type.

Both the first and second types of deposits are very dangerous and can lead not only to loss of engine functionality, but also to a serious accident. For example, "hot" carbon deposits in the combustion chamber can lead to an excessive increase in the compression ratio and, as a result, detonation. "Cold" deposits can cause valves sticking, difficult starting, uneven operation and loss of power.

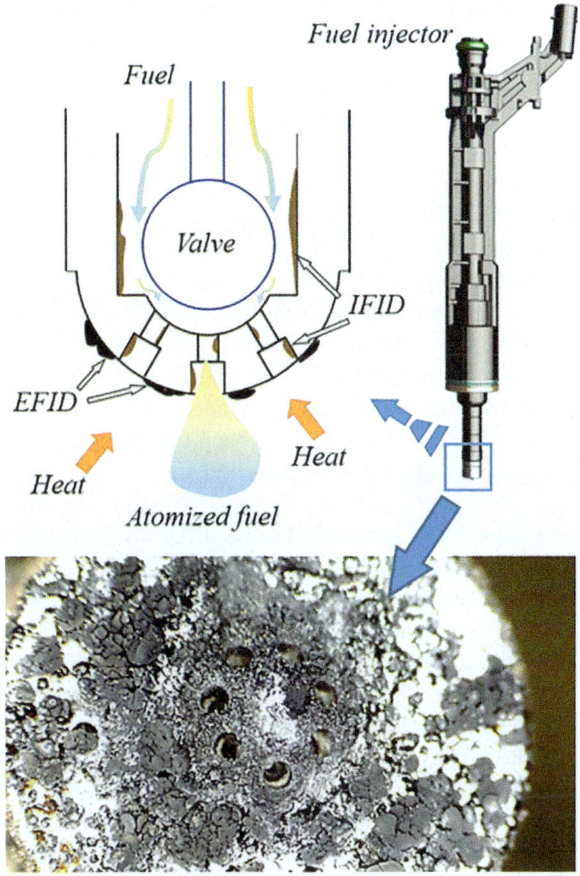

Fig. 3 External (EFID) "hot" and internal (IFID) "cold" deposits of the GDI engine injector nozzle

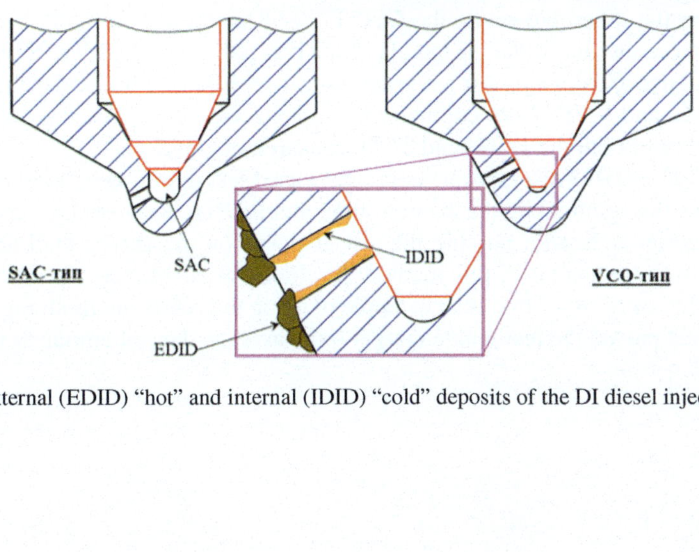

Fig. 4 External (EDID) "hot" and internal (IDID) "cold" deposits of the DI diesel injector nozzle

3 Methods of Control of Fuel Deposits

CEC (Coordinating European Council for the Development of Performance Tests for Transportation Fuels, Lubricants and Other Fluids) has established a number of requirements to limit both "hot" and "cold" fuel deposits approved by the Worldwide Fuel Charter (WWFCh—Worldwide Fuel Charter [5]) as quality standards for automotive fuels. The standards (methods) adopted by the WWFCh at different times are presented in the table. Currently, these standards have different statuses, and some are formally inactive, but are widely used "unofficially" and for research purposes. Green in this table indicates current methods included in the list of WWFCh requirements regarding the tendency of fuels to form deposits.

A common problem in the practice of using these techniques is the significant difficulties in equipping testing laboratories with reference engines. Typically, these engines are engines that have been out of production for a long time, so providing laboratories with consumables and spare parts is becoming an increasingly difficult task. Engine manufacturers refuse to undertake further technical support for CEC activities, explaining this by the difficulties of warehousing and logistics of spare parts. In Table 1, in the "Descriptions" column, the serial production dates of the reference engines are indicated. Only for the VW EA111 engine with direct injection, there is an agreement on technical support for the CEC TDG-F-113 methodology, but the methodology itself, despite the long development period, still remains in the "temporary" status and has not been officially approved. For this reason, the CEC TDG-F-113 test is not included in the list of mandatory WWFCh procedures, although it is quite widely used by both automakers and fuel companies for comparative tests.

The CEC F-20-98 test is listed twice in the table—in the "cold" section and in the "hot" section. This is one of the most popular tests, which evaluates the effectiveness of detergent additives aimed at preventing the formation of the IVTD, on the one hand, and the degree of negative impact of these additives on combustion chamber deposits (TCD), on the other. However, the CEC F-20-98 method is formally "invalid" or "inactive" because the WWFCh has classified it as non-representative. Nevertheless, our experience of using this test while applying a modified method and using Russian VAZ-21116 and VAZ-21126 engines with a displacement of 1.6 L has shown good repeatability of both the IVTD and TCD measurement results.

The CEC F-05-93 test is also a commonly used test to evaluate the effectiveness of gasoline detergent additives in controlling and limiting intake valve deposits. For fuel categories 4–6, after passing the test, deposits on the valves shall not exceed 30 mg/valve. An analog of this method developed at NAMI has also shown good repeatability of results. Here is an example of using the NAMI method: Fig. 5 shows that the fuel passed the test with a margin—the average deposit amount was 10 mg/valve.

Table 1 Fuel quality standards for "hot" and "cold" fuel deposits

Standard designation	Fuel deposit types/name of test method	Descriptions
Cold deposits		
CEC F-05-93 (deposits on intake valves)[b]	Inlet valve cleanliness in the MB M102E engine	For "old" engines with 2-valves/cyl. Ref. engine M102, 2.3 L (1983–1993)
CEC F-20-98[a] (intake valves and comb. chamber deposits)[b]	Deposit forming tendency on intake valves	Ref. engine M111, 2.0 L, with 4-valves/cyl. comb. chamber (1992–2003). Currently not supported by the WWFCh
ASTM D6201 (intake valves and comb. chamber deposits)[b]	Test method for dynamometer evaluation of unleaded spark-ignition engine fuel for intake valve deposit formation	Ref. engine Ford Ranger 2.3 L (1994)
CEC F-16-96 (intake valve sticking)	Assessment of the inlet valve sticking tendency of gasoline fuels	Ref. engine VW Waterboxer (1982–1992)
CEC F-110-16[a] (test for internal contamination of DI engine diesel injectors)	Internal diesel injector deposit test	Ref. engine PSA DW10C (2009–2015). The test is not approved. Currently not supported by the WWFCh
Hot deposits		
CEC F-20-98[a] (comb. chamber deposits)[b]	Combustion chamber deposit test method	Ref. engine M111, 2.0 L, 4-valves/cyl. comb. chamber (1992–200). Currently not supported by the WWFCh
CEC F-23-01 (test for external contamination of diesel engine injectors)[b]	Cylinder indirect injection diesel engine	Ref. engine Peugeot XUD9 A/L (1982–2005). Test for pre-chamber diesel engines of the previous generation
CEC F-98-08 (test for external and internal contaminations of diesel DI engine injectors)[b]	Direct injection, common rail diesel engine nozzle coking test	Ref. engine PSA DW10BTED4 (2004–2015)
CEC TDG-F-113[a] (test for external and internal contaminations of gasoline DI engine injectors)[b]	Fuel injector deposits in direct injection spark-ignition engines	Ref. engine VW EA111, 1.4 L (1985–2015)

[a] Tests that have lost or have not received official status are not supported by the WWFCh, but are actively used in research work and in comparative tests
[b] Alternative Russian proven methods are available

Fig. 5 Intake valves of a PFI engine after testing without a detergent additive (left) and with a detergent additive (right) using an analog method to CEC F-05-93 on the Russian VAZ-21116 engine

The quality suitability of gasoline for use in the engines with direct injection (GDI—gasoline direct injection) is confirmed by a special test CEC TDG-F-113, which has not currently passed the approval procedure, but has been published for the purpose of extended testing in the automotive industry [6]. The essence of the method is as follows. First, the injectors are "contaminated"—the engine runs on base fuel without detergent additives. During this period, gradual contamination of the injector with deposits occurs, as a result of which the duration of the ti signal for opening the injector increases (the injection duration compensates for the loss of fuel pressure). This process is illustrated by the left part of Fig. 6—within 48 h, there is an increase in the ti signal duration from 2.35 to 2.55 ms. Then, the engine is switched to fuel with a detergent additive—DCA (deposits control additive). If the DCA "works", then the injector begins to clean itself and, as a result, the duration of the signal decreases. The "clean-up" procedure is carried out within 24 h (the interval is 48–72 h). It can be seen that during this time, the duration of the ti signal is almost completely restored. For the next 24 h, the engine operates in the "keep-clean" mode, during which it is checked whether the ti signal is maintained at the level of residual contamination. Figure 6 illustrates the passing of the test with a positive assessment—during the "keep-clean" mode, the deviation of the ti signal does not exceed acceptable limits.

For diesel engines, the CEC F-98-08 test is used, which also consists of two stages: first—"dirty-up", artificial contamination of the injector when the engine is running without the DCA (32 h); second—"clean-up", engine operation with the same fuel, but with the addition of the DCA (12 h). To accelerate the pollution process, zinc is added to the diesel fuel in the form of zinc neodecanoate $Zn(C_{10}H_{19}O_2)_2$ at a concentration of 1 ppm. During the first stage, due to contamination of the injectors, there is a natural decrease in the cyclic fuel supply and, as a result, a decrease in the engine power by approximately 5–12% of the original value. At the second stage, the injectors are cleaned and the power is restored. According to the WWFCh

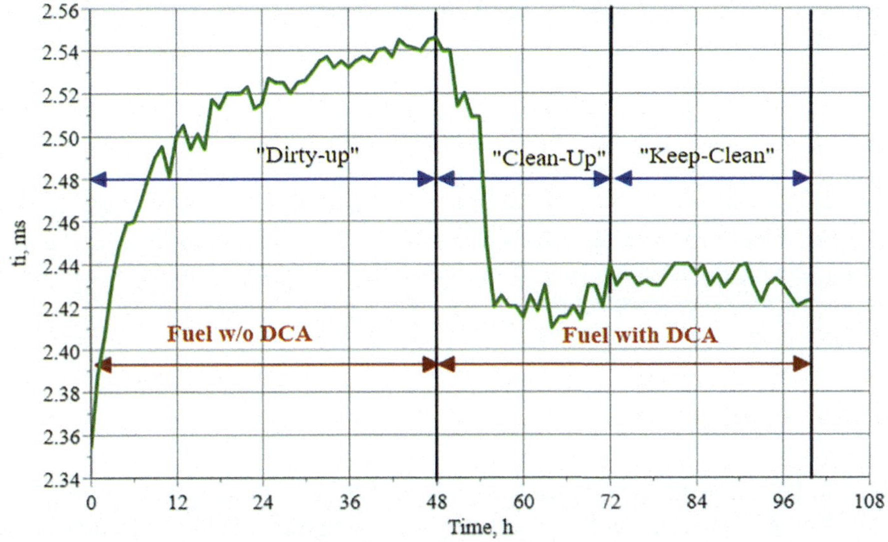

Fig. 6 Result of assessing the cleaning properties of fuel using a method analogous to CEC F-98-08

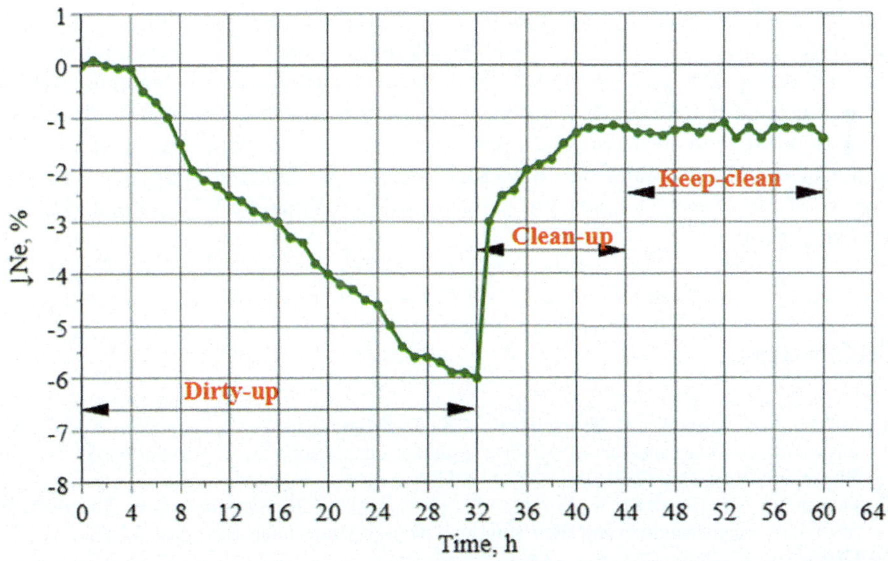

Fig. 7 Result of diesel fuel testing using a method analogous to CEC F-98-08

requirements, if the restored power is no more than 2% lower than the original power, then the test is considered to have been passed. Sometimes, this test is added with the "keep-clean" option, during which the ability of detergent additives to prevent contamination of injectors is assessed. Figure 7 shows, as an example, the result of a test that includes all three stages mentioned. At the end of the first stage, the drop in the engine power was 6%. At the second stage (from 32 to 44 h), when the engine was running on fuel with the additive, the power was almost restored—the decrease in power relative to the initial value was less than 1.5%. At the third stage, the engine power remained almost unchanged, which indicates the ability of the fuel additive to provide the "keep-clean" mode.

4 Conclusion

Monitoring the quality of fuels with respect to their tendency to form deposits is a critical objective of the environmental program. A contaminated or dirty injector creates an asymmetrical fuel spray with a greater depth of penetration into the air charge compared to clean injectors. This leads to wetting of the surfaces of the cylinder, piston and nozzle tip and, as a result, an increase in particulate emissions. During operation, the consequences of injector contamination are not always noticeable—the driver may not feel any change in the unevenness of the ride, nor a decrease in the engine power, or a loss of vehicle acceleration dynamics. However, the hidden environmental damage can be quite significant—suspended particulate matter (PN) emissions can increase by 1–2 orders of magnitude, and mass emissions (PM) can increase more than 10 times [7]. In the table, the "letter b" marks the standard CEC and ASTM methods, which now have Russian analogs, allowing independent quality control of fuels for the tendency to deposits, using completely Russian material and technical support.

References

1. Luksho, V.A., Kozlov, A.V., Terenchenko, A.S., Ter-Mkrtichian, Yu.G., Karpukhin, K.E.: Technical and economic analysis of vehicles pollutant emissions reduction technologies. Biosci. Biotechnol. Res. Asia **12**(2), 1867–1872 (2015)
2. Panchishny, V.I., Luksho, V.A., Kozlov, A.V., Terenchenko, A.S., Karpukhin, K.E.: The problem of toxicity of gas engines and their solutions. Biosci. Biotechnol. Res. Asia **12**(2), 217–224 (2015)
3. Zuev, N., Kozlov, A., Terenchenko, A., Karpukhin, K., Azimov, U.: Detailed injection strategy analysis of a heavy-duty diesel engine running on rape methyl ester. Energies **14**, 3717 (2021)
4. Zlenko, M.A., Terenchenko, A.S.: E-fuels—last hope for internal combustion engine. Trudy NAMI **4**, 15–34 (2022) (in Russ.)

5. World-Wide Fuel Charter. https://www.acea.auto/files/WWFC_19_gasoline_diesel.pdf
6. CEC Activity Report January–June 2023
7. Barker, J., Reid, J., Mulqueen, S., Innospec Langley, G.J., Wilmot, E.M.J., Vadodaria, S., Whitaker, J.: The investigation of the structure and origins of gasoline direct injection (GDI) deposits. JSAE 20199078. SAE 2019-01-2356

Development and Verification of a Mathematical Model of Fuel Consumption Estimation Suitable for Research of Algorithms of Energy-Efficient Control on Digital Roads

Alexander Koudrin and Sergey Shadrin

Abstract The purpose of this work is to develop a multifactor mathematical model in Matlab Simulink environment, taking into account fuel consumption and a digital terrain model with variable inclinations of road surface with intersections and traffic light regulation, to conduct field experiments, as well as computer simulation and verification of the developed model. The paper proposes a mathematical model for calculating the instantaneous fuel consumption of a car with an internal combustion engine. The design of a digital twin of the car is described. A full-scale experimental run on a test road section is carried out, and the conditions of computer simulation are described. The results of calculation of instantaneous fuel consumption of the experimental car when driving on a test road section with elevation differences, controlled intersections, and different speed ranges are compared with the results of computer simulation. The results of computer simulation matched the field experiment by 98%. Conclusions are drawn about the applicability of the proposed approach to the process of development and verification of intelligent speed control algorithms for improving the energy efficiency of road vehicles.

Keywords Road vehicle · Digital twin · Simulation · Digital terrain model · Fuel consumption · Mathematical model verification

A. Koudrin (✉) · S. Shadrin
Advanced Engineering School of Electric Transport, Moscow Polytechnic University, Bolshaya Semyonovskaya Street, 38, 107023 Moscow, Russia
e-mail: a.koudrin@outlook.com

MADI, 64, Leningradsky Prosp., Moscow 125319, Russia

D. H. Tien et al. (eds.), *Proceedings of the 1st International Conference on Sustainability and Emerging Technologies for Smart Manufacturing*, Proceedings in Technology Transfer, https://doi.org/10.1007/978-981-97-7083-0_23

1 Introduction

Computer simulation plays an important role in the design and testing of new vehicles. Mathematical models that simulate real physical processes have been in use for decades and offer a number of advantages over natural experiments [1]. They allow rapid testing of different variants of object characteristics, simulation of changing external conditions, and error tolerance without additional costs [2].

In modern world, the issue of saving resources, including fuel, is becoming more and more urgent. On the one hand, this is due to the constant growth of energy prices, and on the other hand to the need to reduce emissions of toxic substances into the atmosphere [3]. One of the ways to solve this problem is to improve the energy efficiency of the road vehicle [4]. To solve this problem, it is important to estimate the fuel consumption of a car with high accuracy when conducting computer simulations to enable the development of new algorithms for more energy efficient control of modern cars, including highly automated and driverless cars.

There are several scientific solutions that offer mathematical models for calculating vehicle fuel consumption. Some papers do not include individual car parameters: tire structure, compression ratio, and others [5, 6]. This makes it difficult to apply this model on a pro-industrial scale. There are solutions that have a detailed model of engine operation, but do not take into account the physics of the car [7, 8]. For example, drag forces have a significant impact on fuel consumption. Without taking these factors into account, the results of the calculation during the operation of the car will not be complete. It is fair to use such a model for engine bench testing. There are also multifactor models that have not been tested in the real world, which raises doubts about their reliability [9].

The aim of this study is to develop a mathematical model that considers the specific parameters of both the engine and vehicle design, and to test it through a field experiment. The model will be customizable with various factors, making it applicable to different vehicle types. The detailed mathematical model will ensure high accuracy.

2 Description of the Mathematical Model of Fuel Consumption

In the present work, the main equation, which is used for calculation of fuel consumption and optimization of control actions, is formula (1) [4].

$$G_{\mathrm{t}} = \frac{\int \frac{3600 \cdot \rho_{\mathrm{K}} \cdot \eta_{v}}{p_{\mathrm{e}} \cdot \alpha \cdot l_{0}} \cdot v \cdot (P_{\mathrm{r}} + P_{\mathrm{a}} + P_{\mathrm{i}} + \mathrm{Br}) \cdot \frac{1}{\eta_{\mathrm{T}}} \cdot \mathrm{d}t}{\int 755 \cdot v \cdot \mathrm{d}t}, \quad (1)$$

where: G_t—fuel consumption per unit of traveled distance, ρ_k—charge density, η—transmission efficiency, η_υ—fill factor, p_e—average effective pressure, α—excess air factor, l_0—theoretically required amount of air for combustion of 1 kg of fuel, υ—vehicle speed, Br—braking resistance, t—time, P_a—air resistance, and P_i—inertia forces.

The model that takes into account the physics of the car and engine, tested in the article [4], was taken as a basis. The proposed model, in comparison with the previous one, additionally takes into account the following factors (2, 3, 4) [10, 11]:

$$\rho_k = \frac{\rho_0 \cdot 10^6}{R \cdot T_0}, \tag{2}$$

where: R—specific gas constant, ρ_0—ambient pressure, and T_0—ambient temperature.

$$p_e = p_i - \left(0.089 + 0.0118 \cdot \frac{s * n}{30}\right), \tag{3}$$

where: S—piston stroke, n—crankshaft revolutions, and p_i—average indicator pressure.

$$P_r = \sin\alpha \cdot g \cdot G + G \cdot g \cdot \cos\alpha \cdot \frac{k}{1000} \cdot \left(5.1 + \frac{550{,}000 + 90 \cdot G_k}{P_{TP}} + \frac{1100 + 0.0388 \cdot G_k}{P_{TP}} V^2\right), \tag{4}$$

where: k—coefficient accounting for tire structure; G_k—vertical wheel load; P_{TP}—tire pressure; V—vehicle speed, g—free fall acceleration, α—longitudinal slope of roadway profile, and G—vehicle weight.

The general view of the implementation of the mathematical model in Matlab Simulink environment is shown in Fig. 1.

3 Vehicle's Digital Twin

In this research, we used Simcenter Prescan software for vehicle model description. The Prescan program includes a set of actors that can be used in scenario simulations. Actors are objects that differ in their behavior and/or dynamics (e.g., a car or a pedestrian). Each of these objects can be configured with its own behavior and dynamics model through the configuration window.

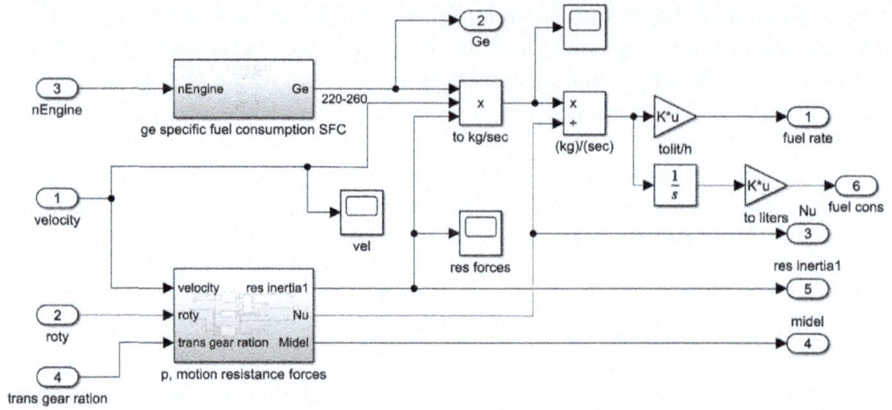

Fig. 1 General view of realization of the mathematical model of car fuel consumption in Matlab Simulink environment

The configuration parameters are as follows:

(1) Movement trajectory.
(2) Assignment of controls.
(3) Dynamics properties.
(4) Customizing the performer animation ports for export to the compilation table, which allow you to set lighting, spinning wheels, etc.
(5) Selecting which object response models should be assigned to the subject.

Within the set experiment, it is necessary to create a virtual copy of Hyundai Getz 1.4 car. There is no such car in the set of standard Prescan actors. Therefore, changes were made in the physical characteristics of the closest car model available. The basis was the model of the car Toyota Yaris Hatchback. The program allows you to make changes in the dynamics of the car, transmission, powertrain, suspension, and steering. We have applied the appropriate characteristics of transmission gear ratios and engine torque map in accordance with characteristics of Hyundai Getz experimental car tested on real road.

Some parameters like overall dimensions were taken from the Internet resource [12]. The moment of inertia of the car was determined according to the method described in [13].

Weight distribution was measured directly, and center of gravity position was estimated after several sets of weight distribution measurement of inclined and horizontally set vehicle with equations described below.

The center of gravity position of the experimental car was determined by the following mathematical Eqs. (5) and (6):

$$H = R_{\text{st\_av}} + \frac{\Delta G \cdot L}{G_{\text{a}} \cdot \text{tg}(\alpha)}, \tag{5}$$

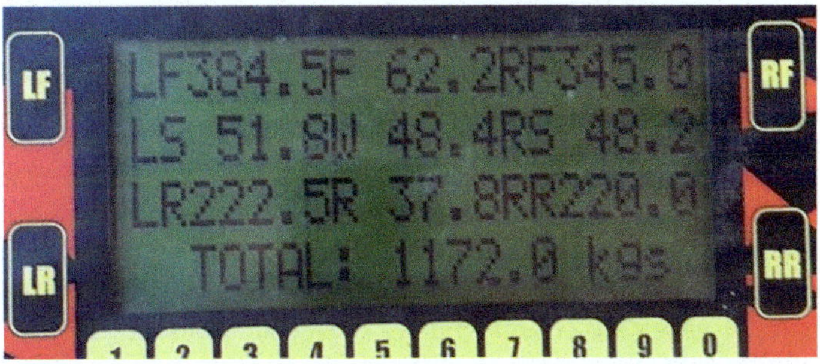

Fig. 2 Weight of the experimental vehicle together with a driver and full fuel tank

$$R_{\mathrm{st\_av}} = R_{\mathrm{st1}} + \frac{G_{\mathrm{ar}}}{G_{\mathrm{a}}}(R_{\mathrm{st2}} - R_{\mathrm{st1}}), \tag{6}$$

where

$R_{\mathrm{st\_av}}$	average static wheel radius, mm;
R_{st1}	static radius of front axle wheels, mm;
R_{st2}	static radius of rear axle wheels, mm;
L	vehicle base, mm;
ΔG	is the difference between the front axle loads in horizontal and inclined positions of the vehicle, kg;
G_{a}	vehicle weight, kg;
G_{ar}	load on the rear axle of the vehicle, kg;
$\mathrm{tg}(\alpha)$	value of the tangent of the angle of inclination of the car during static experiment.

During the measurements, INTERCOMP SW500 E-Z equipment was used to determine the weight distribution under the each wheel as shown in Fig. 2.

4 Experiment Description

Purposes of the experiment:

- Validation of the results of the experiment with virtual simulation and real road tests.
- Evaluation of the effectiveness of the developed mathematical model for fuel consumption estimation.
- Identification of the shortcomings of the developed system in real conditions.

To achieve the objectives, computer simulation and full-scale experiment were carried out.

As the results of the experiment, we will compare the obtained fuel consumption obtained in the field experiment with a similar scenario in the virtual environment.

Computer simulation was carried out on a digital twin of the real road (Fig. 3), taking into account the elevation difference and traffic light regulation. The field experiment was conducted on the same road section to verify the mathematical model. The number of recorded road test on this route is 10. The characteristics of the route are given in Table 1.

The test object was a car of category M1—Hyundai Getz 1.4, technical characteristics of which are given in Table 2. Test conditions are given in Table 3.

Car condition at the time of the experiment: tire pressure was set according to the manufacturer's recommended values, car mileage—78,573 km, fuel tank was filled with A-92 fuel for 45 L, and car load was the same as during weight measurement.

Recorded parameters and indicators:

- driving speed;
- traveled distance;
- instantaneous fuel consumption;
- geolocation of the test object;

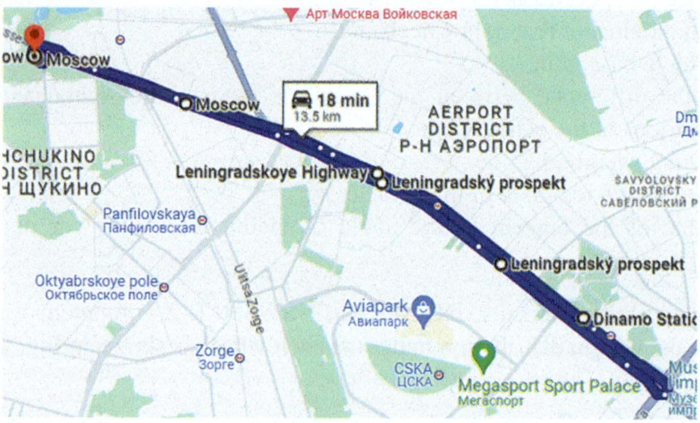

Fig. 3 Plot of a real road from which was made the digital terrain model

Table 1 Travel route characteristics

Characteristics	Value
Length (km)	13.5
Route type	Closed
Average gradient (%)	± 1
Maximum ascent (%)	~ 14
Maximum descent (%)	~ 17
Type of pavement	Asphalt concrete
Surface condition	Clean

Table 2 Technical characteristics of the experimental car Hyundai Getz

Parameter	Value
Engine	
Engine type	Petrol
Arrangement	Front, lateral
Number of valves per cylinder	4
Cylinder diameter (mm)	75.5
Piston stroke (mm)	78.1
Compression ratio	10
Toxicity standard	Euro 4
Fuel tank capacity (L)	45
Powertrain	
Transmission	Manual
Wheel drive	Front

Table 3 Conditions of full-scale experiment

Date and time	October 10, 2023, 23:27
Air temperature	+ 1 °C
Atmospheric pressure	745 mm Hg
Wind speed	3.2 m/s, northwest direction
Air humidity	60%
Road condition	Dry asphalt

- accelerator pedal position;
- crankshaft rpm.

The data was recorded on the Xiaomi Mi Mix 2 cell phone via the OBDLink app connected to the OBDLinkLX diagnostic scanner. Due to possible errors in the calculation of fuel consumption by the diagnostic scanner, it is necessary to determine the correction factor.

The correction factor was determined during the separate experimental run. The fuel tank was filled up to 45 L. After that the car traveled 140 km and returned to the same gasoline station. The ratio of fuel consumption from the diagnostic scanner and the specially calibrated for this purposes fuel column was 0.96.

The computer simulation used the throttle position from the real road experiment and the number of activated gear. The following assumptions were used during simulation:

- the reduction of vehicle speed during maneuvering is not taken into account;
- the traction properties of the wheels and the road surface are unchanged;
- weather conditions (temperature, humidity, air pressure, and wind speed) are set to be stable.

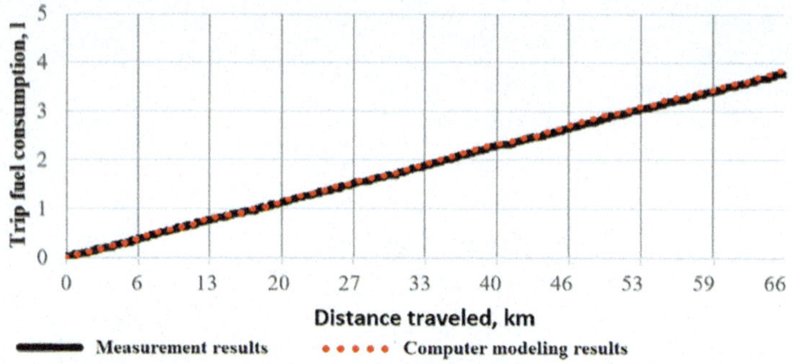

Fig. 4 Comparison of computer simulation and field experiment results

5 Experiment Results

The unknown parameters included in Eq. (1) were obtained by solving the optimization problem by comparing the simulation results with the results of field tests on a road section with a total route length of 66.5 km. The adequacy of the fuel calculation model was performed by comparing the simulation results with the road test results and subsequent statistical analysis of the error functions. Figure 4 shows the comparison of the computer simulation and real road experimental results. The relative error of fuel consumption calculation did not exceed 3%. The 3% difference can be reduced by determining more accurate coefficients for Eq. (1) through additional experimental runs. Additionally, accuracy can be improved by segmenting the road surface of the digital road twin more finely.

6 Conclusion

The main results and conclusions of this study are as follows:

1. A mathematical model for estimating the fuel consumption of a car with an internal combustion engine is proposed.
2. A digital twin of an experimental passenger car Hyundai Getz 1.4 is designed.
3. The virtual tests of the developed algorithms were carried out on a digitized section of a real road, taking into account the height difference, intersections, and traffic light regulation.
4. A full-scale experiments on the developed test road section with measurements of instantaneous fuel consumption and driving conditions were conducted.
5. The correction factor (calibration) of the diagnostic scanner fuel consumption measurement was determined during a separate experimental run on a distance of 140 km.

6. The comparison of the results of fuel consumption simulation and full-scale test on real road was performed, and the relative error of calculations did not exceed 3%.
7. It is proved and established that the proposed mathematical model and the described approach are applicable to the process of development and verification of intelligent speed control algorithms to improve the energy efficiency of modern road vehicles, including highly automated and driverless cars, which is the subject of further research.

Current trends in the development of road transport are reduced to the active development of vehicles with alternative energy sources. Various systems for electric vehicles are being developed or adapted [14]. The next step in this research is to improve the accuracy of the mathematical model to 98% or higher. After that, the proposed model will be adapted to calculate the energy consumption of hybrid cars.

Funding The research was carried out with the financial support of the Ministry of Science and Higher Education of the Russian Federation within the framework of the project "Development of a mathematical model of chassis operation (transmission, chassis, and control mechanisms) in static and dynamic states and creation of a digital twin of a passenger car platform on its basis" (code: FZRR-2023-0007).

References

1. Kulikov, I., Korkin, S., Kozlov, A., Terenchenko, A., Karpukhin, K., Azimov, U.: Component-in-the-loop testing of automotive powertrains featuring all-wheel-drive. Energies **14**, 2017 (2021). https://doi.org/10.3390/en14072017
2. Kurmaev, R.K., Umnitsyn, A.A., Karpukhin, K.E.: Conceptual test rig for real-time analysis of operation of thermostatic system of high-voltage components of electric vehicle traction drive. Int. J. Emerg. Trends Eng. Res. **8**(4), 1208–1211 (2020). https://doi.org/10.30534/ijeter/2020/42842020
3. Vehicle Energy Consumption Calculation Tool (VECTO). Accessed 2024/01/10. https://climate.ec.europa.eu/eu-action/transport/road-transport-reducing-co2-emissions-vehicles/reducing-co2-emissions-heavy-duty-vehicles_en#vehicle-energy-consumption-calculation-tool-vecto
4. Litvinov, A.S., Farobin, Y.A.: Car. Theory of Operational Properties, 240 pp. Moscow (1989)
5. Pantoja Gallegos, J.L., Dominguez Vergara, N., Dominguez Perez, D.N.: Calculation of the fuel consumption, fuel economy and carbon dioxide emissions of a heavy-duty vehicle. J. Phys. Conf. Ser. **2307**(1), 012025 (2022). https://doi.org/10.1088/1742-6596/2307/1/012025
6. Krivoshapov, S.I., Nazarov, A.I., Mysiura, M.I., Marmut, I.A.: Calculation methods for determining of fuel consumption per hour by transport vehicles. IOP Conf. Ser. Mater. Sci. Eng. **977**(1), 012004 (2020). https://doi.org/10.1088/1757-899X/977/1/012004
7. Krivoshapov, S., Gorbik, Y., Kashkanov, V.: Method consumption accounting of fuel consumption for ambulance cars. J. Mech. Eng. Transp. **17**(1), 71–77 (2023). https://doi.org/10.31649/2413-4503-2023-17-1-71-77
8. Yang, X., Zhizhong, L., Min-ye, C., Xiao-min, J.: Research on calculation software of fuel consumption for heavy trucks. In: Third International Conference on Measuring Technology and Mechatronics Automation (2011). https://doi.org/10.1109/ICMTMA.2011.563
9. Pielecha, I., Sidorowicz, M.: The proposal of the calculation algorithm predicting the real vehicle fuel consumption. J. Mech. Transp. Eng. **69**(3), 63–69 (2017). https://doi.org/10.21008/j.2449-920X.2017.69.3.06

10. Koudrin, A.B., Nikiforova, P.A., Sadovskiy, D.A., Shadrin, S.S.: Adaptation of Renault Fluence fuel consumption measurement, taking into account tire pressure. Automob. Road Infrastruct. **1**(27) (2021)
11. Lukanin, V.N., Morozov, K.A., Khachiyan, A.S.: Internal Combustion Engines. Theory of Work Processes, 479 pp. Moscow (2005)
12. Car specs database, automobile-catalog.com. Accessed 2024/01/10
13. Mango, N.: Measurement & calculation of vehicle center of gravity using portable wheel scales. In: SAE World Congress, Detroit, MI, 8–11 Mar 2004
14. Karpukhin, K.: Development of X-in-the-loop testing system for electric vehicles. IOP Conf. Ser. Earth Environ. Sci. **633**, 012008 (2021). https://doi.org/10.1088/1755-1315/633/1/012008

Improve Flutter Stability for Streamlined Sections by Arranging Eccentric Masses

Hong-Son Nguyen, Van-Bao Hoang, Van-Quyen Nguyen, Xuan-Thuan Nguyen, and Ngoc-An Tran

Abstract Airfoils and bridge decks usually have streamlined geometries for efficient aerodynamic analysis. These shapes allow for the use of Theodorsen functions, which offer approximations specifically for thin plates. This research presents a new method to improve the efficiency of streamlined sections by strategically arranging eccentric masses. Unlike traditional use of tuned mass dampers (TMDs), this approach provides a clear benefit by decreasing the need for many springs and viscous dampers. This text elaborates on the development of equations that describe the movement of bridge deck-eccentric mass systems, laying the groundwork for future research. Using the revised step-by-step technique, the research thoroughly analyzes flutter stability and concludes with numerical simulations that highlight the effectiveness of the suggested strategy. This study not only enhances the comprehension of flutter dynamics but also provides a practical approach for enhancing the stability of streamlined structures in real-world engineering applications.

Keywords Flutter stability · Streamlined sections · Eccentric masses · Wind speed · Revised step-by-step

H.-S. Nguyen
Department of Science and Technology, Hanoi University of Industry, No. 298 Cau Dien Street, Bac Tu Liem District, Hanoi, Vietnam

V.-B. Hoang · V.-Q. Nguyen · X.-T. Nguyen
Department of Mechatronics, Hanoi University of Science and Technology, No. 1 Dai Co Viet, Hai Ba Trung, Hanoi, Vietnam

N.-A. Tran (✉)
Faculty of Civil Engineering, Vietnam Maritime University, No. 484 Lach Tray, Haiphong, Vietnam
e-mail: antn.ctt@vimaru.edu.vn

D. H. Tien et al. (eds.), *Proceedings of the 1st International Conference on Sustainability and Emerging Technologies for Smart Manufacturing*, Proceedings in Technology Transfer, https://doi.org/10.1007/978-981-97-7083-0_24

1 Introduction

The cross-sections of airfoils and bridge decks are usually streamlined shapes [4, 6, 7, 14]. The phenomenon of flutter is understood as the coupled bending and torsional vibration of an object under the effect of wind. This phenomenon was initially studied for airfoils, then extended to bridge decks. Flutter stability of airfoils and bridge decks often is studied with the sectional model [4, 7, 11, 13, 15, 17]. The aerodynamic forces acting on the streamlined sections are normally approximated by the Theodorsen functions of thin plate [4, 7, 15, 17].

To improve the critical flutter wind speed for the airfoils, aerodynamic methods are usually used. The airfoils are equipped with flaps to create additional aerodynamic forces [4, 7, 17, 18]. For bridge decks, both aerodynamic methods [1, 2, 16] and mechanical methods (arrangement of TMDs) [5, 8, 9] are used. In some cases, for bridge sections, these two methods are used together to create high efficiency in improving aerodynamic stability [3, 10].

In this paper, the authors present the use of eccentric masses to improve flutter stability for streamlined sections. These masses are arranged inside the section. The advantage of this method compared to using TMDs is that the number of springs and viscous dampers arranged is less. Simulation results show that this method is effective. The optimization method used is the grid search method [12].

2 Research Model

Consider a streamlined section with eccentric masses arranged as shown in Fig. 1. The section has a mass per unit length of m and a mass moment of inertia per unit length of I_a. The heaving stiffness, the coefficient of heaving damper, the torsional stiffness, and the coefficient of torsional damper are denoted by k_w, c_w, k_φ, and c_φ, respectively. The eccentric masses are rigidly connected by a light bar. Each eccentric mas has a weight of m_C. The bar has a joint connection with the section and is equipped with a torsion spring of stiffness k_θ and a torsion damper of coefficient c_θ. The width of the section is B. The distance from the mass to the joint connection is d. The wind speed is U. The self-excited forces are L_{se} and M_{se}.

The displacements of the section consist of two components: vertical displacement w and torsional displacement φ. The torsional displacement of the bar connecting the eccentric masses is θ. Note that φ and θ are chosen as absolute displacements.

Applying the Lagrange method, the equations of motion have the form

$$(m + 2m_C)\ddot{w} + c_w\dot{w} + k_w w = L_{se} \tag{1}$$

$$I_a\ddot{\varphi} + (c_\varphi + c_\theta)\dot{\varphi} - c_\theta\dot{\theta} + (k_\varphi + k_\theta)\varphi - k_\theta\theta = M_{se} \tag{2}$$

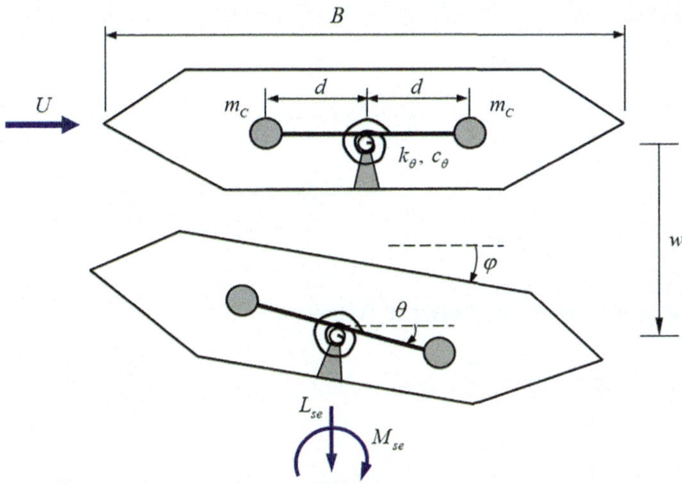

Fig. 1 Sectional model installed with eccentric masses

$$I_b\ddot{\theta} - c_\theta\dot{\varphi} + c_\theta\dot{\theta} - k_\theta\varphi + k_\theta\theta = 0 \tag{3}$$

The mass moment of inertia of the eccentric masses taken with the joint connection is determined

$$I_b = 2m_C d^2 \tag{4}$$

The self-excited forces acting on the section is expressed as the real forms [6, 13]

$$L_{se} = \frac{1}{2}\rho U^2 B\left[KH_1^*(K)\frac{\dot{w}}{U} + KH_2^*(K)\frac{B\dot{\varphi}}{U} + K^2H_3^*(K)\varphi + K^2H_4^*(K)\frac{w}{B}\right] \tag{5}$$

$$M_{se} = \frac{1}{2}\rho U^2 B^2\left[KA_1^*(K)\frac{\dot{w}}{U} + KA_2^*(K)\frac{B\dot{\varphi}}{U} + K^2A_3^*(K)\varphi + K^2A_4^*(K)\frac{w}{B}\right] \tag{6}$$

where K is the reduced frequency and A_i^*, H_i^* are flutter derivatives.

In the case of streamlined section, it is possible to approximate the flutter derivatives according to the Theodorsen functions of thin plate [6, 13]. The expressions of Theodorsen functions are presented in [13, 15, 17].

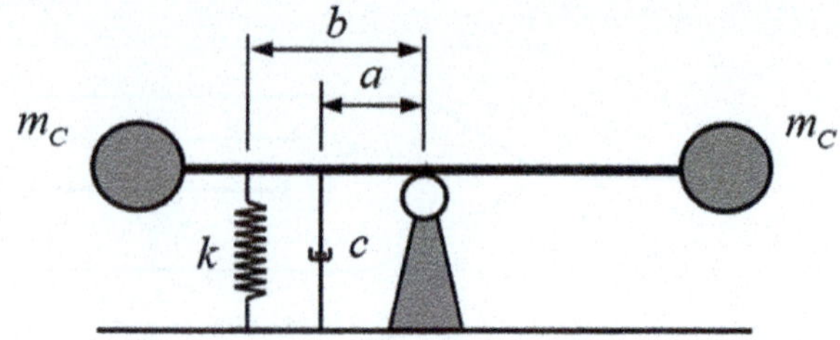

Fig. 2 Arrangement diagram of eccentric masses

3 Arrangement of Eccentric Masses

A vertical spring and a vertical damper of the connecting bar can be arranged as shown in Fig. 2. The distances from this spring and this damper to the joint connection are b and a, respectively.

With the arrangement as shown in the diagram, we have equivalent torsion spring and torsion damper with the coefficients

$$k_\theta = kb; \quad c_\theta = ca \tag{7}$$

where k and c are the stiffness of vertical spring and the coefficient of vertical damper; k_θ and c_θ are the stiffness of torsion spring and the coefficient of equivalent torsion damper.

4 The Revised Step-by-Step Method

The revised step-by-step [11] is very suitable for calculating flutter stability for section models. This method has been developed to calculate flutter stability when the section is installed with TMDs [9] or connected to the wings [1]. The idea of this method is as follows

Step 1: Assume that torsional vibration has the form [11]

$$\varphi = \varphi_0 e^{-\zeta_F \omega_F t} \sin \omega_F t \tag{8}$$

where ζ_F and ω_F are Lehr damping ratio and flutter frequency, respectively.
Step 2: Substitute expression (8) into the equations for heaving vibration of section (1) and torsional vibration of the bar connecting the eccentric masses (3). We find the amplitude ratios and phase differences of the vibrations and then substitute them into the torsional vibration equation of section (2).
Step 3: Express the torsional vibration equation of the section in the standard form, we have two equations with two unknowns: Loga decrement δ_F and flutter frequency ω_F. Thus, for each value of wind speed U, we will determine the corresponding values of δ_F and U_F.

5 Numerical Simulations

Consider the section model (suspension bridge) with the parameters presented in Table 1. Using the revised step-by-step, the critical flutter wind speed and flutter frequency for the bridge section are determined to the 48.8 m/s and the 1.2538 rad/s. The critical flutter wind speed in [7] is 49.38 m/s, and the error is 1.17%.

The variations of flutter frequency and Loga decrement with wind speed are shown in Figs. 3 and 4.

Table 1 Parameters of the bridge section [7]

Parameters	Values	Unit
m	12,879.8	kg/m
I_a	670,056.71	kgm^2/m
B	18.288	m
ω_w	0.880	rad/s
ω_φ	1.552	rad/s
ζ_w	0	
ζ_φ	0	

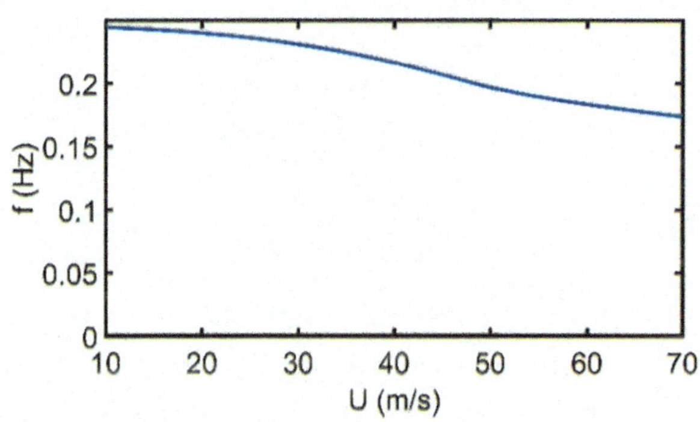

Fig. 3 Flutter frequency–wind speed diagram (without eccentric masses)

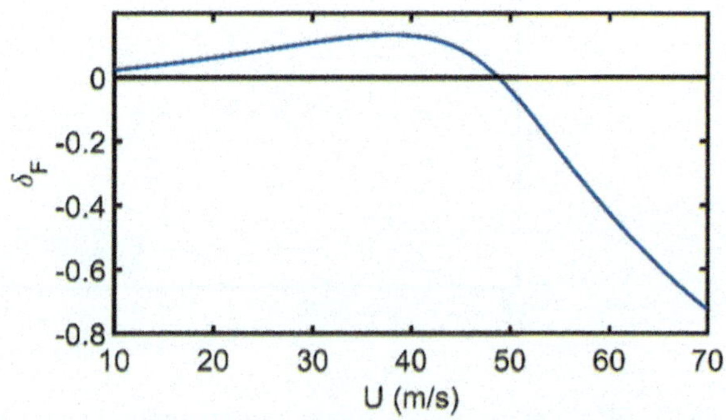

Fig. 4 Loga decrement–wind speed diagram (without eccentric masses)

Consider the case of the section with the eccentric masses, $m_C = 225.4$ kg/m and $d = 8$ m. The mass ratio $2m_C/m = 3.5\%$ and the mass moment of inertia ratio $I_b/I_a = 4.3\%$. Assume that $c_b = 0$.

The frequency ratio is determined

$$\mu_F = \frac{\omega_b}{\omega_F}$$

where ω_F is the flutter frequency of the section without eccentric masses and ω_b is the torsional frequency of the bar connecting eccentric masses.

The stiffness of the torsion spring $k_b = I_b \times \omega_b{}^2$.

To find the optimal stiffness of the torsion spring, the grid search method is used [12]. Consider μ_F in the range [0.9; 1.1] [8]. For each value of μ_F, we can determine a corresponding value of k_b. The critical flutter wind speeds of the section with eccentric masses (denoted by U_{cr}) corresponding to the values of μ_F are presented in Table 2. The maximum critical flutter wind speed is 57 m/s, increased by 16.8% (Fig. 5).

The variations of flutter frequency and Loga decrement with wind speed in the case of μ_F having the optimal value are shown in Figs. 6 and 7.

Table 2 Critical flutter wind speeds when installing eccentric masses

μ_F	0.90	0.91	0.92	0.93	0.94	0.95	0.96
U_{cr} (m/s)	53.2	53.5	53.8	54.2	54.5	55.0	55.4
μ_F	0.97	0.98	0.99	1.0	1.01	1.02	1.03
U_{cr} (m/s)	55.9	56.5	57.0	42.9	43.3	43.7	44.1
μ_F	1.04	1.05	1.06	1.07	1.08	1.09	1.10
U_{cr} (m/s)	44.4	44.7	44.9	45.2	45.4	45.6	45.8

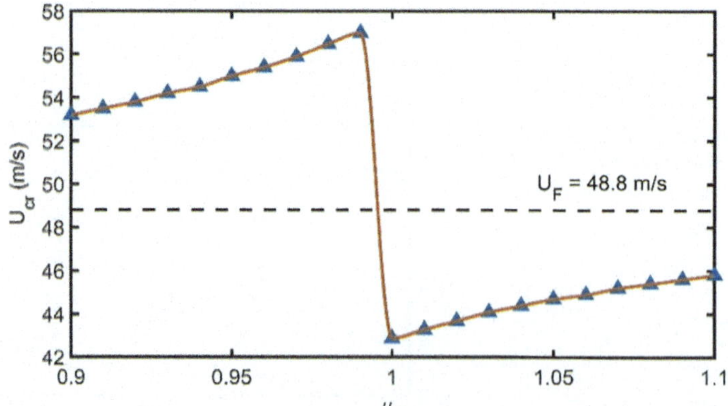

Fig. 5 Effects of the frequency ratio μ_F on the critical flutter wind speed

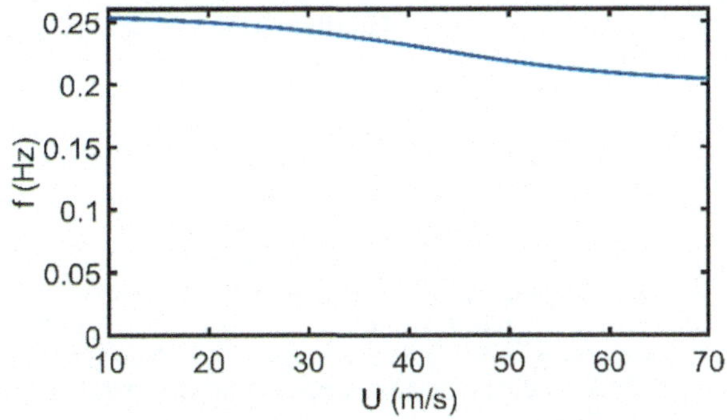

Fig. 6 Flutter frequency–wind speed diagram (with the optimal value of $\mu_F = 0.99$)

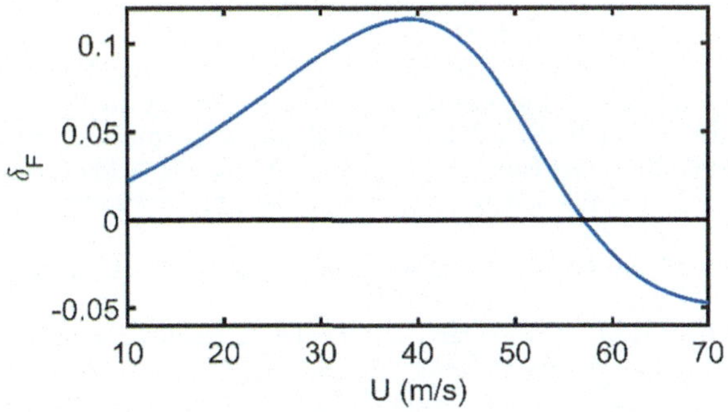

Fig. 7 Loga decrement–wind speed diagram (with the optimal value of $\mu_F = 0.99$)

6 Conclusions

This work suggested a technique for improving flutter stability in streamlined sections by organizing eccentric masses. Numerical simulation results demonstrate the effectiveness of this strategy. Looking forward, our study will focus on performing multi-objective optimization evaluations for bluff body sections using experimentally determined aerodynamic parameter data. Furthermore, the investigation will look at the effect of torsional stiffness and the coefficient of torsion damper on critical flutter wind speed. This comprehensive approach enables a detailed knowledge of the interaction between many elements determining flutter stability, which will guide future research in the topic.

References

1. An, T.N., Khang, L.T.: Increase of critical flutter wind speed of long-span bridges using passive separate control wings. In: Proceedings of the 16th Asia Pacific Vibration Conference, pp. 649–654 (2015)
2. Arco, D.C., Aparicio, Á.C.: Improving suspension bridge wind stability with aerodynamic appendages. J. Struct. Eng. **125**(2), 1367–1375 (1999)

3. Bakis, K.N., Massaro, M., Williams, M.S., Graham, J.M.R.: Passive control of bridge wind-induced instabilities by tuned mass dampers and movable flaps. J. Eng. Mech. **143**(9), 04017078 (2017)
4. Bishlinghoff, R.L., Ashley, H., Halfman, R.L.: Aeroelasticity. Dover Publications, Inc. (1995)
5. Bui, H.-L., Tran, N.-A.: Multi-objective optimal design of TMDs for increasing critical flutter wind speed of bridges. J. Wind Eng. Ind. Aerodyn. **225**, 104992 (2022)
6. Dowell, E.H.: A Modern Course in Aeroelasticity. Kluwer Academic Publisher (2005)
7. Fung, Y.C.: The Theory of Aeroelasticity. Dover, New York (1969)
8. Gu, M., Chang, C.C., Wu, W., Xiang, H.F.: Increase of critical flutter wind speed of long-span bridges using tuned mass dampers. J. Wind Eng. Ind. Aerodyn. **73**, 111–123 (1998)
9. Khang, N.V., Seils, A., An, T.N., Dien, N.P., Nghia, N.T.: An improvement of the step-by-step analysis method for study on passive flutter control of a bridge deck. Arch. Appl. Mech. **86**, 557–573 (2016)
10. Kwon, S.-D., Sungmoon, M.S., Chang, S.-P.: A new passive aerodynamic control method for bridge flutter. J. Wind Eng. Ind. Aerodyn. **86**, 187–202 (2000)
11. Matsumoto, M., Okubo, K., Ito, Y., Matsumiya, H., Kim, G.: The complex branch characteristics of coupled flutter. J. Wind Eng. Ind. Aerodyn. **96**, 1843–1855 (2008)
12. Rao, S.S.: Engineering Optimization—Theory and Practice, 4th edn. Wiley, New Jersey (2009)
13. Simiu, E., Scanlan, R.H.: Wind Effects on Structures, 3rd edn. Wiley (1996)
14. Son, N.H., Ngoc, H.T.B., Phong, D.V., Hung, N.M.: Experiments and numerical calculation to determine aerodynamic characteristics of flows around 3D wings. Vietn. J. Mech. VAST **36**(2), 133–143 (2014)
15. Starossek, U.: Brückendynamik-Winderregte Schwingungen von Seilbrücken. Vieweg, Braunschweig, Wiesbaden (1992)
16. Starossek, U., Ferenczi, T., Priebe, J.: Eccentric-wing flutter stabilizer for bridges—analysis, tests, design, and costs. Eng. Struct. **172**, 1073–1080 (2018)
17. Theodorsen, T.: General Theory of Aerodynamic Instability and the Mechanism of Flutter. NACA Report No. 496, pp. 413–433. Washington (1935)
18. Theodorsen, T., Garrick, I.E.: Nonstationary Flow About a Wing-Aileron-Tab Combination Concluding Aerodynamic Balance. NACA Report No. 736, pp. 129–138. Washington (1942)

Study on Aerodynamic Noise Characteristics of Quadcopter UAV Considering the Influence of Separation Distances Between Rotor Tips and Rotational Speeds

Quang Quyen Le, Thanh Dong Pham, and Quoc Tru Vu

Abstract The interactions between adjacent rotors induce intricate and unsteady airflow in their vicinity, directly influencing both the aerodynamic efficiency and acoustic emissions of the multi-rotor UAVs. This interaction is appreciably influenced by the spatial distance between adjacent rotor tips. In this study, the aerodynamic noise characteristics of APC quadcopter rotors with the different separation distances between neighboring rotor tips and with the different rotational speeds of multi-rotors in hover were investigated. We employed the unsteady panel method and acoustic analogy, solving the FW-H equation using formulation Najafi 1C.

Keywords Aerodynamic noise · Rotor interaction · Overall sound pressure level

1 Introduction

Unmanned aerial vehicles (UAVs) are increasingly developing and becoming useful devices both in military and civil applications. Multi-rotor UAVs have a simple structure, flexibility, maneuverability, and the ability for vertical takeoff and landing. However, the rotor-rotor interactional effect in multi-rotor configurations is more severe than in a single rotor. Interactions between adjacent rotors induce intricate and unsteady airflow in their vicinity, directly influencing both the aerodynamic efficiency and acoustic emissions of the multi-rotor UAVs.

Q. Q. Le (✉) · T. D. Pham · Q. T. Vu
Le Quy Don Technical University, Hanoi, Vietnam
e-mail: le.quang-quyen@lqdtu.edu.vn

D. H. Tien et al. (eds.), *Proceedings of the 1st International Conference on Sustainability and Emerging Technologies for Smart Manufacturing*, Proceedings in Technology Transfer, https://doi.org/10.1007/978-981-97-7083-0_25

Presently, there have been many investigations into multi-rotor UAVs employing experimental methods and numerical simulations. However, the main focus of these studies is still on evaluating UAV's aerodynamic performance. Furthermore, the analysis of rotor interaction and its impact on the aerodynamic noise characteristics of multi-rotor UAVs holds considerable significance. This interaction is appreciably influenced by the spatial distance between adjacent rotor tips [1, 2].

In this paper, the aerodynamic noise characteristics of an advanced precision composites (APC) quadcopter with the different separation distances between neighboring rotor tips and with the different rotational speeds of multi-rotors in hover are investigated.

2 Methodology

2.1 Calculational Method

To ascertain the aerodynamic noise characteristics of rotors, it is imperative to address two principal issues: the aerodynamic problem and the aerodynamic noise problem. In particular, the calculated pressure distribution serves as input parameters for the aerodynamic noise problem [3, 4].

The aerodynamic issue identifies the rotor aerodynamic characteristics and the pressure distributions across blade surfaces. To address this matter, we employ the unsteady panel method to construct a rotor model and compute rotor aerodynamic characteristics.

The aerodynamic noise issue identifies the rotor aerodynamic noise characteristics, utilizing the FW-H equation [5]. Acoustic pressure is defined employing formulation Najafi 1C [6], which is a solution of the FW-H equation.

Utilizing a combination of the unsteady panel method and acoustic analogy rooted in solving the FW-H equation with formulation Najafi 1C, we constructed a computational tool for determining the rotor aerodynamic noise characteristics [3, 4].

2.2 Validation

Validating the obtained results by comparing them with experimental data in published works [7, 8]. The rotor model is adapted from the 1/4 scale UH-1 helicopter rotor. The rotor model included two blades of rectangular shape and used a NACA 0012 airfoil. The rotor blade has a diameter of 3.658 m, a blade chord of 0.1334 m, and a blade pitch angle of 9°. The rotational speed of the rotor is 1296 rpm.

The rotor operates within a chamber at a specific wind velocity of 100 knots. Observers located at microphones 4 and 5 with coordinates (– 3.27, 2.16, – 0.17) m and (– 3.28, – 2.23, – 0.17) m. Following 500 computational steps ($dt = 0.00046$ s), equivalent to five revolutions, the obtained results are presented in Figs. 1, 2 and 3.

Figures 1 and 2 show the vortex wake behind rotors and pressure distributions on rotor blade surfaces in forward flight.

The results depicted in Fig. 3 indicate that the disparity between the average lift coefficient of this computational model and data published in work [8] is small (1.6%).

The results depicted in Figs. 4, 5, 6 and 7 indicate that the fluctuations of the acoustic pressure and the SPL at microphones 4 and 5 from this computational tool are quite similar to data published in works [7, 8].

Fig. 1 Vortex wake behind rotors

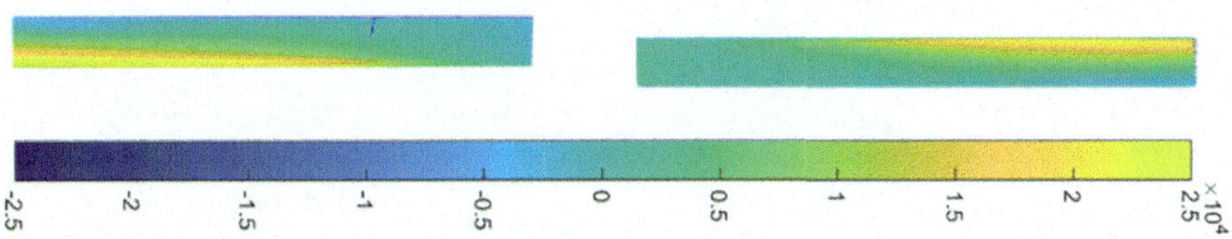

Fig. 2 Pressure distributions on blade surfaces

Fig. 3 Variation of the lift coefficient of rotors

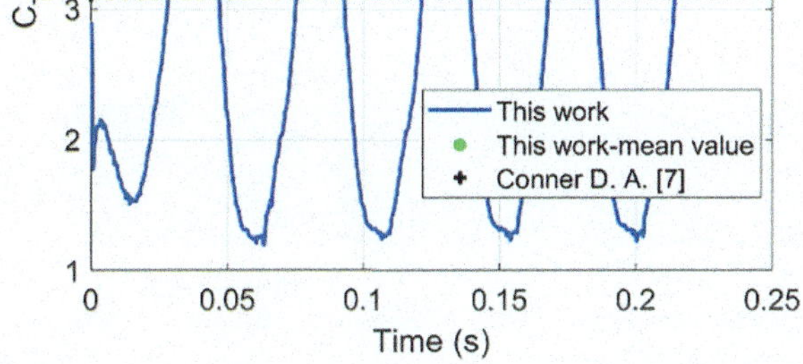

Fig. 4 Temporal fluctuations of the acoustic pressure at mic. 4

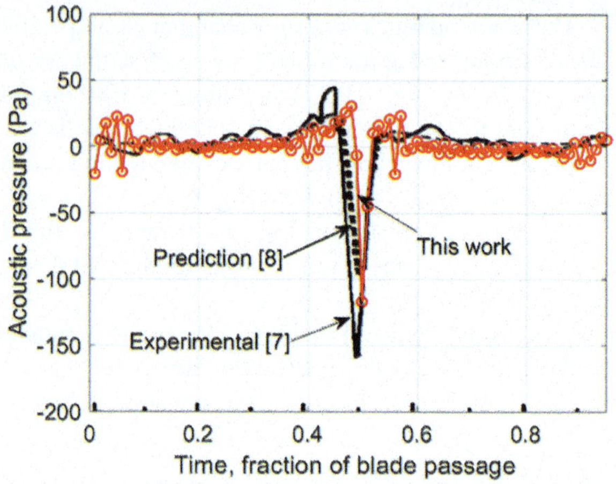

Fig. 5 Temporal fluctuations of the acoustic pressure at mic. 5

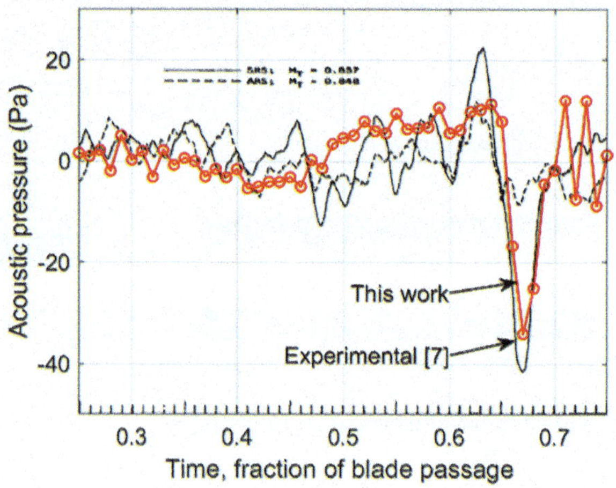

Fig. 6 Sound spectrum at mic. 4

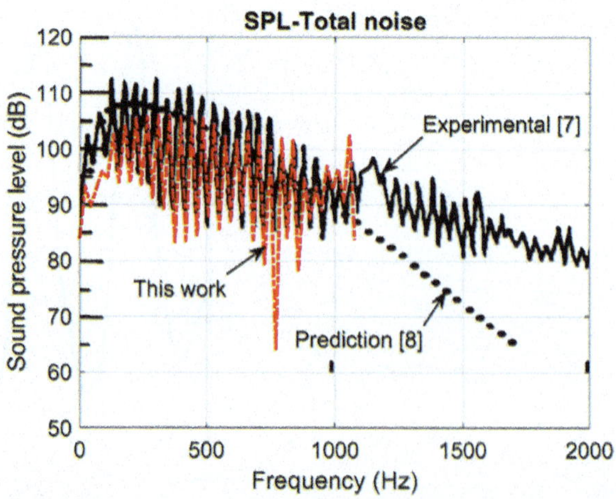

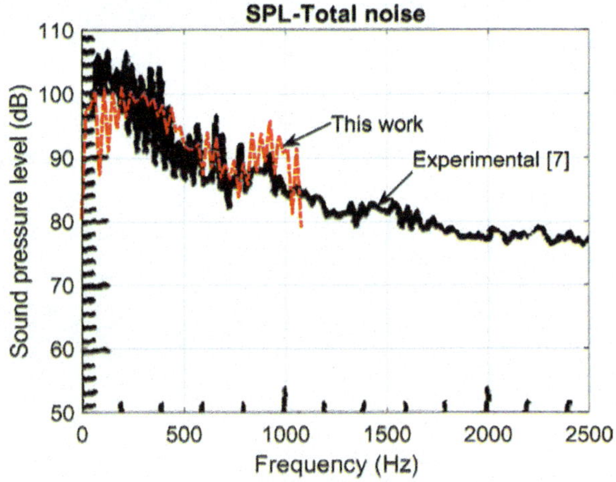

Fig. 7 Sound spectrum at mic. 5

3 Results and Discussion

3.1 Quadcopter Model

The quadcopter model employed in this work resembles an APC 12 × 5.5 MR [1]. Each rotor consists of two blades, with the chord and twist angle along the blade span depicted in Fig. 8.

The APC quadcopter uses a NACA4412 airfoil at the blade root, an E63 airfoil at 20% blade span, and a Clark Y airfoil at the blade tip. The rotor blade has a diameter of 0.1524 m. The adjacent rotors are configured to rotate in opposing directions, thereby balancing the angular moment and maintaining the rotational stability of the quadcopter. When viewed from overhead, rotors 1 and 3 exhibit counterclockwise rotation, while rotors 2 and 4 rotate clockwise (Fig. 9).

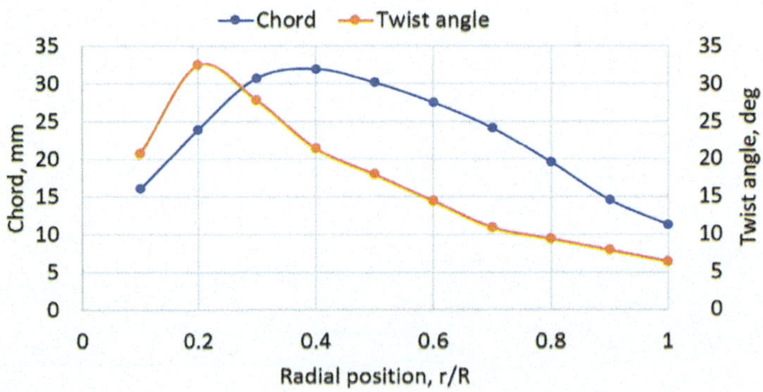

Fig. 8 Chord and twist angle distribution on the APC quadcopter rotor blade

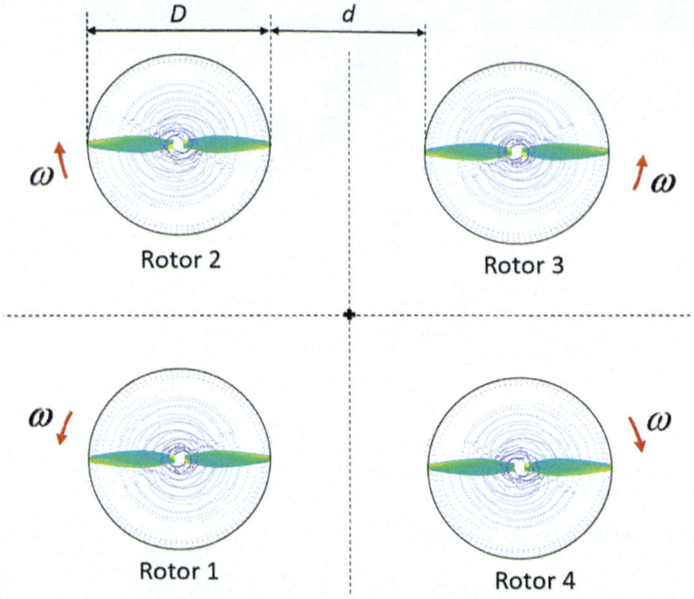

Fig. 9 APC quadcopter configuration

First, the separation distance d has a value $d = 1 \times D$; here, D is the diameter of the rotor. The rotational speeds of rotors ω are 3900, 4500, and 5400 rpm. The obtained results are time variations in the lift values of the APC quadcopter depending on the different rotational speeds (see Fig. 10).

The results in Fig. 10 show that the lift force of multi-rotors is directly proportional to the rotor rotational speed. When the rotational speed increases, the lift force of multi-rotors also increases, and oppositely.

Fig. 10 Temporal variation of the lift force value of the APC quadcopter

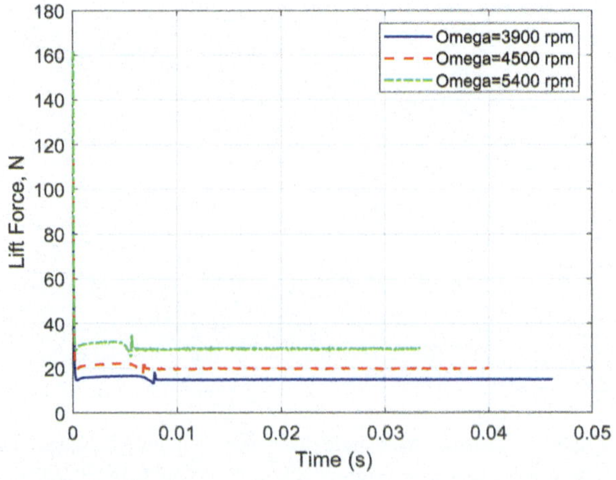

Calculating the aerodynamic noise characteristics of multi-rotor at different observes from micro. 1 ($\beta = 0^o$) to micro. 4 ($\beta = 90^o$) in three cases $\omega =$ 3900; 4500; 5400 rpm. These observers are at a distance of 10*R from the center of the quadcopter (Fig. 11). The obtained results are the variation of sound pressure and sound spectrum diagram at the microphone positions 1 and 4 as given in Figs. 12, 13, 14 and 15.

Upon comparing the aerodynamic noise characteristics of multi-rotors shown in Figs. 12, 13, 14 and 15, it is evident that, at the observers, a rise in rotational speed leads to a corresponding rise in both acoustic pressure and SPL generated by the multi-rotor, and oppositely.

The overall sound pressure level (OASPL) is a single-number value that describes a noise source. The OASPL is the average of all individual noise levels emitted by the noise source. The OASPL is expressed as:

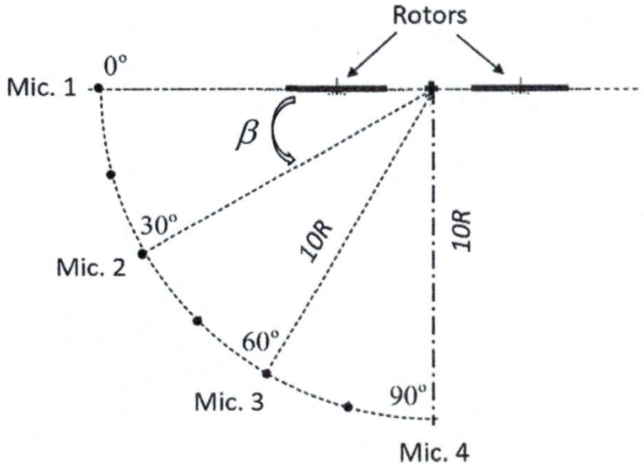

Fig. 11 Microphone positions for measuring aerodynamic noise characteristics

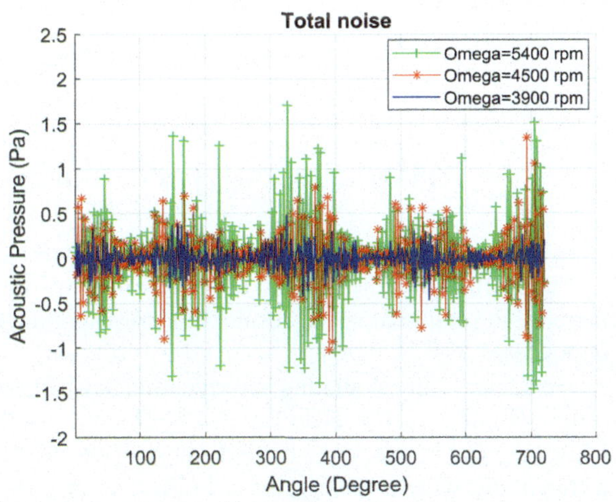

Fig. 12 Variation of the acoustic pressure in 2 periods at mic. 1

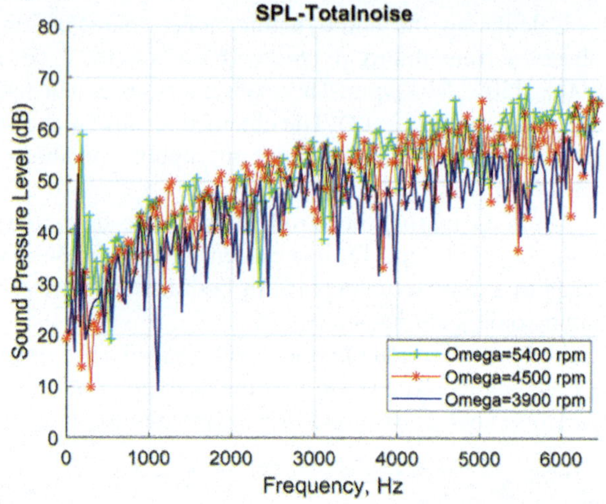

Fig. 13 Sound spectrum diagram at mic. 1

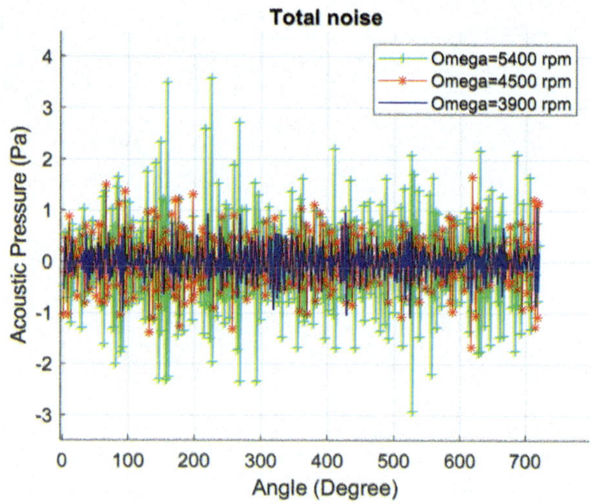

Fig. 14 Variation of the acoustic pressure in 2 periods at mic. 4

$$\mathrm{OASPL} = 10 \times \log_{10}\left[10^{\frac{\mathrm{SPL}_1}{10}} + 10^{\frac{\mathrm{SPL}_2}{10}} + \cdots + 10^{\frac{\mathrm{SPL}_n}{10}}\right] \tag{1}$$

Here SPL_1, SPL_2… SPL_n are individual sound pressure levels in dB.

Figure 16 shows the OASPL at observers below the rotor plane. At observers, an increase in rotational speed results in a corresponding increase in the OASPL, and oppositely.

Next, carry out calculating aerodynamic noise characteristics of APC quadcopter rotors with $d = 0.1 \times D$ and $d = 1 \times D$ at $\omega = 4500$ rpm. Time variation in the lift values and vortex wake behind rotors of the APC quadcopter depending on the different distances d are expressed in Figs. 17 and 18.

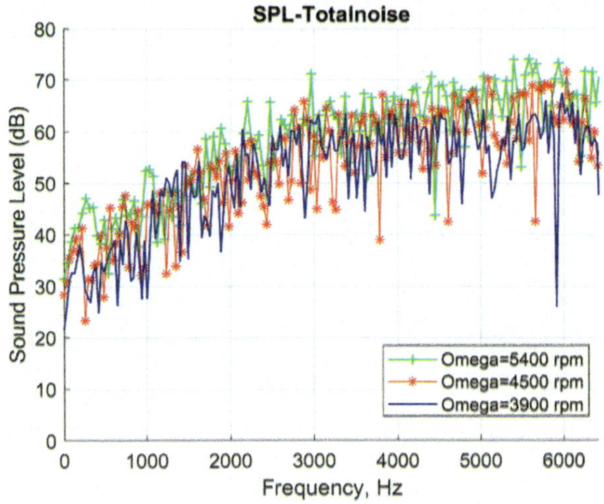

Fig. 15 Sound spectrum diagram at mic. 4

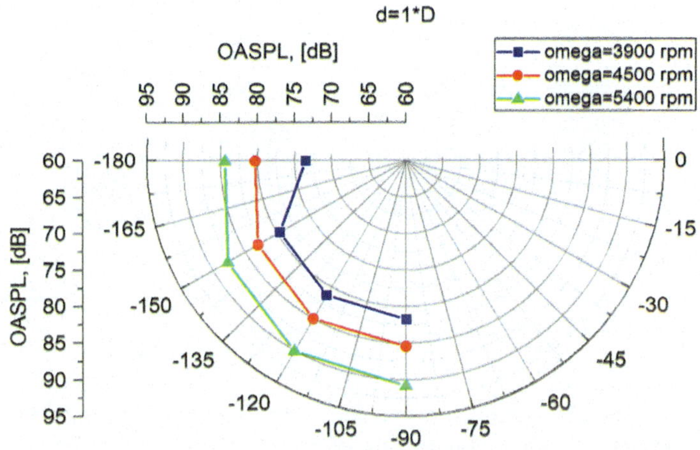

Fig. 16 Comparative analysis of OASPL directionality of the quadcopter according to the rotational speed

The results of comparing the lift force variation in Fig. 17 show that when the spacing between rotor tips was changed from $d = 0.1 \times D$ to $d = 1 \times D$, the lift force of the multi-rotor changed insignificantly. But, when the separation distance for the quadcopter is decreased, the rotor interactional effects become apparent, especially at positions of rotor tips and the normal direction of the rotor plane (Fig. 18). This can affect the aerodynamic noise characteristics of the rotors.

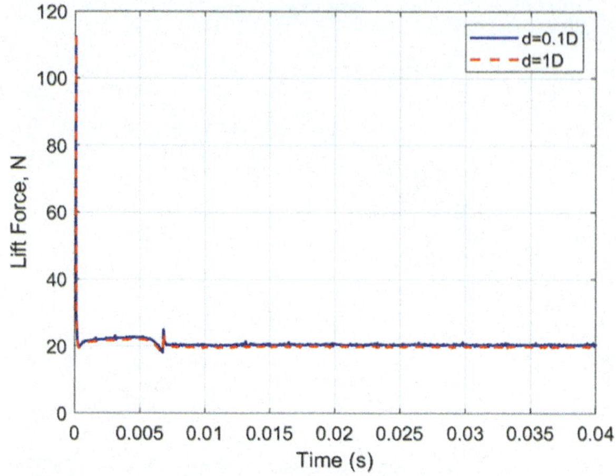

Fig. 17 Temporal variation of the lift force value of the APC quadcopter with $d = 0.1 \times D$ and $d = 1 \times D$ at $\omega = 4500$ rpm

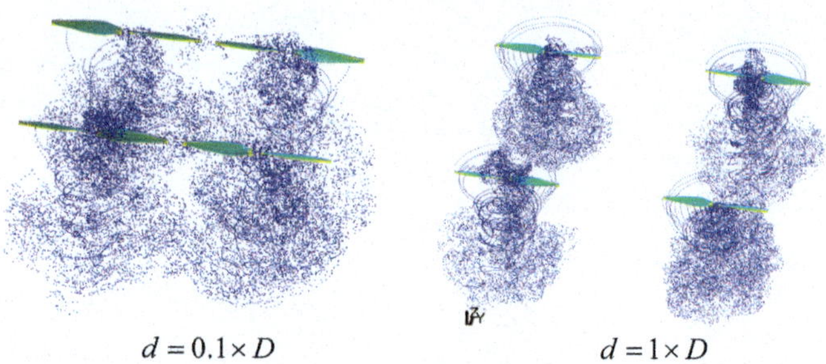

Fig. 18 Vortex wake behind quadrotors $d = 0.1 \times D$ and $d = 1 \times D$

Comparing the aerodynamic noise characteristics of multi-rotors in Figs. 19, 20, 21 and 22, we see that, when changing the distance d from $d = 0.1 \times D$ to $d = 1 \times D$, the acoustic pressure and SPL of multi-rotors at a microphone 1 are insignificant changed. However, the sound signal received at microphone 4 has changed significantly. When $d = 0.1 \times D$, the acoustic pressure and SPL received at the microphone 4 is higher than when $d = 1 \times D$ as a result of the impact of rotor interaction on aerodynamic noise characteristics of multi-rotors.

Figure 23 expresses a comparative analysis of the OASPL directionality of the quadcopter according to the distance between rotor tips. It can be seen that when the rotor spacing is decreased, the OASPL generated by the multi-rotor system increases, and oppositely. At observers below perpendicular to the rotor plane, the OASPL has the most significant variation, whereas, in the plane of the rotors, the change in OASPL is negligible.

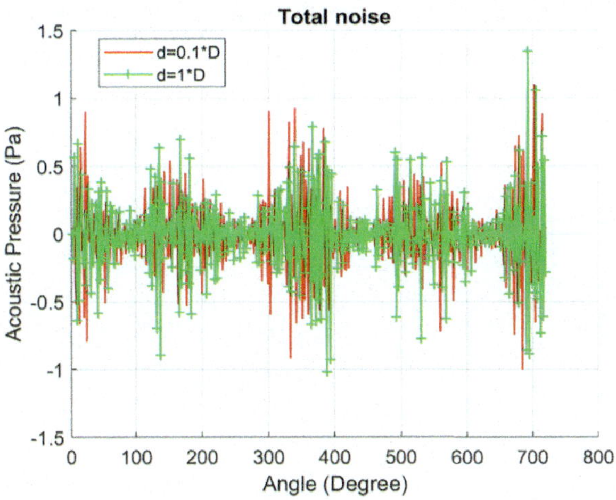

Fig. 19 Variation of the acoustic pressure in 2 periods at mic. 1

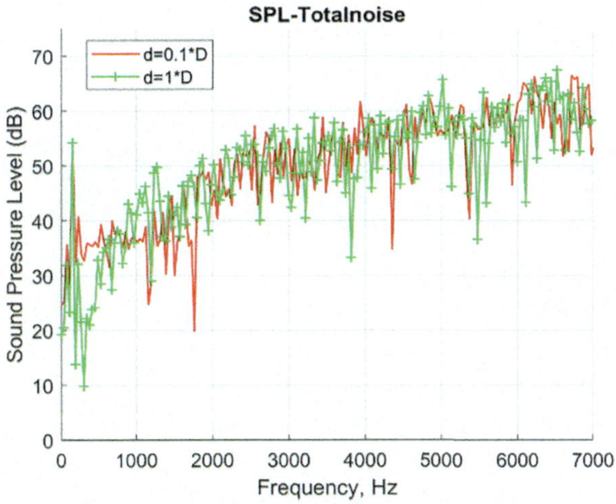

Fig. 20 Sound spectrum diagram at mic. 1

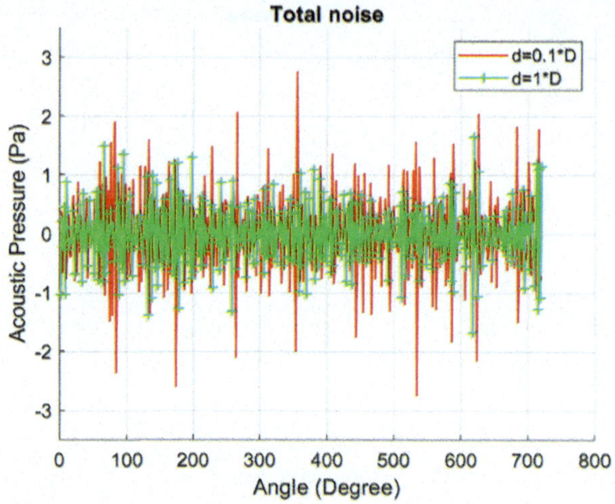

Fig. 21 Variation of the acoustic pressure in 2 periods at mic. 4

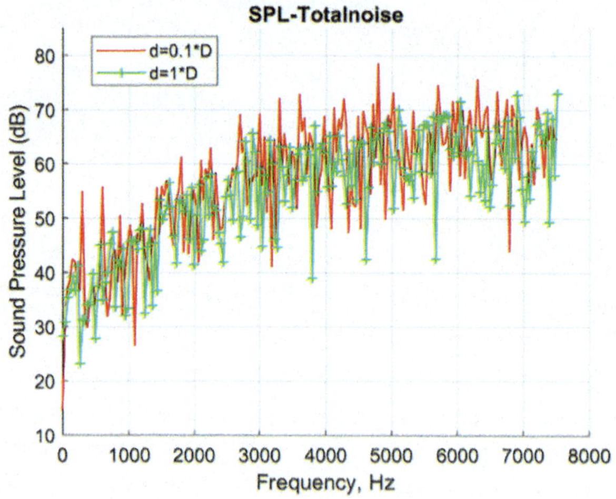

Fig. 22 Sound spectrum diagram at mic. 4

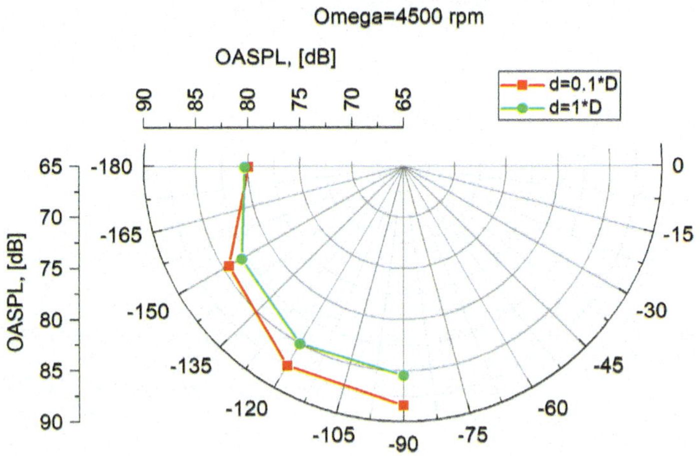

Fig. 23 Comparative analysis of OASPL directionality of quadcopter according to the distance between rotor tips

4 Conclusion

In this paper, the aerodynamic noise characteristics of APC quadcopter rotors with the different separation distances between neighboring rotor tips and with the different rotational speeds of multi-rotors in hover were investigated. Obtained results show that:

- The lift force of multi-rotors is directly proportional to the rotor rotational speed. When the rotational speed increases, the lift force of multi-rotors also increases, and oppositely.
- At observers, a rise in rotational speed leads to a corresponding rise in both acoustic pressure and SPL generated by the multi-rotor, and oppositely. Hence, the OASPL at observers also increases.
- When the spacing between rotor tips was changed, the lift force of the multi-rotor changed insignificantly. However, reducing the distance between adjacent rotor tips resulted in the appearance of rotor interaction effects, particularly pronounced at locations corresponding to rotor tips and the below perpendicular to the rotor plane.
- When the distance between adjacent rotor tips is decreased, the OASPL generated by the multi-rotor system increases, and oppositely. At observers below perpendicular to the rotor plane, the OASPL has the most significant variation, whereas, in the plane of the rotors, the change in OASPL is negligible.

References

1. Misiorowski, M., Gandhi, F., Oberai, A.A.: Computational study on rotor interactional effects for a quadcopter in edgewise flight. AIAA J. **57**(12), 5309–5319 (2019). https://doi.org/10.2514/1.J058369
2. Lee, H., Lee, D.-J.: Rotor interactional effects on aerodynamic and noise characteristics of a small multirotor unmanned aerial vehicle. Phys. Fluids **32**, 047107 (2020)
3. Le, Q.Q., Pham, T.D., Nguyen, A.T., Vu, Q.T.: Building and verifying a tool for calculating the aerodynamic noise of helicopter rotors. J. Sci. Tech. **17**(05) (2022). https://doi.org/10.56651/lqdtu.jst.v17.n05.530
4. Le, Q.Q., Pham, T.D., Vu, Q.T., Nguyen, A.T.: Research on aerodynamic noise characteristics of helicopter rotors considering the influence of blade shapes. J. Sci. Tech. **18**(03) (2023). https://doi.org/10.56651/lqdtu.jst.v18.n03.613
5. Ffowcs, W.J., Hawkings, D.: Sound generation by turbulence and surfaces in arbitrary motion. Math. Phys. Sci. (1934–1990) **264**(1151) (1969)
6. Najafi-Yazdi, A., Brès, G.A., Mongeau, L.: An acoustic analogy formulation for moving sources in uniformly moving media. Proc. R. Soc. A Math. Phys. Eng. Sci. (2011). https://doi.org/10.1098/rspa.2010.0172
7. Conner, D.A., Hoad, D.R.: Reduction of high-speed impulsive noise by Blade planform modification of a model helicopter rotor. NASA Technical Memorandum-84553, AVRADCOM TR 82-B-6 (1982)
8. Brentner, K.S.: Prediction of helicopter rotor discrete frequency noise: a computer program incorporating realistic blade motions and advanced acoustic formulation. NASA Technical Memorandum-87721, NASA Langley Research Center Hampton (1986)

An Approach to the Development of a Methodology for Energy-Efficient Control of Cargo Vehicles

Sergey Shadrin, Daria Makarova, Andrey Keller, and Yury Furletov

Abstract Cargo vehicles play a significant role in the road transport industry and make a major contribution to fuel consumption and air pollution. The authors of the article analyzed and structured the factors affecting fuel consumption in the process of truck cargo transportation. From the point of view of research and optimization, the factors of road parameters and driver influence were taken as priorities, on the basis of which the dependence of fuel consumption was compiled. The procedure of optimizing the speed profile over the predetermined route with respect to specific transport task was described. Tests on public roads were conducted to determine unknown coefficients of the target function and further evaluate the results of optimizing fuel consumption. The authors of the article analyzed various methods used and being developed for energy-efficient vehicle management. It is established that the proposed in the article approach is suitable for analysis and rapid prototyping of intelligent energy-efficient vehicle's control systems.

Keywords Fuel consumption · Energy-efficient control · Cargo vehicles · Digital routs · Simulation

S. Shadrin · D. Makarova (✉) · A. Keller · Y. Furletov
Moscow Polytechnic University, Bolshaya Semyonovskaya Street, 38, 107023 Moscow, Russia
e-mail: makarovadaria224@gmail.com

S. Shadrin
e-mail: s.s.shadrin@mospolytech.ru

A. Keller
e-mail: andreikeller@rambler.ru

Y. Furletov
e-mail: yury.furletov@gmail.com

S. Shadrin · D. Makarova · A. Keller
Sociocenter, Tverskoy Boulevard, 13, Str. 1, 123104 Moscow, Russia

D. H. Tien et al. (eds.), *Proceedings of the 1st International Conference on Sustainability and Emerging Technologies for Smart Manufacturing*, Proceedings in Technology Transfer, https://doi.org/10.1007/978-981-97-7083-0_26

1 Introduction

Energy efficiency and safety traditionally belong to the main directions of development of the automotive industry [1]. The cost of automobile fuel is a major component of operating costs both in automotive enterprises and for private owners of cars and trucks. In addition, the uncontrolled use of non-renewable natural resources has led to their sharp reduction and, as a result, the rise in price and strict control of both the extraction of minerals and the control of environmental damage caused to the planet during extraction, processing, use, and subsequent disposal.

The environmental crisis, along with a dramatic reduction in stocks of refined petroleum products, led to the need for decisive and urgent measures in all affected areas of industry, including cargo transportation [2–4]. Along with the development of alternative energy sources, which is one of the trends in the technological development of the automotive industry in the long term, the most important direction remains the optimization of the consumption characteristics of trucks using existing power systems in gasoline, diesel, and hybrid power plants. According to numerous studies, the use of techniques aimed at saving fuel when driving, as well as choosing a route, allows you to save about a quarter of the fuel used, which makes this direction very promising, both from the point of view of the short-term and from the point of view of the long-term perspective. At the same time, in terms of improving the energy efficiency of cars, researchers are constantly faced with the human factor, which occupies an important share in the issues of inefficient fuel use.

Cargo vehicles are one of the largest sources of greenhouse gas emissions and air pollutants. Today, most vehicles with medium and high load capacity are powered by diesel fuel, which accounts for about 92% of the total energy consumption of heavy-duty vehicles [5–9]. In order to increase fuel efficiency and reduce air pollution, strict standards and regulations have been established for cargo vehicles [10–13]. In addition, recent developments in this field offer additional approaches to improving the energy efficiency of vehicles based on monitoring the dynamic characteristics of the vehicle and special driving techniques. Today, emerging technologies of communication between vehicles and infrastructure further stimulate the development of energy-efficient driving.

2 Factors Affecting Fuel Consumption

In order to identify the algorithm needed to optimize fuel consumption, it is necessary to identify and describe the factors that directly affect fuel consumption. Figure 1 schematically shows the identification of the factors under consideration. Further details about each of them.

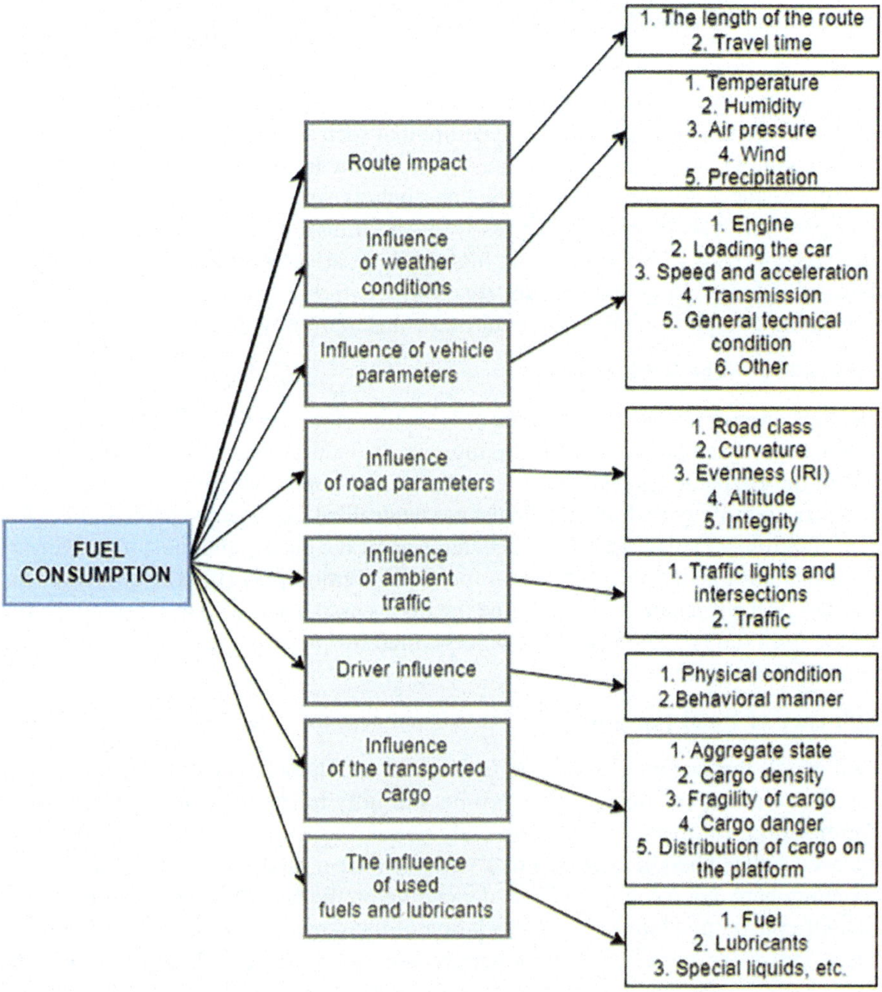

Fig. 1 Structure of factors affecting the fuel consumption of a car

The Impact of the Route

Route influencing factors include distance and travel time on the route. Often, the shortest route by distance may not be the most profitable in terms of total fuel consumption per trip, for example, due to the resulting traffic and low average speed. According to various studies [14], the difference between the minimum and maximum fuel consumption when driving between two destinations can reach up to 41% only depending on the choice of route [15].

Influence of Weather Conditions

Weather factors include temperature, humidity, the presence and speed of wind, air pressure, and the presence of precipitation. These factors affect fuel consumption, on the one hand—through additional equipment such as air conditioning, heater, and cooling system elements, on the other—through the influence on the characteristics of the resistance to movement of the car, such as frontal and lateral aerodynamic drag and removal of pneumatic tires from the transverse wind load. Air pressure can, in turn, be a limiting factor in high-altitude areas with thin air. However, by default, without taking into account zones with extreme environments, the variation of weather conditions affects the final fuel consumption within 1% [16].

Influence of Vehicle Parameters

The most important, however, not the only factors of the car that affect fuel consumption are the characteristics of the engine, weight load, speed, and acceleration. The size of the engine, its power characteristics, the type and quality of fuel used, as well as the availability of additional exhaust gas neutralization systems directly affect fuel consumption [17]. The speed and acceleration of the car are the most intuitive variables that have a significant effect on fuel consumption [18, 19]. The efficiency and features of the transmission determine the losses associated with the transmission of torque from the engine to the drive wheels, thereby contributing an additional factor to fuel consumption.

Influence of Road Parameters

First of all, the factors of road parameters are understood as class, evenness, horizontal curvature, and height of the profile. For hilly roads, there are possibilities for optimization, for example, by adequately gaining speed before climbing and using the accumulated height as energy to accelerate the car on the descent by inertia. Fuel economy here can reach up to 20%, depending on the height of the road profile [20, 21]. Of course, in addition to the gradation of the roadway, its actual condition also has a significant impact with the presence/absence of road defects, repair work, etc., forcing the driver to additional maneuvering and speed reduction.

Influence of the Surrounding Traffic

It is difficult to overestimate how the car receives information about the surrounding traffic, including information about traffic lights at nearby intersections. The availability of such information systems, coupled with computing power, allows us to calculate the most optimal pace of movement in order to preserve the so-called "green corridor" and reduce stops or braking with subsequent accelerations at intersections. Of course, this will also be facilitated by the development of intelligent systems for exchanging information about the intentions and planned maneuvers of surrounding cars. In the future, the development of such intelligent transport systems can potentially provide up to 50% of fuel economy [22].

Driver's Influence

Driver factors are essentially a human factor in a broad sense. First of all, they include his physical condition, that is, how much the driver is focused on the driving process itself. In addition, this includes his aggressiveness when driving and behavioral manners, due, among other things, to accumulated experience both in general and in particular in specific areas. Experienced drivers are able to adjust their driving speed to the rhythm of traffic lights, avoiding unnecessary stops, as well as excessive acceleration or braking. Further development of various assistant and autopilot systems will help to reduce the influence of the driver as a significant factor. An experiment conducted in Japan between different drivers showed that fuel consumption from driver to driver can differ by up to 25% [23, 24].

Influence of the Transported Cargo

The factor of the transported cargo is also a limiting factor due to the requirements for the transportation of certain types of cargo. Some types of cargo are dangerous, and therefore access to certain roads or their sections may be limited, which should be taken into account when laying the route. The aggregate state, as well as the fragility of the cargo, will impose limiting requirements on the speed of the car, as well as accelerations along the entire route. Also, the distribution of cargo on the platform and the position of its center of gravity will have an impact on fuel consumption.

The Influence of Used Fuels and Lubricants

The factor of the fuels and lubricants used and their quality is an important element. Although at first glance, it may seem that this factor has already been taken into account in the influence of the parameters of the car, in fact, the quality of these may vary depending on the place of refueling or the distributor of materials. For example, the desire to save on the fuel used, and refueling with low-quality fuel will increase consumption and reduce the resource of the power unit. The choice of low-quality coolant, engine, and transmission oils will also have a negative impact on the final fuel consumption.

In the current research work aimed at improving the energy efficiency of cargo vehicles based on the synthesis of digital traffic routes and vehicle control algorithms, the factors of road parameters and driver influence are priority from the point of view of research and optimization.

Thus, the main dependence on which the calculation of fuel consumption and optimization of control actions is carried out is the following equation:

$$B_{\mathrm{e}} = \frac{\int b_{\mathrm{e}} \cdot \frac{1}{\eta} \cdot \left[\left(m \cdot f \cdot g \cdot \cos\alpha + \frac{\rho}{2} \cdot c_{\mathrm{W}} \cdot A \cdot \upsilon^2\right) + m \cdot (a + g \cdot \sin\alpha) + B_{\mathrm{r}}\right] \cdot \upsilon \cdot \mathrm{d}t}{\int \upsilon \cdot \mathrm{d}t} \quad (1)$$

where B_{e}—fuel consumption per unit distance traveled, b_{e}—specific fuel consumption, η—is the efficiency of the powertrain, m—is the mass of the car, f—reduced rolling resistance coefficient, g—is the gravitational acceleration, α—longitudinal slope of the profile of the roadway, ρ is the air density, c_{w} is the drag coefficient,

A—frontal area of the car, υ is the speed of the car, a longitudinal acceleration, B_r—resistance braking, and t is time.

This formula takes into account at once many components that affect fuel consumption from the point of view of the car's design: the efficiency of the engine's fuel cycle, losses on transmission of torque through the transmission, tire rolling resistance, aerodynamic drag of the car, reaction to acceleration, resistance to climb, and resistance to deceleration [25]. This dependence was derived when determining the number of emissions during multiple dynamometric tests, for example MNEDC in the EU or FTP75 in the USA. In the course of these studies, exhaust gases are collected into specially designed bags, with the help of which, in addition to emissions of harmful components such as oxides, oxygen dioxides, or nitrogen oxides, fuel consumption is calculated.

3 Digital Routes

One of the most important components of the work of the developed algorithm for energy-efficient vehicle management is to obtain and process data about the location of the car, the road on which it is located, as well as about the upcoming road section along which the route runs.

There are various approaches to creating HD maps, but it all starts with creating a digital terrain model (DTM), which is then added to "information layers" with the corresponding objects (road markings, infrastructure elements, etc.). In turn, geospatial data for DTM can be obtained by satellite imagery, laser scanning (ground and air), radar sensing, and other methods [26].

4 Current Research Results

Based on the analysis of the above fuel consumption formula, the objective function is the following equation:

$$U = b_e \cdot \frac{1}{\eta} \cdot \left[\left(m \cdot f \cdot g \cdot \cos\alpha + \frac{\rho}{2} \cdot c_W \cdot A \cdot \upsilon^2 \right) + m \cdot (a + g \cdot \sin\alpha) + B_r \right] \cdot \upsilon. \tag{2}$$

To determine the unknown coefficients of the objective function and further evaluate the results of optimizing fuel consumption, a full-scale experiment was conducted using a Mercedes–Benz Actros 1845 truck on the Moscow–St. Petersburg route along the M11 highway carrying 15t cargo (S. S. Shadrin). The coefficients of the objective function related to the characteristics of a truck were determined using the numerical least squares method of multicriteria optimization. Further, the

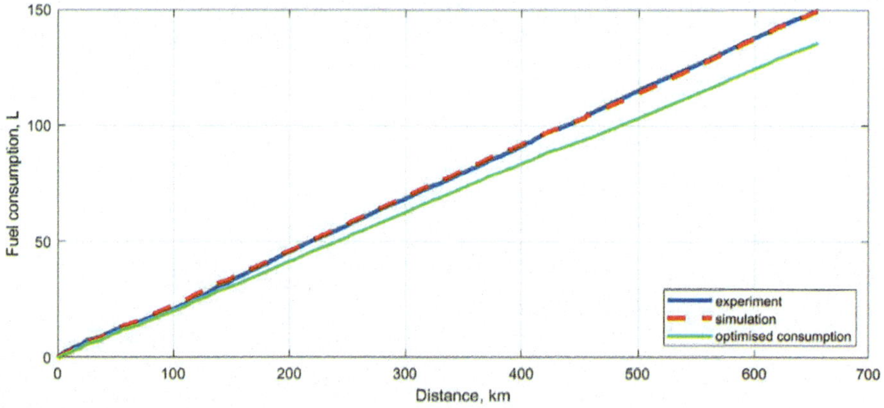

Fig. 2 Comparison of fuel consumption graphs

optimization of speed profiles is carried out by the numerical method of gradient descent using Matlab Simulink tools.

Figure 2 shows the result of comparing a full-scale experiment, simulation results, and predicted fuel consumption when using an optimized speed profile on the route.

In the graph, the optimized speed profile was selected taking into account the task of achieving minimum fuel consumption while driving. However, this profile may differ if there is a different task, for example, delivery to a certain point in time (if there are loading and unloading windows), or if necessary, speed up the total time of the route. As a result, there are several speed profiles for the selected route at the output, between which you can choose the appropriate one (Fig. 3).

5 Discussion

Many pilot projects for energy-efficient driving have been developed on the basis of connected transport systems, an example in this case is the Compass4D project [27] and a project aimed at protecting the environment: "Real-time Information Synthesis Program (AERIS)" [28]. In the urban area, the environmental approach (EAD) shows the greatest effectiveness.

In projects based on connected transport systems, information about planned traffic, as well as information about the distance traveled and the characteristics of the connected vehicle, is used to save energy and reduce emissions in order to determine the most economical high-speed trajectory for the implementation of a dynamic control task. These studies have shown that with correctly calculated trigonometric speed indicators, the driver will drive the vehicle in the most energy-efficient and environmentally friendly way, since this reduces the number of installations and makes it possible to avoid sudden accelerations and decelerations [29].

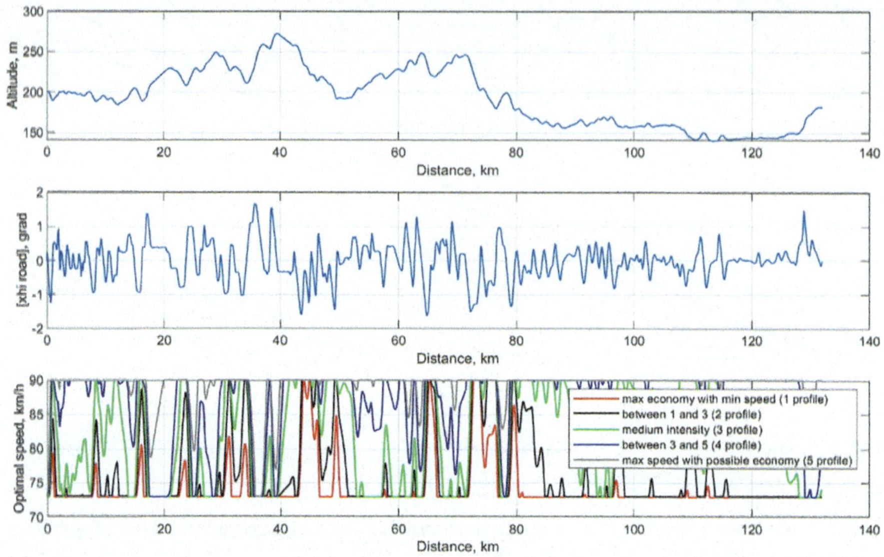

Fig. 3 Selection of optimal speed profiles

The EAD approach was first applied to a scenario with fixed time signals without any other restrictions and tested during simulation and field experiments and showed a 10–20% reduction in fuel consumption [30]. Based on the initial model, many adaptation options have been developed for various conditions of signal transmission and traffic, such as switched-on signals [30, 31], signaled corridor [32], and overloaded transport [33]. Applications have significantly expanded the scope of the EAD approach and have shown significant improvements in fuel efficiency, but trigonometric speed indicators have limited further expansion of the EAD approach, for example, the ability to adjust speed indicators to adapt to a specific vehicle power plant and road terrain.

In these studies, studies [34, 35] have developed energy-efficient and environmentally friendly driving models typical of certain types of powertrains. In this source, the authors developed a strategy to minimize fuel consumption for a gasoline engine with a continuously variable transmission [36]. In this source, the authors developed an integrated model to optimize the dynamics of the vehicle and powertrain for passenger vehicles that move along highways with hilly terrain [37].

6 Conclusion

In this research work aimed at improving the energy efficiency of the movement of cargo vehicles based on the synthesis of digital traffic routes and vehicle control algorithms, the factors affecting fuel consumption were analyzed, described, and structured in details.

The more factors and elements listed above will be taken into account when developing an energy-efficient management methodology, the higher the accuracy of forecasting and optimizing fuel consumption on the existing route will occur. From the point of view of research and optimization, the factors of road parameters and driver influence were taken as priorities, on the basis of which the dependence of fuel consumption was compiled.

The procedure of optimizing the speed profile over the predetermined route with respect to specific transport task (delivery to a certain point in time, maximum economy, fastest time of delivery, etc.) was described.

It is established that the proposed approach is suitable for analysis and rapid prototyping of intelligent energy-efficient vehicle's control systems.

Funding The research was carried out with the financial support of the Ministry of Science and Higher Education of the Russian Federation within the framework of the project "Development of a mathematical model of chassis operation (transmission, chassis and control mechanisms) in static and dynamic states and creation of a digital twin of a passenger car platform on its basis" (code: FZRR-2023-0007).

References

1. Saykin, A.M., Bakhmutov, S.V., Terenchenko, A.S., Endachev, D.V., Karpukhin, K.E., Zarubkin, V.V.: Tendency of creation of "driverless" vehicles abroad. Biosci. Biotechnol. Res. Asia **11** (2014)
2. Nonami, K., et al.: Autonomous control systems and vehicles. In: Intelligent Unmanned Systems. Springer Japan (2013)
3. Cheng, H.: Autonomous intelligent vehicles. In: Theory, Algorithms, and Implementation. Springer (2011)
4. Salvi, B.L., Subramanian, K.A.: Sustainable development of road transportation sector using hydrogen energy system. Renew. Sustain. Energy Rev. **51**, 1132–1155 (2015)
5. Daw, C., Gao, Z., Smith, D., LaClair, T., et al.: Simulated fuel economy and emissions performance during city and interstate driving for a heavy-duty hybrid truck. SAE Int. J. Commer. Veh. **6**(1), 161–182 (2013). https://doi.org/10.4271/2013-01-1033
6. California Air Resources Board (CARB): On-Road Heavy-Duty Vehicle Program, July 2017. Accessed 31 Mar 2020. https://www.arb.ca.gov/msprog/onroadhd/onroadhd.htm
7. Frey, H.C., Zhang, K., Rouphail, N.M.: Fuel use and emissions comparisons for alternative routes, time of day, road grade, and vehicles based on in-use measurements. Environ. Sci. Technol. **42**(7), 2483–2489 (2008)
8. U.S. Energy Information Administration (EIA): Annual Energy Outlook 2021. Accessed 26 Mar 2021. https://www.eia.gov/outlooks/aeo/pdf/05%20AEO2021%20Transportation.pdf
9. U.S. Environmental Protection Agency: https://www.fueleconomy.gov/feg/atv.shtml

10. U.S. Energy Information Agency: Greenhouse Gas Emissions and Fuel Efficiency Standards for Medium- and Heavy-Duty Engines and Vehicles—Phase 2, Oct 2016. Accessed 31 Mar 2020. https://www.gpo.gov/fdsys/pkg/FR-2016-10-25/pdf/2016-21203.pdf
11. Ben, C.M., Shmerling, E., Kuperman, A.: Analytic modeling of vehicle fuel consumption. Energies **6**(1), 117–127 (2013)
12. An, F., Earley, R., Green-Weiskel, L.: Global overview on fuel efficiency and motor vehicle emission standards: policy options and perspective for international
13. Zhou, X., Huang, J., Lv, W., Li, D.: Fuel consumption estimates based on driving pattern recognition. In: Green Computing and Communications (GreenCom), 2013 IEEE and Internet of Things (iThings/CPSCom), IEEE International Conference on and IEEE Cyber, Physical and Social Computing, pp. 496–503. IEEE (2013)
14. Park, S., Rakha, H., Farzaneh, M., Zietsman, J., Lee, D.W.: Development of fuel and emission models for high speed heavy duty trucks, light duty trucks, and light duty vehicles. In: 2010 13th International IEEE Conference on Intelligent Transportation Systems (ITSC), pp. 25–32. IEEE (2010)
15. Wang, J., Yu, Q., Li, S., Duan, N., Li, K.: Eco speed optimization based on real-time information of road gradient. J. Automot. Safe. Energy **5**(3), 257–262 (2014)
16. Carrese, S., Gemma, A., La Spada, S.: Impacts of driving behaviours, slope and vehicle load factor on bus fuel consumption and emissions: a real case study in the city of Rome. Procedia-Soc. Behav. Sci. **87**, 211–221 (2013)
17. Asadi, B., Vahidi, A.: Predictive cruise control: utilizing upcoming traffic signal information for improving fuel economy and reducing trip time. IEEE Trans. Control Syst. Technol. **19**(3), 707–714 (2011)
18. Taniguchi, M.: Eco-driving and fuel economy of passenger cars. In: Proceedings of Annual Meeting of IEE Japan 2008, p. S21 (5–8) (2008)
19. Van Mierlo, J., Maggetto, G., Van de Burgwal, E., Gense, R.: Driving style and traffic measures-influence on vehicle emissions and fuel consumption. Proc. Inst. Mech. Eng. Part D J. Automob. Eng. **218**(1), 43–50 (2004)
20. Robert Bosch GmbH: Bosch Automotive Handbook, 10th edn., 1280 pp. (2018)
21. GPS: Global Positioning System (or Navstar Global Positioning System): Wide Area Augmentation System (WAAS) Performance Standard, Section B.3, Abbreviations and Acronyms. https://web.archive.org/web/20170427033332/http://www.gps.gov/technical/ps/2008-WAAS-performance-standard.pdf
22. Shadrin, S.S.: Energy-efficient management of autonomous wheeled vehicles based on the analysis of high-precision geoinformation environment data. In: Technologies and Components of Intelligent Transport Systems, pp. 100–112 (2018)
23. KML—Keyhole Markup Language. Wikipedia. https://en.wikipedia.org/wiki/Keyhole_Markup_Language
24. Shadrin, S.S.: Geoinformation support for autonomous movement of ground transport. Proc. R.E. Alekseev Natl. Techn. Univ. **3**(122), 172–177 (2018)
25. Robert Bosch GmbH. Bosch Automotive Electrics and Automotive Electronics: Systems and Components, Networking and Hybrid Drive, 5th edn., 530 pp. Springer Vieweg (2014)
26. Digital Elevation Model (DEM) Format. Sustainability of Digital Formats: Planning for Library of Congress Collections. https://www.loc.gov/preservation/digital/formats/fdd/fdd000285.shtml
27. International Road Transport Union (IRU): Compass4D, improving road safety and efficiency by sharing information wirelessly. Accessed 26 Mar 2021. https://www.iru.org/innovation/iru-projects/compass4d
28. U.S. Department of Transportation: Applications for the Environment: Real-Time Information Synthesis (AERIS), June 2017. http://www.its.dot.gov/aeris/
29. Barth, M.J., Mandava, S., Boriboonsomsin, K., Xia, H.: Dynamic eco-driving for arterial corridors. In: 2011 IEEE Forum on Integrated and Sustainable Transportation System (FISTS), Vienna, Austria, June 2011, pp. 182–188

30. Hao, P., Wu, G., Boriboonsomsin, K., Barth, M.: Developing a framework of eco-approach and departure application for actuated signal control. In: IEEE on Intelligent Vehicles Symposium, Seoul, Korea, June 2015
31. Hao, P., Wu, G., Boriboonsomsin, K., Barth, M.: Eco-approach and departure (EAD) application for actuated signals in real-world traffic. IEEE Trans. Intell. Transp. Syst. **20**(1), 30–40 (2017)
32. Xia, H., Boriboonsomsin, K., Barth, M.: Dynamic eco-driving for signalized arterial corridors and its indirect network-wide energy/emissions benefits. J. Intell. Transp. Syst. **17**(1), 31–41 (2013)
33. Ye, F., Hao, P., Qi, X., Wu, G., et al.: Prediction-based eco-approach and departure at signalized intersections with speed forecasting on preceding vehicles. IEEE Trans. Intell. Transp. Syst. **20**(4), 1378–1389 (2019)
34. Li, S., Peng, H.: Strategies to minimize the fuel consumption of passenger cars during car-following scenarios. Proc. Inst. Mech. Eng. Part D J. Automob. Eng. **226**(3), 419–429 (2012)
35. Saykin, A.M., Buznikov, S.E., Endachev, D.V., Karpukhin, K.E., Terenchenko, A.S.: Development of Russian driverless electric vehicle. Int. J. Mech. Eng. Technol. **8**(12), 955–965 (2017)
36. Tuan, N.Kh., Karpukhin, K.E., Terenchenko, A.S., Kolbasov, A.F.: World trends in the development of vehicles with alternative energy sources. ARPN J. Eng. Appl. Sci. **13**(7), 2535–2542 (2018)
37. Hu, J., Shao, Y., Sun, Z., Bared, J.: Integrated vehicle and powertrain optimization for passenger vehicles with vehicle-infrastructure communication. Transp. Res. Part C **79**, 85–102 (2017)

The Effect of Cutting Tool Tilt Angle on Surface Quality in Machining on a Four-Axis Turn-Mill Machine

Hai-Chung Luong, Hoanh-Son Truong, Kien-Trung Nguyen, Van-Huong Tran, Thai-Son Le, and Chuong-Dao Nguyen

Abstract In the process of machining 3D surfaces, the cutting tool (tool) often is an end-ball tool, so the position of the contact point (cutting point) between the edge of the cutting tool and machined surface is always altered resulting in the change of the surface finish. With the machining process performed on a CNC machine with more controlled axes, the wide diversity machining methods will be applied. The problem related to difficult-to-reach cutting positions with thin walls could be controlled. Simultaneously, it is possible to create multiple tool approaches to optimize the machining process and heat dissipation in the machining area. In this paper, the variation in surface roughness and microstructure of the machining surface is studied by changing of contact point of end-ball tool (applying different tilt angles for the tool axis). The experimental cutting process was carried out with C45 steel on a four-axis turn-mill machine (EmcoTurn E65), and the CNC machining program was created on NX Cam software.

Keywords Surface roughness · Eccentricity · Tilt angle · CNC milling · Mill-turn machine · NX

1 Introduction

In the field of metal cutting, especially mold machining, the machining surfaces are mostly curved and complex that requires high capability of software to program for CNC machines. The four-axis turn-mill machine can combine turning and milling to create two simultaneous movements: the rotation of the chuck (rotating workpiece) and three-axis movement of the tool (rotating and translational tool). Therefore, this type of CNC machine can create complex tool movements and simultaneously

H.-C. Luong (✉) · T.-S. Le · C.-D. Nguyen
Mechanical Engineering Department, Vinh University of Technology Education, Vinh, Vietnam

H.-S. Truong · K.-T. Nguyen · V.-H. Tran
School of Mechanical Engineering, Hanoi University of Science and Technology, Hanoi, Vietnam
e-mail: son.truonghoanh@hust.edu.vn

D. H. Tien et al. (eds.), *Proceedings of the 1st International Conference on Sustainability and Emerging Technologies for Smart Manufacturing*, Proceedings in Technology Transfer, https://doi.org/10.1007/978-981-97-7083-0_27

perform few machining strategy such as milling, turning, drilling, and boring with single clamping setup of jig and fixture.

In 1990, the pioneering research projects on turn-mill machines were conducted by Schulz et al. [1], and his research classifies machining methods into two types: coaxial turn-mill and orthogonal turn-mill. The coaxial turn-mill can create the inner and outer surfaces while the orthogonal turn-mill can only mill the outer surface. By cutting experiments on a four-axis turn-mill machine, the milling plain bearing half liners for motors at high speed has demonstrated good chip removal, high surface quality, and low cutting forces. The measurement data shown that the surface roughness will increase as the feed rate (F) is increased. Uysal et al. [2] by machining experiments on AISI 1050 steel performed on a four-axis turn-mill machine according to the orthogonal method with 32 mm Seco Micro-Turbo 217.69-03 milling tool, have shown the relationship of feed rate and tool/workpiece rotation speed ratio (r_n) to the surface roughness. The results are comparative in considering the influence of workpiece radius and tool radius. In addition, with two different types of cutting tools (wiper inserts and conventional inserts), Uysal determined the influence of tool eccentricity on the surface roughness corresponding to different r_n values (100, 150, and 200). They also showed that the influence of machining distance is proportional to tool wear for machining AISI 1050 steel with 50 mm Seco Duratomic milling tool. Savas et al. [3] studied the surface roughness with milling experiments on a turn-mill machine using an end-mill tool and SAE 1050 steel workpiece. By sequentially changing in input parameters such as depth of cut, axial longitudinal feed rate, and workpiece revolutions, they established the relationship of these parameters to the surface roughness.

Kopac and Pogacnik [4] analyzed the influence of tool eccentricity on the surface quality on turn-mill machines. They showed that the roughness value with eccentrically machining will be much better than those without eccentricity on a turn-mill machine. Yuan and Zheng [5] focused on developing geometric models for machining processes on turn-mill machines to effectively predict the surface roughness. By proposed algorithms, they could simulate the influence of machining parameters such as tool radius, workpiece radius, tool rotations, workpiece rotations, and eccentric distance on the surface roughness. Huang and Cai [6] studied the cutting area model of each point on the cutting edge to determine the effects of different machining parameters on tool wear and propose a model to balance tool wear on cutting length. They conducted machining experiments to prove the validation of the simulation model and optimization process. Ozay and Savas [7] studied the optimization of cutting parameters to achieve the best surface roughness results in the tangential cutting on turn-mill machines. They applied the Taguchi method to experimental design and results to build a model that influence model of tool rotation speed, rotation speed of the workpiece, depth of cut, feed rate, and helix angle of the cutting tool on the surface roughness. Boozarpoor et al. [8] comprehensively studied the influence of machining parameters on surface quality in machining process of Inconel 718 alloy by the orthogonal method on a turn-mill machine. The machining experiments is produced with fixed parameters (tool rotation speed of 1000 rpm, feed rate of 0.12 mm/rev, workpiece rotation speed of 300 rpm, and eccentricity of

0.2 mm), and the results showed that feed rate was the most influential and decisive factor on the surface roughness and residual stress. Sun et al. [9] built a mathematical model of cutting layers and cutting forces by the orthogonal cutting method on a turn-mill machine. By experimental machining, based on the geometric shape of the cutting layers, they analyzed and built a corresponding mathematical model of the cutting force. In addition, they determined the trend of tool wear at various locations on the cutting edge corresponding to different shapes of the cutting layer. Mohamad and Zahid [10] built a program to improve programming and simulation of the fourth axis machining process in the NX Cam system. The results showed that the developed programs are capable of optimizing the fourth axis machining simulation process by minimizing the machining programming steps and time. Lazoglu et al. [11] studied the method of optimizing toolpaths in machining free-form surfaces. This research shown a new approach to generate optimized toolpaths for free-form surfaces commonly used in automotive, aerospace, biomedical, mold, and household appliance manufacturing.

The above studies on eccentricity machining in turn-mill machines often focuses on eccentricity research in the *Y*-axis direction with a determined *Y* distances. The contact point of the tool cutting edge and the workpiece is altered, leading to a variation in cutting velocity (V_c) and the surface roughness. In the case of workpiece with a large diameter or 3D curved surface with a small surface inclination, the changing in the eccentric distance along the *Y*-axis does not make significant variation of the position of the contact point, and the effect on the surface roughness is negligible. This is clearly shown in Fig. 1b, with the same eccentricity (*e*) distance, the contact point position will be varied with different diameters of the workpieces. In addition, the machining process of 3D parts with complex surfaces, especially in the field of mold machining, is always carried out on 3-, 4-, or 5-axis machines and use CAM software to support machining programming, so the determination of the surface quality according to eccentricity distance will be significantly difficult.

In this article, the study will focus on the influence on the surface roughness in turn-mill machining with different tool tilt angles (other factors are considered fixed).

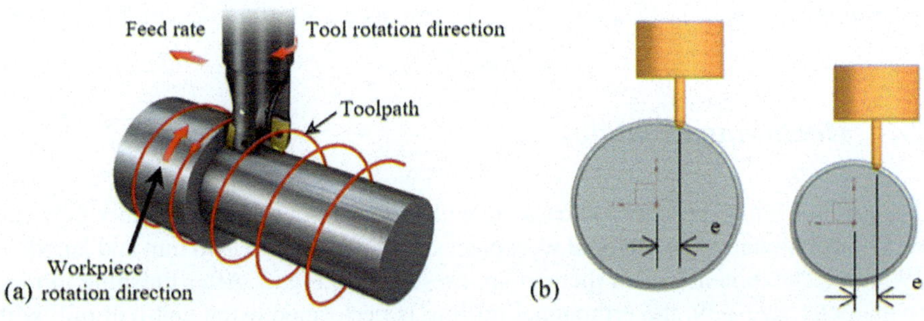

Fig. 1 **a** Movements and toolpaths on the turn-mill machine; **b** eccentricity distance (*e*) for workpieces with different diameter sizes

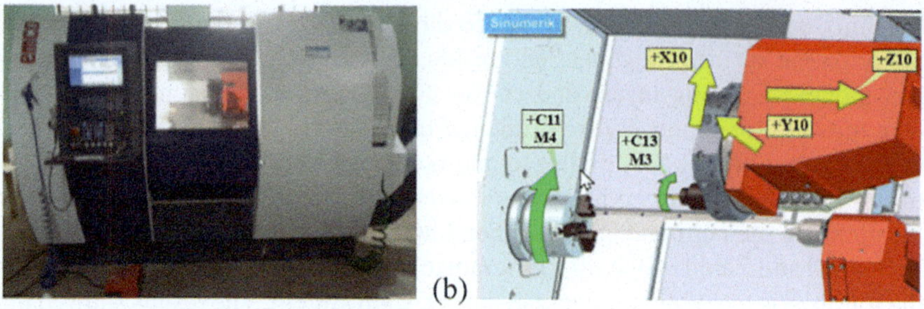

(a) (b)

Fig. 2 **a** Image of EmcoTurn E65 machine; **b** main movements of the machine

Table 1 Technical data of EmcoTurn E65 machine

Total dimensions ($L \times W \times H$) (mm)	3480 × 2080 × 2060	Tailstock travel	500 mm
$X/Y/Z$ travel (mm)	260/± 40/610	C-axis resolution	0.001°
Maximum turning diameter	Ø 500 mm	C-axis motion speed	1000 rpm
Maximum part length	520 mm	Number of tool	12
Main spindle speed range	0–5000 (3500) rpm	Maximum drive power	22 kW
Counter spindle speed range	0–5000 rpm	Total weight (kg)	Approx. 6500

2 Experimental Process Setup

2.1 CNC Machine

The machining tests were conducted on EmcoTurn E65 machine manufactured in Austria (Fig. 2a), and the cutting tool can move in X, Y, Z axes. The machine can perform combined milling and turning (orthogonal milling or coaxial milling). In turning, the chuck rotates according to value $C11$ while the milling tool rotates according to value $C13$ (Fig. 2b). Table 1 presents main parameters of the machine.

2.2 Machining Condition

The part machining process is carried out with the orthogonal milling method (Fig. 1a) on C45 rotationally symmetrical workpieces with diameter of $\phi 40$ mm and length of 190 mm. The dimensions of the part are presented in Fig. 3. After finishing turning to diameter size $\phi 39$, the orthogonal milling is performed using end-ball mill with diameter of $\phi 8$ (Table 2), spindle speed of 3000 rpm, feed rate of 0.05 mm/rev, and depth of cut of 0.15 mm. During the machining process, the cutting tool is set at different tilt angles of 0°, 5°, … to 90°.

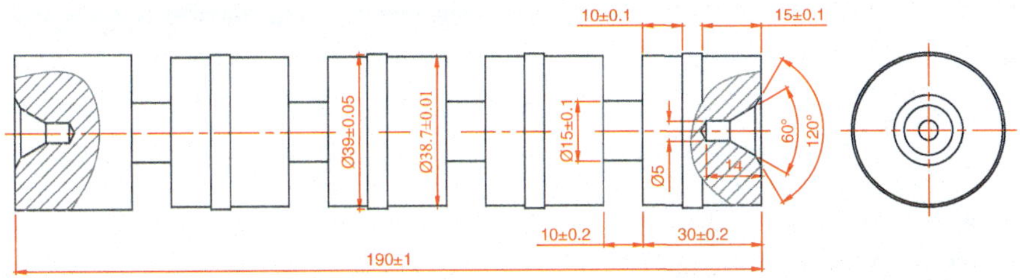

Fig. 3 Dimension of machining part

Table 2 Alloy end-ball milling tool vertex: ball nose end mill (56 HRC)

Type	Diameter (D)	Radius (R) ($\pm$ 0.01)	Length (L)	Cutting length (L_c)	Shank diameter (D_h)
VB263-8	8 (mm)	4 (mm)	100 (mm)	14 (mm)	8 (mm)

2.3 *Setup for the Toolpaths and Machining Tests*

The machining program is performed on NX Cam software. The tool tilt angle is fixed during the machining process (which means the Y coordinate value will not change), the variable contour machining method is most suitable. The cutting speed F is formed from two movements: the tool moves along the Z-axis (F_d) and the workpiece rotates at angle $C13$ (F_t) (Fig. 4a), creating a spiral tool path.

$$\vec{F} = \overrightarrow{F_d} + \overrightarrow{F_t} \quad (1)$$

To create a tilt angle for the tool axis, enter the tilt angle value in the rotation angle box (Fig. 4b), and the contact point will be shifted when the tilt angle changes (Fig. 4c for tilt angle of 40°, Fig. 4d for tilt angle of 80°). Figure 5a shows the position of the experimental machining surfaces according to different tilt angles of the tool, while Fig. 5b presents the clamping process of the workpiece for machining tests.

2.4 *Measurement of Machined Surface Quality*

After completing all machining tests, the roughness machined surfaces is measured with Mitutoyo SJ.201 surface roughness meter (Fig. 6a) and machined surface quality is checked on 2D Arcs Kim450U meter (Fig. 6b). Each machining surface is inspected at three different locations (Table 3).

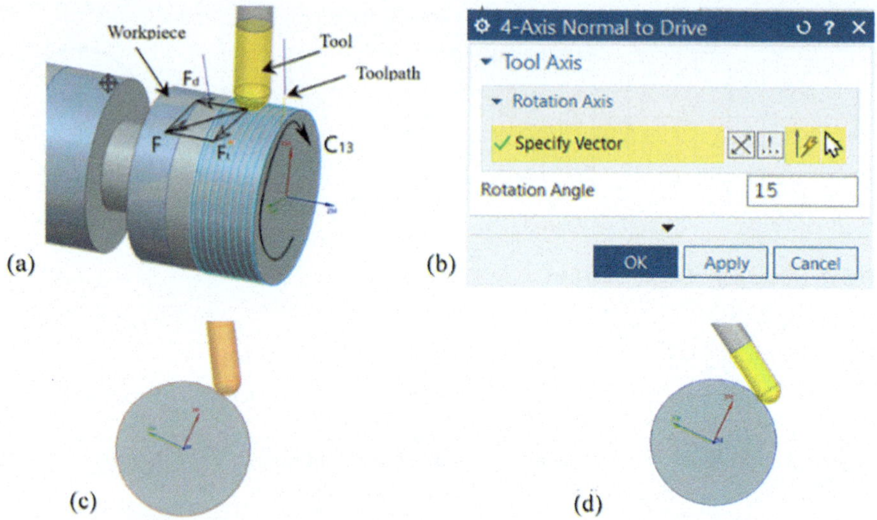

Fig. 4 **a** Tool movements in the machining process; **b** set up the tilt angle of tool; **c** tilt angle of 40°; **d** tilt angle of 80°

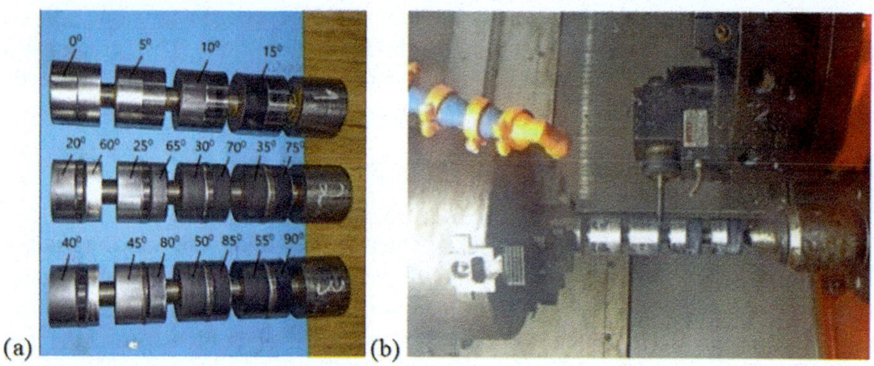

Fig. 5 **a** Diagram of machining locations; **b** images of jigs and machining setup

3 Experimental Results and Discussion

3.1 Experimental Data

The surface roughness graphs in Figs. 7 and 8 are established from the data in Table 4. The experimental results show the change in surface roughness (Ra) as the tool tilt angle (α) changes.

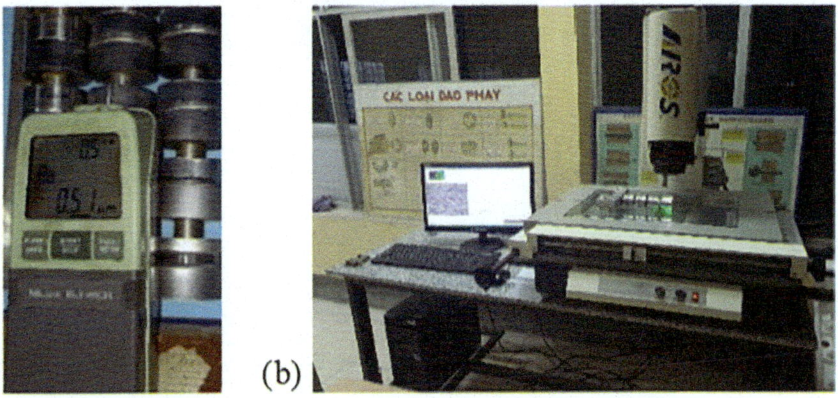

Fig. 6 **a** Mitutoyo SJ.201 surface roughness meter, **b** Arcs Kim450U 2D meter

Table 3 Results of machining surface quality inspection on Arcs Kim450U 2D meter

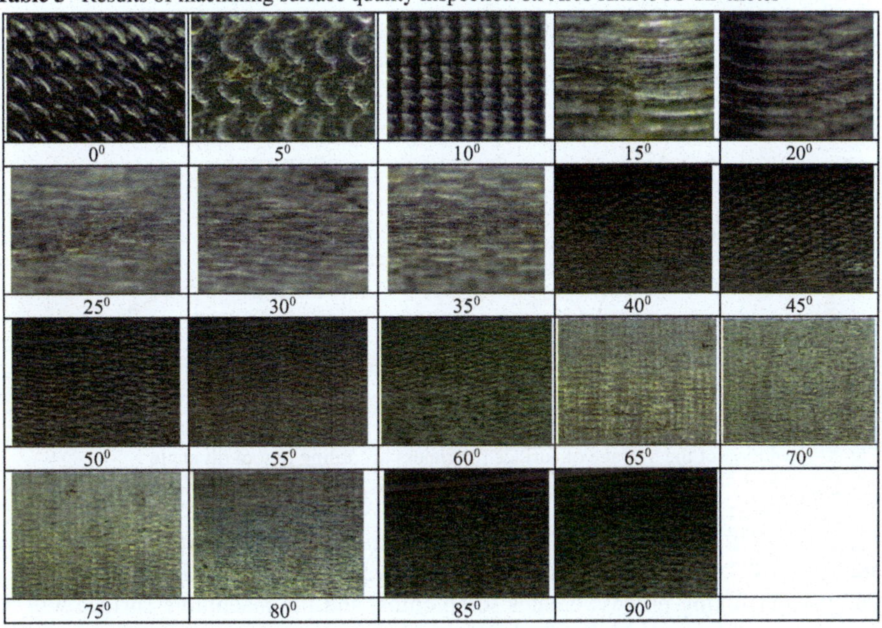

With the graph in Fig. 7, it is clear that when the tilt angle increases, the surface roughness decreases. The tilt angle is in the range of $0° < \alpha < 15°$, the surface roughness value is large. This is consistent with the rough appearance of the machining surface shown in Table 3. Machining test with the tilt angle $\alpha > 15°$, the surface roughness is stable, with small fluctuations in the range of $0.25\ \mu m < Ra < 0.5\ \mu m$. Simultaneously, by observing the machining surface quality (Table 3), it can be seen that the surface has many short and similar scratches with tilt angle in in the range of

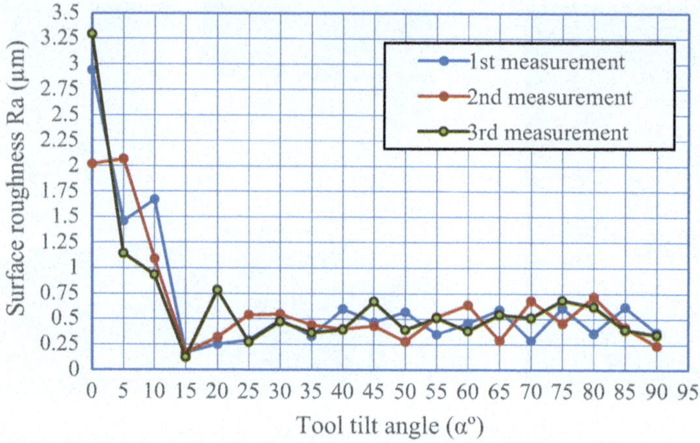

Fig. 7 Evolution of surface roughness according to tool tilt angle of three times measured

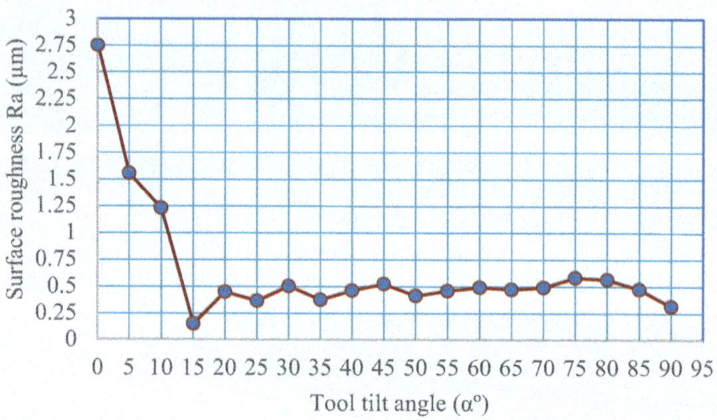

Fig. 8 Evolution of the average of surface roughness according to tool tilt angle

15° ÷ 35°. This proves that the chip shape is mainly short wire chip. As the large tilt angle $\alpha \geq 40°$, the machining surface appears very small undulating peaks, proving that the machining process creates scrap chips, and the machining surface will be smoother which is roughly equivalent to grinding machining methods.

3.2 Surface Roughness Regression Model

Based on the data of the graph in Figs. 7 and 8, it can be seen that the change of surface roughness value to tilt angle could be expressed follows an exponential nonlinear regression model:

Table 4 Results of the surface roughness with three measurements (Ra: μm)

Part No. 1								
	Tilt angle							
Measurement	0°	5°	10°	15°				
1st measurement	2.94	1.46	1.67	0.17				
2nd measurement	2.02	2.07	1.09	0.17				
3rd measurement	3.29	1.14	0.93	0.12				
Part No. 2								
	Tilt angle							
Measurement	20°	25°	30°	35°	60°	65°	70°	75°
1st measurement	0.25	0.29	0.49	0.33	0.46	0.59	0.29	0.61
2nd measurement	0.32	0.54	0.55	0.44	0.64	0.29	0.68	0.46
3rd measurement	0.78	0.27	0.47	0.36	0.38	0.54	0.51	0.68
Part No. 3								
	Tilt angle							
Measurement	40°	45°	50°	55°	80°	85°	90°	
1st measurement	0.6	0.47	0.57	0.35	0.36	0.62	0.37	
2nd measurement	0.4	0.43	0.28	0.52	0.72	0.42	0.24	
3rd measurement	0.39	0.67	0.39	0.51	0.62	0.39	0.34	

$$Y = f(X_i, \beta) + \varepsilon_i \tag{2}$$

By applying testing data with the exponential function $y = a(x)^b$, the results with low reliability (variance $R^2 \leq 0.5$). Therefore, the following dependency function was proposed, where $\boldsymbol{x}$ is the input parameter (tilt angle of the cutting tool); $\boldsymbol{a}$, $\boldsymbol{b}$, $\boldsymbol{c}$, $\boldsymbol{d}$ are constants, and $\boldsymbol{y}$ is the output (surface roughness).

$$y = a(x)^b + c(x) + d \tag{3}$$

Statistical Package for Social Sciences SPSS_26.0 software is used to analyze data and determine the constants $\boldsymbol{a}$, $\boldsymbol{b}$, $\boldsymbol{c}$, $\boldsymbol{d}$. The tilt angle value according to the input variable $\boldsymbol{x}$ and the surface roughness value according to the dependent function $\boldsymbol{y}$ are imported in the SPSS software (Fig. 9a). The constants $\boldsymbol{a}$, $\boldsymbol{b}$, $\boldsymbol{c}$, $\boldsymbol{d}$ are listed into the parameters box, while the dependent function y is set into the dependent box and the formula model are expressed into the model expression box (Fig. 9b). By conducting analysis, the results of constants $\boldsymbol{a}$, $\boldsymbol{b}$, $\boldsymbol{c}$, $\boldsymbol{d}$ are presented in Table 5, and roughness regression model could be expressed by Eq. 4.

$$y = 5.243(x)^{-0.68} + 0.007(x) - 0.288 \tag{4}$$

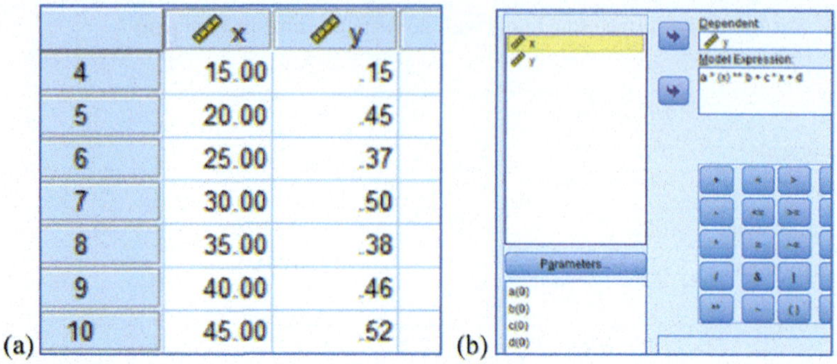

Fig. 9 **a** Input variable values and output parameters; **b** regression formulas and regression constants

Table 5 R^2 variance value and constants

Parameter estimates				
95% confidence interval				
Parameter	Estimate	Std. error	Lower bound	Upper bound
a	5.243	0.655	3.847	6.638
b	− 0.680	0.308	− 1.337	− 0.024
c	0.007	0.005	− 0.004	0.017
d	− 0.288	0.674	− 1.725	1.149

Anova[a]			
Source	Sum of squares	df	Mean squares
Regression	13.013	4	3.253
Residual	0.485	15	0.032
Uncorrected total	13.498	19	
Corrected total	5.423	18	
Dependent variable: *y*			

[a] *R* squared = 1 − (Residual Sum of Squares)/(Corrected Sum of Squares) = 0.911

4 Conclusion

By the experimental orthogonal milling tests on C45 rotationally symmetrical workpieces with ball nose end mill using turn-mill CNC machine, the following conclusion can be drawn:

- The regression equation showing the dependence of surface roughness on tool tilt angle have been established with high accuracy ($R^2 = 0.911$). This results is significantly important for machining parts with complex surfaces using four- or five-axes CNC machines, as the contact point on the cutting edge always changes position with variation in tool tilt angle. The result can be used to select machining methods, so that the surface quality is mostly stabilized.

- Based on the measuring results of surface roughness and inspection of the machining surface quality, the cutting tool tilt angle range can be determined for the best surface quality. This analysis will be the basis for further research on the other fields such as monitoring and controlling of vibration, cutting tool wear, dimensional accuracy, as well as machining surface profile.

References

1. Schulz, H.: High speed turn-milling—a new precision manufacturing technology for the machining of rotationally symmetrical workpieces (1990)
2. Uysal, E., Karaguzel, U., Budak, E., Bakkal, M.: Investigating eccentricity effects in turn-milling operations. In: 6th CIRP International Conference on High Performance Cutting, HPC2014. Procedia CIRP **14**, 176–181 (2014)
3. Savas, V., Ozay, C.: Analysis of the surface roughness of tangential turn-milling for machining with end milling cutter. J. Mater. Process. Technol. **186**, 279–283 (2007)
4. Kopac, J., Pogacnik, M.: Theory and practice of achieving quality surface in turn milling. Int. J. Mach. Tools Manuf. **37**(5), 709–715 (1997)
5. Yuan, S., Zheng, W.: The surface roughness modeling on turn-milling process and analysis of influencing factors. Appl. Mech. Mater. **117–119**, 1614–1620 (2012)
6. Huang, C., Cai, Y.: Turn-milling parameters optimization based on cutter wear. Adv. Mater. Res. **602–604**, 1998–2001 (2013)
7. Ozay, C., Savas, V.: The optimization of cutting parameters for surface roughness in tangential turn-milling using Taguchi method. Adv. Nat. Appl. Sci. **6**(6), 866–874 (2012). ISSN 1995-0772
8. Boozarpoor, M., Teimouri, R., Yazdani, K.: Comprehensive study on effect of orthogonal turn-milling parameters on surface integrity of Inconel 718 considering production rate as constrain. Int. J. Lightw. Mater. Manuf. **4**, 145–155 (2021)
9. Sun, T., Qin, L., Fu, Y., Liu, C., Shi, R.: Mathematical modeling of cutting layer geometry and cutting force in orthogonal turn-milling. J. Mater. Process. Technol. **290**, 116992 (2021)
10. Mohamad, M.H., Zahid, M.N.O.: Simulation program for 4th axis CNC machining in NX CAM system. Int. J. Automot. Mech. Eng. **16**(3), 6971–6986 (2019)
11. Lazoglu, I., Manav, C., Murtezaoglu, Y.: Tool path optimization for free form surface machining. CIRP Ann. Manuf. Technol. **58**(1), 101–104 (2009). https://doi.org/10.1016/j.cirp.2009.03.054

3D Avatar Interactive System on Hologram Fan Projector Using Motion Recognition Control by Mediapipe Holistic

Thai Anh Ngoc Huynh, Hoang Nguyen Pham, Ngoc Tho Thi Nguyen, and Tri Thanh Trang

Abstract This paper presents a novel system for creating and displaying a 3D avatar model on a fan hologram device. The system uses a webcam to capture the gestures and movements of the user and applies an AI model to recognize the body parts and joints of the user. The system then maps the recognized joints to a 3D avatar model and renders it on a fan hologram device using a persistence of vision (POV) technique. The system allows the user to interact with the 3D avatar model in real time and see it from different angles. The system also provides various functions, such as changing the model, using the input as image or video, and displaying the skeleton of the user. The system is based on the mediapipe holistic solution, which combines the pose, face, and hand landmarkers to create a complete landmarker for the human body. The system is tested and evaluated in terms of basic functions, display quality, and user experience. The results show that the system performs well in most conditions and provides an immersive and engaging experience for the user. The accuracy of the recognition function varies from 55 to 100%, depending on the environment and lighting condition. The success rate of the identification function is 82.69% for all body parts and acts. The system has a high potential for further development and application in various domains, such as education, entertainment, art, and communication.

Keywords 3D avatar model · Fan hologram · AI recognition function · Mediapipe holistic · Augmented reality

T. A. N. Huynh (✉) · H. N. Pham · N. T. T. Nguyen · T. T. Trang
Can Tho University, Can Tho City, Vietnam
e-mail: hntanh@ctu.edu.vn

H. N. Pham
e-mail: pnhoang@ctu.edu.vn

N. T. T. Nguyen
e-mail: ntntho@ctu.edu.vn

T. T. Trang
e-mail: tritrang@ctu.edu.vn

D. H. Tien et al. (eds.), *Proceedings of the 1st International Conference on Sustainability and Emerging Technologies for Smart Manufacturing*, Proceedings in Technology Transfer, https://doi.org/10.1007/978-981-97-7083-0_29

1 Overview

1.1 Augmented Reality Technology on Hologram

Augmented reality technology on hologram augmented reality (AR) has become a powerful tool and has been applied in various fields in the twenty-first century [1–3]. AR has achieved remarkable progress, significantly surpassing previous implementations in gaming applications such as Pokémon Go [4]. It is projected that, beginning in 2024, AR technology will continue its trajectory of advancement, emerging as a pivotal support mechanism within the realm of business efficiency [5]. The application of AR is very diverse and rich in many fields because there is no limit to the content of the virtual components when displayed. AR uses various methods to enhance the display of 3D virtual content into the real natural environment that the user is observing [1, 2, 6]. These virtual contents can be accessed through the camera system on smart mobile devices, virtual reality glasses or through 3D hologram forms [1, 2, 7] (Fig. 1).

Modern AR systems can also create different scenarios and provide rich multisensory experiences for users [2–4]. With the development of advanced AR technologies such as integrating computer vision, combining sensors on smartphones, or object recognition, information about the surrounding environment of the user's real world opens up more interaction channels for AR content [2, 3, 6, 7]. The AR content display and how users interact with this content will be limited by the type of device and corresponding AR technology. AR applications on mobile devices bring high personal experience to users, but require high hardware configuration of the device at the present time as well as certain technological capabilities of users. The approach that the authors evaluate as bringing a balance between the three factors of investment cost, 3D content display capability, and the ability to provide AR experience for many users at the same time is 3D hologram or hologram AR systems.

Fig. 1 Character on hologram using OpenPose

Fig. 2 3D holographic fan

1.2 *Fan Hologram*

Fan hologram, also known as 3D LED holofan, is a technology that creates 3D images floating in the air. When the fan starts spinning, the controller will manage the lighting of each LED pixel. The result of this control is to create images or videos on a circular screen due to the fan rotation. The 3D content is designed on a black background and accordingly the LED pixels displaying the black background will not light up. The 3D model presented on the hologram fan generates an effect where the model seems to float in mid-air, facilitated by the use of a transparent background. This feature significantly enhances the viewer's experience beyond that of traditional 2D displays, as the model is not constrained by the boundaries of a screen and appears against a backdrop that many viewers can be seen through at the same time (Fig. 2).

1.3 *Factors Affecting the Display of Hologram*

The display quality of a hologram depends on many factors, including resolution, contrast, brightness, color reproduction, and interactivity. The resolution of the hologram is determined by the number of pixels used to create the image and is the decisive factor for the level of detail of the image. Contrast is the degree of difference between light and dark points and creates the vividness of the image. Brightness, the amount of light reflected from the hologram, is considered as the brightness display of the hologram and high brightness makes the image easier to see. Color reproduction, the accuracy of the displayed color, makes the image more realistic. And the last factor affecting the AR content display is the user's ability to interact with these models. These are the abilities or levels that users can interact with the image from simple as controlling the presentation to more complex as simulating according to the user's gestures. These interactions create a realistic and lively experience for users. In addition, the display quality is also affected by some other factors such as the focal length of the hologram, the contrast ratio of the viewing environment, and the distance between the viewer and the hologram. The appropriate focal length will make the image clearer. The high contrast ratio between the viewing environment and the brightness of the hologram will make the image easier to see.

1.4 Interactive Hologram

Interactive hologram is a potential development direction for future hologram systems, but it is a complex system consisting of many main components, in which the basic components include shaping device, display device, and interactive device. The shaping device can be a computer, projector or laser, LED fan, etc. The display device can be a screen, mirror, or other reflective surface to display 3D images. The interactive device is the most important part and has many different solutions, which can be a controller, sensor or AI camera system, etc. The main task is to create a control mechanism that allows users to interact with 3D images. The high cost of hologram devices and interactive controllers limits accessibility for most users, making affordable, expandable interactive hologram designs a potential solution for future devices.

2 Research Content

2.1 Data Processing

In this research, images from cameras or input videos are utilized for skeleton detection and extraction via the MediaPipe Holistic library. This library, introduced by Google in October 2020, offers the ability for fast, simultaneous recognition of face, body pose, and hand gestures within a single image frame. The movements of the 3D avatar model are precisely replicated to reflect the user's movements captured in the input video, with these synchronized actions subsequently displayed on the hologram device (Fig. 3).

First, the input frames will be identified by BlazePose detector. Then, based on the joint points of the identified pose, we will extract three regions of interest located at the two hands and face, and use a re-crop model to improve the region of interest. After having the cropped regions at the two hands and face with the highest resolution, we will apply models to identify the detailed joint points in those regions and merge them to create a model with more than 540 landmarks. To cater to a broader user base, we utilize a popular asset from Unity's asset store, namely the 3D model of the Unity-Chan character from Unity. Subsequently, we initialize the application's basic environment, including the ground, the wall behind the character, and the 3D model of the character in a T-pose. Figure 4 illustrates the basic environment.

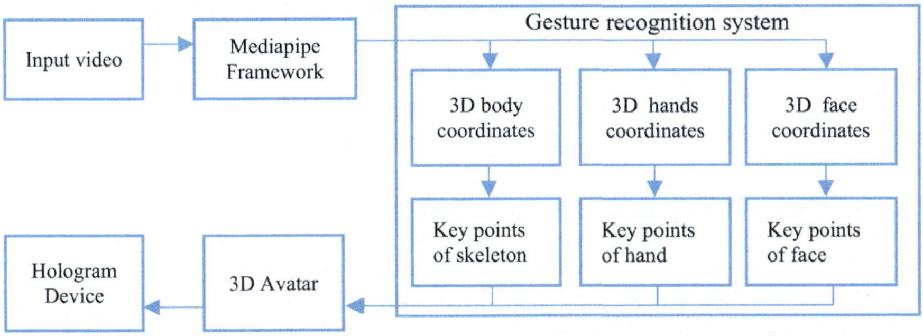

Fig. 3 Data processing procedure

Fig. 4 Initializing the environment for the application

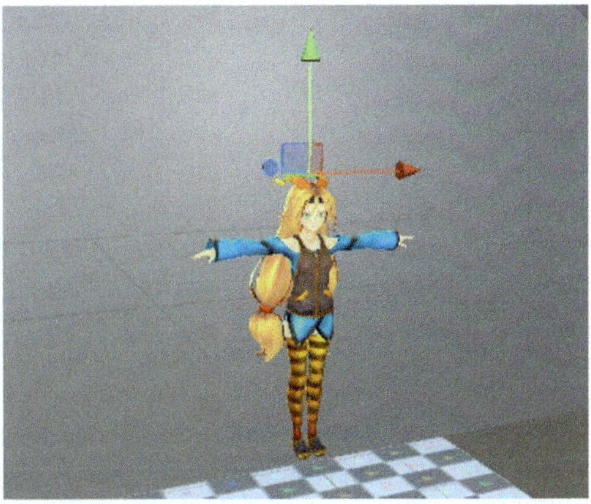

After creating the environment for the application, we proceed to identify the joints for the character model including three types: body joints, hand joints, and anchor points on the face. Finally, we will transfer the coordinates in 3D space obtained from the joints to the positions of the corresponding joints on the 3D model of the character. The joint points are applied to the 3D model by calculating the relative distances between the point we are tracking and a point we choose as the axis. This distance is represented as a vector in 3D space Oxyz. Therefore, to track the movement of the face and two hands of the user, we will store and process on 16 pairs of points on the left hand, 16 pairs of points on the right hand, and 4 pairs of points on the upper body. Additionally, to optimize hand movement tracking performance, we also monitor the palm's direction to ensure that the 3D avatar character's palm direction matches the user's movement direction. To achieve this, we identify the palm plane formed by three joint points (0, 5, and 17), as illustrated in Fig. 4. Subsequently, we calculate and store the normal vector perpendicular to the identified palm plane.

After synchronizing the positions of the joints obtained from the user and the joints on the 3D avatar model, the system will render them into frames and display them on the hologram device.

2.2 Displaying Images on Fan Hologram

Currently, there are many different types of holograms, but we will use a fan-shaped hologram with a LED system to display the image of the 3D avatar character. The operating principle of the fan hologram is based on the persistence of vision (POV) system. Due to the characteristic of human eyes, they cannot perceive movements that occur in less than 50 ms. Therefore, if there is a point in space that flashes and repeats at a frequency of 20 Hz, human eyes will not see this flicker. Based on the above POV principle, the LED strips will be attached to the fan blades of the holofan and will flash when spinning at high speed to create a 3D image effect floating in the air. In 2011, Liong and colleagues [8] designed and developed a rotating LED system to display messages in the form of alphabet characters. This system will define 36 characters including 26 characters in the alphabet and numbers from 0 to 9. Then, the system will turn on and off 7 LEDs on the fan blade to create the desired characters. Although Liong's system was limited to displaying only alphabet characters, this was an important milestone in the development of later holofan systems. Another study by Al-Natsheh et al. [9] in 2019 aimed to design and fabricate a POV display in cylindrical space and allow users to observe from any direction around it. The authors also synchronized the frequency of the control signal and the speed of the motor to create a screen with enough resolution to help viewers distinguish characters, letters, and symbols. In this topic, we will use an algorithm to convert images from pixel matrix to encoded form suitable for the rotating light device of the holofan.

The input data is an image with a width of w pixels and a height of h pixels. The holofan device will display the image in the form of a circle with a radius of $L_{\max}$, where $L_{\max}$ is the number of LEDs on a fan blade. This circle will be divided into n angular regions called "time slots" because the angle of each region corresponds exactly to the rotation time from the reference point to that region. The angle between two adjacent time slots α will be calculated by $\alpha = (2 * \pi)/n$. In the direction of the radius, each time slot will be divided into $L_{\max}$ small regions corresponding to $L_{\max}$ LEDs. Finally, the input pixel matrix will be converted into points on the corresponding time slot by the following formula:

$$P_{x,i} = L_j * \cos(\alpha_i) + c_x \tag{1}$$

$$P_{y,i} = L_j * \sin(\alpha_i) + c_y \tag{2}$$

$$V_{i,j} = A\left[I_x + w * \left(h - 1 - I_y\right)\right] \tag{3}$$

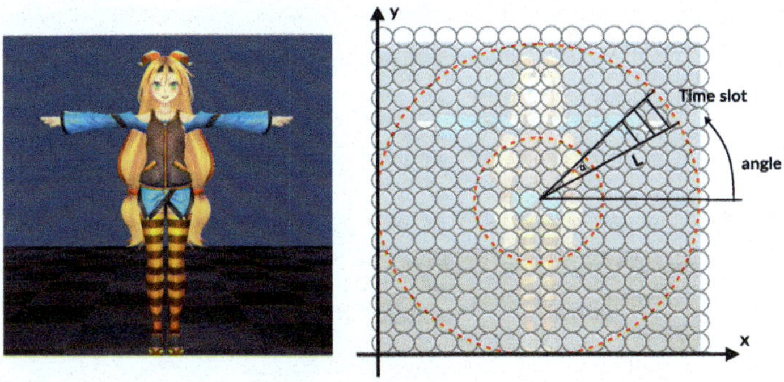

Fig. 5 Process of converting from pixel matrix to display points on time slot

$$I_x = \frac{P_{x,i} * w}{2 * L_{\max}} \quad (4)$$

$$I_y = \frac{P_{y,i} * h}{2 * L_{\max}} \quad (5)$$

where P_i is the pixel coordinate of the slot i, $I(x, y)$ is the coordinate of the pixel on the input image calculated by P_i, and $V(i, j)$ is the display value of the slot i at the position of the LED j. The process of converting pixels on the input image matrix to coordinates of points to display is shown in Fig. 5.

3 Conclusion

3.1 *Testing Environment*

Testing of the system's main functions was conducted. The setup for testing included the use of a high-definition webcam connected to a computer; the computer was positioned on an elevated surface relative to the ground to ensure the webcam could capture the user's entire upper body; the user maintained a distance of approximately 60–80 cm from the webcam. The computer configuration for the tests was as follows:

- Display: 1920 × 1080 Pixels, 120 Hz, 250 nits LED-backlit
- CPU: Intel Core i5, 9300H
- RAM: DDR4, 2666 MHz
- Hard drive: SSD, 512 GB
- Graphics: Intel UHD Graphics, 630
- Operating system: Windows, 10 Home Single Language.

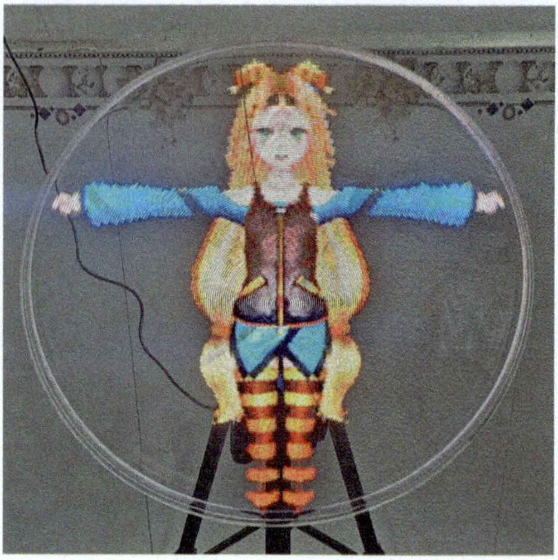

Fig. 6 3D character displayed on fan hologram with resolution is 4.32 PPI

The holofan in use is a device with two blades and a diameter of 42 cm, with a total of 224 LED bulbs. The corresponding display resolution is about 4.32 PPI (Fig. 6).

3.2 Testing Basic Functions

The basic function testing process involves performing the following functions: changing the model, using the input as webcam, image input, and video input. During the function testing process, should the function be executed and the system continues to recognize the object and display the skeleton, it is deemed a successful operation (Table 1).

Thirty operations were performed for each function, and the number of successful occurrences was recorded. The results indicated that the system achieved a 100% success rate in changing the model, utilizing the webcam, and processing video inputs. However, the success rate fell to 33% with image inputs, attributed to the image quality and the proportion of the object within the image.

Table 1 Basic function test results

Function	Change model	Use webcam	Use images	Use video
Successful	30/30	30/30	10/30	30/30

3.3 Testing the Recognition Function of the AI Model

This testing process will test the most important function of the application. The testing process will be performed in different types of environments to evaluate the effectiveness of the AI model. The time of each test in an environment is between 10 and 20 s. Test results with webcam input. The details of the testing process are given in Tables 2 and 3.

Note

- *Environment*: there are two cases, which are the environment with a simple background (behind the user there is only a plain wall with a single color), and the environment with a complex background (behind the user there will be objects or a wall with a complex pattern).
- *Lighting Condition*: there are two cases, which are the condition with adequate lighting and low lighting.
- *Number of Objects*: the number of objects includes the user who is interacting with the application.
- *Action*: create basic gestures such as waving hands, making two-finger gestures, opening and closing hands, and fingers in front of the webcam.

Table 2 Test results of AI model identification function using MediaPipe Holistic

STT	Background	Bright conditions	Objects	Number of tests	Successful	Accuracy (%)
1	Simple	Full-light	2	40	36	90
2	Simple	Low-light	2	40	30	75
3	Simple	Full-light	24	40	32	80
4	Simple	Low-light	4	40	26	65
5	Complex	Full-light	2	40	38	95
6	Complex	Low-light	2	40	26	65
7	Complex	Full-light	4	40	34	85
8	Complex	Low-light	4	40	22	55

Table 3 Test results using 3D avatar character model

STT	Identification	Act	Objects	Tests	Success rate (%)
1	Left hand	Basic gesture	2	40	80
2	Right hand	Basic gesture	2	40	75
3	Eye	Blink of an eye	2	40	100
4	Mouth	Closing/opening	2	40	100
5	Eyebrow	Murmur	2	40	100

The recognition function of the AI model was tested under various environmental and lighting conditions, incorporating two or four objects in each of the 40 tests conducted for each scenario. The accuracy of recognition was assessed by comparing the model's output with the actual gestures and movements of the objects. It was observed that accuracy improved in environments with simple backgrounds and adequate lighting, achieving up to 95%. Conversely, accuracy diminished in settings with complex backgrounds and poor lighting, falling to 55%. The model also successfully recognized the left hand, right hand, eye, mouth, and eyebrow with a success rate of 75% or higher.

The test results indicated that the system accurately performed the basic functions for which it was designed. Nevertheless, accuracy decreased under low-light conditions and when gestures were executed too rapidly. While the environment itself did not impact application accuracy, obstacles in front of the user or additional people entering the webcam frame introduced noise, detracting from the user experience. The first individual to appear in front of the webcam, provided they meet the standard ratio (full upper body visible), was designated as the primary user for the 3D model reporting. Input selection as an image failed when the image did not meet the recognition ratio standard, such as when the user's image was angled or contained more than one person. Body part recognition faced challenges, especially with hands moving too swiftly or exiting the frame, resulting in differing success rates for each hand due to variable movement speeds during testing.

The display quality and user experience of the model on a fan hologram were evaluated based on image quality, response speed, resolution, and viewing angle. The evaluation demonstrated that images on the fan hologram were displayed effectively, with clear, vivid colors, and smooth animations. The response speed was appropriate, enabling real-time tracking and mimicry of user movements. The fan hologram's resolution influenced the 3D model's display quality, where higher resolution enhanced detail and sharpness. The fan hologram provided an effective display for viewers 2–10 m away and in front, offering diverse perspectives of the 3D model. However, the viewing experience diminished for users moving to the sides of the fan hologram, as the depth and volume of the 3D model became less perceptible.

3.4 Potential Development

This topic has a high potential for further development in the field of augmented reality and holographic display. The model on fan hologram can be applied to various domains, such as education, entertainment, art, and communication. For example, the model can be used to create interactive and immersive learning experiences for students, such as simulating historical events, demonstrating scientific phenomena, or exploring different cultures. The model can also be used to create engaging and creative entertainment products, such as games, movies, or performances. The model can also be used to create artistic and expressive works, such as sculptures, paintings, or animations. The model can also be used to enhance communication and collaboration, such as video conferencing, telepresence, or social networking.

To achieve these applications, the model on fan hologram can be improved in several aspects, such as increasing the resolution, contrast, and brightness of the display; adding more features and functions to the model, such as voice, facial expression, and gesture recognition; integrating the model with other devices and platforms, such as smartphones, tablets, computers, or virtual reality headsets; and expanding the scope and scale of the model, such as creating multiple or larger fan holograms, or combining them with other types of holograms. These improvements can enhance the quality and user experience of the fan hologram model, opening up new possibilities and opportunities for applying augmented reality and holographic display in advertising, education, communication, and many other fields.

References

1. Mekni, M., Lemieux, A.: Augmented reality: applications, challenges and future trends. Appl. Comput. Sci. **20**, 205–214 (2014)
2. Michalos, G., et al.: Augmented reality (AR) applications for supporting human-robot interactive cooperation. Procedia CIRP **41**, 370–375 (2016)
3. Furht, B.: Handbook of Augmented Reality. Springer Science & Business Media (2011)
4. Piekarski, W., Thomas, B.H.: ARQuake: the outdoor augmented reality gaming system. Commun. ACM **45**(1), 36–38 (2002)
5. Schmalstieg, D., Höllerer, T.: Augmented Reality: Principles and Practice. Addison-Wesley Professional (2016)
6. Wang, R., You, S., Neumann, U.: Annotating and navigating tourist videos. In: Proceedings of the 2004 ACM SIGMM Workshop on Effective Telepresence, New York, NY, pp. 24–31 (2004)
7. Zhang, Z.: Microsoft kinect sensor and its effect. IEEE Multimed. **19**(2), 4–10 (2012)
8. Liong, B.Y.: Design and development of spinning LED message display system. Dissertation, Universiti Tun Hussein Onn Malaysia, 2011
9. Al-Natsheh, W.H., Hammad, B.K., Abu Zaid, M.A.: Design and implementation of a cylindrical persistence of vision display. In: 2019 6th International Conference on Electrical and Electronics Engineering (ICEEE). IEEE (2019)

The Study of Methods for Measuring and Controlling Oxygen Storage Capacity

Alexey Terenchenko, Andrey Kozlov, Givi Nadareishvili, and Andrey Porsin

Abstract Reducing the toxicity of exhaust gases from internal combustion engines remains an urgent task both now and in the future, despite the widespread development of electric vehicles. The three-way catalyst (TWC) system for reducing the toxicity of gasoline engines is well known. One of the characteristics of the catalytic unit used in after treatment systems is the oxygen storage tank (OSC). OSC can be measured in various ways. The technological oxygen storage capacity is a parameter measured by an electronic control unit on a car by comparing signals from oxygen sensors installed on the hot end. The engine method is an investigative method for obtaining the full oxygen content on the surface of the neutralizer in a wide range of temperatures and calculating OSC values. This article presents a comparison of the OSC values obtained by these different methods. In order to establish a relationship between the true oxygen capacity measured during coating development and block production and the process capacity continuously measured during calibration, certification preparation and block aging, it is necessary to supplement the existing catalyst model. For the first time, the possibility of correlation using a scaling factor between OSC values was established in the practice of system development and calibration.

Keywords Vehicle · Engine toxicity · After treatment system · Oxygen storage capacity

1 Introduction

For engine management system calibrations and environmental certifications, testing catalytic converters [1, 2] mounted in the engine exhaust system is of interest for measuring the oxygen storage capacity (OSC) of primary (starter) catalytic converters.

A. Terenchenko (✉) · A. Kozlov · G. Nadareishvili · A. Porsin
NAMI Russian State Scientific Research Center, Moscow, Russia
e-mail: terenchenko@nami.ru

D. H. Tien et al. (eds.), *Proceedings of the 1st International Conference on Sustainability and Emerging Technologies for Smart Manufacturing*, Proceedings in Technology Transfer, https://doi.org/10.1007/978-981-97-7083-0_30

OSC is the amount of oxygen that can be stored in or released from the catalytic block/converter when the oxygen content in the exhaust gases changes within the range from zero (period when oxygen is released by the catalyst) to its excess, i.e., above the stoichiometric quantity (period when oxygen is absorbed by the catalyst). The process of oxygen accumulation-release is based on physicochemical processes occurring in the catalytic layer. The ability to store oxygen throughout the entire service life is one of the most important properties of a catalytic system.

The correlation between OSC and the efficiency of three-way catalytic converters (TWC) is well-studied.

Technological OSC is a technological parameter measured by comparing signals from oxygen sensors located before and after the test object (catalytic converter) during cyclic changes in the oxygen content in the exhaust gases. The precise composition of the exhaust gases is not taken into account. As a result, the values of technological oxygen capacity and true oxygen capacity can significantly differ from each other.

For the first time, the possibility of correlation using a scaling factor between OSC values was established in the practice of system development and calibration.

2 Main Part

Measurement of technological oxygen capacity is used for diagnosing the functionality of the catalytic converter in accordance with UNECE Regulation No. 83. Demonstrating the functionality (diagnosis) of the catalytic converter is part of the certification tests [3, 4]. Technological OSC is measured continuously during the vehicle's operation and throughout its lifespan according to a specific algorithm, which is an integral part of the control unit software (Fig. 1). This diagram is obtained from the electronic control unit of the car.

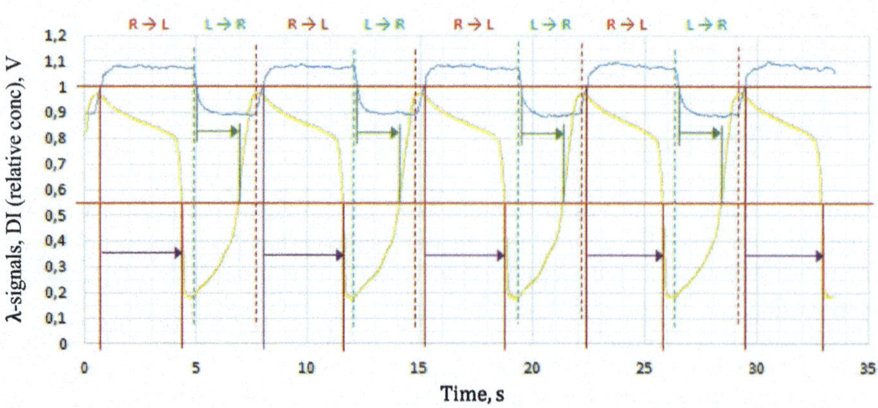

Fig. 1 Duration diagram for the transition from L (poor $\lambda = 0.95$) to R (rich $\lambda = 1.05$) and backward

The OSC parameter is determined using the aging-related formula:

$$\mathrm{OSC} = Z * Q * \Delta\lambda * A, \tag{1}$$

where

- OSC is oxygen storage capacity, mg;
- Z—transition delay time in each oxygen sensor voltage measurement through the threshold of 0.55 V at $1.05 > \lambda > 0.95$;
- $\Delta\lambda$—change in mixture composition—$\Delta\lambda = 0.05$;
- Q—mass of exhaust gas passing through the neutralizer, mg/s;

$$Q = \left(q + \frac{\lambda * q}{14,7}\right) \times \frac{1000000}{36}, \tag{2}$$

where Q—exhaust gas flow, kg/hour; A—conversion factor to units of measurement mg; $A = 0.23$.

Level of concentration of components in exhaust gas:

- CO—600–2000 ppm;
- NS, NOx—units—tens of ppm;
- H_2—concentration is comparable to the concentration of CO.

The duration of the transition from L to R depends on the value of OSC, determined by the amount of cerium oxide in the catalytic block and the mass flow rate of CO, H_2, and HC in the exhaust gas.

To measure true OSC, unlike technological OSC, methods are typically employed based on measuring the composition of the exhaust gases and precise measuring the quantity of oxidized components (e.g., carbon monoxide) by the oxygen stored in the catalytic layer.

In addition, the engine method of measuring OSC is applied as a control for the block's oxygen capacity. This method is used by the catalytic coatings manufacturer to monitor the quality of the blocks.

The engine method is a research technique for exploring the OSC of cerium oxide across a broad temperature range. Traditionally, this method relies on the oxidation of CO by oxygen released from the lattice of cerium-containing oxides. However, the limited activity of cerium in this reaction at lower temperatures hampers the OSC measurement process. An alternative approach involves adding a more active Pt/Al_2O_3 catalyst to the cerium-containing oxide to ease the CO oxidation stage.

Cerium-containing oxides ($Ce_xM_{1-x}O_2$) play a crucial role in three-way catalytic converters (TWC) [5, 6]. They enhance TWC activity while reducing noble metals burden, thus lowering the overall cost of TWC, maintaining it at stoichiometric ratios on the catalyst surface. These conditions ensure effective oxidation of Co and hydrocarbons to Co_2 and H_2O concurrently with the reduction of NO_x to N_2 [6, 7, 9]. The catalysts' activity in soot oxidation is increased [8, 10]. The involvement of crystalline ceria in soot oxidation is demonstrated in experiments with samples:

soot/Ce_xZrxO_2 [11]. Among the series of Ce_xZr1xO_2 oxides, $Ce_{0.69}Zr_{0.31}O_2$ exhibits the most effective soot oxidation due to its high oxygen storage capacity [Battin, 2008] [10]. Additionally, cerium oxide enhances soot oxidation by facilitating the oxidation of NO to NO_2, which serves as a stronger oxidizer than oxygen. The utilization of cerium-containing oxides capitalizes on the redox properties of cerium cations ($Ce_4 + Ce_3 +$), which depend on the oxygen content in the gas phase, thereby enabling oxygen release or absorption. The quantitative determination of stored/released oxygen, or oxygen storage capacity (OSC), has been extensively discussed in various publications. Although several approaches to measuring OSC are presented in the literature [12], a standardized method is yet to be proposed.

OSC measurement entails determining the amount of reducing gas (e.g., H_2, CO) that reacts with oxygen stored in cerium oxide. These measurements are typically conducted under cyclic conditions in the absence of gaseous oxygen. There are two primary types of OSC measurements. continuous reduction involves determining the final OSC, whereas dynamic measurements assess the final OSC during the passage of variable pulses of reducing or oxidizing gas through the sample. Typically, H_2, CO [15 –19], or hydrocarbons [13, 14] are used as reducing gases.

Experimental data confirm that the limiting factor in engine method OSC measurement is the reaction rate between CO and O.

The dynamic oxygen storage capacity model, accepted as the base and based on [6], utilizes a simplified physics-based approach that focuses solely on the chemical species critical for oxygen storage. Specifically, only one type of oxidizing gas, O_2, and one type of reducing gas, CO, are taken into account. The capacity to store oxygen in the converter is elucidated by introducing two surface particles, CeO and Ce, which represent the oxidized and reduced forms of cerium, respectively. Adhering to the Eley–Rideal mechanism, the reaction scheme is outlined as follows:

$$O_2 + 2Ce = 2CeO \tag{3}$$

$$CO + CeO = CO_2 + Ce, \tag{4}$$

where (3) represents oxygen absorption on ceria, and (4) describes the oxidation of CO by previously stored oxygen. Both forward and reverse reactions are accounted for, resulting in the subsequent overall reaction rates, which are contingent upon the concentrations of reactants:

$$r_1 = k_1^f[Ce]^2[O_2] - k_1^b[CeO]^2C_0 \tag{5}$$

$$r_2 = k_2^f[CeO]^2[CO_2] - k_2^b[Ce]^2[CO_2] \tag{6}$$

k_2^f, k_2^b are rate constants for the forward and reverse reactions, respectively. The forward reaction constants are influenced by the catalyst temperature according to the Arrhenius equation.

Consequently, unlike many semi-empirical models, the rates of oxygen adsorption and desorption are influenced not only by the actual level of oxygen accumulation [CeO] but also on the catalyst temperature [Lewis, 1987] [17]. The oxidized and reduced sites on the surface collectively contribute to the overall OSC:

$$[\mathrm{CeO}] + [\mathrm{Ce}] = \mathrm{OSC} \tag{7}$$

The normalized oxygen storage level is defined as:

$$\varphi = \frac{[\mathrm{CeO}]}{\mathrm{OSC}} \tag{8}$$

$$\frac{\mathrm{d}\varphi}{\mathrm{d}t} = \frac{1}{\mathrm{osc}}(2r_1 - r_2) \tag{9}$$

The concentrations of chemical species entering the three-way catalyst (TWC) are estimated based on wide-scale measurements from the lambda sensor upstream of the catalyst, determining the air–fuel ratio, as presented in [18]. This model finds application in real-time diagnostics in vehicle, as the required model input data can be measured by commercially available sensors and readily assessed by the electronic control unit (ECU).

To establish a connection between the true OSC measured during coating development and block production, and the technological OSC measured continuously during calibration, certification preparation, and block aging, the model needs to be supplemented.

Determining the correlation between the two types of oxygen capacity is crucial for shortening the aging period of the block for constant monitoring of its efficiency.

It is important to note that in a real engine, oxidation (oxygen utilization) is associated not only with CO but also with H_2 and CH.

Let us supplement the oxidation model with the following reactions:

$$\mathrm{H_2} + \mathrm{CeO} = \mathrm{H_2O} + \mathrm{Ce} \tag{10}$$

$$2\mathrm{CH_2} + 3\mathrm{CeO} = \mathrm{H_2O} + 2\mathrm{CO} + 2\mathrm{Ce} \tag{11}$$

According to numerous sources, the limiting reaction remains the oxidation of carbon monoxide. The reaction rate constants are approximately the same for all major transformations.

Let us analyze the volumetric concentrations of the main elements in the exhaust gas composition participating in oxidation reactions (Fig. 2). This diagram is obtained from the electronic control unit of the car.

Curves of the CH, NO, O_2 concentration have been plotted in the area of the external speed characteristic of the studied engine, in relation to CO concentration (Fig. 3).

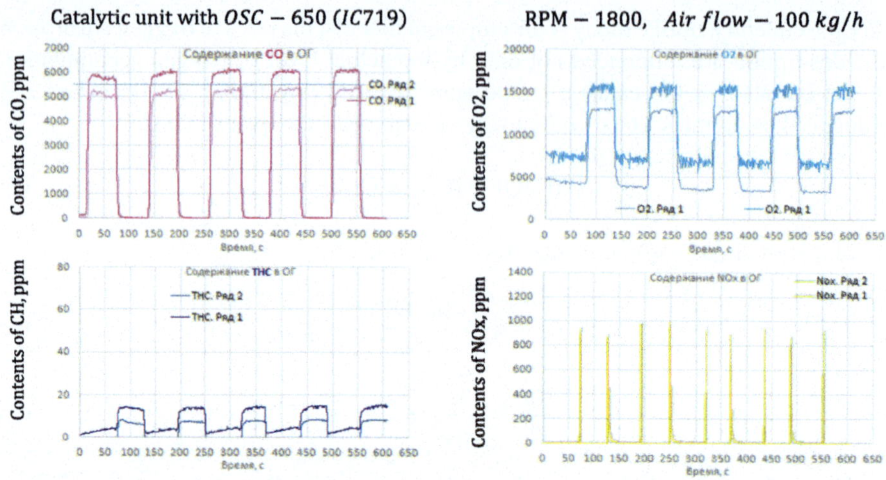

Fig. 2 Change in concentrations of O_2, CO, CH, and NO when measuring oxygen capacity

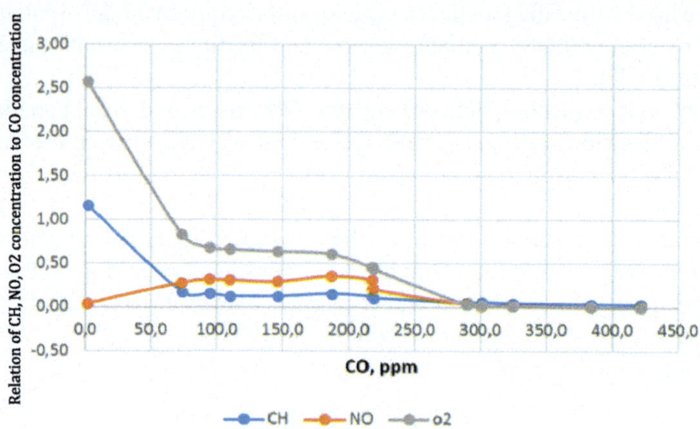

Fig. 3 Ratios of O_2, CH, and NO concentrations in relation to CO concentration

The impact of nitrogen oxides (NO), mentioned earlier, on the readings of both narrow-band and wide-band oxygen sensors is insignificant, as the sensor responds to free oxygen.

The calculation of the mass flows of CO, CO_2, and HC is based on the mass balance of carbon atoms, where the number of carbon atoms entering the engine equals the number of carbon atoms in the exhaust gases:

$$Q_C^T = Q_c^{CO} + Q_c^{HC} + Q_c^{CO_2} + Q_c^{soot}, \tag{12}$$

where

- Q_C^T is the mass flow rate of carbon atoms entering the engine, contained in the fuel (neglecting the content of CO_2 in the air);
- Q_c^{CO} is the mass flow rate of carbon atoms in CO in the exhaust gas;
- Q_c^{HC} is the mass flow rate of carbon atoms in hydrocarbons (CH);
- $Q_c^{CO_2}$ is the mass flow rate of carbon atoms in CO_2 in the exhaust gas;
- Q_c^{soot} is the mass flow rate of carbon atoms in soot in the exhaust gas (neglecting this term further).

$$Q_C^T = \frac{M_C}{M_{CO}} \cdot \nu_{CO} \cdot M_{CO} + \frac{M_C}{M_{CO_2}} \cdot \nu_{CO_2} \cdot M_{CO_2} + \frac{M_C}{M_{HC}} \cdot \nu_{HC} \cdot M_{HC}$$

or

$$Q_C^T = M_C \cdot \left(\nu_{CO} + \nu_{CO_2} + \nu_{HC}\right), \tag{13}$$

where

ν_{CO}—molar consumption of CO in exhaust gases (mol/h);
M_C—molar mass of carbon (kg/mol);
M_{CO}—molar mass of CO;
ν_{CO_2}—molar consumption of CO_2 in exhaust gases;
M_{CO_2}—molar mass of CO_2;
ν_{HC}—molar consumption of hydrocarbons in exhaust gases;
M_{HC}—molar mass of methane.

For the components of a gas mixture, the ratio of their molar flow rates is equal to the ratio of their corresponding volumetric concentrations k_i. From here, we get the relations:

$$\frac{\nu_{CO}}{\nu_{CO_2}} = \frac{k_{CO}}{k_{CO_2}} \tag{14}$$

$$\frac{\nu_{CO}}{\nu_{HC}} = \frac{k_{CO}}{k_{HC}} \tag{15}$$

where k_{CO}, k_{CO_2}, k_{HC}—volumetric concentrations, respectively, of CO, CO_2, and HC in the exhaust gases, measured in experiments.

From Eqs. (14 and 15), we express ν_{CO_2}, ν_{HC} through ν_{CO} and substitute it into Eq. (11). As a result, we get:

$$Q_C^T = M_C \cdot \nu_{CO} \cdot \left(1 + \frac{k_{CO_2}}{k_{CO}} + \frac{k_{HC}}{k_{CO}}\right) \tag{16}$$

where from we get the expression for the molar consumption of CO:

$$\nu_{CO} = \frac{Q_C^T}{M_C \cdot \left(1 + \frac{k_{CO_2}}{k_{CO}} + \frac{k_{HC}}{k_{CO}}\right)} \tag{17}$$

Let us express the mass flow of carbon atoms Q_C^T contained in the fuel through the measured fuel flow Q_{fuel}:

$$\nu_{CO} = \frac{Q_C^T}{M_C \cdot \left(1 + \frac{k_{CO_2}}{k_{CO}} + \frac{k_{HC}}{k_{CO}}\right)} \tag{18}$$

where M_{C8H18} is the molar mass of octane.

Then, expression (18) will take the form:

$$\nu_{CO} = \frac{8 \cdot Q_{\text{fuel}}}{M_{C8H18}\left(1 + \frac{k_{CO_2}}{k_{CO}} + \frac{k_{HC}}{k_{CO}}\right)} \tag{19}$$

From Eqs. (14 and 15), we obtain expressions for calculating the molar consumption of CO_2 and CH through the molar consumption of CO:

$$\nu_{CO_2} = \nu_{CO} \cdot \frac{k_{CO_2}}{k_{CO}} \tag{20}$$

$$\nu_{CH} = \nu_{CO} \cdot \frac{k_{HC}}{k_{CO}} \tag{21}$$

Let us express the mass flow in terms of molar flow:

$$Q_{CO} = \nu_{CO} \cdot M_{CO} = M_{CO} \cdot \frac{8 \cdot Q_{\text{fuel}}}{M_{C8H18}\left(1 + \frac{k_{CO_2}}{k_{CO}} + \frac{k_{HC}}{k_{CO}}\right)}, \tag{22}$$

$$Q_{CO_2} = \nu_{CO_2} \cdot M_{CO_2} = \frac{k_{CO_2}}{k_{CO}} M_{CO_2} \cdot \frac{8 \cdot Q_{\text{fuel}}}{M_{C8H18}\left(1 + \frac{k_{CO_2}}{k_{CO}} + \frac{k_{HC}}{k_{CO}}\right)}, \tag{23}$$

$$Q_{HC} = \nu_{HC} \cdot M_{CH_4} = \frac{k_{HC}}{k_{CO}} M_{CH_4} \cdot \frac{8 \cdot Q_{\text{fuel}}}{M_{C8H18}\left(1 + \frac{k_{CO_2}}{k_{CO}} + \frac{k_{HC}}{k_{CO}}\right)}, \tag{24}$$

The derivation of formulas for calculating H_2 is based on the mass balance of oxygen and hydrogen atoms, when the number of corresponding atoms at the engine inlet is equal to the number of carbon atoms in the exhaust gases. Balance equation for oxygen atoms:

$$Q_{\text{air}}^{O} = Q_{CO}^{O} + Q_{CO_2}^{O} + Q_{H_2O}^{O}, \tag{25}$$

where

Q_{air}^{O}—mass flow rate of oxygen atoms at the inlet of the air sucked into the engine, measured;

$Q^{\mathrm{O}}_{\mathrm{CO}}$—mass flow rate of oxygen atoms in the composition of CO in the exhaust gas;

$Q^{\mathrm{O}}_{\mathrm{CO_2}}$—mass flow rate of oxygen atoms in the composition of CO_2 in the exhaust gas;

$Q^{\mathrm{O}}_{\mathrm{H_2O}}$—mass flow rate of oxygen atoms in the composition of H_2O in the exhaust gas.

$Q^{\mathrm{O}}_{\mathrm{CO}}$ and $Q^{\mathrm{O}}_{\mathrm{CO_2}}$ can be determined from the corresponding mass flow rates of CO (23) and CO_2 (24) using the formulas:

$$Q^{\mathrm{O}}_{\mathrm{CO}} = Q_{\mathrm{CO}} \frac{M_{\mathrm{O}}}{M_{\mathrm{CO}}}. \tag{26}$$

$$Q^{\mathrm{O}}_{\mathrm{CO_2}} = Q_{\mathrm{CO}} \frac{M_{\mathrm{O}}}{M_{\mathrm{CO_2}}}. \tag{27}$$

The mass flow rate of oxygen atoms in water molecules in the exhaust gases from (25) taking into account (20 and 21) will be:

$$Q^{\mathrm{O}}_{\mathrm{H_2O}} = Q^{\mathrm{O}}_{\mathrm{air}} - Q_{\mathrm{CO}} \frac{M_{\mathrm{O}}}{M_{\mathrm{CO}}} - Q_{\mathrm{CO}} \frac{M_{\mathrm{O}}}{M_{\mathrm{CO_2}}}. \tag{28}$$

Then, the mass flow rate of water $Q^{\mathrm{O}}_{\mathrm{H_2O}}$ in the exhaust gases will be:

$$Q^{\mathrm{O}}_{\mathrm{H_2O}} = Q^{\mathrm{O}}_{\mathrm{H_2O}} \cdot \frac{M_{\mathrm{H_2O}}}{M_O}. \tag{29}$$

The volumetric concentration of water $k_{\mathrm{H_2O}}$ in the exhaust gases can be calculated using the formula:

$$k_{\mathrm{H_2O}} = \frac{Q^{\mathrm{O}}_{\mathrm{H_2O}}}{M_{\mathrm{H_2O}}} \cdot \frac{k_{\mathrm{CO}}}{\nu_{\mathrm{CO}}}. \tag{30}$$

Balance equation for hydrogen atoms:

$$Q^{H}_{\mathrm{fuel}} = Q^{H}_{\mathrm{H_2O}} + Q^{H}_{H_2}, \tag{31}$$

where

Q^{H}_{fuel} is the mass flow rate of hydrogen atoms in the fuel;

$Q^{H}_{\mathrm{H_2O}}$ mass flow rate of oxygen atoms in the composition of H_2O in exhaust gases (water flow rate is calculated according to (30);

$Q^{H}_{H_2}$ mass flow rate of hydrogen atoms in gaseous form H_2.

The mass flow rate of hydrogen atoms in the fuel is calculated using the formula:

$$Q^{H}_{\mathrm{fuel}} = Q_{\mathrm{fuel}} \cdot \frac{M_H}{M_{\mathrm{C8H18}}} \tag{32}$$

The mass flow rate of oxygen atoms in the composition of H_2O in the exhaust gases is expressed through the mass flow rate of water (29) according to the formula:

$$Q^{H}_{\mathrm{H_2O}} = Q^{O}_{\mathrm{H_2O}} \cdot \frac{M_H}{M_{\mathrm{H_2O}}} \tag{33}$$

Then, from (31 and 32), the mass flow of hydrogen atoms in the gaseous form of H_2 can be expressed in terms of fuel flow and water flow in the exhaust gases:

$$Q^{H}_{H_2O} = Q^{O}_{H_2O} \cdot \frac{M_H}{M_{\mathrm{H_2O}}} \tag{33}$$

The volumetric concentration of molecular hydrogen kH_2 in the exhaust gases can be calculated using the formula:

$$k_{H_2} = \frac{Q^{H}_{H_2}}{M_{\mathrm{H_2}}} \cdot \frac{k_{\mathrm{CO}}}{\nu_{\mathrm{CO}}} \tag{35}$$

Thus, it is possible to consider the presence of H_2 in the exhaust gas by increasing the CO concentration by $\frac{Q^{H}_{H_2}}{M_{\mathrm{H_2}}} \cdot \frac{k_{\mathrm{CO}}}{\nu_{\mathrm{CO}}}$ times.

Measurements were conducted on the emissions reduce system of a gasoline V8 internal combustion engine with an attached exhaust system in a stand configuration. The engine had a displacement of 4392 cm^3, direct fuel injection with electronic control, and an injection pressure of 20–30 MPa. The nominal power was 440 kW at a rotational speed of 5500 rpm. The engine was installed on an engine test stand based on the dynamometer module for engine testing from FEV Europe GmbH, equipped with the MEXA-ONE-D2 gas analysis system.

The recalculated oxygen capacity, measured by two methods considering the derived formulas, showed good agreement (Table 1).

Table 1 Recalculated oxygen storage capacity (OSC), measured by two methods

Method of measurement	OSC value, mg, for exhaust gas flow					
	50 kg/h		100 kg/h		140 kg/h	
	Row 1	Row 2	Row 1	Row 2	Row 1	Row 2
Technological OSC	453	481	422	436	425	347
Motor OSC	502	502	420	414	429	375
Relative error	9%	4%	0,5%	5%	1%	7%

3 Conclusion

The recalculated oxygen storage capacity (OSC), measured by two methods considering the derived formulas, showed good match (Table 1). An error within 9% indicates the possibility of monitoring the aging of blocks on a vehicle or engine test stand using diagnostic programs and assessing the functionality of the block promptly. Calculations of chemical kinetics show a direct relationship between the values of oxygen storage activity measured in two ways. The technological OSC is constantly measured by the car control unit and calculates indirect values showing the relative oxygen activity of the unit. Using the conversion factor, it is possible to judge the true oxygen content in the catalyst and its change during service life, aging, and calibrations. This allows you to select the optimal block size and calibration.

References

1. Nadareishvili, G.G.: Generalized models of processes occurring in neutralization systems. Int. J. Emerg. Trends Eng. Res. **8**(1), 119–129 (2020)
2. Nadareishvili, G.G.: Methods for developing a system for after-treatment of exhaust gases. Monograph, Ed. "Transport Literature", pp. 192, Moscow (2020)
3. Panchishny, V.I., Luksho, V.A., Kozlov, A.V., Terenchenko, A.S., Karpukhin, K.E.: The problem of toxicity of gas engines and their solutions. Biosciences Biotechnology Research Asia, vol. 12(spl. Edn. 2), pp. 217–224 (2015)
4. Kulikov, I.A., Shorin, A.A., Bakhmutov, S.V., Terenchenko, A.S., Karpukhin, K.E.: A Method of powertrain's components sizing for a range extended electric vehicle. SAE Technical Papers 2016–01–8096
5. Beccari, S., Pipitone, E.: A new simple function for combustion and cyclic variation modeling in supercharged spark ignition engines. Energies **15**, 3796 (2022). https://doi.org/10.3390/en15103796
6. King, J.: The king review of low-carbon cars. Part II: recommendations for action/J. King. HM Tresaruy (2007)
7. Bellis, D.V.: Performance optimization of a spark-ignition turbocharged VVA engine under knock limited operation. Appl. Energy **164**, 162–174 (2016)
8. Shojaeefard, M.H., et al.: Cooled EGR for a turbo charged SI engine to reduce knocking and fuel consumption. Int. J. Autom. Eng. **3**(3), 474–481 (2013)
9. Yi, J., Wooldridge, S., Coulson, G., Hilditch, J., Iyer, C.O., Moilanen, P., Schamel, A.: Development and optimization of the Ford 3.5 L V6 ECOBoost Combustion system. SAE Int. J. Engines **2**(1), 1388–1407 (2009)
10. Li, R., Liu, F., Yang, D., Hu, Q., Hua, Y., Yang, M., & Liu, S.: Study on the formation and ignition law of ignition core in ammonia atmosphere diesel cylinder. Fuel//vol. 365, 1 June 2024, p. 131257
11. Gusakov, S.V., Gusakov, S.V., Markov, V.A., Afanasyeva, I.V.: Improving the operational performance of vehicles when using hybrid power plants. BMSTU J. Mech. Eng. no. 2, pp. 32–41 (2012)
12. Zhang, Y., Shen, W., Zhang, H., Wu, Y., Lu, J.: Effects of inert dilution on the propagation and extinction of lean premixed syngas/air flames. Fuel **157**, 115–121 (2015)
13. Wiemann, S.: Combined production of power and syngas in an internal combustion engine–experiments and simulations in SI and HCCI mode. Fuel **215**, 40–45 (2018)
14. Wierzba, I.: The flammability limits of H_2–CO–CH_4 mixtures in air at elevated temperatures. Int. J. Hydrog. Energy **31**(4), 485–489 (2006)

15. Varghese, R.J., Kolekar, H., Kumar, S.: Demarcation of reaction effects on laminar burning velocities of diluted syngas–air mixtures at elevated temperatures. Int. J. Chem. Kinet. **51**(2), 95–104 (2019)
16. Suzuki, R., Shoji, H., Yoshida, K., Iijima, A.: Analysis of knocking in an SI engine based on in-cylinder: spectroscopic measurements and visualization (No. 2010-32-0092). SAE Technical Paper (2010)
17. Shang, R., Zhang, Y., Zhu, M., Zhang, Z., Zhang, D., Li, G.: Laminar flame speed of CO_2 and N_2 diluted H_2/CO/air flames. Int. J. Hydrog. Energy **41**(33), 15056–15067 (2016)
18. Singh, D., Nishiie, T., Tanvir, S., Qiao, L.: An experimental and kinetic study of syngas/air combustion at elevated temperatures and the effect of water addition. Fuel **94**, 448–456 (2012)
19. Lewis, B., Von Elbe, G.: Combustion, Flames and Explosions of Gases. Elsevier (2012)

Automotive Aerodynamics Analysis: An Implementation of OpenFOAM Software

Huu Khai Nguyen, Duc Khanh Nguyen, Duy Tien Nguyen, and The Truc Nguyen

Abstract This work focuses on the computational study and simulation of aerodynamics around a passenger car using the OpenFOAM software. The problem is defined and calculated using the Tesla Model 3 prototype–an all-electric vehicle highly regarded for its performance and aerodynamic design. Throughout the work, various results were analyzed, including drag force, lift force, drag coefficient, moment coefficient, and front and rear lift coefficients. Additionally, the work pays special attention to pressure distribution results and streamlining force distribution. The results found that the electric car model being studied exhibits high aerodynamic efficiency, allowing for stable operation at both regular and highway speeds. Furthermore, the optimization in the design is shown to improve the car's grip on the road at the front and rear and optimize the side mirrors' design, reducing drag, providing good visibility, and minimizing friction noise between the vehicle and the surrounding airflow.

Keywords Aerodynamics · Computational · OpenFOAM · k-omega-RANS model

1 Introduction

Recently, climate change has become a global concern. The greenhouse effect has accelerated ice melting in the Arctic, leading to rising sea levels. The increasing temperatures along the equator have also created dry regions and expanded deserts. Several main factors contributing to the greenhouse effect include excessive use of fossil fuels for transportation, heating, and deforestation for housing. Major cities

H. K. Nguyen · D. K. Nguyen · D. T. Nguyen · T. T. Nguyen (✉)
Department of Vehicle and Energy Conversion Engineering, School of Mechanical Engineering, Hanoi University of Science and Technology, No.1 Dai Co Viet Street, Hanoi 10000, Vietnam
e-mail: truc.nguyenthe@hust.edu.vn

H. K. Nguyen
Department of Mechanical Engineering, Gyeongsang National University, Jinju, Korea

D. H. Tien et al. (eds.), *Proceedings of the 1st International Conference on Sustainability and Emerging Technologies for Smart Manufacturing*, Proceedings in Technology Transfer, https://doi.org/10.1007/978-981-97-7083-0_31

worldwide, such as Beijing in China, London in the UK, Paris in France, and Washington DC in the USA, have reached alarming levels of air pollution, affecting public health. Therefore, addressing, managing, and reducing environmental pollution have become urgent priorities in many countries. Currently, the transportation sector consumes a sizable portion of fossil fuels in the form of gasoline and diesel. In Europe, the transportation sector accounts for up to 33% of total fossil energy consumption [1]. As a result, automotive manufacturers are focusing on finding solutions to reduce fuel consumption, increase efficiency, and minimize losses. Electric and hybrid vehicles have become top priorities as part of the electrification efforts. While fuel consumption and drivetrain optimization are addressed, other factors influencing the vehicle's performance are also being considered. According to Piechna [2], the energy consumption of a vehicle is significantly affected by aerodynamic drag. For electric vehicles, aerodynamic drag can account for up to 48% of the total driving resistance at high speeds [3]. Improving the aerodynamics of electric vehicles can increase their operating range (the maximum distance the vehicle can travel on a full charge) [4]. As a result, in this study, the influence of aerodynamics on the design of electric vehicles is explored through a simulated analysis of the Tesla Model 3 prototype. Through the observations by using OpenFOAM, it is evident that this vehicle possesses high aerodynamic efficiency and maintains stable performance at both normal and highway speeds. The simulation results also show the optimization design of the vehicle.

2 Model Design and Numerical Method

2.1 *Model Design*

In aerodynamics, the blockage ratio is the ratio between the cross-sectional area of an object placed in the fluid flow (such as air) and the total cross-sectional area of the flow channel. For example, in wind tunnels, the test model represents the object, while the cross-sectional area of the wind tunnel represents the flow channel. The blockage ratio is important because it affects the accuracy of the test results: a higher blockage ratio means that more fluid flow is obstructed by the object, which can cause disturbances and affect the accuracy of measurements. Therefore, it is desirable to keep the blockage ratio as low as possible during testing or experimentation.

$$B = A_{\mathrm{f}}/A_{\mathrm{w}} \tag{1}$$

In which: A_{f} is the frontal area of the model, and A_{w} is the cross-section of the wind tunnel. When the blockage ratio is high, it should be adjusted to minimize the blockage and achieve the best results:

$$C_{d\_c} = wC_{d\_m} \tag{2}$$

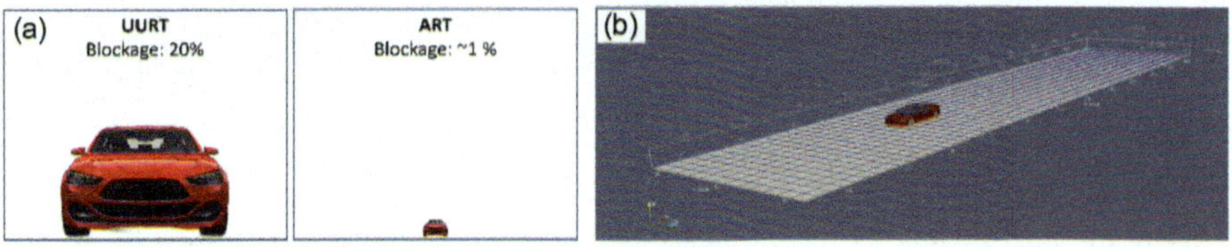

Fig. 1 Two kinds of tunnel models (**a**) [6] and ART Tunnel model used for simulation (**b**)

With: C_{d_c} and C_{d_m} are, respectively, the corrected and measured drag coefficients; w is the correction factor, represented as a function of the blockage ratio [5]. The correction factor (w) is inversely proportional to the square of the blockage ratio:

$$w = f.(1 - B)^2 \tag{3}$$

Typically, aerodynamic studies are performed in wind tunnels. The "UURT" open-jet wind tunnel determines the blockage effect with a cross-section of 0.7 m × 0.6 m and a free-stream turbulence level of 0.05, with a maximum free-stream velocity is 30 m/s. The "ART" closed-circuit wind tunnel has a 3.05 m × 2.44 m cross-section area with a maximum free-stream speed of 90 m/s and a turbulence level 0.005. The main difference between the two wind tunnels is the turbulence level and the magnitude of the blockage in the experiments. UURT has a blockage effect of 20% for the 1/5 scale model, while the ART tunnel has a blockage effect of 1% (Fig. 1a).

Based on the assessment of blockage ratios, the ART wind tunnel model is considered optimal for aerodynamic simulations and studies as it provides results closest to reality without being limited by the survey space. In this study, the authors used the ART wind tunnel model to investigate the aerodynamics of the Tesla Model 3 [7]. Therefore, the model's dimensions are Length × Width × Height = 4690 mm × 1930 mm × 1440 mm. The dimensions of the ART wind tunnel are calculated as follows: L_ART = 3L + L + 8L = 12L; W_ART = 6W; H_ART = 5H.

Additionally, due to the complexity and time-consuming nature of drawing and simulating the original shape of the Tesla Model 3, the authors directly utilized available open-source CFD models of these cars at a lower cost, suitable for research purposes. We used Blender to remove unnecessary parts for the calculations and proceeded with numerical simulations (Fig. 1b).

2.2 Mathematical Model

When operated at any speed, a vehicle will experience interactions with airflow. Because the flow around the vehicle is an incompressible flow, based on common computational fluid dynamics (CFD) assessments, the model used in this study is the Reynolds-Average Navier–Stokes (RANS) model due to its advantages such as low computational cost and the ability to produce accurate and reliable simulation results:

$$\left(\frac{\partial v}{\partial t} + v\nabla v\right) = -\nabla p + \mu\nabla^2 v + f \tag{4}$$

Among various RANS turbulence models, the k-ω model, developed by Wilcox [8], has garnered widespread acceptance due to its high accuracy in drag prediction and its relative computational efficiency. In light of these advantages, the k-ω model has been designated as the priority turbulence model for this investigation.

Additionally, in most aerodynamic problems, aerodynamic drag force is the primary quantity focused. This drag force is generated by the combined effects of frictional forces and pressure on the vehicle's geometric shape. The more streamlined the vehicle's geometric shape is, the less aerodynamic drag force is generated. The variation of aerodynamic drag force is called aerodynamic performance. Several nondimensional parameters are used to compare the aerodynamic performance between different geometric shapes, such as different vehicle designs. These parameters are known as the drag coefficient (C_{D}) and the lift coefficient (C_{L}), which are calculated using the forces acting on the geometry, implied from Eq. 5. Besides, in this work, the other components' coefficient forces, like the pitching moment coefficient, frontal lift coefficient, and rear lift coefficient, are also figured out.

$$F_{\mathrm{Drag/Lift}} = \frac{1}{2}\rho C_{\mathrm{D/L}} A_{\mathrm{f}} v^2 \tag{5}$$

2.3 Numerical Method

Simple FOAM is a non-compressible flow solver within OpenFOAM that addresses simple aerodynamic problems. It is an algorithm based on the finite volume (FVM) method combined with numerical solution methods.

Understanding the interaction between objects and fluid flow is crucial for this study. While Fig. 2a delves into the *k*-omega turbulence model algorithm within a simple FOAM solver, Fig. 2b highlights "libforces.so," a dynamic library empowering OpenFOAM applications to efficiently simulate and analyze such interactions. This library enables the calculation of various aerodynamic effects on objects moving through fluids, including drag, lift, and other relevant forces. It achieves this by enhancing OpenFOAM's ability to accurately compute fluid forces acting on object surfaces governed by the Navier–Stokes equations.

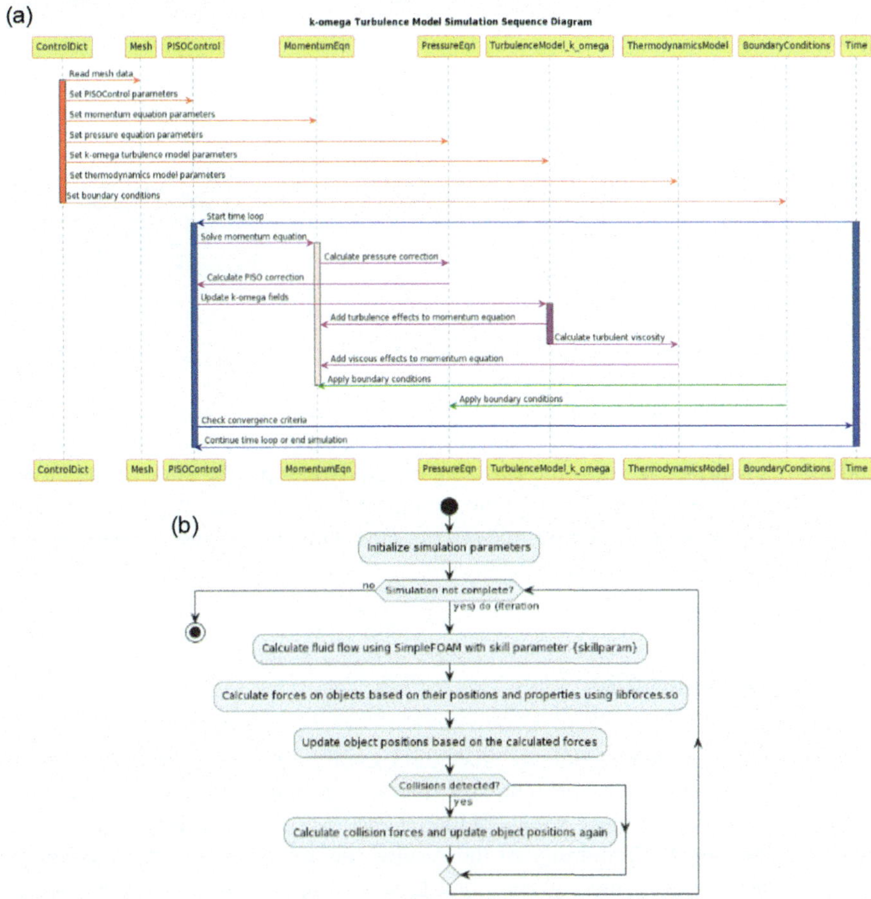

Fig. 2 **a** Algorithm of the turbulence model k-omega in simple FOAM solver and **b** Algorithm of the libforce.so library in OpenFOAM

2.4 Boundary Conditions

To calculate lift force, drag force, and other relevant coefficients, specific parameters such *magUInf* (magnitude velocity infinite), *lRef* (wheelbase), and *Aref* (frontal cross-sectional area) are also set up. As officially announced by the Model 3 Tesla manufacturer, the values of *lRef* and *Aref* are 4.69 m and 2.20 m [7], respectively. The *magUInf* represents the wind tunnel velocity in the investigation, with values of 60 km and 100 km. Regarding boundary conditions, the problem involves two main components: the wind tunnel and the vehicle. The inertial conditions of the surveyed vehicle cause its velocity to be zero, whereas the wind tunnel will have an incoming flow velocity of *magUInf*. This study sets the transport model conditions to *Newtonian* and dynamic viscosity coefficient nu = 42.1e-05 [N^2/m]. Additionally, the

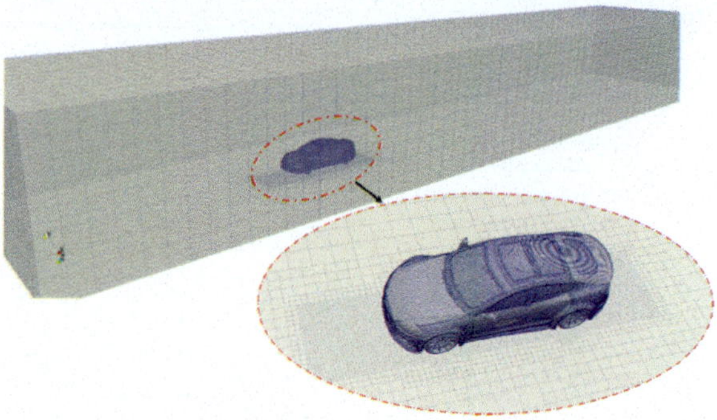

Fig. 3 Calculated mesh for aerodynamics across simulation model

zero-gradient condition is applied to the *pressure* at the wind tunnel inlet boundaries, road surface, and vehicle model. Conversely, the *velocity* at the boundary of the inlet and the road surface is set to the *magUInf* value. Normally, the state of the flow is determined by the value of the Reynolds number (Re = U*L/μ). Since when the car is running, there is a chaotic flow, the mixing flow will be set to the system such that the Reynolds value, Re, is set to 9,500,000.

2.5 Mesh Generation

In simulations, the mesh elements' size significantly affects the results' accuracy, convergence, computation time, and how interactions between surfaces are handled in the problem. A finer mesh results in higher accuracy [9]. Therefore, in this simulation, to increase the reliability of the results, the author divides the model into two separate parts for mesh generation. The wind tunnel is created and meshed using the blockMesh utility in OpenFOAM. The remaining part, the Tesla model, is divided using the snappyHexMesh utility. The mesh is generated with 1,770,207 cells, 5,624,654 faces, and 2,120,525 points (Fig. 3).

3 Results and Discussion

3.1 Drag Force and Drag Coefficient

Aerodynamic drag results from the pressure difference between the vehicle's front and rear. As the velocity increases, aerodynamic drag also increases. Generally, it is influenced by the cross-sectional area, speed, and air density surrounding the vehicle. At different velocities, the pressure difference varies, and this difference is indicated as shown in Fig. 4.

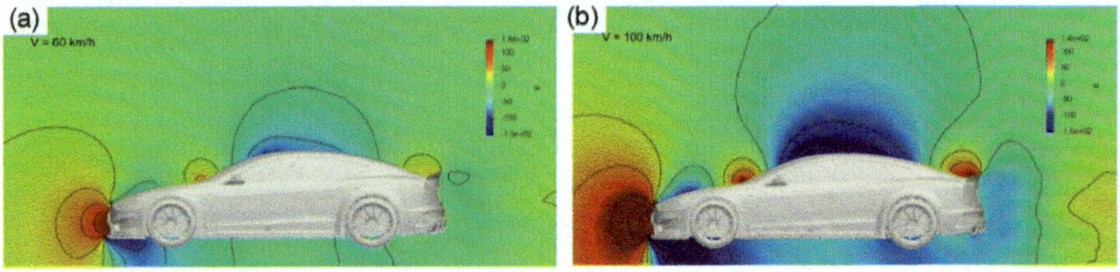

Fig. 4 Simulation result of pressure contour at 60 km/h (**a**) and 100 km/h (**b**)

Figure 4 represents the distribution of pressure around the vehicle at different speeds, showing the areas of high and low pressure, which contribute to the overall aerodynamic drag experienced by the car. As the velocity increases, the pressure difference also increases, leading to an increase in aerodynamic drag. The relationship between flow velocity, pressure, and aerodynamic drag in the problem is described based on the RANS equations (Eq. 4) and drag equation (Eq. 5). In this study, as mentioned earlier, the Tesla Model 3 prototype was investigated at two velocity values of 60 km/h and 100 km/h. These are the two most common velocity values standing for two driving modes of the car: normal driving and high-speed driving. From the aerodynamic study, the drag force of the car was calculated and presented clearly, as shown in Fig. 5.

Clearly, at a regular driving speed of 60 km/h, the aerodynamic drag on the car is not significant, indicating that the car will operate at a high level of stability (Fig. 5a). In contrast, at a highway speed of 100 km/h, the drag force considerably impacts the car, with the average drag force value reaching 311.38 N, which is three times higher than the average drag force at 60 km/h. The influence of aerodynamic drag will affect the stability of the car's operation. In Fig. 5b, points with large drag force values are areas such as the front of the car, windshield, side mirrors, and rear. These

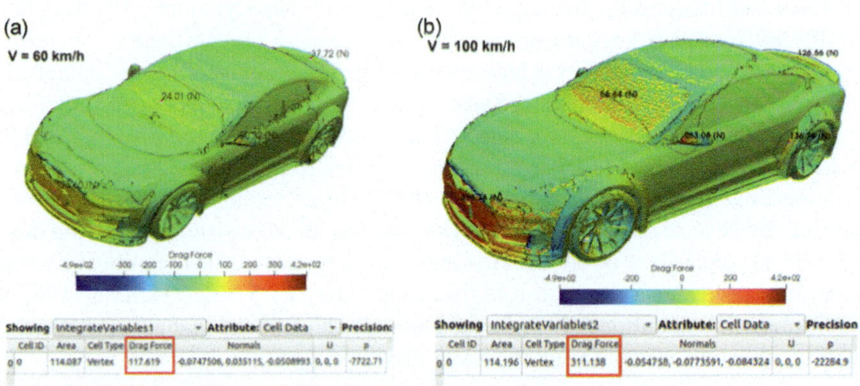

Fig. 5 Simulation results of drag force at 60 km/h (**a**) and 100 km/h (**b**)

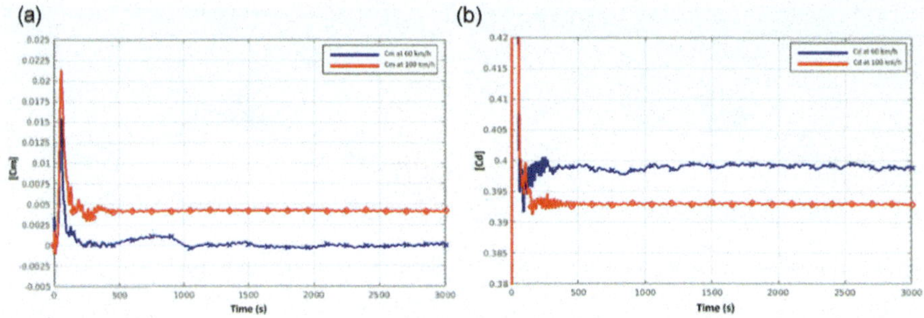

Fig. 6 Simulation results of C_m (**a**) and C_d (**b**) at 60 km/h and 100 km/h

areas cause instability in the flow of air around the car, creating resistance and being the main energy-consuming area of the car.

Parallel to the drag and lift force investigation, the coefficient of moment and drag are also important. Thus, in this section, we analyze two critical parameters: the moment coefficient (C_m) and the drag coefficient (C_d) in Fig. 6a and b. Obviously, both figures showcase an impressive level of stability. The moment coefficient remains negligible throughout the observation period, barely exceeding 0.0042 even at 100 km/h (as shown by the red-dotted line). Similarly, the drag coefficient stays remarkably constant, fluctuating between 0.39 and 0.4, regardless of the operating mode. This remarkable consistency in C_m and C_d points toward a highly stable prototype model under both working conditions.

3.2 *Lift Force and Lift Coefficient*

Lift force and lift coefficient are crucial in automotive aerodynamics, affecting both stability and road grip. For cars, optimal lift force creates a delicate balance, providing enough downforce for stability at high speeds while maintaining efficient contact with the road. This translates to enhanced steering precision, improved control, and a safer and more enjoyable driving experience. This study delves into the lift force and lift coefficient of the Tesla Model 3, presenting insights in Figs. 7 and 8.

Interestingly, at 60 km/h and 100 km/h, the lift force remains negative (downward, opposite the upward axis). This translates to the Model 3 consistently pressing down on the road, regardless of speed. However, the force intensifies significantly with increasing speed. A startling 2.8-fold rise occurs, from 68.5 N at 60 km/h to 194.7 N at 100 km/h. This additional downforce, a boon for high-speed maneuvers, enhances stability and control, translating to a safer driving experience.

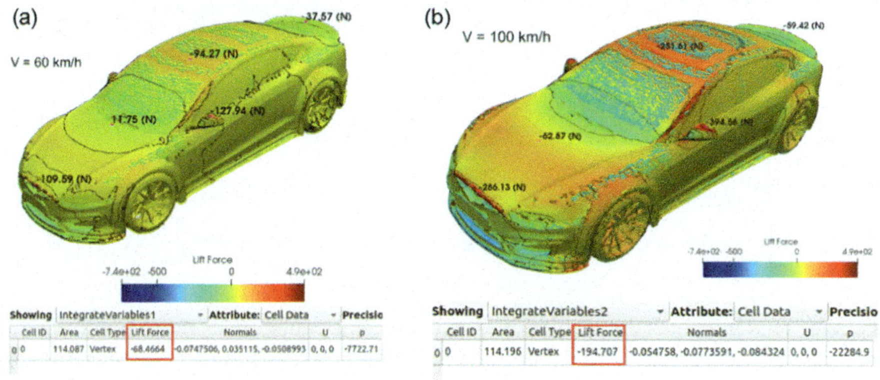

Fig. 7 Estimating of lift force at 60 km/h (**a**) and 100 km/h (**b**)

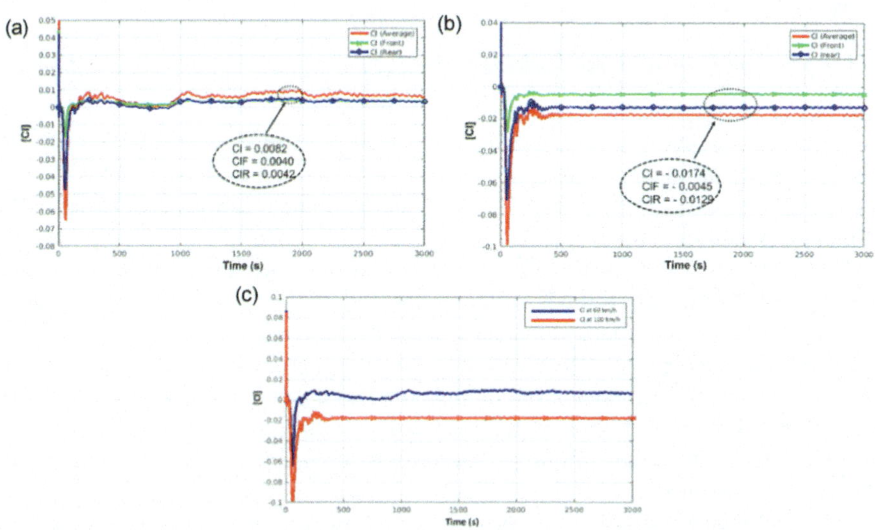

Fig. 8 Estimating of lift coefficients for front-end, rear-end, and total lift coefficient

While the areas experiencing the most lift force are the front bonnet, sunroof region, and rearview mirrors, Fig. 7a and b reveal a surprising uniformity in force distribution across the car. This showcases the design's effectiveness in generating increased grip and distributing it evenly. This even distribution ensures exceptional stability by minimizing the impact of localized lift forces on specific parts of the vehicle.

Further insights come from the lift coefficients. The disparity between the front and rear lift coefficients is speed-dependent. At 60 km/h (Fig. 8a), the car glides smoothly with admirable stability, seemingly unfazed by the lift force. However, at 100 km/h (Fig. 8b), a difference emerges of – 0.0045 for the front and – 0.0129 for

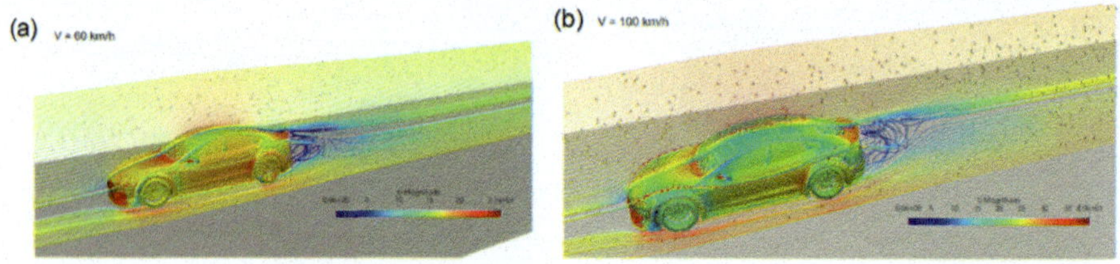

Fig. 9 Streamline across the vehicle's body at 60 km/h (**a**) and 100 km/h (**b**)

the rear. This indicates that at higher speeds, the rear experiences more lift compared to the front. Nevertheless, the overall lift coefficient (Cl) still promotes stability and improved road grip. Although some disparity exists between front and rear Cl values, the car's overall stability remains enhanced.

3.3 Streamline Across Vehicle

In this study, the deformation of streamlines is also a significant focus of analysis. The two most important positions to consider are the deformation of streamlines over the car's body and the side mirrors because these are the locations experiencing the most aerodynamics (Fig. 9).

Generally, there is not a significant difference in the streamline deformation between the two cases studied. A common trend is the flow becoming turbulent and generating vortices behind the car's tail. The main cause is the pressure difference, as mentioned earlier. These vortices also play a crucial role in influencing the vehicle's drag and lift forces. Even though vortices appear behind the car, it can be seen that the velocities of these vortices are low at both studied velocities. Thanks to this, the manufacturer has optimized the level of influence of these vortices on the vehicle's actual values.

By aerodynamic simulations, observing the streamlines over the side mirror provides developers with insights into the streamlines' deformation, enabling them to understand how to design the mirrors most suitable for optimal efficiency. In Fig. 10, this design shows that the streamlines over the side mirrors are quite smooth. This design, Tesla Model 3's side mirrors, has effectively guided the flow smoothly without causing any disruption or instability, leading to the absence of vortices at both low and high velocities. This means that this mirror design has optimized at least three aspects: aerodynamics (smooth flow over the mirrors), optical efficiency (uninterrupted flow, providing no blind spots for rear observation by the driver), and reduced noise (lack of vortices behind the side mirrors, reducing air disturbances and potential collisions with the airflow).

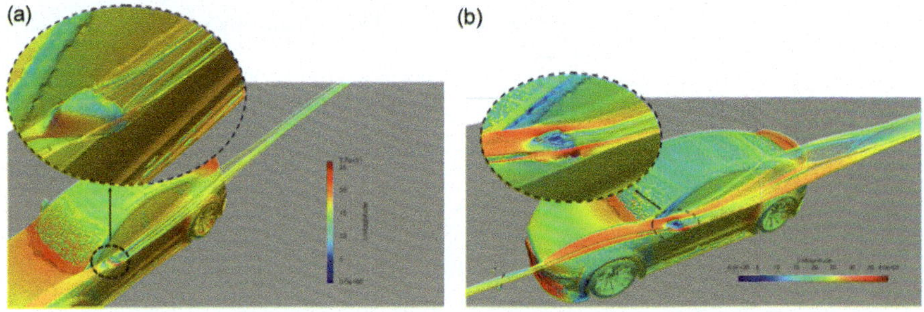

Fig. 10 Streamline across the vehicle body at 60 km/h (**a**) and 100 km/h (**b**)

4 Conclusion

In this work, OpenFOAM simulation software was used to conduct aerodynamic investigations on the Tesla Model 3 prototype. Through the observations, it is evident that this vehicle possesses high aerodynamic efficiency and maintains stable performance at both normal and highway speeds. Additionally, the results show the successful optimization of the design, not only ensuring a balanced downforce at the front and rear of the car but also perfecting the side mirror design to reduce drag, supply clear visibility, and minimize noise generated from the interaction between the vehicle and the surrounding airflow.

References

1. Eurostat.: Energy & transport in figures. Luxembourg: European Union. Available at: ec.europa.eu/transport/factsfundings/statistics/doc/2013/pocketbook2013.pdf (2015). Accessed 15 Dec 2023
2. Piechna, J.: A review of active aerodynamic systems for road vehicles. Energies **14**(23), 7887 (2021). https://doi.org/10.3390/en14237887
3. Lohse-Busch, H., Duoba, M., Rask, E., Stutenberg, K. et al.: Ambient temperature (20°F, 72°F and 95°F) impact on fuel and energy consumption for several conventional vehicles, hybrid and plug-in hybrid electric vehicles and battery electric vehicle. SAE Technical Paper 2013–01–1462 (2013). https://doi.org/10.4271/2013-01-1462
4. Palin, R., Johnston, V., Johnson, S., D'Hooge, A. et al.: The aerodynamic development of the tesla model S—Part 1: overview. SAE Technical Paper 2012–01–0177 (2012). https://doi.org/10.4271/2012-01-0177
5. Giri, F., El Majdoub, K., Ouadi, H.: Accounting for tire effect in longitudinal vehicle control. American Control Conference. IEEE (2009). https://doi.org/10.1109/acc.2009.5159898
6. Altinisik, A., Kutukceken, E., Umur, H.: Experimental and numerical aerodynamic analysis of a passenger car: influence of the blockage ratio on drag coefficient. ASME J. Fluids Eng. **137**(8), 081104 (2015). https://doi.org/10.1115/1.4030183
7. Model 3 | Tesla. (n.d.). Retrieved from https://www.tesla.com/model3. Accessed 15 Dec 2023

8. Wilcox, D.C.: Formulation of the k-w turbulence model revisited. AIAA J. **46**(11), 2823–2838 (2008). https://doi.org/10.2514/1.36541
9. Suresh, K., Regalla, S.P.: Effect of mesh parameters in finite element simulation of single point incremental sheet forming process. Proc. Mater. Sci. **6**, 376–382 (2014). https://doi.org/10.1016/j.mspro.2014.07.048

Achieving FIT Manufacturing Through the Lens of Industry 5.0? A Lean Perspective

Ton Nguyen Trong Hien, Noppadol Amdee, and Adisak Sangsongfa

Abstract There are still gaps in the literature, as there is a lack of comprehensive evaluations regarding whether the fifth industrial revolution contributes to businesses achieving leanness, one of the three components of FIT manufacturing, the production philosophies that aid businesses in optimizing operations, responding to market changes, staying competitive, innovative and fostering sustainable development. The effectiveness of the fourth industrial revolution in delivering clear benefits in this regard remains unproven. To avoid redundancy with previous studies, a combined approach utilizing keyword analysis through preference landscape analysis and content analysis was employed. To this end, extracting all relevant papers and deeply analyzing studies from the Scopus database, the study results reveal that Lean research within the context of the fifth industrial revolution is still in its early stages. While existing research shows promising signs compared to previous revolutions, answering this question requires further time to assess the long-term costs and benefits.

Keywords FIT manufacturing · Industry 5.0 · Lean · Sustainability

1 Introduction

Manufacturing is considered the backbone of many countries, especially those in the process of development. The manufacturing sector's contribution to Thailand's overall domestic output (GDP) has shown fluctuations in recent years. However, according to the World Bank [1], the value added by manufacturing in Thailand has not dropped below 26% over the past three years (2020 to 2022). In the case of Vietnam, the most recent three-year data indicates positive growth, reaching 25% in

---

T. N. T. Hien (✉) · N. Amdee · A. Sangsongfa
Department of Industrial Technology Management, Faculty of Industrial Technology, Muban Chom Bueng Rajabhat University, Chom Bueng 70150, Thailand
e-mail: 66p951003@mcru.ac.th

D. H. Tien et al. (eds.), *Proceedings of the 1st International Conference on Sustainability and Emerging Technologies for Smart Manufacturing*, Proceedings in Technology Transfer, https://doi.org/10.1007/978-981-97-7083-0_32

2022. In the same year, other representative countries such as Malaysia and Singapore had their manufacturing sectors accounting for approximately 23% and 21% of GDP, respectively. In pursuit of growth and lasting success in the global economy, manufacturers seek reliable approaches to enhance their businesses and outperform competitors. In their quest for improvement, manufacturers initially aimed to reduce costs but soon found themselves under pressure for continuous enhancements in cost optimization, quality, and efficiency. In the current scenario, organizations must adapt to evolving circumstances, and to foster economic growth, companies are revisiting their corporate strategy by incorporating and implementing more comprehensive sustainable business practices.

The concept of "FIT manufacturing" (FIT) was arguably introduced in an article by Pham and Thomas [2]. The FIT manufacturing framework, as outlined by Thomas and Pham, plays a crucial role in helping manufacturers achieve economic sustainability and navigate market competition. In their examination of manufacturing trends, Thomas and Pham addressed the subject by introducing the foundational concept of the "FIT" strategy, which integrates three models of manufacturing concepts: Lean management, agile management, and sustainability concepts. FIT is regarded as a manufacturing strategy that converges elements sought after by manufacturers because it suggests factors for manufacturers to embrace and develop sustainability in the dynamic economy. However, achieving FIT is not a straightforward task. The fourth industrial revolution (I4.0) brings about significant changes in the manufacturing revolution. According to Hien's research [3], the positive influence of I4.0 on FIT is clearly evident in a positive direction. Conversely, similar to the findings of some other studies, the impact of I4.0 on Lean manufacturing remains unclear. Looking back at history, each revolution has become the foundation for subsequent revolutions. Therefore, it is expected that the fifth revolution (I5.0) will be built upon I4.0, but it will take a step further. Both I4.0 and I5.0 are crucial stages in technological development, but they exhibit clear differences. I4.0 is technology-driven, aiming for automation and a focus on technology enhancement. In contrast, I5.0 is value-driven, believed to balance efficiency and productivity, promoting collaboration between humans and technologies. The question arises whether I5.0 helps manufacturing enterprises achieve the "FIT" state. To answer this question, we need to clarify the issue that I4.0 has not been able to address, which is the Lean factor.

2 Literature Review

In addition to numerous studies suggesting that I4.0 has a positive impact on Lean implementation, contributing to speed productivity and facilitating autonomous operations [4–7], and enhancing cost-competitiveness [6, 7], some earlier research has proposed worries about influence of I4.0 on Lean manufacturing. The eagerness to embrace I4.0 technologies often overlooks crucial considerations such as cost–benefit analysis, deployment strategies, and the potential disruption to existing manufacturing operations and Lean models [8, 9]. Despite the touted benefits, integrating

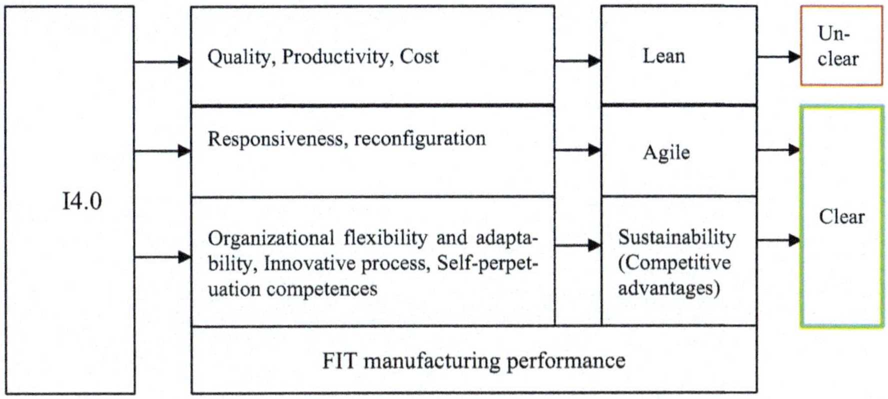

Fig. 1 Previous findings have explored the relationship between I4.0 technology and various aspects of FIT manufacturing

Lean principles with I4.0 experiences high failure rates due to the novelty of the concept and its limited scope. Achieving higher levels of I4.0 adoption requires careful planning, otherwise organizations risk ineffective digital transformations if they blindly follow the I4.0 trend without tailoring solutions to their unique attributes [10, 11]. Gaps persist between Lean, typically operational, and I4.0, which tends to domain of engineering or information technology. Researchers explains that Lean practices are automated without proper integration with I4.0 technologies, it can result in digital waste [12, 13].

The experimental study [3] has demonstrated that I4.0 has a positive impact, enabling manufacturers to achieve what is referred to as FIT manufacturing. This is supported by clear evidence documented through the Production Adaptive Index (agility) and Production Stability Index (sustainability). The research model is further illustrated in Fig. 1.

Considering a revolutionary change, the 5.0 revolution has brought about a breath of fresh air. However, when a new term is introduced, there are often multiple interpretations and different understandings. Therefore, a study [14] has summarized and synthesized 72 documents to provide a multidimensional perspective on the 5.0 revolution (Fig. 2). The I5.0, considered a breakthrough, introduces a fresh perspective. As a new term is often subject to various interpretations, a study [15–17] has synthesized 72 documents to provide a multidimensional view of the 5.0 revolution. The focus of the 5.0 revolution lies in the collaboration between machines and humans, emphasizing management methods that support sustainability in the industrial ecosystem. This collaborative approach optimizes management and production systems through the integration of smart production systems including hardware and software solutions. I5.0 is gradually exploring

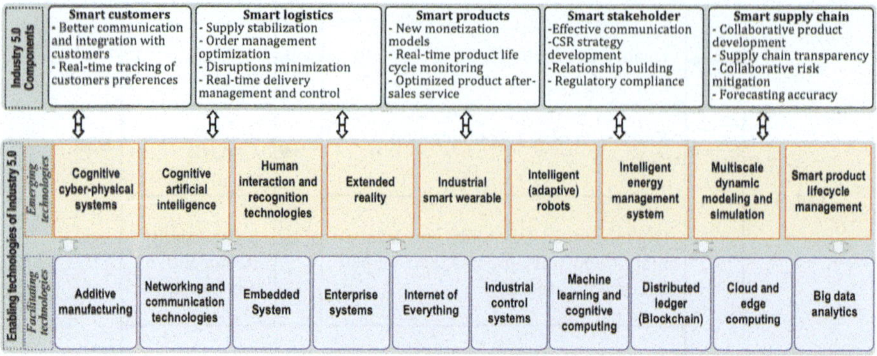

Fig. 2 The abstract for the architectural design of Industry 5.0

new potentials, and unique technologies are emerging in this endeavor. Cyber–Physical systems, capable of self-learning and automatic interaction, merge artificial intelligence with the physical world. This capability not only creates self-sustaining systems but also opens up new prospects in automation and management. In the future, artificial intelligence becomes increasingly "cognitive," with the ability to understand and make decisions akin to humans. Recent studies [15–17] suggest that this not only aids computers in processing complex information but also enhances robust human–machine interactions, optimizing work processes and strategic decision-making.

Technological advancements in human recognition and interaction, including voice, facial, and gesture recognition, have become crucial. These developments not only provide personalized user experiences but also improve communication between humans and machines, especially in areas such as customer service and education. In addition to the mentioned technologies, such as extended reality, industrial smart wearables, intelligent robots, intelligent energy management systems, multi-scale dynamic modeling and simulation, and smart product lifecycle, the 5.0 industrial revolution is witnessing significant changes in how we manage, produce, and interact with technology [15–17]. Examples of these changes include extended reality creating new spaces in user experiences by combining the real and virtual worlds, applicable from education to entertainment, posing challenges in data management and personal information security. Industrial smart wearables not only provide data and information directly at the workplace but also enhance safety and labor efficiency, requiring flexible management to harness the full potential of this technology. Intelligent robots, with the ability to learn and interact with the environment, open doors for automation in various industries. Managing them requires flexibility and interaction between humans and machines. While the 4.0 industrial revolution introduced connectivity and automation, the 5.0 revolution presents challenges in flexible management, integrated information, and closer interaction between humans and technology.

It can be observed that I4.0 has limited relevance and unclear outcomes for businesses applying Lean, as evidenced by the evaluation criteria. Recognizing that I5.0 brings positive changes in management and how individuals interact with technology—something lacking in I4.0. Therefore, this study aims to investigate whether I5.0 assists manufacturing enterprises in achieving a "FIT" state, specifically whether the barriers of Lean have been addressed or not. To the best of our knowledge, the purpose of this research has not been explored.

3 Research Methodology

The study conducts an analysis of previous literature, aiming to be comprehensive and thoroughly analyze the content.

In the initial step, a reference landscape map analysis is carried out to gain a clear understanding of the relationships between reference documents, providing an overall view of how ideas and research works are interconnected. This aids in tracking the development of a research field. This could be a network structure built based on top references in a specific research field. These references may include significant studies within that domain. Preference linkages serve to demonstrate the pertinence of others' works to the subject under discussion. Consequently, if a collection of records receives more frequent citations from other publications within a specific domain, there is a higher likelihood that these records pertain to that same domain [18, 19].

Using general keywords such as "industry 5.0" or "the fifth industrial revolution" (I5.0), "production" or "manufacturing" a significant number of documents were obtained, utilizing the reliable ScienceScape [20] literature exploration software.

In the second step, when specifically searching for articles discussing the relationship between I5.0 and Lean, there were not many pre-existing research documents. Therefore, content analysis is conducted to thoroughly examine the content.

The data was extracted at 15:01 on January 20, 2024 (GMT + 7). To ensure a certain level of reliability, the Scopus database is utilized, and only documents in the English language are considered for convenience in synthesis and comparison. The details are illustrated in Fig. 3.

4 Results and Discussion

Figure 4 illustrates the top references used to construct the structure of the citation network. Information about keywords is mapped onto this map based on relationships and connections with key reference documents. Important keywords that appear frequently may be depicted prominently in the chart, signified by a larger and more prominent bubble compared to others. It can be observed that I5.0 has a close relationship with I4.0 when two network nodes are positioned closely, indicating a

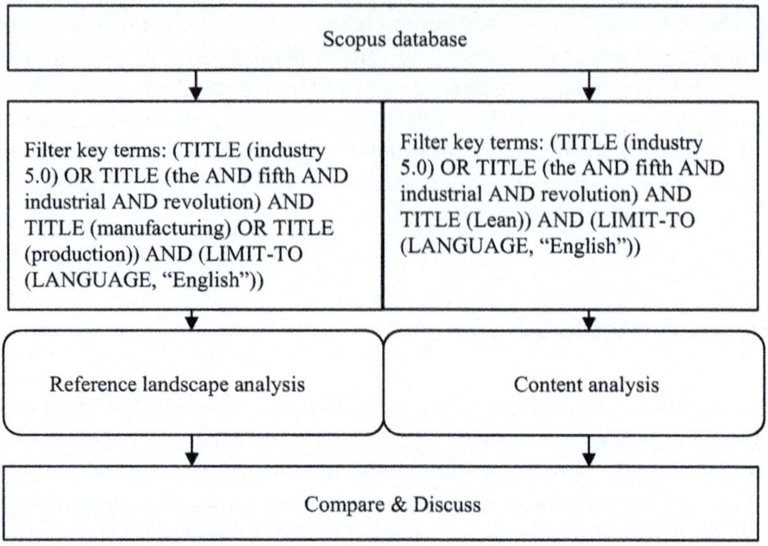

Fig. 3 Research framework model

seamless transition from I4.0 to I5.0. Alongside I5.0, where two crucial keywords are highlighted: "human-centric" and "sustainability" closely interlinked, indicating that the Industry 5.0 production model aims for sustainable development, placing human factors at the core.

Analyzing 79 received documents, "Digital twin," "AI," or "IoT" are indispensable technologies, but overall, these terms primarily signify support (Lying distant from the central network nodes of I4.0 and I5.0). Despite the strong presence of the keyword "augmented reality" (AR), both of these factors aim at optimizing production efficiency. AR assists in monitoring and managing the production process, providing continuous information about machinery status, while ergonomics optimizes job design and work performance. The relationship between ergonomics and AR is particularly close as they both target improving working conditions, safety, and comfort for workers. Through AR, crucial information about the production process, instructions, and machinery status can be directly displayed in the worker's field of view, helping to optimize workflow.

Meanwhile, there is still limited research on Lean and I5.0. Table 1 shows the overall development of I5.0 is clearly based on the advancements of I4.0. There is evidence of a connection between Lean tools and concepts and the design principles of I4.0 and I5.0. Indeed, the application of I5.0 principles in conjunction with Lean tools is highly feasible for enhancing optimistic indicators related to quality, efficiency, and productivity. Despite the prominence of technologies such as AI, IoT, and Digital Twin, it can be argued that expecting too much regarding sustainability in production may be unrealistic. Assessing the impact of 4.0 and I5.0 is a complex process that requires time for accurate evaluation. As mentioned, there are studies indicating

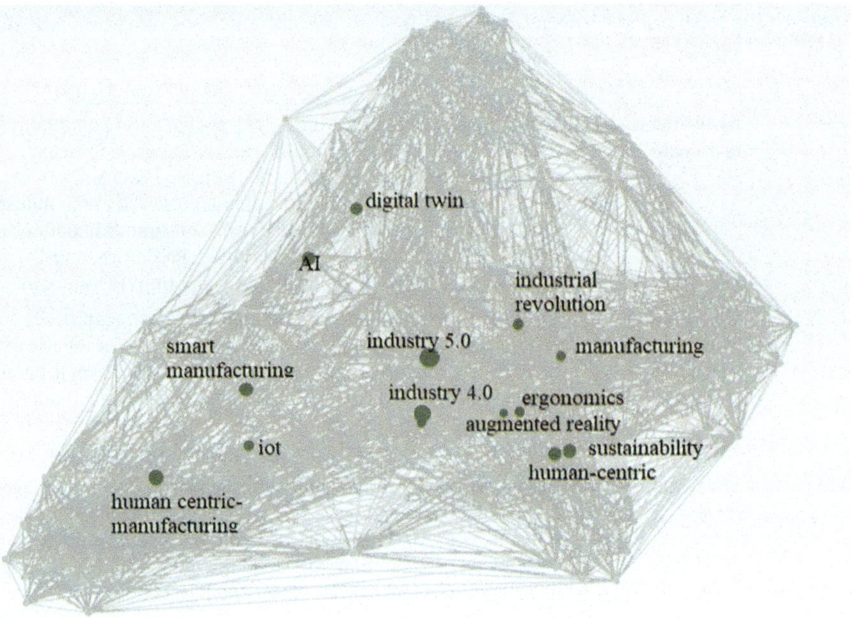

Fig. 4 Reference landscape

ineffective implementations leading to economic setbacks and digital waste. I4.0 has shown potential in improving efficiency and reducing waste through advanced technologies like IoT, AI, and automation, proving its effectiveness remains challenging, while I5.0 presents a new prospect of collaboration between humans and intelligent machines, substantiating its true value requires time and effort. This may entail innovative approaches to technology adoption and deployment, as well as thorough research and analysis to draw precise conclusions regarding the effectiveness and impact of both industries. Future research directions will require time to assess the experimental effectiveness of long-term profitability criteria concerning environmental, social, and occupational safety factors. If successfully addressed, I5.0 could indeed mark a pivotal step for manufacturing enterprises to achieve FIT manufacturing.

Table 1 The relationship between Lean and I5.0

Author (s)	Method	Findings
[21]	Real-time case approach	Traditional Lean manufacturing can evolve into Lean intelligent manufacturing by leveraging I5.0 technologies. Key Lean techniques, including VSM, 5S, Kaizen, and Kanban, can be effectively integrated with I5.0 technologies such as the Industrial Internet of Things (utilizing RFID and sensors), Simulation, Big Data Analytics (comprising Descriptive, Predictive, and Prescriptive Analytics), and Cloud Computing (for storage)
[22]	Game-based learning	The primary components of I5.0 concentrate on employees and activities during assembly. The paper advocates for the use of the Lean management training game methodology, which is based on the POKA YOKE method, 5S method, SMED method, SPC method, and push–pull method. This methodology serves as a tool to equip students with the necessary skills for I5.0
[23]	State-of-the-art review	Lean (specifically Lean manufacturing principles) and I5.0 are related in the application of Lean tools and concepts to the design principles of I5.0. Lean tools are used to implement the design principles of I5.0, aiming to extend the influence of technology to economic, social, and environmental aspects
[24]	Systematic review	Several interactions were identified between Lean tools and concepts and the design principles of I4.0 and I5.0. The study addresses significant potential for improvement by applying technologies focused on human-centric approaches, such as collaborative robots and arm exoskeletons. These technologies have the capability to enhance the strength and productivity of workers while promoting diversity and inclusivity in the workplace. Creating conditions for individuals with different languages, educational backgrounds, and disabilities to participate effectively in work contributes to establishing a diverse and inclusive work environment. This helps foster a workplace where people with unique characteristics and needs can participate and contribute
[25]	Empirical research	I5.0 aims to create a responsible and human-centric working environment. Lean Six Sigma also employs a similar approach by focusing on process optimization and quality, but with the primary goal of improving the experiences of both workers and customers. Both Lean and I5.0 prioritize sustainability at their core
[26]	Empirical research	The emergence of I5.0, in contrast to I4.0, considers sustainable production by aiming to enhance productivity, operability, minimize workplace accidents, and streamline production processes in an environmentally friendly and sustainable manner
[27]	Review	The article discusses the relationship between I5.0 and Lean. I5.0, with a focus on sustainability and human-centric approaches, is seen as aligning well with the Lean management philosophy, a proven method for improving processes and organization. Lean can play a significant role in supporting SMEs in adapting to I5.0 and addressing associated challenges

5 Conclusion

The overall development of I5.0 is evidently built upon the foundation of I4.0. Despite the prominence of AI, IoT, and Digital Twin technologies, it can be acknowledged that expecting significant strides in sustainability within production might be overly optimistic. There is evidence of a correlation between Lean tools and concepts and the design principles of I4.0 and I5.0. The application of I5.0 principles combined with Lean tools is highly feasible for enhancing optimistic indicators related to quality, efficiency, and productivity. However, the future research direction will require time to evaluate the experimental effectiveness of long-term profitability criteria concerning environmental, social, and occupational safety factors. If successfully addressed, this would position I5.0 as a crucial milestone for manufacturing enterprises to achieve FIT manufacturing.

References

1. Worldbank.: https://data.worldbank.org/indicator/nv.ind.manf.zs?locations=th-vn-id-sg-ph-my (2024). Last Accessed 26 Dec 2023
2. Thomas, A.J., Pham, D.T.: Making industry fit: the conceptualization of a generic "fit" manufacturing strategy for industry. In: 2nd IEEE International Conference on Industrial Informatics, pp. 523–528. IEEE, Berlin (2004)
3. Hien, TNT.: The interrelation between Industry 4.0 and FIT manufacturing: a step towards sustainability. In: Proceedings on the 6th International Conference on Contemporary Issues in Economics, Management and Business (6th CIEMB 2023), pp. 2147–2156. National Economics University Publishing house, Ha Noi (2023)
4. Narula, S., Puppala, H., Kumar, A., Luthra, S., Dwivedy, M., Prakash, S., Talwar, V.: Are Industry 4.0 technologies enablers of lean? Evidence from manufacturing industries. Int. J. Lean Six Sigma **14**, 115–138 (2022)
5. Shahin, M., Chen, F.F., Bouzary, H., Krishnaiyer, K.: Integration of Lean practices and Industry 4.0 technologies: smart manufacturing for next-generation enterprises. Int. J. Adv. Manuf. Technol. **107**, 2927–2936 (2020)
6. Yilmaz, A., Dora, M., Hezarkhani, B., Kumar, M.: Lean and Industry 4.0: mapping determinants and barriers from a social, environmental, and operational perspective. Technol. Forecast. Soc. Change **175**, 121320 (2022)
7. Reyes, J., Mula, J., Díaz-Madroñero, M.: Development of a conceptual model for lean supply chain planning in Industry 4.0: multidimensional analysis for operations management. Prod. Plann. Control **34**, 1209–1224 (2021)
8. Buer, S.-V., Strandhagen, J.O., Chan, F.T.S.: The link between Industry 4.0 and lean manufacturing: mapping current research and establishing a research agenda. Int. J. Prod. Res. **56**, 2924–2940 (2018)
9. Bouzary, H., Chen, F.F., Krishnaiyer, K.: A modified discrete invasive weed algorithm for optimal service composition in cloud manufacturing systems. Proc. Manuf. **17**, 403–410 (2018)
10. Rossini, M., Costa, F., Tortorella, G.L., Portioli-Staudacher, A.: The interrelation between Industry 4.0 and lean production: an empirical study on European manufacturers. Int. J. Adv. Manuf. Technol. **102**, 3963–3976 (2019)
11. Alsadi, J., Antony, J., Mezher, T., Jayaraman, R., Maalouf, M.: Lean and Industry 4.0: a bibliometric analysis, opportunities for future research directions. Qual. Manag. J. **30**, 41–63 (2023)

12. Hines, P., Tortorella, G.L., Antony, J., Romero, D.: Lean Industry 40: past, present, and future. Qual. Manag. J. **30**, 64–88 (2023)
13. Alieva, J., Powell, D.: The significance of digital waste in the automation of Lean practices. Qual. Manag. J. **30**, 121–134 (2023)
14. Ghobakhloo, M., Iranmanesh, M., Tseng, M.-L., Grybauskas, A., Stefanini, A., Amran, A.: Behind the definition of Industry 5.0: a systematic review of technologies, principles, components, and values. J. Ind. Prod. Eng. **40**, 432–447 (2023)
15. Leng, J., Sha, W., Wang, B., Zheng, P., Zhuang, C., Liu, Q., Wuest, T., Mourtzis, D., Wang, L.: Industry 5.0: prospect and retrospect. J. Manuf. Syst. **65**, 279–295 (2022)
16. Maddikunta, P.K.R., Pham, Q.V., Prabadevi, B., Deepa, N., Dev, K., Gadekallu, T.R., Ruby, R., Liyanage, M.: Industry 5.0: a survey on enabling technologies and potential applications. J. Ind. Inf. Integr. **26**, 100257 (2022)
17. Golovianko, M., Terziyan, V., Branytskyi, V., Malyk, D.: Industry 4.0 vs. Industry 5.0: co-existence, transition, or a hybrid. Proc. Comput. Sci. **217**, 102–113 (2023)
18. Cooper, H., Hedges, L.V., Valentine, J.C.: The handbook of research synthesis and meta-analysis. Russell Sage Foundation (2009)
19. Huang, Y., Schuehle, J., Porter, A.L., Youtie, J.: A systematic method to create search strategies for emerging technologies based on the Web of Science: illustrated for 'Big Data.' Scientometrics **105**, 2005–2022 (2015)
20. Patra, R.K., Pandey, N., Sudarsan, D.: Bibliometric analysis of fake news indexed in Web of Science and Scopus (2001–2020). Glob. Knowl. Memory Commun. **72**, 628–647 (2022)
21. Faisal, A.M., Karthigeyan, L., Suresh, G.: Development of conceptual model for the introduction stage of lean intelligent manufacturing using simulation model based on the identification of the appropriate technologies of Industry 5.0. Intelligent Manufacturing and Energy Sustainability, pp. 237–246 (2023)
22. Husár, J., Knapčíková, L.: Lean management training game methodology as a tool for preparing students for Industry 5.0. Lecture Notes in Mechanical Engineering, pp. 359–369 (2023)
23. Hadi, A., Cheung, F., Adjei, S., Dulaimi, A.: Evaluation of lean off-site construction literature through the lens of Industry 4.0 and 5.0. J. Constr. Eng. Manag. **149**(12), 03123007 (2023)
24. Moraes, A., Carvalho, A.M., Sampaio, P.: Lean and Industry 4.0: a review of the relationship, its limitations, and the path ahead with Industry 5.0. Machines **11**, 443 (2023)
25. Rahardjo, B., Wang, F.K., Lo, S.C., Chu, T.H.: A sustainable innovation framework based on lean six sigma and Industry 5.0. Arab. J. Sci. Eng. **49**(5), 7625–7642 (2024)
26. Fernandes Thomaz, J.P.C., Bispo, H.I.N.: Lean manufacturing and Industry 4.0/5.0. Increasing Supply Chain Performance in Digital Society, pp. 101–130 (2022)
27. Mladineo, M., Ćubić, M., Gjeldum, N., & Crnjac Žižić, M.: Human-centric approach of the lean management as an enabler of Industry 5.0 in SMEs. 10th International Conference on Mechanical Technologies and Structural Materials, MTSM 2021, pp. 111–117, Split (2021)

Simulation of the Operating Process of a Spark Ignition Engine Powered by Carbon-Free Fuel

Van Dung Nguyen, Thin Quynh Nguyen, Andrey Y. Dunin, and Dinh Thi Luong

Abstract The results of mathematical modeling of the operating process of a spark ignition engine when running on gasoline and two carbon-free fuels: hydrogen and ammonia are presented. To carry out the computational study, a software package developed at MADI was used. In this model, the factor *m* in the Wiebe formula that corresponds to each fuel type is determined. This value selection is based on the condition of coincidence of experimental and calculated data on the maximum cycle pressure and the highest rate of pressure increase. In the second stage of computational studies, the impact on the performance of a spark ignition engine from switching to hydrogen and ammonia was assessed. The results show that, with the transition to hydrogen, the maximum value of the heat release rate increases by 17.5% compared to gasoline, while with ammonia increases by 1.6%. Engine operation on ammonia has the highest hourly fuel consumption, which is 2.6 times higher than that of gasoline and 6.2 times higher than that of hydrogen.

Keywords Spark ignition engine · Carbon-free fuel · Hydrogen · Ammonia · Operating process modeling

V. D. Nguyen · A. Y. Dunin
Department of Heat Engineering and Automotive Engines, Moscow Automobile and Road Construction State Technical University (MADI), Moscow, Russia

T. Q. Nguyen (✉)
Faculty of Mechanical Engineering, University of Transport and Communications, Hanoi, Vietnam
e-mail: thinquynh@utc.edu.vn

D. T. Luong
Institute of Vehicle and Energy Engineering, Military Technical Academy, Hanoi, Vietnam

D. H. Tien et al. (eds.), *Proceedings of the 1st International Conference on Sustainability and Emerging Technologies for Smart Manufacturing*, Proceedings in Technology Transfer, https://doi.org/10.1007/978-981-97-7083-0_33

1 Introduction

Today, the Internal Combustion Engine (ICE) continues to be the main power plant in road transport, which is subject to a whole range of requirements related to power, economic indicators, and exhaust gas toxicity (EG). In this regard, one of the important directions for the development of ICE is the use of fuels alternative to gasoline and diesel fuel obtained from oil. In addition to expanding the oil production facilities, the use of alternative fuels is considered an opportunity to reduce the carbon until it is completely eliminated from the exhaust gas of ICE [1, 2]. The trend toward reducing emissions of carbon and carbon-containing substances is illustrated by the example of the European Union requirements for heavy-duty vehicles. Among the alternative fuels, gas fuels are especially distinguished, providing faster mixture formation and better completeness of mixing with air. Table 1 presents the thermodynamic properties of some gas alternative fuels in comparison with gasoline.

Table 1 shows that hydrogen has the lower heating value almost three times higher than that of gasoline, and ignition energy is required 1.5–1.6 times. However, hydrogen has several disadvantages as a fuel: low density (compared to methane); high fire hazard, and, therefore, the need for storage using cryogenic systems. This increases the weight of the vehicle and requires more space in it to accommodate fuel storage tanks.

Based on the disadvantages of hydrogen, the use of ammonia as a secondary hydrogen carrier is of practical interest. The prospects for using ammonia as a motor fuel are due to several of its positive properties.

Ammonia (NH_3) occurs naturally as salts. Its purest form is anhydrous NH_3. Ammonia is lighter than air and highly soluble in water, allowing leakage to be controlled without the risk of fire or explosion. NH_3 is an effective hydrogen carrier with 1.5 mol of hydrogen per 1 mol of ammonia, and ammonia has a low pour point (195.7 K), making it a promising fuel for cold climate regions [3]. The low viscosity of ammonia improves fuel atomization conditions. With the octane value up to 130, ammonia is very suitable for engines with high compression ratios because it will help reduce engine detonation [5].

Table 1 Thermodynamic properties of different types of fuels [3, 4]

Properties	Petrol	Methane	Hydrogen	Ammonia
Lower heating value (MJ/kg)	44.5	50.05	120	18.8
Density at 0.1 MPa and 298 K (kg/m^3)	736	0.66	0.08	0.73
Auto-ignition temperature (K)	503	859	773–850	930
Octane rating (RON)	90–98	120	> 100	130

2 Description of the Mathematical Model and Its Verification

To compare the performance of a spark ignition (SI) engine when powered by gasoline and carbon-free fuels, computational studies were carried out using the "SI-MADI" calculation element of the "ICE-MADI" PC software package. "SI-MADI" was developed based on a mathematical model proposed at MADI. In the ICE-MADI PC, methods of thermodynamic and gas-dynamic calculation of gas processes are used to simulate gas-dynamic and thermal processes in the intake and exhaust systems of the engine, as well as in its cylinders. During gas exchange, the engine cylinder is considered as an open thermodynamic system of variable volume, and upon completion of the gas-exchange processes as a closed one. The gases in the cylinder are considered ideal and in a quasi-equilibrium state. The processes in the cylinder are described by the equation of the first law of thermodynamics in differential form. For gas-exchange processes, the equations of conservation of mass and the flow of gases through the valve hole were used, and for the finite volume of the gas mixture, the Clapeyron equation was used. The internal energy of a gas mixture is determined taking into account its temperature and composition. The boundary conditions characterizing the exchange of mass and heat of the thermodynamic system under consideration with the environment are expressed by the equations of heat transfer and flow in a quasi-stationary formulation and the distribution of parameters at the boundaries. Heat transfer into the walls is described by Newton's equation. In the processes of gas exchange, compression, and expansion, the heat transfer coefficient is determined by the Woshni. Calculation of heat transfer into the cylinder wall is carried out taking into account the temperature distribution along its height and changes in the heat transfer area due to movement of the piston. Gas-dynamic boundary conditions are considered in the section near the cylinder. It is assumed that the pressure in the inlet and outlet pipelines does not change during gas exchange. The gas flow through a valve opening of a time-varying area is determined by assuming the flow to be one-dimensional and quasi-stationary. The reverse ejection from the cylinder and its recharging are taken into account. In this case, the calculation is carried out using two calculation modules: SI engine operating cycle; joint work of SI engine and turbocharger. The information diagram of the inflatable SI engine model is presented in Fig. 1.

The parameters of the naturally aspirated SI engine 4CH 7.6/7 were set as the initial data for the computational study (Table 2).

The inlet parameters corresponded to standard atmospheric conditions: pressure $p_k = 0.1$ MPa, temperature $T_k = 298$ K. The outlet pressure (p_t) is 0.102 MPa, which corresponds to the values found for the 4CH 7.6/7 engine. The outlet temperature (T_t) was approximately (according to experimental data for the 4CH 7.6/7 engine) set equal to 843 K and then refined by the program through repeated calculations. The average temperatures of the surfaces of engine parts were assumed to be equal temperature of the piston bottom: $T_b = 573$ K; temperature of the internal surface

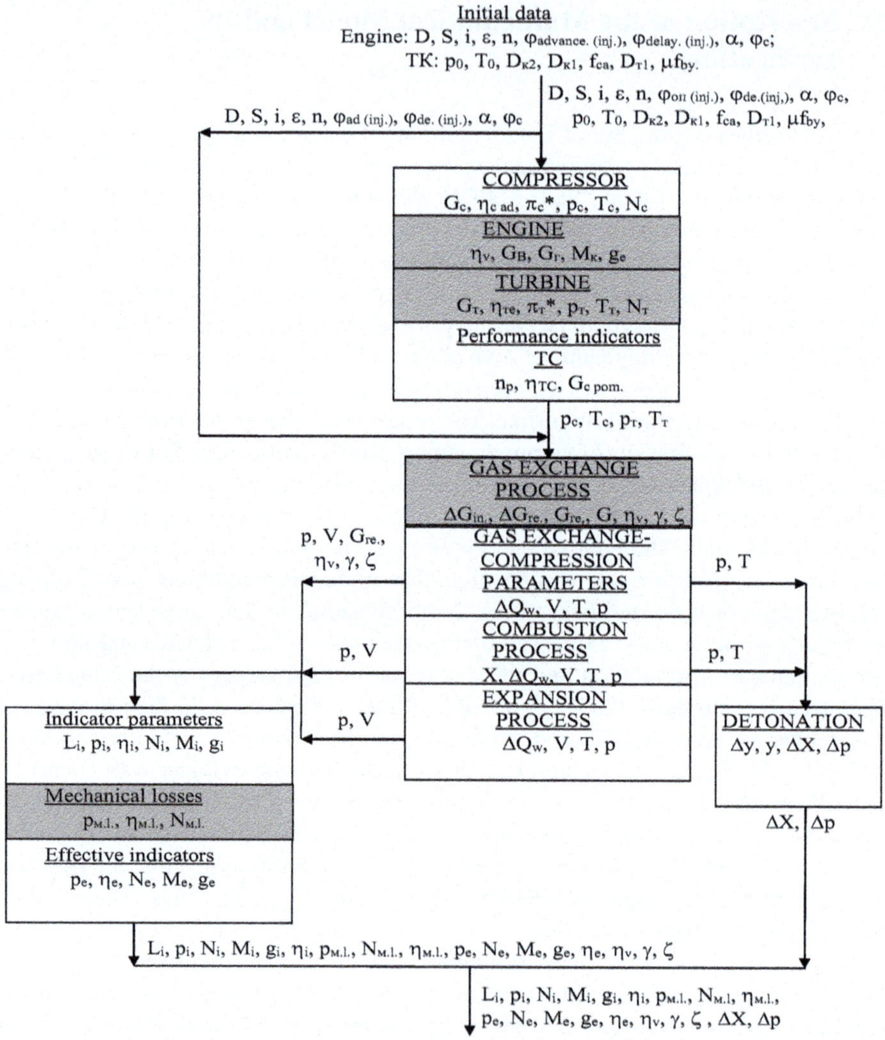

Fig. 1 Information diagram of the mathematical model of SI engine with supercharged

of the combustion chamber located in the cylinder head: $T_{\text{h}} = 548$ K; cylinder wall temperature: $T_{\text{w}} = 518$ K.

To calculate the heat release process in the SI engine combustion chamber, the formula proposed by Wiebe is used as follows:

$$x = 1 - \exp\left[-6.908 \cdot \left(\tau_1/\tau_{\text{cp1}}\right)(m+1)\right], \quad (1)$$

Table 2 Parameters of the 4CH 7.6/7 engine with spark ignition, specified as initial data in the ICE-MADI PC software package

No.	Options	Designation	Magnitude
1	Cylinder diameter (mm)	D	76
2	Piston stroke (–)	S	70
3	Number of cylinders (–)	i	4
4	Compression ratio (–)	ε	10.5
5	Ratio of combustion chamber surface to piston crown area (–)	F_{cc}/F_p	2.3
6	Ratio of crank radius to connecting rod length (–)	λ	0.280
7	Maximum intake valve lift (mm)	$h_{v\ (iv)\ max}$	8.8
8	Intake valve neck diameter (mm)	$d_{r\ (iv)}$	36
9	Intake valve opening advance angle (BTDC)	$\varphi_{o\ (iv)}$	33
10	Duration intake valve opening	$\varphi_{0\ (iv)}$	292
11	Maximum exhaust valve lift (mm)	$H_{v\ (ev)\ max}$	8.65
12	Exhaust valve neck diameter (mm)	$d_{r\ (ev)}$	31
13	Exhaust valve opening advance angle (BBDC)	$\varphi_{o\ (ev)}$	47
14	Duration opening the exhaust valve	$\varphi_{0\ (ev)}$	244

where

m the heat release curve indicator
τ_1 current time of the combustion process
τ_{cp1} duration of the combustion process.

When using the Wiebe formula, it is necessary to set the values of the indicator m and duration τ_{cp1}, which depend on the engine operating mode and the type of fuel used.

At the first stage of the study, in order to determine factor m (Table 3), a series of calculations was carried out and their results were compared with experimental data for various engines. The choice of the m value was determined by comparison of achieving values, which comprise the maximum cycle pressure p_z and the highest rate of pressure increase $(dp/d\varphi)_{max}$ close to the experimental data.

To find the m indicator when operating the on gasoline, we used the data obtained when testing the 4CH 7.6/7 engine at the experimental stand of the Problem Laboratory of Transport Engines of MADI. A comparison of calculated data and experiment

Table 3 The value of the heat release curve index m for the Wiebe formula depending on the type of fuel

Type of fuel	The value of the indicator m
Hydrogen	3.0
Petrol	1.7
Ammonia	1.2

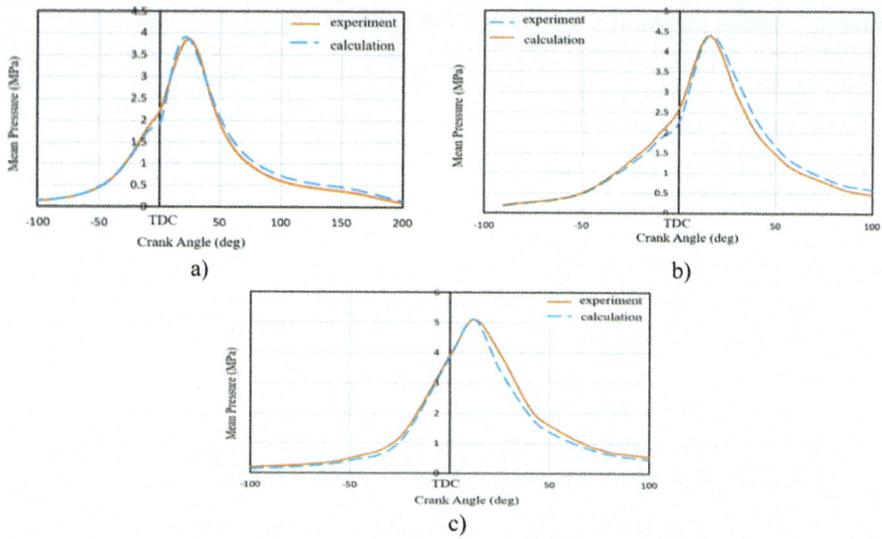

Fig. 2 Comparison of experimental and calculated indicator diagrams. **a** SI engine 4CH 7.6/7 when running on gasoline ($n = 3000$ min^{-1}; $\alpha = 1$; IT = 15° BTDA; $m = 1.7$; $\varphi_z = 70°$). **b** SI engine 4CH 8.6/8.6 when operating on hydrogen ($n = 3000$ min^{-1}; $\alpha = 1.64$; IT = 15° BTDA; $m = 3.0$; $\varphi_z = 40°$). **c** SI engine 4CH 11.5/7.7 when operating on ammonia ($n = 1000$ min^{-1}; $\alpha = 1$; IT = 15° BTDA; $m = 1.2$; $\varphi_z = 52°$)

is shown in Fig. 2 at $n = 3000$ min^{-1}, the excess air coefficient $\alpha = 1$, the ignition timing (IT) angle IT = 15° before TDC.

Work [6] presents experimental data on the operation of SI engine 1CH 11.5/7.7 (Table 4), converted to operate on ammonia [7]. In Fig. 2c, presents the comparison results of calculated and experimental data in the operating mode: $n = 1000$ min^{-1}; $\alpha = 1$; IT = 15° BTDC, $\varphi_z = 52°$.

Experimental data on the operating process of hydrogen-powered SI engine are taken from [8]. The authors conducted research on the 4CH 8.6/8.6 engine model CA20 (Table 4). The results of a comparison of the obtained calculated and experimental data are shown in Fig. 2b. Mode parameters: $n = 3000$ min^{-1}, $\alpha = 1.64$, IT = 15° before TDC, $\varphi_z = 40°$.

Table 4 Parameters of engines, converted to run on hydrogen and ammonia

Parameters	CA20 engine [8]	EP6 engine [7]
Number of cylinders	4	1
Cylinder diameter/piston stroke (mm/mm)	86/86	115/77
Connecting rod length (mm)	142.8	177
Compression ratio	10	11.75

In three cases of comparison between simulation and experiment, the error value between simulation and experiment is from 5 to 10%.

3 Simulation Results of the Operating Process of a Spark Ignition Engine Powered by Carbon-Free Fuel

At the second stage of computational research, the domestic SI engine 4CH 7.6/7 was selected as the object (Table 2). The purpose of the study is to assess the impact on engine operating performance from switching to carbon-free fuel (hydrogen and ammonia). The composition of the mixture in this study was assumed to be stoichiometric: $\alpha = 1$. This value of the excess air coefficient is necessary for the effective operation of a three-way catalyst when gasoline is supplied to the SI engine.

Figure 3 shows a comparison of heat release rates ($dQ/d\varphi$), and calculated indicator diagrams. Operating mode indicators: $n = 2000$ min^{-1}; fully open throttle; the onset angle of heat release is 15° BTDC. As can be seen that, with the transition to hydrogen, the maximum value of the heat release rate $(dQ/d\varphi)_{max}$ increases by 17.5% compared to gasoline, while with ammonia it increases by 1.6%. The highest value of the maximum cycle pressure p_z is achieved with hydrogen, while with gasoline and ammonia its values are almost the same. Despite this, the operation of SI engine on ammonia is characterized by the highest hourly fuel consumption Q_t (Table 5), which is 2.6 times higher than that of gasoline and 6.2 times higher than that of hydrogen. The temperature in the combustion chamber varies as shown in Fig. 3c. The maximum temperature values in the case of using hydrogen are $T_{zmax} = 2958.9$ K at 19° ATDC, when using Petrol (Gasoline) $T_{zmax} = 2578.3$ at 20° ATDC and with Ammonia $T_{zmax} = 2451.3$ at 20° ATDC. In addition to the mode $n = 2000$ min^{-1} under similar conditions, calculations were carried out in other modes of the external speed characteristics of the engine.

The external speed characteristics (ESC) of a spark ignition gasoline engine when operating on various types of fuel are presented in Fig. 4.

The use of carbon-free fuels, provided that $\alpha = 1$ is maintained, provides a higher average effective pressure p_e. Thus, in the maximum torque mode ($n = 3400$ min^{-1}), the increase in p_e when operating on ammonia was 11% relative to operating on gasoline, while on hydrogen it was 22% (Fig. 4a). The effective efficiency changed so that for ammonia, there was a tendency to reduce it by 3.5%, and for hydrogen, it tended to increase by 2.3% (Fig. 4d). Most significantly, operation on ammonia is inferior in specific effective fuel consumption g_e, which in the mode $n = 3000$ min^{-1} is 2.4 times higher than when operating on gasoline (Fig. 4b). For hydrogen, on the contrary, a decrease in g_e by 2.7 times was noted. The results obtained allow us to recommend the use of ammonia in SI engine operating as part of a combined power system. This approach will make it possible not to significantly increase the size of fuel tanks while maintaining the vehicle's range, its dynamic and power characteristics.

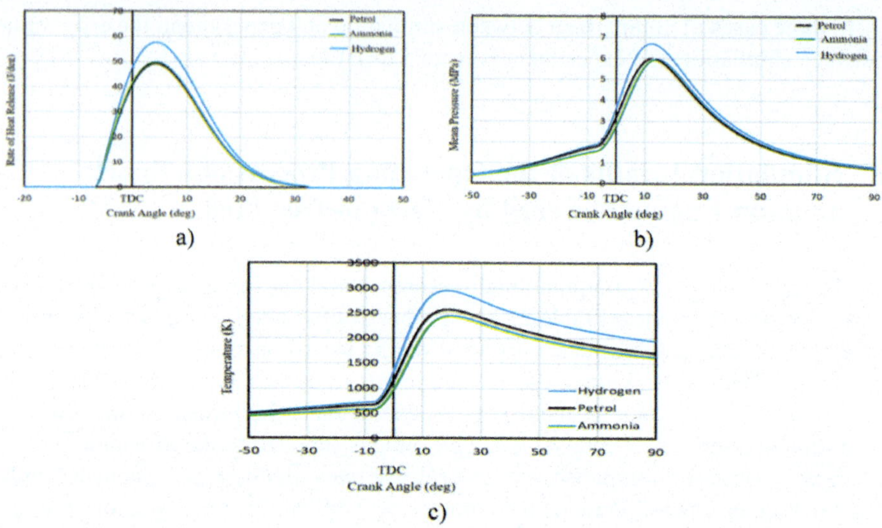

Fig. 3 Variation of values in the combustion chamber of SI engine 4CH 7.6/7 at $n = 2000$ min^{-1} and $\alpha = 1$. **a** Heat release rate $dQ_t/d\varphi$; **b** mean pressure in cylinder; **c** temperate in the cylinder

Table 5 Hourly consumption when operating SI engine on various types of fuel: SI engine 4CH 7.6/7; $n =$ 2000 min^{-1}; $\alpha = 1$

Type of fuel	Hourly flow rate Q_t (kg/h)
Hydrogen	2.15
Petrol	5.17
Ammonia	13.37

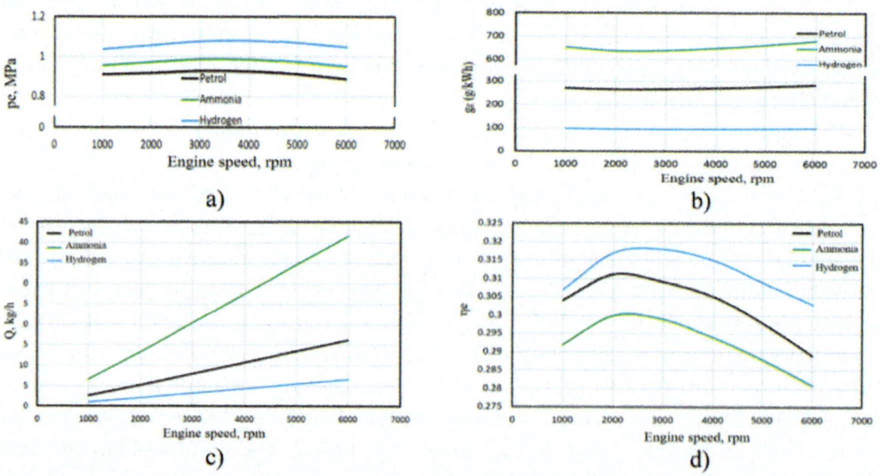

Fig. 4 Variation of the SI engine 4CH 7.6/7 parameters when operating on different types of fuel and $\alpha = 1$. **a** Average effective pressure p_e; **b** specific effective g_e fuel consumption; **c** specific effective hourly Q fuel consumption; **d** effective efficiency η_e

4 Conclusions

1. When converting a spark ignition engine to hydrogen, the maximum heat release rate increases compared to gasoline, while with ammonia it decreases. So, in the case of the calculations carried out for the 4CH 7.6/7 engine, these changes amounted to 17.5% and 1.6%, respectively.
2. The operation of SI engine on ammonia, provided that the specified engine power is provided, is associated with an increase in its specific effective fuel consumption. Thus, in the external speed characteristic mode at $n = 3400\ \text{min}^{-1}$ and $\alpha = 1$, the value of g_e when operating on ammonia is 2.4 times higher than when operating on gasoline. For this reason, based on the condition of maintaining the volume occupied by fuel tanks on a car, it is rational to use a spark ignition engine operating as part of a combined power system.

References

1. Thin Quynh, N., Duc, L.H., Dunin, A.Y.: Influences of injector geometry parameters on fuel injection characteristics and parameters of a diesel engine. Transp. Commun. Sci. J. **74**(4), 530–543 (2023)
2. Thin Quynh, N., Duc, L.H., Dunin, A.Y., Shatrov, M.G.: Engine performance, emissions characteristics of the compression ignition engine using HHO gas (Brown's gas). IOP Conf. Ser. Mater. Sci. Eng. **2021**, 1159 (2021)
3. Kurien, C., Mittal, M.: Review on the production and utilization of green ammonia as an alternate fuel in dual-fuel compression ignition engines. Energy Convers. Manag. **251**, 114990 (2022)
4. Dimitriou, P., Javaid, R.: A review of ammonia as a compression ignition engine fuel. Int. J. Hydrogen Energy **45**(11), 7098–7118 (2020)
5. Hales, J.M., Drewes, D.R.: Solubility of ammonia in water at low concentrations. Atmos. Environ. **13**, 1133–1147 (1979)
6. Pozzana, G., Bonfanti, N., Frigo, S., Doveri, N., Dario, P., Mattoli, V., Ragnoli, M.: A Hybrid Vehicle Powered by Hydrogen and Ammonia. SAE Technical Paper Series (2012)
7. Pelé, R., Brequigny, P., Bellettre, J., Mounaïm-Rousselle, C.: Performances and pollutant emissions of spark ignition engine using direct injection for blends of ethanol/ammonia and pure ammonia. Int. J. Engine Res. (2023)
8. Lukanin, V.N., Morozov, K.A., Khachiyan, A.S., Alekseev, I.V., Golubkov, L.N., Chernyak, B.Ya., Trusov, V.I., Kamfer, G.M., Makhov, V.Z., Prishvin, S.A., Sinyavsky, V.V., Matyukhin, L.M., Nazarov, N.I., Shatrov, M.G.: Dvigateli vnutrennego sgoraniya. Kniga 1 (Internal Combustion Engines. Book 1). Izdatel'stvo "Vysshaya shkola", Moscow (2010)

Evaluating the Impact of Cutting Speed and Feed Rate on Surface Roughness Utilizing a Four-Insert Carbide Face Milling Cutter on CNC Machines

Hoanh-Son Truong, Trong-Thanh Nguyen, and Tuan-Anh Bui

Abstract A machining process with significantly higher cutting speeds than conventional machining can enhance surface finish, machining productivity, and tool life. However, it also results in increased cutting temperatures, which can impact the surface quality of workpiece and the life of cutting tool. This study explores the effects of cutting speed (100–550 m/min), feed rate (0.01–0.2 mm/tooth), and cutting depth (0.2–0.5 mm) on the surface roughness of parts machined on a milling CNC machine. The experiment used carbon steel C45 as the workpiece material, a carbide face milling as the cutting tool, and soluble cutting oil as the coolant. The findings revealed that a rise in both the cutting speed and feed rate leads to an increment in surface roughness. However, these increases also lead to higher cutting temperatures, which can have implications on the life of the cutting tool and the quality of the surface that has been machined. This research is crucial for choosing appropriate cutting parameters to achieve the desired surface roughness and ensure efficiency in high-speed machining applications. Furthermore, the research findings contribute to broadening the understanding of machining and material processing.

Keywords Surface roughness · Cutting speed · Feed rate · Carbide inserts · CNC machines

1 Introduction

The quality of machine parts significantly influences the functionality of component joints within mechanical structures. This intricate relationship warrants further exploration through theoretical and experimental research. Identifying methods to

H.-S. Truong · T.-T. Nguyen (✉) · T.-A. Bui (✉)
School of Mechanical Engineering, Hanoi University of Science and Technology, No 1 Dai Co Viet Rd., Hanoi, Vietnam
e-mail: thanh.nguyentrong@hust.edu.vn

T.-A. Bui
e-mail: anh.buituan@hust.eud.vn

D. H. Tien et al. (eds.), *Proceedings of the 1st International Conference on Sustainability and Emerging Technologies for Smart Manufacturing*, Proceedings in Technology Transfer, https://doi.org/10.1007/978-981-97-7083-0_34

enhance surface quality, especially in high-speed machining, will enhance functionality and ensure the quality of machine component joints. Analyzing the factors affecting part quality during high-speed machining, particularly milling, is essential in manufacturing technology. During machining, part surface quality is a key determinant of process quality, highlighting the need to study technological factors influencing surface roughness in high-speed machining.

In 2017, a study by Mohd Rasidi Ibrahim et al. emphasized the significance of machining parameters in lathe operations, which affect system dynamics and crucial customer requirements such as the quality of the workpiece surface and the wear of the cutting tool. The investigation revealed that the ideal feed rate and cutting speed enhance the smoothness of the surface and diminish tool wear, with the cutting speed playing a pivotal role in affecting surface roughness [1]. In a study, Pranav Singhal et al. investigated how tool overhang, cutting speed, and feed rate influence the surface texture of EN8 material. This was done during the production of SW17 hexagons on a CNC Turning machine with a C-axis, using a 4-flute end mill with a high-speed steel parallel shank. The study used a Taguchi-inspired experimental design to establish an L9 orthogonal array for trial runs. The results, analyzed using Taguchi S/N ratios, suggest that the cutting speed plays a pivotal role in influencing the surface texture, succeeded by tool overhang and feed rate [2]. In 2019, a study was conducted by Jéferson Hilário Wallendorff and his team, where they evaluated the performance and longevity of a mixed ceramic tool featuring S-type edge microgeometry. The study explores the influence of cutting parameters on tool failure, employing comprehensive factorial design methods for experimental design and scrutiny. The research establishes tool life curves and equations for the studied machining conditions, aiding in future process optimization [3]. In a study, Sabi Sabev et al. explore surface roughness in a 45 mm diameter, 50 mm length hole in AISI304 steel employing an anti-vibration bar. They examine the influence of cutting parameters on the surface quality of the component. Through regression modeling, they establish a comprehensive understanding of the correlation between parameters and surface quality, leading to the optimization of surface roughness [4]. Moreover, research on the influence of technological parameters on the machining quality of high-hardness materials is also of interest to scientists [5], especially, machining methods with the support of ultrasonic technology can be applied [6, 7]. In 2001, R. F. Avila et al. presented a study which highlights the significance of cutting fluids in enhancing machining performance. The study contrasts the use of three different fluids and dry cutting on AISI 4340 steel, evaluating aspects such as tool longevity, surface quality, and chip formation. The study finds that the use of emulsion-based fluid, which does not contain mineral oil, and dry cutting methods outperform synthetic fluid and emulsion that includes mineral oil [8]. In 2022, a study led by Azhar Equbal and his colleagues focused on the CNC milling process of AISI 316 stainless steel. The study finds depth of cut impacts material removal rate most, feed rate affects surface roughness, and cutting speed has little effect on surface roughness [9]. A study by Fahri et al. in 2023 investigates the wear age of a 4-flute endmill during CNC milling, analyzing its impact on tool life. Using a feed rate of 770 mm/min and depths of cut ranging from 0.5 to 1.75 mm on AISI 4340 material, the research concludes that the 0.5 mm cutting depth induces less wear compared to the 1 mm and 1.75 mm depths [10].

Surface roughness refers to the texture of a material's surface, affecting characteristics such as wear resistance, corrosion resistance, light reflectance, and adhesion. It plays a crucial role in machined surfaces, impacting factors like friction, wear resistance, and fatigue life. The texture of surfaces on machined components is notably affected by various cutting parameters, such as the speed of cutting, rate of feed, and the cut's depth. This study examines the effect of feed rate and cutting speed on surface roughness during the machining of carbon steel using a face milling cutter with four carbide inserts. The primary objective is to establish the relationship between feed rate, cutting speed, and surface roughness parameters, specifically Ra and Rz. Understanding this relationship allows manufacturers and engineers to adjust cutting parameters to attain the preferred surface texture for various machining applications.

2 Materials and Methods

This section focuses on providing a detailed description of the equipment used in the experiment, including related materials. In addition, the experimental method is also presented specifically, with the aim of evaluating the influence of some technological parameters on the quality factor of the workpiece surface. Among these quality factors, surface gloss is given special attention. By adjusting these technological parameters, we can grasp the relationship between them and the surface gloss, thereby proposing solutions to optimize the production process.

2.1 Experimental Equipment and Materials

The experimental process was set up based on machining at the CNC GV503 vertical machining center, a product of Mori Seiki. The material used in the machining process is common machine-making steel, with a carbon content of 0.45%. Clearly, the impact of technological parameters on surface quality has been evaluated by many researchers. However, under specific working conditions, further research is still necessary, with the aim of finding the most suitable parameter set, ensuring a balance between quality requirements and machining productivity.

In this study, the experimental samples were machined using a face milling cutter fitted with 4 carbide inserts. The cutting tool parameters are detailed as in Fig. 1. In which, the number of teeth (number of chips) of the cutter is 4, the tool face diameter (D_1) is 63 mm, the largest diameter of the cutter (D_2) is 75.9 mm. The height of the cutter base (L_2) is 40 mm, the length of the cutter (L_1) is 125 mm, and the diameter of the cutter shank (D_4) is 32 mm. All of these parameters are crucial in maintaining the efficiency and quality of the machining operation.

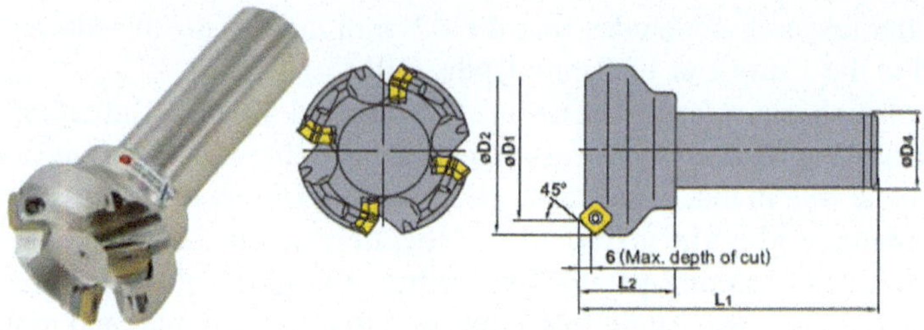

Fig. 1 Basic parameters of the face milling cutter with hard alloy inserts

The experimental samples were rectangular blocks of carbon steel, each with dimensions of 60 mm in length, 50 mm in width, and 40 mm in height. These samples were numbered sequentially to ensure accuracy during the evaluation process. Before conducting the experiment, the samples were cut, cleaned, and basic parameters were measured to ensure that the input conditions of the experimental process were consistent. After machining, the samples were cleaned again, and surface roughness was measured on a Mitutoyo SJ—400 measuring machine. The standard length for sampling measurement results, denoted as *L*, ranged from 0.8 to 5 mm. The measuring head was moved from the outside in during the measurement process. This helped ensure the accuracy and consistency of the measurement results.

2.2 Experimental Methods

As presented above, the scope of this study is limited to investigating the influence of cutting parameters, including cutting speed V_c (m/min) and feed rate S_r (mm/tooth), on the surface roughness of machined parts (Ra & Rz) when performing high-speed machining on CNC machines with different materials. To ensure accuracy and consistency during the study, cutting speeds were selected ranging from 100 to 550 m/min and S_r from 0.01 to 0.2 mm/tooth. Additionally, the cutting depth t was determined with two values of 0.2 mm and 0.5 mm. These parameters were chosen based on previous studies and practical experience to ensure that the experimental process was conducted efficiently and yielded reliable results. In this context, the experiments will be carried out under two scenarios: (1) The cutting speed is held constant at 550 m/min, while the feed rate *Sr* changes from 0.01 to 0.2 mm/tooth, increasing in steps of 0.01 mm/tooth; (2) The cutting speed varies from 100 to 550 m/min in increments of 100 m/min, with the feed rate *Sr* maintained at 0.05 mm/tooth. The cutting parameters for the scenarios are computed and displayed in Tables 1 and 2.

The calculated parameters shown in Tables 1 and 2 will be applied to adjust the cutting parameters in the NC program. This enables precise control of the actuators during the machining of the experimental samples. Additionally, to ensure consistency throughout the machining process, other machining conditions are also fixed,

Table 1 Cutting mode parameters set in the NC program with constant cutting speed

Sample no	*Sr* (mm/tooth)	*n* (rpm)	*F* (mm/min)
1	0.01	2780	111
2	0.02	2780	222
3	0.04	2780	445
4	0.06	2780	667
5	0.08	2780	890
6	0.10	2780	1112
7	0.12	2780	1334
8	0.14	2780	1557
9	0.16	2780	1779
10	0.20	2780	2224

Table 2 Cutting mode parameters set in the NC program with variable cutting speed

Sample no	*n* (rpm)	*F* (mm/min)
1	506	101
2	758	152
3	1011	202
4	1264	253
5	1517	303
6	1769	354
7	2022	404
8	2275	455
9	2528	506
10	2780	556

except for the parameters adjusted according to Tables 1 and 2. In this way, any variations in the results can be attributed to the changes in these parameters, enhancing the reliability of the study. By controlling all other machining conditions and isolating the effects of the varied cutting parameters, the impact of cutting speed and feed rate on the surface roughness of machined components can be assessed with assurance. This systematic approach helps us draw meaningful conclusions and gain valuable insights into the high-speed machining process.

3 Results and Discussion

The experiment was conducted on samples numbered 1 to 10. Each sample was face milled in turn, with the cutting parameters determined according to the experimental procedure sheet. After machining, the machined surface underwent cleaning with an acetone solution. Subsequently, a dedicated roughness measuring device was employed to measure the surface roughness. Each sample was measured four times, and the average was considered as the result. These results underwent statistical analysis to ensure experiment accuracy and reliability. Surface roughness values, represented by Ra and Rz, were measured under two conditions: one with varying cutting speed and a constant feed rate, and another with a constant cutting speed but varying feed rate, as shown in Tables 3 and 4, respectively. Notably, the cutting depth was adjusted in each experiment to 0.2 mm and 0.5 mm, while maintaining other parameters constant, to assess the impact of cutting depth on surface roughness.

Table 3 Surface roughness of workpieces machined with $S_r = 0.05$ mm/tooth

Sample no	*V* (m/min)	Sr (mm/tooth)	*t* = 0.2 mm		*t* = 0.5 mm	
			Ra (μm)	Rz (μm)	Ra (μm)	Rz (μm)
1	100	0.05	0.135	0.760	0.110	0.790
2	150	0.05	0.145	0.718	0.115	0.780
3	200	0.05	0.128	0.721	0.100	0.770
4	250	0.05	0.143	0.790	0.150	0.850
5	300	0.05	0.125	0.820	0.143	0.860
6	350	0.05	0.105	0.890	0.155	0.875
7	400	0.05	0.180	0.811	0.148	0.901
8	450	0.05	0.175	0.923	0.153	0.892
9	500	0.05	0.185	1.050	0.178	0.943
10	550	0.05	0.230	1.150	0.200	1.100

Table 4 Surface roughness of workpieces machined with $V_c = 550$ m/min

Sample No	*V*c (m/min)	Sr (mm/tooth)	*t* = 0.2 mm		*t* = 0.5 mm	
			Ra (μm)	Rz (μm)	Ra (μm)	Rz (μm)
1	550	0.01	0.115	0.770	0.067	0.560
2	550	0.02	0.128	0.820	0.083	0.600
3	550	0.04	0.123	0.855	0.087	0.630
4	550	0.06	0.128	0.870	0.091	0.670
5	550	0.08	0.130	0.900	0.095	0.700
6	550	0.10	0.129	0.890	0.093	0.710
7	550	0.12	0.138	0.948	0.098	0.720
8	550	0.14	0.140	0.943	0.110	0.710
9	550	0.16	0.158	0.948	0.130	0.830
10	550	0.20	0.167	0.960	0.133	0.880

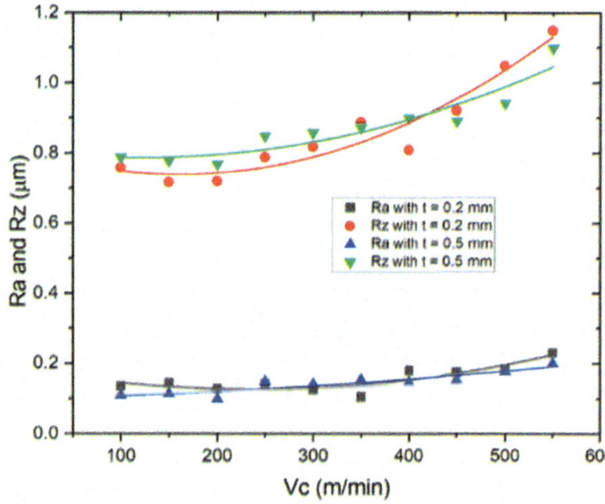

Fig. 2 Surface roughness depending on cutting speed

The graphs shown in Fig. 2 illustrate the direct correlation between cutting speed and surface roughness of the workpiece. A lower feed rate results in reduced micro-roughness height. However, for cutting speeds less than 300 m/min, the surface roughness remains relatively constant compared to speeds greater than 300 m/min. The experimental interpolation curve shows minor variance intervals within the cutting speed range of 100 m/min to 300 m/min. This can be attributed to the following factors: (1) At lower cutting speeds, increased levels of plastic and elastic deformation on the machined surface led to a plowing effect, thereby increasing surface roughness; (2) At higher cutting speeds, the impact of plastic deformation due to plowing becomes negligible. The roughness is primarily influenced by the work hardening of the workpiece's surface layer due to the cutting force.

As work hardening intensifies, roughness decreases, improving the workpiece surface quality. The intensity and depth of work hardening increase with the cutting force, force duration, and level of plastic deformation. Alongside work hardening, a softening phenomenon occurs due to heat in the cutting zone. High temperatures and prolonged heating can cause the workpiece to lose the work hardening achieved during machining. The final state of the metal surface depends on the force and heat generated during the machining process, with the depth of work hardening varying based on the machining method, cutting conditions, and the geometrical shape of the cutting tool.

Modifying the cutting conditions alters the cutting force, increases the level of plastic deformation, and enhances work hardening. Extending the force duration also deepens work hardening. However, changing the cutting conditions raises the temperature in the cutting zone, softening the metal. In practice, the influence of cutting temperature on work hardening is complex due to factors such as friction, heat dissipation conditions from the cutting zone, and changes in the metal structure.

When the cutting depth (t) changes, the machining conditions also vary, consequently impacting surface roughness. In the graph, particularly when $Vc > 450$ m/

min, the discrepancy between the two cutting conditions with different cutting depths becomes most apparent. Given identical machining parameters, the Rz roughness is higher when the cutting depth is 0.2 mm compared to when it is 0.5 mm. At higher cutting speeds, the generated heat decreases, resulting in more stable plastic deformation. However, with smaller cutting depths, the plastic deformation of the surface layer is also reduced, leading to an increased tendency for surface roughness compared to cases with larger cutting depths. The observed outcomes can be attributed to a combination of influential factors. Firstly, at lower cutting speeds, the machined surface experiences heightened levels of plastic and elastic deformation, leading to significant tool deflection and subsequently increasing surface roughness. Conversely, as cutting speed increases, the impact of plastic deformation due to tool deflection becomes less pronounced. Moreover, the relationship between cutting speed and surface roughness is further complicated by the influence of surface hardening. With higher cutting speeds, the surface layer of the workpiece undergoes less hardening due to reduced effects of cutting forces. Consequently, this results in heightened surface roughness and diminished surface gloss. Additionally, elevated cutting speeds lead to increased heat generation, causing a rise in cutting temperatures. This phenomenon softens the metal within the cutting zone, thereby diminishing surface hardness and durability, and consequently elevating surface roughness. Furthermore, the cutting depth plays a crucial role in surface roughness dynamics. Generally, increasing cutting depth results in higher cutting forces and heat generation, leading to greater levels of plastic and hard deformation, and subsequently increasing surface roughness. Conversely, when cutting depth is limited and machining is performed at high cutting speeds, the reduced plastic deformation layer may contribute to an increase in surface roughness.

Figure 3 distinctly illustrates the effect of the feed rate (*S*r) on the surface roughness. It is observed that as *S*r increases, the surface roughness of the machined component also proportionally rises. In both machining modes with different cutting depths, machining with a cutting depth of 0.5 mm yields smaller roughness compared to machining with a cutting depth of 0.2 mm. This can be elucidated as follows: The chip formation process comprises two stages: the cutting process, which generates chips, and the tool's sliding process on the workpiece surface. When *S*r is too small, the cutting process diminishes, leaving primarily the sliding and pushing of chips. Consequently, under the same speed and feed rate, roughness is higher with a cutting depth (t) of 0.2 mm compared to 0.5 mm; with smaller cutting depths, the tool's sliding process on the workpiece surface increases. Thus, *S*r should not be excessively reduced, as doing so may fail to improve surface quality and may even decrease it, reducing machining productivity; Decreasing *S*r results in improved surface finish due to geometric influence. However, further reduction intensifies the impact of plastic deformation on the workpiece surface, slowing down the rate of surface finish improvement.

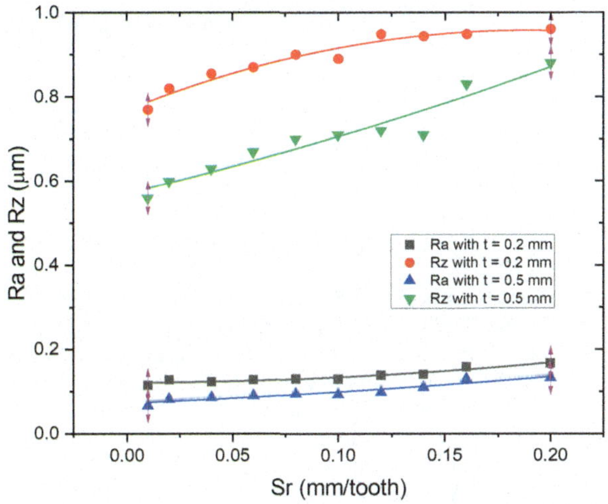

Fig. 3 Surface roughness depending on feed rate

4 Conclusions

The research delved into the effects of machining parameters on the surface texture of the processed parts. The procedure was executed on a CNC milling machine, with carbon steel C45 serving as the material for the workpiece. The face milling technique was employed, utilizing four carbide inserts. The results showed that: (1) Cutting speed significantly impacts surface roughness, with higher cutting speeds leading to reduced surface roughness due to enhanced work hardening; (2) The effect of cutting depth on surface roughness varies depending on cutting speed and feed rate. Smaller cutting depths coupled with high cutting speeds may result in increased surface roughness due to reduced plastic deformation; (3) Higher feed rates generally result in rougher machined surfaces due to increased plastic deformation. However, excessively low feed rates can also intensify plastic deformation, affecting surface finish and machining productivity. An increase in cutting speed and feed rate generally reduces surface roughness but can also elevate cutting temperatures, impacting tool life and surface quality.

References

1. Ibrahim, M.R., Sreedharan, T., Fadhlul Hadi, N.A., Mustapa, M.S., Ismail, A.E., Hassan, M.F., Tajul Arifin, A.M.: The effect of cutting speed and feed rate on surface roughness and tool wear when machining machining D2 steel. Mater. Sci. Forum **909**, 80–85 (2017)
2. Singhal, P., Bunker, O.M.: Effect of overhanging of tool, cutting speed and feed rate on surface roughness of SW17 hexagon produced on CNC turning machine with C-axis. Int. J. Adv. Res. Innov. **8**(4), 127–133 (2020)

3. Nicolodi, J.H., Consalter, L.A., Durán, O., Souza, A.J.: Influence of cutting speed and feed rate on the wear of an S-type mixed ceramic tool in turning of AISI 4140 hardened steel. Tehnički Vjesnik **26**(4), 913–919 (2019)
4. Sabev, S., Kasabov, P.: The influence of feed rate and cutting speed to surface roughness during hole boring of AISI 304 with anti-vibration boring bar. In: 10th International Scientific Conference "TechSys 2021"—Engineering, Technologies and Systems, pp. 060004, (2022)
5. Poplăcenel, M., Iordache, D.M.: A brief review of milling hard metals. University of Pitesti. Sci. Bull. Autom. Ser. 32(1): 1–9 (2022)
6. Kumar, J.: Ultrasonic machining—a comprehensive review. Mach. Sci. Technol. **17**(3), 325–379 (2013)
7. Singh, R., Khamba, J.S.: Ultrasonic machining of titanium and its alloys: a review. J. Mater. Process. Technol. **173**(2), 125–135 (2006)
8. Ávila, R.F., Abrão, A.M.: The effect of cutting fluids on the machining of hardened AISI 4340 steel. J. Mater. Process. Technol. **119**(1–3), 21–26 (2001)
9. Equbal, A., Equbal, M.A., Equbal, M.I., Ravindrannair, P., Khan, Z.A., Badruddin, I.A., et al.: Evaluating CNC milling performance for machining AISI 316 stainless steel with carbide cutting tool insert. Materials **15**(22), 8051 (2022)
10. Fahri, F., Aminy, A.Y., Arsyad, H.: The cutting effect of four flute endmills machining result for AISI 4340 material on vertical mazak CNC mill machine. In: Proceeding of international conference on energy, manufacture, Advanced Material and Mechatronics 2021, pp. 030013, (2023)

Effects of Heat Accumulation on the Part Quality and Methods for Reducing Heat Accumulation in Wire Arc Additive Manufacturing: A Short Review

Quang Huy Hoang, Thuy Duong Nguyen, Van Thao Le, Van Canh Nguyen, and Tat Khoa Doan

Abstract Metal 3D printing technology using an arc energy source—widely known as wire arc additive manufacturing (WAAM)—has a great promising for manufacturing large-sized and medium–complex parts. However, with its very large heat input characteristics, the complex thermal cycle and heat accumulation cause major challenges in WAAM. This paper presents a short review on the influence of heat accumulation on the process stability, geometric accuracy, and material properties. In addition, methods for reducing the heat accumulation phenomenon to improve the product quality are also exported. The findings reveal that the heat buildup is one of the most important causes that leads to low surface quality, high thermal distortion, and high residual stress. The heat accumulation also causes anisotropy in microstructures and materials properties. The paper provides an insight into the WAAM process, especially on the heat buildup phenomenon and several methods to reduce such a phenomenon and improve the quality of WAAMed parts.

Keywords Wire arc additive manufacturing · Heat accumulation · Part quality · Material properties

---

Q. H. Hoang
Weapon Institute, Hanoi, Vietnam

T. D. Nguyen
School of Mechanical Engineering, Hanoi University of Science and Technology, Hanoi, Vietnam

V. T. Le (✉)
Advanced Technology Center, Le Quy Don Technical University, Hanoi, Vietnam
e-mail: vtle@lqdtu.edu.vn

V. C. Nguyen
School of Mechanical and Automotive Engineering, Hanoi University of Industry, Hanoi, Vietnam

T. K. Doan
Faculty of Mechanical Engineering, Le Quy Don Technical University, Hanoi, Vietnam

D. H. Tien et al. (eds.), *Proceedings of the 1st International Conference on Sustainability and Emerging Technologies for Smart Manufacturing*, Proceedings in Technology Transfer, https://doi.org/10.1007/978-981-97-7083-0_35

1 Introduction

Wire arc additive manufacturing (WAAM) is an emerging metal additive manufacturing (MAM) technology and has received a lot of attention in research and industrial applications because the WAAM system can be easily implemented using welding robots or integrating a welding source into a multi-axis CNC milling machine, and various wire materials are available on the market. The key advantages of the WAAM technology are following [1, 2]: First, the WAAM technology allows the fabrication of thin-walled parts with large sizes, up to (1000–3000) mm with very high material deposition speeds of (50–130) g/min than other AM technologies such as PBF (powder bed fusion) and powder-based DED. Second, the use of wire materials also results in higher material efficiency (up to 100% of wire metal is deposited into fabricated components), and they are friendly to operators and environment [3]. Third, the production of wire materials also requires lower costs than the production of metal powders for powder-based AM technologies. Lastly, compared to conventional manufacturing processes, WAAM technologies can reduce the production time by 40–60% and finishing time by 15–20% (depending on the part size) [1, 2].

However, during the WAAM process, the layer's temperature is continuously increased, as heat from the newly deposited layer is continuously added, leading to the local heat accumulation that overflows the melting pool to the layer sides, reducing dimension accuracy and surface quality [2]. In addition, if the temperature is highly accumulated, it can cause the thin-walled part to collapse. The heat accumulation can also reduce the materials' mechanical properties, making the material brittle and prone to defects.

During the deposition, the arc energy also creates uneven heat zones in a sequence of heating and cooling. Typically, the heat is often dispersed by conducting through the substrate, air convection, and radiation to the air, as shown in Fig. 1 [4]. The temperature field in the deposited part is complex and unstable, causing defects and effects on the shape, dimensional accuracy, microstructure, and material properties. Therefore, controlling the heat input will help to control the geometrical shape (e.g., width and height), improve mechanical properties, and reduce residual stress after the deposition.

2 Effects of the Heat Buildup in the Deposition Process

2.1 Effects on the Shape of As-Built Parts

Figure 2 reveals the change in width of the wall according to the height of deposited layers [5]. For the first layer, the heat is transferred quickly through the substrate, leading to the melting pool being cooled quickly. When the layers are increased, the layer temperature increases due to the heat accumulation. The temperature mainly

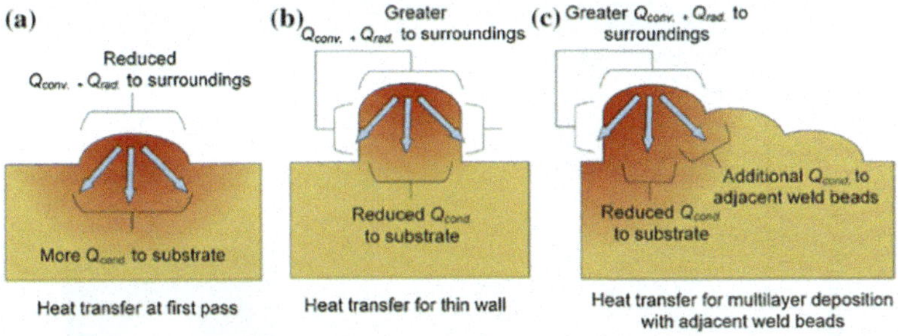

Fig. 1 Illustration of the modes of heat dissipation: **a** in the first layer, **b** after several deposited layers, and **c** in the case of multi-overlapped-bead deposition [4]

Fig. 2 An example of the thin wall built by the WAAM process [5]

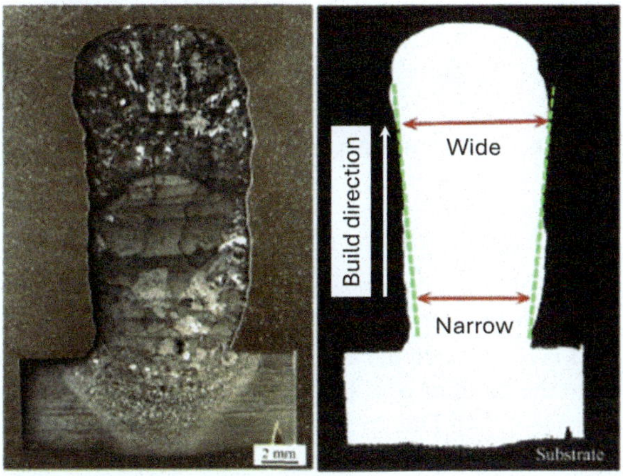

dissipates to the air through convection and radiation, causing the cooling speed to slow down significantly, resulting in a gradual increase in wall width.

To reduce the impact of the heat buildup on the component's shape, a practical solution is to increase the waiting time between two successive layers. By using this method, the next layer is made on a surface with a sufficiently low temperature, making the melting pool size constant and the geometrical shape stable [4]. However, controlling the interlayer temperature in this method is difficult, especially when components are built with multiple overlapping beads in a layer. Determining the dwell time to achieve a reasonable temperature is often through an experimental method of trial and error [6]. Therefore, it is rarely implemented in practical terms.

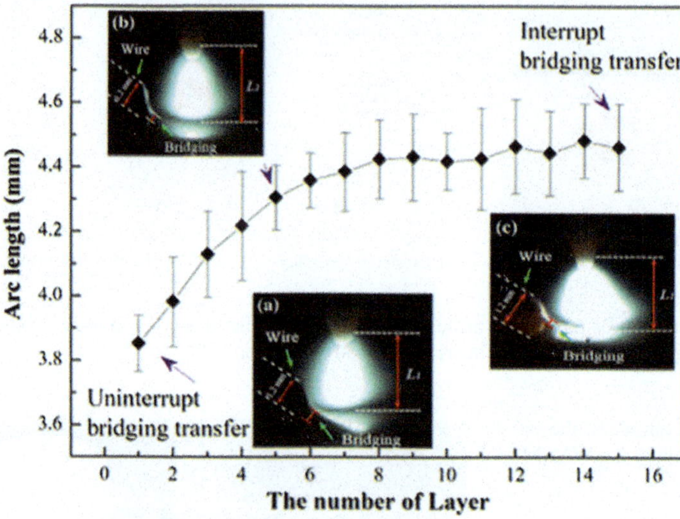

Fig. 3 Relations of the arc length and the layer number [5]

2.2 *Effects on the Process Stability*

The heat buildup in WAAM is often considered as a pre-heating process for the next layers. When the accumulation of heat augments, the vapor of metal generated from the melting pool increases the radiation's loss in the arc region, leading to arc narrowing [7]. Due to the change in the wall shape, especially the first layers, the distance between the welding torch tip and the melting pool increases, causing the arc length to increase (Fig. 3). The variation of arc shape occurred in various layers due to the combination of the arc's constriction and the layer's shape.

Figure 3 depicts the WAAM process using the GTAW source for Ti6Al4V alloy. The arc region expands then stabilizes along the number of deposited layers. The mode of transferring metal changed from the continuously transferring bridge mode with a regular and continuous surface to the free transferring mode, which tends to generate the surface waviness [8]. Therefore, controlling and minimizing the heat accumulation can enhance the process stability.

2.3 *Influence on Microstructures of As-Built Components*

Generally, components fabricated by WAAM experienced complicated thermal cycles because of multiple heating and cooling sequences and the heat accumulation during the deposition process [3]. The complicated thermal cycles and heat accumulation phenomenon generate considerable variation in microstructures and an anisotropy in microstructures with coarse and fine grains.

Figure 4a shows an example of complex thermal cycles in the WAAM of a SS308L thin wall with twenty deposited layers [9]. It is exhibited that the cooling rate reduces with the layer levels. Moreover, because of the heat accumulation, the interlayer temperature also increase with the number of deposited layers. For example, the temperature of the first layer after the deposition is 222 °C, and this temperature value increases to 653 °C and 797 °C for the tenth and the 18th layer, respectively.

The decrease in the cooling rate from the bottom to the top layer leads to the grain size that raises along the bottom-to-top direction of the wall (Fig. 4b). The

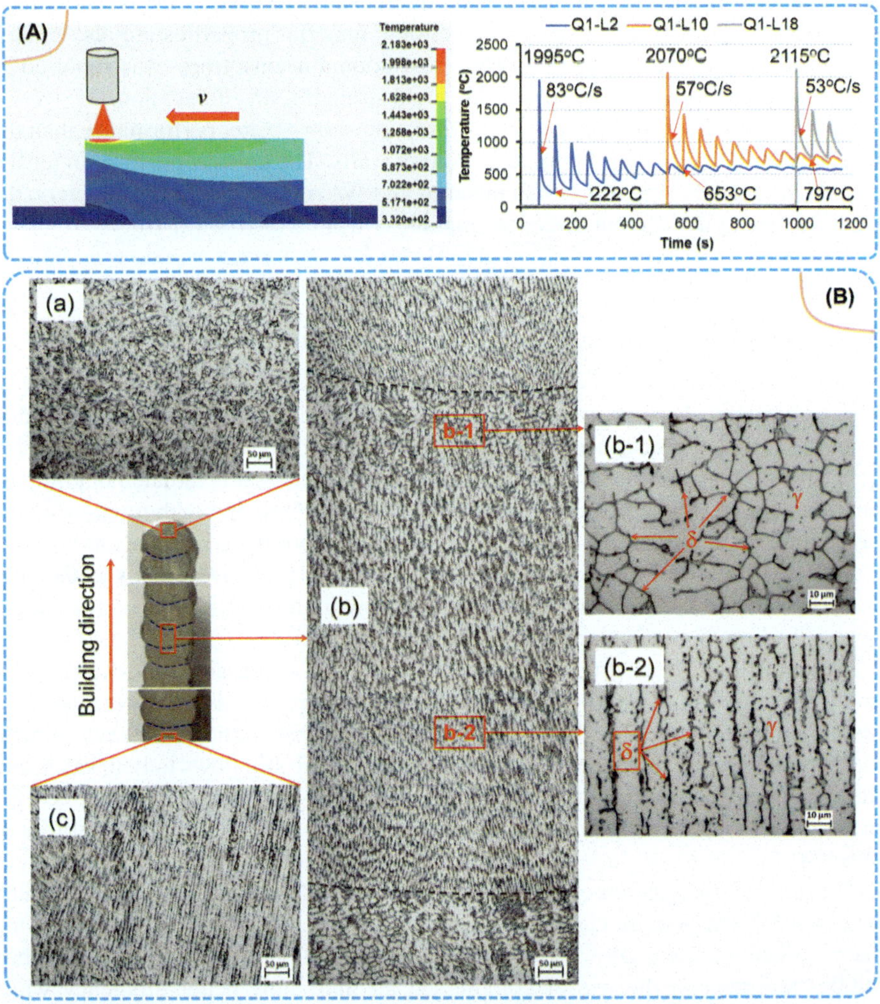

Fig. 4 **A** Thermal cycles in WAAM of a SS 308L thin wall [9] and **B** the anisotropy in microstructures in the SS 308L thin wall fabricated by WAAM [1]: microstructures in **B-a** the upper, **B-b** the middle, and **B-c** the lower region

anisotropy in microstructures and properties of WAAMed components is known as a common concern of AM as-built parts and poses great challenges to engineering applications. This can limit the applicability of the WAAM components when the materials' properties are inferior to those of conventional wrought products [10].

The anisotropic properties of WAAM components are mainly attributed to textures, grains size, and shape. Under rapid heating and cooling speeds, columnar grains generally develop in the vertical direction, in particular in the cases of thin-walled components [1, 11], as shown in Fig. 4B-b.

Normally, coarse columnar grains are not preferred, while the equiaxed grains are beneficial in hindering the expansion of cracks by increasing the density of the grain boundary [12]. However, the anisotropic material properties are also related to defects, for example porosity. The most common anisotropy was reported in mechanical response in terms of tensile, yield strengths, and elongation.

In general, the deposition/horizontal direction shows better performance than the building/vertical direction in the cases of thin walls. The anisotropic level in tensile properties strongly depends on the material and WAAM processing parameters; the tensile strength in different directions can differ from 5 to 15% [2, 10].

3 Methods for Reducing the Heat Accumulation in WAAM

There are many different methods available to reduce the heat buildup. The first solution is that the heat accumulation can be limited by keeping a certain delay between two consecutive layers. Then, the accumulated heat is radiated out into the environment. To implement this technique, it can be combined with monitoring the layer temperature to ensure that the cooling temperature between adjacent layers is similar [1]. Thus, the heat accumulation can be eliminated. However, this method has the disadvantage that the waiting time increases with the height of the embankment layers, leading to a reduction in manufacturing productivity.

Second, the heat accumulation can be minimized using an active water-based cooling method. In this method, the printing table is immersed in a water tank. As a result, the heat is transferred through the workpiece base to the water environment. However, the disadvantage of this method is that the arc still flows to both sides, and the cooling speed is not guaranteed to be continuous throughout the process. This method is also quite cumbersome, complex to deploy and reduces the flexibility of WAAM.

Third, a compressed air cooling system from an independent source is integrated with the WAAM system. Gases such as CO_2, N_2, Ar, and H_2 are considered. Using such a cooling system allows to reduce the heat buildup [13]. Previously published studies recommend the use of compressed air and show better efficiency while reducing costs due to the convenience and availability of compressed air systems. Wu et al. [14] used an active cooling method using cold CO_2 compressed air in the WAAM process (Fig. 5). They indicate that this method remarkably decreased the amplitude value of heat and the dwell time between two successive layers, making

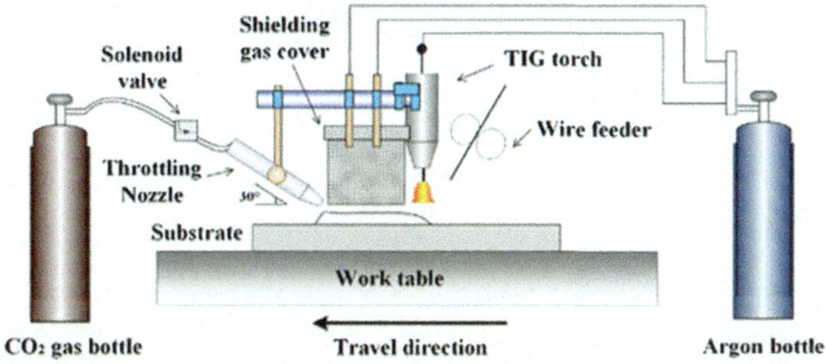

Fig. 5 An active cooling configuration of equipment [5]

the performance improved by up to 80% compared to free-cooled WAAM. Le et al. [15] used dry compressed air cooling for 8 s between layers compared to free cooling with 30 s of rest between layers. They showed that the actively cooled part has higher hardness, layer thickness, and effective width, while the surface undulation is lower than the part made by the free cooling method.

Briefly, the method mentioned above has provided effective strategies to stabilize the process, improve the geometry, and enhance the quality of parts fabricated by WAAM. Other benefits include limiting residual stresses and other defects during metal solidification. However, choosing the appropriate cooling method still needs to be studied in more detail and completely. A too high cooling rate during solidification is also detrimental to the material properties. Therefore, it is necessary to determine the cooling gas, the rate of gas flow, and the cooling time.

4 Conclusions

WAAM technology is increasingly widely applied in practice because of the superiority of an advanced production method that produces high-value parts and large parts in different industry sectors. In the WAAM process, controlling the thermal cycle and eliminating the heat accumulation are the most important tasks to stabilize the deposition process and improve the part quality. In this paper, we provided a comprehensive review on the influences of the heat accumulation on the process stability and the as-built parts' quality. Moreover, different methods that are implemented in the literature to reduce the heat accumulation were reviewed. The results indicate that excessive heat buildup is a primary factor contributing to diminished surface quality, increased thermal distortion, and elevated residual stress. Additionally, the heat accumulation induces variations in microstructures and material properties. The paper offers an understanding of the WAAM process, particularly focusing on the issue of heat buildup and proposing various approaches to mitigate it and enhance the quality of WAAM-produced components. This helps manufacturers and process planners understand and decide an appropriate method for their applications.

Acknowledgements This research was funded by the Vietnam Ministry of Education and Training, grant number B2024-BKA-10.

References

1. Le, V.T., Mai, D.S., Paris, H.: Influences of the compressed dry air-based active cooling on external and internal qualities of wire-arc additive manufactured thin-walled SS308L components. J. Manuf. Process. **62**, 18–27 (2021). https://doi.org/10.1016/j.jmapro.2020.11.046
2. Jafari, D., Vaneker, T.H.J., Gibson, I.: Wire and arc additive manufacturing: opportunities and challenges to control the quality and accuracy of manufactured parts. Mater. Des. **202**, 109471 (2021). https://doi.org/10.1016/j.matdes.2021.109471
3. Le, V.T., Mai, D.S., Bui, M.C., Wasmer, K., Nguyen, V.A., Dinh, D.M., Nguyen, V.C., Vu, D.: Influences of the process parameter and thermal cycles on the quality of 308L stainless steel walls produced by additive manufacturing utilizing an arc welding source. Weld. World **66**, 1565–1580 (2022). https://doi.org/10.1007/s40194-022-01330-4
4. Cunningham, C.R., Flynn, J.M., Shokrani, A., Dhokia, V., Newman, S.T.: Invited review article: strategies and processes for high quality wire arc additive manufacturing. Addit. Manuf. **22**, 672–686 (2018). https://doi.org/10.1016/j.addma.2018.06.020
5. Wu, B., Ding, D., Pan, Z., Cuiuri, D., Li, H., Han, J., Fei, Z.: Effects of heat accumulation on the arc characteristics and metal transfer behavior in wire arc additive manufacturing of Ti6Al4V. J. Mater. Process. Technol. **250**, 304–312 (2017). https://doi.org/10.1016/j.jmatprotec.2017.07.037
6. Geng, H., Li, J., Xiong, J., Lin, X.: Optimisation of interpass temperature and heat input for wire and arc additive manufacturing 5A06 aluminium alloy. Sci. Technol. Weld. Joining **22**, 472–483 (2017). https://doi.org/10.1080/13621718.2016.1259031
7. Zhou, X., Zhang, H., Wang, G., Bai, X.: Three-dimensional numerical simulation of arc and metal transport in arc welding based additive manufacturing. Int. J. Heat Mass Transf. **103**, 521–537 (2016). https://doi.org/10.1016/j.ijheatmasstransfer.2016.06.084
8. Geng, H., Li, J., Xiong, J., Lin, X., Zhang, F.: Optimization of wire feed for GTAW based additive manufacturing. J. Mater. Process. Technol. **243**, 40–47 (2017). https://doi.org/10.1016/j.jmatprotec.2016.11.027
9. Le, V.T., Bui, M.C., Nguyen, T.D., Nguyen, V.A., Nguyen, V.C.: On the connection of the heat input to the forming quality in wire-and-arc additive manufacturing of stainless steels. Vacuum **209**, 111807 (2023). https://doi.org/10.1016/j.vacuum.2023.111807
10. Wu, B., Pan, Z., Ding, D., Cuiuri, D., Li, H., Xu, J., Norrish, J.: A review of the wire arc additive manufacturing of metals: properties, defects and quality improvement. J. Manuf. Process. **35**, 127–139 (2018). https://doi.org/10.1016/j.jmapro.2018.08.001
11. Fu, Y., Zhang, H., Wang, G., Wang, H.: Investigation of mechanical properties for hybrid deposition and micro-rolling of bainite steel. J. Mater. Process. Technol. **250**, 220–227 (2017). https://doi.org/10.1016/j.jmatprotec.2017.07.023
12. Zhang, S., Gong, M., Cen, L., Lu, Y., Gao, M.: Differences in properties between hybrid wire arc additive-milling subtractive manufactured aluminum and magnesium alloys. Appl. Sci. **13**, 2720 (2023). https://doi.org/10.3390/app13042720
13. Henckell, P., Günther, K., Ali, Y., Bergmann, J., Scholz, J., Forêt, P.: The influence of gas cooling in context of wire arc additive manufacturing—a novel strategy of affecting grain structure and size. Presented at the (2017)
14. Wu, B., Pan, Z., van Duin, S., Li, H.: Thermal Behavior in Wire Arc Additive Manufacturing: Characteristics, Effects and Control. Springer, Singapore (2019)

15. Le, V.T., Mai, D.S., Hoang, Q.H.: Effects of cooling conditions on the shape, microstructures, and material properties of SS308L thin walls built by wire arc additive manufacturing. Mater. Lett. **280**, 128580 (2020). https://doi.org/10.1016/j.matlet.2020.128580

Simulate and Optimize the Front Bumper System of the Vehicle with LS-DYNA Software

Hieu Minh Pham, Tien Van Tang, Huu Van Hoang, Quynh Ba Luu, Phuong Thai Do, Hoang Vuong Chu, and Khanh Quoc Pham

Abstract Ensuring the safety of passengers is a top priority in the automobile industry. To achieve this, engineers have designed new vehicles with advanced safety systems. These systems can be classified into two categories: active and passive safety. This article will examine the passive safety system, namely the front bumper. The front bumper is an essential part of the vehicle that protects passengers from injuries during a frontal collision. It works by absorbing part of the vehicle's energy during the collision. This study aims to optimize the front bumper's structure by calculating and analyzing its ability to withstand collisions and absorb energy. The finite element method is used, with LS-DYNA software being the primary research tool. The collision results are evaluated based on NCAP standards, with a speed of 64 km/h. The study focuses on the front bumper's ability to absorb collision energy and its deformation. By increasing the front bumper's thickness from 3 to 5 mm, the results show that a thickness of 5 mm is optimal, with a deformation parameter of more than 0.84%.

Keywords Bumper simulation · LS-DYNA · Optimizing bumper

H. M. Pham · T. V. Tang (✉) · H. V. Hoang · Q. B. Luu · P. T. Do · H. V. Chu
Faculty of Transportation Engineering, School of Mechanical and Automotive Engineering, Hanoi University of Industry, No. 298 Cau Dien Street, Bac Tu Liem District, Hanoi, Vietnam
e-mail: tangtienthoha@gmail.com

K. Q. Pham
Department of Mechanical and Computer-Aided Engineering, National Formosa University, No. 64 Wenhua Rd, Huwei Township, Yunlin County, Taiwan 632

D. H. Tien et al. (eds.), *Proceedings of the 1st International Conference on Sustainability and Emerging Technologies for Smart Manufacturing*, Proceedings in Technology Transfer, https://doi.org/10.1007/978-981-97-7083-0_36

1 Introduction

Designing a car model with the help of available tools is no longer considered a daunting task. Nonetheless, it is of paramount importance to ensure that every model adheres to global safety regulations, which entails meticulous research and consideration of multiple factors. The NCAP standards are the most extensively utilized benchmarks for assessing the safety level of a car model [1, 2].

NCAP assesses car safety features globally. Standards are set in each region based on simulated collisions. Test results evaluate car models, with passenger protection in front collisions being the most significant factor. Front protection systems are crucial for passenger safety during frontal collisions [3, 4]. Our priority is designing and implementing the passive safety system with accuracy to ensure your safety and peace of mind.

Passive safety systems have garnered considerable attention from researchers. In a recent study, Liu's team improved a vehicle's front deformation during a collision by optimizing the front bumper's thickness using HyperWorks and LS-DYNA software. The front beam's deformation parameters were significantly enhanced as a result [5]. By conducting an analytical study, researchers Kim and Lee [6] compared traditional front bumpers with those equipped with "the crash box," following RCAR standards. The front bumper and vehicle frame were analyzed for stress distribution and displacement. LS-DYNA software was used to measure energy absorption, revealing that the crash box absorbed energy up to 7.79E+06 (Jun) while protecting the frame. Shock absorbers were installed behind the bumper. Ramesh Kannan and colleagues used experimental and numerical methods to evaluate collision likelihood. Researchers used LS-DYNA to predict vehicle safety during collisions. They found barrier and bumper configurations that protect passengers. Their paper highlights the importance of vehicle safety and the need for effective measures to reduce collision damage [7]. Besides, there are a few studies that have analyzed the typical parameters of the front bumper system, such as the design and analysis of the front bumper for light passenger vehicles [8] and the design and analysis of an automotive bumper beam in low-speed frontal crashes [9].

Previous vehicle safety research has focused on modifying the structure or material of the front bumper. This article explores increasing bumper thickness and stiffness. LS-DYNA software is used to simulate the bumper's collision resistance after thickness augmentation.

2 Theory

2.1 *Finite Element Method*

The finite element method is a numerical technique used to solve partial differential equations. It breaks down a large system into simpler parts called finite elements, which are created by discretizing an object or system. This method is used in various fields, including structural analysis, heat transfer, and fluid flow. Dynamic analysis determines the structure's response over time and helps find parameters, such as displacement, deformation, and stress. The basic equation of motion is solved using the following equation:

$$[M]\{\ddot{u}\} + [C]\{\dot{u}\} + [K]\{u\} = \{F(t)\} \tag{1}$$

$[M]$ is the mass matrix
$[C]$ is the damping matrix
$[K]$ is the stiffness matrix.

2.2 *Implicit and Explicit Integration Schemes*

Numerical analysis approximates solutions to differential equations using explicit and implicit methods. They are used in computer simulations of physical processes. Explicit predicts the future state based on the current state, while implicit solves an equation that includes both current and future states.

In the explicit method, we determine $X_{(t+\Delta t)}$ by evaluating the current state (X_t). We can derive the state $X_{(t+\Delta t)}$ from X_t by applying the following equation [10]:

$$X_{(t+\Delta t)} = F\left(X_{(t)}\right) \tag{2}$$

Alternatively, in the implicit method, we obtain the state $X_{(t+\Delta t)}$ by solving the subsequent state Eq. (8):

$$F\left(X_{(t)}, X_{(t+\Delta t)}\right) = 0 \tag{3}$$

2.3 *Time Step Calculations*

It is important to note that the choice of time step is a critical factor that can impact the reliability of simulation results. For a more comprehensive explanation of the calculation process, consult the "Theory manual" [11]. Ultimately, the method you select should be tailored to the specific problem you are trying to solve. For explicit simulations, the time step is determined for all deformable elements in the model. The expression varies depending on which type of element is used. The general idea is:

$$\Delta t_e \approx \frac{L_e\sqrt{\rho}}{\sqrt{E}}$$

L_e Charteristic length of an element
ρ Density
E Young's modulus.

The critical time step is determined as the minimum of all calculated time steps:

$$\Delta t_c \approx \min(\Delta t_e)$$

2.4 Displacements, Strains, and Stresses in the Element

In the finite element method, the nodal displacement values are determined through interpolation and approximation based on the displacement of an element's nodal points. Once shape functions have been identified, the displacement field can be expressed using a specific formula [12].

$$U_e = N \cdot q_e \tag{4}$$

U_e Displacement field
q_e Displacement vector at the *e*th element
N Shape function.

The displacement field at the *e*th element can be obtained through the application of formula (4). It follows that the deformation at the *e*th element can be calculated using the following equation:

$$\varepsilon_e = \partial \cdot U_e \cdot q_e = Bq_e \tag{5}$$

B is the deformation matrix.

According to Hooke's law, we have the following expression for calculating stress:

$$\sigma_e = E \cdot B \cdot q_e \tag{6}$$

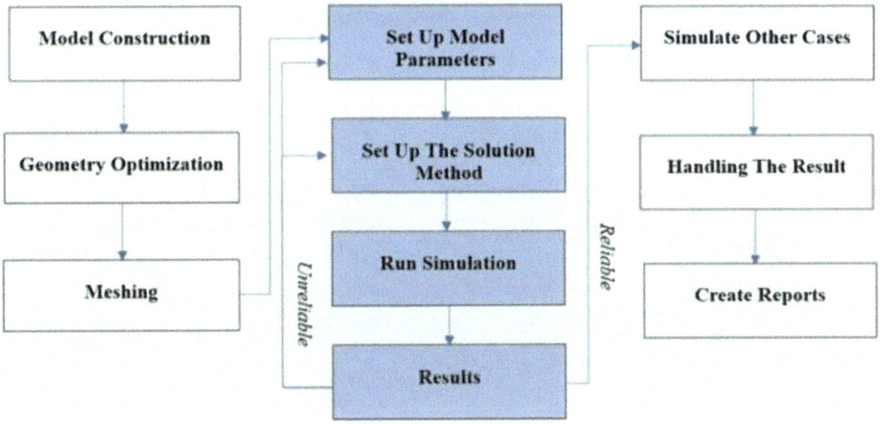

Fig. 1 Block diagram of process of setting up problem

3 Setting Up the Problem

3.1 Block Diagram of the Process of Setting Up the Problem

The finite element method is a widely used approach for solving a diverse range of problems, and it consists of three fundamental stages: preprocessing, problem execution, and results evaluation. In the preprocessing stage, the model undergoes optimization and discretization within the computational domain, utilizing specialized software. This is the phase where the key parameters of the problem are established, which ultimately influence the results. In the subsequent stage, the problem is executed, and the results are generated. Finally, experienced engineers will assess the reliability of the findings and recommend enhancements if required (Fig. 1).

3.2 Material Parameters

In car manufacturing, it is important to know the different types of steel used. The three most common types are soft steel, hard steel, and super hard steel. HSS is the main steel code used for car body production, with a strength range of 210–550 MPa. The bumper material is from the Johnson–Cook Law 2 Elasto-Plastic material line, which has linear elastic behavior and specific parameters. Density: RHO = 7.8E−9 tons/mm^2; Elastic modulus: $E = 210{,}000$ MPa; Poisson's coefficient: PR = 0.3; Plastic stress: SIGY = 280 MPa.

Fig. 2 Problem model after setup is complete

3.3 Model Construction

This research project was partially funded by the guidelines established by NCAP, a global organization that evaluates a vehicle's capacity to safeguard its occupants in the event of an accident. As outlined by these criteria, the vehicle model was subjected to a frontal collision test at a speed of 64 km/h [13]. For your convenience, a simulation configuration utilizing rigid columns within LS-DYNA software is depicted (Fig. 2).

In this write-up, we will assess the outcomes of a simulation model that contrasts two front bumper bars of differing thicknesses—3 and 5 mm. The comparison will take into account various parameters, such as collision energy absorption, deformation, stress, and absorption capacity.

4 Result

4.1 Contact Force

The front bumper absorbs collision energy and prevents damage to the vehicle body. It is designed to deform and dissipate collision energy, adding an extra layer of protection. We aim to compare the effectiveness of two different front bumper designs (Fig. 3).

Upon impact, the force between a barrier and a wall peaks at 61.3 kN and gradually decreases over time, ranging from 20 to 25 kN. A 3-mm thick bar can reduce about 66% of the interaction force between itself and the hard wall. A barrier bar's force parameters depend on its thickness. A 5-mm barrier has a significantly higher maximum force value (90.3 kN) and peaks earlier than a 3-mm barrier. The time for the force to decrease after reaching its maximum is more abrupt for the 5-mm barrier.

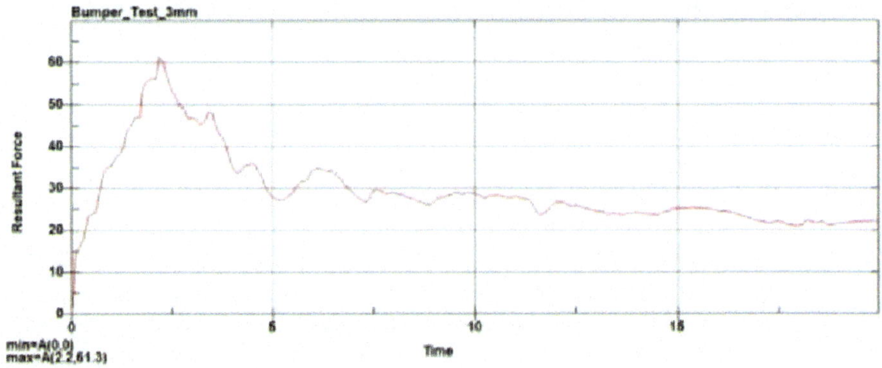

Fig. 3 Graph of relative interaction force of 3-mm model

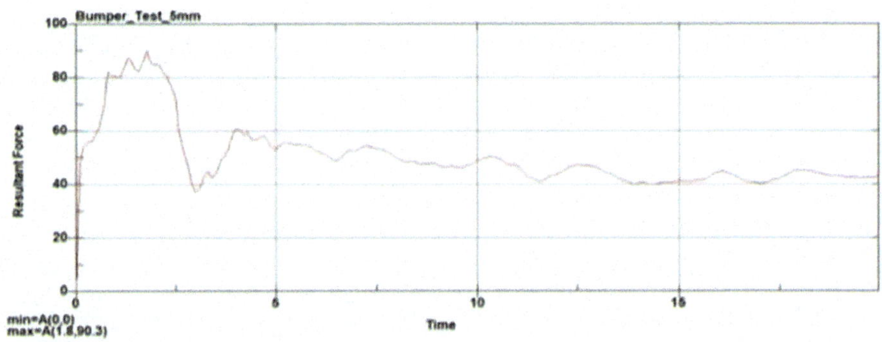

Fig. 4 Graph of relative interaction force of 5-mm model

While this sudden change is not ideal for passenger safety during vehicle travel, it could result in injuries due to the inertial force caused (Fig. 4).

When a bumper bar with a 5-mm thickness is completely deformed, the force of interaction between the bar and a rigid wall decreases to approximately 40 kN. This suggests that the 5-mm barrier reduces the interaction force with the wall by 55%.

4.2 Displacement

Figure 5 depicts a graph of the displacement value when the bar is fully deformed. At the point where the bar comes into contact with the hard field, with a thickness of 3 mm, the displacement value measures 237 mm. Likewise, at the same position with a 5-mm barrier, the displacement parameter reaches 235 mm, as illustrated in Fig. 6.

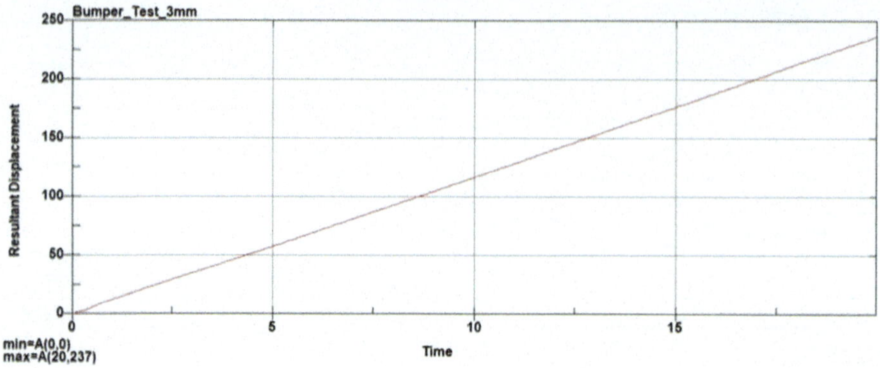

Fig. 5 Displacement of 3-mm model

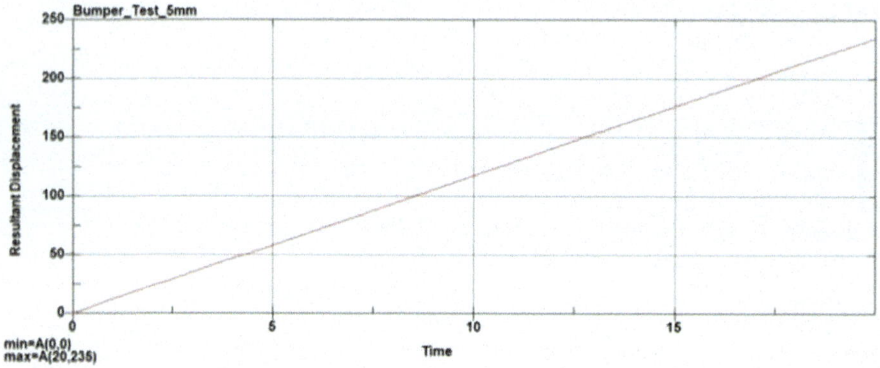

Fig. 6 Displacement of 5-mm model

Based on the data presented in Figs. 5 and 6, there appears to be a subtle shift in displacement parameters as the barrier thickness increases. Nevertheless, this shift is not considered significant. Notably, when the barrier thickness is increased from 3 to 5 mm, the displacement value of the barrier decreases from 237 to 235 mm. As such, one can surmise that a 5-mm barrier is only marginally more optimal than a 3-mm barrier, by a mere 0.84%.

4.3 Stress

Von Mises stress is a widely accepted metric for evaluating the durability of plastic structures. It helps to calculate and assess the deflection of flexible materials by measuring a scalar stress parameter. By analyzing the parameters presented in Fig. 7, it is evident that a 3-mm thick barrier bar does not surpass the allowable plastic stress threshold of the steel material even when subjected to complete deformation, as its standard stress is only 3.07E−1 kJ.

The present study aims to compare the standard stress parameters of two bumper bars, namely the 5 and the 3 mm ones. The results of the analysis are depicted in the form of a simulation image and a corresponding value in Fig. 8. It is important to note that the stress parameter of the 5-mm bumper bar is found to be comparatively larger than that of the 3-mm counterpart. This information could aid in making informed decisions regarding the selection of the appropriate bumper bar for a given application.

The 5-mm bumper bar has a standard stress value of 3.11E−1 kJ, which is the maximum stresses it can withstand a collision without surpassing the allowable

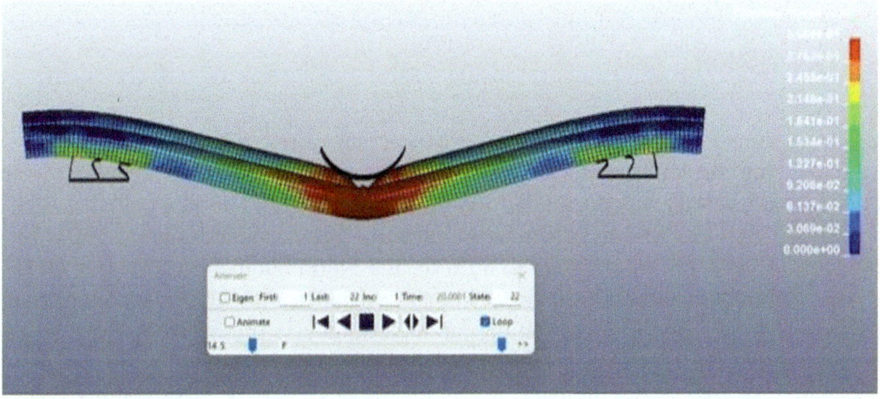

Fig. 7 Stress parameters of 3-mm bumper bar

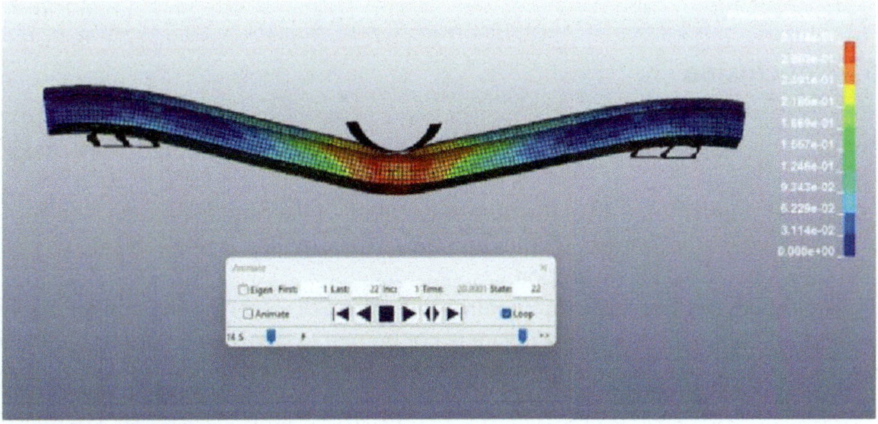

Fig. 8 Stress parameters of 5-mm bumper bar

Table 1 Comparison table of results between 2 cases

Parameter	Thickness		
	3 mm	5 mm	Comparison (%)
Contact force	21 kN	41 kN	95
Absorption ability	66%	55%	– 16.7
Deformation	237 mm	235 mm	– 0.84
Stress	3.07×10^{-1} kJ	3.11×10^{-1} kJ	1.3

threshold for the material. However, when compared to another bumper bar, it experiences 1.3% more stress. As a result, the 3-mm barrier bar has an edge over the 5-mm barrier bar in terms of stress parameters.

5 Discussion

Based on the above result parameters, we can create Table 1.

Increasing the bumper thickness from 3 to 5 mm has pros and cons. A 5-mm bumper bar has 95% greater interaction force than a 3 mm one in a collision with a rigid wall. However, a 3-mm bumper bar absorbs better, reducing up to 66% of the interaction force value compared to 55% for the 5-mm bumper bar.

The 5-mm bumper bar offers better impact resistance compared to its 3-mm counterpart, as shown by the deformation parameter analysis presented in Table 1. The 5-mm bar exhibits a deformation parameter of 235 mm, while the 3-mm bar shows 237 mm, resulting in an efficiency advantage of 0.84% for the 5-mm bumper bar.

The stress on both bumpers during impact is lower than the plastic stress of the steel material. However, the 3-mm bumper performs better than the other.

6 Conclusion

Objective: Analyze the advantages and disadvantages of modifying bumper thickness.

Method: Simulated collision with two bumpers of varying thickness.

Results: A thicker bumper (5 mm) minimizes deformation, while a thinner bumper (3 mm) improves force absorption.

In addition, engineers can now control and minimize errors in problem-solving by utilizing software. The use of software is not limited to small-scale issues, but it can also handle larger-scale problems, such as simulating the behavior of vehicles during and after a collision. Nevertheless, there is room for further research on passive safety systems in vehicles to improve their capacity to absorb collision energy, particularly in enhancing the vehicle's crash box capacity of passive safety systems on vehicles to absorb collision energy; further research can be done to improve the vehicle's crash box.

References

1. Anders, L., Claes, T.: How do Euro NCAP results correlate with real-life injury risks? A paired comparison study of car-to-car crashes. Traffic Inj. Prev. **3**(4) (2002)
2. Anders, K., Anders, L., Claes, T.: Comparison between Euro NCAP test results and real-world crash data. Traffic Inj. Prev. **11**(6) (2010)
3. Euro NCAP: Test Result, BMW 5 Series Standard Safety Equipment (2023)
4. Euro NCAP: Test Result, VinFast VF8 Standard Safety Equipment (2023)
5. Liu, C., Song, X., Wang, J.: Simulation analysis of car front collision based on LS-DYNA and hyper works (2014)
6. Ravinder Reddy, P., Harish, T.: Design and crash analysis of passenger car frontal bumper beam using Hypermesh and Radioss (2015)
7. Ramesh Kannan, C., Padmanabhan, P., Rajkumar, G.: Optimization of Bumper Design by Using Crash Test (2014)
8. Pareschi, L., Russo, G.: Implicit–explicit Runge–Kutta schemes for stiff systems of differential equations (2001)
9. Joshi, N.: Design and analysis of front bumper for light passenger vehicles. Int. J. Rapid Res. Eng. Technol. Appl. Sci. **2**(4) (2016)
10. Marzbanrad, J., Alijanpour, M., Kiasat, M.S.: Design and analysis of an automotive bumper beam in low-speed frontal crashes. Thin-Walled Struct. (2009)
11. LS-DYNA: Theory Manual (2006)
12. Zienkiewicz, O.C., Taylor, R.L., Zhu, J.Z.: The Finite Element Method: Its Basis and Fundamentals (2013)
13. Jawi, Z.M., Kassim, K.A.A., Sadullah, A.F.M.: ASEAN NCAP—the best bet for vehicle safety in the region's safe system (2013)

Performance Analysis of Copper-Coated Aluminum Tool in Electro Discharge Machining of Ti-6Al-4V Alloy

Van Dua Tran, Shailesh S. Shirguppikar, L. Selvarajan, Van Bong Pham, and Anh Nguyen Mai

Abstract Electric Discharge Machining (EDM) is one of the most popular non-traditional machining processes applied for machining very hard conductive materials with desired shape, size and dimensional accuracy. In this paper, copper-coated aluminum alloy is applied as EDM electrode instead of copper electrode. Electrodeposition, simple and economical process, was used for preparation of copper coating on aluminum alloy. The method of anodization was applied in order to get a strong adhesive copper coating on aluminum alloy substrate. Taguchi-based design of experiment technique were used to form relationship of process parameters and response variables. The copper-coated surface was characterized with XRD report. Lower tool wear rate (TWR), surface roughness (SR) and higher material removal irate (MRR) were found for copper-coated aluminum alloy EDM electrode compare to uncoated aluminum alloy electrode.

Keywords EDM · Coated electrode · MRR · TWR

1 Introduction

Electrical discharge machining (EDM) has become one of the most important production technologies to manufacture very accurate three-dimensional complex components on any electrically conductive material Ho and Newman [1]. Examples include precision machining of hardened steels, carbides and ceramic materials Klocke and

V. D. Tran (✉) · V. B. Pham · A. N. Mai
Hanoi University of Industry, No. 298, Cau Dien Street, Bac Tu Liem District, Hanoi, Vietnam
e-mail: duatv@haui.edu.vn

S. S. Shirguppikar
Department of Mechanical Engineering Department, RIT, Rajaramnagar Affiliated to SUK, Walwa, Maharashtra, India

L. Selvarajan
Department of Mechanical Engineering, Mahendra Institute of Technology (Autonomous), Namakkal, Tamilnadu 637503, India

D. H. Tien et al. (eds.), *Proceedings of the 1st International Conference on Sustainability and Emerging Technologies for Smart Manufacturing*, Proceedings in Technology Transfer, https://doi.org/10.1007/978-981-97-7083-0_37

König [2]. The cost of a part manufactured by EDM is essentially determined by the electrode cost, which consists of the raw material cost and the tool production. The contribution of the electrode cost to the total EDM operation cost can be more than 70%. The electrode production is a major cost and time spent in the EDM process cycle, which can account for over 50% of the conventional machining costs Ozgedik and Cogun [3]. Teng studied the electroplating preparation process of the Cr–Cu composite tool. Based on this, the brush plating preparation process of the composite tool was proposed, and the rotating brush plating device was made and carried out in experiments Teng [5]. Author prepared a new copper-based nickel-tungsten alloy composite tool by the nickel-tungsten co-deposition electroplating method and carried out microhole machining experiments. The results showed that the nickel-tungsten alloy coating has good erosion resistance performance, which can effectively reduce the side wear of the tool, especially at the end and edge, maintaining the shape of the tool and improving the shape precision of the machined microhole Lv [6]. Zhang used the gel comprised mixture of TiN–Cu powders sodium alginate dispersant and gelatin binder and concluded that the wear rate of the tool is lower than that of the copper tool Zhang [7]. Researchers were proposed a wear-variation EDM (WV-EDM) approach to fabricate a laminated disc electrode (LDE) with stable-shape microchannels on the outer edge surface. This electrode was subsequently used to machine microgrooves on workpieces by EDM, and it is concluded that this tool could promote the machining efficiency Lei et al. [7]. According to the above research, although the current researches on composite tools have achieved rich results, due to the numerous methods for preparing composite tools, the influence of composite tool wear and shape change remains to be further studied. +e experimental results show that the material removal rate of composite tools increases about 2 times and the processing precision is also improved Cao et al. [8]. Al–Cu–Si–TiC composite tool produced by powder metallurgy technology and processed it for processing analysis. It is found that the Al–Cu–Si–TiC tool is more sensitive to peak current and pulse conduction time than the conventional tool and the percentage of TiC based on the optimal demand of the composite is 18% El-Taweel [9]. In order to prevent tool loss, Hai-Liang and Cao proposed a method of processing small hole machining by EDM using dispersant dielectric fluid. +e results showed that the reasonable choice of the composite tool and dispersant dielectric fluid could effectively improve the processing speed, lower tool wear, and improve depth-to diameter ratio of the drilling hole significantly An and Cao [9]. Influence of coated electrode to improve machining performance were studies by Shirguppikar and Patil [8]. Effect of copper coated electrode and Cu–MWCNT coated electrode were used for machining of AISI 304 SS and it was observed that Cu–MWCNT & copper coated electrode influenced to improve MRR and reduce TWR & Prosun and Mondal [9].

Table 1 Chemical composition of Ti-6Al-4V

Contents	Ti (Ti)	Al	V	Fe	O	C	N	H
Percent %	89.86	5.89	3.93	0.165	0.128	0.017	0.005	0.002

In this paper, in order to reduce the significant problem of tool wear in EDM, thin film-coated copper tools were made by electroplating technique. Experiments were performed as per Taguchi-based design of experiment method. The effects of the thin film coated tools on tool wear, surface roughness, and material removal rate were analyzed. It provides a new tool preparation method for EDM, and the research results have certain guiding significance for the actual production in EDM.

2 Experimental Details

Experiments were performed on titanium alloy (Ti-6Al-4V) and undergone drilling operation using EDM. Two types of electrodes were used; uncoated aluminum electrode and copper coated aluminum electrode.

2.1 Selection of Workpiece

Titanium alloy (Ti6Al4V) alloy was designed for a good balance of characteristics, including: strength, ductility, fracture toughness, high temperature strength, creep characteristics, weldability, workability and thermal process ability. Also it has high strength to weight ratio, excellent erosion resistance and compatibility with composite structure. Chemical composition and physical properties of Ti-6Al-4 V is given in Table 1, respectively.

2.2 Preparation of Tools Electrode

Electroplating technique is the process of forming a layer of coating material on the surface of tool electrode. It takes place in an electrolytic cell where electrolysis which use D.C current to dissolve a rod of desired coating material and transports the ions onto the tool electrode to be coated. Into a container of water are placed a rod of material to be coated and aluminum rod as a substrate. The water contains an ionic solution which allows a direct electric current to flow from the coating material rod to the substrate. The coating material rod is the anode and the substrate is the cathode. This current flow causes the coating material to ionize, become oxidized which means each atom becomes positively charged by losing an electrons. As the

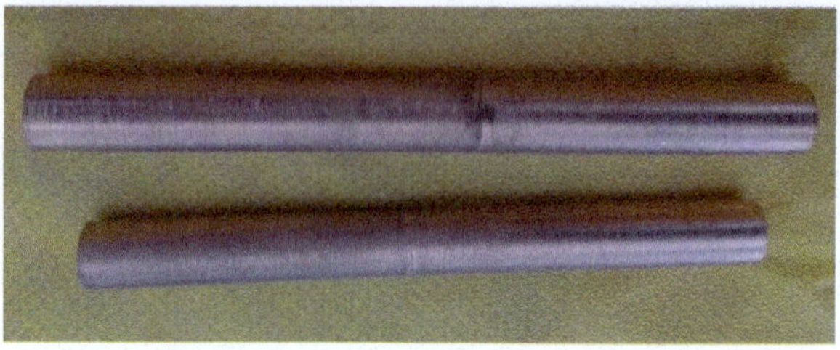

Fig. 1 Thin film-coated copper electrode

coating material ions dissolve into the water, they form a coordination complex with salts already present. The coating material then physically flows to the item, where it is reduced to the metallic state by gaining electrons. This forms a thin, solid, metallic coating material film on the surface of the substrate. Figure 1 shows copper coated tool electrode. Diameter of uncoated aluminum electrode was 9.5 mm and copper coated aluminum electrode was 10 mm.

2.3 *Experimental Work*

Selection of process parameters and its levels were selected as per literature review and machine capabilities, three input process parameters with four levels were selected for the final experimentation, as shown in Table 2.

Taguchi-based orthogonal array were used for experimental work with four levels of Current (I), Gap Voltage (Vg), Pulse-ON Time (TON) and L16 design of experiment were planned and conducted experiments with uncoated aluminum electrode and copper coated aluminum electrode. MRR indicates amount of material removed from workpiece. TWR indicates amount of material removed from tool. Surface roughness of workpiece surface was measured by non-contact type surface roughness instrument (Taylor Hobson machine).

Table 2 Selection of process parameters and levels

Parameters	Symbol	Levels				Unit
Peak current	I	10	20	30	40	A
Gap voltage	Vg	40	45	50	55	V
Pulse-on time	TON	100	500	1000	1500	μS

3 Experimental Results

In this experimental work study has been conducted to investigate the effect of coated and uncoated electrodes on the material removal rate (MRR), tool wear rate (TWR) and surface roughness (SR) by using Ti-6Al-4V as workpiece material (Table 3).

3.1 *Performance Analysis of Material Removal Rate (MRR)*

Main effect plots were plotted with help of Minitab 17 software. These graphs show the behavior of process parameters in relation to response variables. Main effect plot of MRR with uncoated & Cu coated Al electrode is shown in Fig. 2. The MRR chart is a linear graph that increases from 10 to 40 amps in both cases.

High pulse time values, the drop may be explained, due to pulse-OFF values. As the duty cycle was kept constant during the experimentation, at higher values of TON corresponding TOF F values are also higher due to which reduction in the MRR was observed. This phenomenon was been observed by few researchers which support our statement. The calculated MRR for two electrodes is shown in bar chart

Table 3 Experimental results for Ti-6Al-4V using uncoated and copper coated Al electrode

Expt. No	Process parameters			Uncoated Al electrode			Copper coated Al electrode		
	(I)	(Vg)	(TON)	MRR gm/min	TWR gm/min	SR μm	MRR (mg/ min)	TWR (mg/ min)	SR (μm)
1	10	40	100	0.002	0.0023	6.664	0.0033	0.003	6.23
2	10	45	500	0.0022	0.0028	6.683	0.0004	0.0042	6.28
3	10	50	1000	0.0019	0.0041	6.691	0.0019	0.0034	6.34
4	10	55	1500	0.0027	0.0029	6.815	0.0021	0.0045	6.44
5	20	45	100	0.0019	0.005	7.781	0.0062	0.0064	7.112
6	20	40	500	0.0073	0.0058	8.325	0.0088	0.0092	7.354
7	20	55	1000	0.0075	0.0076	8.662	0.0092	0.0126	7.641
8	20	50	1500	0.0111	0.0087	8.981	0.0048	0.0174	7.664
9	30	50	100	0.0053	0.0068	9.116	0.0114	0.011	8.021
10	30	55	500	0.0097	0.0098	9.203	0.0201	0.0137	8.331
11	30	40	1000	0.0141	0.0118	9.412	0.0214	0.0159	8.689
12	30	45	1500	0.0172	0.0124	9.665	0.0204	0.021	8.911
13	40	55	100	0.0093	0.0106	9.706	0.0151	0.0148	8.978
14	40	50	500	0.0164	0.0145	9.783	0.028	0.0214	9.011
15	40	45	1000	0.0229	0.0177	10.112	0.0296	0.0262	9.241
16	10	40	100	0.0282	0.0212	10.391	0.0281	0.0281	9.511

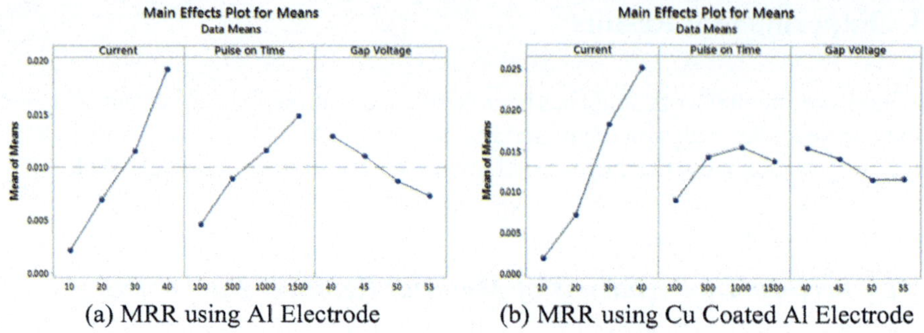

Fig. 2 Main effect plots for MRR with uncoated and Cu-coated Al electrode

Fig. 3 Influence of electrode material on MRR

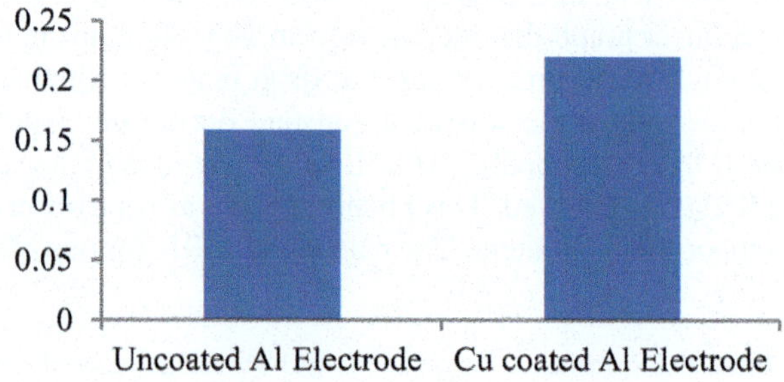

shown in Fig. 3. About 12% higher MRR is achieved for copper-coated Al electrode than uncoated Al electrode. The higher value of electrical conductivity is the main reason for the increasing of material removal rate. The higher value of electrical conductivity can increase more energy transfer into workpiece and it causes more material removal from the workpiece. Though the electrical conductivity of Al is improved by copper coating on it, it does not reach the electrical conductivity as much as copper.

3.2 *Performance Analysis of Tool Wear Irate (TWR)*

The absolute values of tool wear show that there are two phenomena that regulate the wear characteristics of the tools: the erosion of the tool electrode material and the deposition of the workpiece material on the tool electrode. When high MRR conditions occur, large quantities of debris are produced and the quantity of material deposits increases on the tool electrode. In such conditions, the amount of actual material removed from the tool electrode also increases. Main effects plot for TWR is shown in Fig. 4.

Figure 5 shows the calculated electrode wear rate (EWR) for uncoated & Cu copper electrodes. The EWR is decreased by 22% after applying copper coating

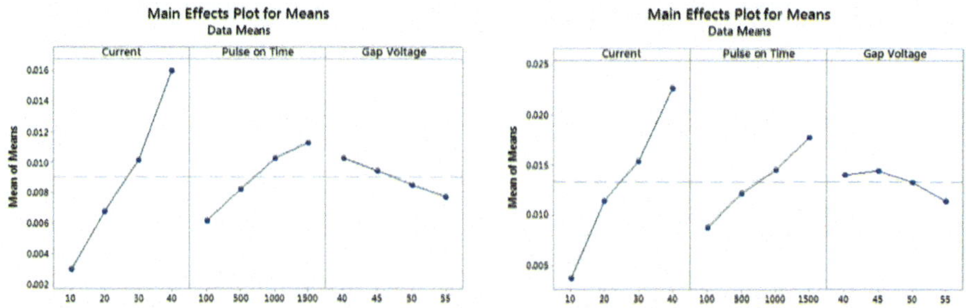

(a) TWR using Al Electrode (b) TWR using Cu Coated Al Electrode

Fig. 4 Main effect plots for MRR with uncoated and Cu-coated Al electrode

Fig. 5 Influence of electrode material on TWR

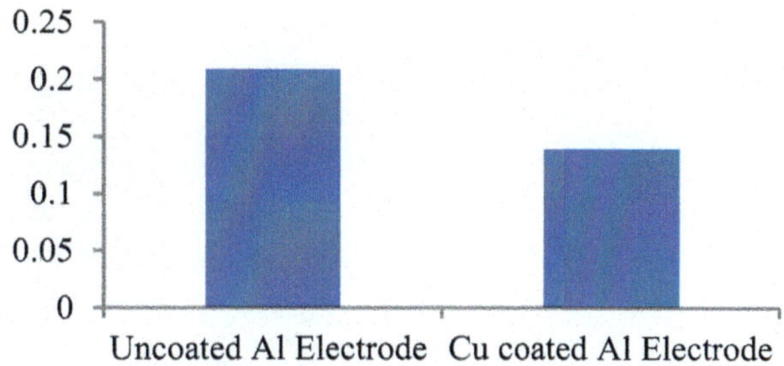

on Al electrode. The thermal conductivity is mainly responsible for electrode wear rate. The higher thermal conductivity of electrode material can be driven out more heat from sparking zone more quickly. This is reduced the electrode erosion. Thermal conductivity of copper-coated Al is higher compared to uncoated Al alloy. The higher thermal conductivity of copper-coated 6061Al electrode can reduce the electrode wear compared to the uncoated Al electrode.

3.3 Performance Analysis of Surface Roughness (SR)

Increase in current and pulse on time shows the increase in the surface roughness (SR) which can be determined from main effects graph, thereby increasing temperature leading to premature melting of the tool electrode. Therefore, as the current level is increased the surface roughness is increased due to increment in the spark energy. Main effects plot for SR is shown in Fig. 6.

Figure 7 shows the influence of three different electrodes on the surface roughness of workpiece. The results show that better surface is achieved using copper-coated Al electrode compared to uncoated Al electrode. The higher wear rate caused the degradation of surface of electrode. The degraded surface of electrode has generated more rough surface.

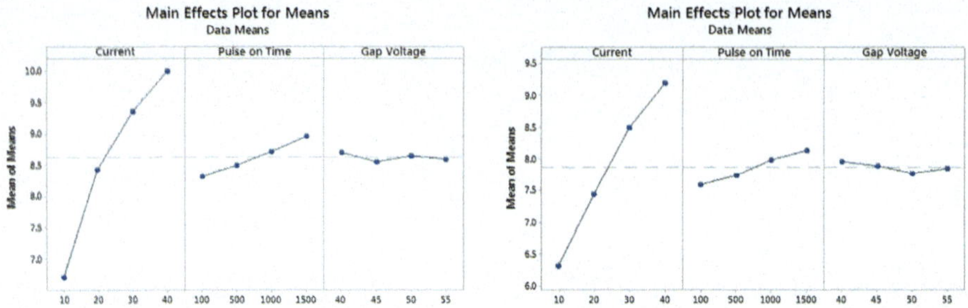

(a) SR using Al Electrode (b) SR using Cu Coated Al Electrode

Fig. 6 Main effect plots for SR with uncoated and Cu coated Al electrode

Fig. 7 Influence of electrode material on TWR

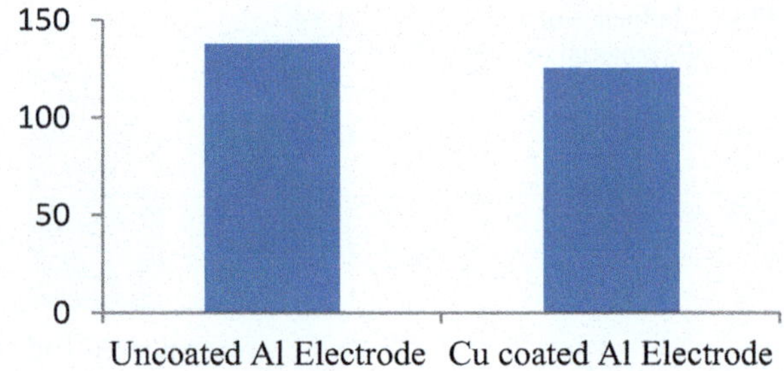

4 Conclusion

Copper-coated aluminum electrode was developed using electrode position process. Prior to electrode position, the surface of aluminum was prepared by anodizing process. The EDM operations were performed with newly developed electrodes. The performance of two electrodes was compared on the basis of MRR, TWR and SR. The results show that copper coating on aluminum has enhanced the performance compared to uncoated aluminum. The result shows that copper coating on aluminum alloy has significantly improved the applicability and sustainability of Al as an EDM electrode. The copper-coated aluminum can be used as EDM tool in a large scale in industry.

References

1. Ho, K.H., Newman, S.T.: State of the art electrical discharge machining (EDM). Int. J. Mach. Tools Manuf **43**, 1287–1300 (2003)
2. Klocke, F., König, W., Fertigungsverfahren—Abtragen, Generieren und Laser material bearbeitung (1997)

3. Ozgedik, A., Cogun, C.: An experimental investigation of tool wear in electric discharge machining. Int. J. Adv. Manuf. Technol. **27**, 488–500 (2005)
4. Jha, B., Ram, K., Rao, M.: An overview of technology and research in electrode design and manufacturing in sinking electrical discharge machining. J. Eng. Sci. Technol. Rev. **4**, 118–130 (2011)
5. Teng, D.: Preparation of Cr–Cu Composite Electrode an Research of Simulation and Experiment. Taiyuan University of Technology, Taiyuan, China (2012). in Chinese
6. Lv, Q.C.: Simulation and Experimental Study on the Fabrication of the Composite Electrode for MEDM. Dalian University of Technology, Dalian, China (2009)
7. Zhang.: Fabrication of TiN-Cu electrical discharge machining electrode by gel injection molding. Chin. J. Vac. Sci. Technol. **39**(2), 169–173 (2019). in Chinese
8. Shirguppikar, S.S., Patil, M.S., Ganachari, V.S., Kolekar, T.V., Jadhav, P.S., Chougule, A.B.: Experimental investigation of CNT coated tools for EDM processes. Mater. Today: Proc. **5**(2), 7131–7140 (2018)
9. Mandal, P., Mondal, S.C.: Development and application of Cu-SWCNT nanocomposite–coated 6061Al electrode for EDM. Int. J. Adv. Manuf. Technol. **103**(5–8), 3067–3076 (2019)

Smart Manufacturing—"Optimizing Energy from Green and Renewable Sources for Sustainability"

Mark Cherukuri, Vijender Kumar Solanki, and Praneet Amul Akash Cherukuri

Abstract The change that is inevitable for future growth is much required to steer the different spheres of social, economic, and human aspects that are required to follow the principles of sustainability so that our progress is harmonious. There is huge pressure on fossil fuels with increasing demand where nations are fighting for their share of fuel on the global map, since there are increasing environmental challenges and also a global commitment to sustainable development. This paper explores the critical integration of green energy, renewable energy sources, energy conservation practices, and smart manufacturing in the pursuit of a sustainable future, as we emphasize the need to shift away from fossil fuels. It highlights the pivotal role of renewable energy sources in mitigating the effects of climate change. This article shall highlight the inexhaustible potential of solar, wind, hydropower, and geothermal energy to revolutionize the different sources of energy, with implementation of smart manufacturing technologies, like automation, AI data analytics, and Internet of Things, playing a crucial role in optimizing resource utilization, reducing environmental impact, and enhancing the overall efficiency of manufacturing processes. It is important to advocate for a holistic approach that intertwines green energy adoption, energy conservation practices, and smart manufacturing solutions through policy changes and proper implementation within the manufacturing facilities and to adhere to the strict norms laid by the governing bodies, and the executive frame work so that it moves the nation toward a sustainable future that harmonizes human activities with the planet, ensuring a resilient, eco-friendly, and sustainable paradigm shift.

Keywords Smart manufacturing · Renewable sources · Sustainable development · IoT · Green energy · Conservation

M. Cherukuri (✉)
University College of Law, Osmania University, Hyderabad, TG, India
e-mail: markson.c9@gmail.com

V. K. Solanki
Stanley College of Engineering and Technology for Women, Hyderabad, TG, India

P. A. A. Cherukuri
Amazon Web Services, Inc, Austin, TX, USA

D. H. Tien et al. (eds.), *Proceedings of the 1st International Conference on Sustainability and Emerging Technologies for Smart Manufacturing*, Proceedings in Technology Transfer, https://doi.org/10.1007/978-981-97-7083-0_38

1 Introduction

The manufacturing space needs a dedication to long-term viability, and ecological responsibility, sustainable design, and production reflect a paradigm change in the industrial sector. Since it requires fundamentally, environmentally conscious design incorporating considerations into all phases of the product life cycle, this all-encompassing method takes into account not just the environmental effect of the finished product but also the procedures involved in its production and disposal. By emphasizing resource efficiency, using eco-friendly materials, and designing goods with end-of-life concerns, such recyclability and disassembly, the overall goal is to minimize the ecological footprint. An essential part of this framework is sustainable manufacturing, which applies these ideas to the actual production processes. Reducing waste, using circular economic principles, and embracing clean energy sources are highlighted. Sustainable design and manufacturing work together to promote innovation that balances environmental care with economic development. This dedication goes above and beyond legal requirements, demonstrating a proactive approach to solving today's environmental issues. Sustainable manufacturing, which supports sustainable design, is the conversion of production processes into ecologically conscious ventures. This entails cutting waste, using energy-efficient technologies, and utilizing sustainable energy sources. Reusing, repurposing, or recycling materials is ensured by the implementation of circular economy concepts, which helps to preserve finite resources.

When it comes to measuring and maximizing the environmental effects connected to products, life cycle assessment (LCA) becomes a crucial instrument. Lean Cycle Assessment (LCA) provides significant insights into opportunities for improvement by methodically evaluating the environmental impacts across a product's life cycle.

Within the framework of sustainable design and manufacturing, green supply chain management also becomes essential. Companies can improve transparency, lower emissions, and advance sustainability throughout the production cycle by incorporating environmental considerations into their supply chains. This article also discusses strategies for developing environmentally responsible supply chains and provides noteworthy case studies of businesses that have effectively incorporated green practices into their daily operations. Adopting sustainable design and manufacturing principles becomes not only a smart economic move but also a moral need as we traverse the complexity of the current industrial period and work toward creating a resilient and ecologically harmonious future [1].

2 Sustainable Design

Sustainable design involves creating products and systems with minimal environmental impact. It encompasses principles and practices aimed at promoting longevity, minimizing waste, and conserving resources. Sustainable design seeks to meet current needs without compromising the ability of future generations to meet their own needs, fostering a balance between environmental, social, and economic considerations.

Sustainable design includes making items and frameworks with negligible natural affect that effects climate. It includes standards and policies pointed at advancing life span, minimizing and preserving assets. The best feasible plan looks for to meet current needs without compromising the capacity of future eras to meet their claim needs, cultivating a compromising and a much adjustable between the natural, social, and economical contemplations.

2.1 Integration of Life Cycle Thinking in Product Design

As life cycle thinking for any product involves considering its whole life, from the basic raw material extraction process to the disposal mechanisms. Life cycle thinking, which stresses longevity, durability, and responsibility for end-of-life management, will guide decisions to reduce environmental impacts at every stage of sustainable development. The principle focuses on the development of products that are easy to break down and recycle at the end of their useful lives. In the context of a circular economy where materials are reused rather than discarded, designing for disassembly contributes to resource recovery, reduces waste, and promotes effective recycling processes. In view of factors, such as revitalization, energy consumption, and toxicity, it is important to select resources that have minimal environmental impact when selecting eco-friendly materials. Sustainable design aims to reduce the total ecological footprint of a product by prioritizing materials which are responsibly grown, environmentally friendly, or made from renewable resources [2] (Fig. 1).

As we can see from the above image depicts the different sources of Energy from where it is clear visible that the products when designed from renewable infinite sources, it forms much less burden on the mother Earth in the form of carbon footprint as there is a clear sources of secondary fuel and even can be reused but in contract when we use finite resources with technical material it always ends in either reuse of product are manufacturing and shall come to an end but will not generate the secondary source of energy. Therefore, it is important for every nation to use renewable resources for their energy consumptions rather burdening the mother earth with higher carbon footprint.

2.2 Sustainable Design Projects

Valuable insights into effective implementation strategies are gained from the examination of successful sustainability design projects. Case studies highlight the practical application of sustainable design principles in everyday life, highlighting innovative

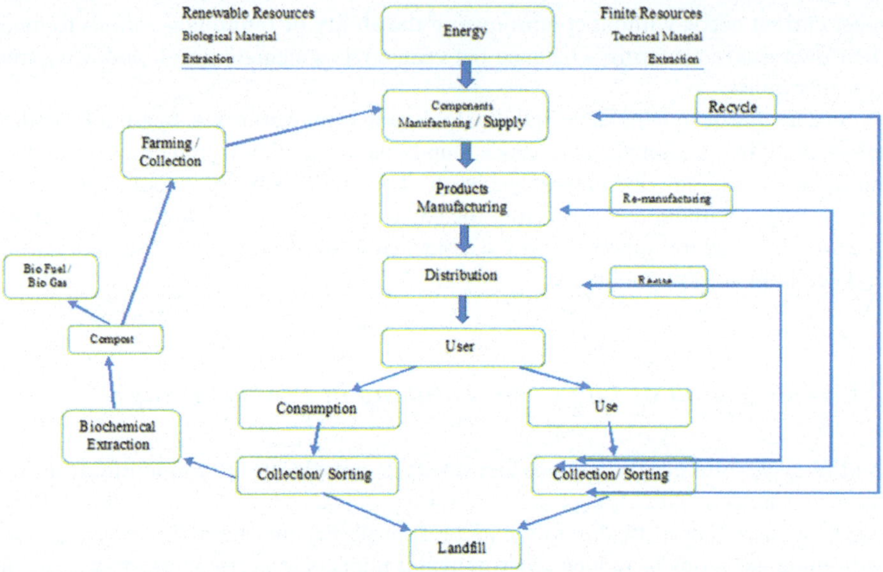

Fig. 1 Sustainable product design practice representation [3]

solutions, beneficial effects on the environment, and financial viability for incorporating sustainability into a planning process. The analysis of these cases will contribute to informing and promoting more widespread adoption of sustainable design principles in the sectors concerned. As we can see that the below graph depicts that from various sources of energies are increased from the previous installed capacities in GW where there is standard to marginal increase in amounts of hydropower between the 2013 to 2022 years but as soon as the solar power, wind power and bioenergy from green sources have started with gradual increases year by year so that the carbon footprint is reduced with production of alternate sources of energies coming up in order to reduce the traditional source of fossil fuels; also there are other sources, such as ocean energy and geothermal energy initiated with gradual increase in the energy production, as the capacity has increase from 1600 GW in 2013 to about 3400 GW in 2022 in about 9 year span the capacity doubled as alternative sources emerged the manufacturing process relied on green and much more sustainable sources of energy for production which led to reduction of carbon footprint (Fig. 2).

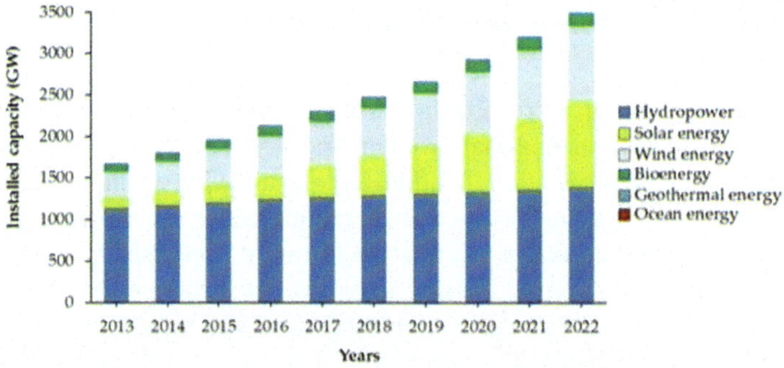

Fig. 2 Represents installed capacity of GW of different sources of energy for continuous reduction of carbon footprint with sustainable sources for manufacturing [4]

3 Sustainable Manufacturing Technologies and Strategies

Sustainable production has become an imperative in the dynamic landscape of today's industry, driven by a pressing need to preserve the environment. To make it more sustainable, the innovative technologies and strategies are changing manufacturing processes to a great extent. The focus is on the development of a roadmap for greener future, which includes adoption of clean energy and circular economy principles that direct the process of sustainability.

3.1 Embracing Clean Energy

In order to reduce the carbon footprint of manufacturing, integrating clean energy sources is a key step. The reliance on fossil fuels is significantly reduced by using renewable energy sources, such as solar and wind power, in production facilities. In doing so, it also helps to mitigate the impact on the environment and puts companies at the forefront of sustainable practice. A key strategy to align economic growth with environmental responsibility is to invest in clean energy infrastructure. The aim of this principle is to produce products that are easy to break down and reuse at the end of their lives. Designing for disassembly promotes resource recovery, reduces waste, and encourages efficient recycling processes, contributing to a circular economy where materials are reused rather than discarded. Always it is the energy that can bring the major impact on the sustainable manufacturing processes [5]. Energy-efficient technologies help industry to optimize energy consumption, reduce operational costs, and minimize GHG emissions. Real-time monitoring and control of energy consumption are enabled by smart manufacturing systems, which incorporate the Internet of Things device as well as artificial intelligence. This data-driven approach enables a high level of precision in resource management which contributes to the sustainability of manufacturing processes.

3.2 Water Conservation and Waste Reduction

As today's grave concern is regarding the prevailing conditions of ground water depletion, water scarcity is a global problem and water conservation is an essential element of sustainable production. In addition to conserving a valuable resource, the adoption of technologies and practices that reduce water use by means of closed-loop systems for drinking water or effective filtration processes also reduces environmental impacts from manufacturing activities. At the same time, sustainability and responsible production are also supported by waste reduction strategies, such as low manufacturing standards or recycling initiatives [6]. It is a paradigm shift in manufacturing that the principles of circular economy are adopted. The circular economy promotes the continuous use and reuse of materials instead of a traditional linear model for "taking, making, and discarding." A closed-loop system that reduces waste and promotes sustainability is created by the design of products for a longer life, ease of repair, or recycling. In addition to environmental protection, companies that embrace the principles of the circular economy are also promoting the resilience of resources.

4 Life Cycle Assessment (LCA)

Sustainable production has become an urgent necessity in the dynamic business environment of today's industry, due to a pressing need for environmental conservation. Innovative technologies and strategies that transform production processes in order to increase their sustainability are examined in this article. The focus is on developing a roadmap for a greener future, from the adoption of clean energy to the principles of circular economy [7]. LCA equips decision-makers with comprehensive information to make informed choices that align with sustainability goals. Also it allows companies to identify areas for improvement and optimize their processes in order to reduce the impact on the environment through identifying bottlenecks and inefficiencies. Products and processes that undergo LCA and demonstrate environmentally friendly attributes can be labeled accordingly, appealing to environmentally conscious consumers and enhancing market competitiveness. In order to ensure that environmental considerations are taken into account at all levels of decision-making, the results of the LCA may inform the development of policies and regulations. Also the challenges of LCA is that it may be difficult to obtain accurate and relevant data for each phase of the life cycle. Accuracy of the input data is a major determining factor for LCA results. In order to isolate the direct impact of a specific product or process, especially in complex systems with several variables, it is difficult to link environmental factors.

5 Environmental Regulations and Policy Measures

The implementation of robust rules and standards has been a key to promoting sustainability across industries during an era marked by growing concerns about the environment. An overview of international standards, such as ISO 14001 and ISO 50001 is given in this article on the world of environmental legislation and standards. In promoting sustainable practices, we will look at the crucial role of government legislation.

5.1 International Environmental Standards

Environmental Management Systems (EMS): The criteria for the establishment of an effective environmental management system are laid down in **ISO 14001**, a globally recognized standard. The framework of this standard is to identify, manage, and continue improving the performance of organizations with regard to their environment. Aspects, such as resource use, waste management, and the impact of activities on ecosystems, are covered. **Energy Management Systems (EnMS)**: Energy management is the focus of **ISO 50001**, which provides a systematic approach to improving energy efficiency, reducing costs, and minimizing carbon emissions for organizations. This standard is intended to help businesses develop energy policies, set objectives, and implement action plans with a view to improving energy efficiency. [8]

6 Challenges and Opportunities

In order to reduce their environmental footprint and contribute to a more environmentally conscious global economy, sustainable design and production have become the basis of businesses. In order to address economic concerns, to find a balance between environmental considerations and technological progress, and to exploit innovation, this article looks at the challenges and opportunities inherent in the pursuit of sustainable practices in design and manufacturing processes.

6.1 Addressing the Economic Concerns

ROI Concerns: The perception of the economic impact is one of the main challenges in sustainable design and manufacturing. Due to concerns about the initial costs, a large number of businesses are reluctant to take on sustainable practices. Investments in environmentally friendly technologies, materials, and processes can be significant leading to doubts as to the return on investment ROI. However, given that the long-term benefits often outweigh the initial costs through cost savings, therefore a shift in perspective is necessary going forward.

6.2 *Balancing Environmental Concerns*

Sustainable Materials and Processes: Integrating sustainable materials and processes into the design and production process is an important step toward achieving a proper balance of environment concerns with advances in technology. It may involve, for example, recycling or biodegradable materials and optimizing the use of resources in order to minimize waste generation. The adoption of the life cycle approach to product development ensures that environmental considerations are taken into consideration at all stages. In order to increase efficiency while maintaining sustainability commitments, it is possible to integrate Industry 4.0 principles, such as the Internet of Things and artificial intelligence. In order to advance sustainable design and manufacturing, cooperation across industries is needed as well as sharing good practices through collaboration and sharing of the technologies with active mediation by the governments.

7 Future Trends in Sustainable Design and Manufacturing

The field of sustainable design and manufacturing is ever changing and evolving to include emerging technologies and innovation, as the world faces its challenges in terms of environment, climate change and also the implications of new technologies, possible advances in materials science, and the key role that artificial intelligence plays to optimize sustainability practices. [7]

7.1 *Advancements in Materials Science*

Bio-based and Recycled Materials: There would be an extensive use of bio-based and recycled materials in the future for sustainable design and production. The development of bioplastics, textiles produced from recycling fibers, and building materials derived from waste products is facilitated by advances in the field of material science. These developments have reduced reliance on scarce resources, minimized environmental impacts, and promoted a sustainable economy [8]. Nanotechnology has the potential to improve the properties of materials used in the production process. Researchers can produce materials with superior strength, conductivity, and durability when they manipulate nanomaterials at the nanoscale. By reducing the need to replace or repair frequently, self-healing materials represent a breakthrough in sustainability. In addition to prolonging product life and reducing overall resource consumption, these materials could repair damage on their own. In infrastructure and transport, applications range from self-healing paints on surfaces to durable materials [9].

7.2 Artificial Intelligence

Energy Management: Artificial intelligence AI is playing a critical role in optimizing energy management within the production process. Real-time energy consumption data can be analyzed by AI algorithms and inefficiencies identified, and recommendations for improving overall energy efficiency can be made. This proactive approach minimizes waste and reduces the environmental impact associated with energy-intensive manufacturing operations [10]. Supply chain: By reducing waste, improving logistics, and minimizing environmental impact, optimization of supply chains based on AI will contribute to sustainability. In order to optimize the level of inventory, simplify transport routes and increase overall efficiency in the supply chain, machine learning algorithms are analyzing large amounts of data. Predictive Maintenance: The optimal performance and longevity of production equipment is ensured through the use of predictive maintenance, which is enabled by artificial intelligence. Artificial intelligence algorithms can predict when equipment is likely to malfunction and schedule maintenance prior to the occurrence of problems, by analyzing sensor data as well as past performance [11]. These trends promise to transform industries for a more sustainable and environmentally friendly future, from 3D printing and Internet of Things to smart manufacturing to bio-based materials and the transformative impact of artificial intelligence.

Rate of Technology Adoption: As per Fig. 3 below which clearly shows that the concept of smart manufacturing is best option to achieve the much required growth along with sustainability through different test technologies which show significant increase in rate of adoption of latest mechanisms, such as geo-special information system, remote sensing, and drone technology, there is great amount of data analysis being used for decision-making processes in smart manufacturing these days with a vision to achieve Smart Factory 4.0.

8 Sustainable Design and Manufacturing: Case Studies

The commitment of **Apple Inc.** to sustainability in design and production has been a point of distinction within the technology sector. By switching to greener sources of energy for its facilities and integrating recycled materials into their product design, the company has achieved significant progress in terms of environmental reduction. For example, Apple's Liam robot disassembles iPhones to recover and recycle valuable

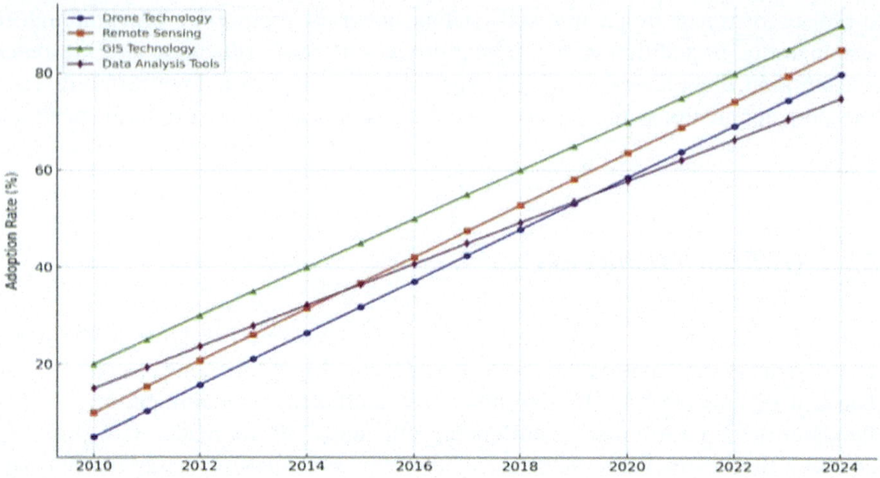

Fig. 3 Represents rate of adoption of various technologies over time in smart manufacturing [11]

materials that promote the circular economy. Implementing closed-loop material management, as demonstrated by Apple's recycling initiatives, is a replicable strategy for other businesses. Companies can reduce waste and encourage the circular use of materials if they design products that are recyclable, as well as establish effective recycling processes. It is a powerful strategy to commit to the transition to renewable energy sources in manufacturing facilities. Also it has committed to bring all of its operations including the supply chain and PLC to net zero by 2030.

With its focus on sustainable design and production, Tesla Inc., a pioneer in electric vehicles, has disrupted the automotive industry. Tesla is integrating sustainability into the whole supply chain, not just production of EVs. It prioritizes energy-efficient production processes, uses recycled materials in its vehicles, and invests in renewable energy sources. With its focus on sustainable design and production, Tesla Inc., a pioneer in electric vehicles, has disrupted the automotive industry. Tesla is integrating sustainability into the whole supply chain, not just production of EVs. It prioritizes energy-efficient production processes, uses recycled materials in its vehicles, and invests in renewable energy sources [12].

9 Conclusion

It is clear that businesses and industries have the power to shape a greener future as we deal with challenges and opportunities related to sustainability in design and manufacturing processes. We derive that the major change makers are the changed thinking moving away from the conventional thinking and processes emphasizing the continued importance of sustainable practices and encouraging further innovative research and implementation of solutions to further us toward a more sustainable, environmentally friendly future [13]. It is not only an environmental necessity, but also a strategic imperative for future growth and resilience for all the nations to come together, to move toward sustainable design and production mechanisms. It is critical time that we act with one goal of integrating renewable energy sources, energy-efficient technologies, and circular economy principles to mitigate and or to stop environmental impacts while fostering economic prosperity for all. Companies that lead the way in sustainability are not just lowering their environmental footprint but also gaining competitive advantages, as it was clearly visible in the case studies. Further innovation for sustainability, with a focus on the need to continue research, collaboration, and investment in green technologies, is expected from emerging trends, such as adopting advanced materials and artificial intelligence for much larger goals. Some of the recommended approaches in this article can contribute toward a proper step of building a sustainable and smart manufacturing, that can help the ecosystem flow in the right direction; also an parallel next version of this paper will also dive deep into each of the approaches and look into the pros and cons of each for every possible situation. Therefore, the path ahead toward more sustainable, efficient, and productive production environments that should highlight the crucial role played by the industry in building smart manufacturing with green energy to shape an environmentally sound future for all.

References

1. Pujari, D., Wright, G.H., Peattie, K.: Green and competitive: influences on environmental new product development performance Research Gate Feb 2003, https://www.researchgate.net/publication/222575837_Green_and_competitive_Influences_on_environmental_new_product_development_performance. Accessed 26 Feb 2024
2. IRENA: Sustainable bioenergy pathways in Latin America: promoting bioenergy investment and sustainability, International Renewable Energy Agency, Abu Dhabi. Author, F.: Contribution title. In: 9th International Proceedings on Proceedings, pp. 1–2. Publisher, Location (2010) (2024)
3. “Sustainable product design practices for a circular economy?|Punctuate Design,” www.punctuatedesign.com. https://www.punctuatedesign.com/insights/sustainableproduct-design-practices. Accessed 13 Mar 2024
4. Lobus, N.V., Knyazeva, M.A., Popova, A.F., Kulikovskiy, M.S.: Carbon footprint reduction and climate change mitigation: a review of the approaches, technologies, and implementation challenges. J. Carbon Res. **9**(4), 120 (2023). https://doi.org/10.3390/c9040120
5. Erkan, Y.: A metamodel for heritage-based urban development: enabling sustainable growth through urban cultural heritage, by Matthias Ripp. Springer Cham, 2022. 209pp. ISBN9783031082375,” Built Heritage, vol. 7, no. 1, Dec. 2023. https://doi.org/10.1186/s43238-023-00107-y
6. Zhang, J., Gao, R.X.: Deep learning-driven data curation and model interpretation for smart manufacturing. Chin. J. Mech. Eng. **34**(1), 71 (2021). https://doi.org/10.1186/s10033-021-00587-y
7. Xu, W., Miao, H., Zhao, Z., Liu, J., Sun, C., Yan, R.: Multi-scale convolutional gated recurrent unit networks for tool wear prediction in smart manufacturing. Chin. J. Mech. Eng. **34**(1), 53 (2021). https://doi.org/10.1186/s10033-021-00565-4

8. Narancic, T., Cerrone, F., Beagan, N., O'Connor, K.E.: Recent advances in bioplastics: application and biodegradation. Polymers **12**(4), 920 (2020)
9. Cherukuri, M., Cherukuri, P.A.A.: Sustainable evolution of business during covid-19 pandemic—innovation a new normal, Annals of Computer Science and Information Systems (2020). https://doi.org/10.15439/2020km21
10. Li, X., Wang, B., Peng, T., Xu, X.: Greentelligence: smart manufacturing for a greener future. Chin. J. Mech. Eng. **34**, 1–6 (2021)
11. Cherukuri, A., Dieu, T., Vududala, S.K., Cherukuri, R.: Keras convolutional neural network model in sequential build of 4 and 6 layer architectures, Lecture notes in networks and systems, pp. 465–472 (2021). https://doi.org/10.1007/978-981-16-20942_56
12. Gholami, H., Abu, F., Lee, J.K.Y., Karganroudi, S.S., Sharif, S.: Sustainable manufacturing 4.0—pathways and practices. Sustainability **13**(24), 13956 (2021). https://doi.org/10.3390/su132413956
13. Sun, J., Sarfraz, M., Turi, J.A., Ivascu, L.: Organizational agility and sustainable manufacturing practices in the context of emerging economy: a mediated moderation model. Processes **10**(12), 2567 (2022). https://doi.org/10.3390/pr10122567

Simulation of Urea Injection in a Closed-Loop Control Using Proteus Software

Huu Tuyen Pham, Huy Chien Nguyen, Xuan Thanh Dinh, Manh Tuan Nguyen, and Thi Huong Tran

Abstract Urea solution plays a crucial role in reducing nitrogen oxide (NO_x) emissions within diesel engine exhausts through selective catalytic reduction (SCR) systems. Control over the injection of urea solution is essential to ensure a complete reaction with the NO_x concentration before the SCR catalyst and that the NO_x concentration does not exceed the allowable threshold. In this study, a closed-loop proportional integral derivative (PID) control method was used using Proteus software. Input data entered into Proteus software includes NO_x concentration after and before the SCR catalyst, exhaust temperature, and exhaust flowrate taken from sensors. Output parameters, including urea injection flowrate and NO_x conversion efficiency, are calculated and displayed on the LED screen. The study has determined the urea injection map when changing NO_x concentration before the SCR catalyst, exhaust flowrate, and exhaust temperature.

Keywords Urea injection control · Closed-loop control

H. T. Pham · H. C. Nguyen (✉)
School of Mechanical Engineering, Hanoi University of Science and Technology, No.1 Dai Co Viet, Hai Ba Trung, Hanoi, Vietnam
e-mail: chiennh@haui.edu.vn

H. C. Nguyen · X. T. Dinh
School of Mechanical Automotive Engineering, Hanoi University of Industry, No.298 Cau Dien Street, Bac Tu Liem District, Hanoi, Vietnam

M. T. Nguyen
FPT Polytechnic, No. 13 Trinh Van Bo Street, Nam Tu Liem District, Hanoi, Vietnam

T. H. Tran
University of Economics—Technology for Industries, No. 456 Minh Khai Street, Hai Ba Trung District, Hanoi, Vietnam

D. H. Tien et al. (eds.), *Proceedings of the 1st International Conference on Sustainability and Emerging Technologies for Smart Manufacturing*, Proceedings in Technology Transfer, https://doi.org/10.1007/978-981-97-7083-0_39

1 Introduction

SCR technology has been widely employed for mitigating nitrogen oxide (NO_x) emissions in diesel engines compliant with EURO 4 standards and beyond. For diesel engines adhering to EURO 3 standards and below, lacking an SCR system, retrofitting SCR stands out as a highly effective solution for NO_x reduction. However, such retrofitting necessitates a bespoke design tailored to each specific engine. Within the SCR setup, urea solution undergoes injection into the hot exhaust, where it chemically decomposes into NH_3, acting as a reducing agent. This ammonia then reacts with NO_x to yield N_2 and H_2O on the catalytic surface, significantly curtailing NO_x levels in the atmosphere. The precise control of urea injection within the SCR system is pivotal, as it directly impacts NO_x reduction efficiency and compliance with emission standards. Various crucial factors must be taken into account for urea injection control, including NO_x concentrations at both the inlet and outlet of the SCR, exhaust flowrate, exhaust temperature, and NH_3 residue. Consequently, the management of urea injection emerges as a vital technology in the advancement of SCR systems.

Researchers worldwide have investigated various urea injection control strategies. Chi JN [1] experimentally studied the effect of opened-loop and closed-loop control on the NO_x conversion efficiency of the SCR systems. The findings revealed that under opened-loop control, NO_x conversion efficiency was contingent upon various exhaust parameters, encompassing NO_x concentration, exhaust flowrate, the amount of urea injection, and NH_3 storage capacity. These disparities led to a deviation of approximately 13.5% in NO_x reduction efficiency from the target value.

Geveci et al. [2] introduced a methodology rooted in noise theory to analyze the sensitivity of NO_x sensors to NH_3. The approach examined the variability characteristics of values obtained from NO_x sensors under varying concentration conditions of both NO_x and NH_3. The proposed methodology has the potential to calculate NO_x and NH_3 concentrations using measurement outcomes.

Shost et al. [3] studied a closed-loop control using information from NH_3 sensors. The author conducted a comparison between closed-loop and open-loop control strategies regarding NO_x conversion efficiency within the U.S. federal test cycle (FTP). The findings revealed a significant surge in NH_3 residuals with the utilization of an opened-loop control strategy. Specifically, the average NH_3 residual reached 19 ppm, while the highest NO_x reduction efficiency attained was 74%. Conversely, upon implementing the closed-loop control, the average residual NH_3 plummeted to 0.2 ppm, accompanied by a remarkable increase in NO_x conversion efficiency, reaching up to 91%.

Willems et al. [4] simulated the effects of closed-loop control on NO_x reduction efficiency and NH_3 residuals using both a European stationary cycle (ESC) and a test cycle comprising 1800 European transient cycles (ETC). Their findings highlighted that closed-loop control is effective primarily under slow-changing operating conditions. Willems et al. [4] came up with block diagram of an opened-loop SCR control shown in Fig. 1.

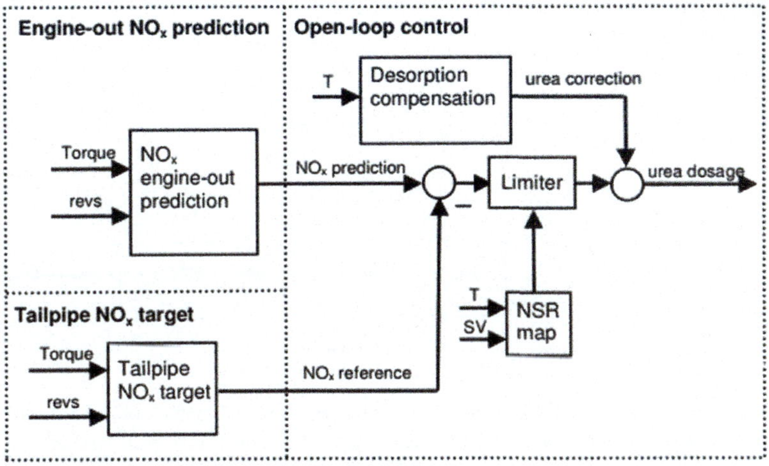

Fig. 1 Block diagram of an opened-loop SCR control [4]

The study results indicate that opened-loop urea supply control can effectively meet EURO 4 standards, achieving a NO_x reduction efficiency of approximately 50–60%. Traditional opened-loop control operates by predicting the engine's NO_x output concentration and leveraging temperature data and exhaust flowrate to reduce NO_x emissions. NO_x conversion efficiency is computed as the disparity between NO_x concentrations in engine exhaust and post-selective catalyst exhaust. The quantity of urea supply necessary for NO_x reduction is constrained by exhaust temperature, catalyst capabilities, and exhaust flowrate to maintain residual NH_3 concentration. However, this opened-loop control system encounters challenges in maximizing NO_x reduction efficiency while minimizing NH_3 residues, particularly during transition modes. One approach to addressing this is by situating the NO_x concentration sensor upstream of the catalyst.

To meet more stringent emission standards, which should be from EURO 4 to EURO 5, NO_x conversion efficiency must also increase, which also increases the NH_3 slip. Although the concentration of excreted NH_3 is not regulated, it is common to apply an average concentration of NH_3 slip of 10 ppm, when the highest allowable is 25 ppm [5] (Fig. 2).

Therefore, the closed-loop control system relies on the control of residual NH_3 and NO_x reduction efficiency through the feedback of the NO_x sensor. Studies developed an SCR control system with a focus on realizing desired NO_x emissions for test cycles with nominal efficiency and, in other words, provide the expected amount of urea for the desire to reduce NO_x.

The closed-loop control structure utilizing feedback from the NO_x sensor is depicted in Fig. 3. The control principle revolves around the nominal urea injection quantity and monitoring signals from the NO_x sensor. Urea injection is alternated between excessive and moderate levels. The impact of urea injection volume variation must be monitored through the NO_x sensor signal. If an increase in urea

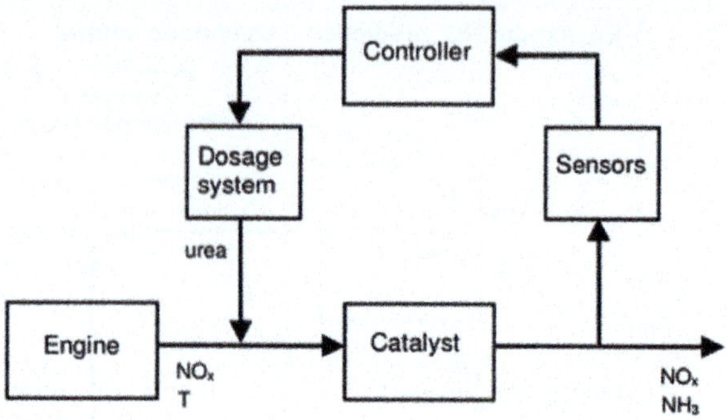

Fig. 2 Automobile urea SCR system structure [5]

injection quantity is reflected in the sensor signal as a decrease in NO_x concentration, indicating effective reduction, further increments in injection quantity are initiated. Conversely, if an increase in urea supply leads to a rise in NO_x concentration, signifying an excess of NH_3, a reduction in urea supply is triggered. The control governed by the NO_x concentration sensor is grounded on a data map featuring nominal urea supply quantities. This sensor ensures corrections during unstable operational modes to align with stable working conditions, facilitated by data mapping derived from opened-loop control.

Upadhyay [6] introduced a method for estimating stored NH_3 utilizing NO_x sensors. They employed a transition control strategy to achieve enhanced NO_x conversion efficiency, especially in situations where significant amounts of NH_3 were initially stored. Furthermore, the study investigated the influence of sensor

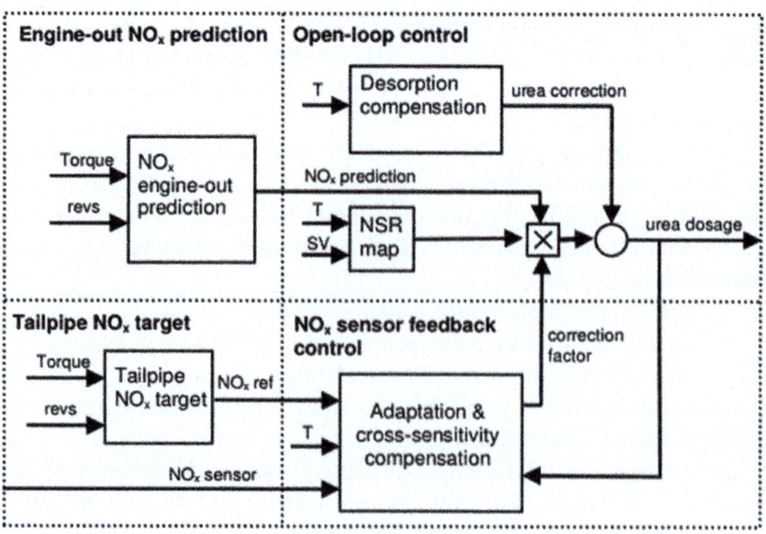

Fig. 3 Block diagram of NO_x sensor-based control strategy [4]

uncertainty on system performance and elucidated the use of NO_x sensors alongside input correlation to distinguish between NO_x residues and NH_3 residues.

Zhao et al. [7] examined the NH_3 storage characteristics within a vanadium-catalyzed SCR system. They discovered that the efficiency of NO_x conversion was highly dependent on the NH_3 storage capacity, particularly when the temperature of the SCR catalyst was below 280 °C. During this transitional process, precise control over NH_3 storage was essential to prevent the accumulation of NH_3 residues.

Mark Shost et al. [8] introduced a method for quantitatively monitoring, providing feedback, and controlling the urea dosage in SCR system. They utilized a model embedded device and a post-SCR NH_3 sensor in a closed-loop setup. A one-dimensional SCR model was developed and integrated into the Simulink/MATLAB software. The integrated model enabled real-time control to precisely measure a urea solution concentration of 32.5% in the exhaust stream. By conducting tests and simulations on engines equipped with embedded SCR models and closed-loop feedback NH_3 sensors, the authors showcased the efficacy of this control method in achieving both NO_x emission standards and reducing NH_3 residuals.

SCR technology has emerged as a promising and extensively employed method in emissions control. It utilizes ammonia (NH_3), typically sourced from urea solution, as a reducing agent [9]. One of the challenges associated with this technology involves accurately dosing the appropriate amount of urea solution to achieve optimal NO_x reduction levels while ensuring that NH_3 slip remains within legal limits. This requires effective control of the injection amount. Closed-loop control of SCR catalysis has been extensively studied in the literature [10–12] and feed-based controllers [13, 14]. In this paper, a closed-loop PID control method using Proteus software is studied. This software is widely used in the field of IC design and simulation. Proteus provides a comprehensive simulation environment, allowing users to build complex electronic circuits from basic electronic components, such as transistors, resistors, capacitors, ICs, and many others. Users can draw and connect these components as they wish, and then test the operation of the circuit before actual deployment. Input data include NO_x concentration after and before the SCR system, exhaust temperature, and exhaust flowrate. Output parameters, including urea injection flowrate and NO_x conversion efficiency, are calculated and displayed on the LED screen. The study has determined the suitable amount of urea injected when NO_x concentration, exhaust flowrate, and exhaust temperature vary.

2 Simulate Urea Injection Control in a Closed Loop Using Proteus Software

2.1 Diagram of the Urea Injection Control Algorithm

With advantages mentioned above, a closed-loop control was selected to study in this paper. The block diagram is shown in Fig. 4.

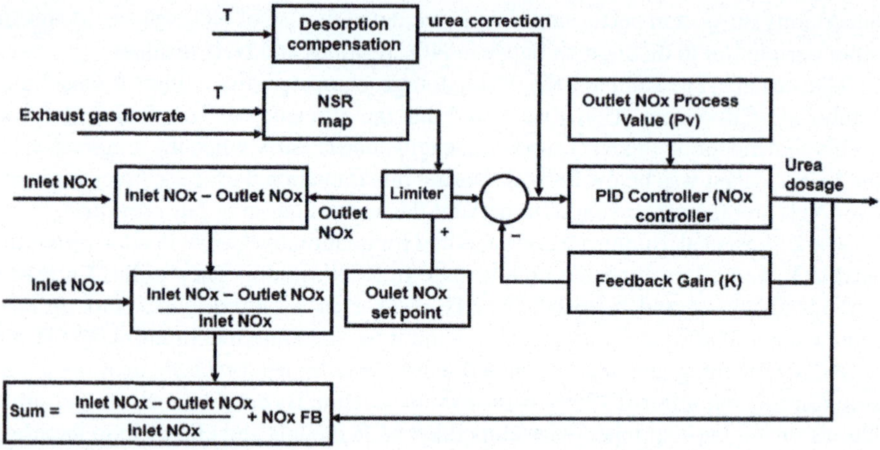

Fig. 4 Block diagram of reducing solution injection system in SCR system

In the proposed system, we enhance the entire process by developing a mathematical model of the system. The NO_x controller functions as a feedback controller, establishing a reference for urea injection to maintain the NO_x emission rate at the desired NO_x set point. This closed-loop controller effectively compensates for calibration errors resulting from NO_x sensor drift. Operating as a Single Input, Single Output (SISO) system, this controller features one input and one output. The input, input NO_x, drives the system, while the output, NO_x feedback, is combined with the percentage reduction output. The controller operates with two states, corresponding to the concentration of NO_x and the quantity of urea injection. The amount of urea injected in basic mode to reduce NO_x is calculated based on Huu Tuyen Pham et al. [15], ensuring the concentration of NH_3 slip is below10 ppm. The control diagram of the reducing solution injection system in the SCR system is fed into the DCU shown in Fig. 5.

2.2 *Simulation of a Urea Injection Control Circuit Using Proteus Software*

The urea injection control simulation circuit is shown in Fig. 5 and Table 1, with input parameters including a signal from the exhaust flowrate sensor, a signal from the temperature sensor, and a signal from the NO_x concentration sensor before and after the SCR system. A control algorithm is loaded into the ATmega328 microcontroller to control the reasonable amount of urea injection. Output parameters include NO_x reduction efficiency and NO_x concentration after the SCR. The urea injection volume is displayed on the LCD screen. The simulation description is shown in Table 2.

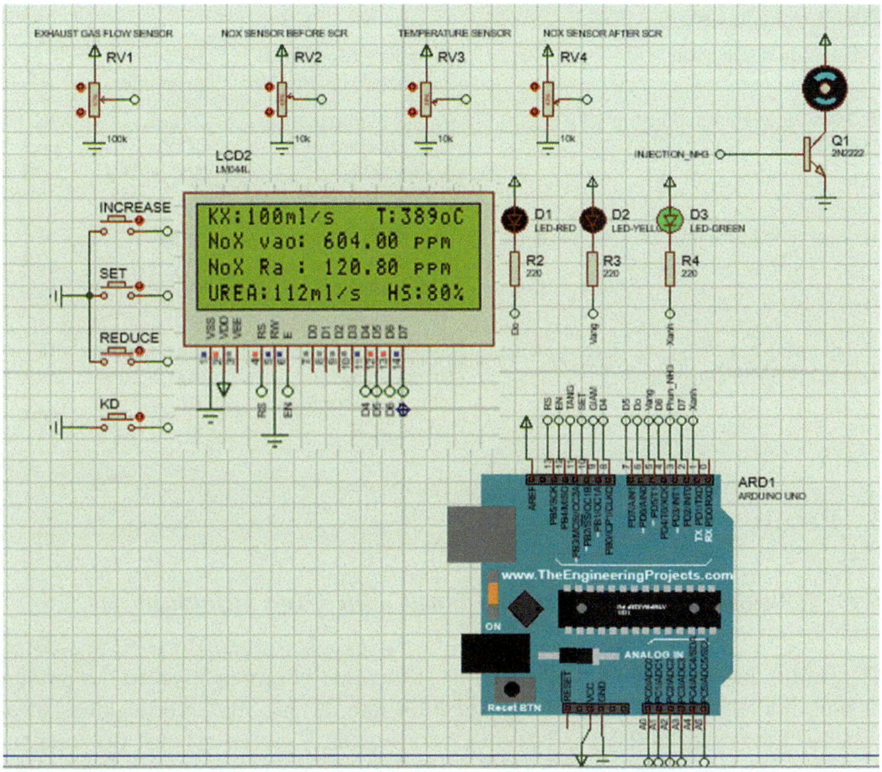

Fig. 5 Simulation of a urea injection control circuit using Proteus software

Table 1 Elements of simulation model

Names of elements	Symbol
Exhaust flowrate sensor	RV1
NO_x sensor before SCR	RV2
Temperature sensor	RV3
NO_x sensor after SCR	RV4
Display screen	LCD
Injection urea	Q1
Microcontroller ATmega328	ARD1

Table 2 Simulation description

Simulation description	Boundary conditions/ Parameters investigated	Exhaust properties	Unit
Exhaust flowrate relationship with urea injection	Exhaust temperature NO_x concentration before SCR Exhaust flowrate	> 200 1000 100 ÷ 500	°C ppm ml/s
Relationship between NO_x concentration before SCR and urea injection	Exhaust temperature Exhaust flowrate NO_x concentration before SCR	> 200 100 100 ÷ 1000	°C ml/s ppm
Relationship between exhaust temperature and NO_x conversion efficiency	Exhaust temperature Exhaust flowrate NO_x concentration before SCR	200 ÷ 600 100 ÷ 1000 100 ÷ 1500	°C ml/s ppm

3 Results and Discussion

3.1 Exhaust Flowrate Relationship with Urea Injection

The results in Fig. 6 show that the exhaust flowrate is proportional to the amount of urea injection. The highest urea injection rate, at 336 ml/s, corresponds to an exhaust flowrate of 500 ml/s. The direct correlation between exhaust flowrate and urea injection may stem from the need to treat NO_x, as an increase in exhaust flowrate also increases the NO_x levels that need to be treated. Therefore, to maintain stable NO_x treatment performance, it is necessary to adjust the urea injection into the system in proportion to the exhaust flowrate. However, this adjustment also depends on many factors, for example the design of the SCR system, the temperature and pressure of the exhaust, and other operational conditions. Properly adjusting the ratio between urea injection and exhaust flowrate is crucial to ensuring the best NO_x treatment performance.

3.2 Relationship Between NO_x Concentration Before SCR and Urea Injection

The results depicted in Fig. 7 shown that the concentration of NO_x before the SCR catalyst is directly proportional to the amount of urea injection. The highest urea injection rate, at 120 ml/s, corresponds to a NO_x concentration before the SCR catalyst of 1000 ppm. The quantity of urea to be injected depends on the NO_x concentration upstream of the SCR catalyst. If the NO_x concentration increases, the urea quantity needs to be increased to ensure that all NO_x is efficiently processed and meets emission standards.

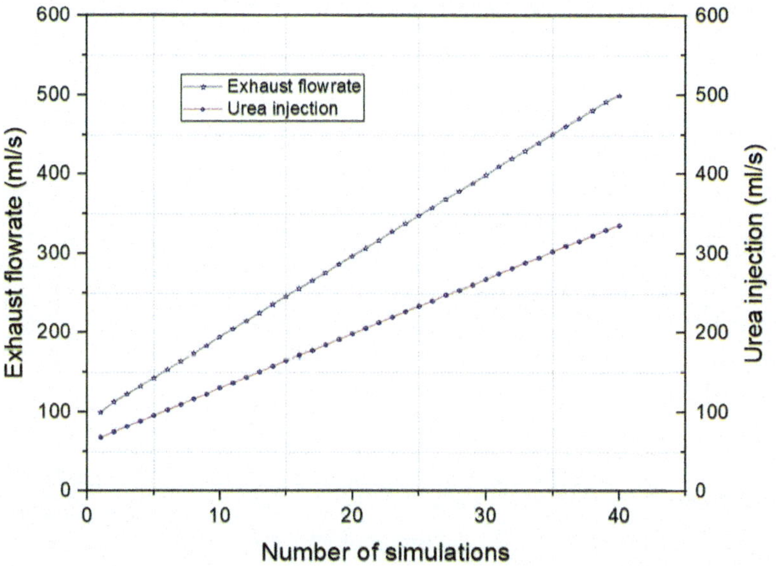

Fig. 6 Exhaust flowrate relationship with urea injection when exhaust temperature > 200 °C, NO_x inlet concentration is 1000 ppm, and exhaust flowrate changes from 100 to 500 ml/s

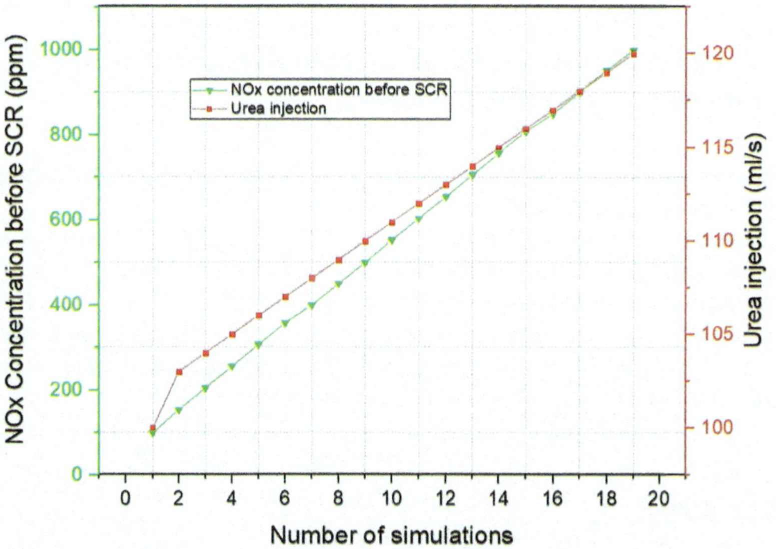

Fig. 7 Relationship between NO_x concentration before SCR catalyst and urea injection is as follows: exhaust temperature > 200 °C, exhaust flowrate is 100 ml/s, and NO_x concentration changes from 100 to 1000 ppm

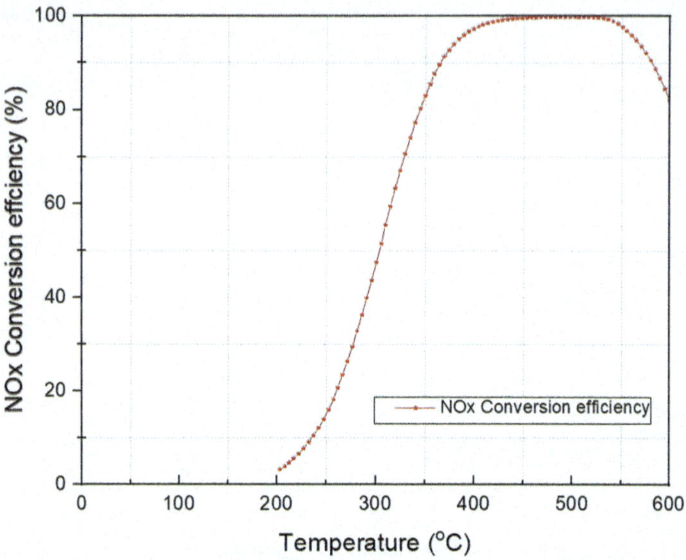

Fig. 8 Relationship between exhaust temperature (200÷ 600 °C) and NO_x conversion efficiency occurs when it changes the exhaust flowrate from 100 to 1000 ml/s and NO_x concentration before SCR from 100 to 1500 ppm

3.3 Evaluate the Relationship Between Exhaust Temperature and NO_x Conversion Efficiency

The exhaust entering the SCR catalyst from the engine will gradually increase the catalyst's temperature. Figure 8 shows the NO_x reduction efficiency of the catalyst versus catalyst temperature with description in Table 2. The conversion efficiency increases as the temperature of the SCR catalyst increases from 200 to 500 °C. The highest efficiency occurs at SCR catalyst temperature of 400–500 °C. When the SCR catalyst temperature is greater than 500 °C, the NO_x conversion efficiency starts to decrease, which can be explained by the high temperature which will cause undesirable oxidation reactions and increase the amount of NO_x.

4 Conclusion

The research has developed a closed-loop PID control system using Proteus software to regulate the optimal urea injection rate based on variations in NO_x concentration, exhaust flowrate, and exhaust temperature. Sustaining consistent NO_x treatment efficacy requires adjusting urea injection in proportion to exhaust flowrate. The amount of urea injected is determined by the pre-SCR catalyst NO_x concentration. An increase in NO_x concentration necessitates a corresponding increase in urea quantity to ensure effective NO_x processing and compliance with emission standards. Concurrently, it is essential to limit NH_3 slip concentration to below 10 ppm. Catalyst conversion efficiency rises within a temperature from 200 to 500 °C. However, conversion efficiency diminishes beyond 500 °C as the priority shifts away from NO_x reduction at higher temperatures.

References

1. JN, C.: Control challenges for optimal NO_x conversion efficiency from SCR aftertreatment systems. SAE Technical Paper 2009-01-0905 (2009)
2. Jones, J.C.P., Geveci, M.: Smart sensing and decomposition of NO_x and NH_3 components from production NO_x sensor signals. SAE Technical Paper 2011-01-1157 (2011)
3. Shost, M., Noetzel, J., Wu, M., Sugiarto, T., Bordewyk, T., Fulks, G., Fisher, G.: Monitoring, feedback and control of urea scr dosing systems for nox reduction, Utilizing an embedded model and ammonia sensing. SAE (2008)
4. Frank Willems, R.C., van den Eijnden, E., van Genderen, M., Verbeek, R.: Is closed-loop SCR control required to meet future emission targets? SAE Technical Paper 2007-01-1574 (2007)
5. Schär, O.: Elsener and Geering, Model-based control of an SCR system for a mobile application. SAE Technical Paper 2004-05-0412 (2004)
6. Van Nieuwstadt, M., Upadhyay, D.: Control of urea SCR systems for US diesel applications. Oil Gas Sci. Technol.-Rev. d'IFP Energies nouvelles **66**(4), 655–665 (2011)
7. Zhao, Y., Hu, J., Hua, L., Shuai, S., Wang, J.: Ammonia storage and slip in a urea selective catalytic reduction catalyst under steady and transient conditions. Ind. Eng. Chem. Res. **50**(21), 11863–11871 (2011)
8. Mark Shost, J.N., Wu, M.C., Sugiarto, T., Bordewyk, T., Fulks, G., Fisher, G.B.: Monitoring, feedback and control of urea SCR dosing systems for NO_x reduction, utilizing an embedded model and ammonia sensing. SAE Int. J. Fuels Lubr. 2008-01-1325 (2008)
9. Gabrielsson, P.L.T.: Urea-SCR in automotive applications. Top. Catal. **28**, 1–4 (2004)
10. Schar, C.M., Onder, C.H., Geering, H.P.: Control of an SCR catalytic converter system for a mobile heavy-duty application. IEEE Trans. Control Syst. Technol. **14**(4), 641–653 (2006)
11. Hu, Y.Z., Zhang, Y., Shuai, S., Wang, J.: Development of closed-loop control strategy for urea-SCR based on NO_x sensors. SAE Technical Paper 2011-01-1324 (2011)
12. Wand, D.Y., Yao, S., Shost, M., Yoo, J., Cabush, D., Racine, D.: Ammonia sensor for closed-loop SCR control. SAE Papers 2008-01-0919 (2008)
13. Patchett, J., V, R., Grimston, K., Rice, G., Calabrese, J., van Genderen, M.: Control system for mobile NO_x SCR applications. U.S. Patent 6 581 374, June (2003)
14. Seher, D.H.E., Wickert, S.: Control strategy for NO_x—emission reduction with SCR. SAE papers 2003-01-3362 (2003)
15. Pham, H.T., Nguyen, T.L., Le, A.T., Do, T.H.: Research on calculating the amount of urea supplied to the selective catalytic NO_x removal unit installed on a bus engine. Vietnam Mech. Eng. J. **10**, 44–49 (2013)

Sustainable Dry Machining of Aluminum Alloy A7075: Utilizing Coral Reefs Optimization and Heatmap Analysis for Impact Assessment and Optimization of Cutting Parameters

Nhu-Trang Le, Van-Canh Nguyen, Duong Van Duc, Hoang Tien Dat, and Van-Tam Ngo

Abstract This comprehensive study centers on optimizing the finishing milling of Aluminum Alloy A7075 material by employing the Coral Reefs Optimization (CRO) algorithm. The main aim is to minimize both surface roughness value (R_a) and cutting force value (F_c). To accomplish this, the research cutting parameters include cutting speed (V_c), feed rate (F), and depth of cut (a_p) within stringent machine and cutting tool constraints. The CRO algorithm, renowned for its effectiveness in complex problem-solving, has been meticulously applied to navigate the trade-offs between these objectives. A set of Pareto-optimal solutions was obtained, with a standout combination featuring a V_c of 335.7291 m/min, an F of 1200 mm/min, and an a_p of 1.3455 mm. This resulted in an F_c of 162.93 N and a R_a of 0.313 μm, signifying an optimal balance between efficiency and quality. These results underscore the potential of bio-inspired algorithms like CRO in optimizing industrial machining processes, offering valuable insights for the advancement of precision manufacturing within the constraints of practical applications. The findings pave the way for improved operational performance in the milling of high-strength aluminum alloys.

Keywords Cutting force · Dry milling · Flood milling · Aluminum alloy · Milling parameters

---

N.-T. Le
University of Economics—Technology for Industry, 456 Minh Khai Str., Hanoi, Vietnam
e-mail: trangln@uneti.edu.vn

V.-C. Nguyen (✉) · D. Van Duc · H. T. Dat
Hanoi University of Industry, 298 Cau Dien Str., Hanoi, Vietnam
e-mail: nguyenvancanh@haui.edu.vn

V.-T. Ngo
Department of Mechanical Engineering, National Cheng Kung University, East District, Tainan 70101, Taiwan
e-mail: n18057047@gs.ncku.edu.tw

D. H. Tien et al. (eds.), *Proceedings of the 1st International Conference on Sustainability and Emerging Technologies for Smart Manufacturing*, Proceedings in Technology Transfer, https://doi.org/10.1007/978-981-97-7083-0_40

1 Introduction

In recent years, a significant increase in research activity focused on bio-inspired metaheuristics has been witnessed. This incorporates innovative algorithms, such as the CRO algorithm, as proposed by Salcedo-Sanz et al. [1], which simulates the dynamic and self-organized process of coral reef formation and reproduction. An interesting study highlighted the adoption of CRO for enhancing service distribution in mobile radio access networks [2], illustrating its applicability in complex, real-world problems [3].

The CRO algorithm is a relatively new bio-inspired optimization algorithm. It draws inspiration from the diverse and self-organized ecosystem of coral reefs. One of the notable advantages of CRO is its ability to effectively combine exploration and exploitation capabilities, making it suitable for a variety of multi-objective optimization problems in engineering [1, 2]. CRO has been increasingly applied in the field of machining optimization. To highlight its applicability, several studies can be referenced where CRO has been effectively utilized.

In this study, CRO is applied to the milling of A7075 material. The machining process involves the optimization of multiple objectives, namely the cutting force (F_c) and surface roughness (R_a). The implementation of CRO in this context aims to find Pareto-optimal solutions for the problem. This approach facilitates the identification of optimal machining parameters that balance between the minimization of cutting force and the improvement of surface finish.

Through this application, contribution is made to the expanding field of machining optimization, demonstrating the effectiveness of CRO in handling complex, multi-objective problems in a real-world manufacturing context.

2 Material and Methods

2.1 Experimental Design

The study applies the experimental design method to develop and evaluate the experimental design model. Through theoretical research on the cutting process and factors affecting the cutting process, while relying on the recommended cutting capability of the cutting tool from the manufacturer and the machine's power to ensure no vibration effects during machining. Alternatively, considering the restricted speed range determined by research for machining A7075 aluminum alloy with titanium nitride-coated end mill cutting tools, cutter diameter $D = 6$ (mm) with number of teeth $Z = 2$. Three adjustable cutting parameters (a_p, F, V_c) in the machining process were selected as input variables in the experimental studies. The selected level values for milling A7075 aluminum alloy with titanium nitride-coated end mill are as shown in Table 1.

Table 1 Parameters and their respective values for cutting

No.	Parameter	Unit	Level		
			− 1	0	1
1	Cutting speed (V_c)	m/min	188	282	376
2	Feed rate (F)	mm/min	800	1200	1600
3	Depth of cut (a_p)	mm	0.5	1.0	1.5

2.2 Workpiece Specimen

The experimental workpiece of Aluminum Alloy A7075 has dimensions of 90 × 60 × 20 mm (Fig. 1). The chemical composition of the workpiece material is presented in Table 2, and its physical properties are outlined in Table 3.

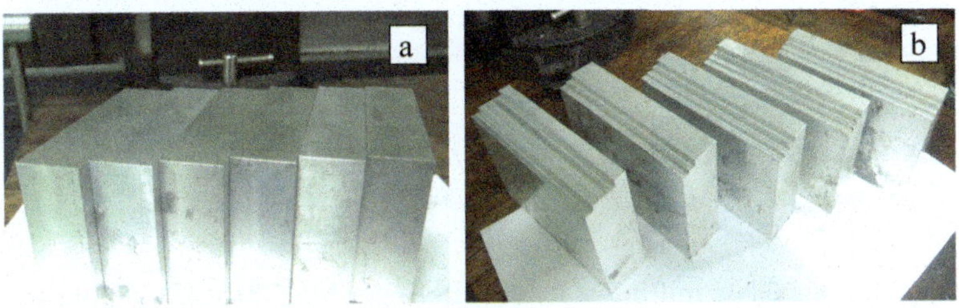

Fig. 1 Workpiece specimen

Table 2 Aluminum alloy A7075's chemical composition

Cu	Mn	Mg	Cr	Zn	Ti	Al
1.2–2.0	0.3	2.1–2.9	0.18–0.28	5.1–6.1	0.2	Balance

Table 3 Mechanical properties of A7075 material

Physical property	Value	Physical property	Value
Melting temperature (°C)	483	Elastic modulus (GPa)	70–80
Thermal conductivity (W/m K)	130	Electrical resistivity (Ω m)	3.99e−006
Specific heat (J/g-°C)	0.89	Flexural stress (MPa)	95
Density (g/cm^3)	2.81	Tensile stress (MPa)	150
Hardness (HB)	60	Poisson's ratio	0.33

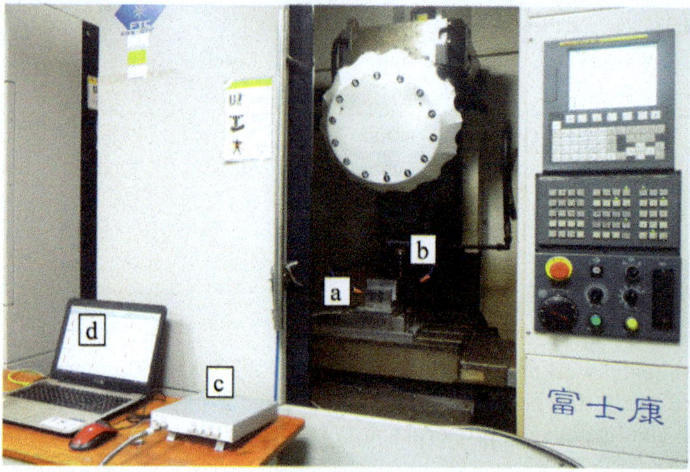

Fig. 2 Experimental setup. **a** Jig and fixture for cutting force measurement; **b** cutting tool; **c** Kistler cutting force dynamometer; **d** computer with DynoWare software

2.3 *Experimental Machine, Cutting Tool, and Measurement Equipment*

The study uses the CNC milling machine HS Super MC500 as shown in Fig. 2 with the spindle rotation speed of 100 –30,000 (rpm); main spindle power of 15 kW, table feed rate of 1–30,000 (mm/min), and rapid traverse rate: 48,000 (mm/min).

The cutting force was measured by the Kistler Type 9139AA dynamometer in 3 directions F_x, F_y, and F_z. The surface roughness value of each workpiece was measured by Surftest SJ-210 (Mitutoyo, Japan) profilometer in accordance with ISO 1997 standards. The experimental results are presented comprehensively in Table 4.

3 Results and Discussion

3.1 *Multivariate Correlation Analysis*

In this study, the heatmap is utilized to analyze the relationships between machining parameters (V_c, F, a_p) and outcomes (F_c, R_a), providing a visual and in-depth insight into how the parameters interact with each other and affect the milling process.

A heatmap is a type of data visualization chart that uses color coding to represent varying values in a matrix or table. Heatmaps are built on a correlation matrix calculated based on correlation coefficients, usually Pearson's correlation coefficients, between pairs of variables. Pearson's correlation coefficient is calculated as follows:

Table 4 Summary of experimental design matrix and results

No.	V_c	F	a_p	R_a	F_c
	m/min	mm/min	mm	μm	N
1	188	800	0.5	0.254	114.95
2	188	800	1.5	0.334	150.14
3	188	1600	0.5	0.361	174.46
4	188	1600	1.5	0.415	216.81
5	376	800	0.5	0.270	101.20
6	376	800	1.5	0.297	135.40
7	376	1600	0.5	0.321	160.33
8	376	1600	1.5	0.373	189.11
9	282	1200	1.0	0.308	162.14
10	282	1200	1.0	0.308	163.79
11	282	1200	1.0	0.313	162.93

$$r_{xy} = \frac{\sum(x_i - \overline{x})(y_i - \overline{y})}{\sqrt{\sum(x_i - \overline{x})^2}\sqrt{\sum(y_i - \overline{y})^2}} \quad (1)$$

where

r_{xy} is the correlation coefficient between x and y.
x_i and x_i are the value of the ith observations in x and y.
$\overline{x}$ and $\overline{y}$ are the mean values of the x and y.

The correlation coefficient ranges from − 1 to 1, where − 1 signifies a perfect negative relationship, 1 signifies a perfect positive relationship, and 0 indicates no linear relationship. In this study, heatmaps are computed and constructed using the Python programming language with the Seaborn library [4].

Correlation analysis serves as a crucial step in comprehending the relationships between process parameters and outcomes. It informs subsequent optimization endeavors utilizing the Coral Reefs Optimization (CRO) algorithm to pinpoint optimal machining conditions.

In the presented heatmap visualization of the correlation matrix (Fig. 3), significant interdependencies between the milling process parameters and outcomes were observed. The cutting force (F_c) and surface roughness (R_a) displayed a strong positive correlation ($r = 0.93$), indicating that these two responses tend to increase or decrease together. The feed rate (F) also positively influenced both F_c and R_a, with correlation coefficients of 0.83 and 0.76, respectively. This suggests that an increase in feed rate is likely to lead to higher cutting forces and rougher surfaces.

Conversely, the cutting speed (V_c) showed a weak negative correlation with both F_c and R_a, hinting that higher cutting speeds may contribute to lower cutting forces

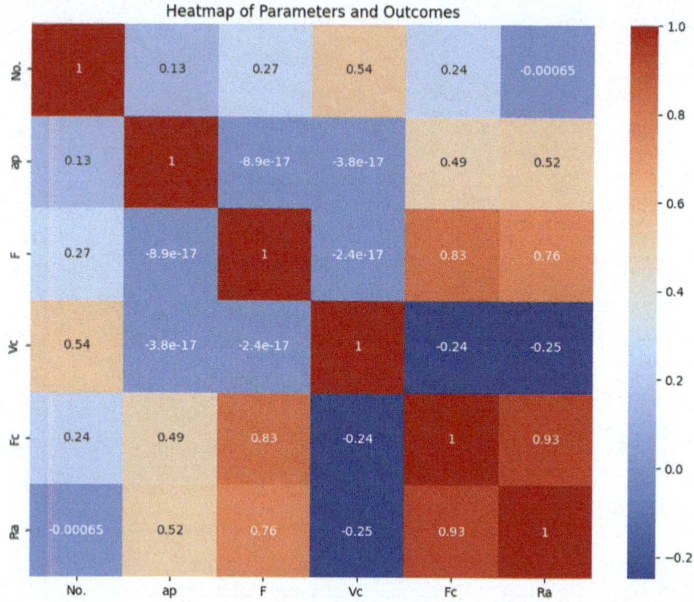

Fig. 3 Correlation Heatmap for milling process parameters and outcomes of A7075 Alloy

and improved surface finishes, although the relationship is not pronounced. Correlations near zero indicate no linear relationship between certain parameters, highlighting the complex nature of the milling process where not all factors interact linearly or significantly.

3.2 *Multiple Optimization of Cutting Parameters in Milling Process*

The research addresses a multiple objective optimization (MOO) problem, aiming to determine the optimal set of machining parameters that concurrently minimize both cutting force (F_c) and surface roughness (R_a) in the milling process of Aluminum Alloy A7075 material.

Multi-objective Optimization Problem Description

Optimization in machining is critical for enhancing productivity and product quality. The multi-objective nature of this problem requires a balancing act between conflicting objectives: the need to reduce the cutting force to prolong tool life and minimize energy consumption, and the need to achieve a superior surface finish for the final product. The optimization aims to discover the optimal combination of cutting $\{V_c, f_z, a_p\}$ while adhering to the operational constraints of the machine and tool capabilities.

Optimization Framework
The solution approach employs the CRO algorithm, a nature-inspired metaheuristic that simulates coral reef formation and the survival of the fittest within the coral population. The specifics of the CRO algorithm are detailed in [5], where the method's robustness in handling complex optimization problems is discussed.

Python programming language is utilized for the computational implementation, alongside scientific libraries, such as Pandas for data manipulation, NumPy for numerical operations, and Scikit-learn for machine learning tasks. The RandomForestRegressor (RFR) from Scikit-learn serves as the predictive model within the CRO framework to evaluate the fitness of potential solutions.

Optimization Results
Through computational analysis, an optimal set of cutting parameters was identified. Remarkably, the optimal solution attained a V_c of 335.7921 m/min, F of 1200 mm/min, and a_p of 1.3455 mm. These parameters correspond to F of 162.93 N and R_a of 0.313 μm, indicating a successful optimization of the milling process.

These findings demonstrate the efficacy of CRO in navigating the complex landscape of multi-objective optimization, providing a valuable strategy for improving industrial milling operations.

4 Conclusion

The research effectively utilized the Coral Reefs Optimization (CRO) algorithm to address the multi-objective optimization problem in the milling process of Aluminum Alloy A7075 material. The outcomes achieved not only met the goals of minimizing cutting force (F_c) and surface roughness (R_a) but also unveiled a new approach in seeking a balance between production efficiency and the quality of the final product. The optimal combination of cutting parameters, which includes a cutting speed (V_c) of 335.7921 m/min, a feed rate of 1200 mm/min, and a depth of cut of 1.3455 mm, highlights the robust applicability of the CRO algorithm in the metal machining sector.

The correlation analysis provided insightful understandings of the relationships between input variables and outcomes, forming a critical basis for setting parameters for the optimization algorithm. The findings of this study not only have practical application value in the manufacturing industry but also contribute to the scientific research field, proving that nature-inspired metaheuristic algorithms like CRO can effectively solve complex problems faced by humans.

Furthermore, detailed programming code along with data and analysis is provided in the appendix of the paper, paving the way for reproduction and verification of the results by the scientific community, while encouraging the widespread development and application of CRO in other optimization problems in industrial manufacturing.

References

1. Salcedo-Sanz, S., Camacho-Gómez, C., Molina, D., Herrera, F.: A coral reefs optimization algorithm with substrate layers and local search for large scale global optimization. In: 2016 IEEE Congress on Evolutionary Computation, pp. 3574–3581 (2016). https://doi.org/10.1109/CEC.2016.7744242
2. Salcedo-Sanz, S., Del Ser, J., Landa-Torres, I., Gil-López, S., Portilla-Figueras, J.A.: The coral reefs optimization algorithm: a novel metaheuristic for efficiently solving optimization problems. Sci. World J. (2014) https://doi.org/10.1155/2014/739768
3. Salcedo-Sanz, S., Casanova-Mateo, C., Pastor-Sánchez, A., Sánchez-Girón, M.: Daily global solar radiation prediction based on a hybrid coral reefs optimization—extreme learning machine approach. Sol. Energy **105**, 91–98 (2014). https://doi.org/10.1016/j.solener.2014.04.009
4. Seaborn.heatmap: https://seaborn.pydata.org/generated/seaborn.heatmap.html (n.d.)
5. Shieh, C.S., Nguyen, T.T., Lin, W.W., Nguyen, D.C., Horng, M.F.: Modified coral reef optimization methods for job shop scheduling problems. Appl. Sci. **12** (2022)

Investigation of Axis Errors of Grinding Tool in Grinding Process for Screw Rotors

Van-Quyet Tran

Abstract In the screw compressor, the screw rotors are the vital component. The machining accuracy for rotors is mainly affected to the screw compressor performance. However, the axis errors of the grinding wheel are often existed in the manufacturing process. To enhance the accuracy for rotor manufacturing, the effecting evaluation of axis errors of the grinding wheel to the tooth profile of the screw rotor is necessary. Therefore, a numerical simulation method is proposed to estimate the grinding wheel axis errors to the rotor profile of screw compressor. It is built to a software by using C# programming language. The validity and advantages of the proposed method are demonstrated via numerical examples.

Keywords Rotor profile · Screw compressor · Grinding wheel

1 Introduction

The screw compressors are widely used to industrial, and the rotor pair is the key component to screw compressor. The rotor geometry and machining precision are mainly affected to the screw compressor efficiency. Most commonly, the forming disk-type is employed to the grinding process (GP) of the SR surfaces. Therefore, several studies related to SR grinding by grinding wheel GW have investigated and presented. Wu [1] proposed a method for simulating of the GP for the compressor SR (SR) based on SW API (SOLIDWORKS Application Programming Interface), and he also presented a mathematical modeling which technologized numerical control form grinder for the form GP of the SR based on vertical multi-axis [2]. A processing method for improving the character and processing performance of the SR was expressed by Shen et al. [3]. Parallel, Lu et al. [4] expressed a calculation method for the GW profile by using engineering science, such as CAD. Besides that, Tao et al.

V.-Q. Tran (✉)
Department of Mechanical Engineering, University of Technology and Education in Hung Yen, 39A Rd., Hung Yen City, Vietnam
e-mail: tranquyetbk86@gmail.com

D. H. Tien et al. (eds.), *Proceedings of the 1st International Conference on Sustainability and Emerging Technologies for Smart Manufacturing*, Proceedings in Technology Transfer, https://doi.org/10.1007/978-981-97-7083-0_41

[5] built a software for calculating the profile of the SR from GW profile by using the GUI graphical interface technics in MATLAB software and also expressed a mathematical model to estimate the effects of installing errors of the GW to the rotor profile in the manufacturing process for the SR of the compressor [6]. In addition, Liu et al. [7] presented a new optimal design algorithm for the form GW to apply to the GP for the SRs. Recently, Shen [8] proposed a fitting processing method for GW profile using a small wave smoothing assumption and curve fitting method. Finally, You et al. [9] expressed a contrivance for generating the GW profile based on the pixel matrix method using the mathematical morphology and spiral motion envelope method.

Numerical simulation of the screw GP is essential before actual machining on CNC machines. It helps save costs and time during the processing process. Based on above studies and the demand of reducing manufacturing errors of the GP for the SR, this study proposed a software for the evaluation of the axis errors of the GW to the rotor profile based on C# programming language. This study will support to install grinding CNC machine and enhance manufacture accuracy for the SR GP.

2 Mathematica Model for Generation of GW Surfaces

This section aims to present the generation method for the GW surfaces. From Fig. 1, the GW surfaces ($\mathbf{r}_c$) can generate the profile of the GW ($\mathbf{r}_w$). The GW profile can be obtained from the discrete points, and these discrete points of the GW profile were fitted by a cubic spline curve.

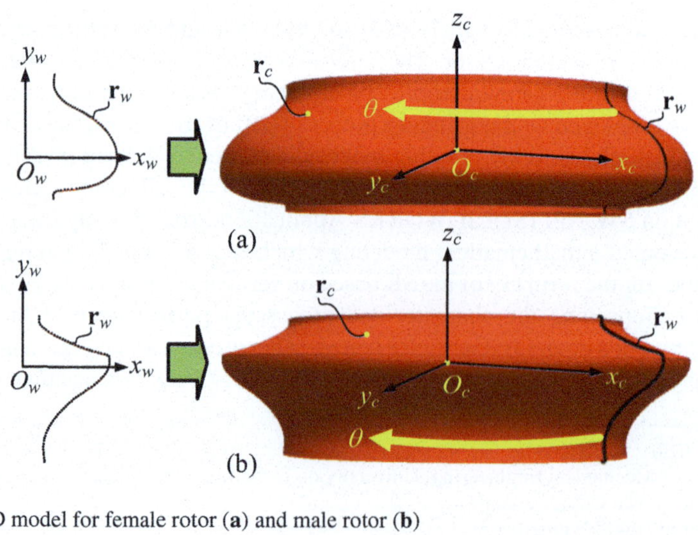

Fig. 1 3D model for female rotor (**a**) and male rotor (**b**)

The equation of each curve fitting section of the GW profile which is fitted by using the cubic spline is represented as:

$$\mathbf{r}_w(u) = [x_w, y_w] = \left[\sum_{k=0}^{3} a_{k,x} q^k, \sum_{k=0}^{3} a_{k,y} q^k, \right], \quad 0 \le q \le q_s \tag{1}$$

where the symbol q denotes the GW profile parameter as fitted curve parameter, $a_{k,y}$ indicates the factors of the curve fitting equations of y_w and $a_{k,x}$ the factors of the curve fitting equations of x_w, k is the sequence of the curve fitting equation with $k =$ 0–3, and q_s denotes the maximum value of the curve fitting parameter. Thus, there are $(m - 1)$ fitted curve equations according to m discrete points of a GW profile.

By using rotating $\mathbf{r}_w$ around the y_w-axis and transformation coordinate system from the coordinate system of $\mathbf{r}_w$ to the coordinate system of $\mathbf{r}_c$, the location vector and unit normal vector can be expressed as:

$$\mathbf{r}_c(u, \theta) = \begin{bmatrix} 1 & 0 & 0 & 0 \\ 0 & \cos\frac{\pi}{2} & -\sin\frac{\pi}{2} & 0 \\ 0 & \sin\gamma & \cos\frac{\pi}{2} & 0 \\ 0 & 0 & 0 & 1 \end{bmatrix} \cdot \begin{bmatrix} \cos\theta & 0 & \sin\theta & 0 \\ 0 & 1 & 0 & 0 \\ -\sin\theta & 0 & \cos\theta & 0 \\ 0 & 0 & 0 & 1 \end{bmatrix} .\mathbf{r}_w(u), \tag{2}$$

$$\mathbf{n}_c(u, \theta) = \frac{\partial_u \mathbf{r}_c \times \partial_\theta \mathbf{r}_c}{|\partial_u \mathbf{r}_c \times \partial_\theta \mathbf{r}_c|}. \tag{3}$$

where θ is the GW surface parameter.

3 Mathematical Model of GP for SR

The relationship of the coordinate systems of the GP for the SR is shown in Fig. 2. Symbols δ_x, δ_y, δ_z, and δ_γ stand for the axis errors of the GW according to x, y, and z axes and crossed axes angle, respectively. By using transformation coordinate system, the location vector $\mathbf{r}_r$ and the normal vector $\mathbf{n}_r$ of the profile of the SR can be obtained and expressed as:

$$\mathbf{r}_r(u, \theta, \phi) = [x_r, y_r, z_r, 1]^T = \mathbf{M}_{rc}(\phi) \cdot \mathbf{r}_c(u, \theta), \tag{4}$$

$$\mathbf{n}_r(u, \theta, \phi) = \mathbf{L}_{rc}(\phi) \cdot \mathbf{n}_c(u, \theta), \tag{5}$$

$$\mathbf{M}_{rc} = \mathbf{M}_1 \cdot \mathbf{M}_2 \cdot \mathbf{M}_3,$$

Fig. 2 Coordinate system of GP for SR

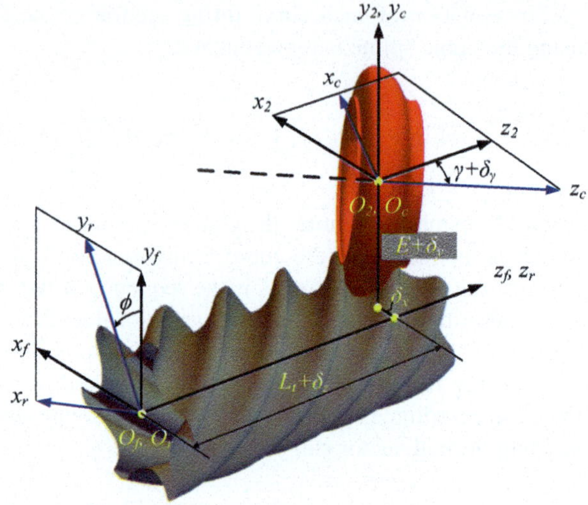

$$\mathbf{M}_1 = \begin{bmatrix} \cos\phi & -\sin\phi & 0 & 0 \\ \sin\phi & \cos\phi & 0 & 0 \\ 0 & 0 & 1 & 0 \\ 0 & 0 & 0 & 1 \end{bmatrix}, \quad \mathbf{M}_2 = \begin{bmatrix} 1 & 0 & 0 & \delta_x \\ 0 & 1 & 0 & E+\delta_y \\ 0 & 0 & 1 & L_t+\delta_z \\ 0 & 0 & 0 & 1 \end{bmatrix}$$

$$\mathbf{M}_3 = \begin{bmatrix} \cos(\gamma+\delta_\gamma) & 0 & \varepsilon\sin(\gamma+\delta_\gamma) & 0 \\ 0 & 1 & 0 & 0 \\ -\varepsilon\sin(\gamma+\delta_\gamma) & 0 & \cos(\gamma+\delta_\gamma) & 0 \\ 0 & 0 & 0 & 1 \end{bmatrix}.$$

where T is defined for the array transpose, coordinate system of the GW S_c transfer to the coordinate system of the SR is performed by the transformation matrix $\mathbf{M}_{rc}$, and $\mathbf{L}_{rc}$ is a sub-matrix 3×3 of the matrix $\mathbf{M}_{rc}$. $L_t = p\phi$ is the GW translation along rotor axis, γ is the crossed axes angle of the SR to GW, and ϕ is the rotational angle of the SR. $E = 0.5(d_a + d_i)$ is the center distant of the GW, and the SR, d_a, and d_i are the wheel inner diameter and rotor outer diameter, respectively. The right and left hands of the SR can be presented as $\varepsilon = 1$ and $\varepsilon = -1$, respectively. p is helix parameter, and it can be calculated as:

$$p = r_p \cot\beta \tag{6}$$

wherein, r_p and β are the pitch radius and helix angle of the SR.

From meshing theory of two surfaces, the meshing condition of the GW to the rotor surfaces can be derived as follows as:

$$f_1 = \mathbf{n}_r^T \cdot \frac{\partial[x_r, y_r, z_r]}{\partial\phi} = 0, \tag{7}$$

Table 1 GWs and SRs input parameters

Items (symbols)	Rotors		GW	
	Male	Female	For male	For female
Crossed-axis angle (γ)	–	–	40°	40°
Center distant (E_r, E_1, E_2)	137		162.200 mm	146.090 mm
Number of teeth (z_1, z_2)	5	6	–	–
Pitch helical angle (β_s, β_h)	50.0° R.H.	50.0° L.H.	–	–
SR Length (L)	284.54 mm	284.54 mm	–	–
Pitch radius (r_{p1}, r_{p2})	62.272 mm	74.727 mm	–	–
Outer radius (r^r_{o1}, r^r_{o2}, r^c_{o1}, r^c_{o2})	90.850 mm	74.720 mm	100 mm	100 mm
Inner radius (r^r_{i1}, r^r_{i2}, r^c_{i1}, r^c_{i2})	62.200 mm	46.090 mm	71.726 mm	72.402 mm
Screw lead (p_1, p_2)	52.253 mm	62.703 mm	–	–

$$f_2 = z_r = 0. \tag{8}$$

The tooth profile of the rotor can be obtained through solving Eqs. (4), (5), (7), and (8) simultaneously with given the GW profile parameter u.

4 Numerical Examples

4.1 SR Tooth Profile Without Axis Errors

In this section, the generated rotor tooth profile of SR by GW can be obtained without considering of axis errors ($\delta_x = \delta_y = \delta_z = \delta_\gamma = 0$). The GWs and SRs input parameters are presented in Table 1.

As presented in Fig. 3, the tooth profiles of the SRs for reference and generated rotor profile are coincided for both of the male and female rotor profiles. It indicates that the truthfulness of the proposed SR grinding mathematical model has been confirmed.

4.2 SR Tooth Profile with Axis Errors

The machining accuracy for rotors is mainly affected to the screw compressor performance. However, the axis errors of the grinder are often existed in the manufacturing process. To enhance the accuracy for rotor manufacturing, the effecting evaluation of axis errors of the GW to the tooth profile of the SR is necessary. In this example, the

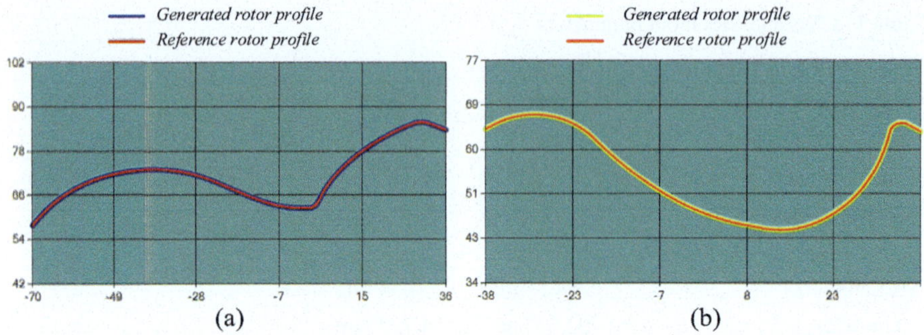

Fig. 3 Rotor tooth profiles of SRs: **a** male rotor, **b** female rotor

influent of the installing errors of the GW is investigated and presented. The interface of the software for the SR grinding process and scale definition are expressed in Fig. 4, and the scale of the tooth profile of the SR is chosen to 0.1 mm. The rotor profile inaccuracy of the simulation rotor profile by GW can be calculated by comparison of the tooth profiles between the simulation and reference of the SRs, and it can be defined by the normal deviation.

In this section, the parameters for the GWs and SRs are shown in Table1. The influent of single axis error to tooth profile of the male rotor is presented in Fig. 5. And the influent of axes errors to the male rotor tooth profile is expressed in Fig. 6. Moreover, the influent of axis errors to female tooth profile is presented in Fig. 7. These results will be supported for position adjusting of the GW to compensate the axis errors for the grinder in the rotor GP.

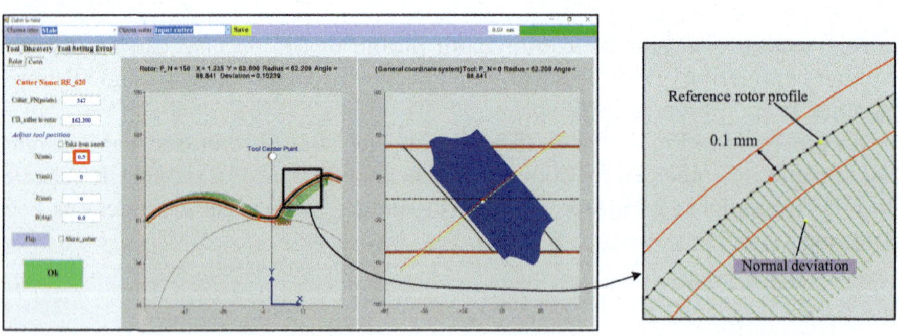

Fig. 4 Interface of SR grinding software

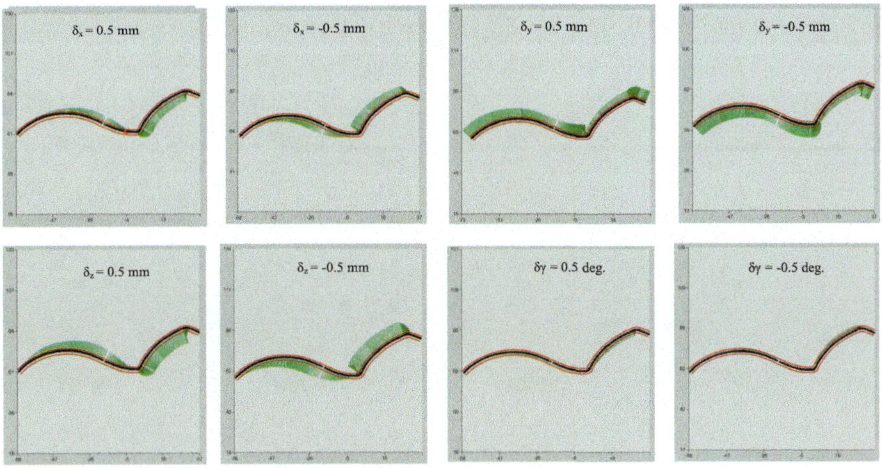

Fig. 5 Influent of single axis error to tooth profile of male rotor

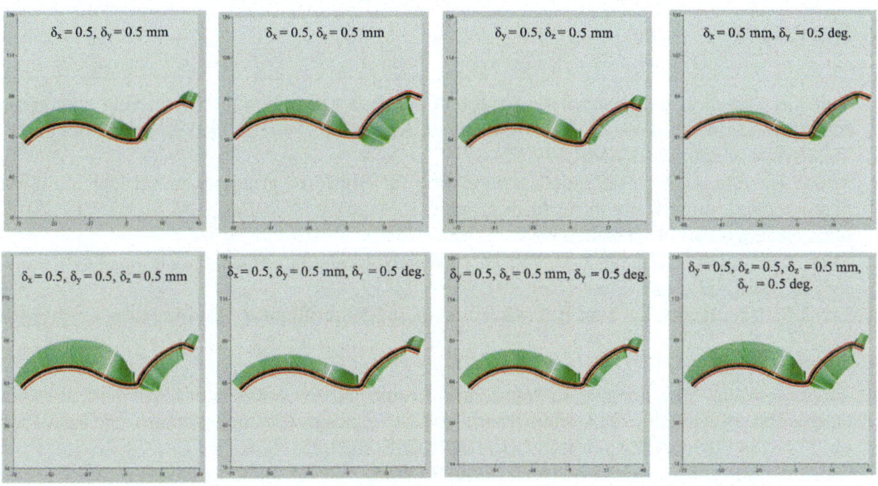

Fig. 6 Influent of axes errors to tooth profile of male rotor

5 Conclusions

A software for screw rotor grinding which created in C# programming language was presented. The grinding process mathematical model for the screw rotor was established. Besides that, the influent of axis errors to of the screw rotors tooth profiles was evaluated and presented. This study will be supported for position adjusting of the grinding tool to compensate the axis errors of the grinder.

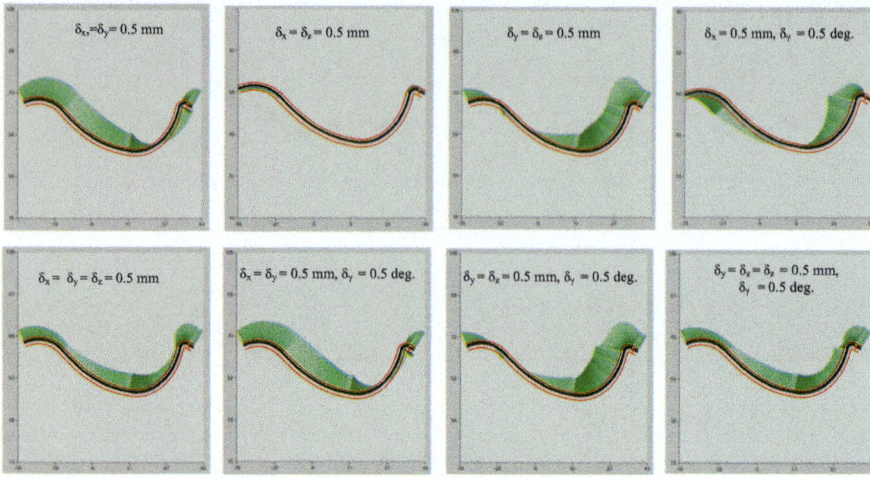

Fig. 7 Influent of axes errors to tooth profile of female rotor

References

1. Wu, Y.R.: Form grinding simulation and tool abrasion estimation for twin screw compressor rotors based on SolidWorks API. Adv. Mater. Res. **579**, 174–183 (2012). https://doi.org/10.4028/www.scientific.net/AMR.579.174
2. Wu, Y.R., Fan, C.W.: Mathematical modeling for SR form grinding on vertical multi-axis computerized numerical control form grinder. J. Manufact. Sci. Eng. **135**(5), 051020 (2013). https://doi.org/10.1115/1.4025339
3. Shen, Z.H., Lu, R.S., Zhang, Z.S., Yao, B.: Research on calculation and smoothing of GW profile for machining SR. Adv. Mater. Res. **291**, 2383–2387 (2011)
4. Lu, R.S., Yao, B., Li, L., Yao, B.S.: Calculation of GW profile and 3D simulation for grinding SR flute. Adv. Mater. Res. **383**, 6769–6775 (2012). https://doi.org/10.4028/www.scientific.net/AMR.383-390.6769
5. Tao, L.J., Wang, Y.L., Feng, H.T., Wang, X.F.: Design and development of SR numerical inverse calculation software. In: 2013 8th International Conference on Computer Science and Education, pp. 579–583 (2013). https://doi.org/10.1109/ICCSE.2013.6553976
6. Tao, L.J., Wang, Y.L., He, Y., Feng, H.T., Ou, Y., Wang, X.F.: A numerical method for evaluating effects of installation errors of GW on rotor profile in SR grinding. Proc. Inst. Mech. Eng. Part B J. Eng. Manufact. **230**(8), 1381–1398 (2016). https://doi.org/10.1177/095440541665
7. Liu, Z.M., Tang, Q., Liu, N., Liang, P.H., Liu, W.: A novel optimization design method of form GW for SR. Appl. Sci. **9**(23), 5079 (2019). https://doi.org/10.3390/app9235079
8. Shen, Z.H.: Novel fitting method of forming tool profile in precision grinding SR. UPB Sci. Bull. Ser. D Mech. Eng. **82**(2), 199–210 (2020)
9. You, M.L., Yao, B.: Application of mathematical morphology in solving the profile of forming GW. Math. Probl. Eng. (2022). https://doi.org/10.1155/2022/5735199

Triboelectric Nanogenerator for the Emerging Technologies and Smart Manufacturing

V. L. Trinh, C. K. Chung, X. C. Nguyen, T. S. Nguyen, N. T. Tran, H. H. Nguyen, and V. N. Hoang

Abstract Triboelectric nanogenerator (TENG) is recently emerging as a sustainable energy source for applications of the emerging technologies and smart manufacturing. TENG is an energy converter that can harvest wasted mechanical energy into electrical energy for many practical applications. The TENG has many precious characteristics for the emerging technologies and smart manufacturing (ETSM), such as green energy, renewable energy, cost effectiveness, lightweight, and self-powered energy. This paper reviews recently TENG technology and its applications for the emerging technology and smart manufacturing. The paper hopes that the TENG energy will be spread for more practical applications of ETSM in the near future.

Keywords Triboelectric nanogenerator · Emerging technologies · Renewable energy · Sustainable energy

1 Introduction

An emerging technology is a technology that will bring benefits of economy and society for our daily living [1]. Emerging technologies show strong impacts on human revolution, such as advanced materials, information and communication,

V. L. Trinh (✉) · T. S. Nguyen · N. T. Tran
School of Mechanical and Automotive Engineering, Hanoi University of Industry, 298 Caudien Street, Hanoi 10000, Vietnam
e-mail: longtv@haui.edu.vn

C. K. Chung (✉)
Department of Mechanical Engineering, National Cheng Kung University, Tainan 701, Taiwan
e-mail: ckchung@mail.ncku.edu.tw

X. C. Nguyen · V. N. Hoang
Vietnam-Japan Center, Hanoi University of Industry, 298 Caudien Street, Hanoi 10000, Vietnam

H. H. Nguyen
Faculty of Natural Sciences, Hungvuong University, Nongtrang Area, Viettri, Phutho Province, Vietnam

D. H. Tien et al. (eds.), *Proceedings of the 1st International Conference on Sustainability and Emerging Technologies for Smart Manufacturing*, Proceedings in Technology Transfer, https://doi.org/10.1007/978-981-97-7083-0_42

robot, biotechnology, energy harvesting technologies, and so on [2, 3]. Figure 1 shows the emerging technologies developed for our social needs including geospatial technology, big data, artificial intelligence, robotics, blockchain, smart applications, additive manufacturing, and Internet of Things (IoT). Triboelectric nanogenerator (TENG), an emerging technology, is an energy harvesting technology that converses the wasted mechanical energy into electrical energy. The TENG's characteristics are precious for the emerging technologies and smart manufacturing (ETSM), such as green energy for more applications of electronics [4], renewable energy addressed in mini-review paper [5], cost effectiveness, lightweight, self-powered energy summarized in review paper [6], and TENG sensors for digital twin applications [7]. TENG generates electrical energy based on the triboelectric effect. In the triboelectric effect, the charge transfer occurs between two triboelectric materials come to contact and separate with each other [8, 9]. TENG has many outstanding characteristics of green, sustainable energy [10, 11]. TENG is the energy generator with non-carbon dioxide emission and friendly environment. By introducing the TENG into the energy production, the lost energy transmission can reduce, and energy cost can degrease [12]. TENG has received many attentions on research and apply for practical applications, such as developing for self-power systems [13], collecting for sustainable and renewable energy applications [14], for Internet of Things (IoT) applications [15], for monitoring systems in the ocean [16], and developing TENGs for industrial applications [6]. TENG can be successfully applied in the emerging technologies and smart manufacturing, such as applying in the IoT systems [17], monitoring the working systems, developing TENG for machine fault detect systems [18], constructing TENG for smart agriculture [19], TENG for human–machine interface [20], and introducing TENG to self-powered robotic system [21]. The challenges involve enhancement of output performance of the TENG [22], cresting vary of weather, developing new technologies, and using new materials [23, 24]. The solutions are related to improve output performance of the TENG by focusing on design and optimization of the TENG [25, 26], get stable working condition [27], and manage effective power output performance [28, 29]. This paper reviews recently TENG technology and its applications for the emerging technologies and smart manufacturing. The paper hopes that the TENG energy will be spread for more practical in the near future.

2 Triboelectric Nanogenerator

Triboelectric nanogenerator (TENG) is one of the latest energy harvesting methods. TENG produces electrical energy upon the triboelectric effect. By the physical phenomena, one triboelectric material donates electric charges, and other triboelectric material accepts electric charges. The triboelectric materials are diverse types of metal, wood, polymer, and eco-materials. In the generation process, two triboelectric materials come to contact and separate with each other to make an electric flow moving via external circuit. Figure 2 shows the working principle of the TENG with four stages of contacting and separating in cycle to produce electricity. Figure 2a

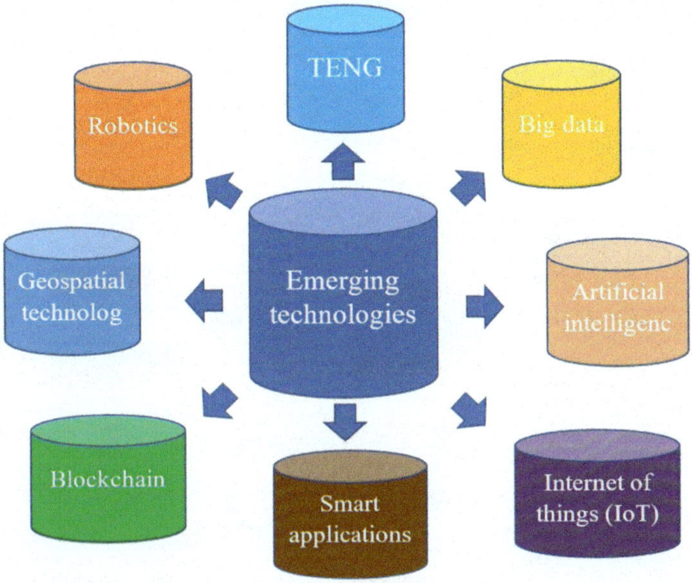

Fig. 1 Emerging technologies developed for our social needs including geospatial technology, big data, artificial intelligence, robotics, blockchain, smart applications, additive manufacturing, and Internet of Things (IoT)

shows the initial state with no charges. Figure 2b shows the two triboelectric surfaces come to contact with each other that produces adverse charges. Figure 2c shows the releasing state that causes the unbalance potentials between the two tribo-materials. The electrical flow will move through the load. Figure 2d shows the released state of the two tribo-materials with the neutral state. Figure 2e shows the pressing state again to make the unbalance potentials between the two tribo-materials. A current will flow via the load. The TENG generates an electricity with the open-circuit voltage (VOC) addressed by following formula [9]:

$$V_{OC} = -\frac{\sigma d}{\varepsilon_0}. \tag{1}$$

where

V_{OC} is the open-circuit voltage.
σ is the triboelectric charge density.
d is the distance between the two contact surfaces.
ε_0 is the vacuum permittivity.

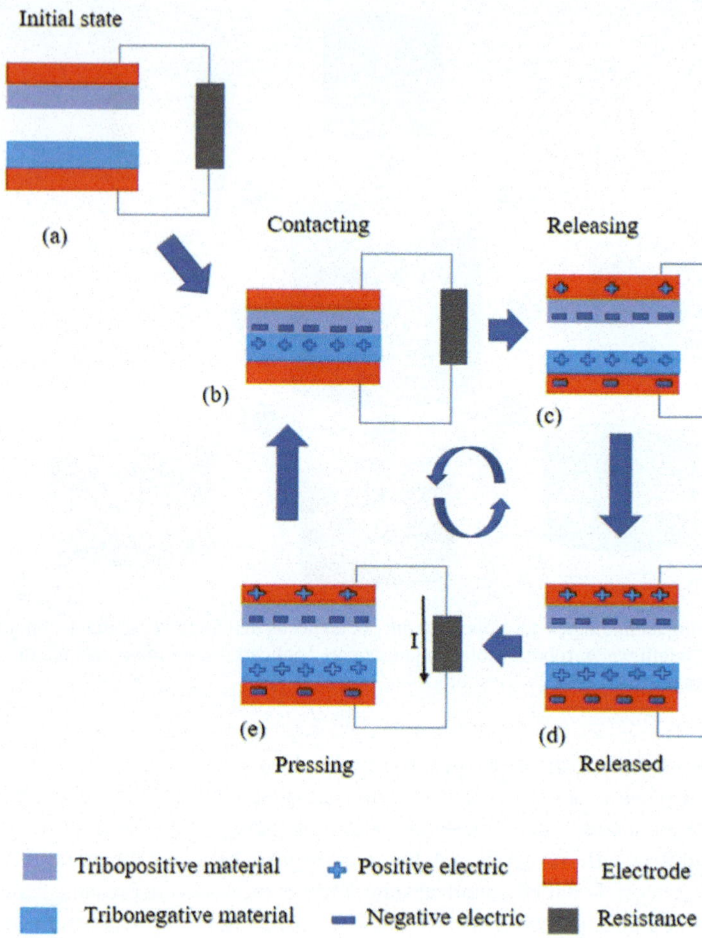

Fig. 2 Working principle of TENG with four stages of contacting and separating in cycle to produce electricity

3 The Emerging Technology and Smart Manufacturing

The emerging technology yields benefits of economy and society for the innovation. There are many emerging technologies applied successfully in our living, such as geospatial technology, big data, artificial intelligence, Internet of Things (IoT), robotics, smart applications, additive manufacturing, and blockchain [30]. Triboelectric nanogenerator, an emerging technology, has been developed for many applications, such as serving as a power supply for sensors to monitor the geological disaster and alarm system [31], developing the self-powered sensing technology for

Table 1 Performances and applications of TENGs that yield advantages of emerging technology

TENG	Output performance	Emerging technology applications	Refs
Au/PTFE-Micro/nanostructure	25.1 V	Marine environmental monitoring	[38]
Microneedle/PDMS/Al	102.8 V	Charging, lighting	[9]
Polyester/Ag nanowire/graphene	7 nW/cm^2	Wearable electronic systems, self-powered clothing	[39]
Fabric structure	3.2 W/m^2	Small electronics	[40]
Graphene sheet/PDMS	4.8 W/m^2	Pressure sensor	[41]
Al/PP-nanowire array	19 mA/m^2	Self-powered temperature sensor	[42]
Micro-pyramid PDMS	1.13 mu A/cm^2	Charging, lighting	[43]
PHFC film	330.6 $\mu mW/cm^2$	Sustainable agriculture	[44]
PVDF nanofiber	10 mW	Water splitting	[45]

the sport sensing system with ability of working in the smart sport facility and equipment [24], a TENG for artificial intelligence (AI) application [32], constructing an emerging TENG for harvesting ocean wave energy[33], and integrating the TENG to the Internet of Thing (IoT) application [15]. Table 1 shows the performances and the applications of the TENGs that yield advantages of the emerging technology.

Smart manufacturing absorbs many advanced technologies of process automation, data storage, prediction technology, digitalization technology, agent technology, and cloud computing technology [34]. In smart manufacturing, sensors are crucial to monitor and manage the work flow. TENG supports many solutions for sensors to serve in smart manufacturing systems, such as developing a FF-TENG to control the fluid flow of the smart factory [35], supporting power for Internet of Things devices in the smart manufacturing system [36], and developing the sustainable triboelectric materials served in the smart manufacturing area [37].

4 Conclusions

The emerging technology brings many advantages for the social innovation of economy and society. The emerging technologies have been successfully developed and applied in practical applications, such as smart applications, geospatial technology, blockchain, artificial intelligence, Internet of Things (IoT), robotics, additive manufacturing, and big data. Triboelectric nanogenerator (TENG) has many potential features for an emerging technology. TENG can harvest ambient mechanical energy into electricity for applications of the emerging technologies. The TENG has diverse characteristics of the emerging technologies and smart manufacturing (ETSM), such as green and renewable energies, low cost in capital investment, lightweight, and self-powered energy for electronics devices. This paper reviews the latest TENG technologies for the emerging technology. The paper hopes that the TENG will apply more and more in the emerging technologies for our daily living in the near future.

Acknowledgements This work was partially sponsored by Department of Mechanical Engineering and Core Facility Center, National Cheng Kung University, Taiwan. It was also supported in part by Hanoi University of Industry, Vietnam. The work was also supported in part by School of Mechanical and Automotive Engineering, Hanoi University of Industry, Vietnam.

References

1. Martin, B.R.: Foresight in science and technology. Technol. Anal. Strategic Manage. **7**, 139–168 (1995)
2. Jones, V., Sampath, H.: Emerging technologies for WLAN. IEEE Commun. Mag. **53**, 141–149 (2015)
3. van der Valk, T., Chappin, M.M.H., Gijsbers, G.W.: Evaluating innovation networks in emerging technologies. Technol. Forecast. Soc. Chang. **78**, 25–39 (2011)
4. Maiti, S., Karan, S.K., Kim, J.K., Khatua, B.B.: Nature driven bio-piezoelectric/triboelectric nanogenerator as next-generation green energy harvester for smart and pollution free society. Adv. Energy Mater. **9**, 1803027 (2019)
5. Walden, R., Kumar, C., Mulvihill, D.M., Pillai, S.C.: Opportunities and challenges in Triboelectric Nanogenerator (TENG) based sustainable energy generation technologies: a mini-review. Chem. Eng. J. Adv. **9**, 100237 (2022)
6. Trinh, V.L., Chung, C.K.: Renewable energy for SDG-7 and sustainable electrical production, integration, industrial application, and globalization: Review. Cleaner Eng. Technol. **15**, 100657 (2023)
7. Jin, T., Sun, Z., Li, L., Zhang, Q., Zhu, M., Zhang, Z., et al.: Triboelectric nanogenerator sensors for soft robotics aiming at digital twin applications. Nat. Commun. **11**, 5381 (2020)
8. Zhu, G., Peng, B., Chen, J., Jing, Q., Lin, W.Z.: Triboelectric nanogenerators as a new energy technology: From fundamentals, devices, to applications. Nano Energy **14**, 126–138 (2015)
9. Trinh, V.L., Chung, C.K.: A facile method and novel mechanism using microneedle-structured PDMS for triboelectric generator applications. Small **13**, 11 (2017)
10. Cho, H., Chung, J., Shin, G., Sim, J.-Y., Kim, D.S., Lee, S., et al.: Toward sustainable output generation of liquid–solid contact triboelectric nanogenerators: the role of hierarchical structures. Nano Energy **56**, 56–64 (2019)
11. Yoon, H.-J., Ryu, H., Kim, S.-W.: Sustainable powering triboelectric nanogenerators: approaches and the path towards efficient use. Nano Energy **51**, 270–285 (2018)
12. Kim, W.-G., Kim, D.-W., Tcho, I.-W., Kim, J.-K., Kim, M.-S., Choi, Y.-K.: Triboelectric nanogenerator: structure, mechanism, and applications. ACS Nano **15**, 258–287 (2021)
13. Wang, Z.L.: Triboelectric nanogenerators as new energy technology for self-powered systems and as active mechanical and chemical sensors. ACS Nano **7**, 9533–9557 (2013)
14. Trinh, V.-L., Chung, C.-K.: Advances in triboelectric nanogenerators for sustainable and renewable energy: working mechanism, tribo-surface structure, energy storage-collection system, and applications. In: Processes (2023)
15. Ahmed, A., Hassan, I., El-Kady, M.F., Radhi, A., Jeong, C.K., Selvaganapathy, P.R., et al.: Integrated triboelectric nanogenerators in the era of the internet of things. Adv. Sci. **6**, 1802230 (2019)

16. Ahn, J., Kim, J.-S., Jeong, Y., Hwang, S., Yoo, H., Jeong, Y., et al.: All-recyclable triboelectric nanogenerator for sustainable ocean monitoring systems. Adv. Energy Mater. **12**, 2201341 (2022)
17. Cao, X., Xiong, Y., Sun, J., Xie, X., Sun, Q., Wang, Z.L.: Multidiscipline applications of triboelectric nanogenerators for the intelligent era of internet of things. Nano-Micro Lett. **15**, 14 (2022)
18. Li, W., Liu, Y., Wang, S., Li, W., Liu, G., Zhao, J., et al.: Vibrational triboelectric nanogenerator-based multinode self-powered sensor network for machine fault detection. IEEE/ASME Trans. Mechatron. **25**, 2188–2196 (2020)
19. Dai, S., Li, X., Jiang, C., Ping, J., Ying, Y.: Triboelectric nanogenerators for smart agriculture. InfoMat **5**, e12391 (2023)
20. Ding, W., Wang, A.C., Wu, C., Guo, H., Wang, Z.L.: Human-machine interfacing enabled by triboelectric nanogenerators and tribotronics. Adv. Mater. Technol. **4**, 1800487 (2019)
21. Hajra, S., Panda, S., Khanberh, H., Vivekananthan, V., Chamanehpour, E., Mishra, Y.K., et al.: Revolutionizing self-powered robotic systems with triboelectric nanogenerators. Nano Energy **115**, 108729 (2023)
22. Chao, S., Ouyang, H., Jiang, D., Fan, Y., Li, Z.: Triboelectric nanogenerator based on degradable materials. EcoMat **3**, e12072 (2021)
23. Wang, H., Cheng, J., Wang, Z., Ji, L., Wang, Z.L.: Triboelectric nanogenerators for human-health care. Sci. Bull. **66**, 490–511 (2021)
24. Luo, J., Gao, W., Wang, Z.L.: The triboelectric nanogenerator as an innovative technology toward intelligent sports. Adv. Mater. **33**, 2004178 (2021)
25. Zhou, L., Liu, D., Zhao, Z., Li, S., Liu, Y., Liu, L., et al.: Simultaneously enhancing power density and durability of sliding-mode triboelectric nanogenerator via interface liquid lubrication. Adv. Energy Mater. **10**, 2002920 (2020)
26. Zhang, H., Quan, L., Chen, J., Xu, C., Zhang, C., Dong, S., et al.: A general optimization approach for contact-separation triboelectric nanogenerator. Nano Energy **56**, 700–707 (2019)
27. Lei, R., Shi, Y., Ding, Y., Nie, J., Li, S., Wang, F., et al.: Sustainable high-voltage source based on triboelectric nanogenerator with a charge accumulation strategy. Energy Environ. Sci. **13**, 2178–2190 (2020)
28. Xi, F., Pang, Y., Li, W., Jiang, T., Zhang, L., Guo, T., et al.: Universal power management strategy for triboelectric nanogenerator. Nano Energy **37**, 168–176 (2017)
29. Cheng, X., Miao, L., Song, Y., Su, Z., Chen, H., Chen, X., et al.: High efficiency power management and charge boosting strategy for a triboelectric nanogenerator. Nano Energy **38**, 438–446 (2017)
30. Mbunge, E., Akinnuwesi, B., Fashoto, S.G., Metfula, A.S., Mashwama, P.: A critical review of emerging technologies for tackling COVID-19 pandemic. Human Behav. Emerging Technol. **3**, 25–39 (2021)
31. Zhou, Q., Deng, S., Gao, A., Wang, B., Lai, J., Pan, J., et al.: Triboelectric nanogenerator with dynamic electrode for geological disaster and fall-down self-powered alarm system. Adv. Funct. Mater. **33**, 2306619 (2023)
32. Jiao, P.: Emerging artificial intelligence in piezoelectric and triboelectric nanogenerators. Nano Energy **88**, 106227 (2021)
33. Rodrigues, C., Nunes, D., Clemente, D., Mathias, N., Correia, J.M., Rosa-Santos, P., et al.: Emerging triboelectric nanogenerators for ocean wave energy harvesting: state of the art and future perspectives. Energy Environ. Sci. **13**, 2657–2683 (2020)
34. Kusiak, A.: Fundamentals of smart manufacturing: a multi-thread perspective. Ann. Rev. Control. **47**, 214–220 (2019)
35. Wang, X., Zhu, C., Wu, M., Zhang, J., Chen, P., Chen, H., et al.: A novel flow sensing and controlling system based on the flapping film triboelectric nanogenerator toward smart factories. Sens. Actuators A **344**, 113727 (2022)
36. Yang, H., Kumara, S., Bukkapatnam, S.T.S., Tsung, F.: The internet of things for smart manufacturing: a review. IISE Trans. **51**, 1190–1216 (2019)

37. Wei, Z., Wang, J., Liu, Y., Yuan, J., Liu, T., Du, G., et al.: Sustainable triboelectric materials for smart active sensing systems. Adv. Funct. Mater. **32**, 2208277 (2022)
38. Chen, H., Wang, J., Ning, A.: Optimization of a rolling triboelectric nanogenerator based on the nano-micro structure for ocean environmental monitoring. ACS Omega **6**, 21059–21065 (2021)
39. Wu, C.X., Kim, T.W., Li, F.S., Guo, T.L.: Wearable electricity generators fabricated utilizing transparent electronic textiles based on polyester/Ag nanowires/graphene core-shell nanocomposites. ACS Nano **10**, 6449–6457 (2016)
40. Feng, P.-Y., Xia, Z., Sun, B., Jing, X., Li, H., Tao, X., et al.: Enhancing the performance of fabric-based triboelectric nanogenerators by structural and chemical modification. ACS Appl. Mater. Interfaces. **13**, 16916–16927 (2021)
41. Xia, X.N., Chen, J., Liu, G.L., Javed, M.S., Wang, X., Hu, C.G.: Aligning graphene sheets in PDMS for improving output performance of triboelectric nanogenerator. Carbon **111**, 569–576 (2017)
42. Feng, Y., Zheng, Y., Ma, S., Wang, D., Zhou, F., Liu, W.: High output polypropylene nanowire array triboelectric nanogenerator through surface structural control and chemical modification. Nano Energy **19**, 48–57 (2016)
43. Xue, C.Y., Li, J.Y., Zhang, Q., Zhang, Z.B., Hai, Z.Y., Gao, L.B., et al.: A novel arch-shape nanogenerator based on piezoelectric and triboelectric mechanism for mechanical energy harvesting. Nanomaterials **5**, 36–46 (2015)
44. Lan, L., Xiong, J., Gao, D., Li, Y., Chen, J., Lv, J., et al.: Breathable nanogenerators for an on-plant self-powered sustainable agriculture system. ACS Nano **15**, 5307–5315 (2021)
45. Ren, X., Fan, H., Wang, C., Ma, J., Li, H., Zhang, M., et al.: Wind energy harvester based on coaxial rotatory freestanding triboelectric nanogenerators for self-powered water splitting. Nano Energy **50**, 562–570 (2018)

Mechanical Properties of Dual-Phase Eutectic High-Entropy Alloys Under Nano-cutting

Duy-Khanh Nguyen, Te-Hua Fang, and Ching-Chien Huang

Abstract This study explores the mechanical characteristics of dual-phase herringbone high-entropy alloys through Molecular Dynamics simulations during the cutting process. Lamellar and herringbone pattern dual-phase workpieces are subjected to multi-directional cutting at varying depths; the behavior is observed in the distribution of shear strain, von Mises stress, temperature, and microstructure evolution with the plastic flow. The findings indicate that structural interfaces influence the uniform directional propagation of shear strain in FCC atoms, with the atomic force being less pronounced on the BCC layer surface but exhibiting high stability. Moreover, atom migration under force occurs concurrently within the FCC layer before gradually redirecting or dispersing in the BCC volume, contingent on the orientation of the structural interface. This results in diverse deformation behaviors in distinct structures and cutting directions.

Keywords Herringbone · Dual-phase · Eutectic high-entropy alloy · Nano-cutting · Molecular dynamics simulation

1 Introduction

In recent years, high-entropy alloys (HEAs) have garnered significant attention from researchers in the material research community [1, 2] due to their outstanding properties compared to conventional alloys, including a remarkable combination of strength, hardness, and fatigue resistance [3, 4]. Most investigated HEAs aim to adopt face-centered cubic (FCC) and/or body-centered cubic (BCC) structures.

D.-K. Nguyen · T.-H. Fang (✉) · C.-C. Huang
Department of Mechanical Engineering, National Kaohsiung University of Science and Technology, Kaohsiung, Taiwan
e-mail: fang.tehua@msa.hinet.net

D.-K. Nguyen
School of Mechanical and Automotive Engineering, Hanoi University of Industry, Hanoi, Vietnam

D. H. Tien et al. (eds.), *Proceedings of the 1st International Conference on Sustainability and Emerging Technologies for Smart Manufacturing*, Proceedings in Technology Transfer, https://doi.org/10.1007/978-981-97-7083-0_43

The synthesis, however, is inevitable in the strength-ductility trade-off because FCC generally offers superior plasticity to BCC but stiffness. Therefore, the creation of materials that combine the advantages is considered to be the introduction of intermetallic compounds or other complex-ordered phases [5] with a growing interest in eutectic high-entropy alloys (EHEAs) due to their multiple-phase content [6, 7]. Hence, recent research efforts have been directed toward designing and optimizing EHEAs, which exhibit tensile fracture strength of around 1000 MPa and uniform elongation of 5–30% [7]. Among these EHEAs, $Al_xCoCrFeNi$ has a microstructure partly controlled by the aluminum content. Precisely, increasing the aluminum content shifts the primary solid phase from FCC, dual-phase (FCC + BCC), to BCC during the solidification of as-cast materials. Notably, the dual-phase homogenized specimen demonstrates significantly higher hardness than FCC and is approximately equal to the BCC structure. Jin et al. [8] present a stable lamellar dual-phase structure by eliminating chromium, which exhibits excellent synthetic mechanical properties and reaches 577 MPa and 1103 MPa in yield and fracture strength, respectively. Shi et al. [9] further enhance the material's elongation from 18.7 to 50% by arranging the phase pattern in a herringbone (HRB) structure through directional solidification. Thus, these scholars have successfully demonstrated the superior properties of EHEAs and significant improvement in material with the employment of micro HRB patterns. On the other hand, HRB structures are also applied to enhance the mechanical properties of conventional alloys, synthesized materials, and nanocomposite foams. Therefore, there is an obligation and necessity to investigate the influence of the HRB on the mechanical properties of EHEAs.

Nano-cutting processes have been conducted through various experiments and computer-aided simulations, offering diverse insights into machining mechanisms and deformation behavior at the nanoscale. In this context, Molecular Dynamics (MD) simulation has emerged as an auspicious approach for exploring nanoscale machining processes [10–12]. Given these considerations, there is a demand for comprehensive studies on the deformation behavior of FCC-BCC dual-phase HRB-EHEAs. Consequently, this study uses MD simulation to elucidate the dual-phase FeCoNiAl HRB-EHEA cutting process. Accordingly, Sect. 3.1 of this work presents the response of HRB workpieces to the cutting process with varying depths and in different directions. Additionally, Sect. 3.2 highlights the distinctions between lamellar and HRB patterns.

2 Computational Method

The simulation model comprises a dual-phase workpiece of FeCoNiAl EHEA and a diamond orthogonal cutting tool (see Fig. 1). The workpiece, intentionally, is constructed from 451,892 atoms in the form of FCC and BCC volumes. Firstly, Atomsk is used to create two distinct monocrystalline specimens in specific lattice constants of 3.54 and 2.86 Å for FCC and BCC, respectively. The process is followed by the equilibrium stage under the isobaric-isothermal constant number of atoms, constant pressure, and continuous temperature (NPT) ensemble. The dual-phase workpiece is established by combining the two monocrystals before relaxation using the thermodynamic calculations model [13]. The final workpiece consists of roundly equal FCC and BCC particles that satisfy the volumic proportion of 3/2 (V_{FCC}/V_{BCC}) [8, 9]. The model uses Morse's potential to describe the tool–workpiece interaction behavior [14] and the interaction between the specimen element and carbon as a diamond tool, as illustrated in Table 1. Meanwhile, interior particles represent their interaction using the embedded-atom method (EAM) [15]. Additionally, during the cut, atoms are dispensed into three distinct layers: boundary, thermostat, and Newtonian (see Fig. 1). The equilibrium and cutting process is conducted in the LAMMPS [16, 17] code environment, with the outputs displayed via OVITO [18]. Crystalline structure is illustrated using Polyhedral Template Matching (PTM) analysis, and dislocation is identified by Dislocation Extraction Algorithms (DXA) with von Mises stress (VMS) used to calculate the local pressure:

$$\sigma_{vM} = \sqrt{3\left(\sigma_{xy}^2 + \sigma_{yz}^2 + \sigma_{xz}^2\right) + \frac{1}{2}\left[\left(\sigma_{xx} - \sigma_{yy}\right)^2 + \left(\sigma_{xx} - \sigma_{zz}\right)^2 + \left(\sigma_{zz} - \sigma_{yy}\right)^2\right]} \tag{1}$$

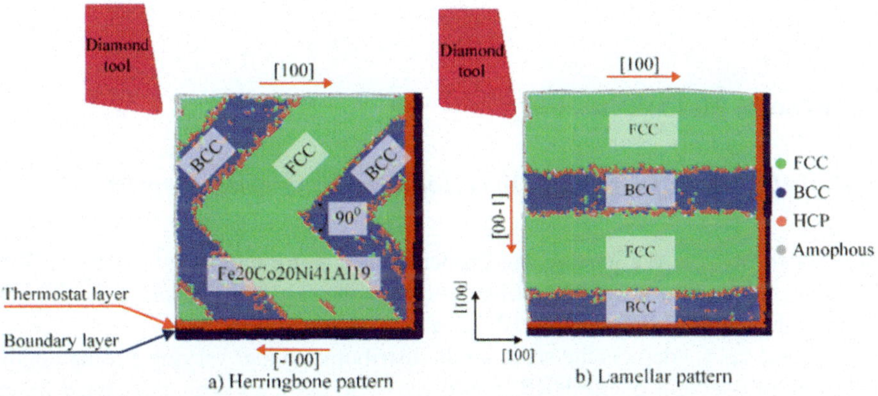

Fig. 1 Schematic of HRB (**a**), lamellar (**b**) workpiece for distinct cut directions

Table 1 Lennard–Jones (L–J) potential parameters

Interaction	ε (eV)	σ (Å)
C–Al	0.108	2.860
C–Fe	0.126	2.711
C–Ni	0.125	2.691
C–Co	0.124	2.703

Table 2 Simulation setup parameters

Parameter	Workpiece	Cutting tool
Workpiece material	$Fe_{20}Co_{20}Ni_{41}Al_{19}$	Carbon (Diamond)
Dimensions (Å) (XxYxZ)	220 × 220 × 108	70 × 40 × 100—R5
Rake/Clearance angle (degree)	–	10/10
Quantity of atoms	451,892	60,193
HRB angle	90^0	
Cutting velocity V_0, (m/s)	100	
Cutting length (Å)	150	
Cutting depth (Å)	10,15, and 20	
Cutting directions	[100], [-100], [00-1]	

where σ_{ij} represents the vital atomic stress, and the process has been investigated in various cutting directions, depths, and dual-structure patterns with simulation parameters shown in Table 2. Accordingly, the HRB workpiece shows distinct deformation behavior in different external force directions [19]. Meanwhile, the surveyed cutting depth is larger than the tool radius to avoid scratching and within the limits of the specimen that exhibits deformation behavior in damaged zone and interior atoms differently [12, 20].

3 Results and Discussion

3.1 Influence of Cutting Direction Under Various Depths

In order to observe the mechanical properties and deformation behavior of HRB-EHEA in the cases of different cutting directions, the workpiece reaction is exhibited in multiple stages of the cut with 10, 15, and 20 Å in depth.

Figure 2 (a_1-f_1) depicts the shear strain distribution under varying cutting depths within a half-section of the HRB workpiece. The particles are color-coded based on shear strain magnitude, with a notable concentration of large shear strain atoms observed on the tool–workpiece contact surface, extending to the underlying area. The penetration of shear strain indicates a prevalent dual microstructure, primarily influenced by intense diffusion in the BCC volume, with limited spread to neighboring areas beneath FCC regions. Significantly, moderate shear bands are formed at the leading edge of the BCC layer [21]. In contrast, the shear strain diffuses indiscriminately into the adjacent damaged area within the BCC layer. The shear strain penetrates deeply along the principal gliding plane {111} of the FCC volume, extending until it undergoes reorientation at the boundary of the underlying layer [22]. The dispersion of VMS in HRB workpieces (see Fig. 2(a_2-f_2)) is examined across various cutting depths. Initially, the central stress core predominantly concentrates beneath the layer, resulting from compression on the topmost atoms during the cutting tool performance. The massive VMS allocated in the chip pile-up volume, the tool–workpiece contact area, and its subsurface while cutting depth unevenly distributes inward interior atoms with more significance in BCC volume. In the cutting along [-100] direction, the stress core, however, focuses on the pile-up separation area at the surface of BCC volumes.

Figure 3 displays the temperature distribution within the interior of the workpiece at different cutting depths and in different directions. The findings indicate that the regions with the highest temperatures are observed in the chip and the contact zone between the workpiece and the tool. This phenomenon can be attributed to the transition from elastic to plastic behavior, which releases kinetic energy [12]. During the cut, more kinetic energy is transferred from the tool to the workpiece, contributing to an elevated local temperature. Furthermore, localized temperature increases are also caused by friction between the workpiece and the cutting tool. Notably, the temperature transmitted inside the sample through the FCC volume is much more significant.

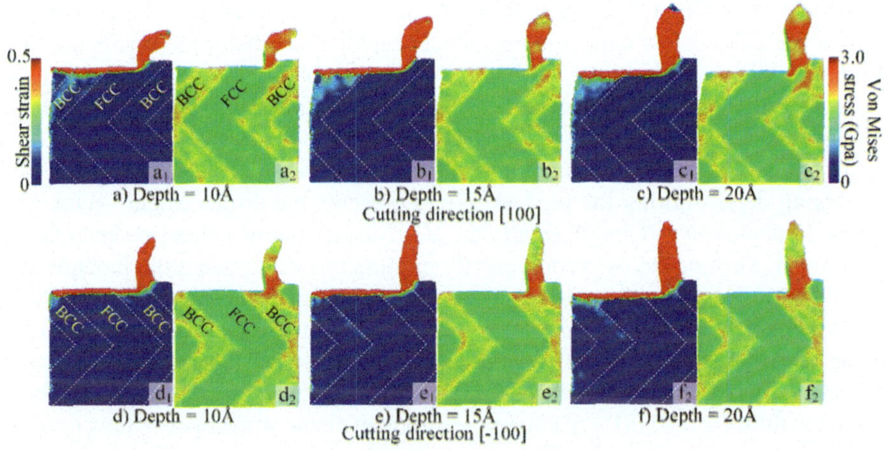

Fig. 2 Shear strain (a_1-f_1) and von Mises stress (a_2-f_2) of HRB-EHEA workpiece in case of different cutting directions with various depths

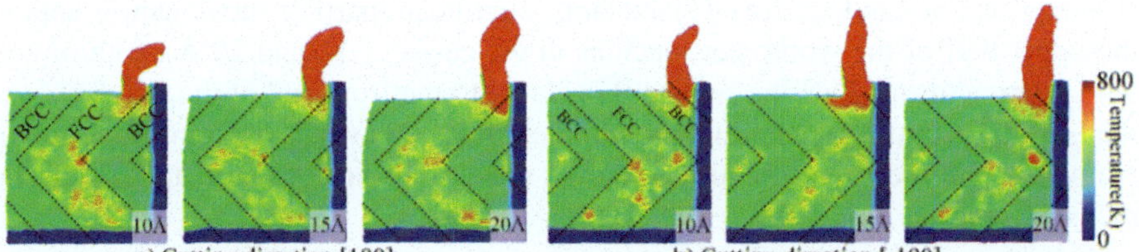

Fig. 3 Temperature distribution on HRB workpiece in [100] (**a**) and [-100] (**b**) cutting direction

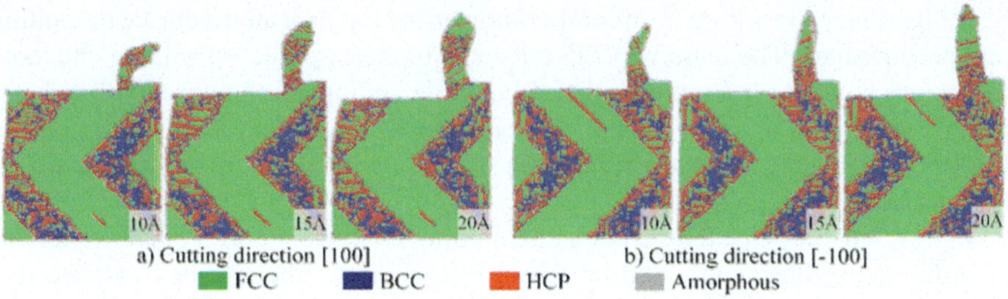

Fig. 4 Physical structure distribution of HRB workpiece in [100] (**a**) and [-100] (**b**) cutting direction

The physical structure distribution of the HRB workpiece FeCoNiAl EHEA during the cutting process is illustrated (see Fig. 4), with particles presented according to the PTM method. Amorphous atoms are prominently present near the direct contact areas and the pile-up volume, experiencing significant stress. Phase transformation occurs regionally from the original BCC to HCP and FCC, with the common plane of nucleation{111}FCC||{12-1}BCC, {-101}FCC||{101}BCC, and {101}FCC||{010}BCC. As an incidental outcome of cutting, the compression of subsurface BCC at low temperatures induces rapid lattice distortion across multiple slipping systems within the BCC structure, promoting phase transformation [22]. The generated FCC seed undergoes partial transformation into HCP under compression through dislocation activity.

To assess workpiece deformation and the manifestation of plastic flow behavior at varying depths during the process, Fig. 5 presents the displacement vector. The arrows' orientation and length signify the atom displacement's direction and relative magnitude. Surface atoms predominantly migrate outward toward the indenter due to the tool movement, illustrated by long vectors in Fig. 5(a1) and (b1). Underlying particles exhibit a consistent orientation to [-101], aligning with the central gliding plane of the crystal in the FCC volume. Intriguingly, BCC atoms partially slide along the {111} plane toward the front end of the layer, while others exhibit a turbulent flow. Figure 5(a1.1) captures the turbulent displacement vector flow forming vortices due to the crystal reaching an orientation where two or more slip systems are equally stressed, leading to high resolution. This phenomenon elucidates the stress core around the front end of the BCC layer in Fig. 2(a2-c2). In Fig. 5a, the majority of displacement is observed in the BCC volume, aligning with the shear strain distribution in Fig. 2(a1-f1). Notably, significant particle movement is kept in the upper half of the workpiece, such as in the cases of 15 and 20 Å depth of cut, with atoms shifting from the second BCC layer to the front end of the first BCC via the [-100] direction, maintaining the turbulent flow. Conversely, a minor fraction of atoms beneath the damaged surface moves upward as part of the material's plastic recovery process.

Figure 6a–c illustrates the variation in force components during the cutting process, showcasing different behaviors at three stages corresponding to the interaction of the diamond tool with the surface of various material microstructures.

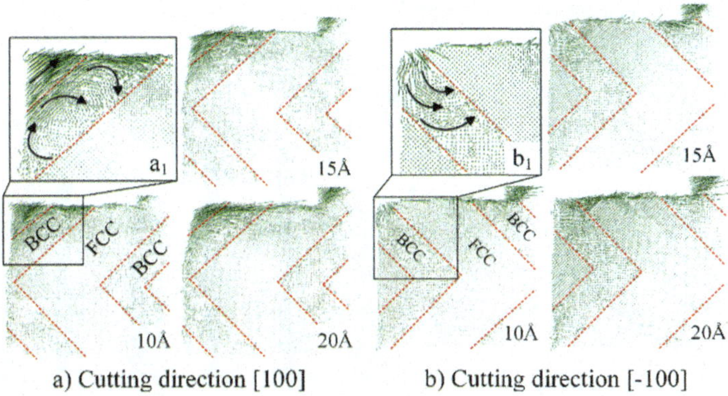

Fig. 5 Displacement vector of HRB workpiece in [100] (**a**) and [-100] (**b**) cutting direction

Initially, both force components gradually increase, reaching their first peak as the tool engages with the workpiece and the number of atoms directly contacting the indenter sharply rises. However, there is a notable difference in the friction and normal force levels, stabilizing after the engagement steps. Additionally, the relative proportion of friction to normal force varies over time for different cutting depths. In the case of a 15 Å cutting depth, the magnitudes of the force components are approximately equal during tool engagement. Conversely, in the shallowest cutting case, the normal component outweighs the friction. Meanwhile, with a cutting depth of 20 Å, the friction component rises significantly faster before stabilizing at its steady level, surpassing the normal force. These differences are primarily attributed to the geometric characteristics of cutting depth and the diameter of the diamond tool. Notably, both force elements gradually decrease while the tool performs on the BCC surface before recovering to a higher value in the next FCC volume. Interestingly, the normal force is greater than friction components, with a better gap shown in the BCC volume.

Figure 7 shows the mean values of the component force and the combined force when cutting in two opposing directions at depths of 10, 15, and 20 Å. Overall, the normal component consistently exhibits smaller values than the other components.

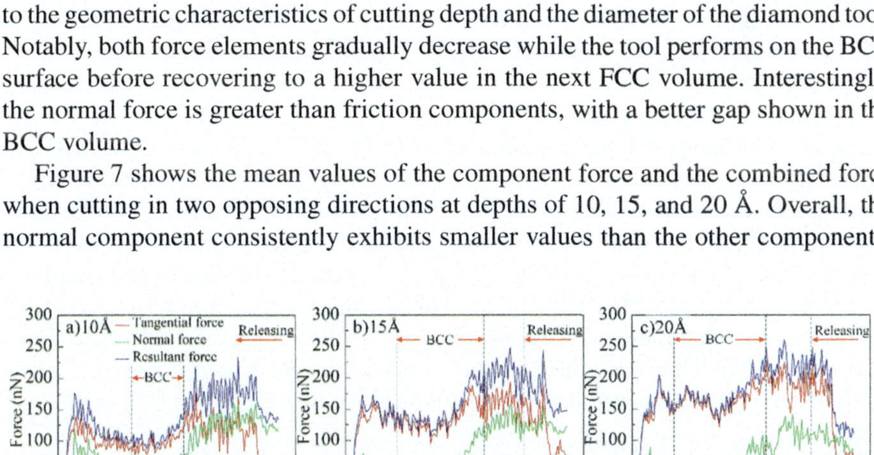

Fig. 6 Force and cutting length relationship under cutting process in [100] direction with various cutting depths

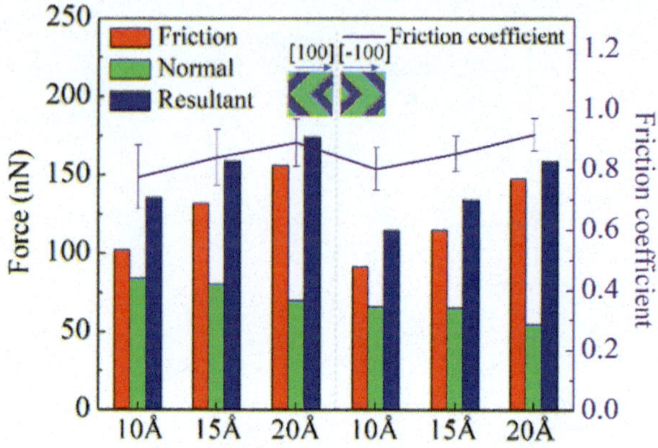

Fig. 7 Variation of average force components, resultant force, and friction coefficient under different cutting depths and cutting directions

At the same depth, the atomic force in the [100] direction surpasses that in the [-100] direction. The disparity between these two force components is more pronounced when cutting in the [100] direction. Notably, the average normal force demonstrates a decreasing trend with an increase in the depth of the cut. The friction coefficient exhibits minor variations when cutting in the [-100] direction and tends to rise with escalating depth of cut.

3.2 *Influence of Dual-Phase Patterns*

In order to compare the effects of different distribution patterns of dual-phase EHEA, this section analyzes the workpiece based on some criteria, such as shear strain, temperature, cutting force, and material removal rate (MRR) of HRB, vertical, and horizontal lamellar specimens.

Figure 8 presents the shear strain of specimens under the cutting process. The initial defection is apparent and focuses on the contact area in the case of horizontal with various depths of 10, 15, and 20 Å. The atoms with high stress are concentrated into a thin strip of constant thickness surrounding the anterior quarter contact of the indenter in the lamellar cases. Although the stress propagating into the workpiece is more significant at the BCC layer of vertical lamellar specimens, the stress distribution evolution remains the strip of concentration. On the other hand, VMS distribution in the case of lamellar workpieces in the secondary layers develops unevenly, especially in the [-110] direction when represented at the BCC layer or [110] in the FCC layers.

Figure 9 depicts the correlation between normal force, friction force, and time during the cutting process for horizontal, vertical lamellar, and HRB microstructure workpieces. Notably, the first case (Fig. 9b) exhibits a distinct trend compared to the others, attributed to the subsurface of the horizontal lamellar case being exclusively FCC. In this case, the force magnitude gradually increases as the cutting tool travels.

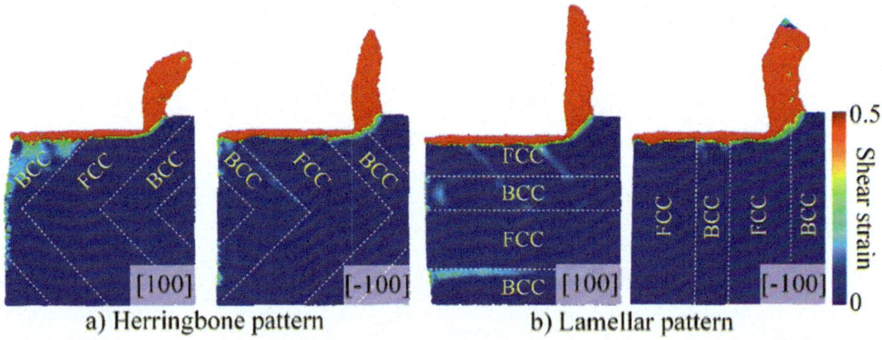

Fig. 8 Shear strain distribution of HRB (**a**) and lamellar (**b**) EHEA workpiece in case of different cutting directions

Conversely, the remaining three instances demonstrate a similar force magnitude trend with a significant reduction in the second stage when interacting with the BCC layer. However, the vertical lamellar cut reveals two nearly leveled peaks at the first and third stages. Notably, asynchronous variation occurs uniquely when cutting across the surface of the BCC layer (Fig. 9b), resulting in nearly equal values of friction force and normal force. The highest force values are approximately similar in specimens with the same cutting depth but varying durations. For the horizontal lamellar case, the force peak is maintained from early to the end of the process. The vertical lamellar case exhibits a decline and subsequent recovery at the BCC layer. Meanwhile, in the HRB case, the force reaches its maximum in the late stage of the process, remaining stable for about half of the peak before experiencing a slight decrease and then rapidly increasing at the BCC layer.

Figure 10 shows the correlation of the number of chip volume particles during the cutting process. In general, the number of worn atoms increases steadily until the end of the cutting stroke at 60 ps before decreasing slightly, as indicated by the elastic recovery phenomenon of the materials. In the case of HRB, the amount of atom generated by the atom tends to increase slightly as the cutting tool moves across the interface from FCC to BCC and an increasing pattern in the opposite direction. With the same cutting depth, the number of atoms in the chip volume does not differ much between the two HRB cutting cases. Meanwhile, the amount of atom plowing of the horizontal lamellar is more significant than in the remaining cases, which is responsible for the lack of direct participation of the BCC volumes.

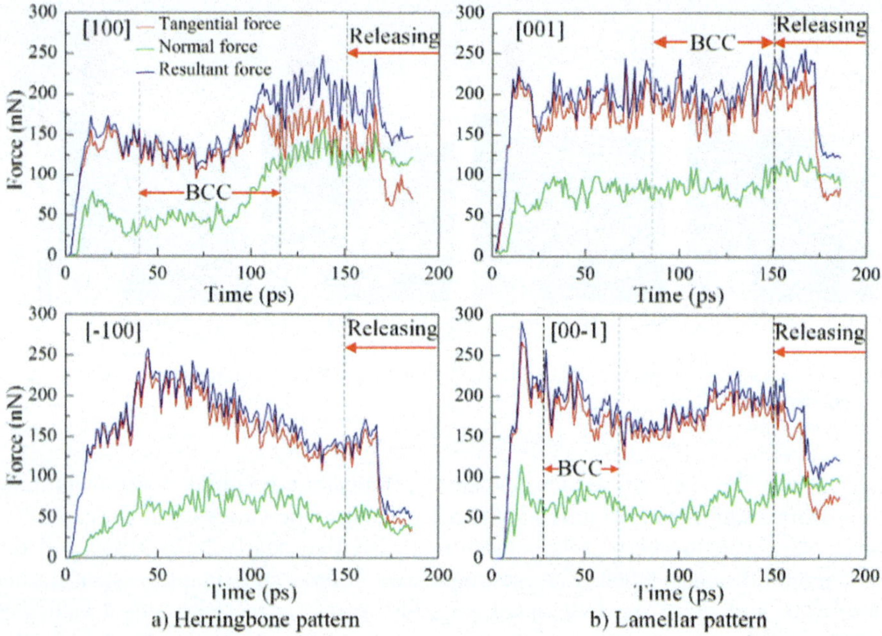

Fig. 9 Normal and resultant forces are involved in evolution of HRB (**a**) and lamellar (**b**) workpieces in different cutting directions

Fig. 10 Number of chip atoms of HRB and lamellar EHEAs under different directions and cutting depths

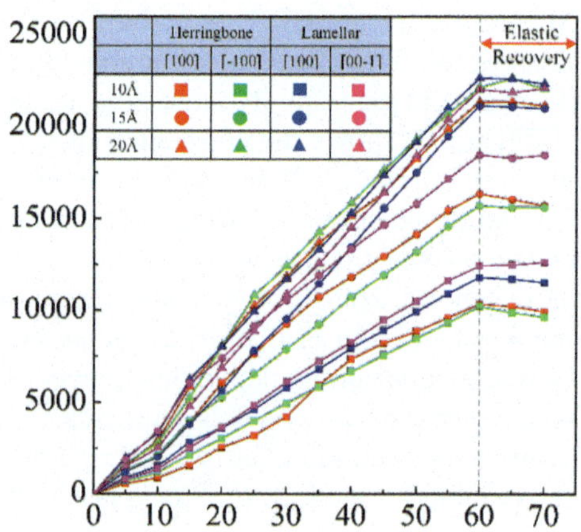

4 Conclusion

This study examines the impact of distinct microstructures, specifically the lamellar and HRB configurations of dual-phase eutectic FeCoNiAl HEA, on the material's response across various directions and depths during the cutting process. Several conclusions are drawn as follows:

(1) The initial state of stress distribution shows an unbalanced distribution with a tendency to stretch in the BCC layers, combined with a lower energy threshold at equilibrium than the FCC, creating sensitivity to mechanical load. The strong phase transformation process creates an interwoven structure between alternating FCC and HCP sheets in FCC {111}FCC and {-101}FCC.

(2) Plastic flow is either immediately reoriented with diffusion in BCC volume by a perpendicular structural interface or gradually shifted direction by a parallel interface. Notably, turbulent atoms flow around the front end of the BCC layer of HRB pattern dual-phase EHEA, creating stress concentration. Therefore, the structure deforms differently based on the BCC layer orientation and shape, resulting in diverse groove depths and dispersion of worn atoms in processed surfaces. Specifically, to cut in the [100] and [001] directions, HRB specimens deform in the shear prevention direction, while the lamellar [00-1] will react upward.

References

1. Lilensten, L., Couzinié, J., Perrière, L., Bourgon, J., Emery, N., Guillot, I.J.M.L.: New structure in refractory high-entropy alloys. Mater. Lett. **132**, 123–125 (2014)
2. Yeh, J.W., Chen, S.K., Lin, S.J., Gan, J.Y., Chin, T.S., Shun, T.T., Tsau, C.H., Chang, S.Y.J.A.: Nanostructured high-entropy alloys with multiple principal elements: novel alloy design concepts and outcomes. Adv. Eng. Mater. **6**(5), 299–303 (2004)
3. Miracle, D.B., Senkov, O.N.: A critical review of high entropy alloys and related concepts. Acta Mater. **122**, 448–511 (2017)
4. Vaidya, M., Muralikrishna, G.M., Murty, B.S.: High-entropy alloys by mechanical alloying: a review. J. Mater. Res. **34**(5), 664–686 (2019)
5. Lu, Y., Dong, Y., Jiang, H., Wang, Z., Cao, Z., Guo, S., Wang, T., Li, T., Liaw, P.K.: Promising properties and future trend of eutectic high entropy alloys. Scripta Mater. **187**, 202–209 (2020)
6. Tang, Z., Gao, M.C., Diao, H., Yang, T., Liu, J., Zuo, T., Zhang, Y., Lu, Z., Cheng, Y., Zhang, Y.J.J.: Aluminum alloying effects on lattice types, microstructures, and mechanical behavior of high-entropy alloys systems. J. Minerals Metals Mater. Soc. **65**, 1848–1858 (2013)
7. Dong, Y., Yao, Z., Huang, X., Du, F., Li, C., Chen, A., Wu, F., Cheng, Y., Zhang, Z.: Microstructure and mechanical properties of $AlCo_xCrFeNi_{3-x}$ eutectic high-entropy-alloy system. J. Alloy. Compd. **823**, 153886 (2020)
8. Jin, X., Zhou, Y., Zhang, L., Du, X., Li, B.: A novel $Fe_{20}Co_{20}Ni_{41}Al_{19}$ eutectic high entropy alloy with excellent tensile properties. Mater. Lett. **216**, 144–146 (2018)
9. Shi, P., Li, R., Li, Y., Wen, Y., Zhong, Y., Ren, W., Shen, Z., Zheng, T., Peng, J., Liang, X.: Hierarchical crack buffering triples ductility in eutectic herringbone high-entropy alloys. Science **373**(6557), 912–918 (2021)

10. Zhang, C., Zhang, F., Chen, S., Cao, W.: Computational thermodynamics aided high-entropy alloy design. J. Minerals Metals Mater. Soc. **64**, 839–845 (2012)
11. Kao, Y.-F., Chen, T.-J., Chen, S.-K., Yeh, J.-W.: Microstructure and mechanical property of as-cast,-homogenized, and-deformed $Al_xCoCrFeNi$ ($0 \leq x \leq 2$) high-entropy alloys. J. Alloy. Compd. **488**(1), 57–64 (2009)
12. Nguyen, D.-K., Fang, T.-H., Cai, Y.-R., Huang, C.-C.: Machining mechanism of polycrystalline nickel-based alloy under ultrasonic elliptical vibration-assisted cutting. Modell. Simul. Mater. Sci. Eng. **31**(8), 085010 (2023)
13. Chaudhary, V., Gwalani, B., Soni, V., Ramanujan, R., Banerjee, R.: Influence of Cr substitution and temperature on hierarchical phase decomposition in the AlCoFeNi high entropy alloy. Sci. Rep. **8**(1), 15578 (2018)
14. Li, J., Fang, Q., Liu, B., Liu, Y., Liu, Y.: Atomic-scale analysis of nanoindentation behavior of high-entropy alloy. J. Micromech. Mol. Phys. **1**(01), 1650001 (2016)
15. Farkas, D., Caro, A.J.J.: Model interatomic potentials for Fe–Ni–Cr–Co–Al high-entropy alloys. J. Mater. Res. **35**(22), 3031–3040 (2020)
16. Junge, T., Molinari, J.-F.: Plastic activity in nanoscratch molecular dynamics simulations of pure aluminium. Int. J. Plast **53**, 90–106 (2014)
17. Goel, S., Luo, X., Agrawal, A., Reuben, R.L.: Diamond machining of silicon: a review of advances in molecular dynamics simulation. Int. J. Mach. Tools Manuf. **88**, 131–164 (2015)
18. Li, J.-Y., Meng, W.-Q., Dong, K., Zhang, X.-M., Zhao, W.-H.: Numerical analysis of solid-liquid two-phase abrasive flow in microcutting polycrystalline materials based on molecular dynamics. Int. J. Precis. Eng. Manuf. **19**, 1597–1610 (2018)
19. Lee, H.S., Kim, B.-S., Cho, C.-W., Oh, M.-W., Min, B.-K., Park, S.-D., Lee, H.-W.: Herringbone structure in GeTe-based thermoelectric materials. Acta Mater. **91**, 83–90 (2015)
20. Lu, Y.-S., Bui, T.X., Fang, T.-H.: Effects of Grain Boundary and Gradient Structure on Machining Property of CoCrFeMnNi Alloys (2024)
21. Smallman, R.E.: Modern Physical Metallurgy. Elsevier (2016)
22. Dieter, G.E., Bacon, D.: Mechanical Metallurgy. McGraw-hill, New York (1976)

A Review of Inverse Methods for Claw-Type Rotor Design

Ngoc-Tuan Phan, Van-The Tran, and Minh-Thuan Hoang

Abstract Due to its efficiency and versatility, claw-type rotors have received significant attention in various engineering applications. The design process of these wheels often involves complex geometry and complex configurations, which makes optimizing performance difficult. In recent years, reverse methods have emerged as promising approaches to dealing with the complexity of design associated with claw rotors. This paper comprehensively reviews the reverse method used to design claw-type rotors. We discuss the use, benefits, and limitations of various reverse design techniques. In addition, we highlight recent advances and future research directions in this field. This study examines the current state of the art of inverse methods for rotor design for claw-type rotors to provide valuable insights for researchers and engineers involved in rotor design and optimization.

Keywords Inverse method · Sealing line · Normal rack · Claw-type rotor

1 Introduction

Vacuum pumps are widely used in various industries to create a vacuum environment through meshed rotors in housing [1–3]. In this case, the claw-type rotor profile has been used in vacuum pump products because it facilitates sealing [4]. Additionally, the rotor profile has a critical effect on pump performance. Therefore, many academic studies seek the inverse method for designing the claw-type rotor profile because this method can predict the pump performance in the design process [2, 5].

The inverse method for designing a rotor profile can be classified into the rack and sealing line generation methods. The first way is firstly provided by Rider [6] to design the screw rotor of a compressor. Stosic and Hanjalic [7, 8] used the rack in a normal plane for generating the screw compressor named "N" profile. However,

N.-T. Phan · V.-T. Tran · M.-T. Hoang (✉)
Faculty of Mechanical Engineering, Hung Yen University of Technology and Education, 39A Rd., Hung Yen, Vietnam
e-mail: Hoangthuan.hut@gmail.com

D. H. Tien et al. (eds.), *Proceedings of the 1st International Conference on Sustainability and Emerging Technologies for Smart Manufacturing*, Proceedings in Technology Transfer, https://doi.org/10.1007/978-981-97-7083-0_44

the equation of a normal rack is not explicitly presented. Therefore, Wu and Fong [9] showed the rack equation in a normal plane named the normal rack, which is combined by nine segments as the tangent continuous. Based on the defined normal rack, Wu and Fong [10] and Xavier [11] presented a mathematical model to design and optimize the screw compressor pair. Hsu et al. [12] used the measured profile data and the method of rack modification to modify the screw rotor profile. Li et al. [13] used two straight lines and one transition curve to define the rack in a normal plane for generating the conical screw rotor. Although the mathematical model for designing the claw-type rotor of the vacuum pump is wholly obtained, the flexibility of the normal rack is not high yet.

In a different approach, Fong et al. [14] presented a numerical method to generate the profile of spur gear by pre-defining the action line combined with the straight line and circle arc. To develop this idea for the screw rotor, Kameya et al. [15], Zaytsev and Ferreira [16, 17] presented the mathematical model for generating the compressor rotor profile using the sealing line and its geometric requirements. Cao et al. [18] used a B-spline curve to determine the sealing line (SL) of the screw compressor rotor. Then, the reverse rotor profile is generated, and the effectiveness of SL parameters on the tooth profile is analyzed. To develop the sealing line generation method on the vacuum pump rotor profile, Hoang et al. [2] presented a mathematical model to design the rotor profile of the fluid machine with the meshing clearance by defining the SL. The straight line and elliptical curve are used to construct it. Based on this idea, Hoang and Tran [5] developed a study on the SL method of asymmetric rotor profiles of vacuum pumps. This article also points out that the reverse rotor profile based on the SL is effective for designing the claw-type rotor of the vacuum pump.

In the inverse design method of the claw-type rotor profile for the vacuum pump, the sealing line generation method is more sufficient and flexible than the rack generation method. Moreover, the geometric performance indices of the rotor profile make it easier to predict the pump performance. Therefore, this article focuses on designing the rotor profile based on the SL.

2 Claw-Type Rotor Profile Generation Method Based on Sealing Line

2.1 *Mathematical Model for Generating Rotor Profile*

The configuration of the claw-rotor, as shown in Fig. 1, can be divided into four different regions: the outer circle designated as A_0B_0, the inner circle designated as C_0D_0, the cyclone curve A_0D_0, and the transition curve B_0C_0. Consequently, the construction of the sealing line is determined by three main areas and four specific points: A_0B_0, B_0C_0, and C_0D_0. Points A_0 and D_0 intersect the outer rotor circle and the y-axis, while points C_0 and B_0 intersect the x and outer axes. The closed line's curves A_0B_0 and C0D0 form the basis for point A_0 and the cycloid curve A_0D_0,

Fig. 1 Comparison of structure between SL and reverse profile

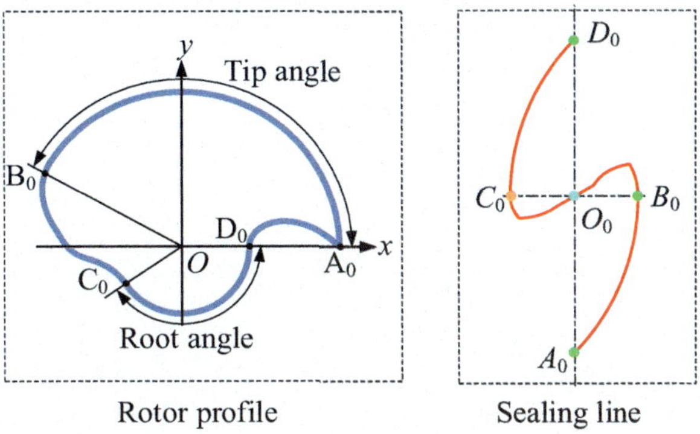

respectively. Meanwhile, the rotor transition curve B_0C_0 comes from the SL curve B_0C_0. Consequently, the SL is only modified by changing the curve B_0C_0. Future research can use the known curve to determine this curve.

The reverse profiles and their unit normal vectors can be defined by transforming from the coordinate system S_s $(O_sx_sy_s)$ to the coordinate system $S_1(O_1x_1y_1)$ and $S_2(O_2x_2y_2)$ shown in Fig. 2. The position vectors of the tooth profile for rotors 1 and 2 can be expressed as follows:

$$\mathbf{r}_1(u,\phi_1) = \begin{bmatrix} \cos\phi_1 & \sin\phi_1 & 0 \\ -\sin\phi_1 & \cos\phi_1 & 0 \\ 0 & 0 & 1 \end{bmatrix} \cdot \begin{bmatrix} 1 & 0 & r_1 \\ 0 & 1 & 0 \\ 0 & 0 & 1 \end{bmatrix} \cdot \mathbf{r}_0(u), \tag{1}$$

$$\mathbf{r}_2(u,\phi_2) = \begin{bmatrix} \cos(\phi_2+\pi) & -\sin(\phi_2+\pi) & 0 \\ \sin(\phi_2+\pi) & \cos(\phi_2+\pi) & 0 \\ 0 & 0 & 1 \end{bmatrix} \cdot \begin{bmatrix} 1 & 0 & -r_2 \\ 0 & 1 & 0 \\ 0 & 0 & 1 \end{bmatrix} \cdot \mathbf{r}_0(u), \tag{2}$$

$$\mathbf{n}_i(u,\phi_i) = \mathbf{M}_{i0}(\phi_i)\mathbf{n}_0(u). \tag{3}$$

The meshing angle $\phi_i (i = 1, 2)$ can be obtained by solving the meshing equation shown as follows:

$$f_i = \mathbf{n}_i \cdot \frac{\partial \mathbf{r}_i}{\partial \phi_i} = 0 \tag{4}$$

To substitute the meshing angle solved by numerical method into the Eqs. (1) and (2), the different shapes and teeth of 2-D reverse profiles are shown in Fig. 3a. The 3-D claw-type rotor results from helical movement for the 2-D profile generated above, including constant lead, variable lead, and conical rotors, as shown in Fig. 3b.

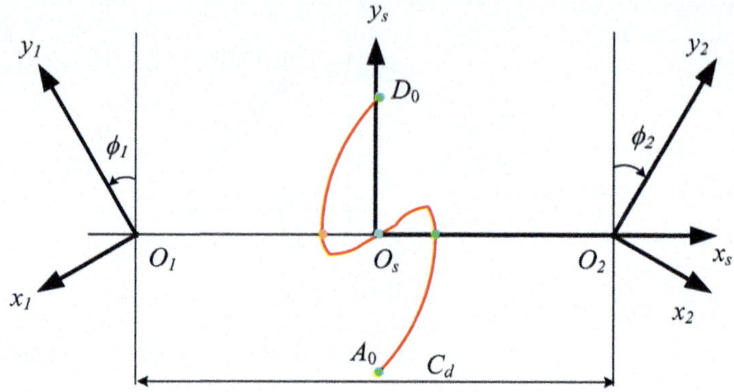

Fig. 2 Relative coordinate system between sealing line and rotor

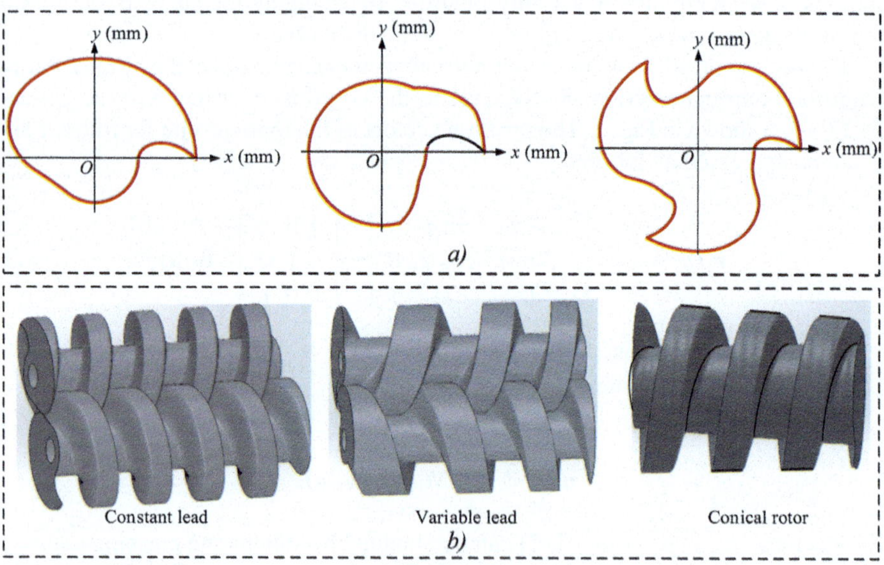

Fig. 3 Some typical forms of reverse profile based on SL: **a** 2-D and **b** 3-D

2.2 *Geometric Constrain of Sealing Line*

The SL must have two constraints for ultimately generating the claw-type rotor of the vacuum pump [2, 5]. First, the SL must be a non-interruption curve. Second, the distance between the points of the SL and the rotor's axis should not be increased, and the distance between its point and the other axis should not be decreased. This condition can be expressed as follows:

$$\begin{cases} \sqrt{\left[x_0(u_{i+1}) + r_1\right]^2 + y_0^2(u_{i+1})} - \sqrt{[x_0(u_i) + r_1]^2 + y_0^2(u_i)} \leq 0 \\ \sqrt{\left[x_0(u_{i+1}) - r_2\right]^2 + y_0^2(u_{i+1})} - \sqrt{[x_0(u_i) - r_2]^2 + y_0^2(u_i)} \geq 0 \end{cases} \quad (5)$$

3 Conclusions

The reverse method plays an important role in the design of the claw-type rotor for vacuum pumps. The paper provides an overview of current state-of-the-art research. It emphasizes mathematical formulation, computational techniques, and the application of inverse methods in the design of claw-type rotors. Despite the challenges, ongoing research has the potential to improve claw rotors' efficiency and reliability and contribute to the progress of various engineering applications.

References

1. Hoang, M.T., Wu, Y.R.: Inter-lobe clearance estimation for twin-screw rotor pairs with point-meshing features. In: ARCHIVE Proceedings of the Institution of Mechanical Engineers Part E Journal of Process Mechanical Engineering 1989–1996, vols. 235 (2020). https://doi.org/10.1177/0954408920971399
2. Hoang, M.T., Wu, Y.R., Nguyen, T.L.: A universal rotor design method for twin-rotor fluid machines with a parameterized sealing line considering inter-lobe clearances. Vacuum **189** (2021). https://doi.org/10.1016/j.vacuum.2021.110276
3. Tang, Q., Zhang, Y.: Screw optimization for performance enhancement of a twin-screw pump. Proc. Inst. Mech. Eng. Part E J. Process Mech. Eng. **228**, 73–84 (2014)
4. Hoang, M.T., Wu, Y.R.: Error compensation method for milling single-threaded screw rotors with end mill tools. Mech. Mach. Theor. **157** (2021). https://doi.org/10.1016/j.mechmachtheory.2020.104170
5. Hoang, M.T., Tran, T.V.: Methodology for generating a general profile of asymmetrical rotors of twin-screw vacuum pumps with point meshing feature by a pre-defined sealing line. Proc. Inst. Mech. Eng. B. J. Eng. Manuf. **237** (2023). https://doi.org/10.1177/09544054221136383
6. Rinder, L.: Screw Rotor Profile and Method for Generating (1987)
7. Stosic, N., Hanjalic, K.: General Method for Screw Compressor Profile Generation (1996)
8. Stosic, N., Hanjalic, K.: Development and Optimization of Screw Machines with a Simulation Model—Part I: Profile Generation (1997)
9. Wu, Y.R., Fong, Z.H.: Rotor Profile Design for the Twin-Screw Compressor Based on the Normal-Rack Generation Method (2008)
10. Wu, Y.R., Fong, Z.H.: Optimization design of an explicitly defined rack for the generation of rotors for twin-screw compressors. Mech. Mach. Theor. **44**, 66–82 (2009). https://doi.org/10.1016/J.MECHMACHTHEORY.2008.02.013
11. Xavier, P.J., Kanthavel, K., Mythili, R.U.: Rotor profile design for twin screw compressor. Int. J. Sci. Eng. Res. **2**, 1–4 (2011)
12. Hsu, R.H., Wu, Y.R., Chen, Z.A.: A numerical method for reversely generating a pair of conjugated rotor profiles of twin-screw compressor from the measured profile data. Proc. CIRP **63**, 483–487 (2017). https://doi.org/10.1016/j.procir.2017.03.137

13. Li, D., He, Z., Wang, C., Guo, Y., Wei, W., Lin, D., Xing, Z.: Design methodology and performance analysis of conical rotors for dry screw vacuum pumps. Vacuum **185**, 110025 (2021). https://doi.org/10.1016/J.VACUUM.2020.110025
14. Fong, Z.H., Chiang, T.W., Tsay, C.W.: Mathematical model for parametric tooth profile of spur gear using a line of action. Math. Comput. Model. **36**, 603–614 (2002). https://doi.org/10.1016/S0895-7177(02)00185-1
15. Kameya, H., Mori, H., Aoki, M.: The shortest seal-line profile for screw compressors. In: International Conference on Compressors and Their Systems, City University, London, UK (2001)
16. Zaytsev, D., Ferreira, C.A.I.: Profile generation method for twin screw compressor rotors based on the meshing line. Int. J. Refrigeration **28** (2005). https://doi.org/10.1016/j.ijrefrig.2004.12.006
17. Zaytsev, D., Ferreira, C.A.I.: Screw compressor rotors profile generation method based on pre-defined meshing line. In: Proceedings of the 3rd International Compressor Technique Conference, August 15–18, Xian Jiaotong University, pp. 117–127 (2001)
18. Cao, S., He, X., Zhang, R., Xiao, J., Shi, G.: Study on the reverse design of screw rotor profiles based on a B-spline curve. Adv. Mech. Eng. **11** (2019). https://doi.org/10.1177/1687814019883782

An Efficient Filter for Topology Optimization of Isotropic Elastic Materials in the Two-Dimensional Design Domain

Quy-Tho Nhu, Ngoc-Tien Tran, Thi-Thao Tran, and Van-Long Trinh

Abstract This study introduces an improved density filter that effectively manages the intricacy of the topology optimization structure. The proposed filter is implemented within a homogeneous isotropic elastic material domain. The findings demonstrate that the utilization of this novel filter effectively mitigates the challenging checkerboard phenomenon commonly encountered in topology optimization problems. The topology optimization approach employed in this study aims to minimize the material based on the minimum compliance objective, while accommodating multiple initial structural constraints. The findings explore the most effective arrangement in terms of the number of iterations and the value of the objective function, supported by various numerical illustrations.

Keywords Topology optimization · Filter · Checkerboard phenomenon · Design domain

1 Introduction

Topology optimization (TO) is an effective method employed in engineering design to attain the most advantageous arrangement of materials within a designated design area. Its objective is to optimize the form and arrangement of a structure, ensuring optimal performance while reducing weight and material consumption. Through exhaustive exploration of various configurations, TO empowers engineers to unveil inventive and resource-efficient designs that might elude conventional design approaches. TO finds extensive applications across diverse industries, including but not limited to aerospace and automotive industries, architectural design, manufacturing and engineering, renewable energy systems, material science, and engineering [1–4]. In recent times, the field of TO has garnered considerable research interest

Q.-T. Nhu · N.-T. Tran (✉) · T.-T. Tran · V.-L. Trinh
School of Mechanical and Automotive Engineering, Hanoi University of Industry, No. 298 Cau Dien Street, Bac Tu Liem District, Hanoi, Vietnam
e-mail: tientn@haui.edu.vn

D. H. Tien et al. (eds.), *Proceedings of the 1st International Conference on Sustainability and Emerging Technologies for Smart Manufacturing*, Proceedings in Technology Transfer, https://doi.org/10.1007/978-981-97-7083-0_45

from numerous scientists. In [5], a methodology was developed for TO specifically for designing second-order phononic topological insulators based on the quantum spin Hall effect. Reference [6] introduced deep learning-based frameworks that align with conventional TO algorithms, enabling high-resolution 3D TO. Ryu et al. [7] proposed a multiobjective TO approach considering diversity and uniformity of optimal solutions. Reference [8] introduced a virtual phase method using structured meshes for TO of complex geometries and nondesign domains. Lopes et al. [9] utilized the BESO algorithm to maximize natural frequency separation in elongated structures. In [10], a method was presented for designing an optimal topological layout with frictionless contact conditions, enabling observation of relative behaviors between deformable parts. These studies highlight the scientific nature of TO research.

To achieve optimal performance with minimal density variation, recent TO studies have investigated and introduced various filters. Bourdin [11] studied a modified version of the conventional minimum compliance formulation, which involved regularizing the density field through a convolution operator. The paper [12] introduced a technique for network TO that utilizes the Kalman consensus filter to minimize communication overhead. Reference [13] introduced a novel advection–diffusion filter for TO, offering computational efficiency and ease of implementation.

In this study, we present a modified density filter for TO. The effectiveness of this method is demonstrated by effectively eliminating checkerboard patterns. Additionally, we apply the proposed method to various design domain boundary conditions to showcase its efficacy.

2 Materials and Methods

The optimization problem involves optimizing the structure of a machine part, a mechanical system, and a sheet of material. The goal is to identify the arrangement of materials within the design domain that minimizes compliance, considering a specific set of loads and boundary conditions. The boundary conditions consist of fixed elements, external forces, and supports.

However, there is a constraint that limits the total volume of materials obtained through optimization to a specified value. Within the Ω design domain, two distinct regions, Ω_m and Ω_e, are present. Region Ω_m represents the area containing the material, while region Ω_e denotes the area without the material. The boundary of Ω encompasses external forces Γ_t and fixed elements Γ_u (as illustrated in Fig. 1). If we denote the design variable as $\mathbf{x}$, then the density of the material within the Ω design domain represented as follows:

$$\rho(\mathbf{x})\colon \Omega \rightarrow \{0, 1\} \text{ where } \rho(\mathbf{x}) = \begin{cases} 1 & \text{for } \mathbf{x} \in \Omega_m \\ 0 & \text{for } \mathbf{x} \in \Omega_m \equiv \Omega \backslash \Omega_m \end{cases} \tag{1}$$

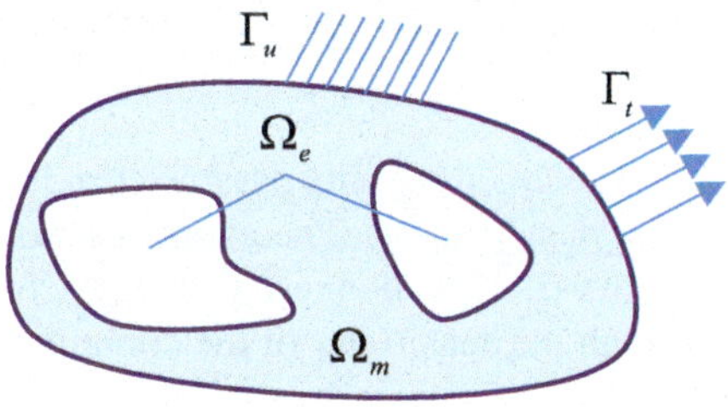

Fig. 1 General design domain [2]

Here, $\boldsymbol{\rho}(\mathbf{x})$ is a vector that serves as an indicator of whether material is present or absent within the design domain. $\boldsymbol{\rho}(\mathbf{x}) = \mathbf{1}$ represents the presence of material, while $\boldsymbol{\rho}(\mathbf{x}) = 0$ signifies the complete absence of material in Ω. However, considering the continuous nature of the material, the material distribution is allowed to vary. This means that the element density can take on multiple values, representing the density of material within the design domain with $\boldsymbol{\rho}(\mathbf{x}) \in [0, 1]$. . In this study, we utilize the SIMP (solid isotropic material with penalization) method [1] to determine the objective function for optimizing the structural configuration of a homogeneous, isotropic, linear elastic material:

$$\begin{aligned} &\min_{\boldsymbol{\rho},\boldsymbol{\psi}} c(\boldsymbol{\rho},\boldsymbol{\psi})= \mathbf{f}^{\mathrm{T}}\boldsymbol{\psi} \\ &\text{s.t.}\mathbf{K}(\boldsymbol{\rho})\psi = \mathbf{f} \\ &\int_{\Omega} \boldsymbol{\rho} d\mathrm{V}=\hat{\mathrm{V}} \end{aligned} \tag{2}$$

where $c(\rho, \psi)$ is the compliance, the global force and displacement vectors are denoted as $\mathbf{f}$ and $\boldsymbol{\psi}$, respectively. $\mathbf{K}$ is the global stiffness matrix, which is dependent on the density distribution. $\hat{\mathrm{V}}$ is the specified maximum volume of design domain. To address the optimization problem described in Eq. (2), we begin by discretizing the design domain using square node elements. The effectiveness of the proposed node elements is assessed by implementing compliance mitigation standards. Let Λ be the overall performance function, we have:

$$\Lambda = c + \boldsymbol{\lambda}^{\mathrm{T}}(\mathbf{K}\psi - \mathbf{f}) \tag{3}$$

where $\boldsymbol{\lambda}$ is the Lagrange multiplier vector derivative of Λ with respect to ρ as:

$$\begin{aligned} \frac{\partial \Lambda}{\partial \boldsymbol{\rho}} &= \frac{\partial c}{\partial \boldsymbol{\rho}} + \frac{\partial c}{\partial \boldsymbol{\psi}}\frac{\partial \boldsymbol{\psi}}{\partial \boldsymbol{\rho}} + \boldsymbol{\lambda}^{\mathrm{T}}\left(\frac{\partial \mathbf{K}}{\partial \boldsymbol{\rho}}\boldsymbol{\psi} + \mathbf{K}\frac{\partial \boldsymbol{\psi}}{\partial \boldsymbol{\rho}}\right) \\ &= \boldsymbol{\psi}^{\mathrm{T}}\frac{\partial \mathbf{K}}{\partial \boldsymbol{\rho}}\boldsymbol{\psi} + \boldsymbol{\lambda}^{\mathrm{T}}\frac{\partial \mathbf{K}}{\partial \boldsymbol{\rho}}\boldsymbol{\psi} + \frac{\partial \boldsymbol{\psi}}{\partial \boldsymbol{\rho}}(2\boldsymbol{\psi}^{\mathrm{T}}\mathbf{K} + \boldsymbol{\lambda}^{\mathrm{T}}\mathbf{K}) \end{aligned} \tag{4}$$

With $2\boldsymbol{\psi}^{\mathrm{T}}\mathbf{K} + \boldsymbol{\lambda}^{\mathrm{T}}\mathbf{K} = 0$ và $\boldsymbol{\lambda} = -2\boldsymbol{\psi}$, Eq. (4) is rewritten as follows:

$$\frac{\partial \Lambda}{\partial \boldsymbol{\rho}} = \boldsymbol{\psi}^{\mathrm{T}} \frac{\partial \mathbf{K}}{\partial \boldsymbol{\rho}} \boldsymbol{\psi} - 2\boldsymbol{\psi}^{\mathrm{T}} \frac{\partial \mathbf{K}}{\partial \boldsymbol{\rho}} \boldsymbol{\psi} = -\boldsymbol{\psi}^{\mathrm{T}} \frac{\partial \mathbf{K}}{\partial \boldsymbol{\rho}} \boldsymbol{\psi} \tag{5}$$

The element density parameter ρ_e is assigned to each displacement of the elements in the design mesh. The element's density becomes the design variable. When the design variables change, the objective function also changes. This change is expressed through the sensitivity of the compliance value:

$$\frac{\partial \Lambda}{\partial \rho_e} = \boldsymbol{\psi}^{\mathrm{T}} \frac{\partial \mathbf{K}_e}{\partial \rho_e} \boldsymbol{\psi}_e = p\rho_e^{p-1} \boldsymbol{\psi}_e^{\mathrm{T}} \mathbf{K}_e \boldsymbol{\psi}_e \tag{6}$$

where p is a penalization factor ($p = 3$), $\boldsymbol{\psi}_e$ represents the displacement vector, while $\mathbf{K}_e$ represents the stiffness matrix of element e.

To enhance the algorithm's performance, a density filter is employed, which incorporates the Gaussian equation. The filter responsible for updating the density elements can be expressed using the following equation:

$$\hat{\rho}_e = \sum_{i \in N_e} \omega(\mathbf{x}_i) v_i \rho_i \left(\sum_{i \in N_e} \omega(\mathbf{x}_i) v_i \right)^{-1} \tag{7}$$

where v_i is the volume of element i. N_e denotes the neighborhood of element e, centered at element e, with a radius of R:

$$N_e \triangleq \{i | \|\mathbf{x}_i - \mathbf{x}_e\| \le R\}, \tag{8}$$

$\omega(\mathbf{x}_i)$ is a weight function given by:

$$\omega(\mathbf{x}_i) = e^{-0.5\left(\frac{2\|\mathbf{x}_i - \mathbf{x}_e\|}{R}\right)^2} \tag{9}$$

Furthermore, we incorporate the sensitivity filter modifications [1] to complement the process:

$$\frac{\widehat{\partial c}}{\partial x_e} = \frac{1}{\max(\varepsilon, x_e) \sum_{i \in N_e} \max(0, r_{\min} - \mathrm{dist}(e, i))} \sum_{i \in N_e} \max(0, r_{\min} - \mathrm{dist}(e, i)) x_i \frac{\partial c}{\partial x_i} \tag{10}$$

where $\varepsilon = 10^{-3}$ is a constant to avoid dividing by zero, $r_{\min}$ is the filter radius, $dist(e, i)$ is the center-to-center distance from element e to element i.

3 Results and Discussion

Based on the proposed equations, we obtained the following results by evaluating the changes made to the improved filter and boundary conditions with multiple forces. In all examples, we assume the material is isotropic and elastic, with Poisson's ratio of 0.3 and a penalty factor $p = 3$ (based on reference [1], a suitable value for p is 3). All applied loads have a unit value of 1 N. The density field is represented using a color scale in the figures, where white indicates void and black represents full material. In this section, we conducted a survey study on the design domain, considering the load and boundary conditions shown in Fig. 2.

We performed link optimization for the design domain in two cases: without using a filter (Fig. 3) and with the implementation of an improved filter (Fig. 4). Figure 5 displays the outcome of the objective function.

From Fig. 3, it is evident that the absence of a filter leads to the presence of a checkerboard phenomenon [1]. This indicates that the optimal structure lacks close connections between mesh elements. The algorithm quickly converges (at the 37th iteration) but yields a high compliance value (c = 30.7456). When using the proposed filter, the results show that the ideal texture eliminates the occurrence of checker-board. The outcome reaches convergence at the 200th iteration with a conformance value 29.86% lower compared to the scenario where no filter is employed (as shown in Fig. 4).

Fig. 2 The initial design domain corresponds to a cantilever beam configuration

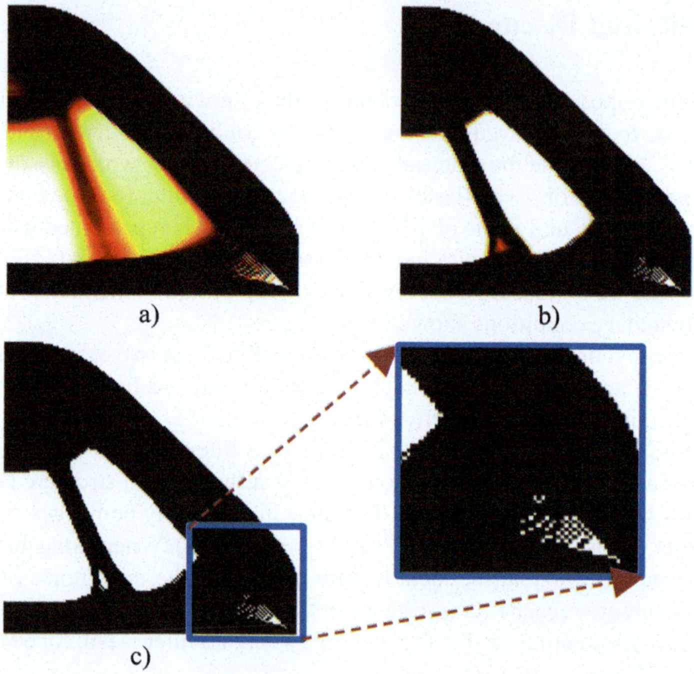

Fig. 3 Optimal design domain without using filters: **a** c = 34.5046; **b** 30.8393; **c** c = 30.7456

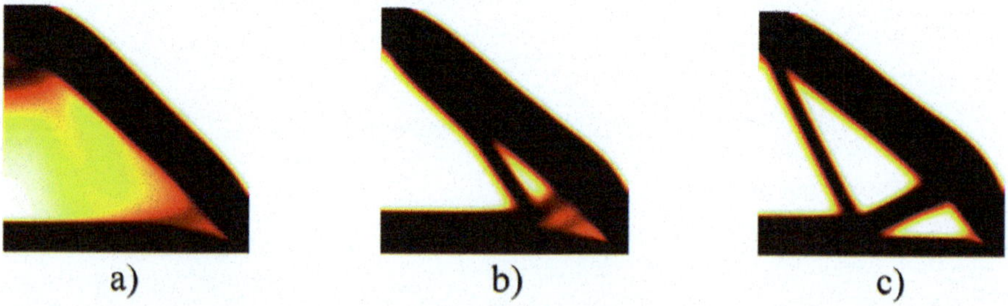

Fig. 4 Optimal design domain using the proposed filter: **a** c = 24.1429; **b** 21.8756; **c** c = 21.5667

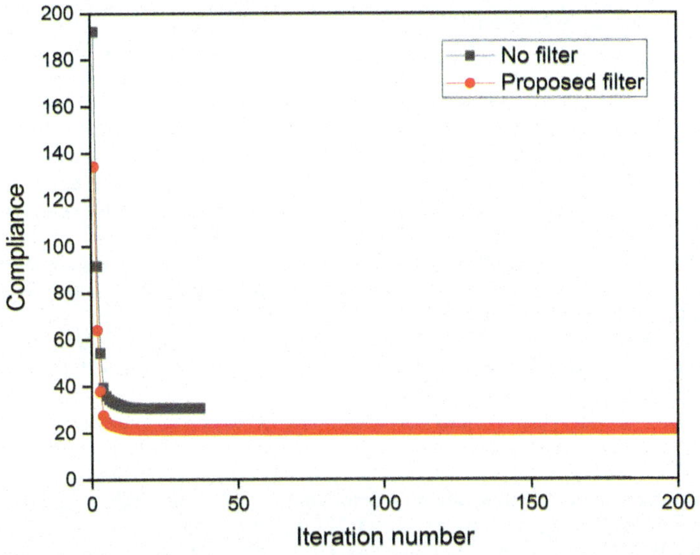

Fig. 5 Compliance results in the optimization process

4 Conclusion

Based on the findings, the article draws the following conclusions:

The absence of filters in the optimization process leads to the appearance of the checkerboard phenomenon, resulting in the loss of material element connections within the structure.

The application of the proposed filter successfully eliminates the checkerboard phenomenon. Convergence is achieved at the 200th iteration, leading to a significantly improved optimal texture with approximately 29.86% lower compliance compared to the scenario without using the filter.

These results hold great significance in establishing optimal structures that exhibit high compliance efficiency, showcasing potential for widespread application in future industrial products.

References

1. Sigmund, O., Maute, K.: Topology optimization approaches: a comparative review. Struct. Multidiscip. Optim. **48**(6), 1031–1055 (2013)
2. Tran, N.T.: Topology optimization of isotropic elastic materials in the two-dimensional design domain with changes in characteristic boundary conditions. EUREKA Phys. Eng. **4**, 133–142 (2023)
3. Tran, N.T.: Topology optimization for isotropic elastic materials using honeycomb tessell. Eastern-Eur. J. Enterprise Technol. **2**(7)(122), 43–49 (2023)

4. Tran, N.T., Tran, T.T, Nguyen, H.T., Phung, X.S., Bui, T.L, Tran, A.S., Luu, V.H., Trinh, V.L., Dinh, G.N.: Investigation on the influence of the sensitivity filter radius on the topology optimization of the 3-dimensional isotropic elastic material domain. In: International Conference on Science, Technology, and Innovation for Sustainable Development (2023)
5. Chen, Y., Chen, Y., Li, J.T.H., Zhu, J.: Topology optimization of quantum spin hall effect-based second-order phononic topological insulator. Mech. Syst. Signal Process. **164**, 108243 (2022)
6. Rade, J., Balu, A., Herron, E., Pathak, J., Ranade, R., Sarkar, S., Krishnamurthy, A.: Algorithmically-consistent deep learning frameworks for structural topology optimization. Eng. Appl. Artif. Intell. **106**, 104483 (2021)
7. Ryu, N., Seo, M., Min, S.: Multi-objective topology optimization incorporating an adaptive weighed-sum method and a configuration-based clustering scheme. Comput. Methods Appl. Mech. Eng. **385**, 114015 (2021)
8. Li., H., Li, H., Gao, L., Li, J., Li, P., Yang, Y.: Topology optimization of arbitrary-shape multi-phase structure with structured meshes based on a virtual phase method. Comput. Methods Appl. Mech. Eng. **387**, 114138 (2021)
9. Lopes, H.N., Cunha, D.C., Pavanello, R., Mahfoud, J.: Numerical and experimental investigation on topology optimization of an elongated dynamic system. Mech. Syst. Signal Process. **165**, 108356 (2022)
10. Jeong, G.E.: Topology optimization of deformable bodies with linear dynamic impact and frictionless contact condition. Appl. Sci. **11**(22), 10518 (2021)
11. Bourdin, B.: Filters in topology optimization. Int. J. Numer. Meth. Eng. **50**, 2143–2158 (2001)
12. Lv, L., Chen, H., Xie, L., Wang, K.: A topology optimization method for reducing communication overhead in the Kalman consensus filter. Appl. Sci. **11**(15), 7107 (2021)
13. Høghøj, L.C., Träff, E.A.: An advection–diffusion based filter for machinable designs in topology optimization. Comput. Methods Appl. Mech. Eng. **391**, 114488 (2022)

Development of Finite Element Tool for the Modeling of Apparent Mechanical Properties of Composites with Random Inclusion Distribution

Tien-Thinh Le, Huan Thanh Duong, Lu Minh Le, Hieu Chi Phan, Quan Tran Quoc, and Van Hai Nguyen

Abstract The general goal of this research is to develop a numerical finite element model in order to calibrate the apparent elasticity tensor of composites with random distribution of inclusions. The statistical fluctuations related to the microstructure are captured through geometrical parameters. The probability density function of the apparent elasticity tensor depends on two factors, the mean value and the fluctuation parameter, both of which are determined by statistical estimates and Monte Carlo realizations. The technique presented in this study can be used to determine the general characteristics of a random elastic microstructure.

Keywords Apparent elasticity tensor · Finite element analysis · Random distribution · Homogenization

T.-T. Le (✉) · V. H. Nguyen
Faculty of Mechanical Engineering and Mechatronics, PHENIKAA University, Ha Dong, 12116 Yen Nghia, Hanoi, Vietnam

A & A Green Phoenix Group JSC, PHENIKAA Research and Technology Institute (PRATI), No.167 Hoang Ngan, Trung Hoa, Cau Giay, 11313 Hanoi, Vietnam
e-mail: thinh.letien@phenikaa-uni.edu.vn

H. T. Duong · L. M. Le
Faculty of Engineering, Vietnam National University of Agriculture, Trau Quy, Gia Lam, 100000 Hanoi, Vietnam

H. C. Phan
Department of Civil Engineering, Faculty of Engineering and Applied Science, Memorial University of Newfoundland, A1B 3X5 St. John's, NL, Canada

Q. T. Quoc
Faculty of Civil Engineering—VNU Hanoi, University of Engineering and Technology, 144—Xuan Thuy, Cau Giay, Hanoi, Vietnam

D. H. Tien et al. (eds.), *Proceedings of the 1st International Conference on Sustainability and Emerging Technologies for Smart Manufacturing*, Proceedings in Technology Transfer, https://doi.org/10.1007/978-981-97-7083-0_46

1 Introduction

Heterogeneous materials are one made up of several phases for which the local mechanical properties vary from one point to another. In general, there are many types of heterogeneous materials. As part of this study, we are only interested in biphasic heterogeneous materials, classified according to their constituents, with a matrix-inclusion type morphology. It is a material composed of a matrix reinforced by inclusions. Particulate polymer composite is a typical example of this type.

Due to various fluctuations, it is not that easy to obtain global properties of the heterogeneous material from the microscopic (local) properties of the phases and the microstructure [1]. For a given representative volume element, several important characteristic sizes should be taken into account. They are the characteristic size of the structural heterogeneities, i.e., d, the characteristic size of the set to be considered, i.e., l, and the characteristic dimension of the structure, i.e., L [2]. Using traditional methods of the continuum mechanics, we generally make the hypothesis of separation of scales to predict the macroscopic properties of the composite materials [3]. However, the characteristic size of the set to be considered, l, must be very large compared to the characteristic size of the structural heterogeneities taken into account, d, and at the same time must be very small compared to the size of the studied structure, L [4]. Concretely, l must be well proportioned with respect to both d and L, in order to respect a double condition [5].

As part of this study, we propose developing a numerical tool for the prediction of the apparent elasticity tensor of the material on the mesoscopic scale, from the local properties of two phases of the material, the matrix and the inclusion and the microstructure of the domain, in linear elasticity. The present work considers a domain of size smaller than the size l of the representative volume element defined above. This is the case of non-separation of scales, for which there are statistical fluctuations that need to be taken into account.

The significance and novelty of the present study are below:

- Calibrate the apparent elasticity tensor: This tensor captures the overall elastic behavior of the composite, but it can vary statistically due to the randomness of the inclusions. This study offers a way to quantify this variation.
- Capture statistical fluctuations through geometrical parameters: By linking the microstructure of the composite (i.e., the arrangement of inclusions) to geometrical parameters, the research allows for a more comprehensive understanding of how randomness affects elasticity.
- Determine the probability density function of the tensor: This function describes the likelihood of encountering different elasticity values within the composite. The study proposes a method to calculate this function using statistical estimates and Monte Carlo simulations.

2 Materials and Methods

2.1 Description of Microstructure

We consider a microstructure domain denoted by Ω of two phases, a matrix and a family of particle reinforcements. The matrix of length $l \times l$ and the inclusions are of diameter d_i. The inclusions are randomly distributed in the matrix. The surface fraction of the inclusions is denoted by f. In terms of mechanical properties, linear elasticity is considered for all phases in the domain. Young's moduli and Poisson's ratios of the matrix and the inclusions are denoted by E_m, E_i and υ_m, υ_i, respectively.

The distribution of inclusion position is random in the domain; therefore, there are two conditions for this distribution, which are represented below:

- No intersection between the edges of the matrix and the inclusions (contact accepted),
- No intersection and contact between the inclusions, so that there is no stress concentration.

2.2 Homogenization Approach

The macroscopic strain tensor is defined such as (see [3]):

$$\mathbf{E} = \langle \varepsilon \rangle = \frac{1}{|\Omega|} \int_{\Omega} \varepsilon(\mathbf{x}) dV \tag{1}$$

According to the mechanics of continuous mediums, under the condition of small deformations, we can deduce:

$$\varepsilon(\mathbf{x}) = \frac{1}{2}(\nabla \mathbf{u}(\mathbf{x}) + (\nabla \mathbf{u}(\mathbf{x}))^T) \tag{2}$$

$$\mathbf{E} = \frac{1}{2|\Omega|} \int_{\Omega} (\nabla \mathbf{u}(\mathbf{x}) + (\nabla \mathbf{u}(\mathbf{x}))^T) dV \tag{3}$$

By applying Gauss's divergence theorem, we have:

$$E_{ij} = \frac{1}{2|\Omega|} \int_{\partial\Omega} (u_i n_j + u_j n_i) dS \tag{4}$$

By definition, the macroscopic stress tensor, Σ is the volume average of the microscopic stress fields $\sigma(\mathbf{x})$ (see [3]):

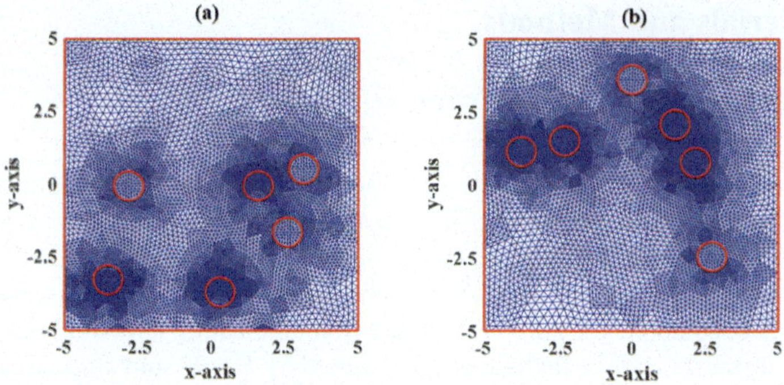

Fig. 1 Illustration of two random realizations (among 10,000) of microstructure for the same inclusion fraction (the length unit is μm)

$$\Sigma = \langle \sigma \rangle = \frac{1}{|\Omega|} \int_{\Omega} \sigma(\mathbf{x}) dV \tag{5}$$

The apparent elasticity tensor is defined by the equation:

$$\mathbb{C}^{app} = \langle \mathbb{C} : \mathbb{A} \rangle \tag{6}$$

where $\mathbb{C}(\mathbf{x})$ is the local elasticity tensor at the point $\mathbf{x}$ and $\mathbb{A}(\mathbf{x})$ is the strain localization tensor at the point $\mathbf{x}$.

2.3 *Finite Element Model*

Figure 1 presents the visualization of the finite element mesh of two random realizations of microstructure for the same inclusion fraction. Mesh refinement is applied in the vicinity between the inclusions and the matrix. A mesh convergence study has been done to determine the appropriate size of the mesh for the considered problem. All the finite element calculations in this study were coded and performed using MATLAB.

2.4 *Mechanical Boundary Conditions*

It is necessary to impose boundary conditions of the domain Ω in order to solve the behavioral equations for the medium considered. In this study, kinematic uniform boundary conditions (KUBCs) are applied. The displacement field $\mathbf{u}$ is imposed at point $\mathbf{x}$ at contour $\partial\Omega$ as follows:

$$\mathbf{u} = \mathbf{E}.\mathbf{x} \qquad \forall \mathbf{x} \in \partial\Omega \tag{7}$$

Here, **E** is the macroscopic deformation tensor.

2.5 Probabilistic Model for Apparent Elasticity Tensor

In this study, Monte Carlo random realizations were applied to propagate the influence of random distribution of inclusions on the apparent elasticity tensor. A set N of realizations of $\mathbb{C}^{app}$ tensor is obtained, and its matrix form is denoted by **C**.

For each realization of $\mathbf{x}$ in Ω, we have $\mathbf{C}(\mathbf{x})$ is a random variable with value in $\mathbb{M}_n^+(\mathbb{R})$, set of positive-definite symmetric real matrices. We take the function $\mathbf{x} \to \underline{\mathbf{C}}$ which is the average of $\mathbf{C}(\mathbf{x})$ and it can be determined by equation below:

$$\underline{\mathbf{C}} = \frac{1}{N}\sum_{i=1}^{N} \mathbf{C}(\theta_i) \tag{8}$$

The function that characterizes the random distribution of the matrix **C** can be written such as

$$P_{\mathbf{C}(\mathbf{x})} = p_{\mathbf{C}(\mathbf{x})}(\mathbf{C})\,\hat{d}\,\mathbf{C} \tag{9}$$

where $p_{\mathbf{C}(\mathbf{x})}(\mathbf{C})$ is the probabilistic density function,

and $\hat{d}\,\mathbf{C} = 2^{n(n-1)/4}\prod_{1\le i\le j\le n} d\mathbf{C}_{ij}$, with $d\mathbf{C}_{ij}$ is the Lebesgue measure in $\mathbb{R}$. The probabilistic model is then written such as [6, 7]:

$$p_{\mathbf{C}(\mathbf{x})}(\mathbf{C}) = \mathbf{1}_{\mathbb{M}_n^+(\mathbb{R})}(\mathbf{C}) \times c_{\mathbf{C}} \times (\det(\mathbf{C}))^{(n+1)\left(1-\delta^2\right)/\left(2\delta^2\right)} \times \exp\left\{-\frac{n+1}{2\delta^2} tr(\mathbf{C})\right\} \tag{10}$$

where $\mathbf{C} \to \mathbf{1}_{\mathbb{M}_n^+(\mathbb{R})}(\mathbf{C}) = 1$ if $\mathbf{C} \in \mathbb{M}_n^+(\mathbb{R})$, and 0 otherwise. $c_{\mathbf{C}}$ is a positive constant which is calculated by:

$$c_{\mathbf{C}} = \frac{(2\pi)^{-n(n-1)/4}\left(\frac{n+1}{2\delta^2}\right)^{n(n+1)/(2\delta^2)}}{\left\{\prod_{j=1}^{n}\Gamma\left(\frac{n+1}{2\delta^2} + \frac{1-j}{2}\right)\right\}} \tag{11}$$

δ is a parameter defining such as

$$\delta = \left\{\frac{1}{\|\underline{\mathbf{C}}\|_F^2} \times \frac{1}{N}\sum_{i=1}^{N}\left(\|\mathbf{C}(\theta_i) - \underline{\mathbf{C}}\|_F^2\right)\right\}^{1/2} \tag{12}$$

The probabilistic density function of $\mathbf{C}$ depends on two parameters such as

- The mean value $\underline{\mathbf{C}}$,
- The fluctuation parameter δ.

Besides, the convergence of the statistical estimator for the number N of realizations of $\mathbf{C}$ can be controlled by the following function (see [7]):

$$conv\,(N) = \left\{ \frac{1}{\left\|\underline{\mathbf{C}}\right\|_F^2} \times \frac{1}{N} \sum_{k=1}^{N} \left(\left\| \mathbf{C}\,(\theta_k) - \underline{\mathbf{C}} \right\|_F^2 \right) \right\}^{1/2} \tag{13}$$

3 Results and Discussion

3.1 Numerical Analysis of Strain and Stress Fields

In this section, numerical application is introduced using the following parameters: $l = 10$, $d_i/l = 1/10$ µm, $f = 0.05$, $E_m = 2.7$ GPa, $E_i = 73$ GPa, $\upsilon_m = 0.37$, $\upsilon_i = 0.22$. The mechanical properties of the matrix and the reinforcement were determined using Molecular Dynamics simulations as introduced in [8]. Figure 2 presents displacement fields (u_{11} and u_{22}) of the microstructure (a) in Fig. 1a in a tensile simulation along x-axis. It is seen that kinematic uniform boundary conditions are well applied at the boundary of the domain in this case of tensile simulation along x-axis. Figure 3 introduces stress and strain fields ($\sigma_{11}, \sigma_{22}, \sigma_{12}, \epsilon_{11}, \epsilon_{22}$, and ϵ_{12}) of the microstructure (a) in Fig. 1a in a tensile simulation along x-axis. The inclusions exhibit Young's modulus much larger than the matrix, so there is almost no stress or strain observed inside the inclusions. However, at the vicinity between the matrix and the inclusions, the stress and strain fields are highly observed as a response to the x-axis loading in this case. Similar results were observed for the tensile test along y-axis and the shear test (but not shown here).

3.2 Identification of Controlled Parameters of Probabilistic Model for Apparent Elasticity Tensor

The statistical convergence of the random samples defined in Eq. (13) is plotted in Fig. 4 over 10000 Monte Carlo random sampling runs. It can be concluded that 10000 Monte Carlo random sampling runs sufficient to obtain a representative convergence estimation, ready for subsequent statistical analysis.

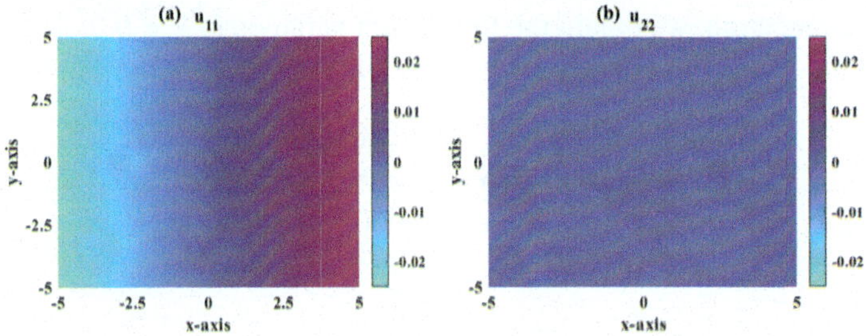

Fig. 2 Displacement fields (u_{11} and u_{22}) of the microstructure (**a**) in Fig. 1a in a tensile simulation along x-axis

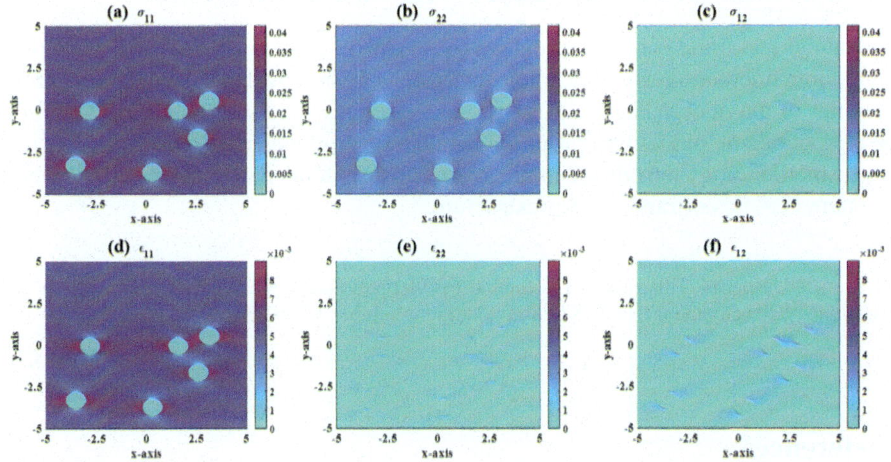

Fig. 3 Stress and strain fields (σ_{11}, σ_{22}, σ_{12}, ϵ_{11}, ϵ_{22} and ϵ_{12}) of the microstructure (**a**) in Fig. 1a in a tensile simulation along x-axis (the length unit is μm)

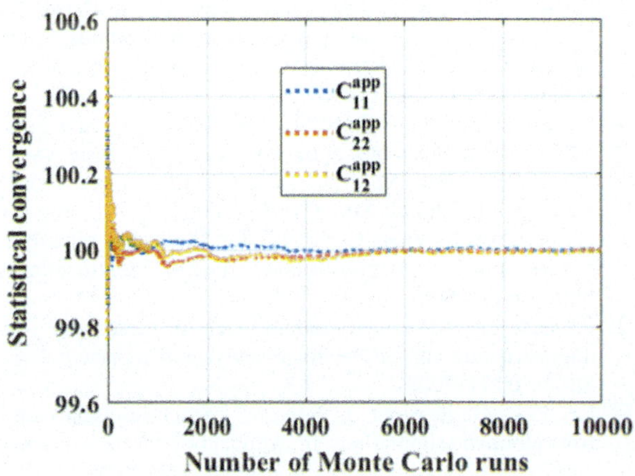

Fig. 4 Statistical convergence for different components of apparent elasticity tensor over 10000 random realizations.

In this application, we obtain the fluctuation parameters $\delta = 0.01$ [(see Eq. (12)] and the components of the $\underline{\mathbf{C}}$ are such as (in GPa):

$$\underline{\mathbf{C}} = \begin{bmatrix} 4.7658 & 2.7819 & -0.0001 \\ 2.7819 & 4.7647 & -0.0003 \\ -0.0001 & -0.0003 & 0.9868 \end{bmatrix} \tag{14}$$

4 Conclusion

The main contributions of the present work are summarized as below:

- a numerical finite element model was developed for homogenization procedure;
- random fluctuations from the microstructure were propagated to the apparent properties of the materials;
- a probabilistic model was proposed for the apparent elasticity tensor;
- the methodology introduced in this work can be extended to determine the general characteristics of random microstructure.

Acknowledgements This research is funded by Vietnam National Foundation for Science and Technology Development (NAFOSTED) under grant number 107.02-2021.62.

References

1. Guilleminot, J., Soize, C.: A stochastic model for elasticity tensors with uncertain material symmetries. Int. J. Solids Struct. **1**(9) (2010)
2. Stefanou, G., Savvas, D., Papadrakakis, M.: Stochastic finite element analysis of composite structures based on material microstructure. Compos. Struct. **132** (2015)
3. Nemat-Nasser, S.: Micromechanics: Overall Properties of Heterogeneous Materials. North-Holland, Netherlands (1998)
4. Karimi, D., Milani, A.S.: SRVE modeling of particulate polymer matrix composites with irregularly shaped inclusions: application to a green stone composite. Compos. Struct. **228** (2019)
5. Ostoja-Starzewski, M.: Material spatial randomnes: From statistical to representative volume element. Int. J. Solids Struct. **1**(17) (2006)
6. Guilleminot, J., Soize, C., Kondo, D.: Mesoscale probabilistic models for the elasticity tensor of fiber reinforced composites: experimental identification and numerical aspects. Int. J. Solids Struct. **1**(1) (2009)
7. Soize, C.: Tensor-valued random fields for meso-scale stochastic model of anisotropic elastic microstructure and probabilistic analysis of representative volume element size. Int. J. Solids Struct. **1**(17) (2008)
8. Le, T.-T., Guilleminot, J., Soize, C.: Stochastic continuum modeling of random interphases from atomistic simulations. In: Application to a polymer nanocomposite. Computer Methods in Applied Mechanics and Engineering, p. 303 (2016)

Optimizing Technological Parameters to Improve the Accuracy of 3D LCD-Printed Products

Huu Nghi Huynh, Bao Khang Quach, and Trong Hieu Bui

Abstract Using resin 3D printing technology to manufacture technical products is a new trend that brings many benefits, especially when using open source or desktop devices due to cost advantages. In order to do this, it is necessary to determine the optimal technological parameter sets according to the intended use, equipment, and materials. In this study, the authors focused on understanding the influence of five technological parameters in the 3D resin printing process as well as the influence of their interaction on dimensional accuracy in each dimension, and all three dimensions L, W, H optimize technological parameters to improve the dimensional accuracy of test samples according to ASTM D5418/ASTM D7028 standards fabricated on desktop LCD devices using standard resin material. The experiment was designed according to the FCCCD method with 32 test samples. Analysis and optimization results on Minitab software and validation results show that the dimensional accuracy of the sample manufactured according to the optimal set of parameters found is close to the predicted value on the software. This demonstrates the process of selecting input technological parameters. The initial values of those parameters are appropriate, and the desktop LCD resin 3D printing device can be used to manufacture technical products for the production process.

Keywords 3D resin · LCD · Dimensional accuracy

H. N. Huynh · B. K. Quach · T. H. Bui (✉)
Ho Chi Minh City University of Technology (HCMUT), 268 Ly Thuong Kiet Street, District 10, Ho Chi Minh City, Vietnam
e-mail: hieubt@hcmut.edu.vn

Vietnam National University Ho Chi Minh City, Linh Trung Ward, Thu Duc City, Ho Chi Minh City, Vietnam

D. H. Tien et al. (eds.), *Proceedings of the 1st International Conference on Sustainability and Emerging Technologies for Smart Manufacturing*, Proceedings in Technology Transfer, https://doi.org/10.1007/978-981-97-7083-0_47

1 Introduction

Resin 3D printing technology, including three main technologies: SLA (stereolithography), DLP (digital light processing), and LCD (liquid crystal display), has been used in many fields including art, dentistry, medicine, rapid prototyping, and other important fields. This technology works on the principle of using UV or laser energy to solidify liquid plastic to create each layer of material. 3D resin printing technology is capable of manufacturing small-sized, complex products with high precision, detail, and surface quality [1, 2]. Currently, scientists are focusing on research to improve many important quality characteristics of resin 3D printing products to manufacture technical products, including size accuracy. Typically, Giudice et al. [3] studied and evaluated the accuracy of dental models manufactured using open-source LCD printers and industrial machines. The results showed that dental samples produced by open-source LCD printers have lower accuracy. However, their accuracy is still satisfactory for a dentally printed sample. Research by Sherman et al. [4] has shown that products in the dental field manufactured using DLP technology with layer thicknesses of 50 and 100 μm have acceptable accuracy and the authors concluded that for 3D Resin printing technology, the best layer thickness is 50 μm. A. Ibrahim and colleagues [5] conducted research on the effects of layer thickness and exposure time on mechanical properties and dimensional accuracy. The authors concluded that products printed at a layer thickness of 50 μm have better mechanical properties than products printed with other layer thicknesses, and when printed at a layer thickness of 50 μm, an exposure time of 9 s will give the highest dimensional accuracy. Zhang et al. [6] studied the influence of layer thickness on the dimensional accuracy of 3D resin-printed products on upper and lower dental models printed with SLA and DLP 3D printing devices. The authors concluded that a layer thickness of 50 μm could be the optimal layer thickness for DLP printing technology. Regarding experimental methods, scientists often use the RSM (Response Surface Method) method to design experiments. For example, Sood et al. [7] studied the influence of technological parameters on the mechanical properties of FDM products, using the RSM method combined with the objective function to determine the optimal set of parameters. The experiment was designed according to the FCCCD method with five input parameters, each of which was divided into three levels. Bansal et al. [8] used the Box–Behnken method with three input parameters, each consisting of three levels, to improve the dimensional accuracy of FDM products. Besides. The authors also use the genetic algorithm GA (Genetic Algorithm) to simultaneously optimize three goals: length size error, width size error, and height size error. Regarding test samples, scientists often choose test samples according to ASTM D5418/ASTM D7028 standards in the shape of a rectangular box with dimensions of 35 × 12.5 × 3.5 mm to improve dimensional accuracy.

A survey of documents shows that experts have used many optimization methods, experimental planning, and testing of many technological parameters to optimize the dimensional accuracy of 3D printing products in general. Resin in particular. There, the RSM method is often used because it has many advantages compared to other methods. In addition, many authors also focus on optimizing technology parameters on open-source LCD devices to improve product quality and replace desktop and industrial devices (DLP, SLA) to reduce production costs.

2 Methodologies and Data

The author's goal is to evaluate the influence of five input technological parameters on dimensional accuracy in L, W, and H dimensions. At the same time, the group also determines the optimal parameter sets for dimensional accuracy in each dimension, and all three dimensions of L, W, H of the test sample serve a specific purpose. The implementation process is carried out according to the flowchart presented in Fig. 1. The experimental sample is a rectangular block selected according to ASTM D5418/ ASTM D7028 standards with dimensions LxWxH in three directions of 35 × 12.5 × 3.5 mm. The model was made on an open-source Voxelab Polaris 3D printer using eSUN's Standard Resin plastic material. The parameters used in the optimization process include two types: fixed parameters and variable parameters, as presented in Table 1. These parameters are selected based on the survey of published works and the practical machine operating experience of the research team.

The experiment was designed according to the FCCCD method with five input parameters, including layer thickness (A), exposure time (B), print bed lifting speed (C), bed lifting distance (D), and printing angle (E). Each parameter is divided into three levels (Table 1). The number of experiments is determined according to the formula (1), where: k is the number of parameters, n is the number of iterations of the mental formula. In this study, k has a value of 5, n has a value of 6, plus the group uses the reduced combination experiment case. From there, the number of experiments is determined as follows the formula (2):

$$N = 2k - p + 2k + n \quad (1)$$

$$N = 25 - 1 + 2.5 + 6 = 32 \quad (2)$$

The experimental plan is presented in Table 2.

Test samples after being fabricated, processed, and stored at room temperature for 24 h (Fig. 2) will be measured in all dimensions and relative errors calculated. For each dimension, to increase reliability, the team measured five times at five different locations and then took the average value as the representative value for that size. The relative errors of the L, W, and H dimensions of 32 samples were calculated according to formulas 3, 4, and 5. The results are presented in Table 2.

$$\%\Delta L_i = \frac{|L_i - L_C|}{L_C} \times 100 \quad (3)$$

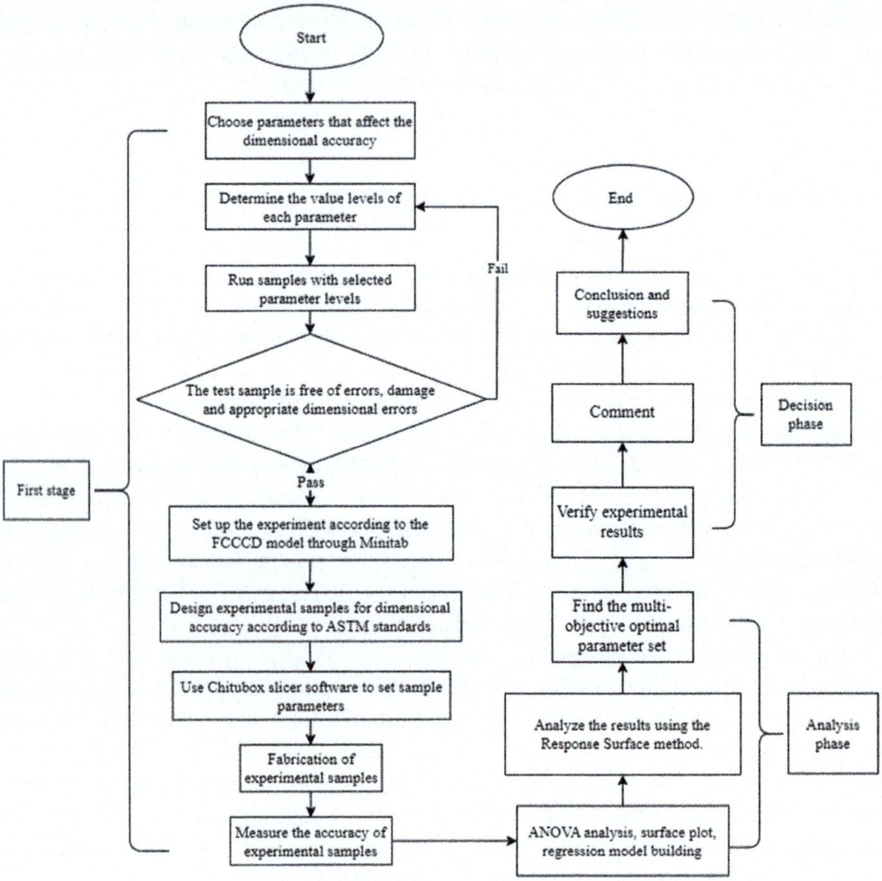

Fig. 1 Experimental procedure flow chart

$$\%\,\Delta W_i = \frac{|W_i - W_c|}{W_C} \times 100 \tag{4}$$

$$\%\,\Delta H_i = \frac{|H_i - H_C|}{H_C} \times 100 \tag{5}$$

In which:

L_i, W_i, H_i: are the measured values on the experimental sample, where i is the serial number of the sample.

L_C, W_C, H_C: is the design dimension of the sample.

$\%\Delta L_i$, $\%\Delta W_i$, $\%\Delta H_i$: is the percentage change in the actual dimension compared to the design dimension of the sample (Fig. 3).

After being collected, the data will be processed on Minitab software to evaluate the influence of each input parameter on the target, determining the optimal parameter

Table 1 Experimental parameters

Parameters	Symbol	Unit	Value		
			Low level (−1)	Normal level (0)	High level (1)
Layer thickness	A	(mm)	0.04	0.05	0.06
Exposure time	B	(s)	10	12	14
Print bed lifting speed	C	(mm/p)	55	60	65
Print bed lift distance	D	(mm)	5	10	15
Raster angle	E	(^{0})	30	45	60
Number of bottom layers	–	–	5		
Build orientation	–	–	Y direction		
Bottom layer exposure time	–	(s)	30		
Light off delay	–	(s)	1		
Bottom light off delay	–	(s)	1		
Lift height	–	(mm)	5		
Lift speed	–	(mm/p)	60		
Retract speed	–	(mm/p)	60		

set for dimensional accuracy in each dimension and the multiobjective optimization parameter set in all three dimensions. Then, re-fabricate the test sample with the optimal set of parameters found. Then the team collects data and compares the achieved dimensional error with the error predicted by the software and the average error value of 32 experimental samples.

3 Results and Discussion

3.1 Optimization Results

For dimension error in length (L). According to the Pareto chart, the layer thickness parameter (A) has the most impact on the target. The parameters lift speed (C), normal layer exposure time (B), and the interaction between parameter A and parameter E also have an influence on the target. The optimal set of parameters found includes layer thickness (A) equal to 0.04 mm, exposure time (B) equal to 10 s, lift speed (C) 55 mm/min, lift height (D) equal to 10.2525 mm, and build angle (E) is 30°. The regression equation representing the relationship between dimensional error in

Table 2 Experimental design according to FCCCD and relative error of experimental samples

Experimental samples	Parameters					$\%\Delta L$	$\%\Delta W$	$\%\Delta H$
	A	B	C	D	E			
1	−1	−1	−1	−1	1	0.2286	0.4140	3.6571
2	1	−1	−1	−1	−1	0.8460	0.9486	3.6120
3	−1	1	−1	−1	−1	0.2016	0.3650	4.0571
4	1	1	−1	−1	1	0.7886	0.8110	5.4286
5	−1	−1	1	−1	−1	0.3528	0.4875	5.0286
6	1	−1	1	−1	1	0.9371	1.1948	3.7714
7	−1	1	1	−1	1	0.4672	0.6413	7.8286
8	1	1	1	−1	−1	1.3486	1.4765	5.4857
9	−1	−1	−1	1	−1	0.0571	0.0880	1.5429
10	1	−1	−1	1	1	0.8229	0.8476	2.0571
11	−1	1	−1	1	1	0.3943	0.5625	4.2857
12	1	1	−1	1	−1	1.4970	1.7564	3.2205
13	−1	−1	1	1	1	0.4457	0.6120	2.0571
14	1	−1	1	1	−1	1.1429	1.3854	1.6571
15	−1	1	1	1	−1	0.5829	0.6425	3.7143
16	1	1	1	1	1	1.1886	1.3125	4.0205
17	−1	0	0	0	0	0.4343	0.6120	4.0571
18	1	0	0	0	0	0.9134	1.0162	3.6571
19	0	−1	0	0	0	0.7771	0.7648	2.6286
20	0	1	0	0	0	0.7314	0.7286	4.2286
21	0	0	−1	0	0	0.5714	0.5260	3.6117
22	0	0	1	0	0	0.7371	0.7380	4.5143
23	0	0	0	−1	0	0.8343	0.8964	5.8286
24	0	0	0	1	0	0.6171	0.6520	2.8571
25	0	0	0	0	−1	0.7486	0.7575	3.0286
26	0	0	0	0	1	0.7314	0.7128	3.8857
27	0	0	0	0	0	0.6286	0.6560	2.9143
28	0	0	0	0	0	0.6143	0.6520	3.3714
29	0	0	0	0	0	0.5886	0.6432	2.8571
30	0	0	0	0	0	0.6405	0.6680	3.2571
31	0	0	0	0	0	0.6814	0.6720	3.8857
32	0	0	0	0	0	0.6286	0.6560	3.6217
Average error value						0.6931	0.7780	3.7384

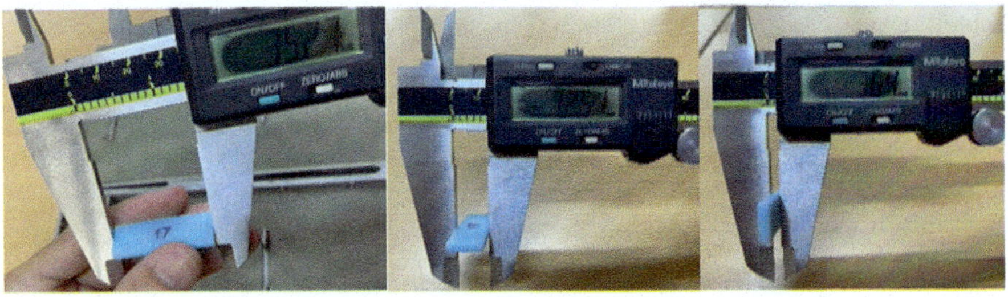

Fig. 2 Measure the dimensions of the sample

Fig. 3 Experimental sample after being fabricated

length (L) and input parameters is shown in Formula 6:

$$\%\Delta L = 0.671 + 0.3511A + 0.0883B + 0.0997C - 0.0899A * E \quad (6)$$

For dimension error in width (W). Also, according to the Pareto chart, the layer thickness parameter (A) has the most impact on the error in width (W). Other parameters such as lift speed (C), normal layer exposure time (B), and the interaction between A * E parameters. B * D · B * E also has a certain degree of influence. The optimal set of parameters found includes layer thickness (A) equal to 0.04 mm, exposure time (B) equal to 10 s, lift speed (C) 55 mm/min, lift height (D) equal to 11.2625 mm, and build angle (E) is 30°. The regression equation representing the relationship between dimensional error in width (W) and input parameters is shown in formula 7:

$$\begin{aligned}\%\Delta W = {} & 0.6864 + 0.3513A + 0.0863B + 0.1206C \\ & -0.1280A * E + 0.0682B * D - 0.0670B * E\end{aligned} \quad (7)$$

For dimension error in height (H) For dimension error in height (H), based on the Pareto chart, the Print table lift distance parameter (D) has the most impact. Factors include lift speed (C), normal layer exposure time (B), build angle (E), and the interaction between B*E, C*D, and A*C parameters also has a certain degree of

Table 3 Optimal parameter set for dimensions in three dimensions L, W, H

Parameters	Layer thickness (A)	Exposure time (B)	Lift speed (C)	Lift height (D)	Build angle (E)
Optimal values	0.04 mm	10 s	55 mm/p	12.879 mm	30^0

influence. The optimal set of parameters found includes layer thickness (A) equal to 0.06 mm, exposure time (B) equal to 10 s, lift speed (C) 63.99 mm/min, lift height (D) equal to 15 mm, build angle (E) equal to 60°. The regression equation representing the relationship between dimensional error in height (H) and input parameters is shown in formula 8:

$$\begin{aligned}\%\Delta H &= 3.576 + 0.902B + 0.368C - 1.073D \\ &+ 0.314E - 0.279A * C + 0.336B * E - 0.314C * D\end{aligned} \quad (8)$$

Multiobjective optimization of all three dimensions: Using the response optimizer tool to analyze the objective function on Minitab software, the optimal set of parameters for dimensional errors in three dimensions L, W, H is determined as follows (Table 3):

3.2 *Verifying the Results*

After finding the optimal set of parameters for dimensional accuracy in all three dimensions L, W, H, the team refabricated three test samples with the above values, measured and analyzed them (Table 4) and compare them with the average value of 32 experimental samples.

Table 4 X, Y, Z dimensions of the samples have been optimized

	Samples	Attempt 1	Attempt 2	Attempt 3	Attempt 4	Attempt 5	Average	Error (%)
Length (L)	1	34.95	35.01	35.06	35.07	35.04	35.024	0.0686
	2	35	35.04	35.03	35.04	35.03		
	3	35.06	35.03	35.01	34.97	35.02		
Width (W)	1	12.53	12.51	12.52	12.49	12.52	12.511	0.0907
	2	12.53	12.51	12.48	12.52	12.52		
	3	12.48	12.51	12.5	12.52	12.53		
Height (H)	1	3.55	3.56	3.57	3.57	3.56	3.563	1.8095
	2	3.55	3.57	3.56	3.57	3.55		
	3	3.56	3.57	3.56	3.58	3.57		

Accordingly, we see that the error values of the three samples after being optimized do not have a large deviation compared to the predicted value. Specifically, the value in length L is 0.0686% compared to 0.0529%, in width W is 0.0907% compared to 0.0715%, and in height H is 1.8095% compared to 1.5114%. This proves that the choice of parameters and initial values is relatively reasonable and that the selected experimental model is appropriate. On the other hand, the experimental validation results also show that the response surface method and the FCCCD model are able to predict the quality characteristic of the LCD printing process quite accurately.

4 Conclusions

Research results showed that the RSM method can be used to evaluate and optimize technological parameters to improve the dimensional accuracy of 3D-printed products using LCD technology. In this study, the authors have identified quadratic regression models to predict dimensional accuracy in each direction and also in the three dimensions L, W, H. At the same time, the team also identified a set of optimal parameters: layer thickness (A) is equal to 0.04 mm, normal layer exposure time (B) is 10 s, lift speed (C) is 55 mm/min, lift height (D) is equal to 12.879 mm, and build angle (E) is equal to 300. The validation results showed that the deviation between the average value of 32 experimental samples and the authentic sample is quite low. Specifically, the error in length L is 0.0157%, in width W is 0.0192%, and in height H is 0.2981%.

Acknowledgements We acknowledge Ho Chi Minh City University of Technology (HCMUT), VNU-HCM for supporting this study.

References

1. Lee, D., Miyoshi, T., Takaya, Y., Ha, T.: 3D micro fabrication of photosensitive resin reinforced with ceramic nanoparticles using LCD microstreolithography. J. Laser Micro/Nano Eng. **1**(2) (2006)
2. Mananes, R.P., Rojo-Manaute, J., Gil, P.: 3D Surgical printing and pre contoured plates for acetabular fractures. J. Elsevier (2016)
3. Lo Giudice, A., Ronsivalle, V., Rustico, L., Aboulazm, K., Isola, G., Palazzo, G.: Evaluation of the Accuracy of Orthodontic Models Prototyped with Entry-Level LCD-based 3D Printers: A Study Using Surface-Based Superimposition and Deviation Analysis (2022)
4. Sherman, S.L., Kadioglu, O., Currier, G.F., Kierl, J.P., Li, J.: Accuracy of digital light processing printing of 3-dimensional dental models. Am. J. Orthod. Dentofacial. Orthop. (2020)
5. Ibrahim, A., Sa'Ude, N., Ibrahim, M.: Optimization of Process Parameter for Digital Light Processing (DLP) 3d Printing (2017)
6. Zhang, Z., Li, P., Chu, F., Shen, G.: Influence of the Three-Dimensional Printing Technique and Printing Layer Thickness on Model Accuracy (2019)

7. Sood, A.K., et al.: Parametric Appraisal of Mechanical Property of Fused Deposition Modeling Processed Parts. Department of Manufacturing Engineering. National Institute of Foundry and Forge Technology, Ranchi 834 003, India (2010)
8. Bansal, R.: Improving Dimensional Accuracy of Fused Deposition Modelling (FDM) Parts using Response Surface Methodology. A thesis submitted in partial fulfillment of the requirements for the degree in Bachelor of Technology in Mechanical Engineering (2011)

A Novel Hierarchical Sliding Mode Controller for Articulated Heavy Vehicles

Huu Viet Nguyen, Tuan Kiet Cao, Ngoc Duy Le, Van Chung Nguyen, Huu Hai Ho, and Tung Lam Nguyen

Abstract Path following and lateral stability remain a major challenge for AHVs with little progress in improving traditional tractor steering. In this paper, a hierarchical sliding mode controller (HSMC) for tractor steering is proposed to achieve path tracking and stability control simultaneously for a tractor-semitrailer. Based on a single-point preview driver reference, an aggregated sliding surface is constructed from three sliding surfaces including yaw rate, slip angle, and articulated angle error, and the control signal is designed to satisfy all sliding surfaces' asymptotic stability. The controller is tested under double lane change conditions at high speed in TruckSim and shows considerable improvement in stability versus the default TruckSim controller.

Keywords Tractor-semitrailers · Preview driver · Hierarchical sliding mode · Lane changing

1 Introduction

Recently, the transportation industry has seen a rise in demand for articulated heavy vehicles (AHVs), offering cost reductions and efficiency improvements in hauling heavy loads. Typically, an AHV comprises a leading unit and a towed unit, connected by mechanical couplings like a dolly or fifth wheel [1]. Due to this configuration, drivers primarily focus on the tractor unit's behavior, often overlooking the towed unit's motion [2]. This setup can lead to sudden occurrences such as trailer swing, jackknifing, and rollover [3].

To replicate the motions and dynamics of AHVs in a simulation environment, researchers formulate many vehicle models and driver steering behavior models [4].

H. V. Nguyen · T. K. Cao · N. D. Le · V. C. Nguyen · H. H. Ho · T. L. Nguyen (✉)
Hanoi University of Science and Technology, Hanoi, Vietnam
e-mail: lam.nguyentung@hust.edu.vn

N. D. Le
Hanoi University of Industry, Hanoi, Vietnam

D. H. Tien et al. (eds.), *Proceedings of the 1st International Conference on Sustainability and Emerging Technologies for Smart Manufacturing*, Proceedings in Technology Transfer, https://doi.org/10.1007/978-981-97-7083-0_48

Ding and He utilized three driver models derived from the dynamic behaviors of the tractor, trailer, and combined tractor/trailer [5]. Szczepaniak examined the multiloop configuration of the drive-articulated model [6]. Similarly, Zhu S and He Y suggested a unified model for multitrailer vehicles, employing closed-loop dynamics [7].

Autonomous control of four-wheeled vehicles has been extensively studied in the last decade with many controllers, such as pure pursuit, Stanley, and PID [8]. For AHVs, those same strategies for tractor steering are being implemented without considering the lateral stability of the towed unit [9]. A tracking longitudinal velocity controller was suggested for multitrailers vehicles [10]. Directional control methods were studied for AHVs [11, 12]. Deng used SMC based on the preview model for the tractor and improved trailer motion with a fuzzy logic compensator [13]. This lack of consideration leads to a potential loss of stability in high-speed maneuvers.

A classic sliding mode controller (SMC) with one sliding surface ensures the convergence of state variables (usually just the tractor states). This paper introduces an extended SMC that employs multiple sliding surfaces following hierarchical principles. Each surface addresses specific subsystems within the articulated heavy vehicle (AHV), encompassing tractor states, and articulated angles as a representation of towed unit behaviors. Controller performance is evaluated using TruckSim software for a trailer-semitrailer system. The study includes (1) developing a model of the tractor-semitrailer, (2) designing the HSMC-based tractor controller, (3) simulating results of the new controllers compared to standard TruckSim's controller in a dual-lane change test, and (4) concluding remarks and future directions.

1.1 Single-Track Articulated Vehicle Model

The AHV dynamic is often studied using a single-track (bicycle) dynamics model in literature [14]. This model simplifies each axle into one equivalent wheel and focuses solely on lateral motion. Assumptions include constant forward velocity, negligible air resistance, small side slip angles, and linear tire force region, while roll and pitch motions are ignored.

Figure 1 shows the configuration of AHV. The global coordinate Oxy is fixed on the ground; the coordinates $O_1x_1y_1$ and $O_2x_2y_2$ move along the CoG of the truck and semitrailer. The motions in the vehicle are described by truck lateral velocity v_{y1}, tractor yaw rate ω_1, and articulated angle θ.

Using the second Newton's law and equivalent momentum, the equations of AHV are

$$\begin{aligned} m_1 a_{y1} &= F_{y1}\cos\delta_f + F_{y2} - F_{hy} \\ m_2 a_{y2} &= F_{y3} + F_{hy}\cos\theta \\ I_{z1}\dot{\omega}_1 &= F_{y1}a_1\cos\delta_f + F_{hy}c - F_{y2}b_1 \\ I_{z2}\dot{\omega}_2 &= F_{hy}a_2\cos\theta - b_2F_{y3} \end{aligned} \tag{1}$$

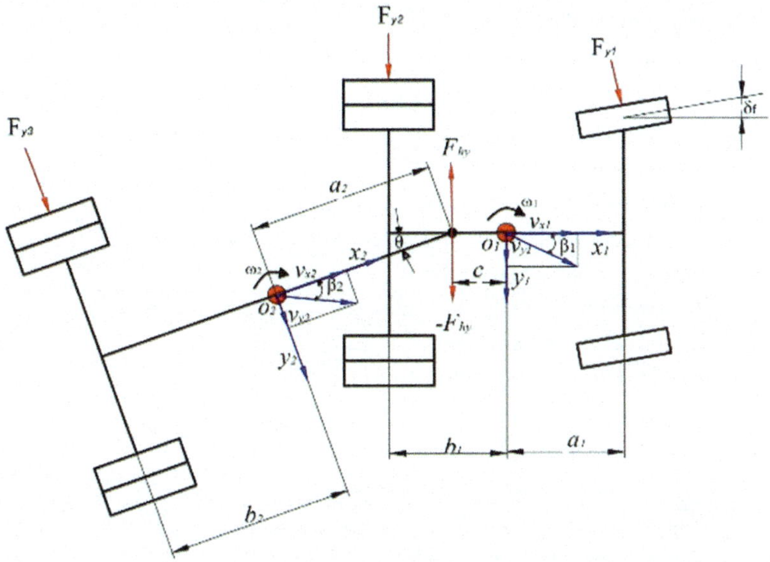

Fig. 1 Dynamic model of 3 axles tractor-semitrailer

where m_1, I_{z1}, m_2, I_{z2} are the mass and yaw inertia of the tractor and its following, respectively. Further, the lateral forces between tires and road surface are proportional to slip angles as in Eq. (2):

$$\begin{aligned} F_{yi} &= C_i \alpha_i, i = 1, 2, 3 \\ \alpha_1 &= \beta_1 + \frac{a_1 \omega_1}{v_{x1}} - \delta_f \\ \alpha_2 &= \beta_1 - \frac{b_1 \omega_1}{v_{x1}} \\ \alpha_3 &= \beta_2 - \frac{b_2 \omega_2}{v_{x2}} \end{aligned} \tag{2}$$

with C_i, α_i is the cornering stiffness and side slip angle of three axles of truck-semitrailers. Besides, assuming that (Table 1):

$$v_{x1} = v_{x2} \tag{3}$$

The physical constraint between the yaw rate of the tractor, trailer, and articulated angle is:

$$\omega_2 - \omega_1 = \dot{\theta} \tag{4}$$

Table 1 Nomenclatures in the system

Symbols	Meaning	Value
a_1	Length between the tractor's center of gravity (CoG) and its front axle	m
b_1	Length between the tractor's CoG and its rear axle	m
a_2	Length between the semi-trailer's CoG and the hitch point	m
b_2	Length between the semi-trailer's CoG and its rear axle	m
c	Length between the tractor's CoG and the hitch point	m
v_{x1}	Forward velocity of the tractor	m/s
v_{x2}	Forward velocity of the semi-trailer	m/s
v_{y1}	Lateral velocity of the tractor	m/s
v_{y2}	Lateral velocity of the semi-trailer	m/s
a_{y1}	Lateral acceleration of the tractor	m/s^2
a_{y2}	Lateral acceleration of the semi-trailer	m/s
F_{hy}	Lateral reaction force at the hitch point	N
β_1	Side-slip angle of the tractor	rad
β_2	Side-slip angle of the semi-trailer	rad
θ	Articulated angle	rad
δ_f	Steering angle	rad
ω_1	Yaw rate of the tractor	rad/s
ω_2	Yaw rate of the semi-trailer	rad/s
ψ	Heading angle of tractor	rad

The slip angle of the trailer is denoted as:

$$\beta_2 = \beta_1 - \theta - \frac{(c + a_2)}{v}\omega_1 - \frac{a_2}{v}\dot{\theta} \tag{5}$$

The state variables are $\mathbf{x} = \left[\omega_1, \beta_1, \dot{\theta}, \theta\right]$. From these above equations, the state-space matrix is as follows:

$$\dot{x} = Mx + N\delta_f \tag{6}$$

where $M = -A^{-1} \cdot B, N = A^{-1} \cdot P$. And A, B, P are the matrix of coefficients, respectively:

$$A = \begin{bmatrix} I_{z1} & m_1cv & 0 & 0 \\ -m_2(a_2 + c) & (m_1 + m_2)v & -m_2a_2 & \frac{C_3(a_2+b_2)}{v} \\ I_{z2} & m_1a_2v & I_{z2} & \frac{C_3a_2b_2-C_3b_2^2}{v} \\ 0 & 0 & 0 & 1 \end{bmatrix} \tag{7}$$

$$B = \begin{bmatrix} -\frac{a_1^2 C_1 + b_1^2 C_2 + C_1 a_1 c - C_2 b_1 c}{v} + m_1 c v & -C_1(a_1 + c) + C_2(b_1 - c) & 0 & 0 \\ m_1 v + m_2 v + \frac{-C_1 a_1 + C_2 b_1 + C_3(a_2 + b_2 + c)}{v} & -(C_1 + C_2 + C_3) & 0 & C_3 \\ m_1 a_2 v - \frac{C_1 a_1 a_2 - C_2 b_1 a_2 + C_3 b_2 (a_2 + b_2 + c)}{v} & -(C_1 a_2 + C_2 a_2 - C_3 b_2) & 0 & -C_3 b_2 \\ 0 & 0 & -1 & 0 \end{bmatrix} \tag{8}$$

$$P = \begin{bmatrix} -C_1(a_1 + c) & -C_1 & -C_1 a_2 & 0 \end{bmatrix}^T \tag{9}$$

2 Proposed Controller Based on Reference Yaw Rate and HSMC

Our proposed controller aims to minimize both the lateral deviation and hitch angle to achieve stable path-tracking. The desired yaw rate is calculated from the reference trajectory and vehicle states. From these values, a steering wheel angle value is computed based on the hierarchical sliding mode (HSMC). The system schematic is in Fig. 2.

2.1 *Single-Point Preview Yaw Rate Reference*

The single-point driver preview model is a simple resemblance of human driver steering behavior based on the looking ahead at the desired path in a fixed preview distance. Adapted from Deng's single-point preview reference model [13], in Fig. 3,

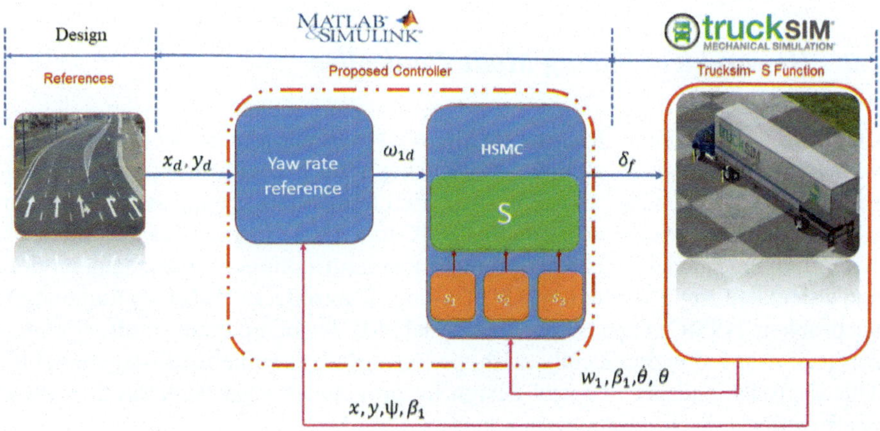

Fig. 2 Proposed controller

Fig. 3 Reference trajectory of tractor

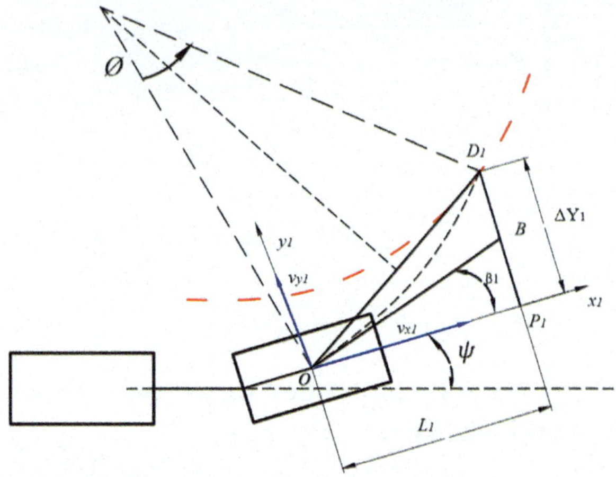

preview lateral displacement error ΔY_1 is denoted as:

$$\Delta Y_1 = \tan\left(\frac{\psi}{2} + \beta_1\right)L_1 \quad (10)$$

Then to follow a circular trajectory from O to D_1, the tractor's yaw rate ω_{1d} is

$$\omega_{1d} = \frac{2\left[\arctan\left(\frac{\Delta Y_1}{vt}\right) - \beta_1\right]}{t} \quad (11)$$

where v is the tractor's forward velocity, t is traveling time from O to P_1, ΔY_1 is calculated from (10).

2.2 *Hierarchical Sliding Mode Controller*

Conventional first-order SMC is a highly robust variable structural control method, which constructs a single sliding surface using a linear combination of all the state variables. It is a classical control method used to maintain driving stability in four-wheel vehicles. However, conventional SMC treats state variables as a whole while AHVs dynamic models illustrate different subsystem motions such as tractor motions and articulated angle oscillation [15]. This article introduces HSMC as a solution to the problem. HSMC expands upon traditional SMC by incorporating multiple sliding surfaces arranged hierarchically, with the top surface containing all state variables. This approach enhances control design by providing a clear structure and direct control effects on each subsystem's motion.

Recall state-space model of AHV in (6):

$$\dot{x} = Mx + N\delta_f$$

where M, N are matrix of known parameters. To control four variables in **x**: $\omega_1, \beta_1, \dot{\theta}, \theta$ three first layer sliding surfaces s_1, s_2, s_3 are constructed. To control all sliding surfaces, a second-order aggregated HSMC layer combining the three first-order sliding surfaces is designed as follows:

$$s = as_1 + bs_2 + cs_3 \tag{12}$$

where a, b, c are the weight for each lower sliding surface. a, b, c are positive constants.

To control all variables' errors with fast convergence, the sliding surfaces are selected as [13].

$$s_i = e_i + c_i \int e_i \tag{13}$$

Here $e_1 = \omega_1 - \omega_{1r}, e_2 = \beta_1 - \beta_{1r}, e_3 = \dot{\theta} - \dot{\theta}_{\text{ref}}$; $\omega_{1r}, \beta_{1r}, \dot{\theta}_{\text{ref}}$ are desired values.

The control law is constructed as follows:

$$u = u_{\text{eq1}} + u_{\text{eq2}} + u_{\text{eq3}} + u_{\text{sw}} \tag{14}$$

where $u_{\text{eq1}}, u_{\text{eq2}}, u_{\text{eq3}}$ are the equivalent control law of s_1, s_2, s_3, respectively. u_{sw} is the switching control signal. To analyze the stability, the Lyapunov function is defined:

$$V = \frac{1}{2}s^2 \tag{15}$$

where s is the second-order surface in (12). Time derivative of V gives:

$$\dot{V} = s(a\dot{s}_1 + b\dot{s}_2 + c\dot{s}_3) \tag{16}$$

Further, substituting the first-order sliding surfaces in, the derivation of V is

$$\dot{V} = s\begin{bmatrix} a(c_1x_2 + f_1(x)) + b(c_2x_4 + f_2(x)) + c(c_3x_6 + f_3(x)) + \\ (aN_1 + bN_2 + cN_3)\delta_f \end{bmatrix} \tag{17}$$

To satisfy Lyapunov theory, $\dot{V} \leq 0$ should exist to ensure asymptotical stability. As a result, the aggregated HSMC's input has the form:

$$\delta_f = \frac{1}{aN_1 + bN_2 + cN_3}\begin{bmatrix} -af_1(x) - ac_1x_2 - bf_2(x) - bc_2x_4 \\ -cf_3(x) - cc_3x_6 - kS - \int sign(S) \end{bmatrix} \tag{18}$$

where $k > 0, \in > 0, f_1(x) = \mathbf{M}(1,:)\mathbf{x} - \dot{\omega}_{1r}, f_2(x) = \mathbf{M}(2,:)\mathbf{x} - \dot{\beta}_{1r}, f_3(x) = \mathbf{M}(3,:)\mathbf{x} - \dot{\theta}_{\text{ref}}$; N_1, N_2, N_3 are $\mathbf{N}(1), \mathbf{N}(2), \mathbf{N}(3)$, respectively. Since $V = \frac{1}{2}s^2 \geq 0$ and $\dot{V} \leq -kS^2 - \in |S| \leq 0$, the second-order surface is stable in terms of Lyapunov.

However, the stability of three first-layer sliding surfaces is not checked. According to Qian and Yi [16], each subsystem state progresses along the initial layer sliding surface, advancing toward its respective subsystem coordinate origin determined by the state variables of the subsystem. Thus, three first-layer sliding surfaces are of asymptotic stability.

3 Simulation

In this part, we compare how well the HSMC controller performs against the default TruckSim controller during a high-speed DLC maneuver. The parameter values of AHV are provided in Table 2. The tractor-semitrailer is cosimulated on the TruckSim and MATLAB/Simulink software.

3.1 *TruckSim Vehicle*

The AHV model in TruckSim includes a 2A Tractor 5.5 T/8.5 T (225 kW) as the leading unit and 1A Flatbed Semi-Trailer connected to tractor with a fifth wheel hitch. The AHV in TruckSim is shown in Fig. 4, and its parameters are included in Table 2.

Table 2 Parameter of truck-semitrailer in TruckSim

Parameter	Value	Parameter	Value
m_1	4665 kg	m_2	16,620 kg
a_1	1.1 m	a_2	5.21 m
b_1	2.39 m	b_2	3.28 m
I_{z1}	34,823 kgm^2	I_{z2}	179,992 kgm^2
c	1.64 m	C_1	−223,281
C_2	−223,281	C_3	−223,281

Fig. 4 3-axles TruckSim model of loaded combine vehicle

3.2 Testing Scenarios

The proposed controller is tested under TruckSim's double lane change test at 72 km/h. The reference path is the dotted black line in Fig. 5. The AHV goes straight and change to a lane 3.5 m away then moves back to original lane. This test assesses a vehicle's dynamic stability during an emergency maneuver [17].

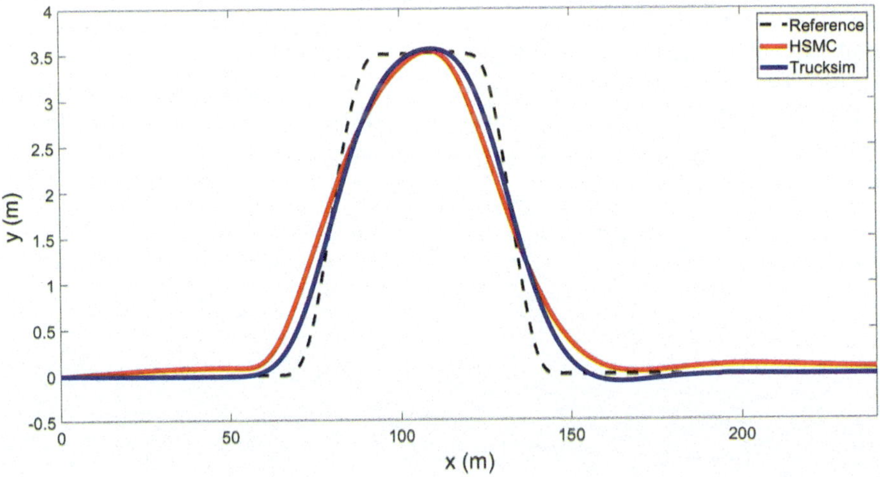

Fig. 5 AHV path trajectory

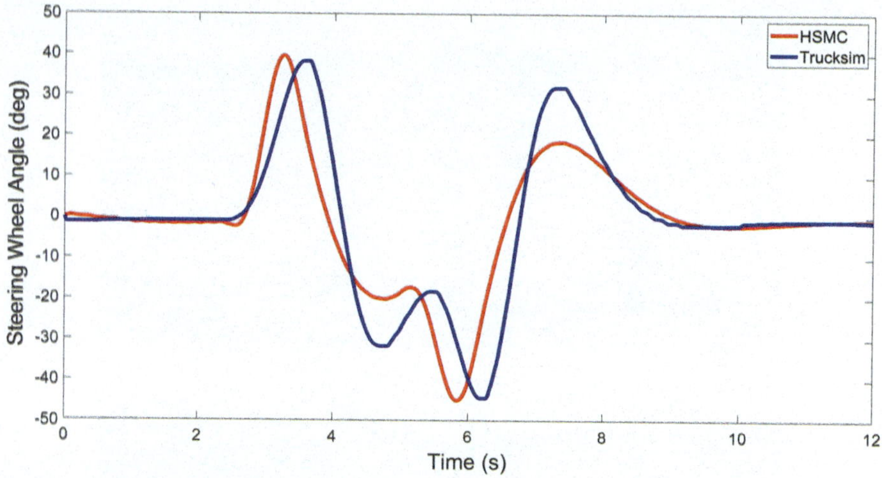

Fig. 6 Steering wheel angle

3.3 *Results and Discussion*

Discussion: In terms of lateral tracking, as depicted in Fig. 5, the HSMC-controlled AHV demonstrates an earlier initiation of steering and finishes the lane change faster than the default controller. This indicates a faster response of the HSMC in this scenario. Both controllers exhibit similar steering behavior in Fig. 6. The first peak steering wheel angle of the HSMC is higher than the default controller by 3.5%. However, the subsequent steering is smoother. The hitch articulated angle is compared to verify stability, as illustrated in Fig. 7. The peaks of the hitch angle of the HSMC-controlled vehicle are smaller, indicating a safer and more stable semi-trailer dynamic. Additionally, in Fig. 8, the semitrailer's lateral acceleration in the HSMC design is always smaller than that of the tractor, implying system stability. Furthermore, the semitrailer acceleration in the HSMC is less than that of TruckSim's default controller. The improvements can be found in Table 3.

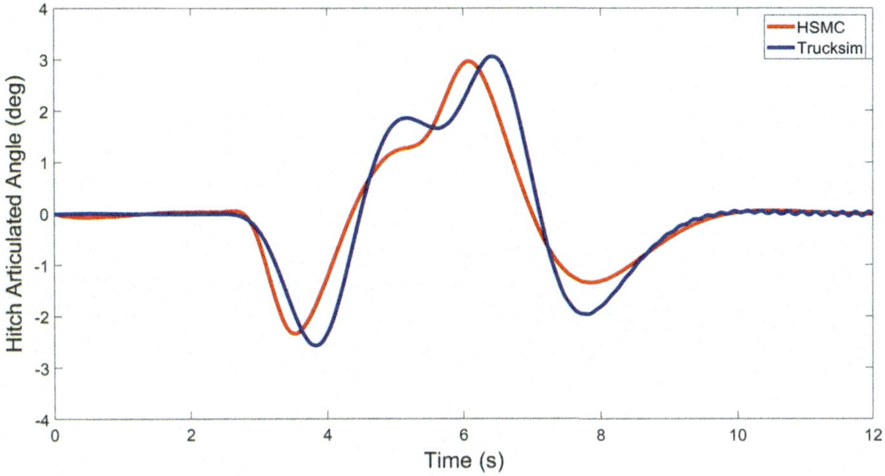

Fig. 7 Hitch articulated angle

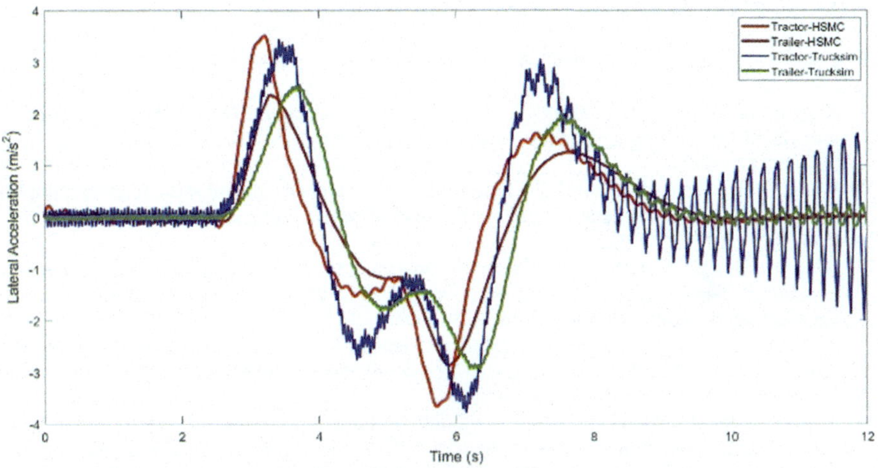

Fig. 8 Lateral acceleration of the tractor and its trailer

Table 3 Performance comparison between the HSMC and TruckSim controller

Parameters	Built-in control	HSMC model	Deviation (%)
Maximum lateral acceleration of semitrailer (m/s^2)	2.35	2.11	↓ 10.2
Articulated angle (°)	3.143	3.05	↓ 2.9

4 Conclusion

This paper addresses the issue of AHV's high-speed lateral stability through a new HSMC design. The dynamic model of a 3-axle truck-semitrailer combination is introduced, followed by a driver preview reference model. Based on the two models, the aggregated HSMC builds a second-order sliding surface comprising three first-order sliding surfaces to control four state variables. Simulation results under TruckSim demonstrate that the HSMC-controlled vehicle exhibits better stability with lower articulated angle while maintaining equivalent lateral tracking performance. Future research may include utilizing more advanced driver preview models with additional visual characteristics, adding state observers, autotuning weight for sliding surfaces, or adopting fuzzy logic to assist drivers with HSMC.

References

1. Xu, X., Zhang, L., Jiang, Y., Chen, N.: Active control on path following and lateral stability for truck–trailer combinations. Arab. J. Sci. Eng. **44**, 1365–1377 (2019)
2. Palkovics, L., Fries, A.: Intelligent electronic systems in commercial vehicles for enhanced traffic safety. Veh. Syst. Dyn. **35**, 227–289 (2001)
3. Kamnik, R., Boettiger, F., Hunt, K.: Roll dynamics and lateral load transfer estimation in articulated heavy freight vehicles. Proc. Inst. Mech. Eng. Part D J. Autom. Eng. **217**(11), 985–997 (2003)
4. Wang, Q., He, Y.: A study on single lane-change maneuvers for determining rearward amplification of multi-trailer articulated heavy vehicles with active trailer steering systems. Veh. Syst. Dyn. **54**(1), 102–123 (2016)
5. Ding, X., He, Y.: Numerical simulation and analysis of closed-loop driver/articulated vehicle dynamic systems. SAE Int. J. Commercial Veh. **5**(2012–01–0244):111–118 (2012)
6. Szczepaniak, J., Tanas, W., Pawlowski, T., Kromulski, J.: Modelling of agricultural combination driver behaviour from the aspect of safety of movement. Ann. Agric. Environ. Med. **21**(2) (2014)
7. Zhu, S., He, Y.: A unified lateral preview driver model for road vehicles. IEEE Trans. Intell. Transp. Syst. **21**(11), 4858–4868 (2019)
8. Yao, Q., Tian, Y., Wang, Q., Wang, S.: Control strategies on path tracking for autonomous vehicle: state of the art and future challenges. IEEE Access **8**, 161211–161222 (2020)
9. Vempaty, S., He, Y., Zhao, L.: An overview of control schemes for improving the lateral stability of car-trailer combinations. Int. J. Veh. Performance **6**, 151 (2020). https://doi.org/10.1504/IJVP.2020.10028995
10. Rahim, A., Huang, W., He, Y.: A longitudinal speed controller for autonomous multi-trailer articulated heavy vehicles. In: Proceedings of the Canadian Society for Mechanical Engineering International Congress 2021, 27–30 June, Charlottetown, PE, Canada (2021)
11. Rahimi, A., Huang, W., Sharma, T., He, Y.: An autonomous driving control strategy for multi-trailer articulated heavy vehicles with enhanced active trailer safety. In: The 27th IAVSD Symposium on Dynamics of Vehicles on Roads and Tracks, Saint-Petersburg, Russia (2021)
12. Liu, L., Wang, B., He, Y.: Research on path-tracking control of articulated vehicle with a trailer based on advanced model prediction control strategy. In: Proceedings of the Canadian Society for Mechanical Engineering International Congress 2021, June 27–30, Charlottetown, PE, Canada (2021)
13. Deng, Z.W., Zhao, Y.Q., Wang, B.H., Gao, W., Kong, X.: A preview driver model based on sliding-mode and fuzzy control for articulated heavy vehicle. Meccanica **57**(8), 1853–1878 (2022)

14. Peng, D., Fang, K., Kuang, J., Hassan, M.A., Tan, G.: Conflict and sensitivity analysis of articulated vehicle lateral stability based on single-track model. Shock Vib. **2021**, 1–18 (2021)
15. Andersson, A.: Road-Tire Friction Estimation for AFS Vehicle Control. MSc Theses (2006)
16. Qian, D., Yi, J.: Hierarchical Sliding Mode Control for Under-Actuated Cranes. Springer, Heidelberg (2016)
17. Bîndac, I.-M., Bradley, S.M., Roşca, P., Pupăză, C., Giurgiu, T.: Theoretical and experimental research on the double lane change maneuver. In: MATEC Web of Conferences, vol. 373, pp. 00054–00054 (2022)

Develop Agents for Autonomous Vehicles Using Reinforcement Learning

Duc-Quang Nguyen, Viet-Anh Le, Trong-Phuoc Le, Phuc-Nam Nguyen-The, and Thanh-Tung Nguyen

Abstract Machine learning and artificial intelligence have been used in various fields, including academic and industries, to make driving safer, more reliable, and improve performance. The objective of this research was to create an agent framework capable of driving a vehicle autonomously in a simulated world. The simulation environment was designed to receive control actions from both agents and humans while providing real-time updates on the state of the vehicle through in-vehicle sensors. The agent framework was developed to learn from past experiences and improve its driving skills. The results showed that the agent can drive the vehicle smoothly and safely. The data collected from the simulation verified the outcomes, exhibiting stable driving behavior and moving speed. This research has potential to accelerate the development of the automotive industry by increasing driving safety without compromising on performance.

Keywords Autonomous vehicle · Machine learning · Reinforcement learning · Driving behavior · Simulation

1 Introduction

The development of computer science and machine learning has enabled computer programs to achieve many achievements in areas that were previously limited to human ability, such as gaming [1], chess [2], and perform complex control operations [3]. Autonomous vehicles are a promising area of technological research and development worldwide. They allow for reducing driver errors, minimizing accidents, and increasing transportation efficiency. Major technology companies, such as Google, Tesla, Uber, Apple, Waymo, GM Cruise, Zoox, Baidu, etc., have significant resources in developing autonomous vehicle technology [4, 5]. Safety and vehicle behavior are critical issues in the processes to ensure public operation. As a result, big

D.-Q. Nguyen · V.-A. Le · T.-P. Le · P.-N. Nguyen-The · T.-T. Nguyen (✉)
School of Mechanical Engineering, Hanoi University of Science and Technology, Hanoi, Vietnam
e-mail: tung.nguyenthanh@hust.edu.vn

D. H. Tien et al. (eds.), *Proceedings of the 1st International Conference on Sustainability and Emerging Technologies for Smart Manufacturing*, Proceedings in Technology Transfer, https://doi.org/10.1007/978-981-97-7083-0_49

data, machine learning applications, and data processing have become the focus of research and development of autonomous vehicles. Machine learning applications, such as deep learning, reinforcement learning, etc., are utilized to enhance automation and improve vehicle system performance [5, 6]. There has been much research in developing virtual environments for general reinforcement learning research such as OpenAI Gym [7], and specialized environments for driving and traffic simulation such as CARLA [8], SUMO [9], and DriverGym [10]. However, simulation environments frequently demand intricate configurations to compute and control complex model parameters, which also require high-performance hardware.

In this research, an autonomous vehicle system was developed that operates in a pre-built simulation environment. The simulation environment needs to balance performance and real operating conditions. The objective of the research is to create an agent that can control a vehicle in the above-simulated environment using reinforcement learning from trial runs and adjustments from past driving experiences. The agent continuously adapts itself to provide appropriate steering angles for different conditions. After receiving data and being trained, the research results allow for the evaluation of the effectiveness of the agent's control. These results will then be compared between the agent and several drivers driving the vehicle on a predetermined simulated road to assess the applicability of this method in the future.

2 Methodology

The reinforcement learning environment built in the research includes the following main components: driving simulation environment and reinforcement learning agent. The interactive information exchange between the driving simulation environment and the agent is summarized in Fig. 1.

2.1 *Driving Environment*

The driving simulation environment includes a closed racetrack with straight and curved sections, the length of the road is about 5 km with two lanes, and each lane is 6 m wide. The condition of the simulation is to control the vehicle to avoid obstacles ahead. Obstacles added to the model are designed by placing barriers along the road, in both the right lane and the left lane. The vehicle is designed based on the multipurpose vehicle (MPV) model, using a simplified bicycle model. In the case of a driver, the test subject will control directly on the computer, and the software will collect data from the driver to provide the best driver data set. In case of automatic control, the agent will send control signals to the environment through a communication server.

In contrast to computer vision-based applications that detect obstacles ahead and often use a lot of resources, the agent model which controls the vehicle in this

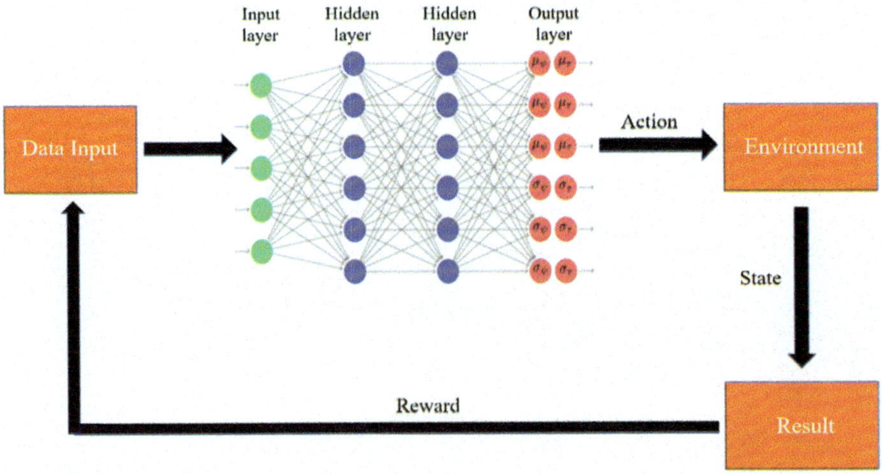

Fig. 1 Structure of the reinforcement learning environment

study employs distance sensors to measure the distance between the vehicle and any obstacles, which is similar to radar or Lidar technology. The range sensor values that measure the distance to the road center, left lane, and right lane. If the vehicle exceeds the road limit, the value becomes unreliable and will be set to −1. Figure 2 illustrates the distance sensor mounted on the simulated vehicle. The grey area that is depicted in the image is a section of the track that is free from any obstacles and contains designated lanes for the vehicle to travel on. The green areas that you see indicate the sections of the track that are considered offtrack and are not designated for the vehicle to travel on. The orange lines that represent the sensor rays, which are being emitted by the distance sensor.

The vehicle's maximum speed is set at 100 km/h. The vehicle's state vectors have different dimensions: the agent sends control signals including steering angle and brake/acceleration control signals: negative braking, positive—acceleration; vehicle status parameters include speed, direction of vehicle movement, distance to the center of the road, distance to the left and right lanes, distance to the nearest front obstacle on the left and right, distance to the next object after the nearest obstacle on the left, right and other distance sensors. So to ensure the learning capability of the agents, the vehicle control or state values have been scaled between [−1, 1] and [0, 1], respectively.

2.2 Driving Agent

In research on reinforcement learning, there are many methods and algorithms to optimize the agent's learning ability. Here the research team uses the proximal policy

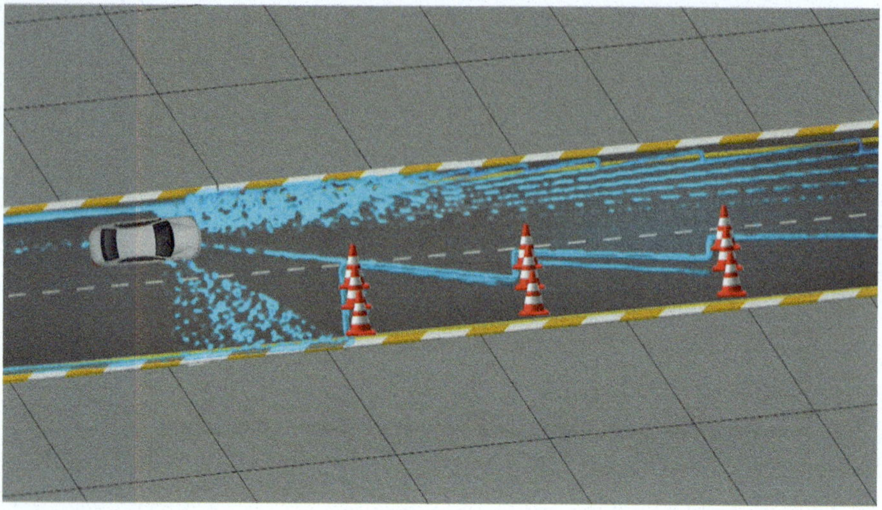

Fig. 2 Distance sensors

optimization (PPO) algorithm. The aim is to replicate the unpredictable actions of a skilled driver while following fundamental traffic regulations. This is achieved by enhancing the traditional PPO algorithm, where the agent's policy (actor) is represented by a mixed density network featuring three Gaussian distributions. The selection of three Gaussians was arbitrary and not employed for hyperparameter tuning. Additionally, another network is employed to estimate the mixing coefficients for each Gaussian distribution.

The actor network architecture is presented as follows: the input layer has 300 units, the next 4 hidden layers have 300 units each, and an output layer of 12 units. The activation function is a ReLU function and is used between different layers except the output layer. The output units are divided into two parts: the first six units represent the average value of steering angle and acceleration, μ_ψ and μ_τ; the latter six represent the variance, σ_ψ and σ_τ, for the three Gaussians in the mixture. The mean values are passed through a SoftSign function to constrain the values between -1 and 1. The variance values are passed a customization function of the form $f_{(x)} = 1/16$ Sigmoid $(3x)$.

The Critic network architecture is presented as follows: The critic shares the identical network structure with the actor, except for the output layers, which comprise only one neuron output, representing the Q value. Throughout all layers, the ReLU activation function is applied except for the final layer, where a linear activation function is utilized.

Lastly, the mixing coefficient network architecture: Likewise, the mixing coefficient network follows the same architecture as the actor and critic, with the exception of the output layer, which comprises three neurons representing the number of Gaussians in the mixture. Throughout the layers, the ReLU activation function is employed, except for the output layer, where a softmax activation function is utilized to ensure the constraint is maintained $\sum_i a_i = 1$.

2.3 Reward Function

The reward function is built to resemble the driving behavior of professional drivers. This reward function is random and is used to recover the expected value and variance of similar distributions.

$$R^j(s_t, a_t) = \begin{cases} \left(V_h^j - \left|V_h^j - V_a\right|\right)_{\sigma(t)} - c_1\left|D_h^j - D_a\right|_{\sigma(t)} - c_2\left|\Psi_t - \Psi_{t-1}\right| - c_3\left|\tau_t - \tau_{t-1}\right| \\ -100,\ if s_{\sigma(t)} = s_T \end{cases}$$

where

s_T represents concluding states as determined by $s_T = \{$obstacle collision, exceeding road boundaries, traveling at velocities lower than 5 km/h, traveling in the reverse direction$\}$.

The indices a and h represent agent and human, respectively, and $\sigma_{(t)}$ denotes the curve's length at time t.

V and D are the value of the vehicle's speed and distance from the center line, respectively. The index j represents the j - th sample, the samples are taken from the GPs model. In each segment, a distinct sample will be chosen randomly to assess the reward function.

The first component in the equation represents the deviation of the agent's speed, V_a, from the speed of the professional driver, V_h. The second part assesses the variance in track positioning between the agent and the professional driver, and c_1 is a weight. The third and fourth components are calculated for steering angle and speed change (brake + gas), respectively.

After calculating the reward function at each control state, the final total reward function is calculated to evaluate the agent's control skill at each learning cycle (episode).

3 Results

The testing process was performed on a computer with the following configuration: CPU: Intel core i7 10750H 2.6 GHz, RAM: 16 GB, and GPU: GeForce GTX 1660 Ti.

It can be seen from the results in Fig. 3 that the amplitude or difference when steering and returning is relatively small and tends to maintain a stable steering angle when the vehicle moves stably. The speed chart shows that the agent is capable of maintaining a speed of between 80 and 90 km/h (around 22.2–25 m/s) when

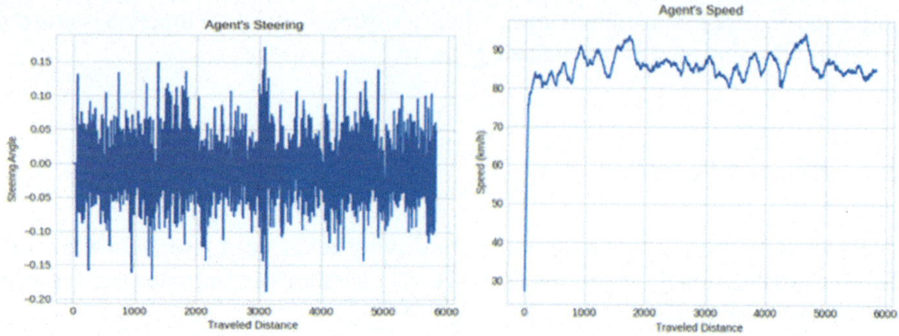

Fig. 3 Steering angle graph (left) and speed graph (right) of the control agent

avoiding obstacles, which shows that the car when controlled by AI is capable of ability to avoid obstacles at high speed.

After 4000 episodes, equivalent to 36 h of driving learning time, the errors of driving trajectory and speed achieved were 0.12 and 6.87, respectively. The control agent can operate well in the simulation environment and almost does not crash into obstacles (corresponding to 10 h of simulation running without hitting obstacles).

4 Conclusion

Advanced technology, such as deep reinforcement learning, combined with numerical simulation, allows self-driving cars to learn and respond to dangerous driving situations and obstacles, just like human drivers. The simulation creates a driving environment with the goal of avoiding static obstacles randomly placed across two lanes. The data collected from these simulations is then used to train learning models to achieve maximum scores or rewards. The results show that this method can help self-driving vehicles react faster than humans in dangerous driving conditions, which enhances the safety of the vehicles and their surroundings. Through deep reinforcement learning, autonomous vehicles can learn from previous experiences and optimize performance, resulting in an improved user experience. However, to make this technology practical, it must be combined with other technologies, such as sensors fusion to determine the vehicle's position in real time and find ways to overcome dynamic obstacles. These challenges present opportunities for future researches and developments.

References

1. Choi, J., Guo, Y., Moczulski, M., Oh, J., Wu, N., Norouzi, M., Lee, H.: Contingency—Aware Exploration in Reinforcement Learning. ICLR 2019 (2019)
2. van Eck, N.J., van Wezel, M.: Application of Reinforcement Learning to the Game of Othello (2006)
3. Tamei, T., Matsubara, T., Rai, A., Shibata, T.: Reinforcement learning of clothing assistance with a dual-arm robot. In: 11th IEEE-RAS International Conference on Humanoid Robots (2011)
4. Liao, J., Liu, T., Tang, X., Mu, X., Huang, B., Cao, D.: Decision-making strategy on highway for autonomous vehicles using deep reinforcement learning. IEEE Access (2020)
5. Ni, J., Chen, Y., Chen, Y., Zhu, J., Ali, D., Cao, W.: A survey on theories and applications for self-driving cars based on deep learning methods. In: Applied Science (2020)
6. Henriksson, J., Borg, M., Englund, C.: Automotive safety and machine learning: initial results from a study on how to adapt the ISO 26262 safety standard. In: 2018 IEEE/ACM 1st International Workshop on Software Engineering for AI in Autonomous Systems (SEFAIAS) (2018)
7. Brockman, G., Cheung, V., Pettersson, L., Schneider, J., Schulman, J., Tang, J., Zaremba, W.: Openai gym (2016). arXiv:1606.01540
8. Dosovitskiy, A., Ros, G., Codevilla, F., Lopez, A., Koltun, V.: CARLA: an open urban driving simulator. In: 1st Annual Conference on Robot Learning (2017)
9. Kheterpal, N., Parvate, K., Wu, C., Kreidieh, A., Vinitsky, E., Bayen, A.M.: Flow: deep reinforcement learning for control in SUMO. In: Simulating Autonomous and Intermodal Transport Systems (2018)
10. Palanisamy, P.: Multi-agent connected autonomous driving using deep reinforcement learning. In: 2020 International Joint Conference on Neural Networks (IJCNN) (2020)

The Compression Behavior of Continuous Carbon Fibers-Based Polylactic Acid (PLA) Resin 3D Printing Materials for Different Infill Strategy

Hoang Tien Dung, Nguyen Ba Thuan, Van Huan-Dau, Nguyen Van Canh, Ngo Quang Tu, Nguyen Dang Viet, Tran Quoc Hung, and Tien-Dat Hoang

Abstract 3D printing, which employs a method of layer-by-layer material deposition, offers several benefits for part manufacturing and includes a reduced number of process steps, increased flexibility for prototyping, particularly for intricate components, and minimal waste of materials. However, the mechanical characteristics of 3D-printed components have a big concern owing to the inherent limitations of interlayer adhesion and surface roughness. The utilization of continuous fiber-reinforced polymer composites in 3D printing involves the incorporation of continuous fiber reinforcements within the polymer matrix, enhancing the mechanical integrity of the printed parts. Compression testing is the primary means of evaluation as this study to explore the mechanical characteristics of continuous carbon fiber (CCF)-based polylactic acid (PLA) in connection with various infill patterns. For demonstration, three models with the same infill pattern but different material layouts are investigated to show the power of CCF-based PLA resin. It can be seen that the results of the case of PLA-CCF achieve excellent mechanical properties that can withstand the peak load of 7 kN with only 31 g of weight. The finite element models are used to predict the stress and strain of the tests which are close to the results of experiments. Furthermore, since 3D printing materials are becoming more and more important in a range of industries, our discovery opens the door for additional research. Finding the ideal features for various applications will require more research into the mechanical properties and behavior of 3D-printed materials.

H. T. Dung · N. Van Canh · N. Q. Tu · T. Q. Hung · T.-D. Hoang (✉)
School of Mechanical and Automotive Engineering (SMAE), Ha Noi University of Industry, Ha Noi, Vietnam
e-mail: hoangdat@haui.edu.vn

N. B. Thuan
Faculty of Mechanical Engineering, Vinh University of Technology Education, Vinh, Vietnam

V. Huan-Dau
Facaulty of Engineering, Dong Nai Technology University, Bien Hoa City, Dong Nai, Vietnam

N. D. Viet
Facaulty of International Training, Thai Nguyen University of Technology, Thai Nguyen, Vietnam

D. H. Tien et al. (eds.), *Proceedings of the 1st International Conference on Sustainability and Emerging Technologies for Smart Manufacturing*, Proceedings in Technology Transfer, https://doi.org/10.1007/978-981-97-7083-0_50

Keywords Composite 3D printing · Printed lattice · Continuous carbon fiber printing

1 Introduction

3D printing, a part of additive manufacturing (AM), is a fabrication technique characterized by the incremental deposition of materials, including metals, polymers, and ceramics, to construct intricate structures layer by layer [1]. This technology has found extensive applications across diverse sectors such as aerospace, automotive, medical devices, construction, and apparel manufacturing. The mechanical attributes of 3D-printed objects significantly influence their functionality, reliability, and performance [2]. The quality of final products is directly affected by many factors including strength, hardness, elasticity, and toughness. The mechanical properties of printed objects are also greatly impacted by infill patterns, material layout, printed material properties in 3D printing [3]. The printing customization can create optimal products with improved mechanical properties and reduced weight.

The evolution of 3D printing technology has seen advancements beyond its foundational methods, including fused deposition modeling (FDM), selective laser sintering (SLS), or stereo lithography (SL), toward specialized printing tailored for different applications [4]. Among these, FDM stands out as the most prevalent due to its cost-effectiveness and operational simplicity, facilitating the printing of diverse materials, in multiple directions, and enabling the fabrication of large-scale objects. However, FDM-produced polymer objects exhibit inadequate interlayer bonding during printing, resulting in weak mechanical properties, stiffness, and functional characteristics. Consequently, 3D-printed polymer products are primarily utilized for conceptual prototyping and research, lacking practical applications because of their weak mechanical properties. To address these limitations, the combination of fiber reinforcement into polymer matrices presents a viable strategy for enhancing the mechanical performance of printed structures [5, 6]. Conventionally, fiber-reinforced plastics are manufactured through molding or lamination techniques, limited manufacturing flexibility, and high production costs. Therefore, the fabrication of fiber-reinforced polymer composites via FDM has garnered increasing interest among researchers. This process can be categorized into two approaches: short or continuous fiber reinforcement with polymers in 3D printing [7, 8]. Short fiber-reinforced plastics demonstrate weaker mechanical properties when contrasted with continuous fiber-reinforced plastics [9].

In this study, polylactic acid (PLA) mixed with short and continuous carbon fiber (CCF) are used to produce porous printed structures via FDM technology. The porous structures were constructed utilizing a single perimeter shell and grid infill pattern type at three different degrees of materials. This study examined the effects of uniaxial compress loading on the compression characteristics of composite printed structures. The optical microscope (OM) is used to characterize the microstructures of carbon-printed composites with the infill scheme. The experimental measurements

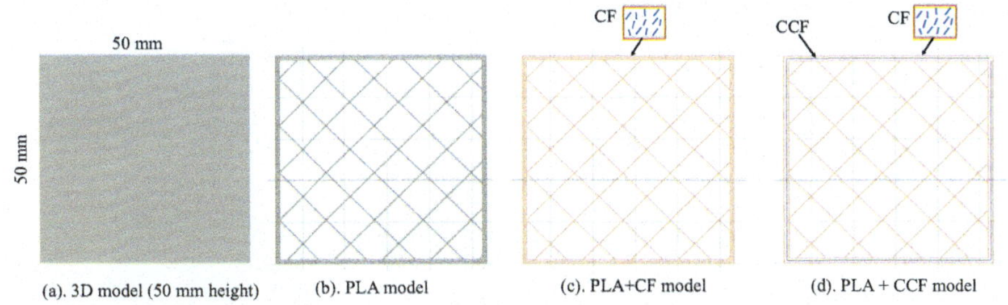

Fig. 1 Printing lattice setting for three case studies

and the estimated stress and strain by Abaqus software meet well. It is important to comprehend these connections in order to optimize material selection and design for efficiency and overall product performance across a wide range of industries and applications.

2 Preparation and Printing Setting

2.1 Model Design

The simulation made use of the case study, which featured cubic models made with Inventor 2023 software that had a volume of 50 × 50 × 50 mm. The printing slicer Aura from Anisoprint company is used to choose the printing settings and generate the lattice patterns in each case study. This application can replicate and show the printed model exactly like the actual printed model. Three different sorts of materials including pure PLA, PLA-short carbon fiber (CF) and CCF were used to begin printing test specimens (see Fig. 1). Every printed model is produced at the same time in order to guarantee a solid connection between the printed models and the glass plate. Samples' 3D models were sliced using Aura program. This program can create a wide range of raster, infill, and fiber patterns, such as solid, triangular, insogrid, anisogrid, etc. Specifically, it can combine any pattern inside layers to create complex lattices. It is possible to arrange fibers at any desired angle. The fiber route is kept clear of the specimen wall even if the fiber volume is still approved.

2.2 Manufacturing Process

The composer A4 printer developed by Anisoprint is employed for the fabrication of three-dimensional models. Equipped with dual nozzles, one dedicated to PLA and the other to CCF, this printer provides numerous opportunities for the creation

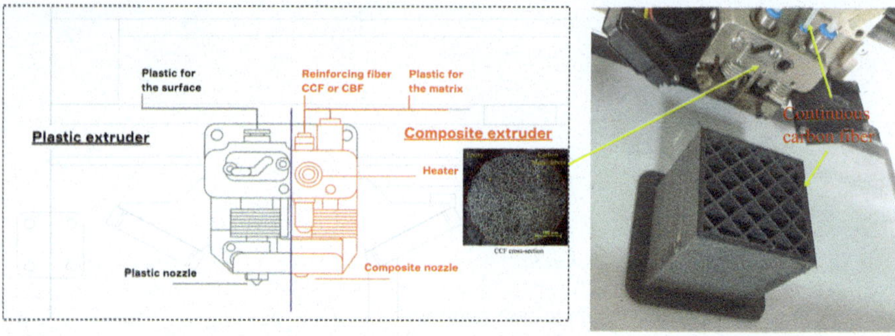

(a). Anisoprint printing extruder (b). Printed model

Fig. 2 Composite printing using short and continuous carbon fibers with co-extruders

of robust and unique products (see Fig. 2). Employing a 3D printer featuring dual nozzles for PLA and CCF enables to manufacture durable and sustainable objects suitable for various applications, ranging from end use parts to prototyping. This high-quality printer has built volume measuring 297 mm × 210 mm × 140 mm and supports dual extrusion capabilities.

The properties of each material used in the article are listed in Table 1. The settings parameters during the printing process recommended by the Anisprint manufacturer are provided in Table 2. The speed of CCF is slower than that of PLA to obtain the best adhesive strength between two layers. The printed surfaces of three models are captured by the OM in Fig. 3. They clear show the mono CFs impregnated to PLA resin in the printing.

Table 1 Printing material properties

Material	Young's modulus (GPa)	Poisson ratio	Density (g/cm^3)	Elongation at break (%)	Tensile strength (MPa)
PLA–ESUN	2.69	0.35	1.24	12	65
Continuous carbon fiber (CCF)–Anisoprint	150	0.26	1.45	–	2200

Table 2 Printing process parameters

CCF layer thickness (mm)	PLA layer thickness (mm)	CCF printing speed (mm/s)	PLA printing speed (mm/s)	CCF temperature (°C)	PLA temperature (°C)
0.36	0.15	20	60	250	200
Building plate temperature (°C)	CCF line width (mm)	PLA line width (mm)	Flowrate (%)	Grid infill (%)	Wall line of model (–)
50	0.8	0.4	100	10	3

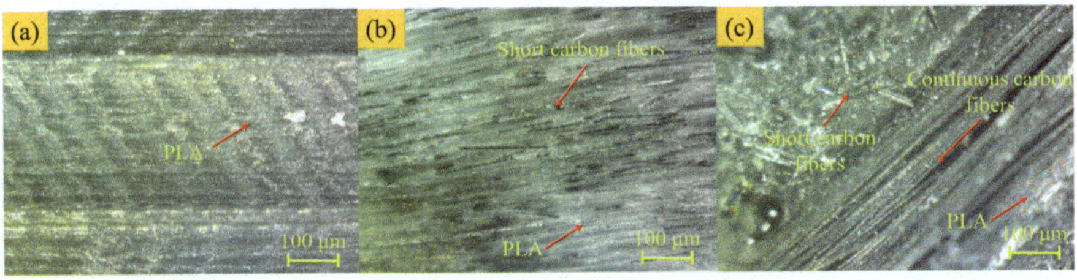

Fig. 3 Printed surfaces of three models: **a** PLA model; **b** PLA-CF model; **c** PLA-CCF model

3 Simulation and Experiment Test

3.1 Compression Simulation

Abaqus, standing out as a finite element analysis (FEA) program renowned for its proficiency in modeling and analyzing solid mechanics and structural behavior, is used in this study. Its exceptional capability to simulate material responses across various loading conditions renders it particularly well-suited for compression testing and simulations. Engineers and researchers apply Abaqus to assess critical variables such as stress, strain, deformation, and failure modes, thereby accurately predicting and visualizing material and structure responses to compression or tension forces during testing. The program's robust material modeling features and adeptness in handling nonlinear and dynamic behaviors make it an optimal choice for compression testing simulations. Furthermore, Abaqus offers robust postprocessing functionalities, facilitating comprehensive analysis, and validation of compression test results. Its flexibility and precision make Abaqus the preferred option for conducting compression test simulations aimed at evaluating the integrity and performance of mechanical components and materials.

The mechanical properties and destructive behavior of 3D-printed models with different materials are clearly shown using element type C3D8R with the number of 9242 elements and 14,966 nodes, models 2 and 3 using the element type C3D8R with the number of 9140 elements and 15,746 nodes. Figure 4 shows the Von Mises stress and deformed shapes of the model under compress load with 5 mm displacement.

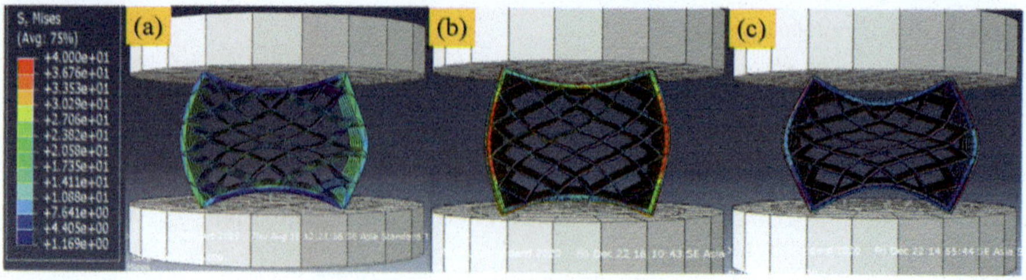

Fig. 4 Von Mises stress and deformed shapes of: **a** PLA model; **b** PLA-CF model; **c** PLA-CCF model

3.2 *Compression Experiment Test*

Based on the layer slicing results and lattice structures configured in the Aura software, a program with a.gcode extension is generated and loaded into the 3D printer composer A4 to produce experimental samples. Figure 5 illustrates the main steps involved in manufacturing samples analogous to Models 1, 2, and 3. The models have been measured and verified to have a dimensional accuracy within ± 0.2 mm compared to the simulated models. The testing speed is 3 mm/min. Figure 6 shows the deformation of the models at the final testing step.

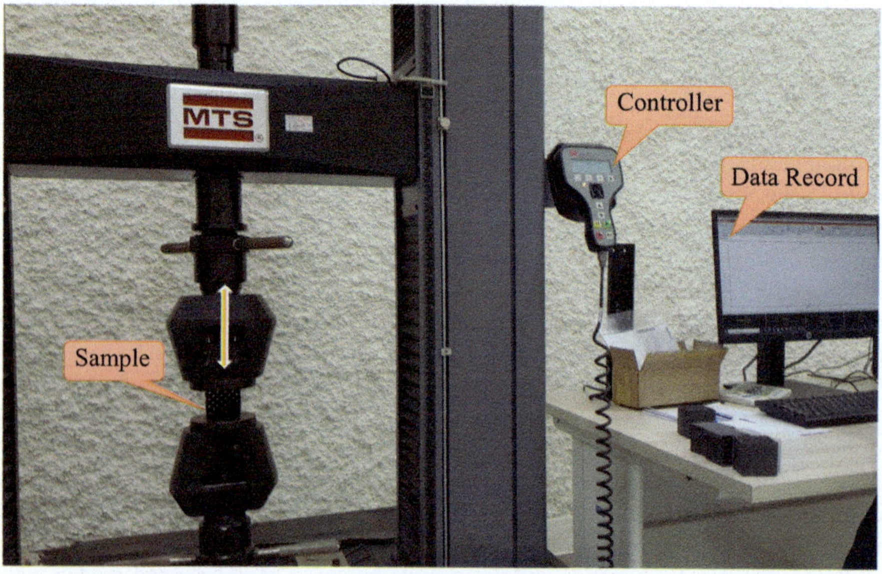

Fig. 5 Compression tests for three models using MTS—45 system (the USA)

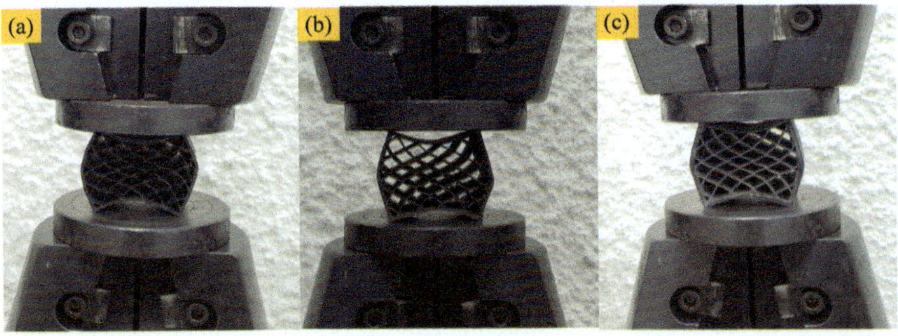

Fig. 6 Deformed models under compression of **a** PLA model; **b** PLA-CF model; **c** PLA-CCF model

3.3 Testing Results

From the load–displacement curve, we can get the result that sample 1 (pure PLA) can withstand a compression force of about 1.6 kN, samples 2 and 3 achieve 5.7 kN and 7 kN, respectively. It can be seen that composite materials play a significant part in improving the stiffness of 3D-printed materials. It can be observed that the specific stiffness (ratio of stiffness over mass) of the printed continuous fiber composite material is noticeably higher than that of the printed pure plastic material. More specimens should be simulated in future studies to highlight the variations in stiffness in each scenario (Fig. 7).

Fig. 7 Load and displacement curve

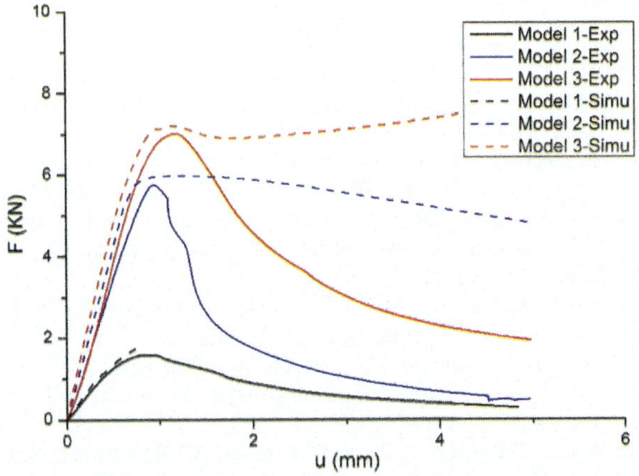

4 Conclusion

Compression testing of the 3D-printed models, all three of which have the same block size of 50 × 50 × 50 mm with variations in insert type and material composition, provided valuable insights into the mechanical properties of these materials. Test results show a clear correlation between the addition of carbon continuous fibers and the compressive strength of the models. Notably, the model incorporating CCF (model 3 with 31 g of weight) exhibits the highest compressive strength, with a peak load of 7 kN. These findings highlight the significant enhancement in compressive strength achieved through the incorporation of CFs, especially in the form of continuous filaments in 3D printing. The observed increase in compressive strength is consistent with the theoretical expectations of CF-reinforced materials. This insight promises to optimize material selection and AM processes, with potential applications in various industries where structural integrity and load-carrying capacity matter. Furthermore, the results of this study also highlight the potential for using advanced composite materials in AM to create lightweight, highly durable components. Compressive strength values obtained from different material compositions provide valuable data to engineers and researchers looking to leverage AM techniques to produce end-user parts.

Acknowledgements This research was funded by the Vietnam Ministry of Education and Training, grantnumber B2022-TNA-27.

References

1. Hu, Y., Lin, Y., Yang, L., Wu, S., Tang, D., Yan, C., Shi, Y.: Additive manufacturing of carbon fiber-reinforced composites: a review. Appl. Compos. Mater. 1–46 (2023). https://doi.org/10.1007/s10443-023-10178-w
2. Parandoush, P., Lin, D.: A review on additive manufacturing of polymer-fiber composites. Compos. Struct. **182**, 36–53 (2017). https://doi.org/10.1016/j.compstruct.2017.08.088
3. Liu, J., Gaynor, A.T., Chen, S., et al.: Current and future trends in topology optimization for additive manufacturing. Struct. Multidiscip. Optim. **57**, 2457–2483 (2018). https://doi.org/10.1007/s00158-018-1994-3
4. Singh, S., Ramakrishna, S., Berto, F.: 3D printing of polymer composites: a short review. Mater. Des. Process. Commun. **2**(2), 97 (2020)
5. Blok, L.G., Longana, M.L., Yu, H., Woods, B.K.S.: An investigation into 3D printing of fibre reinforced thermoplastic composites. Addit. Manuf. **22**, 176–186 (2018). https://doi.org/10.1016/j.addma.2018.04.039
6. Bekas, D.G., Hou, Y., Liu, Y., Panesar, A.: 3D printing to enable multifunctionality in polymer-based composites: a review. Compos. Part B 107540 (2019). https://doi.org/10.1016/j.compositesb.2019.107540
7. Ferreira, R.T.L., Amatte, I.C., Dutra, T.A., Bürger, D.: Experimental characterization and micrography of 3D printed PLA and PLA reinforced with short carbon fibers. Compos. Part B Eng. **124**, 88–100 (2017). https://doi.org/10.1016/j.compositesb.2017.05.013

8. Ferreira, I., Machado, M., Alves, F., Torres Marques, A.: A review on fibre reinforced composite printing via FFF. Rapid Prototyping J. **25**(6), 972–988 (2019)
9. Ho, T.N.T., Nguyen, S.H., Le, V.T., Hoang, T.D.: Coupling design and fabrication of continuous carbon fiber-reinforced composite structures using two-material topology optimization and additive manufacturing. Int. J. Adv. Manuf. Tech. 1–17 (2024). https://doi.org/10.1007/s00170-023-12913-w

The Temperature, Strain, and Strain Rate Dependent Flow Stress of 10B33 Boron Steel Using the Modified Johnson–Cook Model

Thanh-Cong Nguyen, Quang-Cherng Hsu, Jing-Yuan Gao, and Nhu-Tung Nguyen

Abstract The Gleeble-3500 thermomechanical simulator was utilized to perform isothermal compression tests on 10B33 boron steel. These tests obtained reliable stress–strain data at various temperatures (including room temperature, 100, and 200 °C), as well as strain rates (0.1, 1, and 5 s^{-1}). This data served as the foundation for establishing constitutive equations based on the modified Johnson–Cook and power-law models, aiming to estimate the flow stress behavior. The evaluation employed the correlation coefficient (R) as a metric to compare the accuracy of predicted deformation behavior against experimental observations. While the flow stress obtained from the modified Johnson–Cook model demonstrated an excellent correlation with the experimental results across the investigated temperature range except for room temperature, it generally exhibited superior tracking of the deformation behavior compared to the power-law model. This study highlights the enhanced capability of the modified Johnson–Cook model in capturing the stress behavior dependence on strain rate and temperature for 10B33 boron steel under various deformation conditions.

Keywords Constitutive equation · Compression test · Flow stress · 10B33 · Boron steel · Modified Johnson–Cook model

T.-C. Nguyen · Q.-C. Hsu (✉) · J.-Y. Gao
Department of Mechanical Engineering, National Kaohsiung University of Science and Technology, 415 Chien-Kung Road, Kaohsiung City 80778, Taiwan
e-mail: hsuqc@nkust.edu.tw

T.-C. Nguyen
School of Mechanical and Automotive Engineering, Hanoi University of Industry, No. 298, Cau Dien Street, Bac Tu Liem District, Hanoi 11950, Vietnam

N.-T. Nguyen
International School, Vietnam National University, 144 Xuan Thuy Street, Cau Giay District, Hanoi, Vietnam

D. H. Tien et al. (eds.), *Proceedings of the 1st International Conference on Sustainability and Emerging Technologies for Smart Manufacturing*, Proceedings in Technology Transfer, https://doi.org/10.1007/978-981-97-7083-0_51

1 Introduction

The AISI 10B33 steel is a boron-alloyed, medium-carbon steel renowned for its strength and widespread application. The addition of boron to steel improves its hardenability, allowing it to be heat-treated to achieve a higher hardness, despite the fact that the amount of boron in the steel is remarkably low, ranging from 0.002 to 0.005 weight percent [1]. This makes AISI 10B33 steel a good solution for applications where high strength is required. It is a popular choice for a variety of applications, including automotive and construction parts, nuts and bolts, and fasteners [2]. The ability to predict flow stress accurately is essential for commercial metal forming processes, especially extrusion and forging. The constitutive model provides a mathematical relationship between the behavior of stress and deformation by considering temperature, strain rate, and strain [3]. Hence, numerous constitutive equations have been implemented using experimental data to understand the influence of factors such as strain, strain rate, and temperatures on the flow stress of materials during reality hot working processes [4–8]. Li et al. [9] investigate the compressive response of medium-carbon steel samples under varying temperatures and strain rates utilizing the Arrhenius constitutive equation. The data demonstrates that flow stress initially rises with strain and subsequently declines. Li et al. [10] conducted the Gleeble 1500D test of B1500HS, a boron steel, and built the constitutive relationships using the Arrhenius and Johnson–Cook models. It well-predicted the temperature softening, deformation strengthening, and strain rate strengthening effects. Nowadays, integrating the Johnson–Cook model into finite element analysis has been widely attempted to depict the deformation characteristics of materials [11, 12]. However, when considering the combined effects of temperature, strain, and strain rate factors on the flow behavior simultaneously, the original Johnson–Cook model showed its limitations, resulting in reduced accuracy in predictions. To address this issue, plenty of recent studies employ the Modified Johnson–Cook model (MJC) [8, 11, 13, 14], which offers better precision in estimating flow stress. This model demonstrates suitability for real-world situations, offers a high level of reliability, and can be effectively utilized in simulations relating to the plastic deformation of AISI 10B33 boron steel material. This study aims to establish a constitutive equation using the MJC flow stress model for 10B33 boron steel. The study also attempts to compare the predictive ability of that model with the conventional power-law model in terms of flow stress behavior.

2 Experimental Procedure

This study utilized the Gleeble 3500 thermal compression machine to conduct the experiments. Three initial temperatures were set: room temperature (30 °C), 100, and 200 °C. Three strain rates were also set: 0.1, 1.0, and 5.0 s^{-1}. In total, there were nine combinations of conditions for the experiments. To measure and record

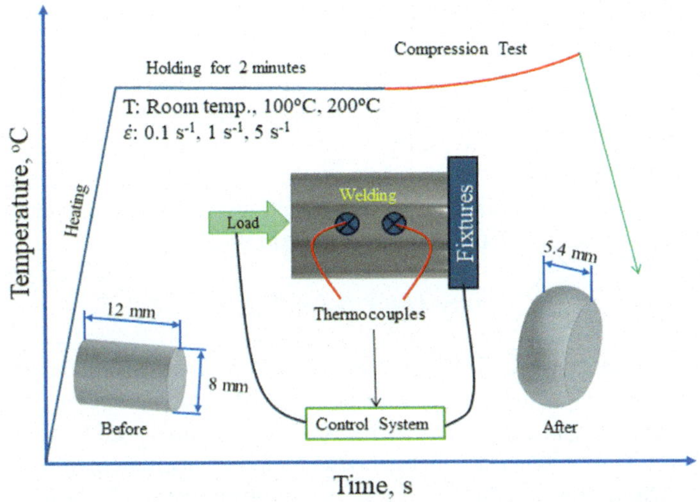

Fig. 1 The compression experimental process

the temperature changes generated in the compression process, thermocouples were used to monitor the rise in temperature of the specimen during compression. The schematic of the compression experimental process and experimental condition are presented in Fig. 1.

3 Results and Discussion

3.1 Modified Johnson–Cook Model

Because of its simplicity and ease of parameter adjustment, the Johnson–Cook (JC) model is commonly applied to investigate the hot deformation characteristics of various materials. However, the JC model maintains a notable limitation in its inability to comprehensively consider the effect of specific temperatures on deformation behavior and the interrelationship between temperature and strain rate. This constraint hinders its precision in capturing the deformation behavior observed. The application of the Modified Johnson–Cook (MJC) model is intended to overcome this remaining problem. The MJC model integrates the influences of strain rate and temperature on the behavior of flow stress, resulting in a more comprehensive and precise depiction of deformation processes. The MJC model can be expressed as:

$$\sigma = \left(A_1 + B_1\varepsilon + B_2\varepsilon^2\right)\left(1 + C_1 \ln \dot{\varepsilon}^*\right) \exp\left[\left(\lambda_1 + \lambda_2 \ln \dot{\varepsilon}^*\right)T^*\right] \quad (1)$$

where the flow stress is presented as σ, and the equivalent plastic strain is marked as ε. Besides, $\dot{\varepsilon}^* = \dot{\varepsilon}/\dot{\varepsilon}_0$ with $\dot{\varepsilon}$ represents the strain rate and $\dot{\varepsilon}_0$ represents the strain rate of reference. The variable T^* denotes the disparity between the present temperature T and the reference temperature T_{ref}. The remaining parameters are material constants. Within the present investigation, the reference temperature and strain rate, which are 100 °C and of 1 s^{-1}, were chosen to evaluate the material constants. This selection effectively normalizes the equation and simplifies the calculations. With these reference values applied, Eq. (1) can be expressed as:

$$\sigma = \left(A_1 + B_1\varepsilon + B_2\varepsilon^2\right) \tag{2}$$

Figure 2 presents the graphical representation of the relationship between experimental flow stress (σ) and strain (ε) observed at the designated reference conditions, which are 100 °C and strain rate 1 s^{-1}. The fitted curve demonstrates a strong correlation with the experimental data, characterized by a high coefficient of determination of 97.07%. Under the specific condition of a deformation temperature of 100 °C, Eq. (1) can be simplified due to the cancellation of temperature-dependent terms with their corresponding reference values. This reduced form of the equation can be expressed as:

$$\frac{\sigma}{\left(A_1 + B_1\varepsilon + B_2\varepsilon^2\right)} = 1 + C_1 \ln \dot{\varepsilon}^* \tag{3}$$

Figure 3 presents the disparity of $\sigma/(A_1 + B_1\varepsilon + B_2\varepsilon^2)$ across different strain rates and strain levels. It is calculated for strain levels ranging from 0.1 to 0.8, with increments of 0.1 for various strain rates. The slope of the fitting curve is utilized to figure out the value of the material constant C_1. Assuming that $\lambda = \lambda_1 + \lambda_2 \ln \dot{\varepsilon}^*$, then Eq. (1) can be expressed as:

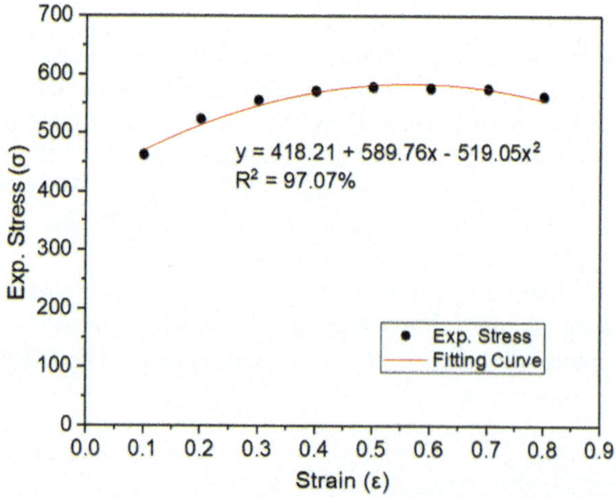

Fig. 2 The plot of relationship between experimental stress and strain at the reference condition

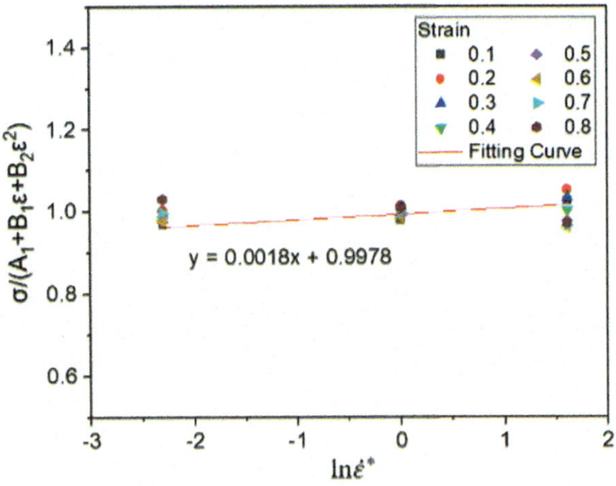

Fig. 3 The graph of $\sigma/(A_1 + B_1\varepsilon + B_2\varepsilon^2)$ and $\ln \dot{\varepsilon}^*$ relationship

$$\frac{\sigma}{\left(A_1 + B_1\varepsilon + B_2\varepsilon^2\right)(1 + C_1 \ln \dot{\varepsilon}^*)} = \exp\left(\lambda T^*\right) \quad (4)$$

Next, applying the natural logarithm to Eq. (4):

$$\ln\left[\frac{\sigma}{\left(A_1 + B_1\varepsilon + B_2\varepsilon^2\right)(1 + C_1 \ln \dot{\varepsilon}^*)}\right] = \lambda\, T^* \quad (5)$$

Subsequently, by performing linear fitting on the left-hand side of Eq. (5) using T^*, the value of λ can be discovered. As illustrated in Fig. 4, the values of that side of the equation for three distinct strain rates can be determined through an examination of the linear fitting curve slope. Then, the parameter λ can be determined based on the plot of λ and $\ln \dot{\varepsilon}^*$ by evaluating λ_1 and λ_2 as the y-intercept and the slope, respectively. Meanwhile, λ is dependent on strain rate, and the variations are plotted in Fig. 5.

Table 1 presents the determined material constants for the MJC model applicable to 10B33 boron steel. These constants were obtained through an appropriate fitting procedure, ensuring that the MJC model accurately captures the behavior of the flow stress of the investigated boron steel under various deformation conditions. By incorporating these material constants into the MJC model, it is possible to form the constitutive equation that characterizes the flow stress (σ) of 10B33 boron steel as follows:

$$\begin{aligned} \sigma = {} & \left(418.21 + 589.76\varepsilon - 519.05\varepsilon^2\right)\left(1 + 0.0018 \ln \dot{\varepsilon}^*\right) \\ & \times \exp\left[\left(-0.0013 - 0.00008 \ln \dot{\varepsilon}^*\right) T^*\right] \end{aligned} \quad (6)$$

Using the constitutive equation that was found, the flow stresses that matched the deformation conditions of the experiments were calculated and shown in Fig. 6.

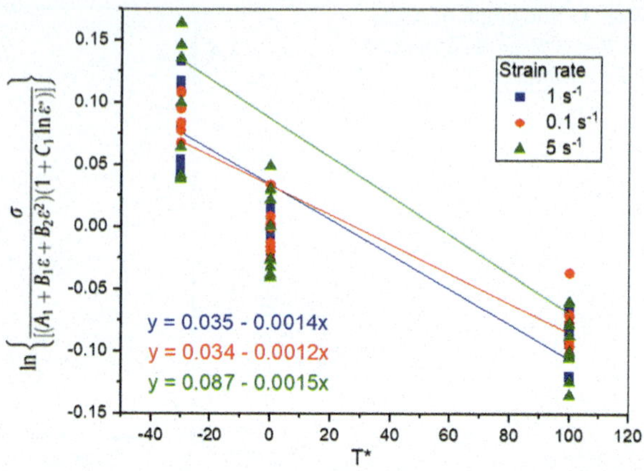

Fig. 4 The correlation between $\ln((\sigma/(A_1 + B_1\varepsilon + B_2\varepsilon^2))/(1 + C_1 \ln \dot{\varepsilon}^*))$ and T^* at different strain rates

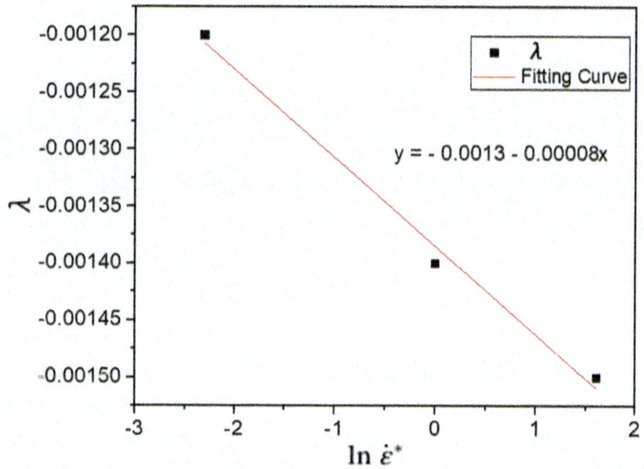

Fig. 5 Relationship between λ and $\ln \dot{\varepsilon}^*$

Table 1 Parameters of MJC model at the designated reference condition

Parameter	λ_1	λ_2	A_1	B_1	B_2	C_1
Value	−0.0013	−0.00008	418.21	589.76	−519.05	0.0018

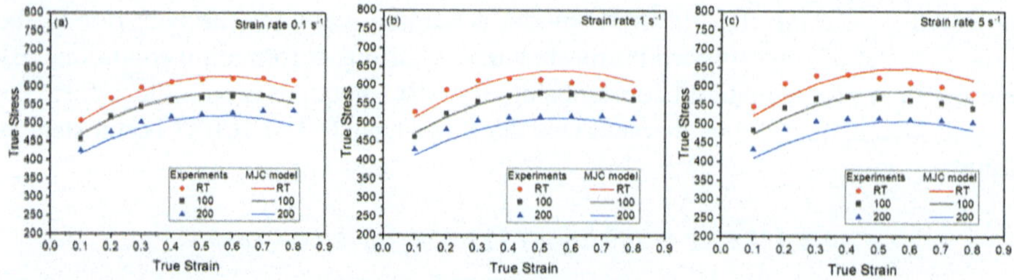

Fig. 6 Comparison graph of estimated flow stress based on MJC model and the experimental data at different strain rates, including **a** 0.1 s^{-1}; **b** 1 s^{-1}, and **c** 5 s^{-1}

Table 2 The parameter of power-law model at investigated conditions

Strain rate	Parameters	RT (°C)	100 (°C)	200 (°C)
$0.1\ s^{-1}$	k	738.18	673.30	602.36
	n	0.1597	0.1658	0.1433
$1\ s^{-1}$	k	744.47	686.71	611.94
	n	0.1615	0.1538	0.1514
$5\ s^{-1}$	k	769.82	687.61	608.47
	n	0.1459	0.1514	0.1434

3.2 *Power-Law Model*

This investigation focused on a test scenario encompassing three distinct temperature and strain rate combinations. Within this context, the Power-Law material model was employed to characterize the material's true stress-true strain behavior.

$$\sigma = k\varepsilon^n \tag{7}$$

Equation (7) formally represents this model, where the key parameters include the yield strain (ε) strain hardening exponent (n), and the strength coefficient (k). To determine the material constants specific to the chosen test conditions, the analysis relied on experimental data extracted from the plastic processing stage, specifically up to a strain of approximately 0.3. Subsequently, a fitting procedure was applied to this data, generating the stress–strain curve equations presented in Table 2.

3.3 *Discussion*

The Modified Johnson–Cook (MJC) constitutive model demonstrated overall effectiveness in predicting the flow stress of 10B33 boron steel. While exhibiting accurate predictions across a variety of deformation conditions, particularly at elevated temperatures, the model faced limitations in representing room temperature flow behavior. Specifically, the calculated flow curve from the MJC model displayed higher accuracy at the reference temperature. A notable deviation was observed at other temperatures, with the model tending to overestimate flow stress at lower temperatures, particularly at room temperature. This sensitivity to temperature variations highlights that 10B33 boron steel flow stress behavior is notably influenced by temperature and suggests model limitations at lower temperatures. Theoretically, flow stress and strain, as well as strain rates, have a proportional relationship with each other, and it is inversely proportional to the deformation temperature. When examining Fig. 6 in a broader context, it is evident that the flow curve indicates a decrease as the strain rate increases from 0.1 to 5 s^{-1} at high strain, while maintaining the same temperature. This phenomenon is even more apparent when observed at room temperature. This can be explained by the fact that the heat generated at lower temperatures appears higher than that at higher initial deformation temperatures, making the softening effect more noticeable. In summary, it is obvious that the influence of temperature on the flow behavior of AISI 10B33 in this investigated case is significant. It also explains why the MJC model has difficulty predicting flow stress at lower temperatures, such as room temperature.

The accuracy of the investigated models was assessed in this study using the correlation coefficient (R). It offers insights into the magnitude of the linear correlation between the measured and estimated values. The aforementioned attribute makes it particularly suitable to evaluate the predictive capacities of models and equations. Mathematically, R is defined as:

$$R = \frac{\sum_{i=1}^{N} \left(\sigma_i^E - \overline{\sigma}^E\right)\left(\sigma_i^P - \overline{\sigma}^P\right)}{\sqrt{\sum_{i=1}^{N} \left(\sigma_i^E - \overline{\sigma}^E\right)^2 \sum_{i=1}^{N} \left(\sigma_i^P - \overline{\sigma}^P\right)^2}} \tag{8}$$

In this context, σ_i^E denotes the experimental value of flow stress, σ_i^P signifies the value of predicted flow stress. $\overline{\sigma}^P$ and $\overline{\sigma}^E$, represent the mean values of the predicted values and the experimental results, respectively, and the totality of data points signifies N. Figure 7 represents the experimental results and the flow stress obtained from the MJC and power-law models. Notably, a large concentration of data points clusters closely around the diagonal line, signifying a strong agreement between model predictions and actual experimental observations.

The calculated correlation coefficient (R) values further support this analysis. The MJC model boasts an R of 0.961, while the power-law model exhibits an R of 0.766. The MJC model outperforms the power-law model in terms of the correlation between experimental data and predicted results.

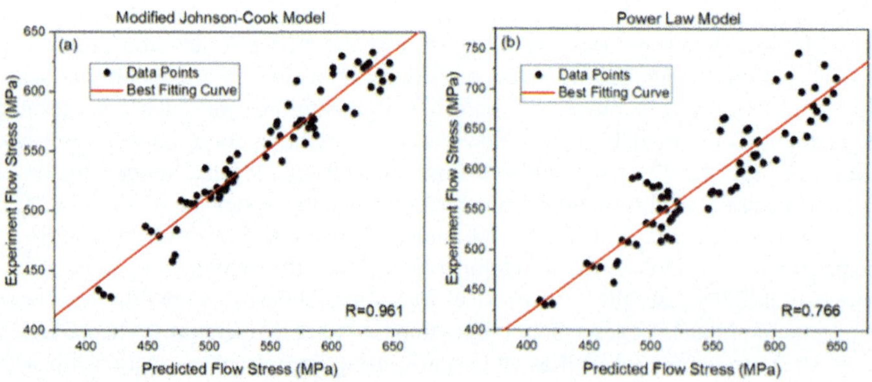

Fig. 7 The correlation plot of experimental results and predicted flow stress given by **a** the MJC model and **b** the power-law model

4 Conclusions

This study explored the flow stress behavior of a cylindrical specimen of 10B33 boron steel during compression testing. The investigation encompassed a wide range of temperatures (room temperature to 200 °C) and strain rates (0.1–5 $s^{-}1$).

1. The relationship between flow stress and temperature as well as strain rate was clearly demonstrated. Notably, the flow stress exhibited a decreasing trend as the deformation temperature increased. This finding aligns with established material behavior in metals, where higher temperatures generally promote easier deformation and lower flow stress.
2. Statistical analysis was employed to assess the dependability of the modified Johnson–Cook constitutive model. The model demonstrated its capacity to deliver accurate stress estimations, suggesting its suitability for computer-aided analysis under the investigated conditions.
3. In comparison to the conventional power-law material model, the research demonstrates that the MJC constitutive model significantly outperforms it when it comes to forecasting flow stress. The modified JC model achieves an R of up to 0.961, exhibiting a significant improvement over the power-law model's R of 0.766. This substantial gap (20.3%) underscores the enhanced accuracy and efficacy of the MJC model for accurate flow stress estimation.

References

1. Campbell, J.: The Mechanisms of Metallurgical Failure. Elsevier Science and Technology Books, Incorporated (2021)
2. Kong, H., Chao, Q., Rolfe, B., Beladi, H.: One-step quenching and partitioning treatment of a tailor welded blank of boron and TRIP steels for automotive applications. Mater. Des. **174**, 107799 (2019)
3. Ottosen, N.S., Ristinmaa, M.: The Mechanics of Constitutive Modeling. Elsevier (2005)
4. Deb, S., Muraleedharan, A., Immanuel, R.J., Panigrahi, S.K., Racineux, G., Marya, S.: Establishing flow stress behaviour of Ti-6Al-4V alloy and development of constitutive models using Johnson-Cook method and Artificial Neural Network for quasi-static and dynamic loading. Theoret. Appl. Fract. Mech. **119**, 103338 (2022)
5. Mirzadeh, H.: Developing constitutive equations of flow stress for hot deformation of AZ31 magnesium alloy under compression, torsion, and tension. Int. J. Mater. Form. **12**(4), 643–648 (2019)
6. Cao, F.R., Bin, Y.I.N., Liu, S.Y., Lu, S.H.I., Wang, S.C., Wen, J.L.: Microstructural evolution, flow stress and constitutive modeling of Al−1.88 Mg−0.18 Sc−0.084 Er alloy during hot compression. Trans. Nonferrous Metals Soc. China **31**(1), 53–73 (2021)
7. Motlagh, Z.S., Tolaminejad, B., Momeni, A.: Prediction of hot deformation flow curves of 1.4542 stainless steel. Met. Mater. Int. **27**, 2512–2529 (2021)
8. Xu, H., Zhao, B., Lu, X., Liu, Z., Li, T., Zhong, N., Yin, X.: A modified Johnson-Cook constitutive model for the compressive flow behaviors of the SnSbCu alloy at high strain rates. J. Mater. Eng. Perform. **28**, 6958–6968 (2019)
9. Li, H.B., Feng, Y.L., Yan, T., Yu, E.L.: Constitutive model of warm deformation behavior of medium carbon steel. J. Iron. Steel Res. Int. **23**(9), 940–948 (2016)

10. Li, H., He, L., Zhao, G., Zhang, L.: Constitutive relationships of hot stamping boron steel B1500HS based on the modified Arrhenius and Johnson-Cook model. Mater. Sci. Eng. A **580**, 330–348 (2013)
11. Sirigiri, V.K.R., Gudiga, V.Y., Gattu, U.S., Suneesh, G., Buddaraju, K.M.: A review on Johnson Cook material model. Mater. Today Proc. **62**, 3450–3456 (2022)
12. Zou, Z., He, L., Zhou, T., Zhang, W., Tian, P., Zhou, X.: Research on inverse identification of Johnson-Cook constitutive parameters for turning 304 stainless steel based on coupling simulation. J. Market. Res. **23**, 2244–2262 (2023)
13. Priest, J., Ghadbeigi, H., Ayvar-Soberanis, S., Liljerehn, A., Way, M.: A modified Johnson-Cook constitutive model for improved thermal softening prediction of machining simulations in C45 steel. Proc. CIRP **108**, 106–111 (2022)
14. Zhu, H., Zhang, C., Chen, S., Wu, J.: A modified Johnson-Cook constitutive model for structural steel after cooling from high temperature. Constr. Build. Mater. **340**, 127746 (2022)

A Systematic Approach to Modeling Crankshaft and Camshaft Signal for the Electronic Control System of the Hyundai D4EA Engine

Tinh Van Tran, Tien Thanh Tran, Thinh Nguyen Cuong, Tri Dinh Quoc, Lam Quang Tran, and Quang T. D. Pham

Abstract Electronic engine control technology stands as a pivotal advancement in modern engine development. The Engine Control Unit (ECU) retrieves operational data from the engine to precisely regulate ignition timing, fuel injection, air–fuel ratio, exhaust emissions, and more, based on signals from various sensors, particularly the crankshaft and camshaft position sensors. This research introduces a methodology for developing a simulation model to analyze the signal of crankshaft and camshaft position sensor in the Diesel D4EA engine using Matlab/Simulink. The waveform model is divided into several segments, each with distinctive parameters represented through sinusoidal, arcsine, and square wave functions to match the characteristic signals of the crankshaft and camshaft within one engine cycle. Both signals display synchronization within one engine cycle, when the crankshaft completes two revolutions, the camshaft completes one revolution during the experimental process. The model is validated through a comparison between simulated and real signals at various engine speeds, demonstrating high precision and reliability. The model is developed with the intention of replacing actual sensors to perform measurements and test the engine ECU in an offline condition, serving educational and research purposes.

Keywords Signal simulator · ECU · Engine speed sensor · Missing teeth

T. V. Tran · T. T. Tran · T. N. Cuong · T. D. Quoc · L. Q. Tran · Q. T. D. Pham (✉)
Faculty of Transportation Engineering, Ho Chi Minh City University of Technology (HCMUT), Ho Chi Minh City, Vietnam
e-mail: quangtdpham@hcmut.edu.vn

Vietnam National University Ho Chi Minh City (VNU-HCM), Ho Chi Minh City, Vietnam

D. H. Tien et al. (eds.), *Proceedings of the 1st International Conference on Sustainability and Emerging Technologies for Smart Manufacturing*, Proceedings in Technology Transfer, https://doi.org/10.1007/978-981-97-7083-0_52

1 Introduction

Nowadays, researchers are looking for ways to control the operating processes to improve engine performance, thus reducing fuel consumption and exhaust emission. To achieve this, the Engine Control Unit (ECU) relies on real-time data obtained from various sensors, such as intake manifold pressure sensors, crankshaft and camshaft position sensors, and oxygen sensor arrays [1, 2]. Through this, the ECU can efficiently control ignition timing, exhaust emissions, air–fuel ratio, and fuel injection. However, employing physical components to test the ECU proves to be highly expensive and time-consuming. Therefore, the construction of signal generators becomes a crucial step in the dynamic testing process of ECU [3, 4]. Utilizing signal generators for testing can slash costs and complexity by up to 90% when contrasted with conventional testing approaches [5].

In the present day, ECUs can be tested using computer tools such as LabVIEW and MATLAB [6, 7], allowing for the simulation of the real operational state of the ECU. Additionally, Hardware-in-the-Loop Simulation (HILS) platforms serve a similar role [8]. The crankshaft (CKPS) and camshaft signals (CMPS) play a crucial role in timing fuel injection and ignition, especially in gasoline engines. Similarly, the camshaft sensor employs a Hall sensor. Some specific research conducted through complex hardware simulation [9] can lead to high costs, lengthy development cycles, and limited flexibility. Additionally, the low accuracy of simulated signals is another common issue, significantly affecting the exactness of following signal processing [10]. Inaccurate signal simulation may hinder successful testing of ECU signal conditioning capabilities, especially considering sensitivity of certain circuits to irregularities in missing teeth waveforms.

This study aims to develop a precise model for synchronizing crankshaft position sensor and camshaft sensor using Matlab/Simulink. The model serves as a substitute for actual sensors to conduct offline measurements and test the ECU. To achieve this, properties of crankshaft and camshaft signals obtained from calibration experiments are mathematically modeled using sine, square wave, and arcsine functions, along with accurately controlled synchronization relationships. Additionally, both sine and arcsine waveforms are employed individually to enhance signal accuracy, especially at positions with two missing teeth.

2 Construction of the Inductive Speed Signal Model

2.1 CKPS Partitioning Model

To construct a segmentation model for the CKPS, it is necessary to determine its characteristic parameters, including the number of teeth N_{tot}, the number of missing teeth N_{mis}, the frequency f of the signal, and the voltage offset deviation V_{offset}. The CKPS can be segmented into two parts: the multiteeth signal $W_{multiteeth}$ and the

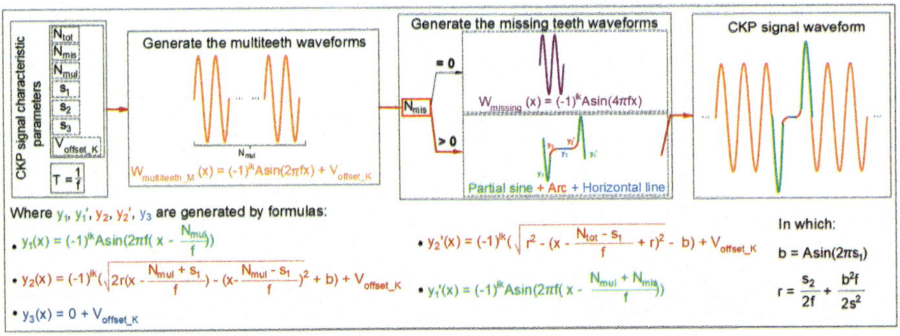

Fig. 1 The process of modeling the CKPS

missing teeth signal W_{missing}. In the case of $N_{\text{mis}} > 0$, the missing teeth waveform is described by combining parts of sinusoidal waves, arcs, and a horizontal line, denoted, respectively, as y_1 and y_1' in green, y_2 and y_2' in red, and y_3 in blue. Finally, the complete CKPS waveform W_{CKP} is shown over one cycle. The process of modeling the CKPS for the D4EA engine follows a series of steps outlined by the red line in Fig. 1, which corresponds to the formulas provided.

The input parameters of the CKPS for the simulation model in this study are presented in Table 1, where specific values are chosen for each parameter.

Furthermore, the voltage amplitude of the signal is a special parameter. This quantity varies linearly with the engine speed through in the following equation:

$$V = 0.00722 \times N \tag{1}$$

The CMPS is simulated as a square wave synchronized with the CKPS. For the Hyundai D4EA engine, the CMPS waveform reverses when teeth numbers 4 and 55 of the crankshaft appear.

Table 1 CKPS characteristic parameters of the D4EA Engine

S. No.	Parameters	Value	Meaning
1	N_{tot}	60	Number of teeth
2	N_{mis}	2	Number of missing teeth
3	N_{mul}	57	Number of multiteeth
4	s_1	0.5	Partial sinusoidal multiple
5	s_2	1	Partial arc multiple
6	s_3	0	Partial line multiple
7	V_{offset}	2.2	Voltage offset of CKPS

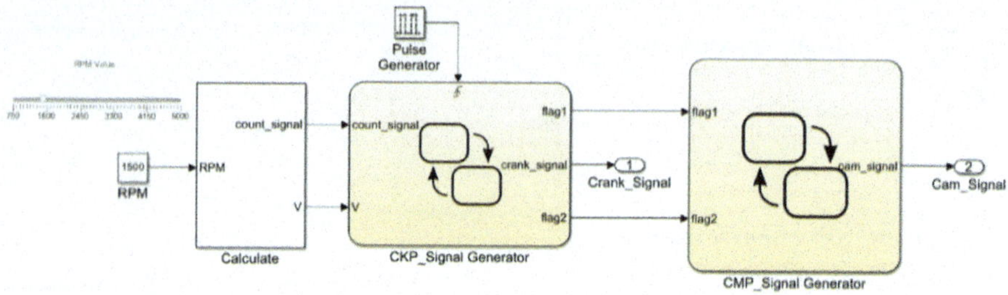

Fig. 2 The simulation model of CKP and CMPS

2.2 *Simulink Model of the CKPS and the CMPS*

In this article, Matlab/Simulink is used as the software interface for simulating the CKPS and CMPS. The developed model consists of an RPM input speed module; a signal generation module cluster, which includes a subsystem used for computing signal amplitude and the number of sine waves; two Stateflow blocks generating signal waves based on algorithms; and two waveform output modules named Crank_Signal and Cam_Signal, as shown in Fig. 2.

The crankshaft model includes two parts as shown in Fig. 3: First_Part and Later_Part. These two states represent the first half and the later half. The activation events timing for each tooth are tracked by a counter variable.

The CKPS includes three states: The CKPS maintains a normal state from teeth 2 to 57, exhibiting a uniform sine wave waveform. As tooth 0 and tooth 59 are missing teeth, the CKPS drops nearly zero, indicating the missing teeth state. This cycle has an arc-shaped waveform and can be represented by the arcs in function. The sudden increase in the amplitude value to a maximum value at the position of the missing teeth is generated by the electromagnetic induction, producing a characteristic waveform. When tooth 1 emerges, the waveform of the signal is segmented into two parts. While it bears resemblance to a sine wave, the CKPS of the latter half surpasses that of the first half, and this is the peak cycle. Tooth 58 resembles tooth 1, but with a greater voltage amplitude in the latter half.

Additionally, the variant flags (1, 2) are crucial in this model. When tooth 4 and tooth 55 of the crankshaft emerge, the CMPS undergoes a reversal (the voltage amplitude of the CMPS changes from high to low or vice versa). Flag variable (1, 2) is employed to notify the CMPS simulation model that it is time to invert the signal state. In this instance, the number of teeth is configured to 60, and this parameter can be adjusted to suit various types of engines.

The camshaft exhibits a square waveform, and its frequency is half that of the CKPS. The CMPS simulation model is depicted in Fig. 4.

The variable flag (1, 2) is a scalar value limited to a maximum of 2. During each revolution of the crankshaft, the flag will increase by 1. This will cause the system to progress to the subsequent state, resulting in an inversion of the CMPS signal.

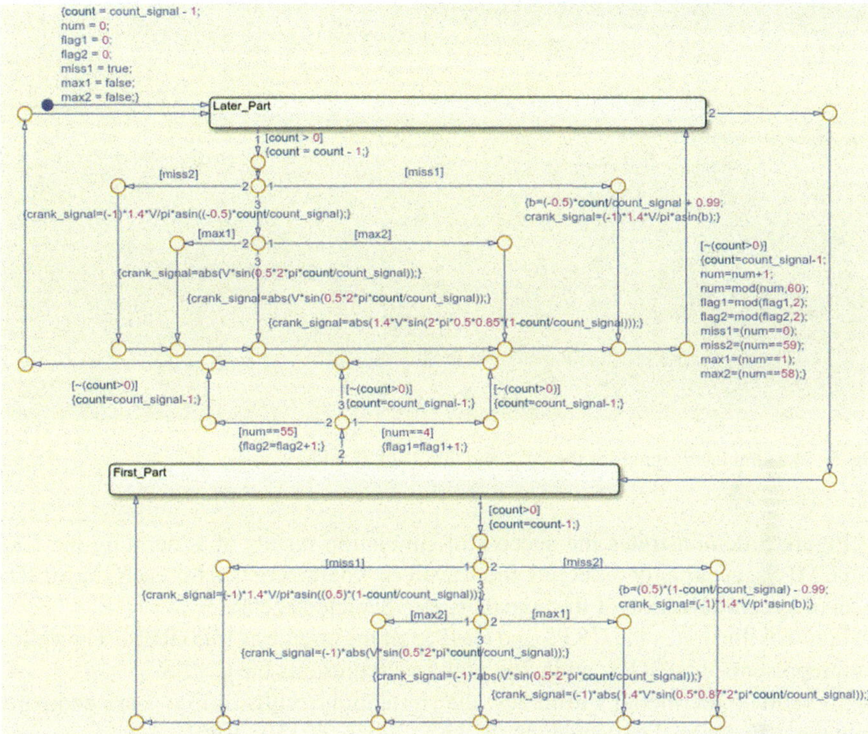

Fig. 3 The CKPS simulation model

Fig. 4 The CMPS simulation model

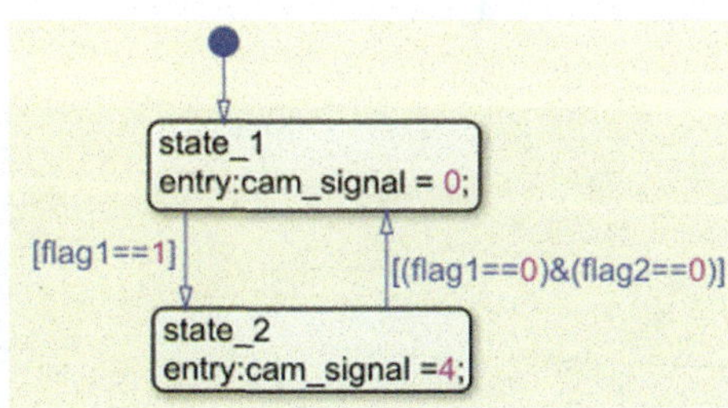

3 Results and Discussions

The simulated signal is displayed in Fig. 5. The engine speed is set at 875 RPM. In the Matlab/Simulink model, the sampling time is configured to 0.5 μs, and the simulation time is set to 0.3 s. The yellow curve depicts the CKPS, while the green curve depicts the CMPS. The time is represented on the horizontal axis, while the voltage amplitude of the signal is depicted on the vertical axis.

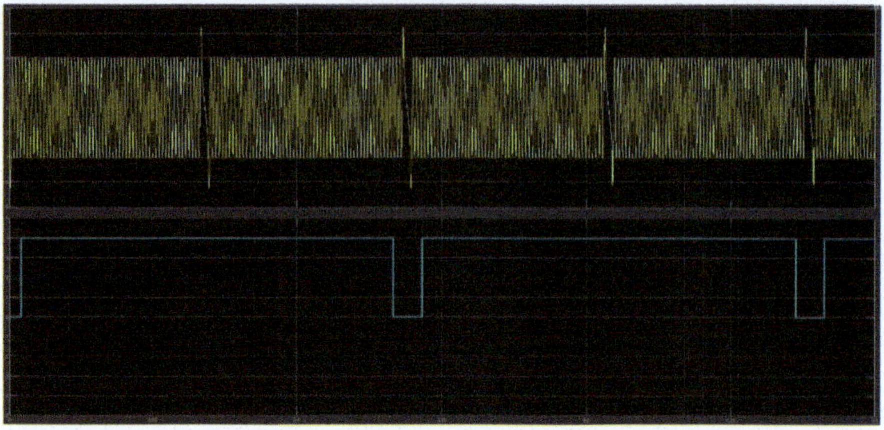

Fig. 5 The simulated signals of the CKP and CMP at 875 RPM

Figure 5 demonstrates the successful simulation results of generating the CKP and CMPSs, accurately reflecting the waveform characteristics for each sensor. The level of synchronization of the signals is also depicted accurately.

Figure 6 illustrates the CKPS and CMPSs of the engine at 1500 RPM. The yellow path represents the CKPS, while the pink line represents the CMPS.

To validate the model's accuracy, the simulation results in Fig. 7 are compared with the actual signal collected, as shown in Fig. 6, at 1500 RPM.

The simulated CKPS (blue curve) closely matches the actual CKPS (red curve) as depicted in Fig. 8, providing evidence for the consistency and high accuracy of the simulation process. Both signals are sampled at a frequency of 2 MHz.

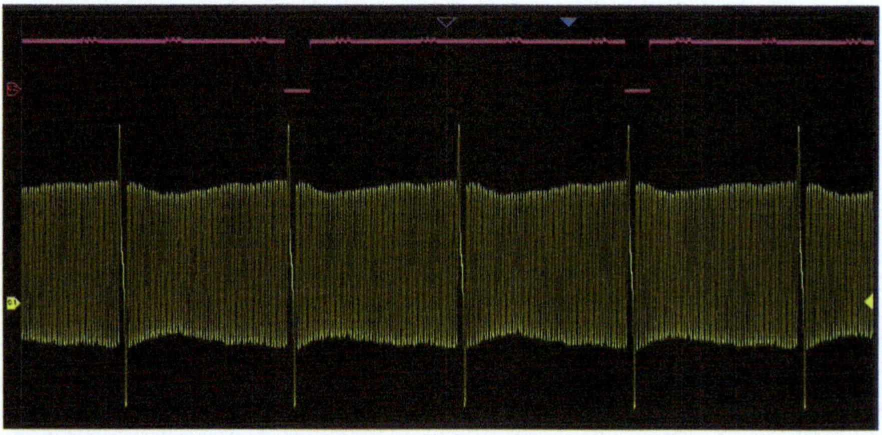

Fig. 6 The collected signal at 1500 RPM

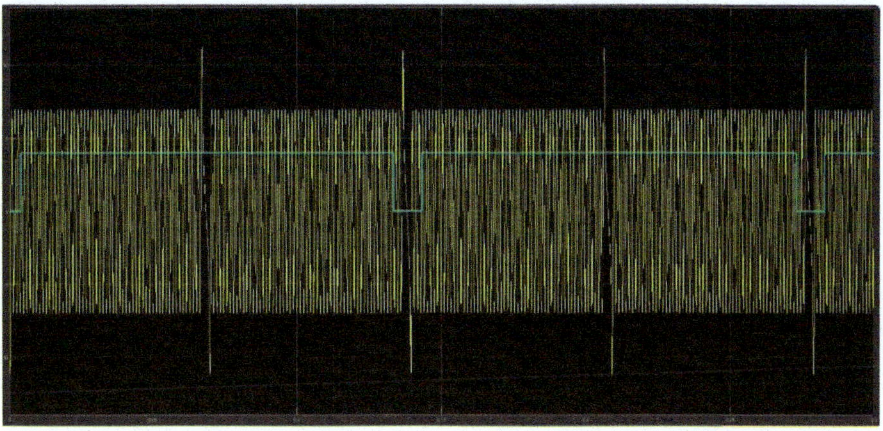

Fig. 7 The simulated signal at 1500 RPM

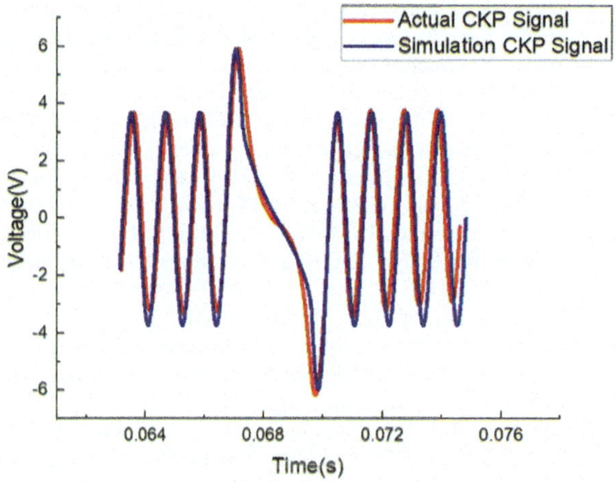

Fig. 8 Comparison between simulated signal and real-world signal

4 Conclusions

The paper has successfully analyzed the characteristic parameters, constructed a segmentation model, and simulated the crankshaft and camshaft signals. The simulated CKP and CMPSs exhibit high accuracy and synchronization. Besides the diesel engine D4EA used for testing, this research model can be further developed for various other engines using electromagnetic speed sensors, by simply adjusting the characteristic parameters and corresponding algorithms. Moreover, the model can be utilized for testing and validating different types of ECUs in real-world scenarios.

The following step of this research is to complete the test bench setup to carry out real-world testing of the ECU in the lab. The simulation program from Matlab/Simulink will be embedded into the DSP C2000 microcontroller, along with an amplification circuit to provide stimulus signals to the ECU. To verify this setup, an oscilloscope will be used to detect pulses from the fuel injector and ignition system to determine whether the CKPS and CMPS are recognized by the ECU. Additionally, an OBD-II diagnostic tool will be connected to the ECU to read operational parameters of simulated sensor signals fed into the ECU via CAN bus communication, such as engine speed, coolant temperature, and ignition timing, ensuring the most accurate simulation of sensor signals.

Acknowledgements We acknowledge the support of time and facilities from Ho Chi Minh City University of Technology (HCMUT) and Vietnam National University Ho Chi Minh City (VNU-HCM) for this study.

References

1. Zheng, T., Zhang, Y., Li, Y., Shi, L.: Real-time combustion torque estimation and dynamic misfire fault diagnosis in gasoline engine. Mech. Syst. Signal Process. **126**, 521–535 (2019)
2. Wang, Y., Wang, G., Yao, G., Shen, L.: Research on the characteristics of operating non-uniformity of a high-pressure common-rail diesel engine based on crankshaft segment signals. IEEE Access: Pract. Innov. Open Sol. **9**, 64906–64917 (2021)
3. Xia, J.L., Sun, Q.D., Chen, B.C., Yuan, X.X., Dong, T.: Development of low cost hardware-in-the-loop test system and a case study for electric vehicle controller. In: International Conference on Applied System Innovation, Okinawa, Japan, pp. 1–4 (2016)
4. Korodi, A., Anitei, D., Boitor, A., Silea, I.: Image-processing-based low-cost fault detection solution for End-of-line ECUs in automotive manufacturing. Sensors (Basel, Switzerland) **20**(12), 3520 (2020)
5. Lattarulo, R., Pérez, J., Dendaluce, M.: A complete framework for developing and testing automated driving controllers. IFAC-PapersOnLine **50**(1), 258–263 (2017)
6. Huang, X., Shen, R.: Engine electronic control sensor signal acquisition system design based on LabVIEW. Automobi. App. Tech. **17**, 65–67 (2016)
7. Cui, Y.Y., Huo, L.S., Zhu, M.Q., Xiang, L.I.: Study and design of the engine ECU test system based on VC++. Electron. Des. Eng **19**, 28–30 (2011)
8. Fang, L.S., Guo, H.Y., Chen, H.: Establishment and application of vehicle Dynamic HIL simulation test platform. Control Eng. China: China **22**(3) (2015)
9. Du, B., Sterpone, L.: An FPGA-based testing platform for the validation of automotive powertrain ECU. In: 2016 IFIP/IEEE International Conference on Very Large Scale Integration (VLSI-SoC), pp. 1–7 (2016)
10. Zheng, X., Wang, B., Ge, Y.: A segmentation model of ECU excitation signal based on characteristic parameters. Sensors (Basel, Switzerland) **21**(12), 4165 (2021)

Research to Evaluate the Effectiveness of a Dehumidifier-Integrated Air Conditioner in Tropical Monsoon Climate

Ta Van Chuong

Abstract A model of a Dehumidifier-Integrated Air Conditioner (DHIAC) with the ability to control air temperature and humidity precisely at the same time has been developed by our research team. In this research, the operation of DHIAC in hot and humid climate conditions for one year was simulated to evaluate the device's performance. The gray box simulation method, combining theory and experiment, gives high accuracy, and a convincing argument is selected for the application. The simulation program is built on Matlab software based on a system of equations to simulate the change of temperature and humidity parameters of the air-conditioned space and regional weather data for a long time. It allows the evaluation of the equipment's operating efficiency in a one-year cycle. Simulation results show that DHIAC can control air temperature and humidity parameters according to national standards with high accuracy. The maximum error of air temperature and humidity compared to the set value is 2.7%. The average EEF energy efficiency factor for the whole year is 2.2. Maximum process time is 1095 s in summer and 607 s in winter. This DHIAC can provide a comfortable environment for people with low energy consumption.

Keywords Air conditioning · Dehumidification · Gray box simulation method · Weather data · COP

1 Introduction

The modernization process in the past increased primary energy consumption and CO_2 emissions in the two decades 1994 to 2014 by 49% and 43%, respectively [1], causing the greenhouse effect and global warming. The Earth's temperature has increased by 0.74 °C over the past century [2]. Climate change threatens humans

T. V. Chuong (✉)
School of Heat Engineering and Refrigeration, Hanoi University of Science and Technology, Hanoi, Vietnam
e-mail: Chuong.tavan@hust.edu.vn

D. H. Tien et al. (eds.), *Proceedings of the 1st International Conference on Sustainability and Emerging Technologies for Smart Manufacturing*, Proceedings in Technology Transfer, https://doi.org/10.1007/978-981-97-7083-0_53

and the entire living environment at different levels across the Earth [3]. According to a 2019 United Nations report, the World's population is expected to increase by 2 billion people between 2019 and 2050. Rapid population growth requires more energy on Earth. In response, most countries have committed to limit global warming to below 2 °C through the Paris Agreement, and a series of measures have been introduced that apply to industry [4], agriculture [5], commercial [6], and residential [7]. Using air conditioning for cooling and heating consumes a lot of energy, accounting for nearly 20% of total electricity consumption in buildings in the World today, and can be up to 80% in high-risk areas in hot and humid climates [8]. Therefore, this is one of the leading causes of global greenhouse gas emissions [9]. Reducing the power consumption of the air conditioning system is one of the leading solutions to reduce CO_2 emissions to zero.

In hot and humid climates, high humidity causes discomfort, fatigue, and many respiratory diseases. It is a favorable environment for viruses and bacteria to grow, potentially leading to mold and rot. Additionally, high humidity can damage electronic devices, so they are often stored in environments with very low controlled air humidity. Air humidity in hot and humid climates can reach over 90% [10], so the requirement for dehumidification is high and has practical significance.

Many dehumidification methods have been researched in industry and civil use, typically including membrane dehumidification technology with integrated air conditioning [11], dehumidification based on the difference in water vapor pressure between the two environments, causing water vapor to move through the separating membrane. Air conditioning integrated liquid dehumidification [12] uses solutions with high water adsorption capacity after being reconstituted to separate water from the air. Combining dehumidification with silica gel and air conditioning [13], the reconstituted silica gel cakes have good water absorption ability to help reduce humidity in the air. Air Handling Unit, Precision Air Conditioning [14], has a dehumidification process based on the principle of cooling the air below the dew point temperature and separating water from the air.

To solve the problem of precise control of air temperature and humidity, a new model of a Dehumidifier-Integrated Air Conditioner (DHIAC) with high operating energy efficiency has been developed [15]. In this study, the processing time and energy efficiency of the equipment at different operating conditions according to actual weather conditions with data from national standards for one year will be researched. According to Afram and Janabi-Sharifi [14], HVAC system modeling methods can be divided into three categories: data-driven methods (also known as black box methods, which monitor input data and the output is correlated through mathematical equations) [16]; physics-based methods (also known as white box methods, built on the laws of physics) [17]; gray box method (basic structure based on physics and some parameters selected from data obtained through tests) [18]. Black box models are developed by measuring data from the system's inputs and outputs and synthesizing mathematical functions based on the obtained data. Black box methods require little understanding of system physics, and they are highly accurate compared to physically based models. However, black box models are built for each specific object, so their generalizability is poor. White box models are more general and can be applied to a group of similar objects, but the accuracy of the method is not high, and when building white box models, a high level of knowledge of system physics is required. The white box method uses equipment specifications given by the manufacturer to model system dynamics. To handle local properties in black box models and large errors in white box models, the gray box method combines system physical equations built on device specifications and data. The simulation program is created based on physical expressions relating to quantities during the processing process, combined with actual survey parameters to help increase accuracy.

2 Materials and Methods

2.1 The Dehumidifier-Integrated Air Conditioner

The Dehumidifier-Integrated Air Conditioner (DHIAC) model based on the cooling effect developed by our research team operates according to the principle diagram shown in Fig. 1 [15]. The process of controlling air temperature and moisture is shown on the *t–w* graph in Fig. 2. DHIAC can regulate room temperature and air humidity, including cooling, heating, dehumidification, and moisture. The air in the room is cooled and dehumidified by the condenser when the air temperature drops below the dew point temperature (I–C process). The heater helps increase the air temperature when the air temperature is lower than required (C–H process). If the room's air humidity is less than the set value, the humidifier will increase the moisture in the air (H–S process).

The DHIAC control algorithm flow chart is introduced, as shown in Fig. 3. The device is operated in four modes: Mode 1—Cooling, Mode 2—Heating, Mode 3—Dehumidification, and Mode 4—Humidification. When the room air temperature

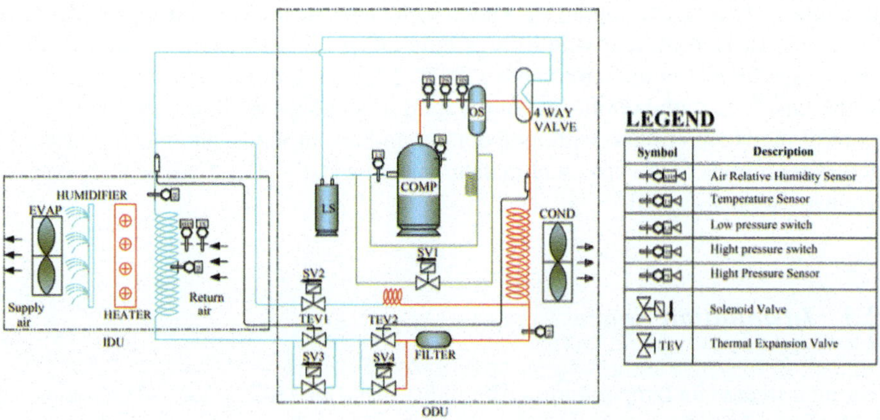

Fig. 1 DHIAC principle diagram

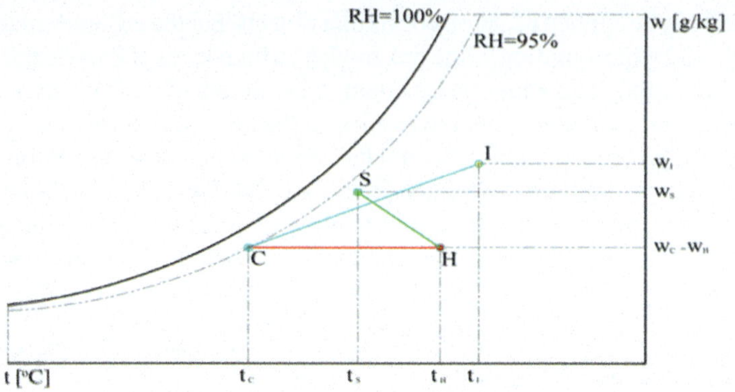

Fig. 2 The process of controlling air temperature and moisture

is greater than the set temperature (T_{sp}) by a preset maximum deviation (ΔT), the device will operate in cooling mode (Mode 1). If the air temperature in the room is greater than the set value by a preset range (ΔT), the device will operate in heating mode (Mode 2). When the room air humidity is higher than the preset humidity level (ΔRH), the device will work in dehumidification mode (Mode 3). If the room's humidity is less than the preset humidity level (ΔRH), the device will operate in humidification mode (Mode 4).

2.2 *Tropical Monsoon Climate*

This type of climate, created by monsoon flows that change direction according to the seasons, is most common in South and Southeast Asia. High temperature is a factor that seriously affects the health of people in these areas [19]. A typical example is Hanoi, the capital of Vietnam, located in the inner tropical region of the Northern Hemisphere with a large amount of radiation, lots of sunshine, and an average annual temperature above 20 °C [20]. In addition, Vietnam is adjacent to the East Sea, with significant annual rainfall and high air humidity. Hanoi has a tropical monsoon climate with temperature and air humidity parameters measured at different times of the year at climate monitoring stations that can be referred to in Vietnam Standards [20].

2.3 *Installation Space*

The air is treated for temperature and humidity in the DHIAC device and then supplied to the room. In addition to outdoor conditions and installation conditions, installation space is also one of the main factors affecting the operation of DHIAC. To determine

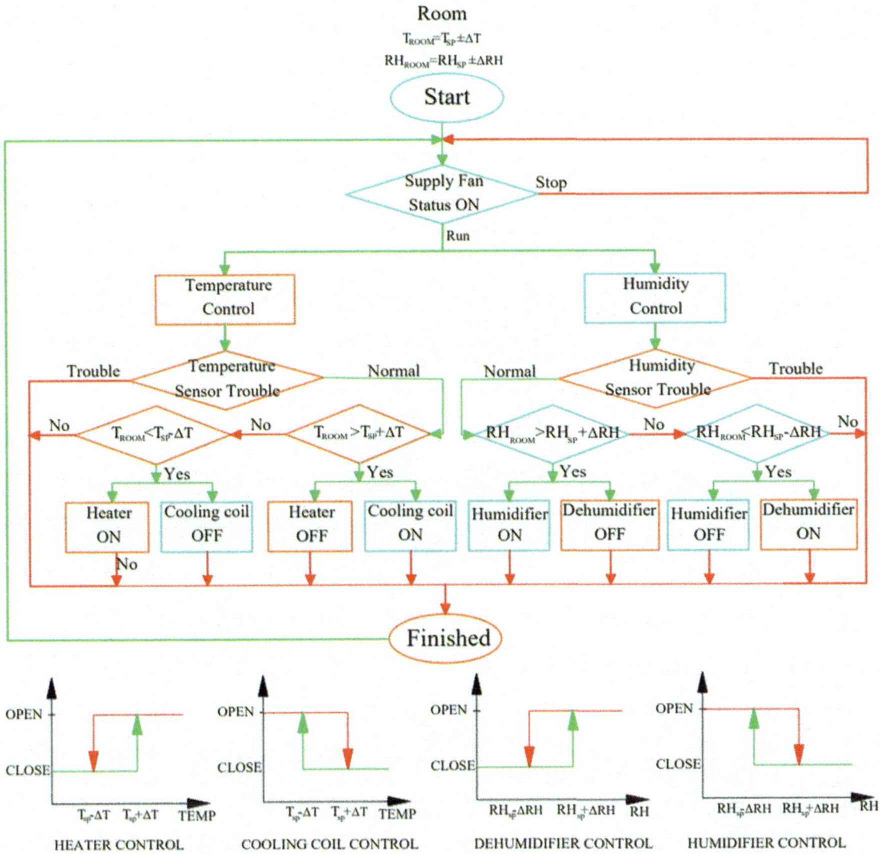

Fig. 3 Flow chart of DHIAC control algorithm

the device's performance, a small dormitory room for two people was used as the installation space. The size of the room is 4 × 3 × 3.5 (m), including one double bunk bed, two computers, doors, and some electrical equipment. Details of the installation space are shown in Fig. 4.

2.4 *Excess Heat and Moisture in the Installation Space*

The total excess heat for the room space is determined by the formula [21]:

$$Q_{\sum} = Q_1 + Q_2 + Q_3 + Q_4 + Q_5 + Q_6 + Q_7 + Q_8 + Q_9 + Q_{10} \quad (1)$$

in which Q_1: heat emitted from machinery and equipment [W]; Q_2: heat emitted from the light [W]; Q_3: heat emitted from humans [W]; Q_4: heat from solar radiation

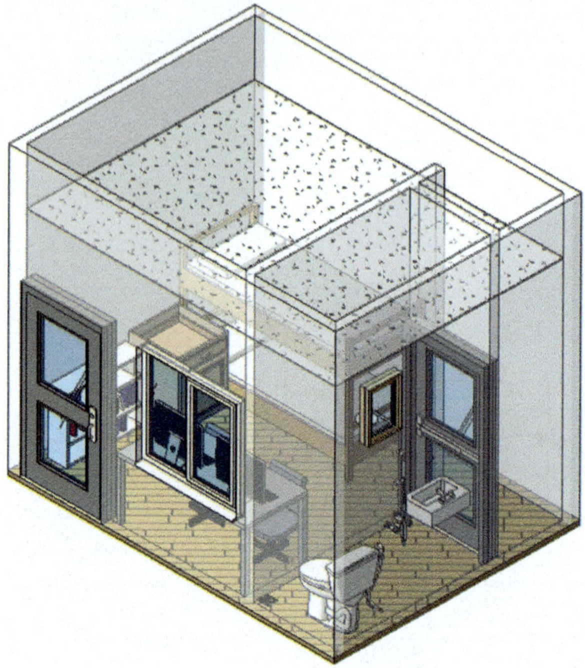

Fig. 4 DHIAC installation space

through glass windows [W]; Q_5: heat from solar radiation through the enclosure [W]; Q_6: heat released due to air leakage through the door [W]; Q_7: heat transmitted through the wall [W]; Q_8: heat transmitted through the ceiling [W]; Q_9: heat transfer through the substrate [W]; and Q_{10}: additional heat due to wind and wall direction [W].

The total excess moisture for the room space is determined by the formula [21]:

$$W_{\sum} = W_1 + W_2 + W_3 + W_4 \tag{2}$$

W_1, W_2, W_3, and W_4 (g/s) are, respectively, excess moisture due to opening the door, excess moisture emitted by people, excess moisture due to door cracks, and excess moisture due to diffusion through walls, floors, and roofs.

Based on the DHIAC installation space and actual climate conditions (temperature, humidity, and radiation intensity) in one year, excess heat and excess moisture in the installation space in one year are shown, respectively, in Figs. 5 and 6.

2.5 Characteristic Equations of DHIAC

Depending on operating conditions and operating modes, the device's capacity will also change. To characterize this, correction factors for cooling capacity and heating

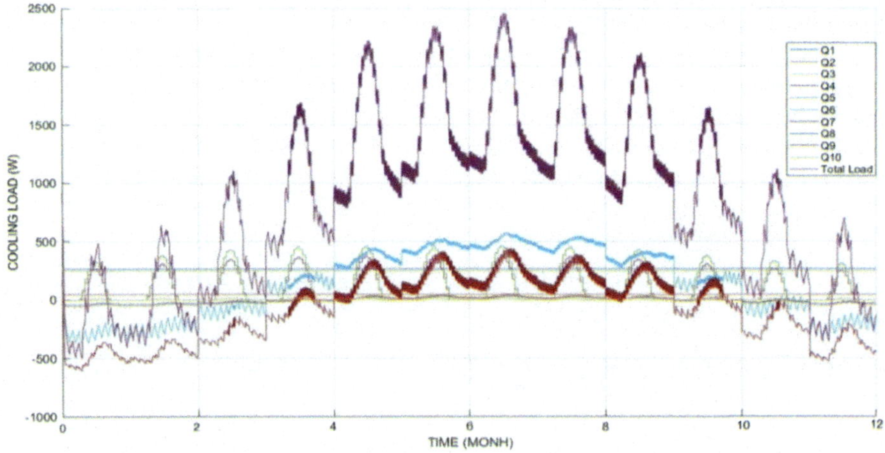

Fig. 5 Excess heat in the installation space

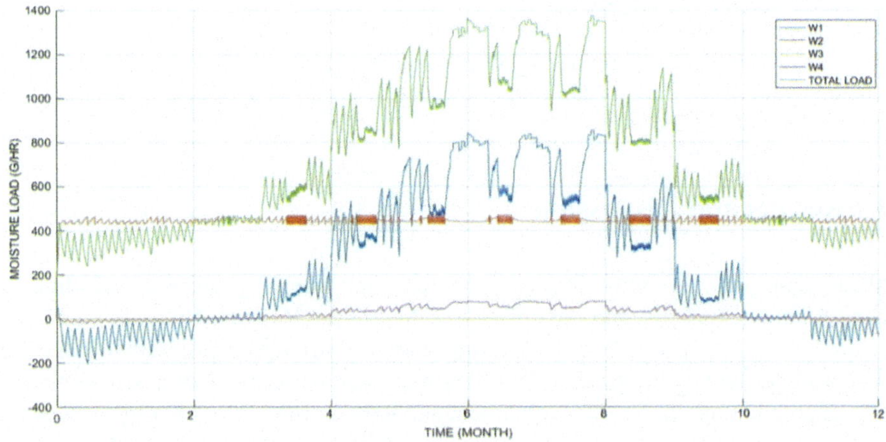

Fig. 6 Excess moisture in the installation space

capacity according to operating conditions have been given in (3) and (4).

$$Q_E = \alpha_{TH}^E.\alpha_v^E.\alpha_{Tod}^E.Q_{Eo} \tag{3}$$

$$Q_C = \alpha_{Tod}^C.\alpha_{Tid}^C.Q_{Co} \tag{4}$$

in which, Q_E: cooling capacity of the evaporator [kW]; Q_C: heating capacity of condenser [kW]; α_{TH}^E: correction factor for cooling capacity according to room temperature and air humidity; α_v^E: correction factor for cooling capacity according to wind speed through the condenser; $\alpha_{T_{od}}^E$: correction factor for cooling capacity according to outdoor temperature; Q_{Eo}: cooling capacity in base conditions [kW]; $\alpha^{\mathrm{C}}_{T_{\mathrm{od}}}$: coefficient to correct heating capacity according to outdoor temperature; $\alpha^{\mathrm{C}}_{T_{\mathrm{id}}}$: heating capacity correction coefficient according to room air temperature; and Q_{CO}: heating heat capacity of the device at original conditions [kW].

The total energy consumption of DHIAC in a year, P [kJ], is determined through the power consumption of DHIAC, heater, humidifier, and fan through the formula:

$$P = \int_{0}^{31536000} \left(Q_{\mathrm{CO}(\tau)} + Q_{\mathrm{HEATER}(\tau)} + N_{\mathrm{HUMIDIFIER}(\tau)} + N_{\mathrm{FAN}(\tau)}\right) \mathrm{d}\tau \quad (5)$$

The energy efficiency factor of DHIAC for one year, EEF, is determined by the ratio of useful energy capacity to the device's power consumption, as shown in the formula:

$$\mathrm{EEF} = \frac{Q_{\mathrm{USE}}}{Q_{\mathrm{E}}} \quad (6)$$

A program to simulate the operation of DHIAC has also been built on Matlab software to perform calculations. After loading the input data, the program will calculate and display important output parameters.

3 Results and Discussion

In the scope of this study, room temperature and humidity are set according to comfortable conditions, and the errors specified in the Vietnam National Standards [20] are 25 ± 1 [°C] and 60 ± 7[%]. Using DHIAC's simulation program to calculate with a signal acquisition time of 10 s for control and display, the output response of air temperature and humidity is obtained, as shown in Figs. 7 and 8, respectively.

Simulation results, as shown in Figs. 7 and 8, show that the room air temperature changes from 24.2 to 25.9 °C, all within the set value range of 25 ± 1 [°C]. The air humidity in the room varied from 54.2 to 68.3%. The humidity is mainly within the set value of 60 ± 7[%]. A few times when the humidity exceeds the set value, but the maximum relative error of humidity is small, reaching a value of 2.17%. The energy efficiency factor of DHIAC in a year is 2.2.

At different conditions of temperature, outdoor air humidity, and setting parameters, DHIAC takes different times to control the temperature and air humidity as set. Outside air parameters used to design air conditioners must be selected according to the number of hours m, hours per year, allowing for uncertainty about the set temperature and humidity or according to the guarantee coefficient K_{g} [20]. Use design outdoor parameters at conditions corresponding to the number of uncertain hours in a year of 0, 100, 200, and 300 [h/year] for Hanoi city and installation room parameters, respectively set to 25 ± 1 [°C] and 60 ± 7[%]. When the room's initial

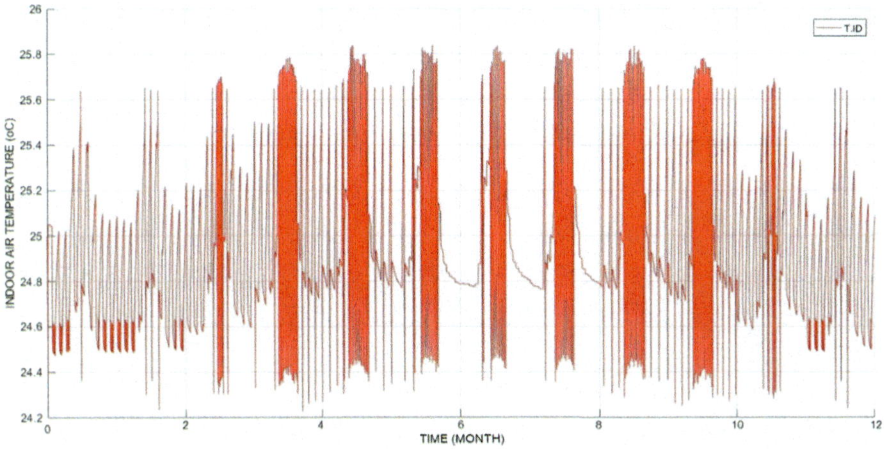

Fig. 7 The air temperature in the room is controlled by DHIAC

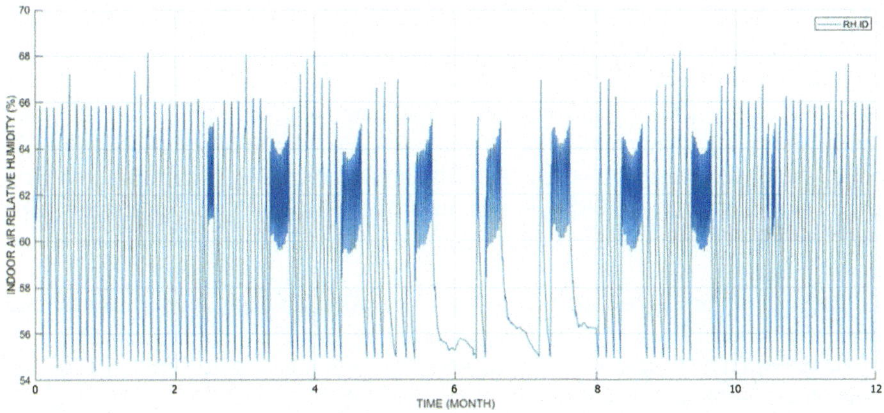

Fig. 8 The air humidity in the room is controlled by DHIAC

temperature and humidity parameters are equal to the outdoor parameters, the process time, τ_{Process}, for the air temperature and humidity in the room to reach the set value range is shown in Table 1.

According to the simulation results (see Table 1), depending on different guarantee coefficients, the time DHIAC takes to control the temperature and humidity of the air to the set range is also other. The larger the guarantee coefficient, the smaller the control time will be. When the guarantee coefficient is 1, this process time reaches its maximum value at 1095 s in summer and 607 s in winter.

Table 1 Process time of DHIAC

Season	m [h/year]	K_g	T_{OD} [°C]	RH_{OD} [%]	T_{SP} [°C]	RH_{SP} [%]	$\tau_{Process}$ [s]
Summer	0	1	40	58.4	25 ± 1	60 ± 7	1095
	100	0.989	36.70	54.80	25 ± 1	60 ± 7	731
	200	0.977	36.10	55.10	25 ± 1	60 ± 7	687
	300	0.966	35.60	56.00	25 ± 1	60 ± 7	672
Winter	0	1.000	5.00	96.30	25 ± 1	60 ± 7	607
	100	0.989	9.60	85.80	25 ± 1	60 ± 7	500
	200	0.977	10.60	85.50	25 ± 1	60 ± 7	468
	300	0.966	11.40	85.40	25 ± 1	60 ± 7	457

4 Conclusions

The operation of DHIAC in hot and humid climate conditions was simulated to evaluate the device's performance. The simulation program is built on Matlab software, which allows the evaluation of the equipment's operating efficiency in a one-year cycle. Simulation results show that DHIAC can accurately control air temperature and humidity parameters according to national standards. The air temperature changes from 24.2 to 25.9 [°C], all within the set value range of 25 ± 1 [°C]. The maximum error of humidity compared to the set value is 2.7%. The average EEF energy efficiency factor for the whole year is 2.2. Maximum processing time, when the guarantee coefficient is 1, is 1095 s in summer and 607 s in winter. The results show that this DHIAC can provide a comfortable environment for people with low energy consumption.

The program has simulated the operation of DHIAC in many different air temperature and humidity settings and compared the air treatment efficiency and energy efficiency of DHIAC with a similar device popular on the market, the Air Handling Unit. However, due to page limitations, the research results will be presented in later articles.

References

1. Akram, M.W, Hasannuzaman, M., Cuce, E., Cuce, P.M.: Global technological advancement and challenges of glazed window, facade system and vertical greenery-based energy savings in buildings: a comprehensive review. Energy Built Environ. (2021)
2. Aysha, S., Mani, M.: Adaptation of buildings to climate change. In: Abraham, M.A. (ed.) Encyclopedia of Sustainable Technologies, pp. 331–349. Elsevier, Oxford (2017)
3. Lee, J.M., Hong, S.H., Seo, B.M., Lee, K.H.: Application of artificial neural networks for optimized AHU discharge air temperature set-point and minimized cooling energy in VAV system. Appl. Therm. Eng. **153**, 726–738 (2019)

4. Sovacool, B.K., Griffiths, S., Kim, J., Bazilian, M.: Climate change and industrial F-gases: a critical and systematic review of developments, sociotechnical systems and policy options for reducing synthetic greenhouse gas emissions. Renew. Sustain. Energy Rev. **141**, 110759 (2021)
5. Elsoragaby, S., Yahya, A., Mahadi, M.R., Nawi, N.M., Mairghany, M., M Elhassan, S.M., Kheiralla, A.F.: Applying multi-objective genetic algorithm (MOGA) to optimize the energy inputs and greenhouse gas emissions (GHG) in wetland rice production. Energy Rep. **6**, 2988–2998 (2020)
6. Kompatscher, K., Seuren, S., Kramer, R., van Schijndel, J., Schellen, H.: Energy efficient HVAC control in historical buildings: a case study for the Amsterdam Museum. Energy Procedia **132**, 891–896 (2017)
7. Rafique, A., Williams, A.P.: Reducing household greenhouse gas emissions from space and water heating through low-carbon technology: identifying cost-effective approaches. Energy Build **248**, 111162 (2021)
8. Opoku, R., Mensah-Darkwa, K., Samed Muntaka, A.: Techno-economic analysis of a hybrid solar PV-grid powered air-conditioner for daytime office use in hot humid climates—a case study in Kumasi city, Ghana. Sol. Energy **165**, 65–74 (2018)
9. Malik, A., Bongers, C., McBain, B., Rey-Lescure, O., Dear, Rd., Capon, A., Lenzen, M., Jay, O.: The potential for indoor fans to change air conditioning use while maintaining human thermal comfort during hot weather: an analysis of energy demand and associated greenhouse gas emissions. Lancet Planet Health **6**(4), e301–e309 (2022)
10. Opaev, A., Gogoleva, S., Palko, I., Nguyen, V.T., Rozhnov, V.: Annual acoustic dynamics are associated with seasonality in a monsoon tropical forest in South Vietnam. Ecol. Ind. **122**, 107269 (2021)
11. Gurubalan, A., Maiya, M.P., Tiwari, S.: Experiments on a novel membrane-based liquid desiccant dehumidifier for hybrid air conditioner. Int. J. Refrig. **108**, 271–282 (2019)
12. Guan, B., Liu, X., Zhang, T.: Investigation of a compact hybrid liquid-desiccant air-conditioning system for return air dehumidification. Build. Environ. **187**, 107420 (2021)
13. Elzahzby, A.M., Kabeel, A.E., Bassuoni, M.M., Abdelgaied, M.: Effect of inter-cooling on the performance and economics of a solar energy assisted hybrid air conditioning system with six stages one-rotor desiccant wheel. Energy Convers. Manage. **78**, 882–896 (2014)
14. Afram, A., Janabi-Sharifi, F.: Review of modeling methods for HVAC systems. Appl. Therm. Eng. **67**(1), 507–519 (2014)
15. Đuc, N.M., Van Chuong, T.: Designation and simulation of dehumidification integrated air conditioning. In: The AUN/SEED-Net Joint Regional Conference in Transportation, Energy, and Mechanical Manufacturing Engineering: 2022, pp. 1200–1214. Springer Nature, Singapore (2022)
16. Afram, A., Janabi-Sharifi, F.: Black-box modeling of residential HVAC system and comparison of gray-box and black-box modeling methods. Energy Build. **94**, 121–149 (2015)
17. Beghi, A., Cecchinato, L., Paggiaro, F., Rampazzo, M.: VAVAC systems modeling and simulation for FDD applications. In: 2011 9th IEEE International Conference on Control and Automation (ICCA): 19–21 Dec 2011, pp. 800–805 (2011)
18. Li, Y., O'Neill, Z., Zhang, L., Chen, J., Im, P., DeGraw, J.: Grey-box modeling and application for building energy simulations—a critical review. Renew. Sustain. Energy Rev. **146**, 111174 (2021)
19. Guo, X., Huang, G., Tu, X., Wu, J.: Effects of urban greenspace and socioeconomic factors on air conditioner use: a multilevel analysis in Beijing, China. Build. Environ. **211**, 108752 (2022)
20. TCVN: TCVN 4088:1997 Climatic data for building design. In: TCVN, Vietnam (1997)
21. Loi, N.D.: Giao trinh thiet ke he thong dieu hoa khong khi. Nha xuat ban Giao duc (2009)

Vision-Based Automatic Hand Measurement System for Ring Selection

Minh-Dung Nguyen, Tieu-Thanh Le, Thi Ngoc Anh La, and Thai-Viet Dang

Abstract Image processing has been significantly transformed in recent years by technological advances that provide solutions that are exceptionally efficient and demand few resources. Utilizing 2D images exclusively eliminates the necessity for expensive and occasionally unattainable scanning technologies such as 3D or 2D scanners, thereby facilitating the development of rapid measurement systems. Using cameras and image processing technology, this article describes an automated technique for determining the size of fingers. After generating a binary image and extracting finger contours via OpenCV image processing and camera calibration, the procedure proceeds to calculate actual sizes via a sampling technique. Comparing hand data from several subjects to manual measurements taken with a ruler, the OpenCV library-implemented Python system was evaluated. The findings highlight the effectiveness of the system, which completes ring size measurements in the jewelry and customer service sectors in a mere 0.5 s, as opposed to the manual method's mean of 11 s.

Keywords Camera calibration · Computer vision · Finger measure · Semantic segmentation · 3D scanner

1 Introduction

The importance of ensuring customer convenience is growing in significance for both companies and customers. At present, there is a demand for the creation of an application that can remotely measure finger size using solely the camera on a mobile device. This development is necessary to expand the market for remote

M.-D. Nguyen · T.-T. Le · T. N. A. La · T.-V. Dang (✉)
Hanoi University of Science and Technology, Hanoi, Vietnam
e-mail: viet.dangthai@hust.edu.vn

D. H. Tien et al. (eds.), *Proceedings of the 1st International Conference on Sustainability and Emerging Technologies for Smart Manufacturing*, Proceedings in Technology Transfer, https://doi.org/10.1007/978-981-97-7083-0_54

purchasing. Furthermore, numerous international standard systems for ring dimensions are available on the market, including those of the United States, Europe, and Japan. Nonetheless, self-measuring finger size by users and comparing it to standard measurements continues to present numerous obstacles and lack user-friendly tools [1].

The fact that several commercially available techniques exist for collecting data on hands and digits demonstrates the continuous development of 3D data retrieval methods, which began in the 1990s and include handheld scanners and mechanical scanning [2]. The ongoing utilization of the 2D scanning CanoScan LiDE 110 and the 3D Nexhand H-100 [3] demonstrates that numerous organizations continue to employ both 2D and 3D scanning techniques. Nevertheless, lackluster performance and increased costs are common disadvantages associated with these devices. Additionally, they are not user-friendly.

Numerous scholarly articles have been validated regarding 3D measurement methods [4]. There are also numerous recently developed methods for utilizing 2D scanners [5]. Nevertheless, the technique of utilizing camera images has yet to be extensively implemented. Concurrently, there is a growing emphasis on the utilization of cameras, face reduction algorithms, and flight tables in research endeavors that demand exceptional precision and rapid computation. Particularly when applied to deep learning methods, models such as MTCNN, three primary network layers P-Net, and R-Net, produce results with a remarkable accuracy rate of 99.6% [6, 7]. An illustrative instance of the implementation of deep learning in the domain of hand recognition is the Google Mediapipe model, which, unlike conventional models, does not incorporate a dedicated deep learning model. In lieu of this, numerous technologies, and deep learning models such as Hand Tracking and BlazeFace are implemented. Mediapipe facilitates the construction of vision pipelines by means of computing devices that are interconnected across data streams. The organization of data and components into a data flow graph facilitates application testing and enhancement. Specifically, performance can be optimized through the implementation of parallel computations in the graph, which does not necessitate an extensive understanding of multithreading. Mediapipe's central concept is to identify the hand skeleton using images and deep learning and to rely on landmark points to monitor movements, hand gestures, and calculate distance [8]. The research team utilized this information to develop a hand point identifier that utilized landmark points to determine the size of digits for the purpose of size measurement.

The paper proposes a straightforward and automatic method for measuring digit size that utilizes hand image analysis. Section 2 provides an overview of current dimensioning methodologies. The contour extraction method, the manual landmark extraction method, and the sample method are detailed in Sect. 3. The procedures for automated finger size measurement utilizing extracted feature points are described in Sect. 4. The experiments are conducted to evaluate the system's efficacy in Sect. 5. Finally, the experimental results with emphasized importance are along with the main conclusions.

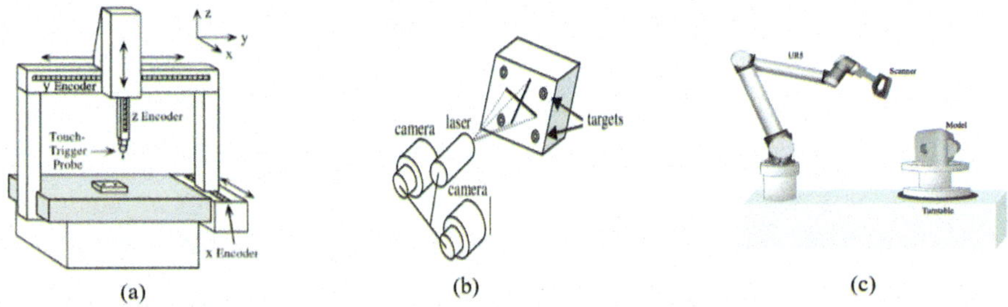

Fig. 1 3D scanning techniques with **a** mechanical scanner, **b** handheld scanner, and **c** scanners with arms

2 Related Work

2.1 3D Scanning Techniques

Mechanical scanner: High-precision 3D measurements are typically done using Coordinate Measuring Machines (CMM), which is known for their expense and slow operation. Portable 3D scanners, like Konica Minolta and Brueuckmann Optical systems.

Handheld scanner: Though lightweight and portable, have limitations such as reduced accuracy and inability to scan metallic objects. They use a wand for 3D surface information, employing a magnetic tracker for position and orientation.

Scanners with arms: Like those from 3D Scanners, Faro and Cimcore integrate laser-based technology into a digitizing arm for efficient 3D spatial awareness and accurate point capture (Fig. 1).

2.2 2D Scanning Techniques

Flatbed Scanners: These scanners have a glass surface where documents or images are placed for scanning. A moving arm or light source captures the image as it moves across the surface.

Contact Image Sensor (CIS) Scanners: These scanners use a compact image sensor to capture light reflected from the document's surface, providing a more compact and energy-efficient scanning solution.

Photo Scanners: Specifically designed for scanning photographs, these devices often include features to enhance image quality and color reproduction. Hence, the authors use this method to measure the hand in this paper (Fig. 2).

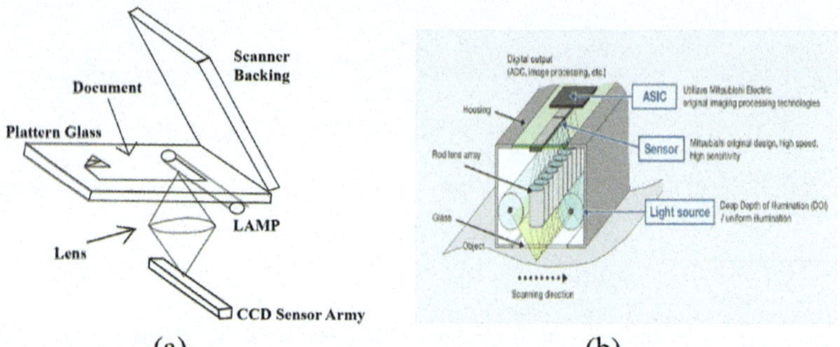

Fig. 2 2D scanning techniques with **a** flatbed scanners, and **b** CIS scanners

3 Proposed Method

For uncomplicated problems with tolerable precision requirements, such as measuring finger dimensions for applications like ring or glove sizing, the measurement methods exhibit excessive complexity and unnecessary cost implications. In the current landscape, propelled by the development of image recognition technologies like AI Mediapipe and Imutils, image processing has attained notable swiftness and resource efficiency. The prospect of seamless integration into diverse systems is also markedly promising. This contributes to a reduction in both cost and time expenditure, particularly for problem domains where stringent precision is not imperative.

3.1 Camera Calibration Using OpenCV

Through correction and remapping, we can correct this distortion. Calibration also allows to determine the relationship between the camera's pixel units and real-world units (e.g., millimeters) (Fig. 3).

First, to use the radial distortion formula:

$$\begin{cases} x_{\text{distorted}} = x\left(1 + k_1 r^2 + k_2 r^4 + k_3 r^6\right) \\ y_{\text{distorted}} = y\left(1 + k_1 r^2 + k_2 r^4 + k_3 r^6\right) \end{cases} \quad (1)$$

Based on (1), to yield the tangential distortion:

$$\begin{cases} x_{\text{distorted}} = x + \left[2p_1 xy + p_2\left(r^2 + 2x^2\right)\right] \\ y_{\text{distorted}} = y + \left[p_1\left(r^2 + 2y^2\right) + 2p_2 xy\right] \end{cases} \quad (2)$$

Finally, the conversion of (x, y) is involved by the matrix equation:

(a) (b)

Fig. 3 2D image calibration with **a** the camera's pixel units, and **b** the real-world units

Fig. 4 The conversion of (*x*, *y*) based on chess board

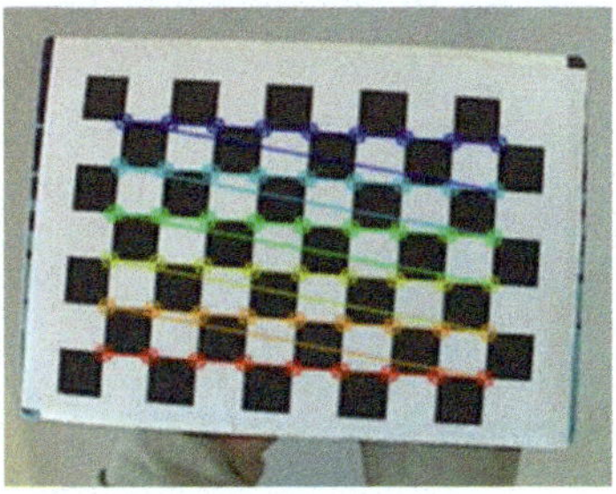

$$\begin{bmatrix} x \\ y \\ \omega \end{bmatrix} = \begin{bmatrix} f_x & 0 & c_x \\ 0 & f_y & c_y \\ 0 & 0 & f_z \end{bmatrix} \begin{bmatrix} X \\ Y \\ Z \end{bmatrix} \quad (3)$$

Use chess board to completely converse to 2D real-world image in Fig. 4.

3.2 Hand Scanning

3.2.1 Preprocessing

In image processing, the process of converting an image to grayscale entails a conversion from its original color format, which is commonly RGB or BGR, to a representation in grayscale. To achieve grayscale conversion of an image, the weighted average of the red, green, and blue color channels is commonly computed. In Eq. (4), the weighted average assigns greater significance to specific color channels in accordance with human perception.

$$\text{Gray} = 0.229R + 0.587G + 0.114B \quad (4)$$

where R, G, and B are the weights of channels red, blue, and green, respectively.

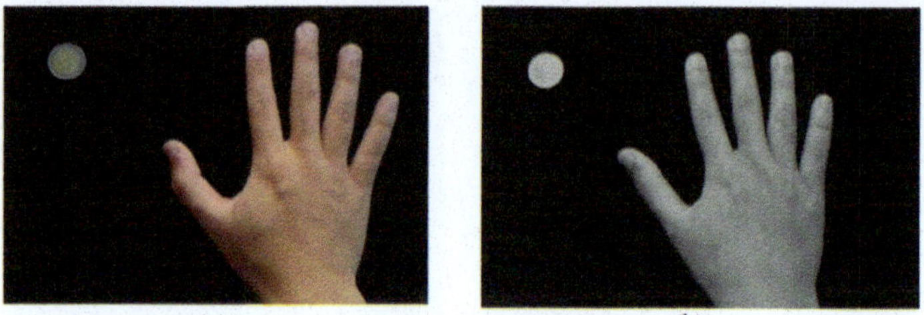

(a) (b)

Fig. 5 Hand image preprocessing with **a** color hand image, and **b** RGB hand image

After Binary threshold, to continue the noise reduction (Fig. 5).

To filter the noise, using lowpass Gaussian filter Kernels, in Eq. (5):

$$W(s,t) = G(s,t) = Ke^{-\frac{s^2+t^2}{2\sigma^2}}$$
$$r = \left[s^2 + t^2\right]^{1/2}$$
$$r_{\max}^2 = \left[\frac{(m-1)}{2}\sqrt{2}\right]^2 = \frac{(m-1)^2}{2} \tag{5}$$

where $G(s,t)$ is the Gaussian value at point (s,t), and σ is the standard deviation. Next, extract the hand contour (Fig. 6).

Then, to segment the binary image including the following steps:

- Convex Hull: Define 2 contour area
- Classify Convex Hull
- Conculate the size of the Concen Hull

to classify the object's size based on size of the contour. Finally, to define the center coordinates:

$$x_{\text{center}} = \frac{1}{n}\sum_{i=1}^{n} x_i; \text{ and } y_{\text{center}} = \frac{1}{n}\sum_{i=1}^{n} y_i \tag{6}$$

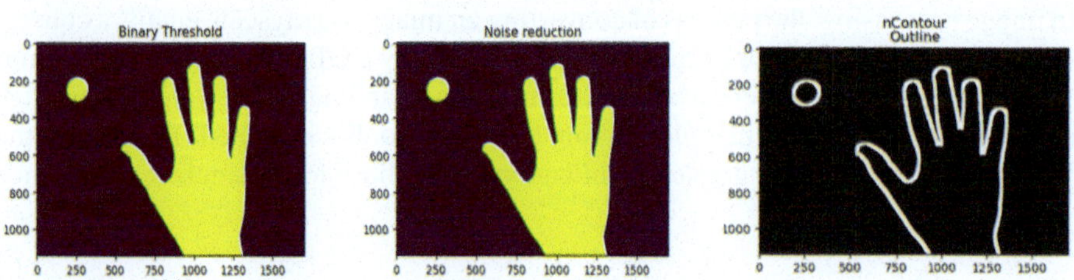

(a) (b) (c)

Fig. 6 Segmented image preprocessing with **a** binary threshold, **b** noise reduction, and **c** hand contour outline where (x_{center}, y_{center}) is the coordinate of the Convex Hull center.

3.2.2 Landmark Determination Method

Applying the landmark determination method from hand outline tracking involves tracking the hand silhouette, extracting features, identifying key points of landmarks, and ensuring temporal consistency across frames, in Fig. 7.

By connecting points in the landmark. The width of a finger is the distance between the two finger slots. The length of the finger is from midpoint of the middle crease line at palmar side of the metacarpalphalangeal joint to tip of the fingers, in Fig. 8. Based on the length of the fingers on the hand, we can calculate the diameter of the ring finger. From there, you can automatically make recommendations about ring styles in the online consultation option.

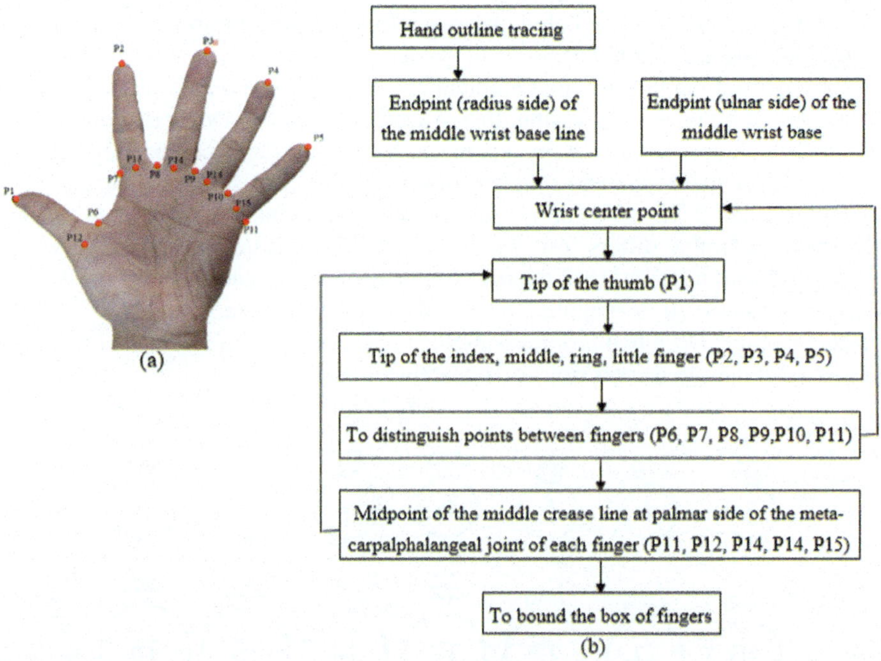

Fig. 7 The landmark determination method with **a** the landmark of hand, and **b** the flow chart of landmark determination

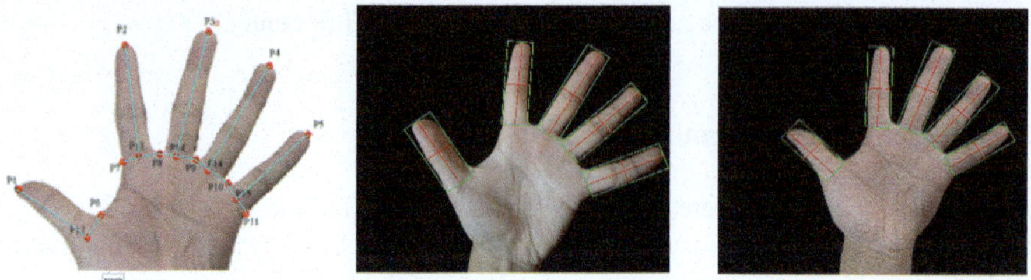

(a) (b) (c)

Fig. 8 The finger's length measurement with **a** the original hand, **b** the student Duy's hand, and **c** the student Dung's hand

4 Experimental Results

Consistent with the width of the inner diameter of the ring, ring manufacturers frequently refer to the table below when determining the appropriate measurement at www.ringsizes.co, in Fig. 9.

An algorithm application that automatically identifies and measures the size of fingers based on image processing. In two scenarios, student Duy's random hand in Fig. 8b and student Dung's in Fig. 8c, the authors conducted a comparison between actual measurement data and standard measurement data using a caliper by hand, in Tables 1 and 2. It can be seen from the results presented above that the manually measured value differs from the electronically measured value. However, the outcomes regarding the size of the ring remain unchanged, and a substantial amount of time is conserved, averaging 0.5 s for the automatic method and 10.7 s for the manual method. Despite minor measurement inaccuracies, the automatic measuring system continues to deliver rings of a guaranteed size.

Fig. 9 The ring size table

Table 1 The comparison between auto-measuring-based vision and manual measuring of Duy

Finger	Auto measure (mm)	Automatic time (s)	Ring size	Manual measure (mm)	Manual time (s)	Ring size
Thumb	21.6	0.5	12	21.3	15	12
Index finger	19.2	0.5	9	18.8	10	9
Middle finger	17.4	0.5	7	17.3	11	7
Ring finger	18.3	0.5	8	18.1	10	8
Little finger	16.6	0.5	6	16.3	8	6

Table 2 The comparison between auto-measuring-based vision and manual measuring of Dung

Finger	Auto measure (mm)	Automatic time (s)	Ring size	Manual measure (mm)	Manual time (s)	Ring size
Thumb	20.6	0.5	11	20.3	16	11
Index finger	20.1	0.5	10	19.9	10	10
Middle finger	20.9	0.5	11	20.8	12	11
Ring finger	19.8	0.5	10	19.4	11	10
Little finger	16.9	0.35	6	16.8	8	6

5 Conclusions

By exclusively employing 2D images, the need for costly and sometimes unaffordable scanning technologies (e.g., 3D or 2D scanners) is eliminated, which expedites the development of measurement systems. The article presented an automatic finger size measurement system based on semantic segmentation network and OpenCV library. The results demonstrate the efficacy of the system, which expedites ring size measurements in the customer service and jewelry industries to 0.5 s from the mean of 11 s required by the manual method.

Acknowledgements This research is funded by Hanoi University of Science and Technology (HUST) under project number T2023-PC-008. Lab-506: Computer Vision and Autonomous Mobile Robot (CVAMS), VJIIST, HUST is gratefully acknowledged for providing funding, guidance, and expertise.

References

1. Bettencourt, K., et al.: Comparison of different methods of measuring finger range of motion via telehealth. J. Hand Surg. (2023)
2. Kus, A., Unver, E., Taylor, A.: A comparative study of 3D scanning in engineering, product and transport design and fashion design education. Comput. Appl. Eng. Educ. **17**(3), 263–271 (2009)
3. Müller, J., et al.: How to measure children's feet: 3D foot scanning compared with established 2D manual or digital methods. J. Foot Ankle Res. **16**(1) (2023)
4. Hu, Z., Lau, A., Frøseth, G.T.: Fast non-contact measurement and alignment method for rail profile using a 3D scanner. In: Analytical and Experimental Methods in Mechanical and Civil Engineering, OES 2023. Structural Integrity, vol. 28 (2023)
5. Ma, K., Yang, Q., Zhang, J., Dang, X., Hu, P.: A new 2D displacement measurement method based on an eddy current sensor and absolute encoding. Machines **10**(11), 1077 (2022)
6. Dang, T.V.: Smart attendance system based on improved facial recognition. J. Robot. Control **4**(1), 46–53 (2023)
7. Dang, T.V., Tran, H.L.: A secured, multilevel face recognition based on head pose estimation, MTCNN and FaceNet. J. Robot. Control **4**(4), 431–437 (2023)
8. Dang, T.V.: Smart home management system with face recognition based on ArcFace model in deep convolutional neural network. J. Robot. Control **3**(6), 754–761 (2022)

Fuzzy Nonlinear Computed Torque Controller for Robot with Fault

Van-Truong Nguyen, Hai-Binh Giap, Dinh-Tan Nguyen, Ngoc-Binh Nguyen, and Van-Tuan Nguyen

Abstract In our study, we provide a Fuzzy Nonlinear Computed Torque Controller (FNCTC) for robot. The Nonlinear Computed Torque Controller (NCTC) is formed by altering the linear proportional-derivative algorithm in the conventional computed torque controller with the nonlinear Proportional-Derivative (PD) method. Fuzzy Logic Control (FLC) is used to approximate uncertainties inside the system. By using the Lyapunov stability theorem, it is demonstrated that the system is stable when configured with the FNCTC. The efficacy of the FNCTC is validated by simulations conducted using the pendubot and by making comparisons with the NCTC.

Keywords Nonlinear computed torque · Fuzzy logic control · Robot

1 Introduction

A Computed Torque Controller (CTC) is a control methodology utilized in the field of robotics and control systems. It is commonly applied to control the movement of robotic manipulators [1, 2]. The core principle of CTC is to calculate the torque needed for each individual joint of a robot to achieve a specific trajectory or motion. When the dynamic model of manipulators is fully corrected, it enables the individual control of each joint's motion through the use of nonlinear feedback. Nevertheless, the change of parameters throughout the trajectory tracking, as well as uncertainties, hinder the controllers' ability to provide effective compensating performance [3, 4].

V.-T. Nguyen · H.-B. Giap · N.-B. Nguyen · V.-T. Nguyen
School of Mechanical and Automotive Engineering, Hanoi University of Industry, Hanoi, Vietnam

D.-T. Nguyen (✉)
Faculty of Mechanical Engineering, East Asia University of Technology, Hanoi, Vietnam
e-mail: Tannd@eaut.edu.vn

D. H. Tien et al. (eds.), *Proceedings of the 1st International Conference on Sustainability and Emerging Technologies for Smart Manufacturing*, Proceedings in Technology Transfer, https://doi.org/10.1007/978-981-97-7083-0_55

In order to contend with this problem, a number of studies have concentrated on the method of adjusting PD gains in the CTC to enhance the precision of robot manipulators' trajectory tracking [5–7]. In [5], FLC is used to determine the nonlinear PD parameters of the CTC. In [6], a study used a neural network-based online self-gain tuning approach to a two-dof parallel manipulator's nonlinear PD computed torque controller. In [7], NCTC is designed around the concept of substituting nonlinear PD for the linear PD component in traditional CTC laws. However, these methods still have drawbacks of not considering disturbance and uncertainties in dynamic models.

In our paper, the system is created with NCTC and FLC for the purpose of controlling the system under the condition of uncertainties. The Lagrange–Euler formulation is employed in our approach to derive the dynamic model of the robot system, which takes the form of typical equations of motion in robotics. FLC is used to deal with the uncertainties in the system. The performance of the FNCTC controller is also evaluated in comparison to NCTC. The suggested control technique enables precise tracking performance, as evidenced by simulations.

2 The Proposed Method

Using the Euler–Lagrange equation, the dynamic model of robot manipulator is described as:

$$M(\phi)\ddot{\phi} + C(\phi, \dot{\phi})\dot{\phi} + G(\phi, \dot{\phi}) + \Xi = U \quad (1)$$

with ϕ standing for the joint angle, $\dot{\phi}$ for the angular velocity, and $\ddot{\phi}$ for the angular acceleration; U is actuator joint torques; $M(\phi)$, $C(\phi, \dot{\phi})$ and $G(\phi, \dot{\phi})$ are, respectively, the moment of inertia, the Centripedal-Coriolis matrix, and the gravitational acceleration vector, Ξ is uncertain component.

By defining $\overline{\phi}$ the desired trajectory, the actual tracking error is determined as

$$\varepsilon = \overline{\phi} - \phi \quad (2)$$

The NCTC law can be written as [8]

$$U = M(\phi)(\ddot{\overline{\phi}} + P(\varepsilon)\varepsilon + D(\dot{\varepsilon})\dot{\varepsilon}) + C(\phi, \dot{\phi})\dot{\phi} + G(\phi, \dot{\phi}) + \Xi \quad (3)$$

where

$$P_i(\varepsilon) = \begin{cases} P_i|\varepsilon|^{\beta_{1i}-1}, & |\varepsilon| > \lambda_i \\ P_i\lambda_i^{\beta_{1i}-1}, & |\varepsilon| \le \lambda_i \end{cases} \quad (4)$$

$$D_i(\dot{e}) = \begin{cases} D_i|\dot{\varepsilon}|^{\beta_{2i}-1}, & |\dot{\varepsilon}| > \lambda_i \\ D_i\lambda_i^{\beta_{2i}-1}, & |\dot{\varepsilon}| \le \lambda_i \end{cases} \quad (5)$$

β_{1i} and β_{2i} is chosen in the interval [0.5, 1.0] and [1.0, 1.5], respectively.

However, Ξ is difficult to determine. In our method, Ξ is approximated by an ideal FLC [9] as

$$F = D(\Phi^* \varsigma + \delta) \tag{6}$$

where ς is the fuzzy basis function vector, Φ^* the parameter vector, δ is an error of the FLC bounded by a small positive constant as $|\delta| \leq \Delta$. As the optimal matrix Φ^* is unknown, we will instead consider its estimate denoted by $\hat{\Phi}$ in the analysis.

The proposed control law is designed as

$$U_{\text{FNCT}} = u_{\text{NCT}} + u_{\text{f}} + u_{\text{n}} \tag{7}$$

where

$$u_{\text{f}} = D\hat{\Phi}^T \varsigma \tag{8}$$

$$u_{\text{NCT}} = D(\phi)(\ddot{\overline{\phi}} + P(\varepsilon)\varepsilon + D(\dot{\varepsilon})\dot{\varepsilon}) + C(\phi)\dot{\phi} + G(\phi) \tag{9}$$

$$u_{\text{n}} = D.\Delta.\text{sign}(\dot{e}) \tag{10}$$

Theorem The system (1) controlled by the FNCTC law (7) is stable if the nonlinear gains D and P are defined by (4) and (5), and the updated law is

$$\dot{\hat{\Phi}} = a.\dot{e}^T \varsigma \tag{11}$$

Prove Selecting the Lyapunov function as

$$L = \frac{1}{2}\dot{\varepsilon}^T \dot{\varepsilon} + \int\limits_0^{\varepsilon} |\sigma|^T P(\sigma)\text{d}\sigma + \frac{1}{2a}\tilde{\Phi}^T \tilde{\Phi} \tag{12}$$

According to [7], $\int_0^{\varepsilon} |\sigma|^{\text{T}} P(\sigma)\text{d}\sigma$ can be proved as a radially unbounded positive definite function. Therefore, L function is positive definite.

Differentiating (12), it is obtained as

$$\dot{L} = \dot{\varepsilon}^T \ddot{\varepsilon} + \varepsilon^T P(\varepsilon)\dot{\varepsilon} + \frac{1}{a}\tilde{\Phi}^T \dot{\hat{\Phi}} \tag{13}$$

Replacing (7) into (1) yields:

$$M(\phi)(\ddot{\varepsilon} + P(\varepsilon)\varepsilon + D(\dot{\varepsilon})\dot{\varepsilon} + \tilde{\Phi}^T \varsigma - \Delta.\text{sign}(\dot{\varepsilon})) = 0 \tag{14}$$

Given that $M(\phi)$ is a uniformly non-negative definite matrix, it follows that $M(\phi)^{-1}$ exists and is limited. Consequently, the closed-loop system may be expressed as:

$$\ddot{\varepsilon} + P(\varepsilon)\varepsilon + D(\dot{\varepsilon})\dot{\varepsilon} + \tilde{\Phi}^T \varsigma + \delta - \Delta.\mathrm{sign}(\dot{\varepsilon}) = 0 \tag{15}$$

By applying the operation of multiplication of $\dot{\varepsilon}$ to both sides of Eq. (15), and subsequently replacing the obtained equation with Eq. (13), we obtain:

$$\begin{aligned} \dot{L} &= -\dot{\varepsilon}^T D(\dot{\varepsilon})\dot{\varepsilon} + \tilde{\Phi}^T\left(\frac{1}{a}\dot{\hat{\theta}} - \dot{\varepsilon}^T \xi\right) + \delta.\dot{\varepsilon} - \Delta.\dot{\varepsilon}.\mathrm{sign}(\dot{\varepsilon}) \\ &\le -\dot{\varepsilon}^T D(\dot{\varepsilon})\dot{\varepsilon} + |\delta|.\|\dot{\varepsilon}\| - \Delta\|\dot{\varepsilon}\| \le 0 \end{aligned} \tag{16}$$

Theorem is proved successfully.

Remark In order to prevent the unintended operation of the FLC and the compensated controller in normal conditions, Eqs. (10) and (11) are adjusted by incorporating a supervised signal ρ as

$$\dot{\hat{\Phi}} = \rho a.\dot{\varepsilon}^T \varsigma \tag{17}$$

$$u_n = \rho D.\Delta.\mathrm{sign}(\dot{\varepsilon}) \tag{18}$$

where

$$\begin{cases} \rho = 0 \text{ if } \phi = \overline{\phi} \\ \rho = 1 \text{ if } \phi \neq \overline{\phi} \end{cases} \tag{19}$$

3 Results

In this section, some simulations conducted with a pendubot system are used to demonstrate the efficiency of suggested method. Parameters of pendubot are as in [10]. Three circumstances are considered in simulation, which are nominal cases, suffered by an external disturbance defined by $\psi(t) = \frac{16\pi}{180} * \exp\left(-\frac{(t-3)^2}{0.025}\right)$ and trajectory tracking case. The reference parameters ϕ_1, ϕ_2 in trajectory tracking case are described as follows:

Table 1 Parameters of controllers

Controller	Symbols	Value
NCTC	P	diag(60, 60)
	D	diag(500, 500)
	λ_1, λ_2	0.002, 0.0002
	β_1, β_2	0.7, 1.25
FNCTC	a	0.1
	$\sigma_1, \sigma_2, \sigma_3$	0.225, 0.225, 0.225
	$\gamma_1, \gamma_2, \gamma_3$	− 1, 0, 1

$$\overline{\phi}_1 = \frac{\pi}{2} + 0.35\sin(1.8t) \tag{20}$$

$$\overline{\phi}_2 = -0.35\sin(1.8t) \tag{21}$$

The input vector of the FLC of proposed controller is defined as $x_i = \{\phi_1, \dot{\phi}_1, \phi_2, \dot{\phi}_2\}$. Each fuzzy input's universe of discourse is subdivided into three fuzzy rules: Positive, Zero, and Negative. Here are the definitions of the three membership functions:

$$w_i(x_i) = \exp\left(-\frac{(x_i - \gamma_i)^2}{2\sigma_i^2}\right) \tag{22}$$

In (17), zero matrices are used to provide the starting values of the weight parameters. NCTC, whose law is as (9), is chosen to compare with proposed control. The parameters of controllers are chosen as in Table 1.

The mean absolute error index is defined to compare the performance of controllers as:

$$\text{MAE} = \frac{1}{z}\sum_{i=1}^{z}\left|\overline{\phi}_i - \phi_i\right| \tag{23}$$

where z is the quantity of samples.

The simulation outcomes are depicted in Figs. 1, 2 and 3. Table 2 provides the values for the mean absolute error of two controllers across three different scenarios. Upon comparing the simulation results of the two controllers, it is evident that both controllers exhibit favorable reactions under three different scenarios. Nevertheless, the suggested controller exhibits superior performance, characterized by reduced error-type values and a more accurate response to the reference signal.

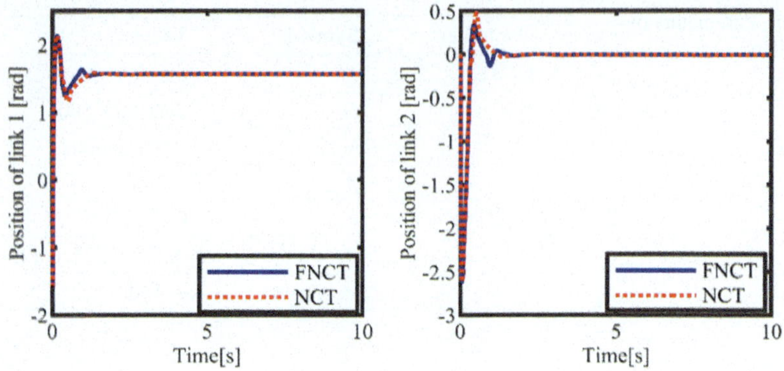

Fig. 1 Position of links' comparison under nominal condition

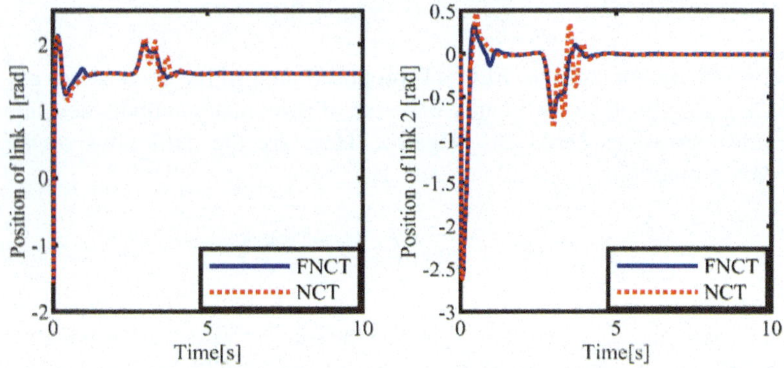

Fig. 2 Position of links' comparison under external disturbance

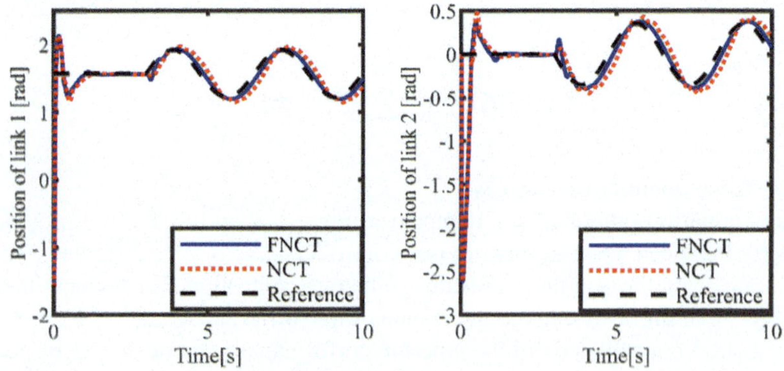

Fig. 3 Position of links' comparison under tracking case

Table 2 Comparison of mean absolute error

Case	Controller	ϕ_1 mean absolute error	ϕ_2 mean absolute error
Nominal	FNCT	0.165	0.174
	NCT	0.201	0.226
Disturbance	FNCT	0.321	0.387
	NCT	0.467	0.51
Tracking	FNCT	0.702	0.779
	NCT	0.916	0.969

4 Conclusion

In our paper, the system incorporates NCTC and FLC, which is designed to govern the system in the face of uncertainty. Using FLC, the uncertainties in the dynamic model of the system are compensated. We also compare the FNCTC controller's performance to that of the NCTC controller. The proposed method of control allows for accurate tracking performance even when faced with uncertainties, as shown in the simulations. It will be possible to enhance NCT controller performance in the future by using various AI approaches.

Acknowledgements This research was funded by Vingroup Innovation Foundation (VINIF) under project code VINIF.2023.DA089.

References

1. Middleton, R.H., Goodwin, G.C.: Adaptive computed torque control for rigid link manipulations. Syst. Control Lett. **10**, 9–16 (1988)
2. Yang, Z., Wu, J., Mei, J., Gao, J., Huang, T.: Mechatronic model based computed torque control of a parallel manipulator. Int. J. Adv. Rob. Syst. **5**, 14 (2008)
3. Nguyen, V.T., Su, S.F., Nguyen, A.T., Nguyen, V.T.: Adaptive nonsingular fast terminal sliding mode tracking control for parallel manipulators with uncertainties. In: 2019 International Conference on System Science and Engineering (ICSSE), pp. 522–525 (2019)
4. Nguyen, V.-T.: Non-negative adaptive mechanism-based sliding mode control for parallel manipulators with uncertainties. Comput. Mater. Continua **74**, 2771–2787 (2023)
5. Llama, M.A., Kelly, R., Santibanez, V.: Stable computed-torque control of robot manipulators via fuzzy self-tuning. IEEE Trans. Syst. Man Cybern. Part B (Cybern.) **30**, 143–150 (2000)
6. Le, T.D., Kang, H.-J., Suh, Y.-S., Ro, Y.-S.: An online self-gain tuning method using neural networks for nonlinear PD computed torque controller of a 2-dof parallel manipulator. Neurocomputing **116**, 53–61 (2013)
7. Shang, W., Cong, S.: Nonlinear computed torque control for a high-speed planar parallel manipulator. Mechatronics **19**, 987–992 (2009)
8. Sancak, K.V., Bayraktaroglu, Z.Y.: Nonlinear computed torque control of 6-dof parallel manipulators. Int. J. Control. Autom. Syst. **20**, 2297–2311 (2022)

9. Chen, Y., Ma, G., Lin, S., Gao, J.: Adaptive fuzzy computed-torque control for robot manipulator with uncertain dynamics. Int. J. Adv. Rob. Syst. **9**, 237 (2012)
10. Nguyen, V.T., Giap, H.B., Tran, N.T., Manh, N.H., Nguyen, V.A.: A study on fuzzy nonsingular fast terminal sliding mode control for pendubot with uncertainties. In: The International Conference on Intelligent Systems & Networks, pp. 12–18 (2023)

Research on Compressive Force on Piles, Driven in Two-Layer Foundation, the Resistance at the Pile's Bottom Remains Constant

T. C. Cong and N. V. Tai

Abstract This paper considers the collision of hammer to pile driven into two layers of foundation base. The pile side friction in each layer is different, pile bottom is subjected to constant resistance. This paper uses the D'alembert wave propagation method and longitudinal collision theory of elastic bars to determine the compressive force of the buffering on the pile; thus, it will be easy to drive piles safely and effectively.

Keywords Pile · Hammer · Two layers · Buffering · Compression force

1 Setting Problem

The method to reinforce foundation by driving reinforced concrete piles is considered as the best method because it is simply implemented as well as ensures the stability of the works when lateral load collides with the works and reduces the volume of materials for excavating and filling soil and shortens the construction time. In reality, to determine the bearing strength of piles, it is defined on the basis of the experiment in the field. Through many figures in other works, statistical method is used to determine the bearing strength of piles by static load in the field. The result of this method is reliable but costly and waste of time. Therefore, this method is only applied to the important works. The equation and experimental method are based on the static theory to calculate the bearing strength of piles following pile material and physico-mechanical property of engineering geology. The experiment of the bearing strength of piles by dynamic load will be simple and less costly than the experiment of the static load; however, the given formula is inappropriate with reality because it is based on Newton's ancient collision theory. Before the theoretical equation

T. C. Cong (✉) · N. V. Tai
School of Mechanical and Automotive Engineering, Hanoi University of Industry, Hanoi, Vietnam
e-mail: truongchicong@haui.edu.vn

D. H. Tien et al. (eds.), *Proceedings of the 1st International Conference on Sustainability and Emerging Technologies for Smart Manufacturing*, Proceedings in Technology Transfer, https://doi.org/10.1007/978-981-97-7083-0_56

had the wrong results with the reality; however, through long researches conducted by many scientists, these equations have been improved and are close to practice. At present, the modern collision theory is born to overcome the shortcomings of Newton's ancient collision theory [1–3].

Calculate for real construction, applying the method of D'Alember experience propagation to solve problems, determine compressive force and activity of elastic compressive force on the pile cap driven into two layers foundation base, the pile side friction in each layer is different, pile bottom encounters constant resistance, which is urgent and meaningful in science and practice.

2 The Problem Establishment

See Fig. 1.

2.1 Set Up the Problem

- **Differential equation of motion of the pile part subjected to friction f_1**

$$\frac{\partial^2 \Pi_1}{\partial t^2} = \left(\frac{\partial^2 \Pi_1}{\partial y^2} - k_1\right) a^2 \quad \text{with } 0 \le y \le L_1; \; t > 0 \tag{2.1}$$

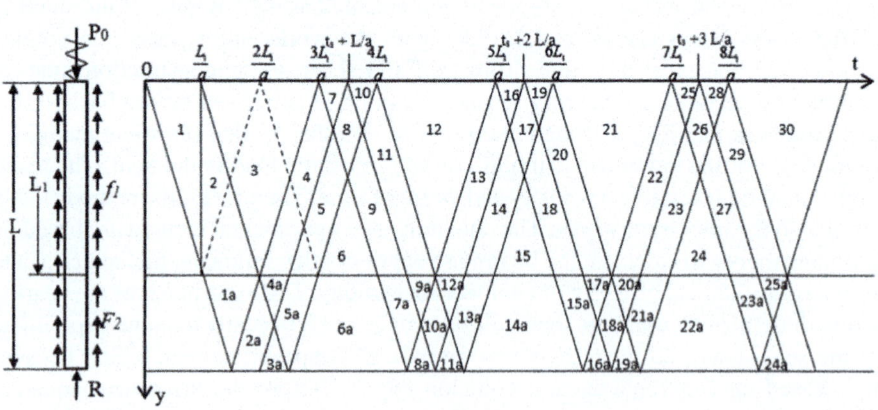

Fig. 1 Problem diagram

where Π_1: Displacement of pile head; $k_1 = r_1 q_1/EF$; $k_1 \neq 0$ when $at - y > 0$. f_1, f_2: The pile side friction. E: Young's module; F: area of pile cross-section; r_1: perimeter of pile cross-section; $a = \sqrt{E/\rho}$: wave transmission velocity; ρ: pile's density.

The solution of Eq. (2.1) corresponds to domain 1:

$$\Pi_1(t, y) = \Phi_1(a.t - y) + k_1.y^2/2 - k_1.a.t.y \tag{2.2}$$

The solution of Eq. (2.1) corresponds to domains 2 and 3:

$$\Pi_1(t, y) = \Phi_1(a.t - y) + k_1(L_1 - y)^2/2 \tag{2.3}$$

The solution of Eq. (2.1) corresponds to other domains:

$$\Pi_1(t, y) = \Phi_1(a.t - y) + \Psi_1(a.t + y) + k_1(L_1 - y)^2/2 \tag{2.4}$$

- **Differential equation of motion of the pile part subjected to friction f_2**

$$\frac{\partial^2 \Pi_2}{\partial t^2} = \left(\frac{\partial^2 \Pi_2}{\partial y^2} - k_2\right)a^2 \quad \text{with } L1 \leq y \leq L; \quad t > L1/a \tag{2.5}$$

where $k_2 = r_1 q_2/EF$; $k_2 \neq 0 \Leftrightarrow at - y > 0$

The solution of Eq. (2.5) corresponds to domain 1a:

$$\Pi_2(t, y) = \Phi_2(a.t - y) + \frac{1}{2}k_2(L_2 - y)^2 \tag{2.6}$$

The solution of Eq. (2.5) corresponds to other domains:

$$\Pi_2(t, y) = \Phi_2(a.t - y) + \Psi_2(a.t + y) + \frac{1}{2}k_2(y - L_1)^2 \tag{2.7}$$

2.2 *The Conditions of Problem [1, 4–6]*

- **Initial conditions and boundary conditions of the problem**

Indicates that time ($t = 0$) is the moment the hammer begins to touch the pile.

With $t = 0$, thus:

$$\begin{cases} \Pi_1 = 0; \ \partial \Pi_1/\partial t = 0 \\ \Pi_2 = 0; \ \partial \Pi_2/\partial t = 0 \end{cases} \tag{2.8}$$

Pile head position $y = 0$:

$$\partial \Pi_1/\partial y = -P(t)/EF \tag{2.9}$$

Corresponding cross-sectional location $y = L_1$:

$$\frac{\partial \Pi_1}{\partial y} = \frac{\partial \Pi_2}{\partial y}; \frac{\partial \Pi_1}{\partial t} = \frac{\partial \Pi_2}{\partial t} \tag{2.10}$$

Solve Eq. (2.10), we have

$$\Psi_1'(a.t + L) = \Psi_2'(at + L_1); \quad \Phi_1'(a.t - L_1) = \Phi_2'(a.t - L_1) \tag{2.11}$$

where Φ_1', Φ_2', Ψ_1', and Ψ_2': are the forward and reverse waves running in the pile corresponding to the soil layers, respectively.

Pile bottom position $y = L$.

In case: the pile has not subsided:

$$EF(\partial \Pi_2/\partial y) + R < 0; \quad \underset{2}{\dot{\Pi}} = 0 \tag{2.12}$$

The pile begins to sink:

$$EF(\partial \Pi_2/\partial y) + R = 0 \text{ and } \underset{2}{\dot{\Pi}} \neq 0 \tag{2.13}$$

The pile stops sinking:

$$EF(\partial \Pi_2/\partial y) + R < 0; \quad \underset{2}{\dot{\Pi}} = 0 \tag{2.14}$$

R is the soil resistance force acting on the pile head, considered constant.

3 Basis for Calculating Compression Force on Piles

Here, we determine the compression force on the pile head in the case ($\omega^2 > 0$) and the case ($\omega^2 < 0$) we also have the same.

Expression to determine the compressive force on the pile's head.

$$P(t) = C(P_h - P_p) \tag{3.1}$$

where Π_h, Π_p and K are the displacement of the hammer's head, pile's head, and the hardness of the elastic buffering, respectively.

The differential equation for displacement of the hammer head:

$$M \ddot{\underset{h}{\Pi}} = -P(t) \tag{3.2}$$

In which, M: mass of the hammerhead.
From Eq. (3.1), we deduce:

$$\ddot{P}(t) = C(\ddot{\underset{h}{\Pi}} - \ddot{\underset{p}{\Pi}}) \tag{3.3}$$

From Eq. (3.2), we deduce:

$$\ddot{\underset{h}{\Pi}} = -P(t)/M \tag{3.4}$$

$$\Pi_p = \Phi(a.t) + \Psi(a.t) \Rightarrow \ddot{\underset{p}{\Pi}} = \left[\Phi''(a.t) + \Psi''(a.t)\right]a^2 \tag{3.5}$$

From Eq. (2.9), we deduce:

$$\partial\Pi_1/\partial y|_{y=0} = -\Phi'(a.t) + \Psi'(a.t) = -P(t)/(EF) \tag{3.6}$$

$$\Phi''(a.t) = \Psi''(a.t) + \dot{P}(t)/(aEF) \tag{3.7}$$

From Eqs. (3.1), (3.2), and (3.3), we obtain the differential equation:

$$\ddot{P}(t) + 2n\dot{P}(t) + (n^2 + \omega^2)P(t) = -2a^2C\Psi''(a.t) \tag{3.8}$$

with

$$n = \frac{aC}{2EF}; \quad \omega^2 = \frac{C}{M} - \left(\frac{1}{2}\frac{aC}{EF}\right)^2 \tag{3.9}$$

4 Establish an Expression to Define Wave Functions

Case ($0 \leq t \leq L_1/a$)

Π_p: Displacement of pile head:

$$\Pi_p = \Phi(a.t) \Rightarrow \ddot{\Pi}_p = a^2\ddot{\Phi}(a.t) \tag{4.1}$$

From Eq. (2.9), we deduce:

$$\frac{\partial \Pi_p}{\partial y} = -\dot{\Phi}(a.t) - k_1.a.t = -\frac{P(t)}{EF} \Rightarrow \ddot{\Phi}(at) = \frac{\dot{\Pi}(t)}{EFa} - k_1 \quad (4.2)$$

Inserting Eq. (4.2) into Eq. (4.1), we have:

$$\dot{\underset{p}{\Pi}} = \left[\dot{\Pi}(t)/(aEF) - k_1\right]a^2 \quad (4.3)$$

The compressive force on pile's head during the time period ($0 \leq t \leq L_1/a$) is $P_0(t)$. During this period of time, the reverse wave $\Psi_2''(at) = 0$, $P_0(t)$ is determined from the equation:

$$\ddot{P}_0(t) + 2n\,\dot{\underset{0}{\Pi}}(t) + \left(\omega^2 + n^2\right)P_0(t) = Ca^2k_1 \quad (4.4)$$

The general solution of Eq. (4.4) is:

$$P_0 = e^{-nt}(C_1 \cos \omega t + C_2 \sin \omega t) + \frac{a^2Ck_1}{n^2 + \omega^2} \quad (4.5)$$

From the initial conditions, we have:

$$C_1 = -\frac{k_1a^2C}{n^2 + \omega^2} \quad C_2 = \frac{1}{\omega}\left(V.C - \frac{k_1a^2Cn}{n^2 + \omega^2}\right) \quad (4.6)$$

From Eq. (2.9), we have an expression to calculate forward waves in domain 1, which is:

$$\Phi_1'(a.t - y) = \frac{1}{E.F}P_0(t - y/a) - k_1at + k_1y \quad (4.7)$$

At ($t = L_1/a$), the boundary between domains 1 and 2, the displacement function is continuous, we have:

$$\Pi_1(L_1/a - 0, y) = \Pi_1(L_1/a + 0, y) \quad (4.8)$$

Hence inferred:

$$\Phi_{1-1}'(a.t - y) = \Phi_{1-2}'(a.t - y) \quad (4.9)$$

Thus, the positive wave in domain 2:

$$\Phi'(a.t - y) = \frac{1}{EF}P_0(t - y/a) - k_1a.t + k_1y \quad (4.10)$$

Positive waves in domains 1a, 2a, 3a:

$$\Phi'(a.t - y) = \frac{1}{EF}P_0(t - y/a) - k_1 at + k_1 y \tag{4.11}$$

Let t_s be the time when the pile begins to sink ($L/a \le t_s \le L/a + L_1/a$). When $t < t_s$, then $EF(\partial\Pi_2/\partial y) + R < 0$ the pile has not yet sunk. According to Eq. (2.12), we can determine reverse waves in domains 2a, 4a as:

$$\Psi'(at + y) = \frac{1}{EF}P_0\left(\frac{y - 2L}{a} + t\right) - k_1(y - 2L + at) \tag{4.12}$$

Positive waves in domains 4, 7:

$$\Psi'(at + y) = \frac{1}{EF}P_0\left(\frac{y - 2L}{a} + t\right) - k_1(y - 2L + at) \tag{4.13}$$

At the time, $t = t_s$, $E.F.(\partial\Pi_2/\partial y) = -R$ and pile bottom begins sinking. Positive waves in domains 3a, 5a:

$$\Psi'(at + y) = \frac{1}{E.F}\left[P_0\left(\frac{y - 2L}{a} + t\right) - R\right] - k_1(y - 2L + at) - k_2(L - L_1) \tag{4.14}$$

Reverse wave in domains 5, 8, 10:

$$\Psi'(at + y) = \frac{1}{EF}\left[P_0\left(\frac{y - 2L}{a} + t\right) - R\right] - k_1(y - 2L + at) - k_2(L - L_1) \tag{4.15}$$

Positive and reverse waves in the remaining domains are calculated similarly.

Actual calculated data

gama $= KL/EF$ with: E: Young's module of pile; $\frac{1}{K} = \sum_{i=1}^{3} \frac{1}{K_i}$, where $K_i = \frac{FE_i}{d_i}$ is the hardness of the buffering ith. In which, corresponding K_i, E_i, d_i is stiffness, Young's module of materials, and thickness of buffering layer i ($i = 1$, 2, and 3) (Tables 1 and 2).

- The ground has two layers and the friction force acting on the side of the pile corresponds to each layer: $f_1 = 2.5$ N/cm^2; $f_2 = 6.5$ N/cm^2. The resistance force of the ground on the pile's bottom: $R = 784{,}800$ N. Hammer's weight = 1800 kg; the falling height of hammer = 180 cm.

Table 1 Pile parameter table

Material	Pile dimension (cm)	Young's module (E) (N/cm^2)	Density (ρ) (kg/cm^3)	Velocity of wave a (cm/s)
Concrete pile M300	40 × 40 × 1200	3.11 × 106	0.024	337 × 10^5

Table 2 Table of buffering stiffness

No.	Type of buffering	E_i (N/cm^2)	Thickness d(cm)	F (cm^2)	K_i (N/cm)	K (N/cm)
1	Sackcloth	364.932	0.8	1600	729,846	
2	Ironwood	65,727	14	1600	7,511,657	
3	Steel	20,601,000	1	1600	32,961,600,000	665,199.32
With above data, we have the non-dimensional stiffness of the buffering: gama = 0.16						
1	Sackcloth	364.932	0,7	1600	834,109,7	
2	Ironwood	65,727	15	1600	7,010,880	
3	Steel	20,601,000	1	1600	32,961,600,000	745,407.1
With above data, we have the non-dimensional stiffness of the buffering: gama = 0.18						
1	Sackcloth	364.932	0.6	1600	973,128	
2	Ironwood	65,727	18	1600	5,842,400	
3	Steel	20,601,000	1	1600	32,961,600,000	834,162.7

With above data, we have the non-dimensional stiffness of the buffering: gama = 0.20

Simulating calculations on the computer, we can calculate the final time of impact $t_E = 0.01419$ (s).

5 Conclusions

- Accuratelty determine the impact time and evolution of impact force during the pile driving process (Fig. 2).

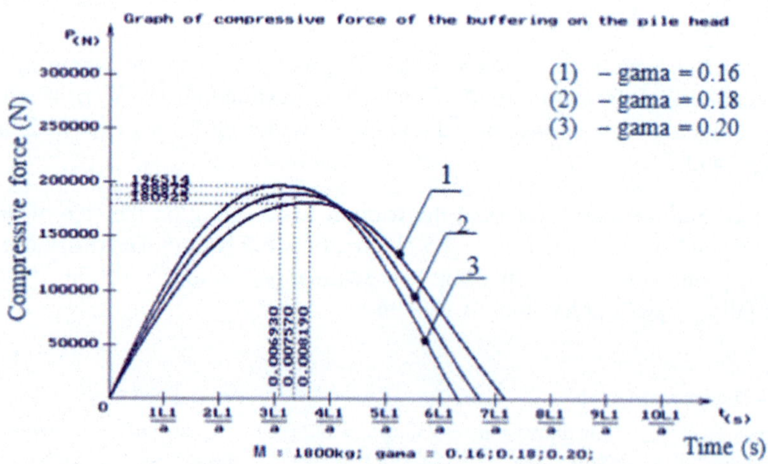

Fig. 2 Graph of compressive force of the buffering on the pile head

Table 3 The result of compressive force on the pile head

gama = 0.16		gama = 0.18		gama = 0.20	
The time pile starts sinking is: 0.0050486749 s		The time pile starts sinking is: 0.0047586749 s		The time pile starts sinking is: 0.0045556749 s	
The final time of impact t_E = 0.0160700000 s		The final time of impact t_E = 0.0150500000 s		The final time of impact t_E = 0.0141900000 s	
Time (s)	Compressive force (N)	Time (s)	Compressive force (N)	Time (s)	Compressive force (N)
0.0000	0.00	0.0000	0.00	0.0000	0.00
0.0020	74,721.67	0.0020	83,323.35	0.0020	91,768.91
0.0040	134,028.65	0.0040	147,235.60	0.0040	159,745.64
0.0060	169,704.49	0.0060	182,270.69	0.0060	193,304.37
0.0080	180,837.57	0.0080	188,446.61	0.0080	193,761.00
0.0100	168,495.69	0.0100	165,792.65	0.0100	159,796.99
0.0120	125,192.23	0.0120	108,931.02	0.0120	89,623.35
0.0140	63,365.18	0.0140	35,595.23	0.0140	6602.17
0.0150	32,062.26	0.0150	1251.97	0.0150	0.00
0.0160	1715.32	0.0160	0.00		
0.0170	0.00				
The maximum value of P is: 180,925 N At the time t = 0.0082 s		The maximum value of P is: 188,875 N At the time t = 0.0076 s		The maximum value of P is: 196,514 N At the time t = 0.0069 s	

- When the hammer weight is fixed = 1800 kg and change non-dimensional stiffness (gama) of the buffering is 0.16; 0.18, or 0.2. When: gama = 0.16, the maximum value of P is: 180,925 N at the time t = 0.0082 s; gama = 0.18, the maximum value of P is: 188,875 N at the time t = 0.0076 s; gama = 0.2, the maximum value of P is: 196,514 N at the time t = 0.0069 s. We choose the buffering with gama = 0.2 because of the highest pile driving efficiency (Table 3).
- These remarks also are suitable to the reference [7] and maybe the reference for people executing the construction to choose the pile head buffering to drive piles safely and efficiently. The results of this method are almost identical to the method within [8].
- This is a new computational theories method to evaluate and design pile. There are many methods of pile driving equipment options; however, this method helps us choose piling equipment quickly, efficiently, and safely. The construction is calculated using a simple programming language; the program is neat.

References

1. An, N.T.: The Theory of Bar Collision and Applying the Wave Theory into the Pile Problem. Water Resource University, Hanoi (1991)
2. Gerxevanopva, H.M.: Guidance in the Construction of Pile Foundation. Construction Publisher, Hanoi (1993)

3. Cuong, N.D., Cong, T.C., Huyen, N.N.: Determining Oscillating energy of piles. Collection of Scientific Works in National Conference of Mechanics, Hanoi (2002)
4. Cong, T.C.: Study the coefficient of energy transfer from hammer to pile driven in two layers foundation base, pile bearing side friction q1#q2 pile bottom encounters constant resistance. Appl. Mech. Mater. **427–429**, 195–199 (2013)
5. Cong, T.C.: Study of vibration energy of the pile driven in two layers foundation base, pile bearing side friction q1#q2 pile bottom encounters constant resistance. Appl. Mech. Mater. **427–429**, 204–208 (2013)
6. Min, Z.Y., Cong, T.C.: Study the stress in the pile driven in two layers foundation base. Energy Educ. Sci. Technol. Part A: Energy Sci. Res. V. (Issues) **32**(5), 3937–3948 (2014)
7. Minh, C.N., Ninh, D.V., Luc, L.H.: Experiment and Test the Pile Quality. Construction Publisher, Hanoi (1999)
8. Goble, G.G., Rausche, F.: Pile capacity testing means, United States Patents, US3960008 A (1976)

Digital Transformation Database for Material Inventory Management in Garment Industry

Tieu-Thanh Le, Thi-Hanh Nguyen, and Thi Ngoc Anh La

Abstract Inventory management is an important part of the manufacturing operations in the garment industry. This study presents an initial proposal for a database system for managing material inventory in the garment industry. It allows for control over the receiving process by managing information related to production plans, suppliers, orders, and order details, as well as addressing the problem of managing the quality of input materials through specified criteria for material inspections. Additionally, it provides a basis for identifying materials with a unique code, establishing a link between the material of the location and the shelf position to optimize storage volume or determine routing for transportation, arrangement, and retrieval of materials. This data system serves the purpose of designing automated management software for inventory warehouses in garment manufacturing businesses. Furthermore, empirical results suggest that there is a near future potential for cost reductions to be realized in the domains of quality control, energy utilization, productivity, and operations and administration. Given the Vietnamese textile industry's endeavor to achieve international competitiveness, this research offers practical implications for policymakers, scholars, and industry experts.

Keywords Database system · Inventory material · MS access · Smart garment factory

1 Introduction

Warehouses play a dual role in the operations of manufacturing organizations, serving as both storage facilities for products and as catalysts for enhancing overall operational efficiency [1]. Raw material warehouses are of utmost importance in textile companies. The following elucidates the underlying processes: Initially, the materials are received and systematically recorded in terms of both amount and quality.

T.-T. Le · T.-H. Nguyen · T. N. A. La (✉)
Hanoi University of Science and Technology, Hanoi, Vietnam
e-mail: anh.lathingoc@hust.edu.vn

D. H. Tien et al. (eds.), *Proceedings of the 1st International Conference on Sustainability and Emerging Technologies for Smart Manufacturing*, Proceedings in Technology Transfer, https://doi.org/10.1007/978-981-97-7083-0_57

Subsequently, they are stored in accordance with precise regulations to uphold their integrity. Warehouse personnel generate delivery notes to ensure accurate documentation of material allocation when production departments want them. Regular inventory checks are essential for ensuring both the quantity and quality of goods [2]. The garment sector has specific demands for a dependable, prompt, and efficient supply of essential materials. The effective administration of the raw material facility is a crucial component in the manufacturing processes of the garment sector. The warehouse management process often encounters issues related to monitoring changes in raw materials within the warehouse and the lack of reliable data on the quantity and condition of these commodities [3, 4]. The "digital transformation" trend has brought forth numerous intricate issues in raw material management, including the optimization of warehouse layouts, warehousing planning, quality control, route routing, and storage planning [5–7]. The components of the text encompass a compilation of raw materials, practices for storage management, an allocation strategy, the flow of products within the warehouse, inventory management, and the process for liquidating items in the warehouse. Further challenges need to be overcome in order to conduct a comprehensive inspection or restore the entirety of the facility. Upon doing a comprehensive assessment of various warehouse management software options such as Bravo, Ecount, and StockPile, it is evident that these programs are not ideal for cloth manufacturers in Vietnam. The inability to fully meet the administrative and computational demands of a smart warehouse during the era of the "fourth industrial revolution" has been observed [8].

The authors suggest developing raw material warehouse management software for Vietnamese garment manufacturers [2, 9, 10] based on an examination of the current state of warehouse management for small and medium-sized organizations and warehouse management software available on the market [8]. Initially, the research team resolved to examine a number of concerns, such as allocation strategy, material identification, storage administration, quality control, and storage optimization. The aforementioned concerns hold paramount significance for clothing enterprises of modest and medium scale in Vietnam. Moreover, resolving challenges is straightforward and applicable to the contexts and circumstances of businesses. As a result, information regarding production plans, suppliers, orders, and order particulars can be efficiently managed through the program, which facilitates control over the reception procedure.

The present study is organized into four distinct sections. Initially, it delineates the precise obstacles linked to efficient and astute warehouse administration. Furthermore, it centers on the procedure of building a database. Moreover, it provides an overview of the results obtained from the process of constructing the database. Finally, it culminates with the presentation of the research findings.

2 The Process of Building a Database for Material Inventory Management

The data fields requiring management are constructed according to the operational procedure of the warehouse, spanning from the input of raw materials to the distribution of goods. The objective of constructing data fields is to provide autonomy, uniformity, and absence of redundancy. The detailed execution steps in the diagram are presented specifically in the following sections.

2.1 Collecting Information to Serve the Construction of the Data Field

Warehouse work records and forms, including technical documents, production plans, orders, packing lists, customers' records, color charts, fabric inspection reports, fabric error classification tables, warehouse cards, warehouse release request forms, lists, balance sheets, import slips, and export slips, are the primary sources of the required information [2]. Practical surveys will yield additional data pertaining to facility conditions, including warehouse shelf sizes and warehouse layout information.

2.2 Dividing the Information into Tables

The primary information objects are chosen during this stage. The information objects comprising warehouse management software will consist of the following: production plans, suppliers, orders, order details, materials, criteria for quality control, shelf information, export request, export detail request, and export. Every individual information object will be correspondingly represented in the MS Access software through a data table.

2.3 Establishing the Data Fields in the MS Access Software

After determining the tables and the information to be stored in each table, proceed to build the data fields. Each data field is displayed as a column in the data table. The applied principle is to store information into the smallest logical parts (breaking down information into logical components) to facilitate the retrieval of specific data when needed. Besides creating data fields for the tables, the selection of data types and setting properties for each field should be carried out.

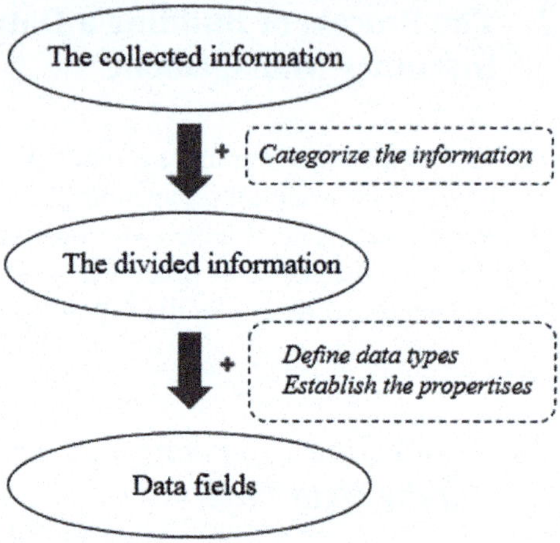

Fig. 1 The diagram illustrates the transformation of information into data field

As such, along with the process of constructing a material warehouse management database, the collected information will undergo transformations, which are represented in the diagram, in Fig. 1.

2.4 Establishing Primary Keys

The primary key is the unique identifier for each table. The primary key is chosen as a column in the table (a data field). Additionally, it serves to link data from multiple tables and combine them. The list of primary keys for each table is listed in Table 1.

Table 1 The list of primary keys for the data tables

No.	Table name	Primary key field
1	Production plan	ProCode
2	Supplier	SupplierID
3	Order	OrderCode
4	Detail order	TypeCode
5	Material	MaterialCode
6	Check result	ID
7	Shelf information	NoteNo

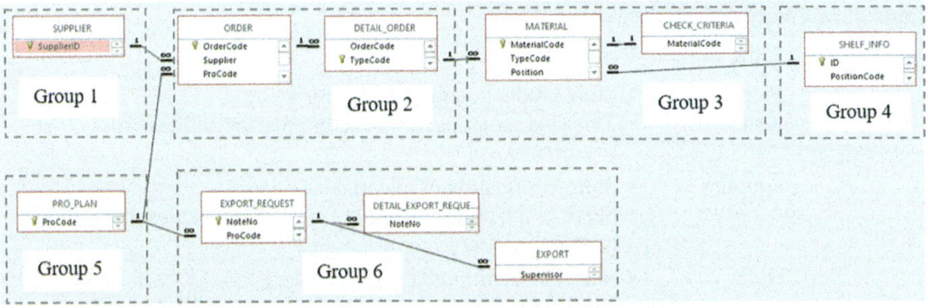

Fig. 2 The relationship between the six groups of data

2.5 *Establishing Relationships for Tables*

The purpose of this task is to establish data relationships within a relational data system. It involves the following steps: Examine the tables and determine the data relationships within each table with the data in other tables; Add fields to the table to establish one-to-one, one-to-many relationships, or create intermediate tables to establish many-to-many relationships; Create links between the tables; The results of establishing relationships for the tables are presented in the diagram shown in Fig. 2.

2.6 *Applying Standard Normalization Rules*

Apply normalization rules to check the suitability of the table structure. The purpose of normalization is to eliminate data redundancy and minimize errors associated with redundant and inconsistent data operations. After applying the normalization rules, a comprehensive inventory management database is obtained.

3 Material Inventory Management Database

After classifying information, defining data types, setting field properties, and establishing relationships between data tables, the author has generalized the data into six groups that are interconnected as shown in Fig. 2 and Table 2.

Table 2 Inventory management data fields

Group	No.	Table name	Purpose
5	1	Production plan	• Store production plan information • This data serves as the basis for ordering and exporting requisitions
1	2	Supplier	• Store information of material suppliers • Serve as the point of contact for ordering and resolving issues related to orders
2	3	Order	• Store basic information about orders • Establish connections with production plans and suppliers
	4	Detail order	• Store specific information about the types of materials for each order • Establish the connection with orders
3	5	Material	• Store all information of each unit of material from the stage of input to storage and output • Establish connections with the "Detail order" table and the "Shelf information" table
	6	Check result	• Store information on inspection criteria and test results for the materials • Establish the connection with the "Material" table
4	7	Shelf information	• Store information about the position, dimensions, and storage capacity of all shelves
6	8	Export Request	• Store the basic information of the export request • Establish the connection with production plans
	9	Detail export request	• Store the specific information of the type of materials to be exported and the quantity • Establish the connection with the "Export Request" table
	10	Export	• Storing information on material distribution • Establish the connection with the "Export Request" table

Building on initial research, a warehouse management database system has been designed specifically for the garment industry (Fig. 3). This system addresses key warehousing challenges by providing centralized management of information critical to the production process. This includes production plans, suppliers, orders, detail orders, materials, check results, etc. Each roll of fabric has a unique code that serves as the primary key in our system. This ensures easy identification of any specific fabric. Additionally, we track the connection between the material's original location and its current shelf location. This two-pronged approach allows us to identify fabric not just by its code, but also by its physical placement within the storage system (Fig. 4).

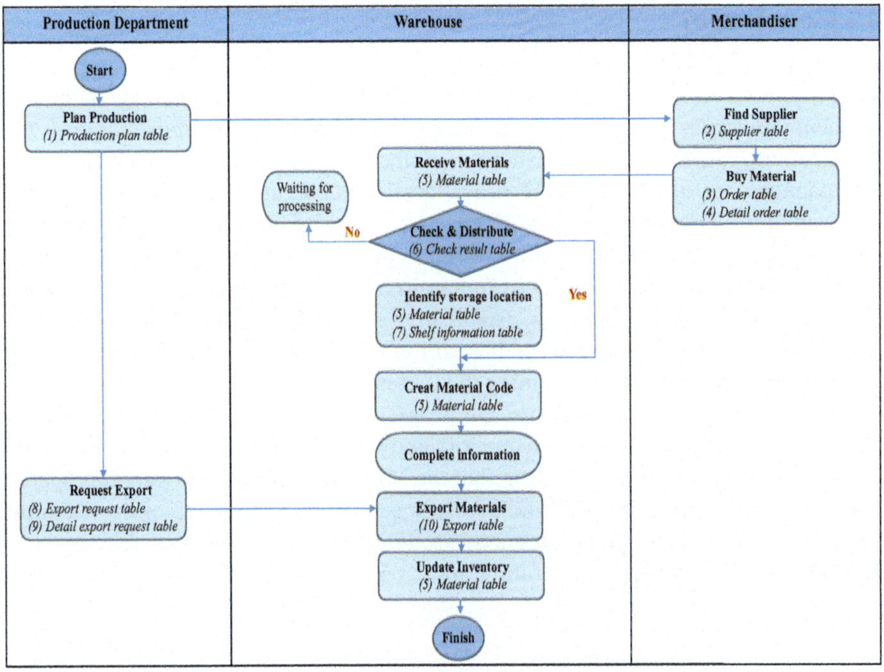

Fig. 3 The diagram for material inventory management in garment industry

Fig. 4 "Storage location information" window interface

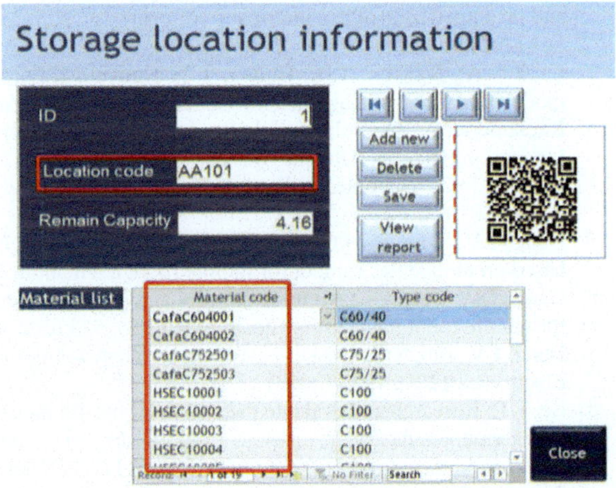

4 Conclusion

The purpose of creating the database system for material inventory management in the garment industry is to facilitate the development of automated management software for inventory warehouses in garment manufacturing firms. This system facilitates the management of the receiving procedure by managing data related to production schedules, vendors, orders, and order particulars. Furthermore, it addresses the challenge of supervising the quality of input materials by setting certain criteria for material inspection. Additionally, it functions as a fundamental basis for the allocation of unique codes to materials, establishing a relationship between the material's location and its corresponding shelf position to enhance the efficiency of storage volume calculations, and determining the routes for material transportation, arrangement, and retrieval. Following this, the program enables the estimation of material allocation, thereby optimizing the quantity and position of materials within the warehouse. However, it is crucial to note that the Microsoft Access software has specific restrictions in terms of file size and access speed. Within the scope of this examination, the MS Access program has successfully accomplished the research objectives and can serve as a blueprint for future, more comprehensive studies.

References

1. Girsang, A.S., Purwanto, A.: Controlling system for stock raw material for production planning and inventory control in a pharmacy company. Int. Rev. Mech. Eng. (I.RE.M.E.) **11**(11), 855–861 (2017)
2. La, T.N.A., Nguyen, T.H., Hoang, T.T., Nguyen, K.O.: Research the digital transformation model of material management in Vietnam Garment companies. MM Sci. J., 6753–6762 (2023)
3. Kaplanidou, A.: Digitalization in the apparel manufacturing process. Master's Thesis-Master Innovation Sciences, Utrecht University (2018)
4. Pascual, D.G., Daponte, P., Kumar, U.: Handbook of Industry 4.0 and Smart Systems. Boca Raton (2019)
5. Dang, T.V., Tran, D.M.C., Tan, P.X.: IRDC-Net: lightweight semantic segmentation network based on monocular camera for mobile robot navigation. Sensors **23**(15), 6907 (2023)
6. Dang, T.V., Bui, N.T.: Multi-scale fully convolutional network-based semantic segmentation for mobile robot navigation. Electronics **12**(3), 533 (2023)
7. Dang, T.V., Bui, N.T.: Obstacle avoidance strategy for mobile robot based on monocular camera. Electronics **12**(8), 1932 (2023)
8. Le, T.T.: Research and evaluate the impact of the 4th industrial revolution on Vietnam's textile and garment industry in order to propose strategic orientations, policies and development solutions in the period of 2019–2030. Report DTDLXH.13/18. Summary Report of Scientific Research Project, Vietnam (2019)
9. The University of Nottingham: Microsoft Access 2016—An Introduction. Watsonia Publishing, Melbourne (2017)
10. La, T.N.A., Vu, T.D., Le, T.T., Bui, T.D.: Research MS access software for applications in smart garment factory machine management. MM Sci. J., 6561–6567 (2023)

An Investigation of High Injection Pressures on Palm Biodiesel Combustion Characteristics Using a Constant Volume Combustion Chamber

Chau Tan Vo, Tai The Mai, Au Hai Ho, and Long Dang Tran

Abstract This study examines the effects of varying fuel injection pressures on the combustion characteristics of diesel (B0) and palm biodiesel (B100) using combustion pressure and heat release rate analysis. Experiments were carried out in a constant volume vessel system, simulating the combustion environment of a CI engine, prior to fuel injection, with precise control to maintain a 21% oxygen concentration and an ambient pressure of 43 bar. Findings demonstrated a 21% decrease in ignition delay for B0, while B100 displayed a noteworthy 33% reduction, matching the surge injection pressure from 400 to 1200 bar. Meanwhile, the peak combustion pressure speed rose from 1.69 to 4.02 bar/ms for B0 and from 1.53 to 3.52 bar/ms for B100, followed by a large integral heat release augmentation of 793 J for B0 and 779 J for B100. At injection pressures exceeding 800 bar, palm biodiesel showcased a shorter ignition delay than diesel, whereas an inverse pattern emerged at pressures of 400 and 600 bar. Moreover, the palm biodiesel exhibited lower integral heat release and maximum combustion pressure compared to diesel at all injection pressures.

Keywords Palm biodiesel · Diesel fuel · High injection pressure · Constant volume combustion chamber · Combustion characteristics

C. T. Vo (✉)
Industrial University of Ho Chi Minh City (IUH), Ho Chi Minh City, Vietnam
e-mail: votanchau@iuh.edu.vn

T. T. Mai · A. H. Ho · L. D. Tran
Ho Chi Minh City University of Technology (HCMUT), 268 Ly Thuong Kiet Street, District 10, Ho Chi Minh City, Vietnam

Vietnam National University Ho Chi Minh City, Linh Trung Ward, Thu Duc City, Ho Chi Minh City, Vietnam

D. H. Tien et al. (eds.), *Proceedings of the 1st International Conference on Sustainability and Emerging Technologies for Smart Manufacturing*, Proceedings in Technology Transfer, https://doi.org/10.1007/978-981-97-7083-0_58

1 Introduction

Addressing emission concerns, especially in NO_x and PM related to diesel engines, presents a formidable challenge when viewed strictly through the lens of emission standards [1]. Optimizing diesel engine combustion typically requires costly adjustments to variables such as exhaust gas recirculation, fuel/air ratios, compression ratios, or high injection pressures. However, adopting a Constant Volume Combustion Chamber (CVCC) facilitates the incorporation of these alterations in research, especially with the addition of a common-rail injection system. Numerous studies have investigated combustion characteristics within a CVCC [2–4].

Escalating oil costs and finite fossil fuel resources have spurred interest in biofuels, in addition to worries regarding energy security and ecological harm. Studies have focused on how biodiesel characteristics affect combustion behavior in internal combustion engines [5–7]. Compared to fossil diesel, biodiesel has unique physical and chemical characteristics, including density, viscosity, and cetane number. These features are important factors in spray formation, combustion, and emissions. Palm biodiesel's excellent lubricating properties promise to extend engine life. Furthermore, biodiesel not only diminishes NO_x and HC emissions but also emits minimal SO_2, making it environmentally friendly with a neutral impact on global CO_2 levels [8]. Fuel injection control includes the manipulation of variables such as fuel injection pressure and the use of strategic methods, which are crucial components of improving combustion. The injection pressure is pivotal in the formation of the fuel–air mixture, which directly affects the combustion process. Higher injection pressures improve atomization and vaporization, reducing CO, HC, and PM emissions while also minimizing ignition delay for enhanced engine efficiency [9].

In research undertaken by Gumus et al. [10], a direct injection (DI) engine was employed to assess the influence of biodiesel on exhaust emissions and brake-specific fuel consumption (BSFC) across an interval of 180–240 bar for fuel injection pressures. The findings indicated a reduction in BSFC for biodiesel-diesel blends with higher concentrations (B50 and B100) as the injection velocity rose. Another experiment by Selvaraj et al. [11] investigated the effects of various fuel injection pressures, ranging from 220 to 260 bar, utilizing frying oil methyl ester (FOME) as the fuel. The findings highlighted notable reductions in carbon monoxide and hydrocarbons at an injection pressure of 260 bar, suggesting an enhancement in combustion efficiency. Notably, these studies did not explore the effects of high injection pressures with the diesel common-rail system commonly found in contemporary diesel engines.

Kuti et al. [12] studied the soot production and combustion features of palm biodiesel in a CVCC, utilizing a common-rail direct injection system at 1000 bar and 3000 bar, coupled with nozzle sizes of 0.08 mm and 0.16 mm. The findings demonstrated that palm biodiesel, possessing a greater cetane number and richer oxygen content, considerably shortened the ignition delay (ID). Moreover, increasing injection pressure improves fuel–air mixing, decreasing ID [12]. Agarwal et al. [13] found that an increased fuel injection pressure of Karanja biodiesel blends led to a greater heat release rate (HRR), resulting in an increase in the highest cylinder pressure due to enhanced fuel–air mixing. Limited research has been carried out regarding the combustion behavior of palm biodiesel under different fuel injection pressures, ranging from low-pressure to high-pressure injection conditions. The combustion mechanism of palm biodiesel and diesel fuels, considering the concurrent impacts of increasing fuel spray velocity and the unique attributes of biofuels, remains unexplored. The purpose of this study is to evaluate how various fuel injection pressures affect palm biodiesel combustion in a constant volume combustion chamber. The analysis encompasses a thorough examination of combustion characteristics, including ignition delay, combustion pressure, heat release rate, and integral heat release. The experimental setting included a 21% oxygen content, an ambient temperature of 1000 K, and pressure variations of 400–1200 bar, typical for conventional diesel engines.

2 Research Methods

2.1 Pre-combustion Technique

The pre-combustion approach is used in this research to generate a combustion environment similar to the ambient conditions inside a diesel engine before the injection starts. Initially, a combustible gas mixture containing acetylene (C_2H_2), oxygen (O_2), and nitrogen (N_2) is introduced into the CVCC vessel, which is then ignited by the spark plug. Subsequently, the pressure in the CVCC quickly climbs to 70 bar, representing the pre-combustion phase in Fig. 1. However, this pressure is excessively high compared to typical diesel engine pressure conditions. The cool-down phase, induced by the heat transfer via the vessel and the two quartz panes, was determined from the time of peak pressure to the desired pressure.

2.2 Combustion Analysis

HRR was determined from pressure rise data recorded using a piezoelectric sensor, adhering to the First Law of Thermodynamics [14] and simplified due to the constant volume condition as shown in Eq. (1) [15, 16].

$$\frac{dQ}{dt} = \frac{1}{\gamma - 1} V \frac{dP}{dt} \tag{1}$$

where dQ/dt represents HRR (J/s), γ denotes the specific heat ratio (1.35), P signifies cylinder pressure (Pa), and V represents vessel volume (546.1×10^{-6} m^3).

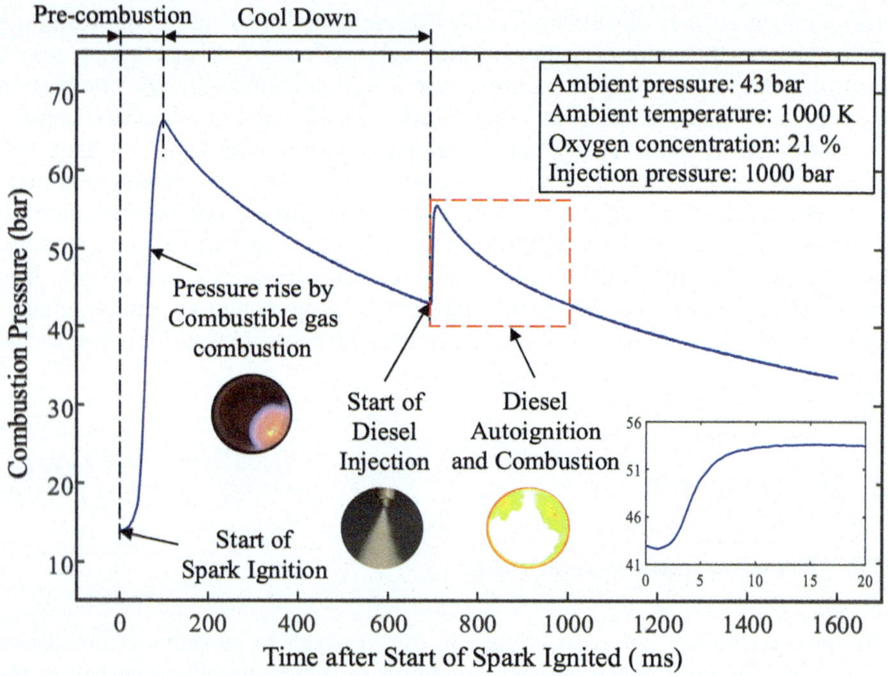

Fig. 1 Principle of the pre-combustion technique and pressure history inside the CVCC

3 Experimental Setup and Test Conditions

3.1 Experimental Setup and Procedure

The CVCC test rig comprises four components, as illustrated in Fig. 2: combustion vessel, gas supply system, fuel supply system, and data acquisition-control (DAC) system. First, a combustible mixture of 3.5% acetylene, 29.38% oxygen, and 67.12% nitrogen by volume is fed into the CVCC through the solenoid valve for a predetermined duration. This gas blend is homogenized by a stirrer fan directly attached to the chamber, as seen in Fig. 3. A single ignition coil was used to ignite the gas combustible mixture. Then, high-pressure fuel is injected into the CVCC chamber after the start of ignition, approximately at 847 ms. The high fuel pressure, created by a high-pressure pump (HP3) and stored in the fuel rail, is delivered to the single-hole solenoid injector (Bosch CRIN2) with a nozzle orifice diameter of 330 μm. A piezoelectric transducer sensor (AVL GU12P) is responsible for measuring pressure data during diesel and palm combustion, subsequently recorded and displayed on an oscilloscope (Tektronix TDS2024) through the charge amplifier (Kistler 5010).

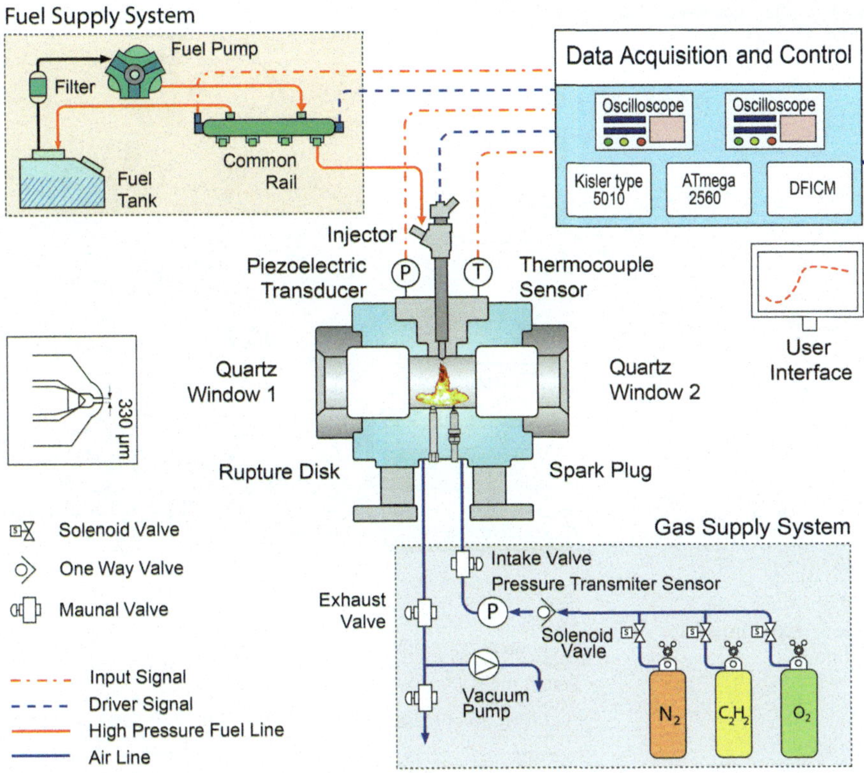

Fig. 2 The setup configuration for the CVCC experiments

The data obtained from the oscilloscope is then utilized to compute the combustion parameters, including combustion pressure, maximum pressure rise, heat release rate, and ignition delay.

3.2 *Experimental Conditions and Fuel Properties*

Table 1 shows the experimental operating conditions in this research. Diesel fuel (B0) and palm biodiesel (B100) were employed to investigate the combustion characteristics. The properties of B0 and B100 fuels were measured according to ASTM methods and are detailed in Table 2. A range of fuel pressures, from 400 to 1200 bar, was chosen to examine how different fuel pressures affect combustion parameters. The injection duration of 2.0 ms was used for all tests with diesel fuel. In this study, the oxygen concentrations of 21% and the ambient pressure of 43 bar, corresponding to ambient temperatures of 1000 K at the moment of fuel injection, were established. Experiments under identical conditions were repeated no less than 10

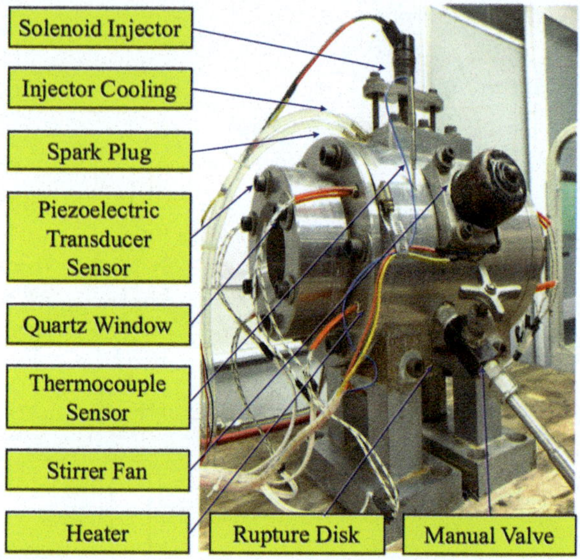

Fig. 3 Components attached on CVCC chamber

Table 1 Experimental conditions

Parameters	Value
Tested fuels	B0; B100
Fuel injection pressure (bar)	400; 600; 800; 1000; 1200
Injector	Solenoid; 1 hole × 330 μm
Injector duration (ms)	2.0
Oxygen concentration (%)	21
Ambient pressure (bar)	43

times to enhance the reliability and precision of the collected data, and the displayed results are the average of these tests. The injector's energizing start time of 847 ms was determined after conducting 20 trials involving burning of gas mixture without injection.

4 Results and Discussion

4.1 Ignition Delay

Ignition delay (ID) was determined from the point of initiation of the injector driver signal until the heat release rate curve exhibited a positive trend [17, 18]. Figure 4 illustrates the ignition delay period with varying fuel injection pressures for both diesel (B0) and Palm biodiesel (B100). With the rise in injection pressure from 400

Table 2 Properties of test fuels

Properties	B0	B100	Test method
Cetane number	51.8	61.5	ASTM D4737
Density @ 15 °C (kg/m^3)	831.9	874.2	ASTM D941
Viscosity @ 40 °C (mm^2/s)	3.24	4.27	ASTM D445
Heating value (MJ/kg)	45.68	39.75	ASTM D 240-19
Distillation T10 (°C)	209	316	ASTM D86-20b
Distillation T50 (°C)	284.3	327.4	ASTM D86-20b
Distillation T90 (°C)	353.6	333.7	ASTM D86-20b

to 1200 bar, the ignition delay shortens by 21%, decreasing from 1.47 ms to 1.16 ms of B0 and by 33%, reducing from 1.52 ms to 1.01 ms of B100, respectively. This phenomenon can be attributed to higher injection velocity and faster breakup with smaller droplet sizes, facilitating spray atomization and vaporization [19, 20]. The ID of B100 is briefer compared to that of B0 at high injection pressure as a result of the greater cetane number of B100 [12, 21]. However, at low injection pressures of 400 bar and 600 bar, the autoignition delay for B100 exceeds that of B0, measuring 1.52 ms for B100 compared to 1.47 ms for B0 at 400 bar and 1.42 ms for B100 versus 1.4 ms for B0 at 600 bar. The substantial impact of B100's higher viscosity and density characteristics becomes more pronounced at low injection pressure, significantly affecting vaporization and atomization processes, consequently resulting in prolonged ignition delay.

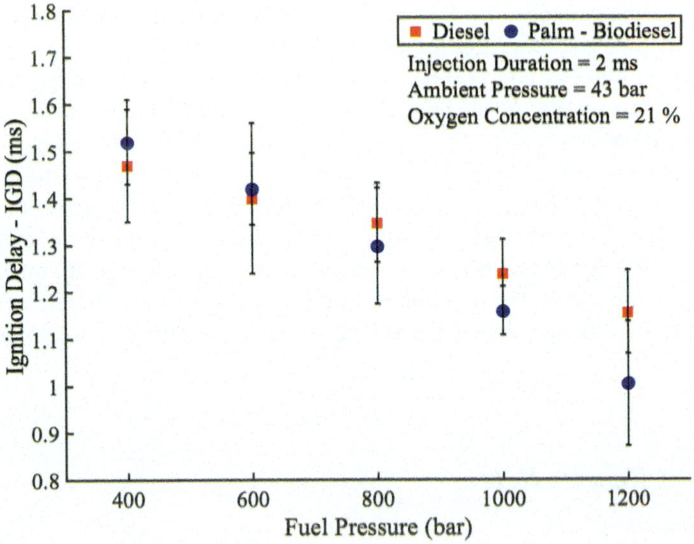

Fig. 4 Ignition delay period of B100 and B0 under various injection pressures

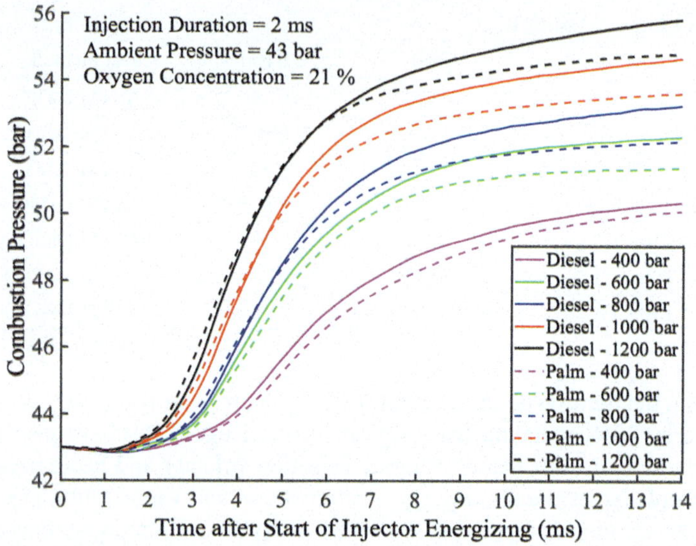

Fig. 5 Combustion pressure of B0 and B100 at various injection pressures

4.2 Combustion Pressure Rise

Figures 5 and 6 illustrate the variations in combustion pressure and the maximum combustion pressure rise over a range of injection pressures from 400 to 1200 bar for both B0 and B100 while maintaining a constant injection duration. At higher injection pressures exceeding 800 bar, the combustion pressure of B100 initiates a slightly earlier rise than that of B0, attributed to the shortened ignition delay. Conversely, at 400 bar and 600 bar, the combustion pressure increase for B100 is slower than that of B0. At all injection pressure conditions, the peak combustion pressure of B100 is consistently lower than that of B0. Because palm biodiesel is more viscous and possesses a reduced calorific value compared to diesel, it burns fuel more slowly and has a lower peak combustion pressure. This phenomenon can be elucidated by the growing dimension of spray drops and a decrease in the amount of fuel drops engaged in combustion, especially in comparison to B0 under varying injection pressures [11]. The maximum combustion pressure rise increased from 1.69 to 4.02 bar/ms for B0 and from 1.53 to 3.52 bar/ms for B100 as the injection pressure increased.

4.3 Heat Release Rate

Figure 7 compares the impact of injection pressures on the heat release rate (HRR) of palm biodiesel (B100) and diesel (B0). Initially, upon fuel injection into the chamber,

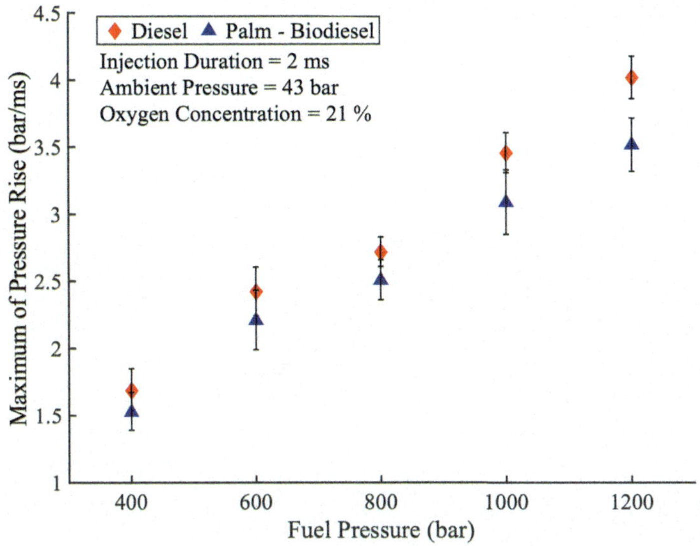

Fig. 6 Maximum combustion pressure rise of B0 and B100 at various injection pressures

the HRR curve declines to negative values as the fuel droplets absorb heat for vaporization. Subsequently, as combustion initiates, the HRR begins to rise. However, a step shoulder appears around 1.7 ms, observed at an injection pressure of 1200 bar. This phenomenon is likely a result of cooler wall regions quenching a portion of the flame upon fuel jet impingement on the chamber [6]. The peak HRR for B0 is 630.6 J/ms at an injection pressure of 1200 bar, while for B100, it is 550.5 J/ms. Additionally, the peak HRR of B100 is lower than that of B0 at all injection pressures. This is because B100 has a greater viscosity and a shorter ignition delay, resulting in the formation of a less accumulative air–fuel mixture and consequently reducing the peak HRR [22].

4.4 Integral Heat Release

Figure 8 depicts the variation in IHR between diesel (B0) and palm biodiesel (B100) across different injection pressures. Throughout all injection pressures, the IHR of B100 consistently remains lower than that of B0, with reductions of 15.6% at 400 bar, 11.1% at 600 bar, 12.7% at 800 bar, 10.2% at 1000 bar, and 9.7% at 1200 bar. The premixed combustion process significantly impacts the IHR, as a higher peak HRR facilitates greater IHR. The lower cetane number, viscosity, and distillation temperatures (T10 and T50) of B0 result in a longer ignition delay and easier droplet atomization, resulting in enhancing the premixed combustion. Additionally, the lower heating value of B100—39.75 MJ/kg versus 45.68 MJ/kg for B0 affects the total

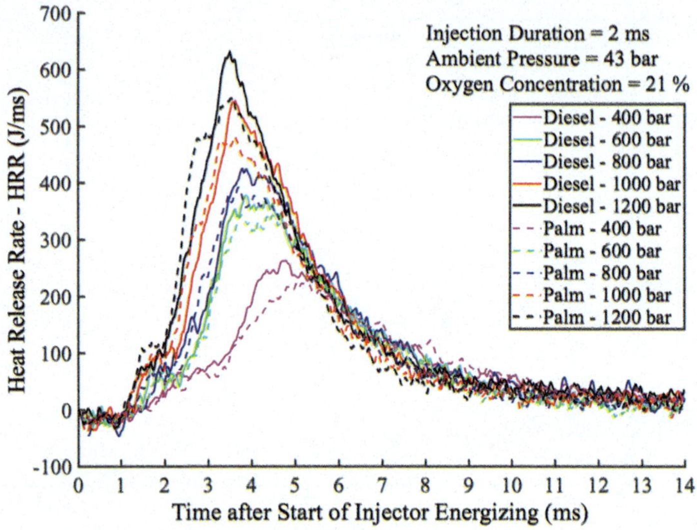

Fig. 7 The heat release rate for various injection pressures

heat supplied to the combustion chamber. As injection pressure decreases, the IHR decreases as a result of the decrease in fuel quantity [11, 23].

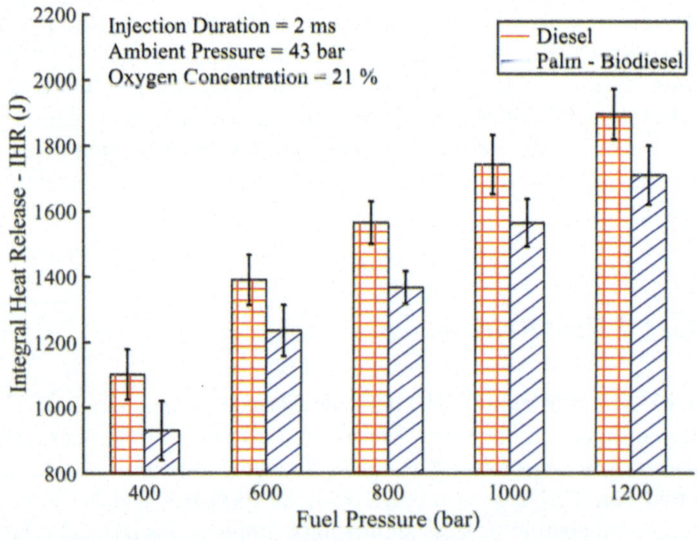

Fig. 8 Integral heat release of B0 and B100 at various experimental conditions

5 Conclusions

This work aims to explore how injection pressure affects the combustion characteristics of palm biodiesel (B100) compared to diesel (B0) within a constant volume combustion chamber. The ignition delay period decreases with higher fuel injection pressure for both B100 (33%) and B0 (21%). B100's higher cetane number contributes to a shorter ignition delay, but at 400 bar and 600 bar, B100's ignition delay exceeds that of B0. Palm biodiesel exhibits lower combustion pressure and peak heat release rate than diesel across all injection pressures. However, both parameters increase with higher injection pressure, enhancing the combustion rate. B100's maximum combustion pressure rise is consistently lower than that of B0 across all injection pressures. This difference can be attributed to the greater influence of higher kinematic viscosity and higher distillation temperatures (T10 and T50) of B100. When the injection pressure increased, the integral heat release (IHR) of both B100 and B0 fuels increased as more fuel was burned. Despite palm biodiesel's higher density, its IHR remains inferior to that of B0 at each injection pressure, primarily attributable to its reduced heating value.

Acknowledgements We acknowledge Ho Chi Minh City University of Technology (HCMUT), VNU-HCM for supporting this study.

References

1. Kadir, W.N.A., et al.: Harvesting and pre-treatment of microalgae cultivated in wastewater for biodiesel production: a review. Energy Convers. Manage. **171**, 1416–1429 (2018). https://doi.org/10.1016/j.enconman.2018.06.074
2. Lee, S.W., Lee, H.S., Park, Y.J., Cho, Y.S.: Combustion and emission characteristics of HCNG in a constant volume chamber. J. Mech. Sci. Technol. **25**(2), 489–494 (2011). https://doi.org/10.1007/s12206-010-1231-5
3. Zhang, J., Jing, W., Roberts, W.L., Fang, T.: Effects of ambient oxygen concentration on biodiesel and diesel spray combustion under simulated engine conditions. Energy **57**, 722–732 (2013). https://doi.org/10.1016/j.energy.2013.05.063
4. Guo, B., Ichiyanagi, M., Ohashi, T., Zheng, Q., Suzuki, T.: Effect of equivalence ratio and mixing time on combustion of ammonia/oxygen/argon mixture using a constant volume combustion chamber with sub-chamber. J. Mech. Sci. Technol. **37**(7), 3829–3840 (2023). https://doi.org/10.1007/s12206-023-0645-9
5. Qi, D.H., Chen, B., Zhang, D., Lee, C.F.: Optical study on the combustion characteristics and soot emissions of diesel–soybean biodiesel–butanol blends in a constant volume chamber. J. Energy Inst. **89**(4), 807–820 (2016). https://doi.org/10.1016/j.joei.2015.03.007
6. Zhang, J., Jing, W., Roberts, W.L., Fang, T.: Soot temperature and KL factor for biodiesel and diesel spray combustion in a constant volume combustion chamber. Appl. Energy **107**, 52–65 (2013). https://doi.org/10.1016/j.apenergy.2013.02.023
7. Ogunkunle, O., Ahmed, N.A.: Overview of biodiesel combustion in mitigating the adverse impacts of engine emissions on the sustainable human–environment scenario. Sustainability **13**(10) (2021). https://doi.org/10.3390/su13105465.
8. Shahid, E.M., et al.: A review of biodiesel as vehicular fuel. Renew. Sustain. Energy Rev. **12**(9), 2484–2494 (2008). https://doi.org/10.1016/j.rser.2007.06.001

9. Kuti, O.A., et al.: Effect of injection pressure on ignition, flame development and soot formation processes of biodiesel fuel spray (2010)
10. Gumus, M., Sayin, C., Canakci, M.: The impact of fuel injection pressure on the exhaust emissions of a direct injection diesel engine fueled with biodiesel-diesel fuel blends. Fuel **95**(x), 486–494 (2012). https://doi.org/10.1016/j.fuel.2011.11.020
11. Selvaraj, K., Thangavel, M.: The effects of fuel injection pressure on combustion and emission characteristics of a diesel engine using frying oil methyl ester. Energy Sources, Part A Recover. Util. Environ. Effects, 1–17 (2019). https://doi.org/10.1080/15567036.2019.1675818
12. Kuti, O.A., et al.: Characterization of spray and combustion processes of biodiesel fuel injected by diesel engine common rail system. Fuel **104**, 838–846 (2013). https://doi.org/10.1016/j.fuel.2012.05.014
13. Agarwal, A.K., et al.: Effect of fuel injection pressure and injection timing of Karanja biodiesel blends on fuel spray, engine performance, emissions and combustion characteristics. Energy Convers. Manag. **91**, 302–314 (2015). https://doi.org/10.1016/j.enconman.2014.12.004
14. Heywood, J.B.: Internal Combustion Engine Fundamentals, 2nd edn. (2018)
15. Wang, Y., Han, J., Maes, N., Cuijpers, M., Somers, B.: Ignition and combustion characteristics of N-butanol and FPBO/N-butanol blends with addition of ignition improver. Front. Energy Res. **10** (2022). https://doi.org/10.3389/fenrg.2022.832509
16. Kook, S., Pickett, L.M.: Effect of fuel volatility and ignition quality on combustion and soot formation at fixed premixing conditions. SAE Int. J. Engines **2**(2), 11–23 (2010). https://doi.org/10.4271/2009-01-2643
17. Vu, D.N., Lim, O.: Experimental study on ignition characteristic of gasoline-biodiesel blended fuel in a constant-volume chamber. J. Mech. Sci. Technol. **33**(10), 5073–5083 (2019). https://doi.org/10.1007/s12206-019-0946-1
18. Ghojel, J., et al.: Ignition characteristics of diesel—water emulsion sprays in a constant volume vessel effect of injection pressure and water content. Energy Fuels (2010)
19. Vo, C., et al.: Effects of variable O_2 concentrations and injection pressures on the combustion and emissions characteristics of the petro-diesel and hydrotreated vegetable oil-based fuels under the simulated diesel engine condition. J. Energy Inst. **91**(6), 1071–1084 (2018). https://doi.org/10.1016/j.joei.2017.07.002
20. Du, W., Zhang, Q., Zhang, Z., Lou, J., Bao, W.: Effects of injection pressure on ignition and combustion characteristics of impinging diesel spray. Appl. Energy **226**(February), 1163–1168 (2018). https://doi.org/10.1016/j.apenergy.2018.06.032
21. Ibrahim, A., El-Adawy, M., El-Kassaby, M.M.: The impact of changing the compression ratio on the performance of an engine fueled by biodiesel blends. Energy Technol. **1**(7), 395–404 (2013). https://doi.org/10.1002/ente.201300048
22. Marasri, S., et al.: Combustion characteristics of hydrotreated vegetable oil-diesel blends under EGR and low temperature. Int. J. Autom. Technol. **20**(3), 569–578 (2019). https://doi.org/10.1007/s12239
23. Ashok, B., et al.: An investigation on CI engine characteristics using pork lard methyl ester at various injection pressures and injection timings. Int. J. Green Energy **16**(11), 834–846 (2019). https://doi.org/10.1080/15435075.2019.1641107

Optimizing Turning Processes for SUS430C Steel: A Comparative Study of RSM and DFA Approaches

Nhat-Tan Nguyen, Anh-Thang Nguyen, Ruan Meng Yue, Viet-Thanh Pham, Van-Long Trinh, and Van-Canh Nguyen

Abstract This study addresses the optimization challenge of the finishing turning of *SUS430C* alloy material. The primary focus is on minimizing surface roughness (R_a) while simultaneously maximizing material removal rate (*MRR*). This approach is critical for enhancing machine efficiency and achieving superior surface quality. Utilizing the Desirability Function Approach (*DFA*) and Response Surface Methodology (*RSM*), the study delved into the effects of cutting parameters, including cutting speed (V_c), feed rate (F), and depth of cut (a_p), on surface roughness (R_a) and material removal rate (*MRR*). The optimization process revealed that a cutting speed of 200 m/min, feed rate of 0.119 mm/min, and depth of cut of 0.200 mm predict optimal values of R_a and MRR with a high desirability index, indicating a significant improvement over the traditional method. This study not only demonstrates the efficacy of combining *RSM* and *DFA* for multi-objective optimization in machining processes but also provides a validated set of parameters that align with industrial application requirements, marking a significant contribution to the field of machining optimization.

Keywords *SUS430C* · RSM-DA · Multiple objective optimization

N.-T. Nguyen · A.-T. Nguyen
Vietnam-Japan Center, Hanoi University of Industry, 298 Cau Dien Str., Bac Tu Liem Dist., Hanoi, Vietnam

R. M. Yue
9F-2, No.1 Taishun Street, Lingya District, Kaohsiung, Taiwan

V.-T. Pham · V.-L. Trinh · V.-C. Nguyen (✉)
School of Mechanical and Automotive Engineering, Hanoi University of Industry, 298 Cau Dien Str., Bac Tu Liem Dist., Hanoi, Vietnam
e-mail: canhnv@haui.edu.vn

D. H. Tien et al. (eds.), *Proceedings of the 1st International Conference on Sustainability and Emerging Technologies for Smart Manufacturing*, Proceedings in Technology Transfer, https://doi.org/10.1007/978-981-97-7083-0_59

1 Introduction

In the machining industry, turning is pivotal for fabricating high-precision components. Turning hard alloys like *SUS430C* presents considerable challenges due to the material's inherent hardness and tendency to adhere, thereby reducing tool life and productivity while escalating machining costs [1]. This underscores the imperative of optimizing cutting parameters (*CPs*) to augment the machining efficiency of *SUS430C* stainless steel, a material distinguished by its corrosion resistance and mechanical properties. This quality makes it applicable across a diverse array of industrial contexts.

Historically, the machining of *SUS430* and similar materials has predominantly focused on minimizing surface roughness (R_a) as a proxy for machining quality. For instance, Sahithi et al. [2] employed regression analysis based on the Taguchi experimental matrix to identify optimal *CPs* for minimizing R_a. However, this study, and others like it, often overlooked the *MRR*, a critical efficiency indicator for production economy. This oversight suggests a gap in comprehensive optimization approaches that simultaneously address both R_a and *MRR*, reflecting the industry's evolving needs for balancing quality and efficiency [2].

In response to this gap, various multi-objective optimization methods, including the *MCDM* method, artificial intelligence (*AI*) algorithms, genetic algorithms (*GA*) [3], etc., have been applied with notable success. Among these, Response Surface Methodology (*RSM*) [4] stands out for its utility in optimizing cutting processes due to its robust statistical framework for model development and optimization. This study aims to leverage RSM, coupled with Desirability Function Approach (*DFA*), to explore the influence of cutting parameters including depth of cut (a_p), cutting speed (V_c), feed rate (f_z) on surface roughness (R_a), and material removal rate (*MRR*). By doing so, it seeks to establish a balanced and empirically validated set of *CPs* that enhance both the machining efficiency and surface quality of *SUS430C*, addressing the dual objectives often faced in industrial machining operations [5–7].

Additionally, the utilization of *RSM* in this context is strategic, not merely a methodological choice. Its aim is to demonstrate the versatility and efficacy of the method across diverse machining scenarios. By focusing on *SUS430C*, this study not only contributes to the specific body of knowledge around this material but also provides a methodological blueprint for optimizing machining processes for hard-to-machine alloys more broadly.

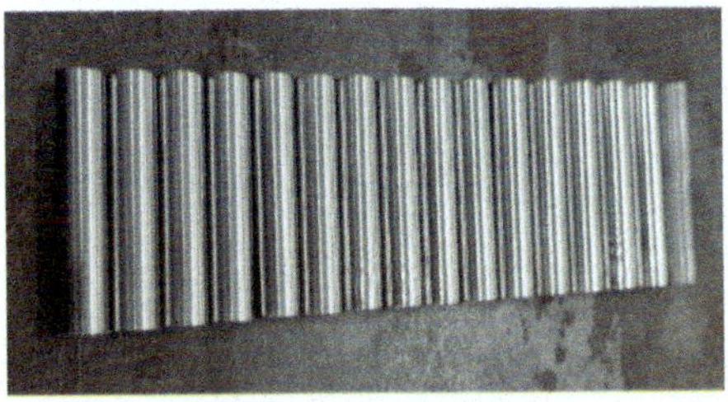

a. Workpiece specimen

b. Experimental machine

Fig. 1 Workpiece and machine

2 Material and Methods

2.1 Materials and Experimental Design

The adoption of a 16-experiment Box–Behnken design, despite its concise scope, is strategically tailored to efficiently investigate the interplay among cutting parameter $\{V_c, f_z, a_p\}$. This choice leverages the design's inherent strength in balancing comprehensive data collection with resource optimization. By focusing on three levels for each parameter, the study ensures a thorough exploration of the parameter space while maintaining a manageable number of experiments. This approach not only aligns with the study's goals but also underscores the practicality of applying *RSM* in industrial research, where efficiency and effectiveness are paramount.

The material used in the experimental study is heat-treated *SUS430* stainless steel, achieving a hardness of 56HRC; the workpieces are prepared to dimension of D16xL70 mm (Fig. 1a). Each experiment is conducted on a Mori Seiki *CNC* lathe (Fig. 1b), and the cutting tool is equipped with Protech inserts to ensure high stability and precision during machining. To eliminate the influence of tool wear on the experimental results, a new insert is used for each experiment, ensuring the objectivity and reliability of the collected data.

The R_a value of each machined surface part was measured in accordance with ISO-4287 standards [8] using a Mitutoyo Surftest JS-201 roughness tester. The *MRR* values were calculated using a formula provided by Sandvik [9]. A summary of experimental results is presented in Table 1, offering a concise overview of the experimental setup and findings in a structured manner suitable for scientific communication.

2.2 Multiple Response Optimization

The multi-objective optimization problem in this study is defined as follows: to find the set of parameters find $X = \{V_c, f_z, a_p\}$ to minimize (R_a) and maximize (*MRR*) within the constraints $\{60 \le V_c \le 200;\ 0.05 \le f_z \le 0.20;\ 0.06 \le a_p \le 0.20\}$. To

Table 1 Design of experiment and acquired data

Run	V_c (m/min)	f_z (mm/rev)	a_p (mm)	R_a (μm)	MRR (mm^3/min)
1	80	0.125	0.2	1.761	2
2	200	0.125	0.06	1.616	1.5
3	200	0.2	0.13	3.25	5.2
4	80	0.05	0.13	0.474	0.52
5	140	0.125	0.13	1.446	2.275
6	80	0.125	0.06	1.358	0.6
7	140	0.125	0.13	1.271	2.275
8	140	0.2	0.06	3.35	1.68
9	80	0.2	0.13	3.271	2.08
10	140	0.125	0.13	1.653	2.275
11	200	0.05	0.13	0.708	1.3
12	140	0.125	0.13	1.373	2.275
13	200	0.125	0.2	1.706	5
14	140	0.2	0.2	3.228	5.6
15	140	0.05	0.2	0.334	1.4
16	140	0.05	0.06	0.604	0.42

address this optimization challenge, a combination of *RSM* and *DA* is employed. *RSM* (Response Surface Methodology) is a statistical technique used for developing, improving, and optimizing processes, was developed by Box and Wilson [10]; while *DA* (*Desirability Function Approach*) is a multi-criteria decision-making method that transforms multiple responses into a single composite desirability function [11]. The efficacy of the *RSM–DA* approach has been validated in previous studies, such as the optimization of milling processes for *SKD11* tool steel under minimum quantity lubrication conditions [12].

In this study, the *RSM–DA* is utilized to pinpoint the optimal set of cutting parameters that achieve a balance between minimizing R_a and maximizing *MRR*.

3 Results and Discussion

3.1 *Effects of Cutting Parameters on R_a and MRR*

The perturbation plot provided indicates the sensitivity of the R_a to variations in the machining parameters from a fixed reference point (Fig. 2). It is observed that the f_z significantly affects R_a, as evidenced by the steep slope of the corresponding line, suggesting that R_a increases considerably with an increase in f_z. In contrast, both

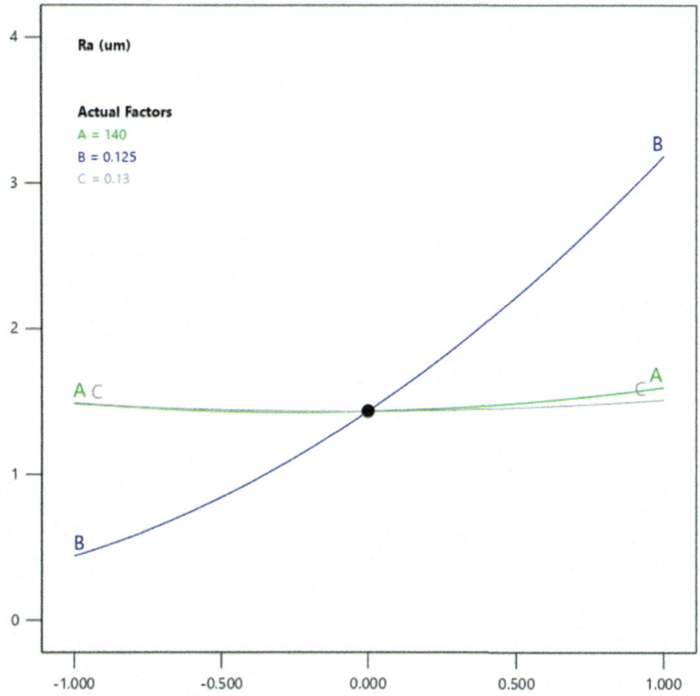

Fig. 2 Perturbation plot for less R_a

the V_c and a_p show minimal effect on R_a, as indicated by their relatively flat lines. This analysis suggests that among the machining parameters studied, f_z is the most critical factor influencing R_a in turning *SUS430C* alloy.

The perturbation plots reveal that both f_z and V_c significantly enhance *MRR* when increased, with f_z showing a slightly greater influence than V_c (Fig. 3). Meanwhile, the a_p has a less pronounced but still positive effect on *MRR*. These insights suggest that optimizing f_z and V_c should be a priority to improve *MRR* in the turning process of *SUS430C* stainless steel.

3.2 Optimization Results

In this study, utilizing the Desirability Function Approach (*DFA*), the impact of each parameter V_c, f_z, and a_p on R_a and *MRR* was independently assessed and synthesized to determine their optimal combination.

The comprehensive analysis revealed that while each individual parameter could achieve a high level of optimization, their combination requires a certain degree of compromise. The combined desirability index, reaching a value of 0.676, reflects a good overall level of optimization, with corresponding values for R_a and *MRR*

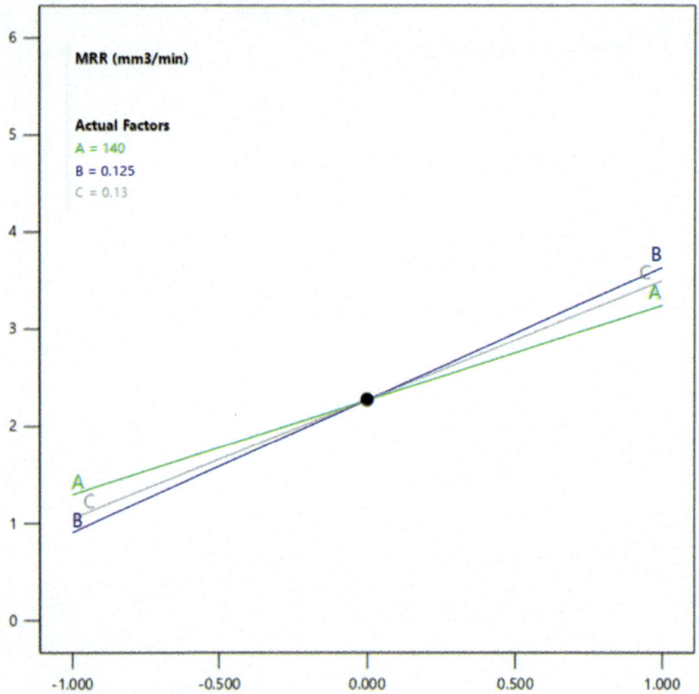

Fig. 3 Perturbation plot for *MRR*

being 0.614 and 0.744, respectively (Figs. 4 and 5). The discrepancy between these values highlights the crucial balance between minimizing surface roughness (R_a) and maximizing material removal rate (*MRR*).

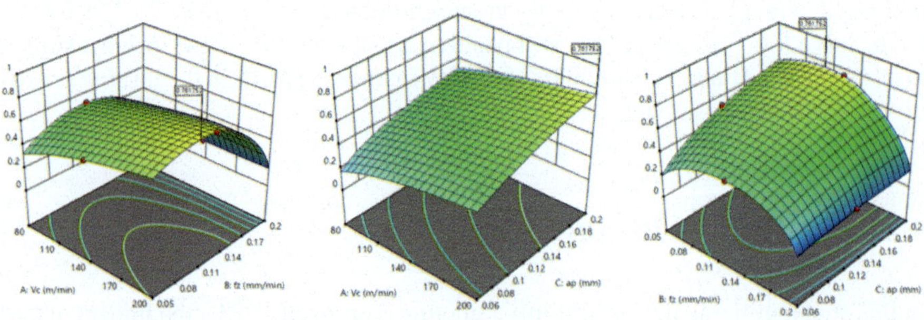

Fig. 4 RSM analysis for R_a and *MRR*

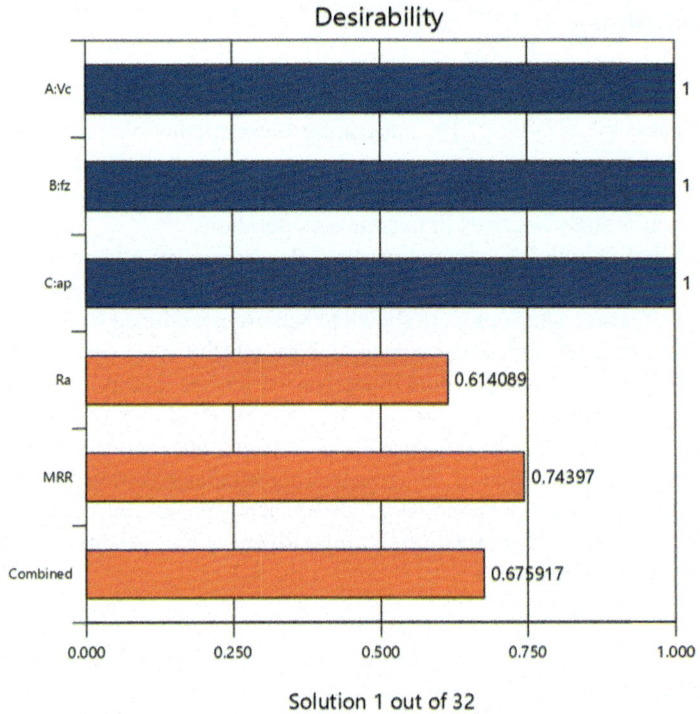

Fig. 5 Desirability profile for optimization solutions

Based on the optimization results obtained through the *DFA*, the multi-objective optimization yielded a set of parameters with a V_c of 200 m/min, a f_z of 0.119 mm/min, and an a_p of 0.200 mm. This combination of parameters predicts a R_a of 1.49895 μm and an *MRR* of 4.27595 mm^3/min, with the observed standard deviations indicating the variability within the experimental results.

The inclusion of this optimized parameter set into the conclusion of the study enhances the practical relevance of the findings. It demonstrates a concrete application of the theoretical analysis by providing a specific, validated parameter configuration that achieves a balance between achieving a minimal surface roughness and maximizing the material removal rate, which are critical factors for the efficiency and quality of machining processes. The high desirability index associated with these results confirms their suitability for industrial application, making this study a valuable reference for future endeavors in machining optimization.

4 Conclusion

The research presented in this paper successfully applied *RSM–DA* to optimize the turning process of *SUS430C*. By integrating these methodologies, the study was able to identify a balanced set of machining parameters that effectively minimized surface roughness (R_a) while maximizing material removal rate (*MRR*), addressing the often-conflicting objectives in machining operations.

The optimal machining parameters were determined to be cutting speed (V_c), feed rate (f_z), and depth of cut (a_p) of 200 m/min, 0.119 mm/rev, and 0.200 mm, respectively. These conditions were shown to achieve a predicted R_a of 1.499 μm and an *MRR* of 4.276 mm^3/min, with a substantial desirability index of 0.676, evidencing a practical reconciliation of the dual objectives of surface quality and productivity.

This study's findings underscore the potential of *RSM–DA* as a robust tool for enhancing machining processes, providing a systematic approach for achieving optimal machining performance. The research contributes significant practical implications for the manufacturing industry, where such precise control over machining parameters is crucial for maintaining competitiveness and producing high-quality machined components.

References

1. Lacerda Amorim, F., Torres, R., Rocha Machado, Á., Sampaio, M.A.: Hard turning of induction hardened SAE 1045 steel with PCBN tools under dry machining and Minimum Quantity of Lubrication (MQL), pp. 959–968 (2018). https://doi.org/10.26678/abcm.cobem2017.cob17-5910
2. Sahithi, V.V.D., Malayadrib, T., Srilatha, N.: Optimization of turning parameters on surface roughness based on Taguchi technique. Mater. Today Proc. **18**, 3657–3666 (2019). https://doi.org/10.1016/j.matpr.2019.07.299
3. Nguyen, V.C., Nguyen, B.N., Tien, D.H., Nguyen, V.Q., Nguyen, X.T., Nguyen, T.D.: Using support vector regression and non-dominated sorting genetic algorithm in multi-objective optimization of milling of S50C steel under MQL condition. J. Appl. Eng. Sci. **20**(1), 123–130 (2022). https://doi.org/10.5937/jaes0-31366
4. Pham, V.H., Nguyen, T.D., Le, V.T., Tien, D.H., Nguyen, V.C.: Optimization of cutting parameters in MQL flat surface milling of SKD11 steel. In: Proceedings of the International Conference on Advanced Mechanical Engineering, Automation, and Sustainable Development 2021 (AMAS2021), pp. 261–266. Springer (2022). https://doi.org/10.1007/978-3-030-99666-6_40
5. Agrawal, S., kumar Gaur, M., kumar Kasdekar, D., Agrawal, S.: Optimization of machining parameters of hard porcelain on a CNC machine by Taguchi-and RSM method. Int. J. Eng. Sci. Technol. **10**(1), 13–22 (2018). https://doi.org/10.4314/ijest.v10i1.2
6. Bhaumik, M., Maity, K., Mohapatra, K.D.A.S.: Multi-objective optimization of EDM process parameters using RSM-based GRA and TOPSIS method for grade 6 titanium alloy. Surf. Rev. Lett. **28**(07), 2150062 (2021). https://doi.org/10.1142/S0218625X21500621
7. Sahoo, P.: Optimization of turning parameters for surface roughness using RSM and GA. Adv. Prod. Eng. Manag. **6**, 197–208 (2011)
8. Mitutoyo: Quick Guide to Surface Roughness Measurement: Reference Guide for Laboratory and Workshop. [Online]. Available: https://www.mitutoyo.com/wp-content/uploads/2012/11/1984_Surf_Roughness_PG.pdf

9. General turning formulas and definitions. [Online]. Available: https://cdn.sandvik.coromant.com/files/sitecollectiondocuments/services/metal-cutting-e-learning/formulas-and-definitions/formulas-and-deinitions-for-turning-metric-enu.pdf
10. Box, G.E.P., Wilson, K.B.: On the experimental attainment of optimum conditions. J. R. Stat. Soc. Ser. B **13**(1), 1–38 (1951). https://doi.org/10.1111/j.2517-6161.1951.tb00067.x
11. Costa, N.R., Lourenço, J., Pereira, Z.L.: Desirability function approach: a review and performance evaluation in adverse conditions. Chemom. Intell. Lab. Syst. **107**(2), 234–244 (2011). https://doi.org/10.1016/j.chemolab.2011.04.004
12. Canh Nguyen, V., Duong Nguyen, T., Thao Le, V.: Optimization of sustainable milling of SKD11 steel under minimum quantity lubrication. Proc. IMechE Part E J. Process. Mech. Eng. **1**(10) (2022). https://doi.org/10.1177/09544089221110978

A Honey Dehydrator Working at Atmospheric Pressure Using a Heat Pump Incorporated with a Falling Film Evaporator

Son Thai Pham and Tho Tran Dang

Abstract This study aims to evaluate the performance of a honey dehydrator system that combines a heat pump with a falling film evaporator, with the specific objectives of achieving enhanced energy efficiency and reducing the moisture content of honey to meet export standards. Various temperature modes of drying air, ranging from 40 to 55 °C with a 5 °C interval, were systematically examined, and the optimal values were determined based on considerations of the moisture extraction capacity with respect to energy consumption. The optimized temperature mode of 50 °C was subsequently employed for the dehydration process of a batch of honey. The moisture content of honey has been reduced from initial value of 22.0–15.5% in batch and the last drops of honey have a moisture content of 12.0%, thereby satisfying the desired specifications required for exportation.

Keywords Honey dehydrator · Atmospheric pressure · Heat pump · Honey moisture content · Energy efficiency

1 Introduction

The elevated moisture content inherent in honey poses challenges for its long-term storage due to its susceptibility to fermentation. Naturally harvested honey exhibits variability in moisture content, as it is influenced by multiple factors including climatic conditions, weather patterns, and extraction methodologies. In order to adhere to prescribed standards, honey producers employ a honey dehydration process. This indispensable procedure facilitates the attainment of optimal moisture levels in honey, thereby enabling its preservation over extended periods

S. T. Pham · T. T. Dang (✉)
School of Mechanical Engineering, Hanoi University of Science and Technology, Dai Co Viet Rd. 01st, Hanoi, Vietnam
e-mail: tho.dangtran@hust.edu.vn

Institute of Heat Engineering and Refrigeration, Hanoi University of Science and Technology, Dai Co Viet Rd. 01st, Hanoi, Vietnam

D. H. Tien et al. (eds.), *Proceedings of the 1st International Conference on Sustainability and Emerging Technologies for Smart Manufacturing*, Proceedings in Technology Transfer, https://doi.org/10.1007/978-981-97-7083-0_60

while simultaneously safeguarding its intrinsic nutritional and medicinal properties. According to Sanz et al. [1], as long as the moisture content of honey remains below 17%, fermentation will not take place, regardless of the yeast count present.

Irrespective of the scale or operational techniques employed, honey dehydration apparatus is consistently engaged in the process of heating honey and extracting its moisture. This can be accomplished by either direct heating utilizing a drying agent, such as air [2], or indirect heating facilitated by a water jacket [3]. The removal of moisture from honey is achieved by means of direct contact between the drying agent, typically air, and the honey itself. Heat pump dryer is commonly employed to address the requirements of both heating and moisture controlling of the drying air.

As reported by Morawski et al. [2], a honey dehydration system utilizes a heat pump system combined with a spraying and stirring mechanism, along with a closed-loop circulation of the drying agent (air). The drying time is minimized by selecting the optimal value of the evaporating temperature of the evaporation unit. Maintaining constant mixing speed and airflow, the optimal evaporating temperature value significantly reduces honey drying time by 8–13% while concurrently enhancing the specific moisture extraction rate (SMER) by 4–32%.

Another investigation on conducting honey moisture reduction processes under atmospheric pressure conditions is reported by Gill et al. [3]. This small-scale honey dehydration system is capable of producing honey with a final moisture content below 17%. In this setup, air serves as the drying agent, operating at atmospheric pressure conditions and temperatures ranging from 30 to 40 °C. Concurrently, a water jacket surrounds the honey container, employing a circulation of heated water with temperatures set at 35, 40, and 45 °C. The honey undergoes heating and is subsequently pumped onto a sieve, allowing it to flow through the perforations of the sieve. This honey stream subsequently encounters the airflow passing underneath the sieve. As a result of this process, the moisture content of the honey is reduced from an initial value of 25.2% to a final value of 16.4%.

As reported in [4], 1468 g of honey was dehydrated, in a batch run, from initial moisture content of 26.6–15.2% in 2 h. The drying agent (humidified air) was heated to 50 °C. The air was in contact with honey under nonturbulent conditions, i.e. the mean velocity of the air was less than 5 m/s and preferably about 2 m/s. Using the air heated to 50 °C but in a continuous run, honey, at supply rate of 15.3–16.0 ml/min, had moisture content reduced from 29.6% to the equilibrium value of 16.9%. The same experiment conditions were repeated but the air temperature was 62–63 °C, which resulted in the moisture content of honey product fluctuating around 18%.

The present study does not extensively explore the intricate aspects concerning the design of the falling film evaporator or the control methodology employed for the heat pump. These specific areas of investigation fall beyond the intended scope of this report. The primary objective of this study is to investigate the dehydration capacity of an established system. Initially, a series of experiments were conducted using water as an artificial load to assess the drying performance of the system. Subsequently, the optimal set of parameters is identified. Finally, the system was tested using honey as the load, and the obtained results were reported.

2 Methodology

2.1 Description of the Established Honey Dehydrator System

The honey dehydrator system, as graphically shown in Fig. 1, consists of a falling film evaporator that is interconnected with a heat pump system. The falling film evaporator is a tower consisting of multiple stages. Each stage has two cones. The expanded section of the upper cone is oriented toward the expanded section of the lower cone. The conical structures facilitate the dispersion of honey into thin films as it descends from the top to the bottom of the tower, thereby enhancing both the contact time and surface area between the drying agent (air) and the honey. The honey is fed from the top of the tower by using a supply tank located on top of the tower. Once the supply valve is opened, honey will be poured slowly into the tower.

The incoming air undergoes a series of sequential treatments, starting with cooling and dehumidification (at the cooling coils of the heat pump), followed by preheating (at the heating coils of the heat pump) prior to its entry into the lower section of the tower. Within the falling film evaporator, the air engages in a moisture exchange

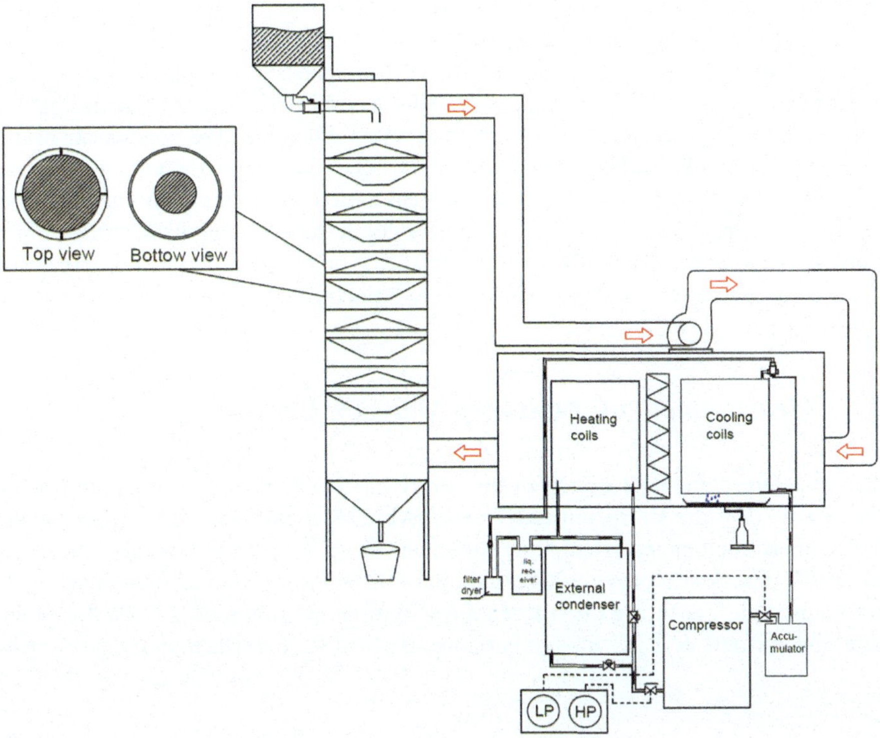

Fig. 1 A schematic representation of the honey dehydrator system

process. Subsequently, it exits through the upper opening of the tower and is circulated back to the air handling box by means of a dedicated fan. Once the air has been cooled, dehumidified, and reheated, it is reintroduced into the lower section of the tower to commence the subsequent cycle of moisture extraction within the evaporator tower.

2.2 *Experiment Using Water as Artificial Load of the Dehydrator System*

Prior to commencing the honey dehydration process in the system, a preliminary phase involves utilizing water as an artificial load. The water undergoes a closed loop circulation, pumped from the bottom section of the evaporator tower to the uppermost level, thus establishing an uninterrupted downward flow of water throughout the entire height of the tower, while the air flows upward internally within the tower. The ingress of air transpires at the lower section of the tower, where it undergoes moisture extraction from the water stream, subsequently emerging from the top of the tower. Subsequent to this, the air proceeds to an air handling unit, where it undergoes cooling and dehumidification. The quantity of condensed water accumulated within this unit serves as an indirect metric for evaluating the moisture extraction efficiency of the air stream within the evaporator tower.

The condensed water is directed, from the condensate tray positioned directly beneath the cooling coil, along the condensate drainpipe, toward the designated location housing a receptacle for condensed water collection. The condensate drainpipe is intentionally installed with an inclined slope to facilitate the unhindered flow of condensate, preventing any potential accumulation within the pipe. Furthermore, during the operational phase of the system, the condensate drainpipe undergoes vibrations, effectively mitigating the risk of condensate buildup within the drainpipe. The amount of condensed water is weighed by a digital scale.

2.3 *Measurement of the Honey Moisture Content*

The moisture content of honey in this study has been accurately measured using the Honey Moisture Refractometer, model PAL-22S, manufactured by ATAGO Co., Ltd. This refractometer offers a measurement range for honey moisture from 12.0 to 30.0%. The device provides a resolution of 0.1% for moisture content and 0.1 °C for temperature, ensuring precise readings. With an accuracy of ± 0.2% for moisture content and ± 1 °C for temperature, the PAL-22S refractometer guarantees reliable and consistent results. It is equipped with automatic temperature compensation, which operates within the range of 10–40 °C, ensuring accurate measurements regardless of temperature fluctuations. The refractometer is designed to operate within the ambient temperature range of 10–40 °C, making it suitable for various laboratory and production environments.

3 Result and Discussion

The experimental results involving the utilization of water as the drying load are depicted in Figs. 2 and 3. The system was subjected to continuous operation throughout an extended period of the day, about 400–540 min. This prolonged experimental period was chosen to encompass and consider the influence of external factors such as variations in ambient temperature, ambient humidity, or even heat dissipation, and other relevant parameters. Subsequent investigations in future study will delve more comprehensively into these parameters, enabling a deeper understanding and analysis of the experimental findings.

The main purpose of the current research is to find an optimized mode for the honey dehydration batch. Within the investigated cases, encompassing four distinct temperature modes ranging from 40 to 55 °C with a 5 °C interval, it was observed that the 50 °C mode exhibited the highest quantity of condensate water, as illustrated in Fig. 2. Conversely, the power consumption at the 50 °C mode, as demonstrated in Fig. 3, did not exhibit a significant deviation from that of the 45 and 55 °C modes.

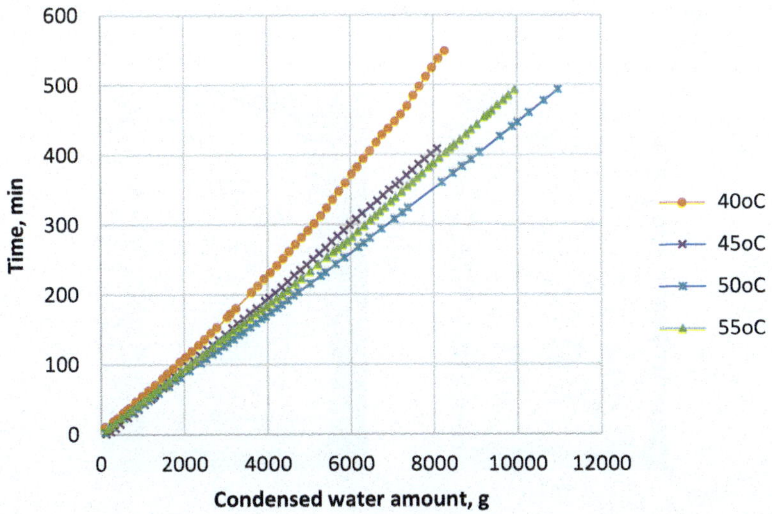

Fig. 2 The accumulated amount of condensed water at different modes of drying air temperatures (lines connecting the data points are guide for the eye)

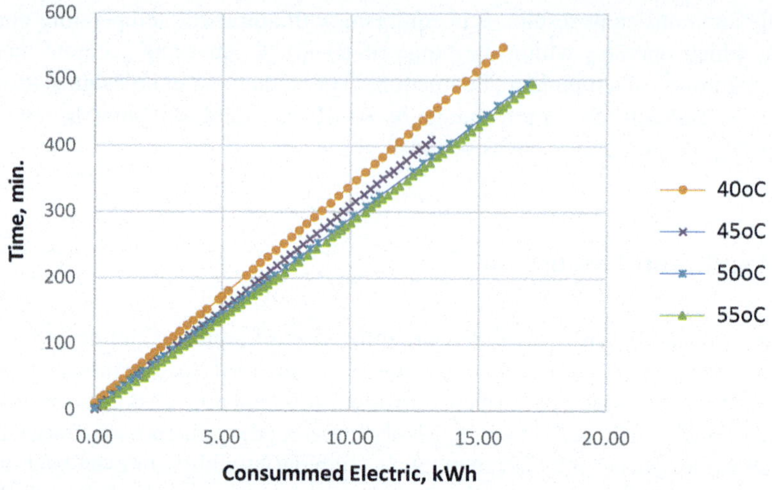

Fig. 3 The accumulated amount of consumed electrical energy at different modes of the drying air temperatures (lines connecting the data points are guide for the eye)

Notably, the 40 °C mode exhibited the lowest power consumption, as indicated in Fig. 4, but it was accompanied by a considerably reduced quantity of condensate water in comparison to the remaining three temperature modes. Consequently, the 50 °C mode has been selected for conducting the dehydrator process of honey.

Following an approximate duration of one hour, a quantity of approximately 64 g of honey is collected at the lowermost section of the evaporator tower. This honey undergoes a moisture reduction process, resulting in a decrease from its initial

(a) (b)

Fig. 4 Image illustrating the application of a refractometer for the measurement of honey moisture content at temperature mode of 50 °C, **a** in batch and **b** last drops of the honey coming from the bottom of the evaporator tower moisture content of approximately 22.0% to a moisture content level of 15.5%. Notably, there are residual drops of honey that exhibit high stickiness and elasticity, requiring a significant amount of time to descend from the bottom of the evaporator tower. The moisture content of these final drops is approximately 12.0%.

4 Conclusion

A honey dehydration system, incorporating heat pumps and a falling film evaporator, was experimentally implemented. Four distinct experimental modes utilizing water as the artificial load were employed to determine the optimal mode for the moisture extraction capacity of the investigated system. The experimental results with honey demonstrated a significant decrease in the moisture content of the output honey, surpassing the established export standards. These encouraging outcomes highlight the potential of the system and drive further endeavors to optimize its operation for enhanced energy efficiency while satisfying the moisture content specifications of the output honey. This optimization aims to ensure the realization of both high-quality and high-volume production. This research demonstrates the potential of heat pump technology as an energy-saving and efficient method for honey dehydration, providing valuable insights for the honey industry and promoting sustainable practices in food processing.

Acknowledgements This research is funded by Hanoi University of Science and Technology (HUST) under project number T2023-TĐ-018. Support and contribution from various individuals, including students and services, are sincerely appreciated.

References

1. Sanz, S., Gradillas, G., Jimeno, F., Perez, C., Juan, T.: Fermentation problem in Spanish north-coast honey. J. Food Prot. **58**(5), 473–578 (1995). https://doi.org/10.4315/0362-028X-58.5.515
2. Morawski, M., Malec, M., Niezgoda-Żelasko, B.: Effective condensing dehumidification in a rotary-spray honey dehydrator. Energies **15**, 100 (2022). https://doi.org/10.3390/en15010100
3. Gill, R.S., Hans, V.S., Singh, S., et al.: A small scale honey dehydrator. J. Food Sci. Technol. **52**, 6695–6702 (2015). https://doi.org/10.1007/s13197-015-1760-0
4. Platt Jr., J.L., Ellis, J.R.B.: Removing water from honey at ambient pressure. U.S. Patent No. 4472450, 18 September 1984

Preparation of Yttria-Stabilized Zirconia (YSZ) Nano-scale Powder by Sol–Gel Method Apply for Preparing Electrolyte of Solid Oxide Fuel Cell

Manh Quang Vu, Duc Dung Le, Ba-Canh Bui, and Cao Ha Hong

Abstract In this study, a method for synthesizing nano-sized 8 mol% Yttria-Stabilized Zirconia (YSZ-8) materials via the sol–gel approach using polyamino-carboxylic acid as a gel-forming aid is developed. Specifically, ethylenediaminetetraacetate disodium salt (EDTA-Na_2) and diethylenetriaminepentaacetic acid (DTPA) are utilized to form the sol–gel in the synthesis process of YSZ-8 materials. DTPA is demonstrated to be more effective than EDTA-Na_2 in the gel formation process due to its strong chelating ability with zirconium and yttrium ions. Consequently, the synthesis yields good results with YSZ-8 particle sizes reaching about 20 nm at low calcinated temperatures of 800–900 °C. These results are validated through analytical techniques such as TGA-DSC, SEM–EDS, and XRD. The successful synthesis of YSZ-8 serves as a crucial foundation in the fabrication of solid electrolytes used in solid oxide fuel cells.

Keywords 8 mol% YSZ · Solid electrolyte · Solid oxide fuel cells · Sol–gel method

---

M. Q. Vu
Petro Vietnam, No. 18 Lang Ha, Dong Da, Hanoi, Viet Nam
e-mail: Vumanh.quang@yahoo.com

M. Q. Vu · D. D. Le · B.-C. Bui · H. H. Cao (✉)
Hanoi University of Science and Technology, No. 1 Dai Co Viet, Hai Ba Trung, Hanoi, Viet Nam
e-mail: ha.caohong@hust.edu.vn

D. D. Le
e-mail: dung.leduc@hust.edu.vn

B.-C. Bui
e-mail: cank.buiba@gmail.com

D. H. Tien et al. (eds.), *Proceedings of the 1st International Conference on Sustainability and Emerging Technologies for Smart Manufacturing*, Proceedings in Technology Transfer, https://doi.org/10.1007/978-981-97-7083-0_61

1 Introduction

Solid oxide fuel cells (SOFC) represent highly efficient electrochemical devices capable of converting chemical energy into electrical power, with demonstrated efficiencies of up to approximately 51% for standalone SOFC systems [1, 2]. These cells typically consist of an anode, a cathode electrode, and an electrolyte material situated between the electrodes. Within this setup, the electrolyte layer serves as a conduit for oxygen ions, which are generated at the cathode, to migrate to the anode where fuel oxidation occurs electrochemically at the triple phase boundary (TPB). The entire process is governed by thermodynamic principles, involving the conversion of reactant enthalpy into useful electrical power as Gibbs free energy.

Zirconia-based oxide exhibits relevant properties desired for SOFC electrolyte material in which yttria-doped zirconia had been extensively studied that possess sufficient conductivity for oxygen ion conducting and proven excellent mechanical and chemical stability [3] even at highly oxidizing and reducing environment and elevated temperature during SOFC operation and fabrication [4].

However, in pure state, zirconia has negligible conductivity [4, 5]. ZrO_2 is monoclinic structure at room temperature but changed to a tetragonal form above 1170 °C and a cubic fluorite structure above 2370 °C [4]. Zirconia structure was improved by doping with yttria that stabilizes the fluorite structure from room temperature to melting point [6–8]. In addition, doping zirconia creates a large concentration of oxygen vacancies, enabling the mobilization of electrons in the fluorite structure, which, as a result, increases conductivity.

It is reported that voltage loss is about 75 mV for each 10 μm thickness of YSZ-8 layer at 600 °C [9]. Therefore, a thickened layer of YSZ electrolyte was not recommended to SOFC application due to highly ohmic loss of YSZ material. In parallel with researches to new material having high conductivity, many efforts have been made to have thin film as much as possible of the electrolyte layer [10–12] to reduce the ohmic loss, in which using ultrafine powder of YSZ material is an alternative selection to obtain thin layer of the electrolyte.

Various methods have been explored to produce nano-sized YSZ-8 particles, including the sol–gel approach pioneered by Penichi [13]. In this method, inorganic metal salts are dissolved in a citric acid solution, followed by the addition of ethylene glycol to facilitate esterification. The resulting intermediate resin is calcinated at elevated temperatures to yield nano-sized particles [12, 14, 15]. Modifications to Penichi's method have been proposed, such as the use of citric acid and EDTA to achieve similar results [16].

In this paper, a novel synthesis approach for YSZ nanopowder utilizing polyamine-polycarboxylic acids is proposed, specifically Diethylenetriaminepentaacetic acid (DTPA), in combination with Disodium ethylenediaminetetraacetate salt (EDTA-Na_2), as alternatives to citric acid. This method offers the potential for improved control over particle size and morphology, thereby enabling the fabrication of thin electrolyte layers with reduced ohmic losses in SOFCs. By elucidating the synthesis methodology and underlying mechanisms, this study aims to contribute to the advancement of YSZ-8 electrolyte materials for high-performance SOFC applications. Additionally, the use of DTPA and EDTA-Na_2 presents an innovative approach to enhance the efficiency and reliability of SOFCs, paving the way for future developments in clean energy technology.

DTPA and EDTA-Na_2, known for their strong chelating properties, form multidentate complexes with a wide range of metal ions [17]. Consequently, YSZ-8 nanopowder can be synthesized via the sol–gel method employing either DTPA or EDTA-Na_2. Samples are made by dissolving zirconyl chloride ($ZrOCl_2$) and yttrium nitrate ($Y(NO_3)_3$) in distilled water and the resulting solution is subjected to chelation after adding DTPA and EDTA-Na_2. The robust chelation provided by DTPA and EDTA-Na_2 not only resists hydrolysis but also inhibits crystallization during the calcination process, leading to the formation of nano-sized particles.

2 Materials and Methods

2.1 Materials

Zirconyl chloride hydrate $ZrOCl_2 \cdot 8H_2O$ (99.9%), yttrium nitrate hydrate ($Y(NO_3)_3 \cdot 6H_2O$ (99.99%), Diethylenetriaminepentaacetic acid (DTPA, 99.99%), Ethylenediaminetetraacetic acid disodium salt (EDTA-Na_2, 99.99%) were purchased from Merck. Distilled water is used for dissolving chemicals.

2.2 Experimental Synthesis

To prepare YSZ-8, $ZrOCl_2 \cdot 8H_2O$ and $Y(NO_3)_3 \cdot 6H_2O$ were dissolved in distilled water with the stoichiometric weight to obtain 8 mol% of yttria-doped zirconia. Diagram of the experimental steps is depicted in Fig. 1. Two experimental solid electrolyte samples S_1 and S_2 (using different complexing agents) are prepared: sample S_1 is prepared by adding EDTA-Na_2 solution into the solution of $ZrOCl_2 \cdot 8H_2O$ and $Y(NO_3)_3 \cdot 6H_2O$; and sample S_2 is prepared by adding DTPA solution into the solution of $ZrOCl_2 \cdot 8H_2O$ and $Y(NO_3)_3 \cdot 6H_2O$. These samples after adding EDTA and DTPA are stabilized at the pH in range of 9–10 to avoid precipitation.

The solution ($ZrOCl_2 \cdot 8H_2O$, $Y(NO_3)_3 \cdot 6H_2O$, and EDTA-Na_2 or DTPA) was stirred and heated up to promote the polymerization process, then was removed the solvents to form the gel state. This gel is dried at 300 °C until it is dried and then is sintered at different temperatures from 600 to 1200 °C for 2 h to obtain YSZ-8 cubic fluorite structure. The calcination process is proven by thermal evolution measurement of drying gels, characterized by TGA–DSC analysis (NETZSCH STA 449F5 with range 10 °C min^{-1} in the air), in this study, sample S_1 (using EDTA-Na_2) is subjected to thermal analysis to find the optimal calcinating program. Sample S_2

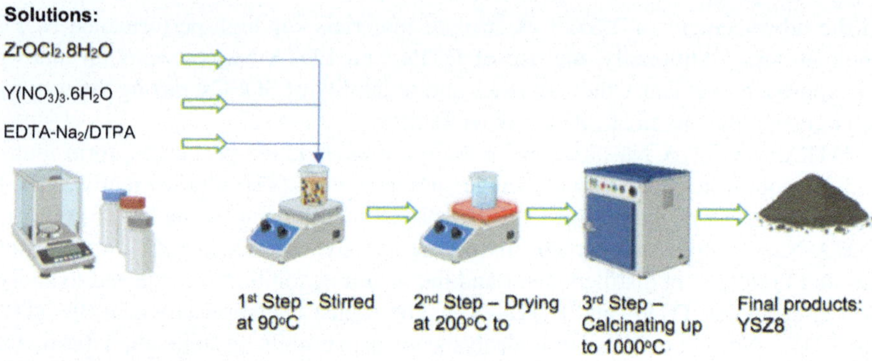

Fig. 1 Diagram of synthesizing YSZ-8 by sol–gel method

(using DTPA) will have similar calcinating program because the thermal decomposition of these two samples will be similar. The YSZ-8 material is then analyzed by the phase existing by X-ray diffractometer (Bruker D8 Advance) with a CuKα radiation source (1.540598 Å) at an accelerated voltage of 40 kV, a selection current of 35 mA, a step size of 0.1°, and a scan rate of $12°\text{min}^{-1}$, and the diffractometer system is used, XPERT-PRO. The morphology by SEM (S-4800, HITACHI, Japan), and elemental analysis by EDS mapping (Bruker Nano Berlin, Germany).

To fabricate SOFC single cell zirconia-based electrolyte and Ni-YSZ anode substrate, this anode substrate is synthesized via solid state reaction by mixing NiO and YSZ-8 with ratio 1:1 by mol (Ni:Y). Then, this mixture is pressed into a circular plate of anode substrate (4.0 cm diameter and 0.1 cm thickness) by hydraulic compressor with about 1000 kgf/cm^2. This plate is sintered at 1350 °C for 3 h. After that, YSZ-8 (previously prepared) is coated on the anode substrate by spin-coating technique to obtain an anode functional layer (AFL) and electrolyte layer. Finally, AFL and electrolyte layers are sintered at various temperatures up to 1350 °C to obtain full densification of the electrolyte. Regarding YSZ-8 spin-coating, the process facilitated a reduction in coating thickness to less than 10 μm for powder-based materials and less than 2 μm for sol-gel-based materials, as recorded in Menzler, N.H's report [18]. Several parameters were theoretically investigated for the impact of dry thickness of the film formation such as speed, viscosity, and mass fraction [19, 20]. Thereafter, the spin-coating was applied to produce such thin layer of the YSZ-8 electrolyte as expected [21–23]. In this study, the nano-powder synthesized as raw material was mixed with sufficient organic binder, solvent, and dispersant for preparing a low viscosity of mixing slurry. The anode substrate placed by movable double-surface sticker on the top of spin-coating machine with minimum speed (200–300 rpm) was adjusted, then the mixing compound was dropped, ensuring that the whole surface of the substrate was covered.

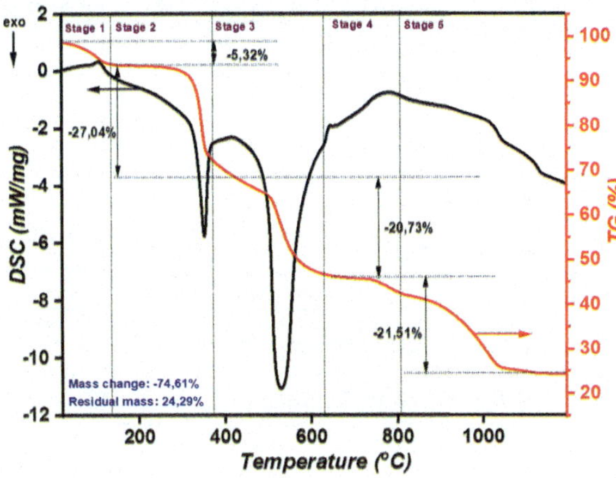

Fig. 2 The TGA/DSC curves obtained for thermal-sprayed YSZ-8 using EDTA-Na_2 as complexing agent (in air with 5 °C/min)

3 Results and Discussion

3.1 Phase Transformation During Calcination

In case of EDTA-Na_2 salt used, shown in Fig. 2. In the case where EDTA-Na_2 salt was used, the TGA/DSC analysis results indicated a total weight loss of around 74.4%, accompanied by exothermic reactions, over the entire heat-treatment range, as shown in Fig. 2

In the initial stage, increasing temperature to 150 °C, the weight loss was observed about 5.3% (Fig. 2) due to moisture evaporation. In the second stage, from 150°C to 350°C, the weight loss remained unchanged, which can be explained by the decomposition of C-H organic components and the evacuation of amino groups during heating, resulting in the formation of an intermediate product. Increasing temperature up to 800 °C, the fluorite structure of YSZ-8 was established with DTPA as the final state. However, using EDTA-Na_2, the presence of Na+ and Cl- caused the crystallization of NaCl. Then, under the effect of high temperature and oxygen-available environment, NaCl transformed into Sodium oxide at high-temperature range from 800 °C to 1000 °C, which caused the weight loss in stage 4 and stage 5, which is also consistent with another report [24].

3.2 Microstructure Analysis

More details were measured by XRD at various temperatures from 600 to 1200 °C to confirm the suggestion for using the chelating agent DTPA and EDTA-Na_2 to the formation of fluorite structure YSZ-8 (Fig. 3a) and formation of intermediate

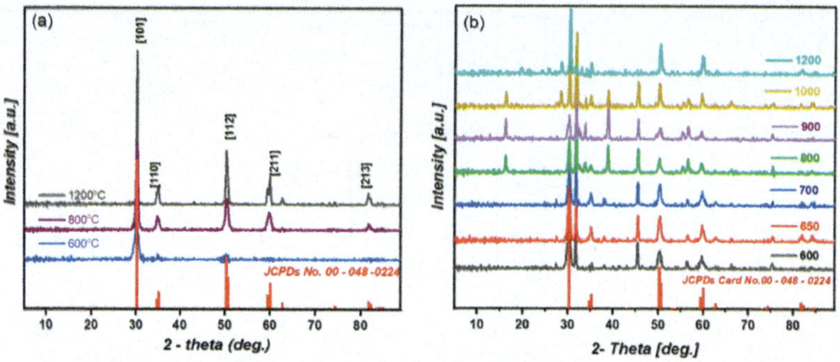

Fig. 3 XRD pattern of YSZ-8 using DPTA (**a**) and EDTA-Na_2 (**b**) as complexing agents

compounds such as Na_2O_2 at temperature above 800–1000 °C (in case of using EDTA-Na_2) as indicated in Fig. 3b.

As shown in Fig. 3a, the basic crystal structure can be identified as the YSZ-8 phase (Ref. JCPDs 00-048-0224). In addition, when using DPTA as a chelating agent, it seems that the crystal structure formation of YSZ-8 is better than using EDTA-Na_2. This can be explained by the formation of Na_2O_2 by using EDTA salt (containing Na) at high temperatures; residual NaCl; and even Na_2O_2 in low-temperature regions (< 1200 °C). Therefore, on the XRD spectrum (Fig. 3b), peaks of these substances appear. At temperatures as high as 1200 °C, these substances evaporate and are no longer part of YSZ-8.

3.3 *Morphology Analysis*

Several morphological characteristics of YSZ-8 were obtained via scanning electron microscopy (SEM) coupled with energy-dispersive X-ray spectroscopy (EDS) elemental analysis (Fig. 4). SEM images reveal that YSZ-8 powder exhibits particle sizes around 15–20 nm (Fig. 5a); these particles exhibit a similar structure.

EDS X-ray mapping data indicates the elemental percentages of Y and Zr to be 2.25% and 26.96%, respectively. Calculations suggest the Y mol ratio in YSZ to be approximately 8%, closely aligning with initial theoretical estimations. This outcome signifies the successful synthesis of YSZ-8. Furthermore, combined with the analysis from Fig. 3 (XRD spectrum), it demonstrates the adequacy of the initial mixing ratio for synthesis, providing accurate insights into the relative proportions of Y and Zr in the electrolyte material YSZ-8. Additionally, the average particle size derived from XRD spectra is approximately 14.8 nm, closely matching the nanoparticle size determined from SEM images.

Finally, in order to reducing the operation temperature of SOFC less than 650 °C, the thin film of YSZ-8 was obtained by the spin-coating technique. YSZ-8 powder was mixed with sufficient solvent and binder to achieve a slurry with suitable

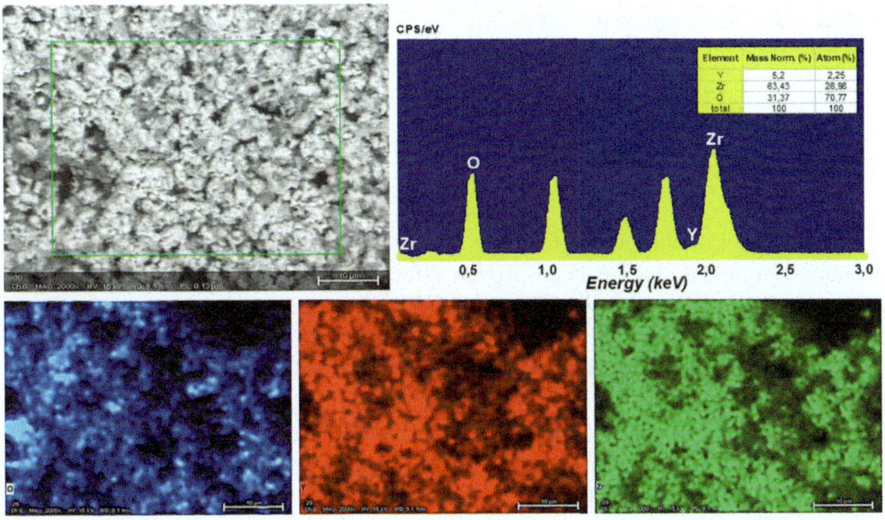

Fig. 4 EDS mapping of YSZ-8 powder using EDTA-Na_2 as complexing agents

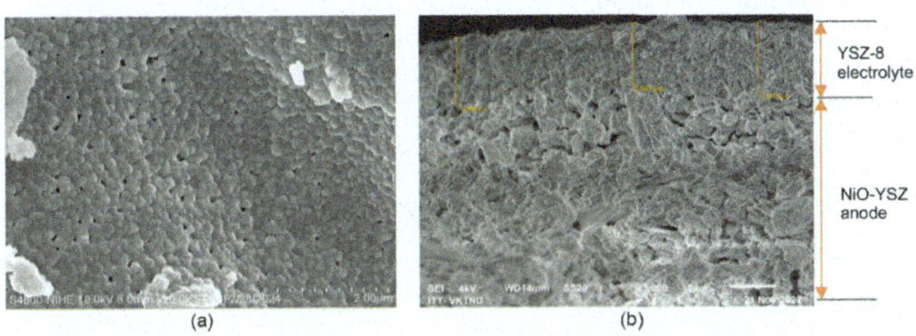

Fig. 5 SEM image of YSZ power being synthesized by sol–gel (**a**) and cross-sectional image of a SOFC showing NiO-YSZ anode support and thin YSZ-8 electrolyte layer (**b**)

viscosity. After that, this layer was sintered with the anode substrate at temperature of 1350 °C for 2 h.

The layered structure and thickness of the anode electrode were examined using cross-sectional SEM images, showing clearly separated layers. The top layer is the electrolyte (YSZ-8), fabricated using the spin-coating technique, and sintering, the thin film reached approximately 7 μm.

It is noted that the anode electrode contains Ni as a catalyst for the conversion reaction of natural gas (fuel of SOFC), but the NiO has not been reduced after preparing NiO-YSZ-8. Theoretically, NiO must be reduced to Ni before coating with YSZ-8. However, Ni will be re-oxidized to NiO if it is exposed to oxygen or heated to 650 °C during the SOFC system operation. Therefore, after completing the cell stack (including anode, cathode, and electrolyte), before working, the hydrogen stream will be brought in to reduce NiO to Ni before injecting a natural gas flow as fuel for the SOFC system to operate.

4 Conclusion

A novel sol–gel method using DTPA acid as a chelating agent instead of citric acid, presented in this study, has been proven effective in synthesizing Yttria Stabilized Zirconia (YSZ) nanoparticles. This method enables the synthesis of material with a high purity of up to 98%, a homogeneous particle distribution, and nanopowders smaller than 20 nm. That is not only DTPA but also different kinds of polyamino carboxylic acids (EDTA, HEDTA, …) could be used with the same result. The YSZ-8 particles were used for spin coating to obtain a thin electrolyte layer up to 7 μm thickness that is sufficient thickness to reduce ohmic loss of the electrolyte and lower working temperature of SOFC.

Acknowledgements This research is funded by the themes of Development of key technologies and processes for the preparation of light aromatics and hydrogen rich gases by in situ catalytic reforming of volatiles from biomass pyrolysis; code: NĐT.94.CHN/20.

The authors would like to thank the RoHan Project funded by the German Academic Exchange Service (DAAD, No. 57315854) and the Federal Ministry for Economic Cooperation and Development (BMZ) inside the framework "SDG Bilateral Graduate School program" for supporting measurements of the TGA/DSC of materials.

References

1. van Biert, L., Woudstra, T., Godjevac, M., Visser, K., Aravind, P.V.: A thermodynamic comparison of solid oxide fuel cell-combined cycles. J. Power. Sources **379**, 382–396 (2018)
2. Rokni, M.: Thermodynamic analysis of an integrated solid oxide fuel cell cycle with a Rankine cycle. Energy Convers. Manage. **51**, 2724–2732 (2010)
3. Weber, A.: Materials and concepts for solid oxide fuel cells (SOFCs) in stationary and mobile applications. J. Power. Sources **127**, 273–283 (2004)
4. Minh, N.Q., Takahashi, T.: Electrolyte. In: Science and Technology of Ceramic Fuel Cells, pp. 74–78. Elsevier, Amsterdam (1995)
5. Yamamoto, O., Arachi, Y., Sakai, H., Takeda, Y., Imanishi, N., Mizutani, Y., Kawai, M., Nakamura, Y.: Zirconia based oxide ion conductors for solid oxide fuel cells. Ionics **4**, 403–408 (1998)
6. Duwez, P., Brown, F.H., Odell, F.: The zirconia-yttria system. J. Electrochem. Soc. **98**, 356–362 (1951)
7. Scott, H.: Phase relationships in the zirconia-yttria system. J. Mater. Sci., 15271535 (1975)
8. Chevalier, J., Grémillard, L., Virkar, V., Clarke, D.: The tetragonal-monoclinic transformation in zirconia: lessons learned and future trends. J. Am. Ceram. Soc. **92**, 1901–1920 (2009)
9. Weber, A., Ivers-Tiffée, E.: Materials and concepts for solid oxide fuel cells (SOFCs) in stationary and mobile applications. J. Power. Sources **127**(1–2), 273–283 (2004)
10. Ried, P., Lorenz, C., Brönstrup, A., Graule, T., Menzler, N.H., Sitte, W., Holtappels, P.: Processing of YSZ screen printing pastes and the characterization of the electrolyte layers for anode supported SOFC. J. Eur. Ceramic Soc. **28**(9), 1801–1808 (2008)

11. Hansch, R., Chowdhury, M.R.R., Menzler, N.H.: Screen printing of sol–gel-derived electrolytes for solid oxide fuel cell (SOFC) application. Ceram. Int. **35**(2), 803–811 (2009)
12. Van Gestel, T., Sebold, D., Meulenberg, W.A., Buchkremer, H.P.: Development of thin-film nano-structured electrolyte layers for application in anode-supported solid oxide fuel cells. Solid State Ionics **179**, 428–437 (2008)
13. Penichi, M.P.: Method of preparing lead and alkaline earths titanates and niobates and coating method using the same to form a capacitor. USA Patent 3,330,697, 11 July 1967
14. Chen, C.C., Nasrallah, M.M., Anderson, H.U.: Synthesis and characterization of YSZ thin film electrolytes. Solid State Ionics **70**(71), 101–108 (1994)
15. Farhikhteh, S., Maghsoudipour, A., Raissi, B.: Synthesis of nanocrystalline YSZ (ZrO_2–$8Y_2O_3$) powder by polymerized complex method. J. Alloy. Compd. **491**, 402–405 (2010)
16. Abdullah, N.A., Hasan, S., Osman, N.: Role of CA-EDTA on the synthesizing process of cerate-zirconate ceramics electrolyte. J. Chem., 1–7 (2013)
17. Friend, M.T., Wall, N.A.: Stability constants for zirconium(IV) complexes with EDTA, CDTA, and DTPA in perchloric acid solutions. Inorg. Chim. Acta **484**, 357–367 (2019)
18. Menzler, N.H., Tietz, F., Uhlenbruck, S., Buchkremer, H.P., Stöver, D.: Materials and manufacturing technologies for solid oxide fuel cells. J. Mater. Sci. **45**, 3109–3135 (2010)
19. Meyerhofer, D.: Characteristics of resist films produced by spinning. J. Appl. Phys. **49**, 1993–1997 (1978)
20. Bornside, D.E., Macosko, C.W., Scriven, L.E.: Spin coating: one dimensional model. J. Appl. Phys. **66**, 5185–5193 (1989)
21. Van Gestel, T., Han, F., Sebold, D., Buchkremer, H.P., Stöver, D.: Nano-structured solid oxide fuel cell design with superior power output at high and intermediate operation temperatures. Microsyst. Technol. **17**, 233–242 (2011)
22. Van Gestel, T., Sebold, D., Buchkremer, H.P., Stöver, D.: Assembly of 8YSZ nanoparticles into gas-tight 1–2 μm thick 8YSZ electrolyte layers using wet coating methods. J. Eur. Ceram. Soc. **32**, 9–26 (2012)
23. Van Gestel, T., Sebold, D., Buchkremer, H.P.: Processing of 8YSZ and CGO thin film electrolyte layers for intermediate-and low-temperature SOFCs. J. Eur. Ceram. Soc. **35**, 105–115 (2015)
24. Mobin, M., Malik, A.U., Ahmad, S.: High temperature interactions of metal oxides with NaCl. J. Less Common Met. **160**(1), 1–14 (1990)

Toward an Open-Source Universal Controller System for Industrial Robots in Industry 4.0

Son Tien Bui and Lac Van Duong

Abstract This article presents the development of a control system for a 3-degree-of-freedom industrial robot, incorporating a user-friendly programming language. The system utilizes the Mitsubishi Q02HCPU PLC controller and the QD75MH4 positioning module. The software offers fundamental functions and enables users to compose programming code for robot control. Experimental results demonstrate the accurate, effective, and stable functioning of the robot controller. The findings suggest that the developed system can be extended to more complex robot configurations with additional degrees of freedom by integrating motion control modules such as the QD77MS16. We open-source the project at https://github.com/lacduong/RobotController.

Keywords Industrial robot · Control system design · Programmable logic controller (PLC)

1 Introduction

Industry 4.0 has brought about significant transformations across various industrial sectors, ushering in new digital technologies that have reshaped manufacturing plants. Industrial robots have seen increasing adoption in the industry over the past three decades, particularly in smart factories, where they perform a diverse range of tasks

Bui Tien Son and Duong Van Lac: The authors contributed equally to this work and designated as co-first authors.

S. T. Bui (✉)
Hanoi University of Industry (HaUI), Hanoi, Vietnam
e-mail: sonbt@haui.edu.vn

L. Van Duong
Hanoi University of Science and Technology (HUST), Hanoi, Vietnam
e-mail: lac.duongvan@hust.edu.vn

D. H. Tien et al. (eds.), *Proceedings of the 1st International Conference on Sustainability and Emerging Technologies for Smart Manufacturing*, Proceedings in Technology Transfer, https://doi.org/10.1007/978-981-97-7083-0_62

including material handling, product packing, picking and placing, palletization, inspection, precise assembly, machining, and additive manufacturing operations.

The majority of industrial robots in use today rely heavily on their manufacturers' proprietary software platforms, with robot programming being based on specific proprietary robot languages provided by each vendor. These characteristics pose challenges for updating, maintaining, or creating new functionalities to meet current production needs, as access to the control system's low levels is restricted. While vendors provide promising robot controllers, they often come with high prices, closed-source code, and limited intervention capabilities [1, 2]. To overcome these limitations and broaden robotic applications, academic research has introduced new technologies and control alternatives [3, 4]. For example, Claudio et al. developed a control structure utilizing two dsPIC30F4013 microcontrollers [5]. However, sustaining this system in an industrial environment over the long term may pose challenges.

This study aims to construct a universal robot controller characterized by flexibility, modularity, manufacturer independence, and an open architecture, making it compatible with most commercially available industrial robot arms, both current and future models. We leverage industrial PLCs to develop controllers for robots and create user-friendly and efficient interactive software interfaces between users and robots. Mitsubishi's Q PLC has been chosen for its robustness and widespread utilization in various automated production settings worldwide. To evaluate the developed controller, a three-axis translational robot, commonly known as a Cartesian robot, has been selected. The results demonstrate that the robot operates with precision and stability.

2 Modeling of the Robot

2.1 Kinematics

For the kinematics problem, we can employ the method of translating and rotating coordinate axes using the familiar Denavit–Hartenberg parameters. The coordinate axis system for the robot is established as depicted in Fig. 1. This reveals that the control values of the motion joints align with the trajectory of the end-effector point. The coordinates of the end-effector point can be determined easily as follows:

$$\begin{cases} x = q_1 \\ y = q_2 \\ z = q_3 \end{cases} \tag{1}$$

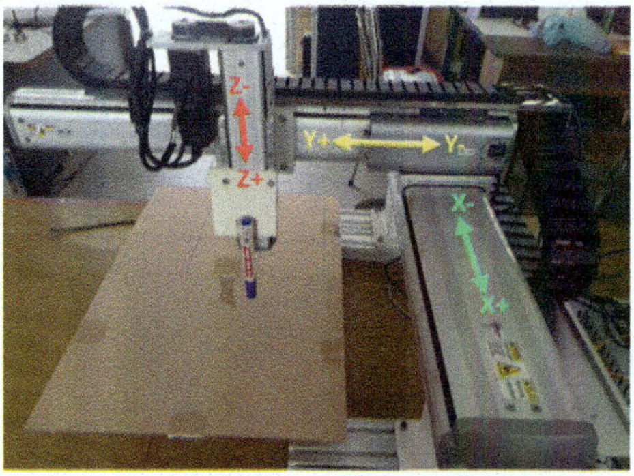

Fig. 1 The coordinate system of a three-degrees-of-freedom robot

2.2 Dynamics

Robot dynamics explores how forces or torques affect a robot's motion. Lagrange's equations are a common method for determining the dynamic equation of a robot. By applying these equations, engineers can systematically derive the equation governing the robot's behavior. Here, using Lagrange's equations, the dynamic equation of the robot is determined as follows:

$$\tau = \begin{bmatrix} \tau_1 \\ \tau_2 \\ \tau_3 \end{bmatrix} = \begin{bmatrix} (m_1 + m_2 + m_3)\ddot{q}_1 \\ (m_2 + m_3)\ddot{q}_2 \\ m_3\ddot{q}_3 - m_3 g \end{bmatrix} \tag{2}$$

3 Design of the Controller System

3.1 Hardware

Figure 2a illustrates the structural diagram of the control system. Signals representing velocity and position will originate from the computer as inputs. These signals will then be scanned and transferred to the Programmable Logic Controller (PLC). Upon receiving the data, the PLC will write it into the functional registers on the QD75MH4 module [6]. Subsequently, the PLC program will execute the commands transmitted from the computer, utilizing the speed and position parameters previously recorded. Once executed, the PLC will receive signals indicating joint positions and transmit them back to the computer for display.

Controlling the servo motor through the QD75MH4 module is straightforward due to its well-organized memory structure, consisting of distinct memory registers with specific functions. Let's consider the example of controlling axis 1, which corresponds to the X-axis motion in this project, denoted as axis X. To achieve

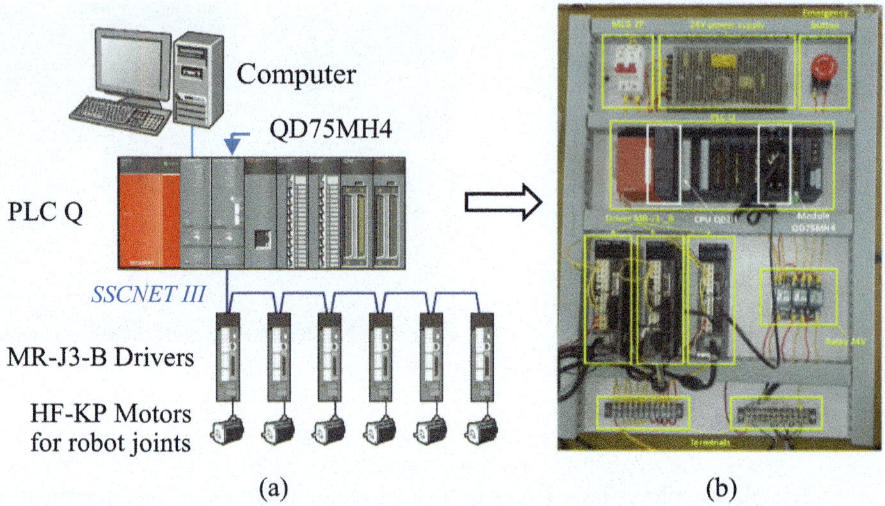

Fig. 2 **a** The structural diagram of the control system. **b** Electrical control cabinet system

movement to a designated position at a specified velocity, two essential parameters are needed: position and velocity. The manufacturer defines a list of 600 movement points for axis 1, ranging from 2000 to 7999. Below, we provide detailed information on the data structure and function of the memory registers (Table 1).

Each point is allocated 10 memory registers, each of which consists of 16 bits. The function of each memory register is as follows:

(1) Data No: The point's sequential number; to move to a specific point, inputting its corresponding sequence value into the G1500 register.
(2) Positioning identifier: to signify the address of the point, the memory cell contains details about control type, movement type, acceleration as well as deceleration times, etc.
(3) M code: Optionally provides an M code value if necessary.
(4) Dwell time: The waiting period until moving to the next point.
(5) Command speed: speed register, composed of two 16-bit registers, equivalent to a 32-bit register and an integer value.
(6) Positioning address: address register with the same structure as command speed.
(7) Arc data: An additional address register, employed when controlling circular interpolation.

By strategically coordinating motor controls, the robot gains the capability to execute movement requests according to the desired trajectories by the operator.

Table 1 Memory structure for the list of points on axis X

Data No.	Positioning identifier	M code	Dwell time	Command speed		Positioning address		Arc data	
				Low-order	High-order	Low-order	High-order	Low-order	High-order
1	2000	2001	2002	2004	2005	2006	2007	2008	2009
2	2010	2011	2012	2014	2015	2016	2017	2018	2019
3	2020	2021	2022	2024	2025	2026	2027	2028	2029
4	2030	2031	2032	2034	2035	2036	2037	2038	2039
5	2040	2041	2042	2044	2045	2046	2047	2048	2049

3.2 PLC Programming

The PLCQ control program is developed using Mitsubishi's GX Works 2 software, which provides all the necessary functions for robot control purposes. These functions include JOG mode for fast movement, OPR mode for home detection, FAST OPR mode for rapid homing, and interpolation mode. JOG mode is utilized when there's a requirement to move the robot to a target position swiftly. It involves inputting the JOG speed and commanding the axis to move accordingly. Regarding the Home position, which represents the origin of the robot, it's crucial to return the robot's axes to this position upon system startup. This process is facilitated using the OPR function of the QD75MH4 module. The robot searches for the origin using the Home sensor, and the Home position is automatically stored as a reference point for subsequent robot movements.

In interpolation mode, the robot adheres to a rule where the Y-axis interpolates along the X-axis, with the Y-axis velocity contingent on the X-axis, thereby generating the desired trajectory. Furthermore, the Z-axis coordinates its movement in conjunction with both the X and Y axes, thereby shaping the robot's motion path. Vital parameters that require specification include the starting point (P1) and endpoint (P2) addresses for linear interpolation. For circular interpolation, an intermediate point (Pa) needs to be added. Other crucial parameters to establish encompass the starting and ending points for the Z-axis to trace the plane trajectory, interpolation velocities for the X and Y axes, and the Z-axis approach speed.

In linear interpolation mode, the process begins by inputting the coordinates of two points, P1 and P2. The robot then moves to point P1, lowers the machining Z-axis, and subsequently moves to point P2, as illustrated in Fig. 3a. In subsequent iterations, if the Z-axis remains unaltered, two additional points, P1′ and P2′, are inputted. The robot proceeds to move along the two straight segments P2P1′ and P1′P2′, with the machining axis still operational, as depicted in Fig. 3b. Should movement be required along a single segment, the coordinates of P1′ are inputted to coincide with point P2, as demonstrated in Fig. 3c. This prompts the robot to move exclusively along the P1′P2′ segment. This process is repeated for subsequent trajectories.

In circular interpolation mode, the process begins with the input of coordinates for P1, P2, and an intermediate point Pa to establish a circular arc. The robot initiates its movement toward P1, lowers the machining axis, passes through Pa, and

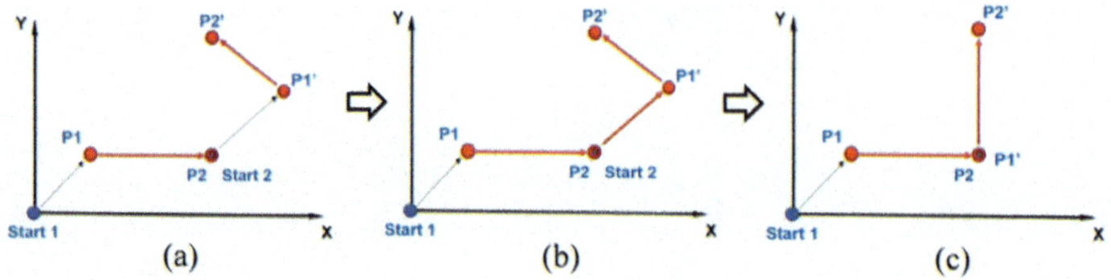

Fig. 3 Line interpolation mode

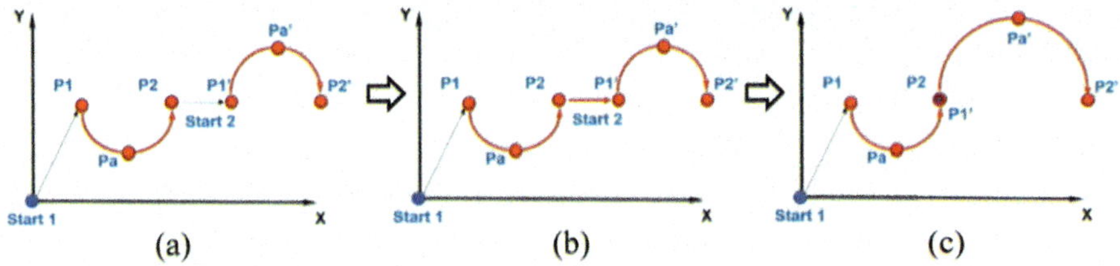

Fig. 4 Arc interpolation mode

subsequently arrives at P2 following a curved trajectory, as depicted in Fig. 4a. In subsequent iterations, if the machining axis remains unchanged, points P1′, Pa′, and P2′ are inputted. The robot continues its machining trajectory: moving linearly on the P2P1′ segment and following the curved arc P1′Pa′P2′, as shown in Fig. 4b. When entering coordinates, where P1′ aligns with P2, the robot exclusively follows the curved arc P1′Pa′P2′, as illustrated in Fig. 4c.

3.3 *Software*

Mitsubishi Electric provides a diverse range of programming software and comprehensive support for working with their devices. We utilized the MX Component software [7] to establish a seamless connection between the programming software on the computer and Mitsubishi Electric's PLC. The communication protocol between the computer and the PLC is facilitated through the USB standard. The software interface, depicted in Fig. 5, is crafted on Windows Form using Visual Studio 2022. Our developed control software incorporates all essential functions for robot control, including independent JOG motion of axes, homing, error code indication, interpolation execution, robot trajectory visualization, and speed adjustment. One notable aspect of the software, setting it apart from many other robot control software solutions, is its integration of a programming language for users during the control process, simultaneously providing ongoing task progress updates.

4 Experimental Results

4.1 *Error Assessment*

To evaluate the accuracy and stability of the robot, measuring the motion errors is crucial. For this, a comparator clock is horizontally positioned with its needle aligned with the axis to be measured. The robot moves along the axis while keeping the other two axes fixed. Upon contact, the clock's needle points to a value. The clock face is

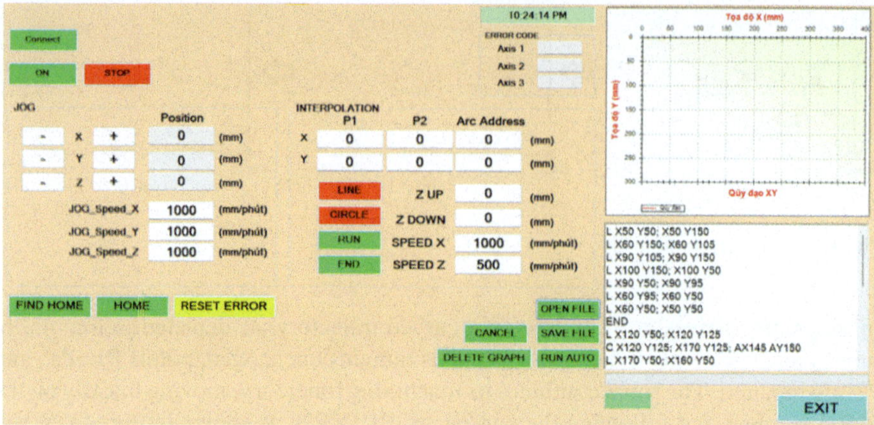

Fig. 5 Interface of the developed control software on the computer

Table 2 Error results on axes X, Y, and Z

	X	Y	Z
Standard deviation (σ)	0.01	0.00607	0.007255

then rotated until the needle aligns with "0" and the robot's current position is noted. After returning the robot to its initial position, the previously recorded position is entered. The robot moves to the position where the needle initially indicated "0". Any deviation from the predetermined position during the same movement command is reflected on the comparator clock. Table 2 presents the outcomes of 20 experimental samples. Bessel's correction formula is applied to calculate the standard deviation as

$$\sigma = \sqrt{\frac{1}{n-1}\sum_{i=1}^{n}(x_i - \overline{x})^2} \tag{3}$$

Using Eq. (3), the local errors of the motion segments are shown in Table 2:

The results demonstrate that the control system effectively manages the robot's operations with a considerable level of accuracy. It's important to note that the reported error encompasses both the inherent mechanical error of the axes.

4.2 Demonstration

To evaluate the performance of the robot using the developed control software, an experimental trajectory was formulated and compared to the actual results, as depicted in Fig. 6. The experimental trajectory was created through the following

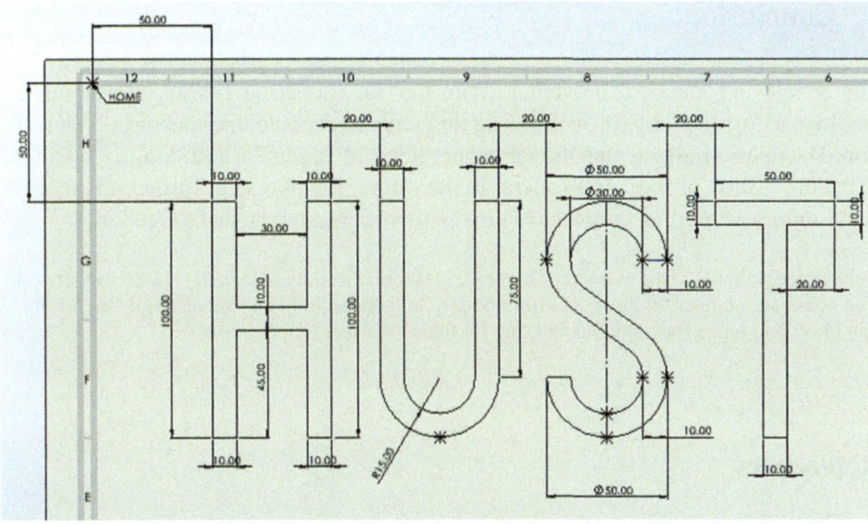

Fig. 6 The desired robot trajectory

Fig. 7 Actual results of the robot running along the desired trajectory

steps: First, collect essential coordinates from the designed trajectory, which are then utilized to control the robot. Figure 7 illustrates the real-world outcome after the control process is completed.

The results indicate that the developed control system functions effectively and maintains stability, accurately following the intended trajectory. However, during operation, it is noted that the robot's speed of response is not yet optimal, and the command processing time requires further optimization. This highlights the ongoing need for the authors to continue their research, focusing on devising control algorithms and optimizing these algorithms to enhance accuracy and the smoothness of the robot's motion.

5 Conclusion

The article introduces a universal controller for industrial robots, which initially employs a customizable programming language to execute line and circle interpolation. The results indicate that the robot operates with precision and stability, attributed to the utilization of PLC controllers. In the future, the plan is to further advance and develop more complex control systems with increased degrees of freedom.

Acknowledgements This research is funded by Hanoi University of Industry (HaUI) under project with constract number 06-2024-RD/HĐ-ĐHCN. Authors would like to thank Bui Duc Minh, Vu Van Phuc, Bui Quoc Bao, Do Quang Long for their technical support.

References

1. DELTA, Robot controller with servo drive integrated. https://www.deltaww.com/en-us/Products/Robot-Controller-with-Servo-Drive-Integrated/ALL/
2. Kondo, D., Kusakabe, K.: Development of Robot Controller MRC01 and Example for Introducing Robot to Equipment, Oriental Motor (2022). https://www.orientalmotor.com/network-products/technology/pdf/MRC01.pdf
3. Nakamura, A., Ohyama, Y., Ito, K., Saito, K.: Proceedings. 1986 IEEE international conference on robotics and automation. Controller Ind. Robots **3**(0), 254–259 (1986). https://doi.org/10.1109/robot.1986.1087714
4. Bilancia, P., Schmidt, J., Raffaeli, R., Peruzzini, M., Pellicciari, M.: An overview of industrial robots control and programming approaches. Appl. Sci. **13**(4), 2582 (2023). https://doi.org/10.3390/app13042582
5. Urrea, C., Kern, J.: Development of an electronic controller applied to a robotized manipulator. Comput. Electr. Eng. **56**, 648–658 (2016)
6. MELSOFT, MELSEC-Q QD75MH Positioning Module User's Manual. http://dl.mitsubishielectric.com/dl/fa/document/manual/ssc/ib0300117/ib0300117b.pdf
7. MELSOFT.: MX Component Version 4 Operating Manual. https://dl.mitsubishielectric.com/dl/fa/document/manual/plc/sh081084eng/sh081084engt.pdf

A Design Review for Generating Involute Tooth Profile of Non-circular Gear Pair

Hong-Phong Nguyen, Van-Quyet Tran, Minh-Thuan Hoang, Banh Tien-Long, and Van-The Tran

Abstract Non-circular gear transmissions are increasingly used in equipment serving daily life and industrial production that the traditional cylindrical gears with constant transmission ratios cannot meet. Significantly, the machines or its part that require a change in speed according to a defined rule. The research and applications of non-circular gears have attracted the attention of many researchers and designers. In this paper, a design review for generating the involute tooth profile of the non-circular gears is presented. A numerical analysis is performed to verify the merit of non-circular gears.

Keywords Involute profile · Non-circular gears · Gear design

1 Introduction

The widely used application of non-circular gears is the step-less speed change boxes often used in mechanical machines, mechanical engineering and continuously variable transmission (CVT) automatic speed changers used in the automobile industry or the defense sector. The non-circular gears are of great significance in applying to transmission systems and solving industrial problems. Along with the development of gears with circular paths with fixed transmission ratios, such as gears with open involute profiles, cycloid gears, Novikov gears, hypoid gears, etc., non-circular gears with variable gear ratios have also been researched and applied by many scientists since the fourteenth century.

The non-circular gear design dates back to the fourteenth-century Italian watch designer Giovanni Dondi, who found the concept of non-circular gears in his design manuscript, which was the first time that the idea of non-circular gears was known. In his work in the 1910s, Dunkerly mentioned the design and application of elliptical

H.-P. Nguyen · V.-Q. Tran · M.-T. Hoang · B. Tien-Long · V.-T. Tran (✉)
Department of Mechanical Engineering, Hung Yen University of Technology and Education, 39A Rd., Hung Yen, Vietnam
e-mail: vanct4.hut@gmail.com

D. H. Tien et al. (eds.), *Proceedings of the 1st International Conference on Sustainability and Emerging Technologies for Smart Manufacturing*, Proceedings in Technology Transfer, https://doi.org/10.1007/978-981-97-7083-0_63

gears. In the 1940s and 1960s, there was a research boom in non-circular gears and even a special study on non-circular gears, but limited by the technical and processing limitations of the industry at the time, the processing of non-circular gears. Accuracy and productivity are not high, and the research boom is also extinguished. Until the 1980s, with the rapid development of computer technology and digital control, the boom of non-circular gears once again flourished. Scientists have gradually modeled, researched, and developed the non-circular gear design theory to put into production machines serving production, people's daily life, as well as national security and defense such as instrumentation, pumps, flow meters, mechanisms in components, and accessories of defense equipment [1–3].

The non-circular gear can replace the cam and linkage to determine the relationship between the input and output of purely mechanical control [4]. The nonlinear relationship between the rotation angle of the driving and the driven gear is used to generate the non-circular gear with different shapes, such as elliptical, squared, oval... etc. [5]. By a different approach, Dooner [6] presented the mechanism of design and optimization of non-circular gear based on the relative displacement of eight links. Doege [7] proposed the design and optimization method for non-circular gear based on symmetrical ram kinematics. Kowalczyk and Urbanek [8] presented the equations for describing non-circular pinion's shape and kinematic parameters. To replace anti-parallel four-bar and suitable manufacturing by grinding, Liu et al. [9] proposed the method for designing the elliptical gear with the velocity of the coupler point as constant. Mundo and Gatti [10] presented the design method for non-circular gear to trace the prescribed trajectory of a linkage. To the same research objective, Modler et al. [11] proposed a method for determining the centrodes and transmission function of non-circular gear based on the geared linkage. Tong and Yang [12] used the critical for a pair of rigid bodies in a rolling motion to design the non-circular gear with various profiles and numbers of teeth. To develop the previous idea, Yang et al. [13] proposed the method named deviation-function to generate the conjugate kinematic pairs by resharing the pitch pair. Lozzi [14] used graphic construction to develop the involute profile of non-circular gear and evaluated the generated profile by implementing computer algorithms, while Riaza et al. [15] generated the non-circular gear based on the geometrical locus of singular points in the involute profile. By another approach, Tsay et al. [16] and Liu et al. [17] proposed a method for approximating the pitch curve of non-circular gear using the Fourier series. Jing [18] defined the pressure angle function to describe the pair geometries of planar gear, especially for non-circular gear. Yan et al. [19] designed the non-circular multilobe internal pitch curves by adding the monotonic function into the pitch curve of the initial profile. Xia et al. [20] used the polar coordinate system and crown rack to design the non-circular bevel gear. Bair et al. [21–24] proposed a method for designing circular-arc elliptical gears using the rack and shaper cutters, considering the undercutting problem. Danieli et al. [25, 26] presented the methodology for enhancing the contact ratio of non-circular gear using the pressure angle constant.

Although the original designs of non-circular gears were proposed in the fifteenth century, methods for machining them were not proposed until the late nineteenth and early twentieth centuries. The presented non-circular gear machining solutions approached the capabilities of machining technology at that time. Studies have been conducted to manufacture specialized non-circular gear machining machines using the forming method and improve conventional gear machining machines to process non-circular gears using the enveloping method [27, 28].

2 Designing of Non-circular Gears

2.1 *Involute Tooth Profile of Non-circular Gears*

The shaper cutters are used to design the involute part of the tooth profile of non-circular gears. The cutters consist of five generating segments, as shown in Fig. 1. The segments *ab* and *eg* are the involute curves generating the involute tooth profile of non-circular gears, segments *bc* and *de* are the circular arcs with centers at O_1 and O_2 generating the fillet profile, and segment *cd* of the shaper cutter is used to generate the root profile. As shown in Fig. 1, the position vector of tooth profile *i* can be obtained by transformation matrix from reference coordinate system $S_s(O_s - x_s y_s)$ to shaper cutter coordinate system $S_c(O_c - x_c y_c)$ as follows:

$$r_c^i = \begin{Bmatrix} x_c^i \\ y_c^i \end{Bmatrix} = \begin{bmatrix} \sin\psi & -\cos\psi \\ \cos\psi & \sin\psi \end{bmatrix} \begin{Bmatrix} x_s^i \\ y_s^i \end{Bmatrix} \tag{1}$$

where $\psi = \frac{\pi}{2z_s} + \tan\alpha - \alpha$, z_s, and α are the tooth number and pressure angle of shapper cutter. Super-cript *i* represents curves *ab*, *bc*, *cd*, *de*, and *eg*.

Fig. 1 The defined structure of the shaper cutter

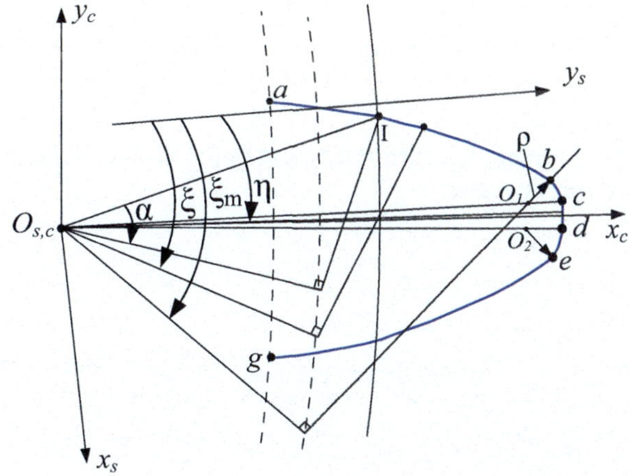

The position vector of involute curves *ab* and *eg* is presented in the coordinate system S_s as follows:

$$r_S^1 = \begin{Bmatrix} x_s^1 \\ y_s^1 \end{Bmatrix} = \begin{Bmatrix} r_b \sin\xi - r_b\xi \cos\xi \\ r_b \cos\xi + r_b\xi \sin\xi \end{Bmatrix} \tag{2}$$

The position vector of segments *bc* and *de* is presented in the reference coordinate system S_s as follows:

$$r_S^2 = \begin{Bmatrix} x_s^2 \\ y_s^2 \end{Bmatrix} = \begin{Bmatrix} r_b \sin\xi_m - r_b\xi_m \cos\xi_m + \rho_1 \cos\xi_m + \rho_1 \cos(\theta + \xi_m) \\ r_b \cos\xi_m + r_b\xi_m \sin\xi_m - \rho_1 \sin\xi_m + \rho_1 \sin(\theta + \xi_m) \end{Bmatrix} \tag{3}$$

The position vector of the segment *cd* is presented in the reference coordinate system S_s as follows:

$$r_S^3 = \begin{Bmatrix} x_s^3 \\ y_s^3 \end{Bmatrix} = \begin{Bmatrix} r_a \sin\eta \\ r_a \cos\eta \end{Bmatrix} \tag{4}$$

where r_b is the base circle radius, ξ is the cutter surface parameter defined as the location of points on the involute region and its range is defined as $0 \le \xi \le \xi_m$, ξ_m is the maximum extension angle of the involute curve at point *b*, θ is the location of points on the fillet region and its range is defined as $0 \le \theta \le {}^{\pi}/_{2} - \tan^{-1}(\xi_m - (\rho_1/r_b))$, ρ_1 is the tip radius of generating cutter on fillet surface. η is a shaper cutter design with its range of $\xi_m + \beta - {}^{\pi}/_{2} \le \eta \le {}^{\pi}/_{2z_s} + \tan\alpha - \alpha$, $r_a = \sqrt{r_b^2 + (r_b\xi_m - \rho_1)^2} + \rho_1$ is the radius of the tip circle's the cutter and $\beta = {}^{\pi}/_{2} - \tan^{-1}(\xi_m - (\rho_1/r_b))$.

The unit normal vectors of shaper cutter surface can be represented as follows:

$$n_v = \frac{\frac{dr_c^i}{dl_i} \times k_c}{\left|\frac{dr_c^i}{dl_i} \times k_c\right|} \tag{5}$$

In which, k_c is the unit vector of the Z_c-axis, Parameter l_i is the design parameters of shaper cutter.

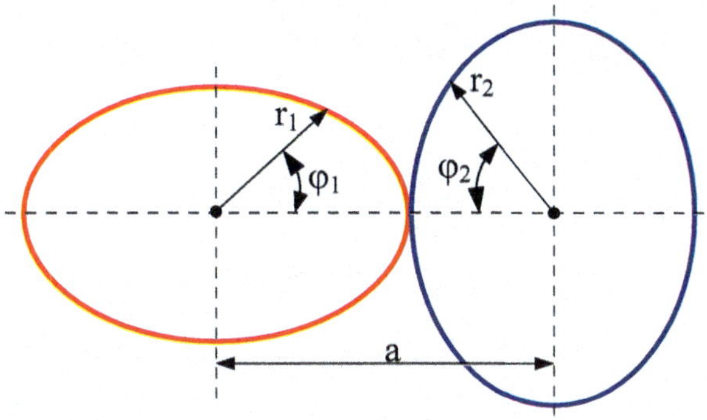

Fig. 2 The operating pitch lines for non-circular gear pairs

2.2 *The Operating Pitch Curves*

The non-circular gearbox can be expressed simply by two operating range lines. According to the law of gear boxing, the operational pitch line must roll together without sliding. The operating line of the uncircular gearbox, which represents the non-circular corresponding gearbox, is shown in Figs 1 and 2.

The corresponding transmission ratio function, denoted as *i*, between the pitch line in the polar coordinate system of the gear pair is formulated as

$$i(\varphi_1) = \frac{\omega_1(\varphi_1)}{\omega_2(\varphi_2)} = \frac{r_2(\varphi_2)}{r_1(\varphi_1)} \tag{6}$$

where $\omega_i(i = 1 \sim 2)$ is the angular velocity of gears.

The center distance between two gear centers can be calculated as follows:

$$r_1(\varphi_1) + r_2(\varphi_2) = a \tag{7}$$

To substitute Eq. (7) into Eq. (6), the transmission ratio function is re-presented as the function of the gear's rotation angle and center distance as follows:

$$i(\varphi_1) = \frac{a - r_1(\varphi_1)}{r_1(\varphi_1)} \tag{8}$$

Based on the no-slip rolling requirement, the relationship between the drive and driven gear angular angle can be presented as follows:

$$r_1(\varphi_1)d\varphi_1 = r_2(\varphi_2)d\varphi_2 \tag{9}$$

The solved rotation angle φ_2 can be separated into two cases:

Case 1: If $r_1(\varphi_1)$ is given, the rotation angle φ_2 is calculated by solving Eq. (8) and (9) together. The form of φ_2 is shown as follows:

$$\varphi_2 = \int_0^{\varphi_1} \frac{r_1(\varphi_1)}{a - r_1(\varphi_1)} d\varphi_1 \tag{10}$$

Case 2: If $i(\varphi_1)$ is given ($r_1(\varphi_1)$ and $r_2(\varphi_2)$ unknown), Eq. (9) (using Eq. (6)) can be rewritten as

$$\varphi_2 = \int_0^{\varphi_1} \frac{1}{i(\varphi_1)} d\varphi_1 \tag{11}$$

By using either Eq. (10) or Eq. (11) , a relation between φ_1 and $\varphi_2(\varphi_1)$can be calculated.

The transmission ratio function $i(\varphi)$ needs to exhibit periodic behavior, with its period T being intricately linked to the periods of revolution of each individual gear, denoted as T_1 and T_2, respectively:

$$T = \frac{T_1}{n_2} = \frac{T_2}{n_1} \tag{12}$$

where n_1 and n_2 are the driving and driven gears' tooth number.

An example is presented to show how to generate the tooth profile of non-circular gears. The input parameters are given in Table 1. The tooth profiles of the non-circular gear pair are generated, as shown in Fig. 3.

Table 1 Input parameters for generating the non-circular gears

Parameters	Gear 1	Gear 2
Number of teeth (n_1, n_2)	25	50
Center distance a	44.82 mm	
Normal module (m_{pn})	1.2	
Normal circular-tooth thickness (s_{pn2})	4.712 mm	
Normal pressure angle (α_{pn})	22°	
Tip rounding (ρ_1, ρ_2)	0.2 mm	0.05 mm
Tooth thickness allowance (mm)	−0.030	
Number of teeth of shaper cutter (z_s)	30	

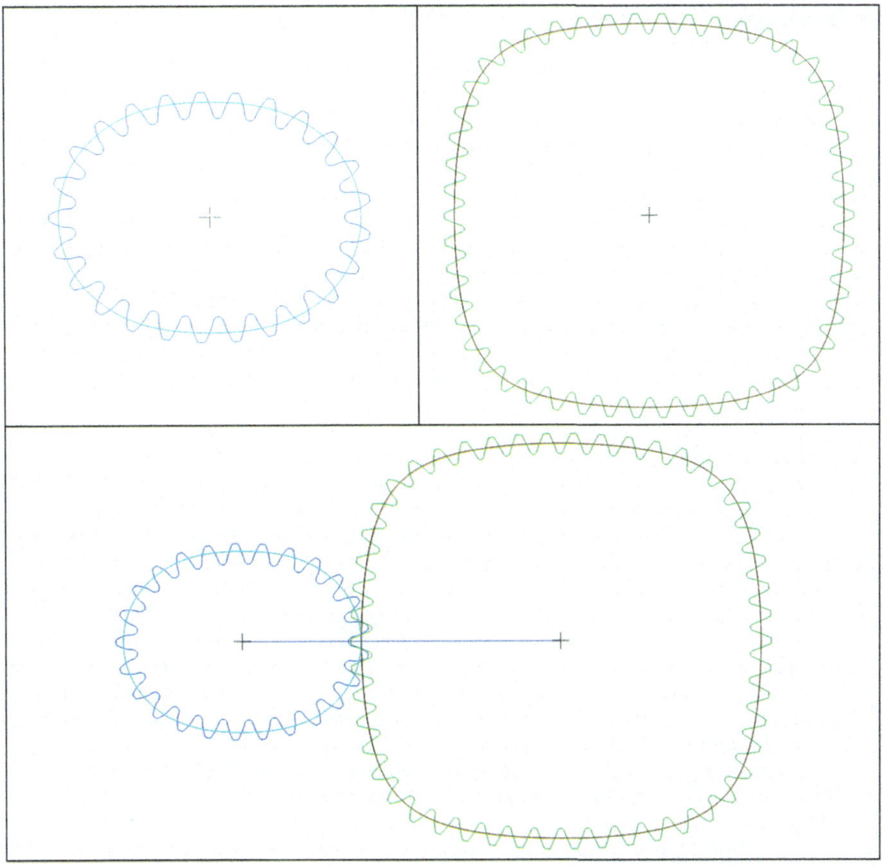

Fig. 3 A non-circular gear pair

3 Conclusion

This article summarizes the creation of non-circular gears. Retrieval into the design, optimization, and experimental production of non-circular gears is essential for synthesis. When designing non-circular gearboxes, calculating the operating range is the main step. According to gearing principles, these gearing pitch lines must roll without sliding. Non-circular gearboxes provide continuous transmission of torque or transmission of torque at specific rotation angles.

References

1. Freudenstein, F., Chen, C.-K.: Variable-ratio Chain Drives with Noncircular Sprockets and Minimum Slack Theory and Application (1991)
2. Emura, T., Arakawa, A.: A new steering mechanism using noncircular gears. JSME Int. J. Ser. 3, Vibr., Control Eng., Eng. Ind. **35**, 604–610 (1992)
3. Dooner, D.B.: Use of Noncircular Gears to Reduce Torque and Speed Fluctuations in Rotating Shafts (1997)
4. Mundo, D.: Geometric design of a planetary gear train with non-circular gears. Mech. Mach. Theory **41**, 456–472 (2006)
5. Vanin, V.A., Kolodin, A.N.: Kinematic structure of gear-cutting machine tools based on a hydraulic step drive for the production of noncircular gears. Russ. Eng. Res. **29**, 291–297 (2009)
6. Dooner, D.B.: Function generation utilizing an eight-link mechanism and optimized non-circular gear elements with application to automotive steering. Proc. Inst. Mech. Eng. C J. Mech. Eng. Sci. **215**, 847–857 (2001)
7. Doege, E., Meinen, J., Neumaier, T., Schaprian, M.: Numerical design of a new forging press drive incorporating non-circular gears. Proc. Inst. Mech. Eng. B J. Eng. Manuf. **215**, 465–471 (2001)
8. Kowalczyk, L., Urbanek, S.: The geometry and kinematics of a toothed gear of variable motion. Fibres Text. Eastern Eur. **11**, 60–62 (2003)
9. Liu, J.Y., Chang, S.L., Mundo, D.: Study on the use of a non-circular gear train for the generation of Figure-8 patterns. Proc. Inst. Mech. Eng. C J. Mech. Eng. Sci. **220**, 1229–1236 (2006)
10. Mundo, D., Gatti, G.: A graphical-analytical technique for the synthesis of non-circular gears in path-generating geared five-bar mechanisms. Trans. Can. Soc. Mech. Eng. **32**, 487–498 (2008)
11. Modler, K.-H., Lovasz, E.-C., Bär, G.F., Neumann, R., Perju, D., Perner, M., Mărgineanu, D.: General method for the synthesis of geared linkages with non-circular gears. Mech. Mach. Theory **44**, 726–738 (2009)
12. Tong, S.H., Yang, D.C.H.: Generation of Identical Noncircular Pitch Curves (1998)
13. Yang, D.C.H., Tong, S.-H., Lin, J.: Deviation-Function Based Pitch Curve Modification for Conjugate Pair Design (1999)
14. Lozzi, A.: Non-circular gears—graphic generation of involutes and base outlines. Proc. Inst. Mech. Eng. C J. Mech. Eng. Sci. **214**, 411–422 (2000)
15. Riaza, H.F.Q., Cardona i Foix, S., Nebot, L.J.: Study of the base curve and formation of singular points on the tooth profile of noncircular gears (2007)
16. Tsay, M.-F., Fong, Z.-H.: Study on the generalized mathematical model of noncircular gears. Math. Comput. Model. **41**, 555–569 (2005)
17. Liu, J.-Y., Chen, Y.-C.: A design for the pitch curve of noncircular gears with function generation, in: Proceedings of the International Multiconference of Engineers and Computer Scientists (2008)
18. Jing, L.: A pressure angle function method for describing tooth profiles of planar gears (2009)
19. Yan, J., Yang, D.C.H., Tong, S.-H.: On the generation of analytical noncircular multilobe internal pitch curves. J. Mech. Des. **130**, 092601 (2008)
20. Xia, J., Liu, Y., Geng, C., Song, J.: Noncircular bevel gear transmission with intersecting axes (2008)
21. Bair, B.-W.: Computer aided design of elliptical gears. J. Mech. Des. **124**, 787–793 (2002)
22. Bair, B.W.: Computer aided design of elliptical gears with circular-arc teeth. Mech. Mach. Theory **39**, 153–168 (2004)
23. Bair, B.-W., Chen, C.-F., Chen, S.-F., Chou, C.-Y.: Mathematical model and characteristic analysis of elliptical gears manufactured by circular-arc shaper cutters (2007)
24. Bair, B.-W., Sung, M.-H., Wang, J.-S., Chen, C.-F.: Tooth profile generation and analysis of oval gears with circular-arc teeth. Mech. Mach. Theory **44**, 1306–1317 (2009)
25. Danieli, G.A.: Analytical description of meshing of constant pressure angle teeth profiles on a variable radius gear and its applications. J. Mech. Des. **122**, 123–129 (2000)

26. Danieli, G.A., Mundo, D.: New developments in variable radius gears using constant pressure angle teeth. Mech. Mach. Theory **40**, 203–217 (2005)
27. Li, J., Wu, X., Mao, S.: Numerical computing method of noncircular gear tooth profiles generated by shaper cutters. Int. J. Adv. Manuf. Technol. **33**, 1098–1105 (2007)
28. Litvin, F.L., Gonzalez-Perez, I., Fuentes, A., Hayasaka, K.: Design and investigation of gear drives with non-circular gears applied for speed variation and generation of functions. Comput. Methods Appl. Mech. Eng. **197**, 3783–3802 (2008)

Static Bending Behavior of Two-Directional Functionally Graded Porous Microshell

Khuat Duc Duong, Nguyen Tuan Linh, Tran Thi Thu Thuy, and Dao Nhu Mai

Abstract This work investigates the static bending of a bi-directional functionally graded porous (BFGP) microshell using Kirchhoff–Love's shell hypothesis, the modified couple stress hypothesis, and the isogeometric analysis technique (IGA). An exponential rule governs the mechanical characteristics of the entire shell, which is supported by an elastic medium. There are two forms of porosity: even and uneven. This approach is unique in that it takes into account the change length-scale parameter as the mechanical characteristics of the material. The validity and effectiveness of the established model are shown by contrasting the numerical results produced by the present formulations with data that have been published in a number of particular circumstances. Furthermore, the influence of properties like porosity, length-scale factor, elastic medium stiffness, and other parameters on the microshell static bending is studied numerically.

Keywords IGA method · Static bending · Bi-directional functionally graded porous

1 Introduction

Functionally graded materials (FGMs) have been widely used in various engineering fields due to their unique mechanical, thermal, and electrical characteristics. The characteristics of materials change over space through the thickness and length of the material. Spatial variation in these characteristics can be used to optimize the performance of engineering structures. These materials find applications in diverse fields such as aerospace, mechanical, and biomedical engineering. In particular, microsheets with bi-directionally variable mechanical properties have attracted

K. D. Duong (✉) · N. T. Linh · T. T. T. Thuy
School of Mechanical and Automotive Engineering, Hanoi University of Industry, Hanoi, Vietnam
e-mail: Duongkd@haui.edu.vn

D. N. Mai
VNU University of Engineering and Technology—VNU-UET, Hanoi, Vietnam

D. H. Tien et al. (eds.), *Proceedings of the 1st International Conference on Sustainability and Emerging Technologies for Smart Manufacturing*, Proceedings in Technology Transfer, https://doi.org/10.1007/978-981-97-7083-0_64

considerable attention due to their unique characteristics and potential applications in micro- and nano-electromechanical systems (MEMS and NEMS). Research theories for micro structures can be mentioned as deformation gradient hypothesis, non-local hypothesis, non-local strain gradient hypothesis, etc. However, the modified couple stress hypothesis is the most widely used because it meets the requirements for results that are close to experimental studies. Thai et al.'s [1] typical study analyzes static bending, particular buckling, and oscillations of functionally graded sheet via modified couple stress hypothesis employing Kirchhoff and Mindlin plate models. Nguyen and associates [2] examined the static nonlinearity of a microplate partially lying on elastic base using Von Karman theory and modified couple stress hypothesis. You can find additional research on the mechanical behavior of microplates in papers [3–7].

Isogeometric analysis is a computational technique that has gained popularity recently. With the use of this technique, finite element analysis (FEA) can be included into standard NURBS-based CAD design tools. Consequently, a single dataset may now be used for all stages of model construction, verification, and improvement [8, 9]. The finite element method (FEM) framework has undergone extensive development, and B-splines are now commonly utilized as standard functions. Thomas Hughes created a paradigm that was consolidated and presented as isogeometric analysis [10–12] for the study of isogeometric lines. Since then, computer approaches have made extensive use of isogeometric analysis methods, particularly for simulating the mechanical behavior of validation and uplift. The shortcomings of the FEM approach are actually overcome by the isogeometric analysis method.

However, to the best of my own knowledge, there are currently not many works on modeling and analyzing the mechanical behavior of microshells with mechanical characteristics that vary in two directions and with porous using numerical methods such as isogeometry. Therefore, in this study, by using the IGA method, modified couple stress hypothesis, and classical shell hypothesis, the static bending characteristics of the BFGP microshell have mechanical properties that variable in both thickness and length direction is discovered. These results can be applied in the process of calculating and designing micrometer-sized structures and variable mechanical characteristics, making a small contribution to understanding the mechanical properties of these structures.

2 Mathematical Foundations

2.1 Describe BFGP Microshell Structure

Let us examine a double-curved shallow microshell made of a include of ceramic and metal phases in a bi-directional FGP. The shell's primary radius in relation to x and y axes is R_1, R_2, respectively. The microshell's geometric parameters are displayed in Fig. 1 with and denoting the lengths of the curve a and b and h denoting the structure's thickness. The two parameters (k_w, k_s) define an elastic substrate on which the structure rests. Here, it shows the stiffness of the Winkler and shear layers. The following mathematical models are used to the porosity laws of shell materials:

Even porosity law (Type E)

$$\vartheta = \frac{\xi}{2} \tag{1a}$$

Uneven porosity law (Type U)

$$\vartheta = \frac{\xi}{2}\left(1 - \frac{2|z|}{h}\right). \tag{1b}$$

The following formula represents the composition of the mechanical characteristics of materials having characteristics: Young's modulus E, length-scale factor l, and mass density ρ [13].

$$E(x,z) = E_m + (E_c - E_m)\left(\frac{1}{2} + \frac{z}{h}\right)^{\alpha_z}\left(\frac{x}{a}\right)^{\alpha_x} - \vartheta(E_c + E_m). \tag{2a}$$

$$\rho(x,z) = \rho_m + (\rho_c - \rho_m)\left(\frac{1}{2} + \frac{z}{h}\right)^{\alpha_z}\left(\frac{x}{a}\right)^{\alpha_x} - \vartheta(\rho_c + \rho_m) \tag{2b}$$

$$l(x,z) = l_m + (l_c - l_m)\left(\frac{1}{2} + \frac{z}{h}\right)^{\alpha_z}\left(\frac{x}{a}\right)^{\alpha_x} - \vartheta(l_c + l_m). \tag{2c}$$

where the subscripts m and c stand for ceramic and metallic parts, respectively; α_z and α_x represent the material's power-law indices through the thickness and x-directions. The field of displacement of every point at any distance z from the middle surface can be found using Kirchhoff–Love shell deformation theory [14]. To do this, the center surface displacement of a two-curved shallow microshell is represented in the

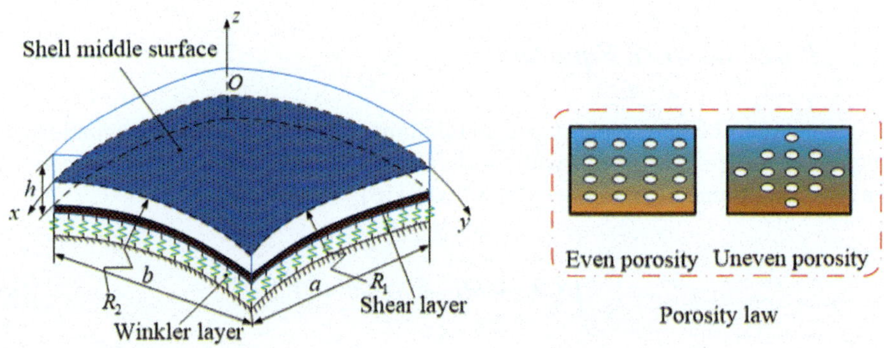

Fig. 1 The BFGP double-curved microshell model, with a rectangular planform

$x-$, $y-$, and $z-$ directions, respectively.

$$\begin{cases} u_x = \left(1 + \frac{z}{R_x}\right) u_0(x, y) - z w_{0,x}(x, y) \\ u_y = \left(1 + \frac{z}{R_y}\right) v_0(x, y) - z w_{0,y}(x, y) \\ u_z = w_0(x, y) \end{cases} \tag{3}$$

A linear elastic isotropic material's constitutive equations are as follows, under modified couple stress hypothesis [15]:

$$\sigma_{ij} = \lambda \varepsilon_{kk} \delta_{ij} + 2\mu \varepsilon_{ij}; \theta_{ij} = 2\mu l^2 \chi_{ij}. \tag{4}$$

where σ_{ij} is components of the local stress vector, θ_{ij} is components of the deviatoric portion, are the components of the deflection of the symmetric stress tensor; where λ and μ are Lamé's constants; χ_{ij} is the symmetric curvature vector; and ε_{ij} is the local strain vector.

The following are the components of strain and curvature:

$$\varepsilon_{ii} = u_{i,i}; i = x, y, z; \tag{5a}$$

$$\varepsilon_{ij} = u_{i,j} + u_{j,i}; j = x, y, z; j \neq i \tag{5b}$$

$$\chi_{ij} = \frac{1}{2}(\Phi_{i,j} + \Phi_{j,i}); i = x, y, z; j = x, y, z; \tag{6}$$

The following is a description of the rotation vector's Φ_i components:

$$\Phi_x = \frac{1}{2}(u_{z,y} - u_{y,z}); \Phi_y = \frac{1}{2}(u_{x,z} - u_{z,x}); \Phi_z = \frac{1}{2}(u_{y,x} - u_{x,y}) \tag{7}$$

2.2 *Force Balance Equation*

Using the Hamiltonian potential energy minimum principle, the general equations of motion of a microphone shell with mechanical properties varying in two directions on an elastic foundation are given as follows [16].

$$\int_{t_1}^{t_2} (\delta U + \delta U_f + \delta V) dt = 0 \tag{8}$$

The formula can be used to display the variation in deformation energy of the two-curved BFGP micro-panel.

$$\delta U = \int_{S_e} \left(\boldsymbol{\varepsilon}^T \mathbf{D}_b \delta\boldsymbol{\varepsilon} + \boldsymbol{\psi}^T \mathbf{D}_{mb} \delta\boldsymbol{\psi} + \boldsymbol{\chi}^T \mathbf{D}_{ms} \delta\boldsymbol{\chi}\right) dxdy \tag{9}$$

where the following is how the components $\boldsymbol{\varepsilon}$, $\boldsymbol{\psi}$, $\boldsymbol{\chi}$, $\mathbf{D}_b$, $\mathbf{D}_{mb}$, and $\mathbf{D}_{ms}$ are identified using the known elements:

$$\boldsymbol{\varepsilon} = \begin{Bmatrix} \boldsymbol{\varepsilon}^0 \\ \boldsymbol{\varepsilon}^1 \end{Bmatrix}; \boldsymbol{\varepsilon}^0 = \begin{Bmatrix} u_{0,x} + \frac{w_0}{R_1} \\ v_{0,y} + \frac{w_0}{R_2} \\ u_{0,y} + v_{0,x} \end{Bmatrix}; \boldsymbol{\varepsilon}^1 = -\begin{Bmatrix} w_{0,xx} - \frac{u_{0,x}}{R_1} \\ w_{0,yy} - \frac{v_{0,y}}{R_2} \\ 2w_{0,xy} - \frac{u_{0,y}}{R_1} - \frac{v_{0,x}}{R_2} \end{Bmatrix}; \tag{10}$$

$$\boldsymbol{\psi} = \begin{Bmatrix} w_{0,xy} - \frac{v_{0,x}}{2R_2} \\ -w_{0,xy} + \frac{u_{0,y}}{2R_1} \\ \frac{u_{0,y}}{2R_1} - \frac{v_{0,x}}{2R_2} \\ \frac{1}{2}\left(\frac{u_{0,x}}{2R_1} - w_{0,xx}\right) - \frac{1}{2}\left(\frac{v_{0,y}}{2R_2} - w_{0,yy}\right) \end{Bmatrix};$$

$$\boldsymbol{\chi} = \begin{Bmatrix} \boldsymbol{\chi}^0 \\ \boldsymbol{\chi}^1 \end{Bmatrix}; \boldsymbol{\chi}^0 = -\frac{1}{4}\begin{Bmatrix} u_{0,xy} - v_{0,xx} \\ u_{0,yy} - v_{0,xy} \end{Bmatrix}; \boldsymbol{\chi}^1 = -\frac{1}{4}\begin{Bmatrix} \frac{u_{0,xy}}{R_1} - \frac{v_{0,xx}}{R_2} \\ \frac{u_{0,yy}}{R_1} - \frac{v_{0,xy}}{R_2} \end{Bmatrix}; \tag{11}$$

$$\mathbf{D}_b = \begin{bmatrix} \boldsymbol{A} & \boldsymbol{B} \\ \boldsymbol{B} & \boldsymbol{F} \end{bmatrix}; \mathbf{D}_{mb} = \mathbf{A}_{mb}; \mathbf{D}_{ms} = \begin{bmatrix} \mathbf{A}_{ms} & \mathbf{B}_{ms} \\ \mathbf{B}_{ms} & \mathbf{F}_{ms} \end{bmatrix}; \tag{12}$$

where $\boldsymbol{A}, \boldsymbol{B}, \boldsymbol{F}, \mathbf{A}_{mb}, \mathbf{A}_{ms}, \mathbf{B}_{ms}$, and $\mathbf{F}_{ms}$ represent the stiffness of the material and may be found using the following formula:

$$(\boldsymbol{A}, \boldsymbol{B}, \boldsymbol{F}) = \int_{-h/2}^{+h/2} Q_b\left(1, \bar{z}, \bar{z}^2\right) \mathrm{d}z; \tag{13}$$

$$\mathbf{A}_{mb} = \int_{-h/2}^{+h/2} Q_{mb} \mathrm{d}z; \left(\mathbf{A}_{ms}, \mathbf{B}_{ms}, \mathbf{F}_{ms}\right) = \int_{-h/2}^{+h/2} Q_{ms}\left(1, \bar{z}, \bar{z}^2\right) \mathrm{d}z \tag{14}$$

where $\bar{z} = z - t_0$ is the separation between the neutral and mean surfaces of the microshell, and Q_{ms} are the tensors of the material. Here's how these factors are calculated:

$$Q_b = \frac{E(x,z)}{1 - \nu(x,z)^2} \begin{bmatrix} 1 & \nu(x,z) & 0 \\ \nu(x,z) & 1 & 0 \\ 0 & 0 & \frac{1-\nu(x,z)}{2} \end{bmatrix};$$

$$Q_{mb} = \frac{E(x,z)l^2}{1+\nu(x,z)}\begin{bmatrix} 1\,0\,0\,0 \\ 0\,1\,0\,0 \\ 0\,0\,1\,0 \\ 0\,0\,0\,2 \end{bmatrix};$$

$$Q_{ms} = \frac{2E(x,z)l^2}{1+\nu(x,z)}\begin{bmatrix} 1\,0 \\ 0\,1 \end{bmatrix};\ t_0 = \frac{\int_{-h/2}^{+h/2} E(x,z)zdz}{\int_{-h/2}^{+h/2} Edz}. \tag{15}$$

The strain energy of the elastic foundation varies depending on the following factors:

$$\delta U_f = \int\limits_{S_e} \left(k_w w_0 \delta w_0 - k_s\left(w_{0,xx} + w_{0,yy}\right)\delta w_0\right) dxdy \tag{16}$$

The following is the calculation for the change in potential energy:

$$\delta V = \int\limits_{S_e} \left(q_0 - N_x^0 w_{0,xx} - N_y^0 w_{0,yy}\right)\delta w_0 dxdy \tag{17}$$

The force balance equation of the shell is obtained as follows, which can be obtained by entering Eqs. (9), (16), and (17) into (8) into the previous Eq. (8):

$$\begin{aligned} &\int\limits_{S_e} \left(\boldsymbol{\varepsilon}^T \mathbf{D}_b \delta\boldsymbol{\varepsilon} + \boldsymbol{\psi}^T \mathbf{D}_{mb} \delta\boldsymbol{\psi} + \boldsymbol{\chi}^T \mathbf{D}_{ms} \delta\boldsymbol{\chi}\right) dxdy \\ &\quad + \int\limits_{S_e} \left(k_w w_0 \delta w_0 + \left(k_s - N^0\right)\left(w_{0,x}\delta w_{0,x} + w_{0,y}\delta w_{0,y}\right)\right) dxdy \\ &= \int\limits_{S_e} (q_0)\delta w_0 dxdy \end{aligned} \tag{18}$$

2.3 Isogeometric Analysis Approach

Use the following equations to compute the displacements of two-curved microshells on the middle surface [17]:

$$\boldsymbol{d}^h = \sum_{I=1}^{Ne} \boldsymbol{C}_I \boldsymbol{d}_I. \tag{19}$$

with

$$\boldsymbol{d}_I = [u_{0e}, v_{0e}, w_{0e}] \tag{20}$$

In which $\boldsymbol{C}_e$ and $\boldsymbol{d}_e$ are the shape-related functions and the unknown displacement tensor at points of control I, respectively, and $Ne = (p + 1)(q + 1)$ is the number of points of control per physical element.

The displacement components u_0, v_0, and w_0 can be calculated as follows:

$$u_0 = \sum_{I=1}^{Ne} [\boldsymbol{C}_I \; 0 \; 0] \boldsymbol{d}_I = \sum_{I=1}^{Ne} \boldsymbol{H}_I^u \boldsymbol{d}_I;$$
$$v_0 = \sum_{I=1}^{Ne} [0 \; \boldsymbol{C}_I \; 0] \boldsymbol{d}_I = \sum_{I=1}^{Ne} \boldsymbol{H}_I^v \boldsymbol{d}_I;$$
$$w_0 = \sum_{I=1}^{Ne} [0 \; 0 \; \boldsymbol{C}_I] \boldsymbol{d}_I = \sum_{I=1}^{Ne} \boldsymbol{H}_I^w \boldsymbol{d}_I; \tag{21}$$

The characteristic equation for the microshell's static bending can be obtained by replacing Eq. (21) into Eq. (18) and using the ensuing formulas:

$$\boldsymbol{K}.\boldsymbol{d} = \boldsymbol{F} \tag{22}$$

The boundary conditions of the structure are determined as follows.

- Clamp-clamped (C)

  $u_0 = v_0 = w_0 = w_{0,x} = w_{0,y} = 0$ at all edge.

- Simply supported (S):

  $u_0 = w_0 = w_{0,x} = 0$ at $y = 0, b$ and $v_0 = w_0 = w_{0,y} = 0$ at $x = 0, a$.

3 Numerical Results

With the mechanical and mathematical model implemented in part 2, a calculation program is implemented on Matlab's software to determine the vertical flection value at the center of the microshell whose mechanical characteristics vary in two directions with any BCs. The convergence and accuracy of the program are verified through numerical comparison with findings that are published. To facilitate the process of providing numerical results, the following several dimensionless formulas are used

$$w_1 = w\left(\frac{a}{2}, \frac{b}{2}\right).\frac{100E_c h^3}{q_0 a^4}; K_w = \frac{k_w a^4}{D_c}; K_s = \frac{k_s a^2}{D_c}; D_c = \frac{E_c h^3}{12(1 - v_c^2)};$$

Table 1 Convergence of vertical displacement values of microshells via various length-scale coefficients (Type E)

		Mesh					
l_c/h	p	3 × 3	5 × 5	7 × 7	9 × 9	11 × 11	15 × 15
0	2	1.5150	1.5191	1.5104	1.5057	1.5030	1.5004
	3	1.5669	1.5002	1.4981	1.4974	1.4972	1.4971
	4	1.4955	1.4968	1.4970	1.4970	1.4970	1.4970
0.5	2	1.4440	1.4522	1.4464	1.4431	1.4412	1.4392
	3	1.4906	1.4392	1.4375	1.4370	1.4368	1.4367
	4	1.4364	1.4366	1.4367	1.4367	1.4367	1.4367
1	2	1.2461	1.2581	1.2563	1.2549	1.2540	1.2531
	3	1.2850	1.2534	1.2524	1.2521	1.2520	1.2519
	4	1.2524	1.2518	1.2519	1.2519	1.2519	1.2519

The input parameters are given as follows: $\frac{a}{h} = 50$, b = a, $R_1 = 5a$, $R_2 = 5b$, $\alpha_z = 0.5, \alpha_x = 1, \xi = 0.1, l_c = h, l_m = l_c/2, K_w = 10, K_s = 1$. The following are the mechanical characteristics of two-curved BFGP microshells:

$$\text{Ceramic:}\quad Al_2O_3, E_c = 380\text{ GPa}, \rho_c = 3800\text{ kgm}^{-3}, v_c = 0.3.$$

$$\text{Metallic:}\quad Al, E_m = 70\text{ GPa}, \rho_m = 2700\text{ kgm}^{-3}, v_m = 0.3.$$

3.1 *Verify Convergence*

The convergence characteristics for the displacement results of mechanical microshell that vary along the thickness direction and are subjected to four-sided single support connection (SSSS) are performed and presented as shown in Table 1. Each shell case's six distinct finite element meshes are examined to determine the suggested IGA method's rate of convergence. Table 1 illustrates that solutions with polynomial degrees of $p = 2$ show comparatively sluggish convergence rates for IGA, while quick convergence is noted for $p = 3$ and $p = 4$. Subsequent surveys employ an 11 × 11 ($p = 4$) cubic NURBS element mesh based on the above results.

3.2 *Study Parameters*

In this subsection, the impacts of material exponent factor, different boundary conditions, and elastic medium stiffness coefficient on the vertical displacement value of

Table 2 Dimensionless vertical displacement value due to the influence of boundary conditions and elastic foundation stiffness coefficient (Type U)

		Boundary condition			
K_w, K_s	$\alpha_x = \alpha_z$	CCCC	SSSS	CSCS	CFCF
0, 0	0	0.2308	0.4230	0.2975	0.4636
	1	1.0598	1.5456	1.2579	2.0713
	10	2.9715	4.0275	3.3898	5.1798
100, 0	0	0.2276	0.4122	0.2922	0.4486
	1	0.9918	1.3998	1.1611	1.7871
	10	2.5012	3.1657	2.7735	3.7210
0, 10	0	0.2233	0.4024	0.2862	0.4446
	1	0.9114	1.2920	1.0682	1.7337
	10	2.0450	2.6732	2.2949	3.4995
100, 10	0	0.2204	0.3926	0.2812	0.4308
	1	0.8605	1.1888	0.9975	1.5308
	10	1.8056	2.2614	1.9908	2.7666

the shell are investigated as shown in Table 2. The results show that the material exponent increases, causing the shell's turnover to increase, which can be explained by the fact that the hardness of the microshell decreases. In addition, the stiffness coefficient of the elastic foundation increases, causing the displacement of the shell to decrease. Furthermore, different boundary conditions will cause the displacement value to change differently, here the boundary condition CCCC gives the largest stiffness and CFCF boundary condition gives the smallest stiffness.

4 Conclusions

In the framework of the current study, an IGA method is developed to analyze static bending of a microshell with mechanical properties changing in two directions via the modified couple stress hypothesis with the new point being the length ratio changes according to the material's mechanical properties. Studies show that vertical displacement tends to decrease with the increase of the length scaling factor, it is not dependent on geometric factors or different boundary conditions. Meanwhile, the increase in the volume and pore exponent always reduces the stiffness of the microshell structure, leading to a rapid increase in vertical displacement. Varying different boundary conditions will give the resulting ranges of vertical displacements. In addition, the elastic foundation stiffness plays an important role in the working process of the microshell, it helps the microshell become significantly stronger. These results are intended to contribute part of our understanding to the general understanding of microshell structures with mechanical characteristics that vary in bi-directions.

References

1. Thai, H.-T., Choi, D.-H.: Size-dependent functionally graded Kirchhoff and Mindlin plate models based on a modified couple stress theory. Compos. Struct. **95**, 142–153
2. Dung, N.T., Tran, V.K., Huyen, T.T.H., Van Minh, P.: Nonlinear static bending analysis of microplates resting on imperfect two-parameter elastic foundations using modified couple stress theory. Comptes. Rendus. Mécanique **350**(G1), 121–141 (2022)
3. Tsiatas, G.C.: A new Kirchhoff plate model based on a modified couple stress theory. Int. J. Solids Struct. **46**(13), 2757–2764 (2009)
4. Yin, L., Qian, Q., Wang, L., Xia, W.: Vibration analysis of microscale plates based on modified couple stress theory. Acta Mech. Solida Sin. **23**(5), 386–393 (2010)
5. Ma, H.M., Gao, X.-L., Reddy, J.N.: A non-classical Mindlin plate model based on a modified couple stress theory. Acta Mech. **220**(1–4), 217–235 (2011)
6. Reddy, J.N., Berry, J.: Nonlinear theories of axisymmetric bending of functionally graded circular plates with modified couple stress. Compos. Struct. **94**(12), 3664–3668 (2012)
7. Kim, J., Reddy, J.N.: A general third-order theory of functionally graded plates with modified couple stress effect and the von Karman nonlinearity: theory and finite element analysis. Acta Mech. **226**(9), 2973–2998 (2015)
8. Nguyen, V.P., Anitescu, C., Bordas, S.P.A., Rabczuk, T.: Isogeometric analysis: an overview and computer implementation aspects. Math. Comput. Simul **117**, 89–116 (2015). https://doi.org/10.1016/j.matcom.2015.05.008
9. Jüttler, B., Langer, U., Mantzaflaris, A., Moore, S.E., Zulehner, W.: Geometry + simulation modules: implementing isogeometric analysis. PAMM **14**(1), 961–962 (2014). https://doi.org/10.1002/pamm.201410461
10. Marussig, B., Hughes, T.J.R.: A review of trimming in isogeometric analysis: challenges, data exchange and simulation aspects. Arch. Comput. Methods Eng. **25**(4), 1059–1127 (2018). https://doi.org/10.1007/s11831-017-9220-9
11. Hughes, T.J.R., Cottrell, J.A., Bazilevs, Y.: Isogeometric analysis: CAD, finite elements, NURBS, exact geometry and mesh refinement. Comput. Methods Appl. Mech. Eng. **194**(39–41), 4135–4195 (2005). https://doi.org/10.1016/j.cma.2004.10.008
12. Hughes, T.J.R., Reali, A., Sangalli, G.: Duality and unified analysis of discrete approximations in structural dynamics and wave propagation: comparison of p-method finite elements with k-method NURBS. Comput. Methods Appl. Mech. Eng. **197**(49–50), 4104–4124 (2008). https://doi.org/10.1016/j.cma.2008.04.006
13. Karamanlı, A., Vo, T.P.: Size dependent bending analysis of two directional functionally graded microbeams via a quasi-3D theory and finite element method. Compos. B Eng. **144**, 171–183 (2018)
14. Shahsavari, D., Karami, B., Fahham, H.R., Li, L.: On the shear buckling of porous nanoplates using a new size-dependent quasi-3D shear deformation theory. Acta Mech. **229**(11), 4549–4573 (2018). https://doi.org/10.1007/s00707-018-2247-7
15. Yang, F., Chong, A.C.M., Lam, D.C.C., Tong, P.: Couple stress based strain gradient theory for elasticity. Int. J. Solids Struct. **39**, 2731–2743 (2002)
16. Tran, V.-K., Pham, Q.-H., Nguyen-Thoi, T.: A finite element formulation using four-unknown incorporating nonlocal theory for bending and free vibration analysis of functionally graded nanoplates resting on elastic medium foundations. Eng. Comput. 1–26 (2020)
17. Cottrell, J.A., Reali, A., Bazilevs, Y., Hughes, T.J.R.: Isogeometric analysis of structural vibrations. Comput. Methods Appl. Mech. Eng. **195**(41–43), 5257–5296 (2006)

Static Analysis of Artificial Bone Pin Made by Using Nanocomposite

Dong Pham Van, Mahesh Thorat, Santosh Patil, Rajanikant Kurane, Shailesh Shirguppikar, Thinh Hoang Xuan, Thanh Pham Viet, and L. Selvarajan

Abstract This paper investigates the fabrication and analysis of artificial bone pins using various nanocomposite materials. The study explores the mechanical properties and biocompatibility of these artificial bone pins through Finite Element Method (FEM) analysis, comparing them with natural bone. Three distinct samples of artificial bone pins were synthesized, each with a unique chemical composition. The analysis includes directional deformation and equivalent stress measurements for each sample, revealing their performance relative to natural bone. Results indicate that Sample 1 and Sample 2 nanocomposite materials closely resemble the properties of natural bone, suggesting their potential for bone pin implants. However, Sample 3 exhibits inferior performance compared to the other samples and natural bone. Overall, this study contributes to the development of more effective artificial bone structures and materials, offering insights into potential implant materials with properties akin to natural bone.

Keywords Artificial bone · Nanocomposite materials · Finite element method (FEM) analysis · Directional deformation · Equivalent stress · Biocompatibility

D. P. Van · T. H. Xuan (✉) · T. P. Viet
Hanoi University of Industry, No. 298, CauDien Street, Bac TuLiem District, Hanoi, Vietnam
e-mail: hoangxuanthinh@haui.edu.vn

M. Thorat
Department of Automobile Engineering (Diploma), Rajarambapu Institute of Technology, Shivaji University, Sakharale, MS 414415, India

S. Patil · S. Shirguppikar
Department of Mechatronics Engineering, Rajarambapu Institute of Technology, Shivaji University, Sakharale, MS 414415, India

R. Kurane
Department of Science and Humanities, Rajarambapu Institute of Technology, Shivaji University, Sakharale, MS 414415, India

L. Selvarajan
Department of Mechanical Engineering, Mahendra Institute of Technology (Autonomous), Namakkal, TamilNadu 637503, India
e-mail: selvalakshmanan86@gmail.com

D. H. Tien et al. (eds.), *Proceedings of the 1st International Conference on Sustainability and Emerging Technologies for Smart Manufacturing*, Proceedings in Technology Transfer, https://doi.org/10.1007/978-981-97-7083-0_65

1 Introduction

The femur is the lengthiest bone within the human anatomy, bears the weight, and facilitates leg movement. Fractures in the femur can result from significant force, accidents, or bone diseases, among other causes. With over three million cases annually in the US, bone fractures are common. The body's natural ability to regenerate bone involves a cycle of resorption by osteoclasts and formation by osteoblasts. However, severe damage or disease may hinder this process, necessitating surgical intervention. Surgeons often use autografts, allografts, or synthetic grafts to replace lost bone tissue. Synthetic grafts, including artificial bone, offer advantages such as reduced invasiveness and improved biocompatibility compared to autografts and allografts. Criteria for designing implanted biomaterials include biocompatibility, osteoconductivity, porosity, and biomechanical compatibility. Early artificial bones, made of rigid materials like metals and ceramics, posed challenges due to poor biocompatibility and excessive rigidity. Engineers strive to develop more comfortable and effective artificial bone structures and materials to enhance quality of life.

Human bones exhibit a sophisticated amalgamation of organic and inorganic elements, each fulfilling pivotal roles in their form and function. The skeletal framework is predominantly constituted of minerals, accounting for approximately 65–70% of bone composition, with calcium and phosphate as the primary components, intricately constructing hydroxyapatite crystals. Meanwhile, organic constituents, comprising roughly 25–30% of bone composition, predominantly feature collagen fibers, which serve as the backbone of bone tissue, imparting both flexibility and resilience. Additionally, water, constituting about 5–10% of bone, permeates the organic matrix, crucially facilitating metabolic processes, nutrient transportation, waste disposal, and sustaining optimal hydration levels within bone tissue.

These research gaps highlight the need for ongoing studies into synthetic bone materials, focusing on biocompatibility, biomechanical compatibility, osteoconductivity, tailored solutions for specific anatomical scenarios, and safety. Addressing these gaps will contribute to the development of more effective and patient-friendly bone replacement solutions (Fig. 1).

2 Selection of Materials

Despite significant advancements in nanomaterials research over the years, our work distinguishes itself through its innovative approach and the use of novel materials. This study explores uncharted territory by employing a unique selection of nanomaterials that have not been extensively investigated in previous research. Our decision to use these specific nanomaterials is based on their exceptional properties, which we believe hold immense potential for addressing key challenges in artificial human bone.

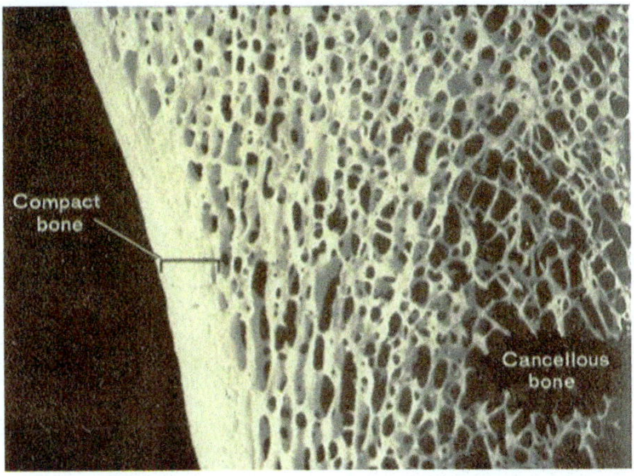

Fig. 1 Macro structure of bone

Materials selected for the fabrication of artificial bone from nanocomposites include PMMA, graphene, zirconia, hydroxyapatite, and ferrocene. Three nanocomposite samples have been synthesized, exhibiting favorable biocompatibility, osteoconductivity, and mechanical properties.

2.1 Sample of Artificial Bone Pin

Artificial bone pins have been produced using the aforementioned biocompatible materials, resulting in three distinct samples.

Sample 1:

The chemical composition consists of polymethylmethacrylate (PMMA) comprising 95%, graphene comprising 3.5%, and ferrocene comprising 1.5%.

PMMA (95%) + Graphene (3.5%) + Ferrocene (1.5%).

Sample 2:

The chemical composition consists of polymethyl methacrylate (PMMA) comprising 85%, zirconia accounting for 10%, graphene making up 3.5%, and ferrocene comprising 1.5%.

PMMA (85%) + Zirconia (10%) + Graphene (3.5%) + Ferrocene (1.5%).

Sample 3:

The chemical composition consists of 85% polymethylmethacrylate (PMMA), 3.5% zirconia, 1.5% graphene, and 10% hydroxyapatite.

PMMA (85%) + Zirconia (3.5%) + Graphene (1.5%) + Hydroxyapatite (10%).

Table 1 Properties of sample 1, sample 2, sample 2, and human bone

Parameters	Sample 1	Sample 2	Sample 3	Human bone
Young's modulus (GPa)	18.87	19.48	8.70	17
Poisson's ratio	0.1694	0.1122	0.08468	0.3
Density (g/cm^3)	0.4106	0.42	0.39	0.3

2.2 Material Properties

See Table 1.

3 Modeling and Analysis of Artificial Bone Pin

3.1 Modeling of Solid Cylinder

To create an artificial bone pin for this study, a straightforward design was employed, which can be implemented using Ansys Workbench software. Specifically, the pin was fashioned to have a diameter of 10 mm and a length of 30 mm.

The design process involved following steps (Fig. 2):

1. Sketching a circle with a diameter of 10 mm.
2. This circle was extruded along the normal direction to achieve a length of 30 mm.

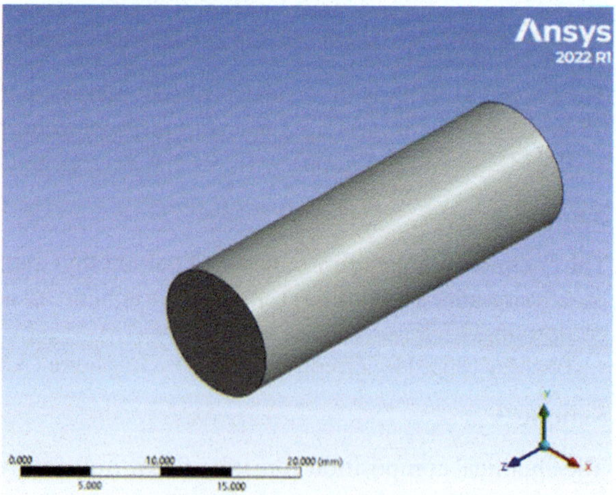

Fig. 2 Model of solid cylinder

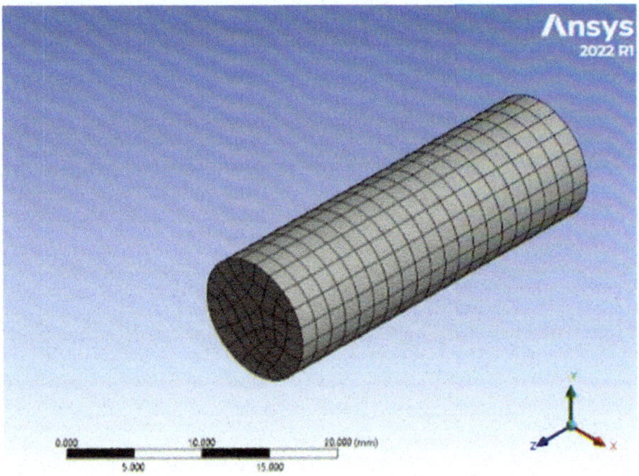

Fig. 3 Meshed model

3.2 *Meshing*

Once the model was constructed, meshing was performed throughout the cylinder's volume for subsequent Finite Element Analysis (FEA). This process utilized a mesh tool, resulting in 4479 nodes and 920 elements. The meshed representation of the artificial bone pin in ANSYS can be observed in the accompanying figure (Fig. 3).

3.3 *Boundary Conditions*

The cylinder is anchored at one end, while a compressive force of 750 pascals, the force exerted by a person weighing 75 kg, is applied to the other end. This compressive load mimics the forces experienced by an artificial bone pin during activities like walking or standing, where the predominant force is compressive in nature (Figs. 4 and 5).

4 Results and Discussion

(1) PMMA (95%) + Graphene (3.5%) + Ferrocene (1.5%)

According to the structural examination conducted on Sample 1 bone, it was observed that the directional deformation measured 1.1898 mm, with the maximum stress reaching 871 MPa and the minimum stress at 541.89 MPa (Figs. 6 and 7).

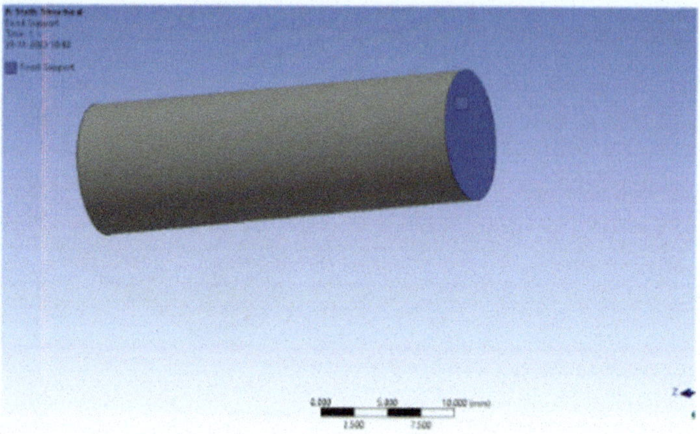

Fig. 4 One end fixed

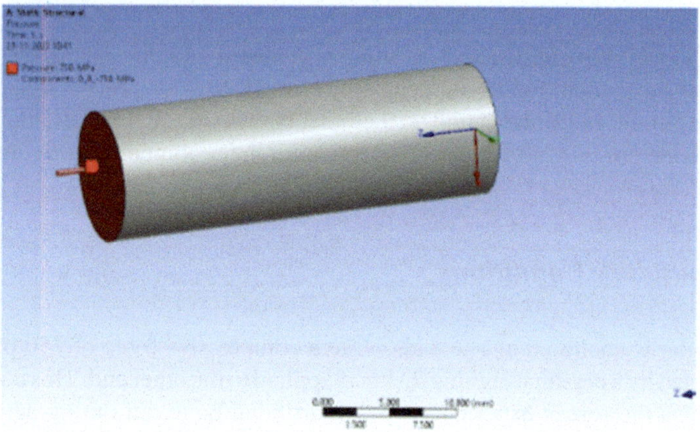

Fig. 5 Load applied on another end

(B) PMMA (85%) + Zirconia (10%) + Graphene (3.5%) + Ferrocene (1.5%)

Based on the structural examination conducted on Sample 2 bone, it was determined that the directional deformation measures 1.298 mm, while the stress levels reached a maximum of 830.38 MPa and a minimum of 611.35 MPa (Figs. 8 and 9).

(C) PMMA (85%) + Zirconia (3.5%) + Graphene (1.5%) + Hydroxyapatite (10%)

According to the structural examination conducted on Sample 3 bone, it was observed that the directional deformation measured 2.2979 mm, with maximum stress recorded at 807.5 MPa and minimum stress at 650.85 MPa (Figs. 10 and 11).

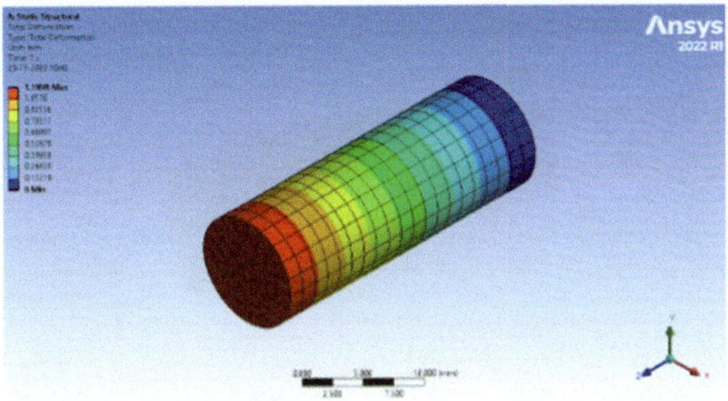

Fig. 6 Sample 1 bone pin directional deformation

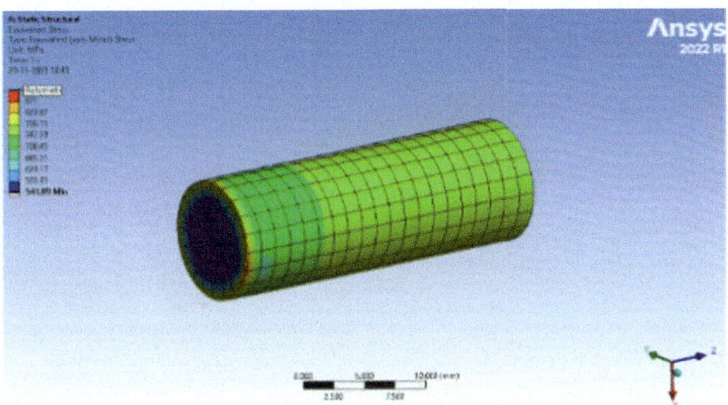

Fig. 7 Sample 1 bone pin equivalent stress

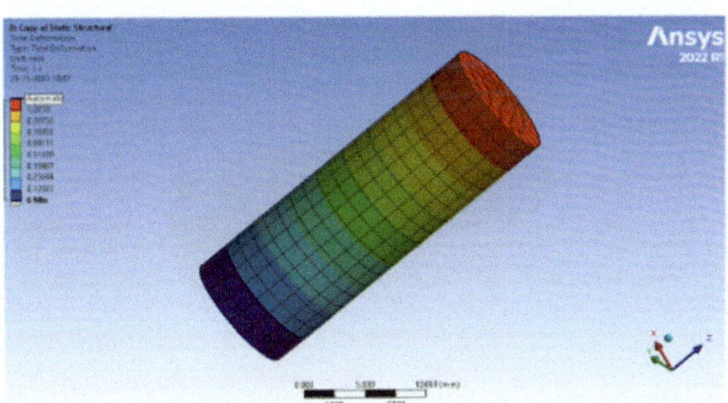

Fig. 8 Sample 2 bone pin directional deformation

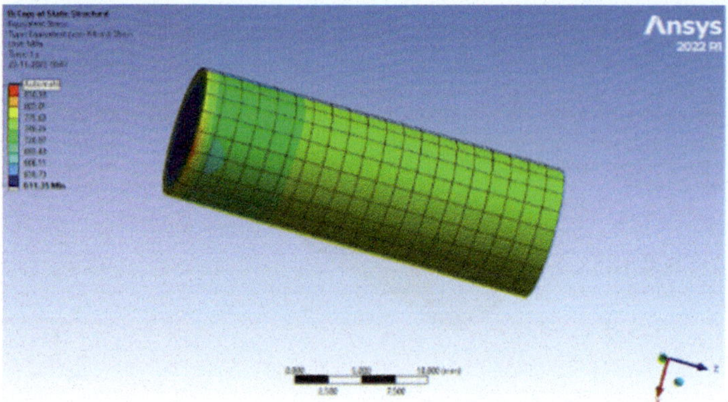

Fig. 9 Sample 2 bone pin equivalent stress

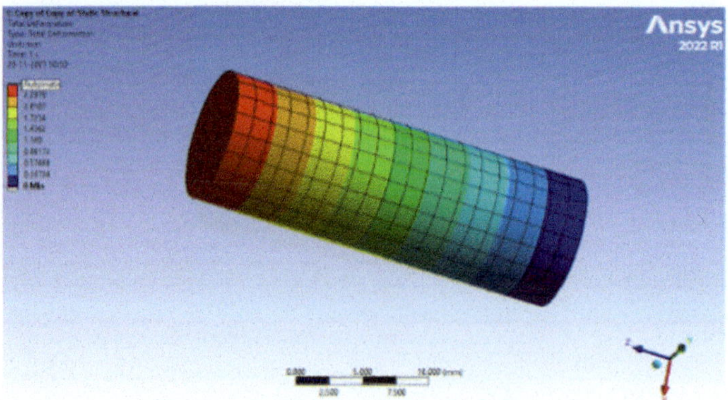

Fig. 10 Sample 3 bone pin directional deformation

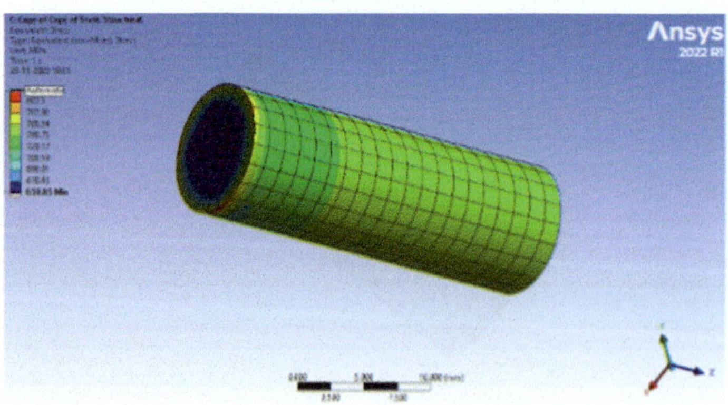

Fig. 11 Sample 3 bone pin equivalent stress

(D) Natural Bone

The structural examination of natural bone pin reveals a directional deformation of 1.3134 mm, with maximum stress reaching 967.1 MPa and minimum stress at 374.24 MPa (Figs. 12 and 13).

Below is a tabulated comparison of the bone pin directional deformation and bone pin equivalent stress for the aforementioned results (Table 2).

Our analysis of directional deformation revealed interesting insights into the mechanical properties of the tested samples compared to natural bone. While both Sample 1 (1.1898 mm) and Sample 2 (1.29 mm) exhibited deformation values closer to natural bone (1.3134 mm), Sample 3 (2.2979 mm) displayed significantly higher deformation, indicating inferior mechanical performance under the applied stress.

Samples 1 and 2: These samples demonstrated deformation values comparable to natural bone, suggesting they possess mechanical properties similar to bone tissue.

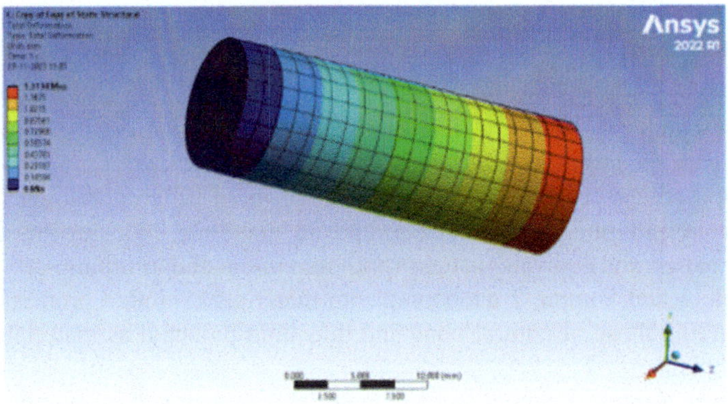

Fig. 12 Human bone pin directional deformation

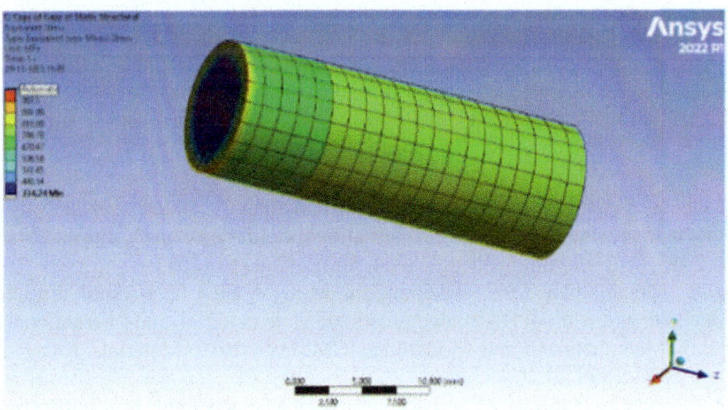

Fig. 13 Human bone pin equivalent stress

Table 2 Result comparison of sample 1, sample 2, sample 2, and human bone

Parameters	Sample 1	Sample 2	Sample 3	Human bone
Directional deformation (mm)	1.1898	1.29	2.2979	1.3134
Equivalent stress (MPa) max	871	830.38	807.5	967.1
Equivalent stress (MPa) min	541.89	611.35	650.85	374.24

This is a promising finding, as these materials could potentially be suitable for applications requiring bone-like stiffness and strength.

Sample 3: This sample exhibited significantly higher deformation compared to both natural bone and the other samples. This indicates a greater susceptibility to bending or displacement under stress, making it potentially less suitable for applications requiring robust mechanical performance.

5 Conclusion

1. The research involved modeling and analyzing artificial bone pins made from various nanocomposite materials.
2. The researchers compared natural bone to three different samples of artificial bone pins made from nanocomposite materials, focusing on directional deformation and equivalent stress (both maximum and minimum).
3. Sample 1 and Sample 2 nanocomposite materials exhibited properties closely resembling those of natural bone and thus hold potential as materials for bone pin implants.
4. However, Sample 3 nanocomposite material demonstrated inferior performance compared to the other samples when compared to natural bone.
5. In summary, Sample 1 nanocomposite material demonstrated results closer to natural bone when compared to Sample 2 nanocomposite material.

Bibliography

1. Senthil Maharaj, P.S.R., Maheswaran, R., Vasanthanathan, A.: Numerical analysis of fractured femur bone with prosthetic bone plates. International Conference on Design and Manufacturing, IConDM 2013, Procedia Engineering vol 64, pp. 1242–1251 (2013)
2. Kirthana, S., Nagamallika, J.V.L., Nizamuddin, M.K.: A FEA on assembled fractured human femur bone with and without HA coated prosthetic plate material. 2nd International Conference on Materials Manufacturing and Modelling, ICMMM—2019. Materials Today: Proceedings vol. 22, pp. 2890–2898 (2020)
3. Kirthana, S., Bindhu Supraja, M., Vishwa, A.S.N., Mahalakshmi, N.: Static structural analysis on femur bone using different plate material. 2nd International Conference on Materials Manufacturing and Modelling, ICMMM—2019. Proceedings vol. 22, pp. 2324–2333 (2020)

4. Kaushlendra, K., Prasad, R.B.: Stress analysis of cortical bone of human femur. Materials Today: Proceedings, vol 44, Part 1, pp. 2054–2060 (2021)
5. Gangwar, A.K.S., Rao, P.S., Kumar, A.: Bio-mechanical design and analysis of femur bone. Materials Today: Proceedings, vol. 44, Part 1, pp. 2179–2187 (2021)
6. Kumar, K.C.N., Griya, N., Shaikh, A., Chaudhry, V., Chavadaki, S.: Structural analysis of femur bone to predict the suitable alternative material. Materials Today: Proceedings, vol. 26, Part 2, pp. 364–368 (2020)
7. Kalaiyarasan, A., Sankar, K., Sundaram, S.: Finite element analysis and modeling of fractured femur bone. Materials Today: Proceedings, vol. 22, Part 3, pp. 649–653 (2020)

Unveiling the Thermal Fingerprint of Magnetorheological Brakes: A Simulation Approach

Hoang Quang Tuan, Trinh Minh Hoang, Nguyen Anh Ngoc, and Ock Taeck Lim

Abstract Magnetorheological brake (MRB) is a promising new braking technology with several potential advantages over conventional hydraulic brakes. However, MRB has not yet been widely commercialized due to a number of challenges, including high operating temperatures. This paper presents a thermal analysis method for the structure of a trapezoidal tooth-shaped disk MR brake. The temperature of the MRF and the main components of the MRB is simulated using Altair Flux software. The simulation results for a period of 60 s show that the temperature of the MRF, rotor, and stator of the trapezoidal brake device is still within the allowable operating range of the material (50–100 °C). The trapezoidal MRB is capable of meeting thermal requirements. The results of this study provide a basis for further research on energy, cooling, and optimization of the trapezoidal MRB structure.

Keywords Magnetorheological Brakes (MRB) · Thermal analysis · Trapezoidal tooth-shaped disk · Magnetorheological fluids · Altair flux

1 Introduction

Magnetorheological fluid (MRF) is a novel smart material composed of micron-sized magnetic particles suspended in a carrier fluid with the help of stabilizers. When exposed to a magnetic field, these particles form chain-like structures, dramatically increasing the fluid's viscosity (shear stress) within milliseconds. This unique property allows MRF to transition from a free-flowing Newtonian fluid to a Bingham

H. Q. Tuan · N. A. Ngoc (✉)
Hanoi University of Industry, Hanoi, Vietnam
e-mail: ngocnguyencnoto@haui.edu.vn

H. Q. Tuan · T. M. Hoang
Hanoi University of Science and Technology, Hanoi, Vietnam

O. T. Lim
School of Mechanical Engineering, University of Ulsan, 93 Daehak-ro, Nam-gu, Ulsan 44610, South Korea

D. H. Tien et al. (eds.), *Proceedings of the 1st International Conference on Sustainability and Emerging Technologies for Smart Manufacturing*, Proceedings in Technology Transfer, https://doi.org/10.1007/978-981-97-7083-0_66

fluid with controllable yield stress [1]. However, elevated temperatures can negatively impact MRF performance. Increased temperature intensifies Brownian motion, causing the magnetic particles to clump together more readily. This reduces the ability of the particles to form the chain-like structures necessary for high shear stress under a constant magnetic field. Furthermore, excessively high temperatures can lead to carrier fluid evaporation and severe wear of the magnetic particles, potentially rendering the MRF inoperable [2].

Karakoc's research [3] simulated the temperature distribution in MRBs during repeated brake applications. The results showed that the MRB temperature could exceed the recommended range for the MR fluid. Therefore, a cooling system is necessary to ensure the lifespan and performance of the MRB. Satyajit [4] analyzed the thermal behavior of MRBs for light electric vehicles like e-bikes under real-world driving cycles for Indian urban conditions. The results showed that the MR fluid temperature in MRBs always remained within the permissible operating range (−40 to 125 °C). Therefore, the proposed MRB can operate safely without any special cooling requirements. WANG Daoming [2] analyzed the simulation and experimental verification of the thermal characteristics of MRBs. The study employed an analytical and computational method based on: heat sources and heat generation rate of MRBs and establishment of a mathematical model for the temperature field of MRBs. The simulation method was conducted on the transient temperature field of MRBs for three braking conditions: normal braking, emergency braking, and intermittent braking. The results showed that the MRB exhibited good continuous deceleration performance under the same coil current. Additionally, it was also observed that the overall temperature of the MRB tended to increase rapidly during the first braking cycle, then increased slowly, and even decreased gradually.

From the aforementioned studies, it can be seen that most of the research on MRBs for automotive applications still has limitations and lacks experimental studies on real vehicles. Most studies are limited to simulation and investigation on 1/4 vehicle models, which may not yield completely accurate results when applied to real vehicles. Therefore, more research is needed to validate the effectiveness of MRBs under various automotive conditions according to current standards. In addition, there is also a need for in-depth research on MRB thermal management to maximize the potential of MRBs and ensure stable operation under all conditions.

This paper presents a study on the simulation of heat generation in different working areas of a MRB disc for auxiliary purposes. After the first overview section, the second section presents the basic structure of the MRB in the form of a trapezoidal disc. The third section describes the steps involved in building a thermal simulation model of the MRB. The fourth section presents the simulation results of the temperature distribution on the brake disc and the magnetic fluid layer in the MRB at different simulation times. The final section is the conclusion of the study.

2 Methodology

2.1 Magnetorheological Brake Structure

In this study, a disc brake with trapezoidal teeth grooves on the surface was used. This vehicle-mounted structure has an additional bolt-on bracket with the active axle housing and the magnetic brake, the purpose of which is to keep the magnetic brake stable when the vehicle is running and when braking (Fig. 1).

Figure 2 shows the main structure of the MRB. The main structure of the MRB assembly includes [5]:

- The rotor of the MRB brake disc is connected to the axle shaft. Trapezoidal grooves with a 75-degree angle are machined on the brake disc.
- The brake housing is assembled using steel plates with round holes for bolts and holes for the coil. The output shaft of the MRB brake is connected to the main transmission.

A small gap of 1–2 mm filled with magnetic fluid is maintained between the disc and the housing. At the braking moment, the electromagnet integrated in the housing is energized. The MRF shear-thickening fluid is then placed in a magnetic field generated by the coil to create a magnetorheological effect. Due to this MRB effect, the fluid becomes semi-solid as the viscosity increases within a few milliseconds and shear friction occurs along the disc surface, resulting in braking and thus deceleration [6–8].

However, this effect is reversible, meaning that the viscosity of the MRF decreases and it returns to a fluid state.

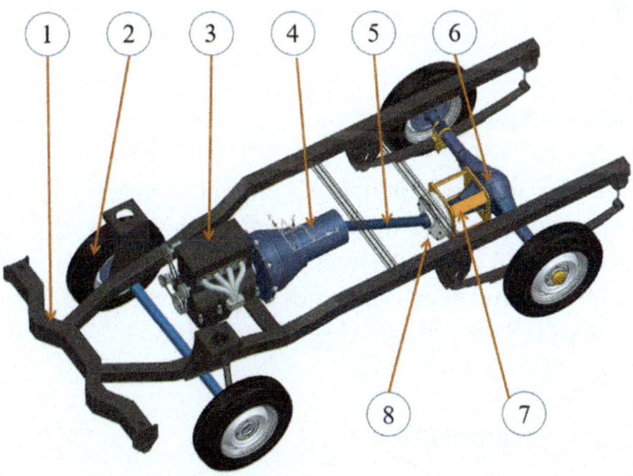

Fig. 1 MRB layout on a car. 1. Chassis; 2. Front axle assembly; 3. Engine; 4. Transmission; 5. Driveshaft; 6. Rear axle assembly; 7. Brake caliper; 8. MR brake

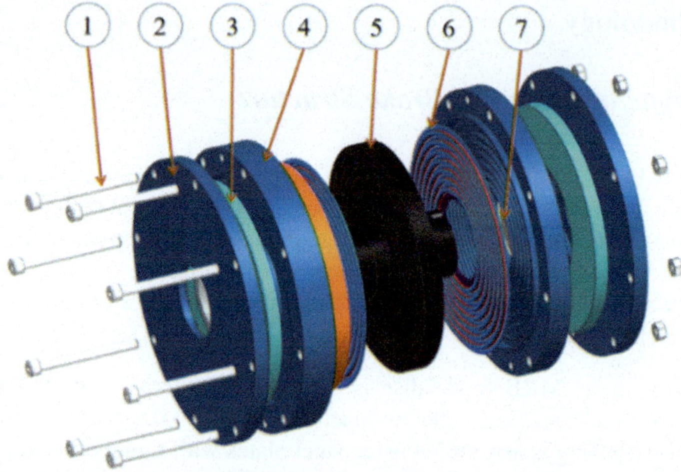

Fig. 2 MR brake construction: 1. Bolts; 2. Outer stopper cover; 3. Coil; 4. Inner stopper cover; 5. Rotary disc; 6. Oil layer; 7. Ball-seal bearing assembly

2.2 *Thermal Simulation Model of Magnetorheological Brake*

Altair Flux tackles the widest range of electromagnetic, electrical, and thermal simulations, seamlessly integrated with Altair's global solutions for multiphysics design exploration and optimization. It empowers innovators to fast-track the electrification of their devices while achieving higher performance levels, improved time-to-market, cost-effective solutions, and enhanced overall efficiency. In the event of thermal issues, it provides a virtual environment to analyze and address the problem. Temperature variations occur over time and with changes in input current.

2.2.1 Meshing

For this meshing model, it's crucial to carefully consider the mesh size to ensure accurate magnetic field interactions between different surfaces. The optimal size will depend on the desired level of detail. For faster meshing, use a value closer to 0, while a value near 1 indicates a sparser mesh. In this specific case, an average mesh size of 0.5 was chosen (Fig. 3 and Table 1).

2.2.2 Material

In this study, the selected materials are MRF-140CG magnetic fluid and AISI 1045 steel. Their relevant thermal properties, including specific heat capacity (C), heat capacity (C_p), and volumetric heat capacity (ρC_p), are determined based on

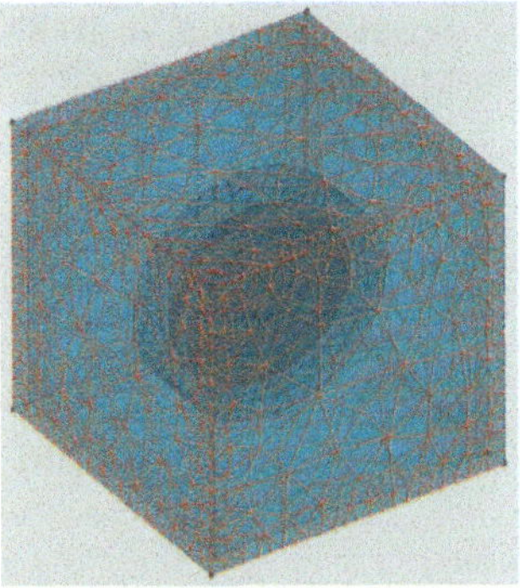

Fig. 3 Meshing the model

Table 1 Meshing parameter

Number of element	
Number of node	487,157
Number of line element	48,996
Number of surface element	433,402
Number of volume element	2,909,637
Element quality	
Number of parts not evaluated	0%
Part numbers of good quality	98.59%
Average number of quality elements	1.1%

the design dimensions and measurements provided by the manufacturer's material specifications.

- Steel AISI 1045:

  Specific heat: $C = 486(\text{J/kg.K})$.
  Mass: $m = 10.07$ kg.
  Volume: $1279 \times 10^6(\text{mm}^3) = 1279 \times 10^{-3}(\text{m}^3)$.
  Specific heat at constant pressure: C_p

$$C_p = C.m = 486 \times 10.07 = 4894\left(\frac{\text{J}}{\text{K}}\right)$$

  Volumetric heat capacity:

$$pCp = \frac{C_p}{V} = \frac{4894}{1279 \times 10^{-3}} = 3826442\left(\frac{\text{J}}{\text{m}^3\text{K}}\right)$$

- MRF140CG

Specific heat: $C = 1000(\mathrm{J/kg.K})$.
Mass: $m = 0.2897$ kg.
Volume: $8.1837 \times 10^5 \left(\mathrm{mm}^3\right) = 8.18 \times 10^{-4} \left(\mathrm{m}^3\right)$.
Specific heat at constant pressure: C_p

$$C_p = C.m = 1000 \times 0.2891 = 289.1 \left(\frac{\mathrm{J}}{\mathrm{K}}\right)$$

Volumetric heat capacity:

$$pC_p = C_p : V = 289.1 : \left(8.18 \times 10^{-4}\right) = 353423 \left(\frac{\mathrm{J}}{\mathrm{m^3K}}\right)$$

2.2.3 Boundary Conditions

Magnetorheological (MR) brakes are a modern type of brake that utilizes MR fluid as a medium for transmitting force. To accurately simulate the heat transfer process in MR brakes, proper definition of boundary conditions is crucial. This paper elaborates on both thermal and current boundary conditions applicable to MR brakes.

Thermal Boundary Conditions: Thermal boundary conditions are defined by the temperature values at the boundaries of the MR brake system. In this case, the boundaries include:

- Outer surface of the brake component: The temperature here depends on the surrounding ambient temperature.
- Inner surface of the brake component: The temperature here depends on the temperature of the MR fluid.

For the outer surface of brake components, the temperature is often determined by a statistical probability method. This method is based on measuring the ambient temperature at different times. The measurement results are used to build a probability distribution function of the ambient temperature. In the simulation problem, the ambient temperature at a specific time is determined by taking a random value from the probability distribution function. For the inner surface of brake components, the temperature of the magnetic fluid is determined by a method that combines analytical and experimental methods. The analytical method is used to calculate the temperature of the magnetic fluid under static conditions. The experimental method is used to calibrate the results of the analytical method.

Current boundary conditions: Current boundary conditions are defined by the current values flowing through the coils in the brake. In this case, the current flowing through the coils is determined by the following method:

- For the magnetic field analysis problem, the current values considered are in the range from 0 to 3 A, with a step of 0.25 A. To increase the magnetic field strength,

the two coils on the two sides will be supplied with currents in the opposite direction.
- For the thermal simulation problem, the current is determined according to the results of the magnetic field analysis problem. In the thermal simulation problem, the current is considered as an input parameter of the problem. The current value is used to calculate the magnetic field strength in the brake. The magnetic field strength is then used to calculate the temperature of the magnetic fluid.

Simulation time: The simulation time is usually determined by considering the time required for the temperature of the magnetic fluid to reach a steady state. In this case, the simulation time is about 60 s.

3 Results

Results of the MR brake thermal simulation show that when supplied with a current of 3A at a rotational speed of 1500 rpm, the temperature of the entire brake assembly, including the rotor and stator, increases gradually over time (from Figs. 4, 5, and 6). The highest temperature region is concentrated mainly in the center along the radial direction. The rotor temperature reaches a maximum value of 326 K at 60 s, while the stator temperature reaches a maximum value of 328^0 K at the same time (Fig. 4). Based on Fig. 4, we can see that (Fig. 7):

- The temperature of the entire MR brake increases gradually over time.
- The hottest region is concentrated mainly in the center along the radial direction.

The increase in temperature of the MR brake is due to the current flowing through the coils in the rotor and stator creating a magnetic field. This magnetic field acts on the iron particles in the magnetic fluid, causing the particles to move and create

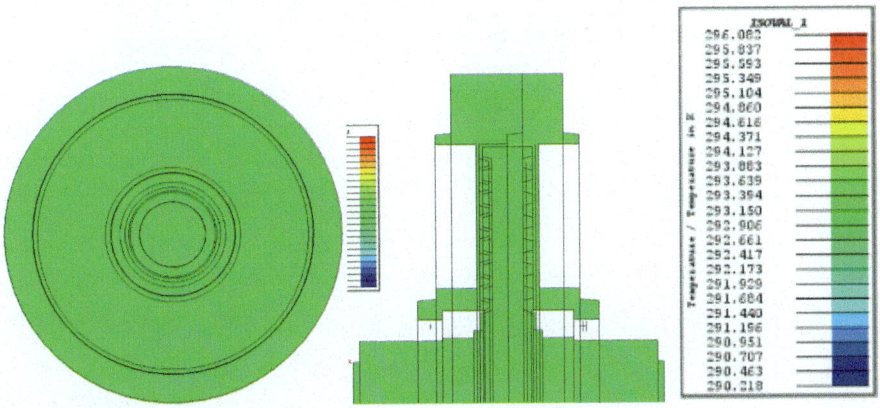

Fig. 4 Temperature at brake disc and oil layer in the time of 0 s

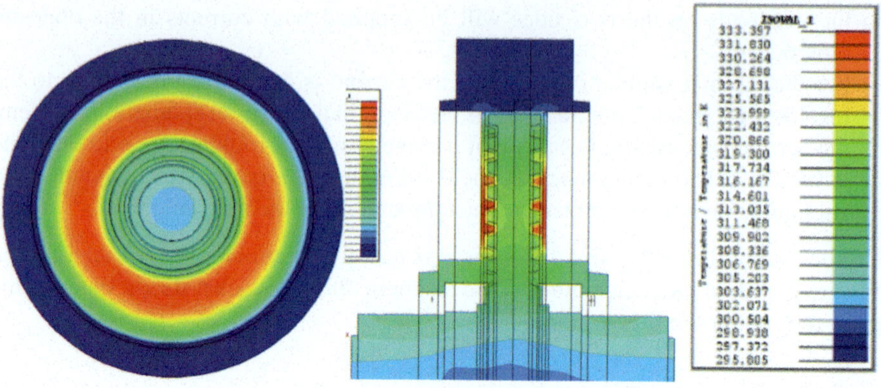

Fig. 5 Temperature at brake disc and oil layer in the time of 30 s

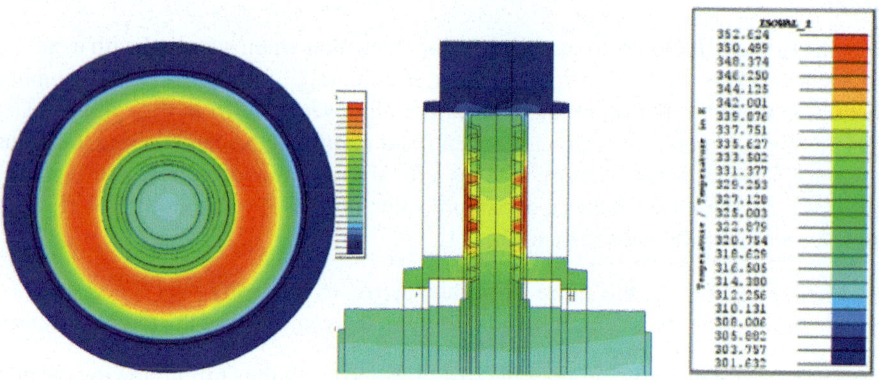

Fig. 6 Temperature at brake disc and oil layer in the time of 60 s

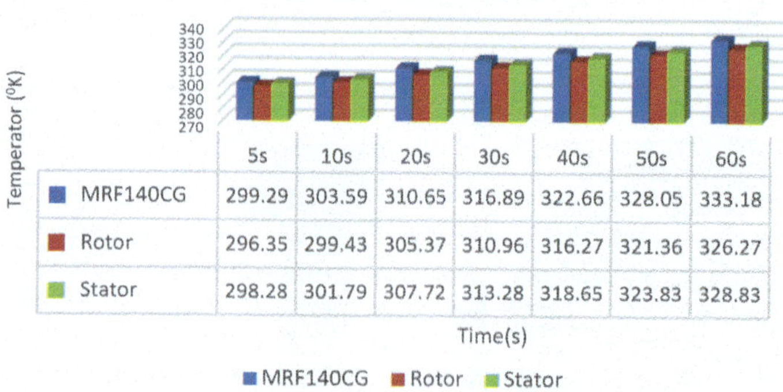

Fig. 7 Graph showing temperature change over time of MRB

friction. This friction generates heat, increasing the temperature of the MR brake. The hottest region is concentrated mainly in the center along the radial direction because the magnetic field is strongest in this region. A stronger magnetic field will create more friction, leading to higher temperatures. The rotor temperature is higher than the stator temperature because the rotor is in direct contact with the coils, where the magnetic field is strongest. The results of the MR brake thermal simulation show that the temperature of the MR brake increases gradually over time and reaches a maximum value at 60 s. The hottest region is concentrated mainly in the center along the radial direction. The rotor temperature is higher than the stator temperature.

4 Conclusion

The paper presents the initial results of a thermal analysis of a magnetically controlled brake (MRB) that is currently under development for automotive applications. The simulation results show that the MRB operates within the allowable temperature range of the material. These results can be used as a basis for the selection of MRB configurations for future automotive auxiliary braking systems. The simulation results are also the basis for the design and manufacturing of a complete brake model that can be used as an auxiliary braking system for the main automotive braking system. However, the issue of heat dissipation in the system needs to be addressed. This is because it is one of the factors that has a great impact on the power and efficiency of the braking system, especially when the MRB operates at high engine speeds in a vehicle.

Acknowledgements This research was supported and funded by Hanoi University of Science and Technology (HUST) under project number T2022-PC-035.

References

1. Jolly, M.R., Bender, J.W., Carison, J.D.: Properties and Applications of Commercial Magnetorheological Fluids (1998)
2. Wang, D., Hou, Y., Tian, Z., Meng, Q.: Temperature rise characteristic of MR fluid in a multi-disc MR clutch under slip condition. Industrial Lubrication and Tribology (2015)
3. Karakoc, K., Park, E.J., Suleman, A.: Design considerations for an automotive magnetorheological brake. Mechatronics **18**, 434–447 (2008)
4. Patila, S.R., Powarb, K.P., Sawanta, S.M.: Thermal analysis of magnetorheological brake for automotive application. Appl. Therm. Eng. https://doi.org/10.1016/j.applthermaleng.2015.11.128 (2015)
5. Hoang, Q.T., Bui, H.D., Tran, T.T.: Design and simulation of magnetorheological brake supporting downhill Van Truck. JST: Engineering and Technology for Sustainable Development (2023)
6. Rosensweig, R.E.: Heating magnetic fluid with alternating magnetic field. J. Magn. Magn. Mater. **252**, 370–374 (2002)

7. Tuấn, H.Q., Hoàng, T.M., Tùng, T.T., Ngọc, N.A.: Nghiên cứu ảnh hưởng của góc nghiêng răng đến mô men phanh sinh ra trên phanh dầu từ trường bằng phương pháp mô phỏng. Tạp chí Cơ khí Việt Nam, số 296, trang 696 (2022)
8. Tuan, H.Q., Hoang, T.M.: A study the effect of key parameters on the magneto-rheological brake torque based on the sim method. ARPN J. Eng. Appl. Sci. **18**(13), (2023). ISSN 1819-6608. https://doi.org/10.59018/0723192

Sliding Mode Control Using Genetic Algorithm for Twin Rotor MIMO System

Dinh-Hieu Phan, Duy-Minh Nguyen, Van-Truong Nguyen, and Van-Anh Nguyen

Abstract In this paper, the purpose of using genetic algorithms (GA) is to enhance the efficiency of sliding mode control (SMC) for a twin rotor system (TRMS). The proposed control strategy, named SMC-GA, combines the traditional sliding controller with GA. Initially, the SMC controller was designed using Lyapunov theory. Then, GA is employed to optimize the parameters of the SMC, ensuring that the TRMS achieves excellent performance. The simulation results demonstrate that the SMC-GA has ability to effectively track different reference trajectories of TRMS.

Keywords Sliding mode control (SMC) · Genetic algorithm (GA) · Pitch angle · Yaw angle · Control parameter

1 Introduction

The TRMS consists of a beam that rotates freely in horizontal planes (pitch angle) and in vertical planes (yaw angle) [1]. These angle are driven by DC motors, with the main rotor generating a lifting force for vertical motion, the tail rotor controlling left or right turning. The aerodynamic force is regulated by adjusting the rotor speeds. As a result, from a control engineering perspective, the TRMS is noticed as a complex, nonlinear system with uncertain parameters [2]. The control objective for the TRMS is to stabilize it under coupled conditions and guarantee precisely and rapid trajectory tracking. Various control strategies can be employed to achieve this goal. Sliding mode control (SMC) methods are particularly appealing in this context, as they enhance robustness and invariance of the system against matched uncertainties and disturbances. Several SMC schemes have been proposed in previous studies.

D.-H. Phan · D.-M. Nguyen · V.-T. Nguyen (✉)
Faculty of Mechatronics Engineering, Hanoi University of Industry, Hanoi, Vietnam
e-mail: nguyenvantruong@haui.edu.vn

V.-A. Nguyen
School of Aerospace, Transport and Manufacturing, Cranfield University, Bedford, UK

D. H. Tien et al. (eds.), *Proceedings of the 1st International Conference on Sustainability and Emerging Technologies for Smart Manufacturing*, Proceedings in Technology Transfer, https://doi.org/10.1007/978-981-97-7083-0_67

For example, in [3], the SMC using super-twisting algorithm was introduced to control the TRMS. Through simulation results, this controller demonstrated robustness against uncertainties and external disturbances, while it also reduces control input chattering. Mustafa and other researchers [4] compared adaptive sliding and adaptive integral sliding control for the TRMS system. Their proposed controller utilized an adaptive coefficient adjustment law to minimize chattering by accurately estimating the upper bound of uncertainty. In [5, 6], a terminal sliding control was proposed. The simulation results showcased the sustainability and effectiveness of this approach. In [7], the focus was on implementing a nonlinear observer to estimate the unmeasured TRMS variables. In [8], the authors explored a TRMS system model and designed an SMC controller without an observer to regulate the pitch and yaw angle. The results underscored the effectiveness of SMC compared to the classic PID control strategy.

In this paper, the SMC-GA control method is proposed for the nonlinear TRMS. The SMC-GA controller offers several benefits, including a straightforward construction, clear physical interpretations for each control parameter. The parameters of the SMC-GA controller are optimized using the GA [9]. To demonstrate its efficiency, the SMC-GA controller quality is compared to the traditional sliding controllers. The simulation results indicate that the SMC-GA controller outperforms the traditional sliding controller in terms of high tracking performance for the TRMS system.

2 TRMS Dynamics

The TRMS is designed for experimental purposes, this system includes the mechanical part and the electrical part. The mechanical part of TRMS includes two rotors (these are DC electric motors independently excited by the excitation source using permanent magnets) combined with a counterweight. Both the rotor and the counterweight are placed on the free-free beam. These parts are attached to the tower. The electrical part located under the tower plays a role in controlling TRMS. The electrical part measures the signals and transmits them to the PC. It also transmits control signals to the actuator. The physical structure of TRMS is described in Fig. 1.

This paper using the results of mathematical model in [9] is presented as:

$$\begin{bmatrix} J_1\cos^2\varphi_v + J_2\sin^2\varphi_v + h^2(m_{T_1}+m_{T_2}) & \\ \quad + J_3 \quad h(m_{T_1}l_{T_1}\sin\varphi_v - m_{T_2}l_{T_2}\cos\varphi_v) & \\ h(m_{T_1}l_{T_1}\sin\varphi_v - m_{T_2}l_{T_2}\cos\varphi_v) & J_1+J_2 \end{bmatrix}\begin{bmatrix} \ddot{\varphi}_h \\ \ddot{\varphi}_h \end{bmatrix} + \begin{bmatrix} h(m_{T_1}l_{T_1}\cos\varphi_v + m_{T_2}l_{T_2}\sin\varphi_v)\dot{\varphi}_v^2 \\ \quad + 2(J_2-J_1)\,\dot{\varphi}_v\,\dot{\varphi}_h\sin\varphi_v\cos\varphi_v \\ (J_1-J_2)\sin\varphi_v\cos\varphi_v\dot{\varphi}_h^2 + g(m_{T_1}l_{T_1}\cos\varphi_v + m_{T_2}l_{T_2}\sin\varphi_v) \end{bmatrix} = \begin{bmatrix} M_{ih} \\ M_{iv} \end{bmatrix} \quad (1)$$

Fig. 1 Physical structure of TRMS

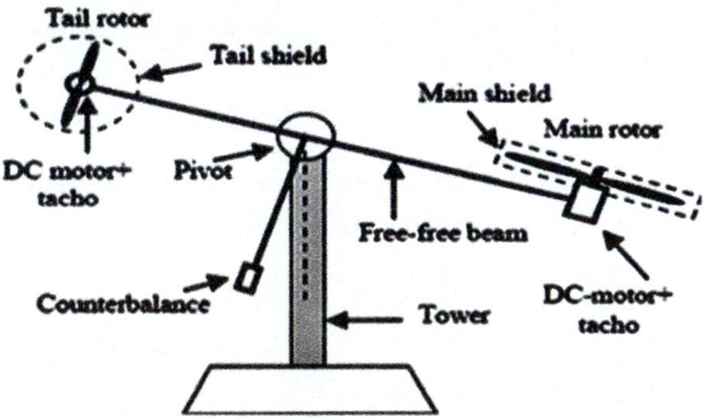

where φ_h, φ_v denote the yaw and pitch angle; m_{T_1}, m_{T_2} denote the mass total of counterweight bar and the mass total of free beam, respectively; l_{T_1}, l_{T_2} are the centroid free-free beam and counter weight bar; J_1, J_2, J_3 are the constant moments of inertia of free-free beam, counterweight bar, and rotary joint; M_{ih}, M_{iv} denote the total of torques.

Let variable $q = \begin{bmatrix} \varphi_h \\ \varphi_v \end{bmatrix}$, from Eq. (1), the general dynamic of TRMS is rewritten as:

$$M(q)\ddot{q} + C(q, \dot{q})\dot{q} + G(q) = \tau + n(t) \tag{2}$$

τ is the input torques; *n(t)* is the uncertain model and unwanted disturbance affecting the system.

3 SMC-GA Controller Design

3.1 SMC Controller

Equation (2) is rewritten as:

$$\ddot{q} = M(q)^{-1}(n(t) - C(q, \dot{q})\dot{q} - G(q)) + M(q)^{-1}\tau = f + gu \tag{3}$$

where $f = M(q)^{-1}(n(t) - C(q, \dot{q})\dot{q} - G(q))$; $g = M(q)^{-1}$; $\tau = u$

We define as: $x_1 = \varphi_h$; $x_2 = \dot{\varphi}_h$; $x_3 = \varphi_v$; $x_4 = \dot{\varphi}_v$; $q = \begin{bmatrix} x_1 \\ x_3 \end{bmatrix}$.

From the Eq. (3), the state equation can be expressed:

$$\begin{bmatrix} \ddot{x}_1 \\ \ddot{x}_3 \end{bmatrix} = \begin{bmatrix} f_1 \\ f_2 \end{bmatrix} + \begin{bmatrix} g_{11} & g_{12} \\ g_{21} & g_{22} \end{bmatrix} \begin{bmatrix} u_1 \\ u_2 \end{bmatrix} \tag{4}$$

The deviation of the yaw angle and pitch angle is modified as:

$$e_1 = x_1 - x_{1d};\ e_3 = x_3 - x_{3d};\ e = [e_1 \quad e_3]^T$$

where $x_{1d}; x_{3d}$ is the desired reference of yaw angle and pitch angle.
The sliding variables are selected as:

$$\begin{aligned} s_1 &= c_1 e_1 + c_2 \dot{e}_1 \\ s_2 &= c_3 e_3 + c_4 \dot{e}_3 \end{aligned} \tag{5}$$

Differentiating $s_1;\ s_2$ in Eq. (5) then combined with the Eq. (4), we have

$$\begin{aligned} \dot{s}_1 &= c_1\dot{e}_1 + c_2(f_1 + g_{11}u_1 + g_{12}u_2) - c_2\ddot{x}_{1d} \\ \dot{s}_2 &= c_3\dot{e}_3 + c_4(f_2 + g_{21}u_1 + g_{22}u_2) - c_4\ddot{x}_{3d} \end{aligned} \tag{6}$$

In order to ensure the TRMS stability, the sliding variable equations must be satisfied:

$$s_1\dot{s}_1 < 0 \ and \ s_2\dot{s}_{.2} < 0$$

Chosen $\dot{s}_1 = -\eta_1 \text{sgn}(s_1);\ \dot{s}_2 = -\eta_2 \text{sgn}(s_2)\ ;\ \eta_1 > 0;\ \eta_2 > 0;$ we have

$$\begin{aligned} s_1\dot{s}_1 &= -s_1\eta_1 \text{sgn}(s_1) = -\eta_1|s_1| < 0 \text{ and } s_2\dot{s}_2 = -s_2\eta_2 \text{sgn}(s_2) \\ &= -\eta_2|s_2| < 0 \end{aligned}$$

From Eq. (6), the control input of SMC can be calculated:

$$\begin{aligned} u_1 &= \frac{(\eta_2 \text{sgn}(s_2) + c_3\dot{e}_3 + c_4 f_2)c_2 g_{22} - (\eta_1 \text{sgn}(s_1) + c_1\dot{e}_1 + c_2 f_1)c_4 g_{22}}{c_2 c_4 g_{11} g_{22} - c_2 c_4 g_{21} g_{12}} \\ u_2 &= \frac{(\eta_1 \text{sgn}(s_1) + c_1\dot{e}_1 + c_2 f_1)c_4 g_{21} - (\eta_2 \text{sgn}(s_2) + c_3\dot{e}_3 + c_4 f_2)c_2 g_{11}}{c_2 c_4 g_{11} g_{22} - c_2 c_4 g_{21} g_{12}} \end{aligned} \tag{7}$$

$c_1, c_2, c_3, c_4, \mu_1, \mu_2$.are the parameters of the SMC controller. It is not easy to determine accurate parameters. To overcome this problems, the GA-SMC controller is used. The SMC controller parameters are determined by the GA.

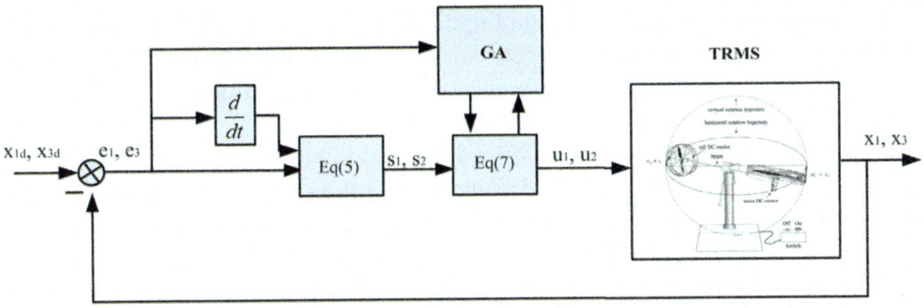

Fig. 2 SMC-GA control system diagram for TRMS

3.2 Parameters Optimization

To improve control quality of TRMS, the selection of appropriate $c_1, c_2, c_3, c_4, \mu_1, \mu_2$ parameters of the SMC controller is extremely important. These parameters are optimally selected by using the GA. The GA parameters can be selected as: $J = \min(e_1^2 + e_2^2)$; max_gen = 150; pop_size = 16; npar = 6; cross_prob = 0.7; mutate_prob = 0.3. The SMC-GA controller diagram for the TRMS is shown in Fig. 2.

4 Simulation Results

Matlab/Simulinks software is used to simulate the TRMS control system. The TRMS dynamics parameters are given in below Table 1.

For the SMC controller, the parameters are chosen experimentally $c_1 = 4, c_2 = 2, c_3 = 6, c_4 = 2.5, \mu_1 = 30, \mu_2 = 40$; For the SMC-GA controller, the parameters are optimized by $J = 24.496$ and SMC-GA controller parameters are optimized $c_1 = 19.789, c_2 = 15.625, c_3 = 40.467, c_4 = 11.789, \mu_1 = 26.474, \mu_2 = 46.847$.

Table 1 Physical parameters of TRMS

Symbol	Definition	Value
m_{T_1}	Mass total of free beam	0.825 kg
m_{T_2}	Mass total of counterweight bar	0.0908 kg
l_{T_1}	Centroid free–free beam	0.0186 m
l_{T_2}	Centroid counter weight bar	0.2443 m
J_1	Moments of inertia of free–free beam	0.0591 kgm^2
J_2	Moments of inertia of counterweight bar	0.0059 kgm^2
J_3	Moments of inertia of rotary joint	0.0000168 kgm^2
H	The length of rotary joint	0.06 m

We choose fix-step time with sampling time $\Delta t_c = 10$ ms in simulink to simulate the TRMS. There are two source types which generate to the TRMS. The first source type includes a sinusoidal wave reference with 1 Hz, amplitude value of 1 and 0.6. The second source type is related to tracking the step function with final value of 1 and 0.6. The simulation results could be described in the below figures.

For the sinusoidal wave reference source, Fig. 3 shows actual yaw and pitch angle tracking more closely their reference using GA-SMC controller. However, it exists the small error between their actual and reference in Fig. 4 using SMC controller. So, it is easily realized that the yaw and pitch angle of TRMS using GA-SMC has better tracking than that using SMC controller. For the step function sourse, the step response of yaw and pitch angle is indicated in Figs. 5 and 6. It is easy to see that the errors of actual and reference angle in Fig. 5 tend to zero while their errors in Fig. 6 do not tend to zero.

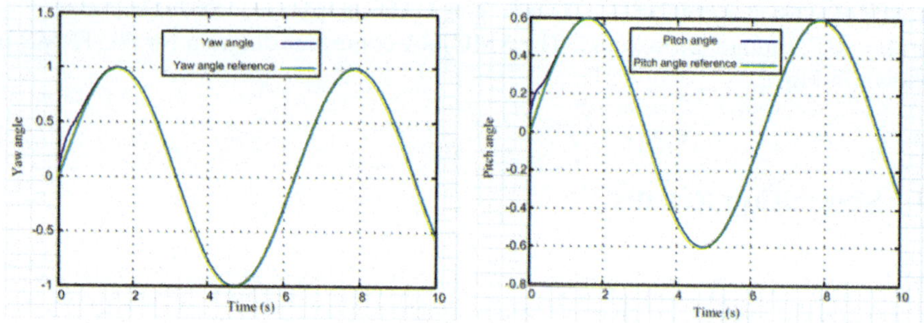

(a) (b)

Fig. 3 Sine wave response for GA-SMC control

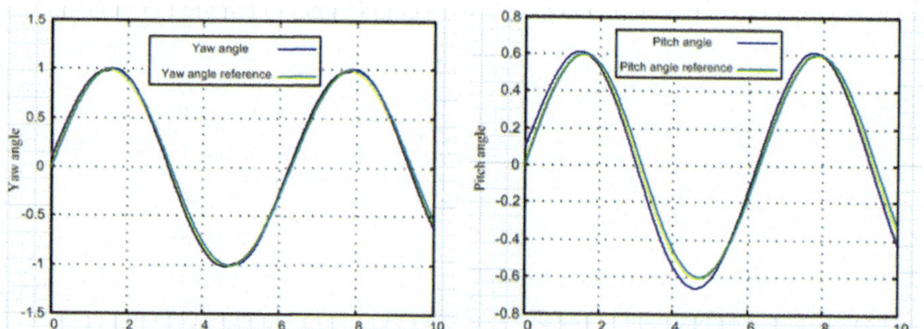

(a) (b)

Fig. 4 Sine wave response for the SMC control

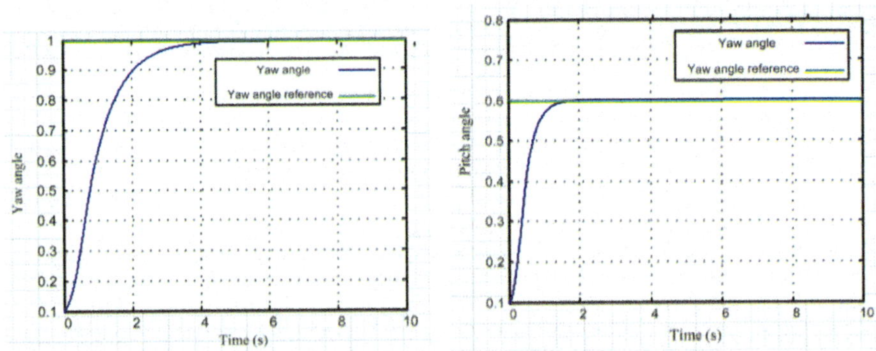

(a) (b)

Fig. 5 Step response for the GA-SMC control

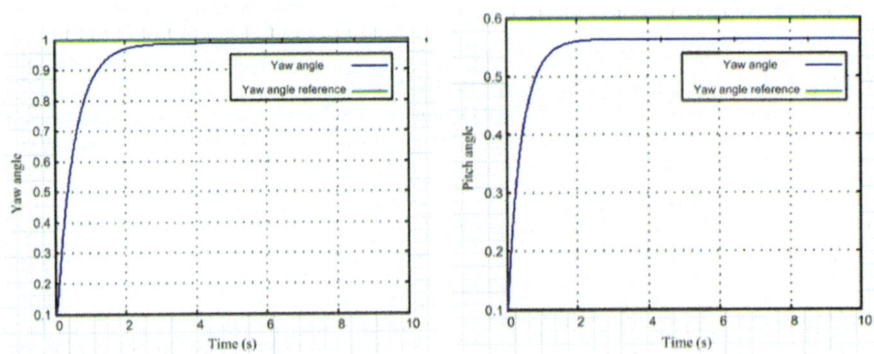

(a) (b)

Fig. 6 Step response for the SMC control

5 Conclusion

This paper has addressed the SMC-GA controller design method for the TRMS. The SMC-GA controller is designed based on traditional SMC controller. SMC-GA control stradegy advantage is that the SMC controller parameters are optimally chosen by using the GA. The parameters c_1, c_2, c_3, c_4, μ_1, μ_2 of the proposed controller are optimally determined in this paper. The simulation results show that the SMC-GA controller has better tracking performance than traditional SMC controllers.

Acknowledgements This research was funded by Vingroup Innovation Foundation (VINIF) under project code VINIF.2023.DA089.

References

1. Juang, J.G., Huang, M.T., Liu, W.L.: PID control using presearched genetic algorithms for a MIMO system. IEEE Trans. Syst. Man Cybern. Part C Appl. Rev. **38**(5), 716–727 (2008)
2. Wen, P., Lu, T.W.: Decoupling control of a twin rotor MIMO system using robust deadbeat control technique. IET Control Theory Appl. **2**(11), 999–1007 (2008)
3. Mondal, S., Mahanta, C.: Second order sliding mode controller for twin rotor MIMO system. India Conference (INDICON), pp 1–5. 2011 Annual IEEE (2011)
4. Mustafa, S., Khan, Q., Khan, I.: Comparative analysis of robust and adaptive control strategies for twin rotor MIMO system. 13th International Conference on Emerging Technologies (ICET), pp. 1–6 (2017)
5. Nguyen, V.T., et al.: Adaptive nonsingular fast terminal sliding mode tracking control for parallel manipulators with uncertainties. 2019 IEEE International Conference on System Science and Engineering, pp. 522–525 (2019)
6. Nguyen, V.T.: Non-negative adaptive mechanism-based sliding mode control for parallel manipulators with uncertainties. Comput., Mater. Continua **74**(2), (2023)
7. Shaikh, P.U., Karvekar, S.: Implementation of sliding mode controller for twin rotor multi input multi output system without state observer. 1st IEEE International Conference on Power Electronics, pp. 1–6. Intelligent Control and Energy Systems (2016)
8. Katoch, S., Chauhan, S.S., Kumar, V.: A review on genetic algorithm: past, present, and future. Multimed Tools Appl. **80**(5), 8091–8126 (2020)
9. Rahideh, A., Shaheed, M.H.: Mathematical dynamic modelling of a twin rotor multiple input-multiple output system. Proc. Inst. Mech. Eng. Part I—J. Syst. Control Eng. 89–101 (2007)

Multi-services Digital Twin for Modular Production System Based on ISO 23247 and Web Server

Giang-Nam Le, Thi Tra Giang Nguyen, Viet-Dzung Tran, Ngoc-Truong Nguyen, Xuan-Hoang Tang, and Thien-Hue Hoang

Abstract In the context of Industry 4.0, Digital Twin is representative of real-world system in the digital world, which relies on physical system data for optimization, manipulation, discovery and interaction with actual system. This paper presents the process of developing and implementing the Digital Twin architecture applied on a modular production system (MPS). The proposed architecture is based on the current Digital Twin framework for manufacturing normalized by ISO 23247. The Digital Twin architecture shows an approach between web services, databases, and key observable production elements to create a bi-directional connection and multi-services Digital Twin system. This approach has been validated on real systems and can be applied to other machines and manufacturing processes.

Keywords Digital twin · Behavior monitoring · ISO 23247 · MPS

1 Introduction

The concept of a Digital Twin (DT) entails creating a digital counterpart that mirrors a physical entity in behavior and appearance, enabling tailored services for monitoring, maintenance, management, optimization, and safety. The concept of the DT gained widespread recognition after its inclusion in NASA's initial draft of the technology roadmap [1]. Since then, numerous research projects across various fields have explored the application of DT technology, with successful instances such as Anna Billey and Thorsten Wuest's reduction of energy consumption in a heat tunnel by up to 40% [2]. There have been few attempts to use DT technology for aquaculture applications [3]. And the DT model also promises to revolutionize biodiversity research [4]. With Industry 4.0's emergence, manufacturing and smart products take center stage, exemplified by proposals like Wilma and her coworker's DT for composite assembly optimization [5]. There is also a DT-based lean design method

G.-N. Le (✉) · T. T. G. Nguyen · V.-D. Tran · N.-T. Nguyen · X.-H. Tang · T.-H. Hoang
Hanoi University of Science and Technology, Hanoi, Vietnam
e-mail: nam.legiang@hust.edu.vn

D. H. Tien et al. (eds.), *Proceedings of the 1st International Conference on Sustainability and Emerging Technologies for Smart Manufacturing*, Proceedings in Technology Transfer, https://doi.org/10.1007/978-981-97-7083-0_68

for Computer Numerical Control Machine Tool [6] and a DT model enabling real-time monitoring and evaluation of service status in intelligent PCB kit-box build assembly process [7]. So, DT offers numerous benefits and has been widely applied across manufacturing and other fields.

Tao et al. [8] outline the purposes of DT services and classify DTs based on implementation features [8]. And an important feature of DT is how the virtual system interacts with the physical system. However, according to [9], two disadvantages of DT in 2019 were the limited quantity of services offered on DT platforms and the lack of bi-directional connectivity between the virtual and physical systems. In 2022, Psarommatis and May published a review of 128 papers on DT and zero-defect manufacturing, highlighting shortcomings such as the absence of standardized processes and Key Performance Indicators (KPIs) for DT development, emphasizing the need for standardized methods and architectures for DT [10].

ISO 23247 addresses these challenges by issuing a series of standards that define a DT framework for manufacturing. ISO 23247 partitions a DT system into layers defined by standards, which are (1) Physical components in real space, (2) Communication devices, (3) Digital copies of real models in space virtual, and (4) User components.

The article proposes an innovative implementation that integrates DT technology with a monitoring control system using Webserver technology, based on the standardized framework for DT in manufacturing (ISO 23247). Conducted at VIMES lab, Hanoi University of Science and Technology, the project successfully built a DT for a modular flexible production system (MPS) known as the "Supply, check, and store" system, and monitored this system. The implementation is based on the integration of component DTs on functional software, SIL solutions (software in the loop), and a web server. It also involves real-time monitoring of sensor data and actuator status through a Web Client, alongside the creation of a synchronized 3D model using the three.js library. Data collected is stored for analyzing past system operations, aiding in problem identification.

2 Theoretical Basis

The ISO 23247 standard, developed by ISO committee TC 184/SC 4 and published in 2021, serves as a comprehensive guide for standardizing Digital Twin implementation architecture in industrial production applications within the context of Industry 4.0, specifying protocols for system interactions and data integration.

ISO 23247 outlines a Digital Twin system across four layers: the base layer encompasses tangible manufacturing components, the second layer involves device communication entities, the third layer integrates spatial, functional, and logical data management functions, and the fourth layer encompasses user, application, and system entities utilizing services from the Digital Twin entity in the third layer (as Fig. 1).

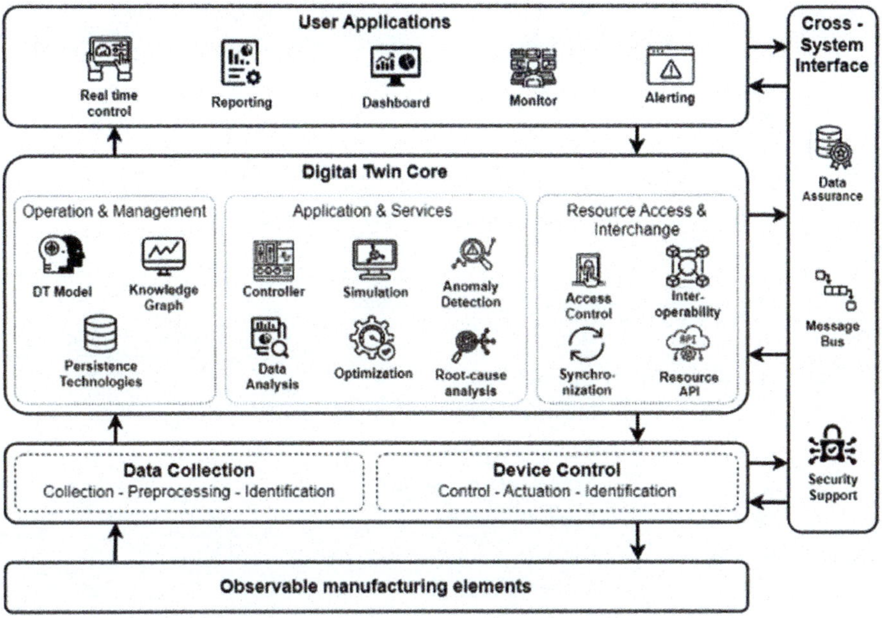

Fig. 1 Concept of DT for manufacturing

3 Environments of Micro-smart Factory for Validation

The study utilizes the MPS at VIMES Lab, Hanoi University of Science and Technology, featuring three stations: supply, inspection, and storage (as Fig. 2), each equipped with field devices including data collection tools and actuators. Workpieces are categorized by height and color, with two options 24 mm and 25 mm in red, silver (metallic), and black (as Fig. 3 and Table 1).

From left to right (as Fig. 3), the first image depicts the types of workpieces, followed by the workpiece feeding station, which is responsible for transferring the workpieces to the next station. This is achieved through two cylinders: one pushes the bar from the feeding tube to the waiting position, while the other is a rotary cylinder

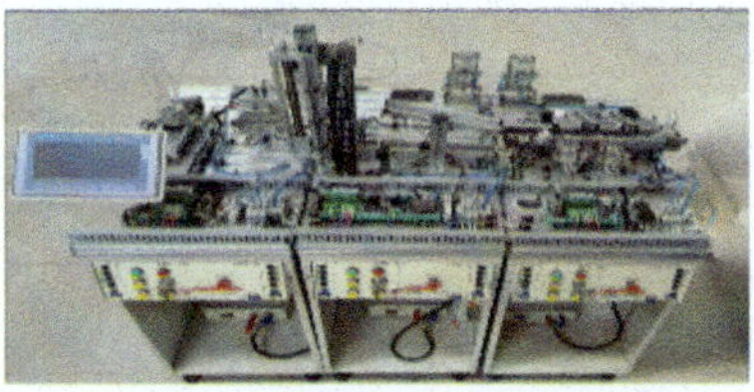

(a) Fixture model

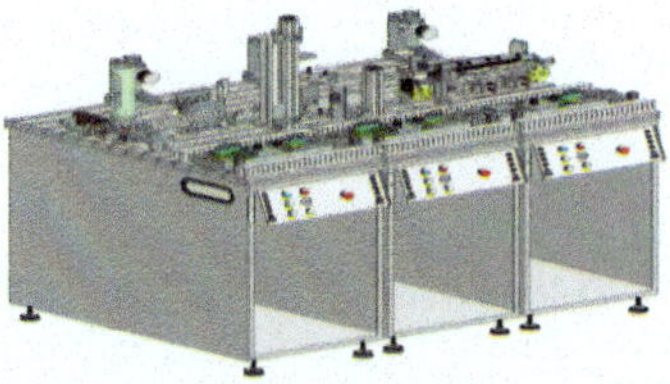

(b) 3D model

Fig. 2 System at VIMES lab and DT entity of production system

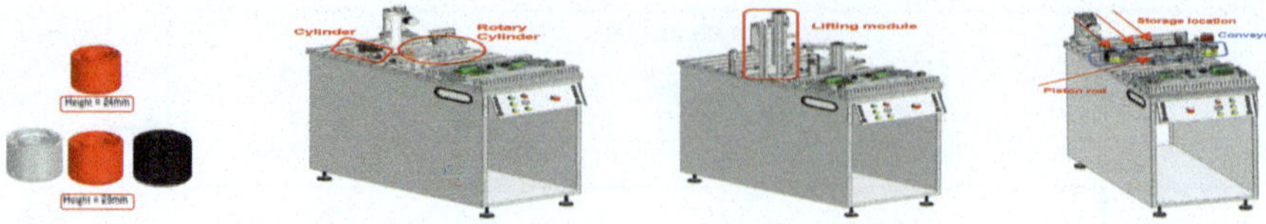

(a) Types of workpieces (b) Workpiece feeding station (c) Inspection station (d) Storage station

Fig. 3 Base models of the system DT entity produced at VIMES lab

Table 1 Data collection equipment and actuator structure

Type	Item	Station 1	Station 2	Station 3
Data collection equipment	Optical sensor	X	X	X
	Inductive sensor	X	X	X
	Small basic switch	X		
	Digital pressure Switch	X		
	Capacitive sensor		X	
Actuator structure	Cylinder	X	X	X
	Vacuum ejector	X		
	Solenoid valve	X	X	X
	DC motor			X

that rotates to advance the workpiece to the next station. The second picture is of the inspection station. At this station, the workpiece is brought to the height measurement position by a shaftless cylinder, and then the workpiece that reaches the height and color will be brought to the next station. Finally, picture three shows station number three, the storage station. After station two, products proceed to storage locations on the conveyor belt, passing through metal and color detection sensors. Each storage position is equipped with a piston rod. Black products are directed to the first storage position, metal products to the second, and red or white products to the third.

4 Digital Twin Architecture Proposal Based on ISO 23247

See Fig. 4.

4.1 Observable Manufacturing Entity–OME

Digital Twin Architecture based on ISO 23247 can be depicted by Fig. 4. At the bottom layer, the manufacturing domain comprises sensors, actuators, and SIMATIC

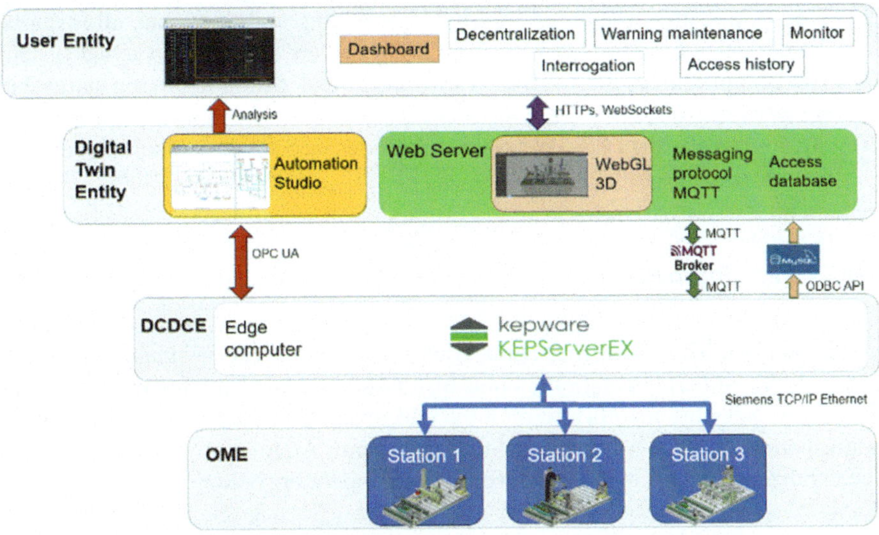

Fig. 4 DT architecture based on ISO 23247

S7-1200 controllers at production stations. These controllers use Siemens TCP/IP via KEPServerEX (DCDCE layer) for data exchange with applications.

4.2 Device Control and Data Collection Entity–DCDCE

In the second architectural layer, the Edge Computer, integrated with KEPServerEX, collects data from field PLCs via Siemens Ethernet TCP/IP. KEPServerEX acts as an OPC-UA Server, ensuring connectivity in line with ISO 23247 standards. The system employs MQTT, OPC-UA, and ODBC API for efficient communication and storage.

4.3 Digital Twin Domain

The third layer encompasses the DT domain, serving as the system's core. It offers various application services such as monitoring, alerting, 3D visualization (simulation) of production stations on a web platform, and historical data retrieval. These services are backed by four distinct data streams representing:

(1) Stream based on OPC-UA and Automation Studio software allows analysis and monitoring of the system pneumatic model.

(2) The second stream is based on MQTT providing real-time data, allowing bi-directional connection between from DT based on webserver to actual system.
(3) The stream utilizes the ODBC API to access real station operating status data stored in the MySQL Cloud Service, enabling historical monitoring services.
(4) Stream based on HTTPs, WebSocket provides web services (station operation, historical monitoring).

4.4 User Domain

In this domain, we offer a web application service with features including system status, dashboard, 3D operations, status history tracking, and maintenance information. To address the inability to monitor real object properties, we utilize simulation capabilities from Automation Studio software. Additionally, we implement a decentralized function with three levels: admin, operator, visitor.

5 System Setup

The production station system connects to an Edge Computer running Windows 10 via Ethernet using KEPServerEX's IoT Gateway and communication driver. Communication driver with the Siemens TCP/IP Ethernet driver for two-way communication. The system operates within a LAN network. Data updates occur every 100 ms, with customizable sampling frequencies of up to 10 ms, depending on the capabilities of the Edge Computer. Application streams depicted in Fig. 5, numbered 1 through 4.

Stream 1: Stream Based on OPC-UA and Automation Studio Software Allowing Analysis and Monitoring of the System Semantic Model.

Automation Studio serves as an integrated solution for developing and deploying complex automation applications across various industries. System components like cylinders, valves, motors, etc., are created and inherited from the research and development stage. KEPServerEX acts as an OPC-UA Server, with Automation Studio serving as the OPC-UA Client, facilitating communication via AS's "Communication Manager" tool. 2D characteristic settings include descriptions of system elements (e.g., cylinder diameter, length, control valve type, and system pressure configuration in AS to closely mimic real-world conditions. The resulting 2D model can be run independently or in parallel to analyze ongoing energy flow in the physical system (Fig. 6).

Stream 2: Based on MQTT Providing Real-Time Data to the DT Entity.

This stream enables real-time communication between the Edge Computer and the user entity layer using MQTT protocol via Keepware's IoT Gateway Plug-In. MQTT facilitates efficient data transmission through a lightweight publisher/subscriber

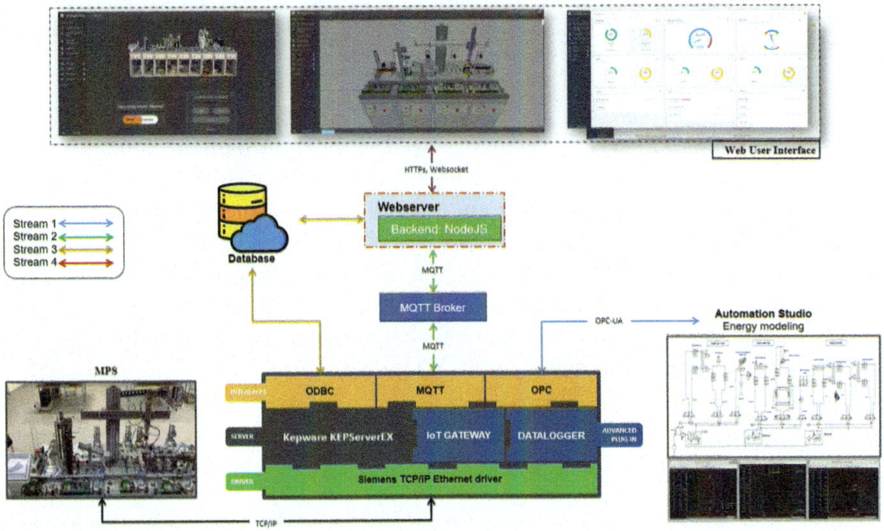

Fig. 5 Base models of the DT entity produced at VIMES lab

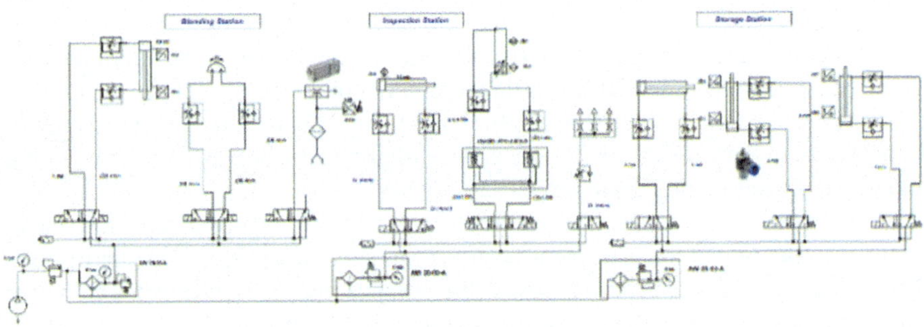

Fig. 6 Pneumatic circuit built in automation studio

architecture, with field data formatted according to Keepware's IOT gateway plug-in and sent to an MQTT Broker. Data transmission occurs at 50ms intervals, with TLS certificates securing connections for MQTT Cloud access. This setup allows other models to access real-time data with provided credentials.

Stream 3: Machine Operating Status Data is Stored at MySQL Cloud Service.

This stream stores operational status data in the cloud using Amazon RDS for MySQL. KeepwareEX encapsulates the data with the Data Logger Plug-In and connects to Amazon RDS using the ODBC API. Configuration involves setting up a Data Source (DSN) and installing the MySQL ODBC driver. ODBC acts as an intermediary layer between the application and database, allowing for easy database switching without modifying the application. Status variables updated every 100 ms are selectively stored in the cloud, mapped into database columns, including: "id, name, value, time_stamp, and quality".

Stream 4: Web Service Provision Stream.

This stream provides a service system on a web application, built on Node.js with the Express framework. Real-time activity data is received via MQTT broker subscriptions using mqtt.js, and websockets are utilized for real-time communication with clients via socket.io. The web client recreates system behavior through a 3D model acting as a DT, developed with three.js and WebGL technology. The 3D model is converted to "*.glb" format for efficient uploading, and mysql.js facilitates retrieval of activity data from MySQL Cloud. Users can monitor data history through tables or simulate operational processes on the 3D model.

6 Result

6.1 Analysis Based on the Digital Twin of the Pneumatic System

In Fig. 7a, within the system analyzed with Automation Software, the operating cycle of the MY1M20-200L-M9BL3 shaftless cylinder commences with the reception of the workpiece from station 1 and the measurement of its height. This measurement forms the basis for evaluating the consistency of properties between the actual model and the deployed Digital Twin model.

The evaluation revealed that the average cylinder transport speed from the starting position to the height measurement position was 75 mm/s over a 200 mm distance. According to Fig. 8, the cylinder moved this distance in 2.7s, resulting in a calculated velocity of 74.1 mm/s and an error of 1.2%.

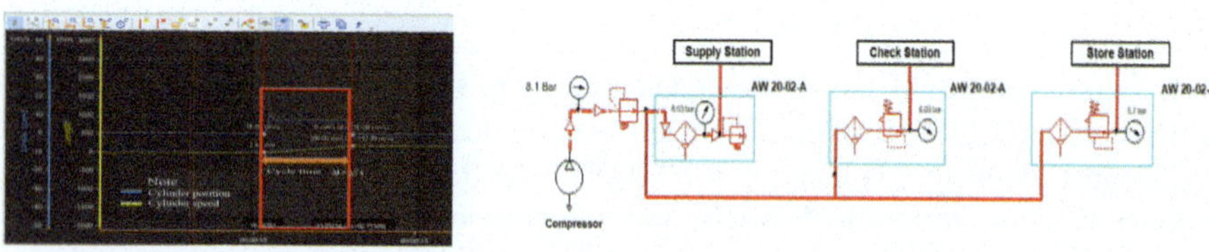

(a) Graph of cylinder operation in AS (b) Pressure regulator valve obtained from digital copy

Fig. 7 Pressure regulator valve obtained from digital copy in automation studio

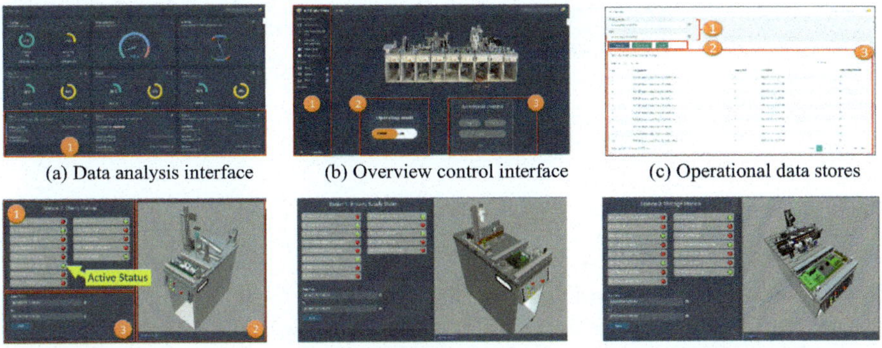

(a) Data analysis interface (b) Overview control interface (c) Operational data stores

(d) Station 1 UI (e) Station 2 UI (f) Station 3 UI

Fig. 8 Interfaces for the service system are available within the web application

6.2 Evaluate the Energy Model in the System

The real system parameters include P_{source} at 8.5 bar for the air compressor, P_1 at 6 bars for the bar feeding station, P_2 at 5 bars for the workpiece inspection station, and P_3 at 6 bar for the workpiece storage station (see Fig. 7b).

Parameters for the digital copy are created using Automation Studio software: P_{source} at 8.1 bar, P_1 at 6.4 bar, P_2 at 5.33 bar, and P_3 at 5.75 bar.

The errors of P_{source}, P_1, P_2, P_3 are 5.1%, 1.1%, 5.5%, and 5.3%, respectively. Through that, the team has obtained the total system pressure error within less than 5% of the real system error, confirming the correctness, accuracy, and fidelity of the digital copy of the energy model built by the team on Automation Studio.

6.3 Digital Twin Base on Webserver Technology

The DT base on webserver technology can be accessed remotely online because the physical and DT system is deployed as network resources. Below are the results of building services and features of the web application:

System control: Users can control system operations by selecting modes like automatic or manual (no. 2 of Fig. 8b) and performing actions such as initiating, halting, or resetting the system (no. 3 of Fig. 8b).

Monitoring operational behavior: Interface (no. 1 of Fig. 8d) enables users to monitor sensor and actuator signals, with green denoting activation and red denoting deactivation. Users can monitor the system's behavior via 3D models, providing real-time simulation of its dynamics, as illustrated in (no. 2 of Fig. 8d).

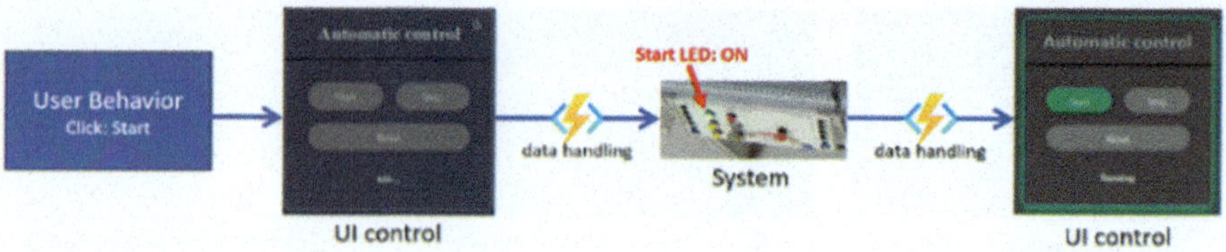

Fig. 9 System's initiating operation flow on the web interface

Role-Based Access Control: Admin role, responsible for managing and operating the entire system. Operator role, restricted to operating and monitoring a specific station. Visitor role, only has the right to observe the entire system's operation.

Database Monitoring: Users can easily monitor the operational status of sensors and the system's executing mechanism through the interface (Fig. 8c). No. 3 of Fig. 8c corresponds to the corresponding activity data within the specified time interval.

Operational Analytics: The data analysis interface (Fig. 8a) provides insights into the system's energy consumption, oil levels in tanks, temperature, and humidity. The interface displays maintenance parameter values (no. 1 of Fig. 8a) that are calculated and directly retrieved from the stored database (Fig. 9).

7 Conclusion

Our work exemplifies an application within the manufacturing domain, bearing relevance to the evolving landscape of Industry 4.0 and the Internet of Things, aligning with newly established standards for DT architecture development. The visualization capabilities offered through the 3D simulation interface, dashboard, and DT of each component present a comprehensive overview of machine states, thereby facilitating enhanced asset management efficiency for process supervisors. Furthermore, the integration of a DT database on a web server enriches the user domain with multifaceted services, including remote control functionalities, behavior monitoring through 3D models, sensor data analysis, access to historical records, and the ability to replay past operational behaviors. Our research successfully realizes a DT conforming to the ISO 23247 architecture, featuring bi-directional connections and a diverse range of services.

References

1. Shafto, M., Conroy, M., Doyle, R., Glaessgen, E., Kemp, C., LeMoigne, J., Wang, L.: Draft modeling, simulation, information technology and processing roadmap. Technol. Area **11**, 1–32 (2010)

2. Billey, A., Wuest, T.: Energy digital twins in smart manufacturing systems: a case study. Robot. Comput.-Integr. Manuf. **88**, 12–23 (2024)
3. Føre, M., Alver, M.O., Alfredsen, J.A., Rasheed, A., Hukkelås, T., Bjelland, H.V., Su, B., Ohrem, S.J., Kelasidi, E., Norton, T., Papandroulakis, N.: Digital Twins in intensive aquaculture—challenges, opportunities and future prospects. Comput. Electron. Agric. **218**, 108676 (2024)
4. Trantas, A., Plug, R., Pileggi, P., Lazovik, E.: Digital twin challenges in biodiversity modelling. Ecol. Inform. 102357 (2023)
5. Polini, W., Corrado, A.: Digital twin of composite assembly manufacturing process. Int. J. Prod. Res. **58**, 5238–5252 (2020)
6. Sheng, J., Zhang, Q., Li, H., Shen, S., Ming, R., Jiang, J., Li, Q., Guoxiong, S., Sun, B., Wang, J., Yang, J., Huang, C.: Digital twin-driven manufacturing equipment development. Robot. Comput.-Integr. Manuf. **83**, 102557 (2023)
7. Wei, Y., Tianliang, H., Dong, L., Ma, S.: Digital twin driven intelligent manufacturing for FPCB etching production line. Comput. Ind. Eng. **186**, 109763 (2023)
8. Tao, F., Zhang, H., Liu, A., Nee, A.Y.C.: Digital twin in industry: state-of-the-art. IEEE Trans. Ind. Inf. **4**, 2405–2415 (2018)
9. Cimino, C., Negri, E., Fumagalli, L.: Review of digital twin applications in manufacturing. Comput. Ind. **113**, 103130 (2019)
10. Psarommatis, F., May, G., Dreyfus, P.-A., Kiritsis, D.: Zero defect manufacturing: state-of-the-art review, shortcomings and future directions in research. Int. J. Prod. Res. **58**, 1–17 (2020)

Shortest Path Planning for Rectangular Holonomic Omnidirectional Mobile Robot Using Improved PRM Algorithm

Li Jie Yew, Muhammad Juhairi Aziz Safar, Khairul Salleh Basaruddin, Mohd Hanafi Mat Som, and Muhamad Khairul Ali Hassan

Abstract An improved and modified Probabilistic Roadmap (PRM) algorithm for non-circular holonomic mobile robot is proposed based on the consideration the shape and its kinematic constraint of non-circular holonomic mobile robot while moving from the initial point to the target point, especially to pass through the narrow passage. The path planning algorithm is also being modified to generate a shorter path distance in the shortest duration of time with the least number of turning points. The comparison of the conventional path planning algorithm will be carried out to determine the most optimal algorithm for modification to achieve the objectives. Then, the chosen path planning algorithm is PRM are improved by obstacle expansion and automatic selection for region of interest (AutoROI). Software simulation and analysis are performed on three different environment maps by a single holonomic mobile robot. The analysis of simulation revealed the performance of the modified path planning algorithm able to reduce elapsed time of PRM and increase the percentage of success path planning.

Keywords Path planning · Probabilistic roadmap (PRM) · Region of interest

L. J. Yew · M. J. A. Safar (✉) · M. K. Ali Hassan
Faculty of Electrical Engineering and Technology, Universiti Malaysia Perlis, Arau, Malaysia
e-mail: juhairi@unimap.edu.my

M. J. A. Safar
Centre of Excellence for Intelligent Robotics and Autonomous Systems (CIRAS), Universiti Malaysia Perlis, Arau, Malaysia

K. S. Basaruddin
Faculty of Mechanical Engineering and Technology, Universiti Malaysia Perlis, Arau, Malaysia

M. H. Mat Som
Faculty of Electronic Engineering and Technology, Universiti Malaysia Perlis, Arau, Malaysia

D. H. Tien et al. (eds.), *Proceedings of the 1st International Conference on Sustainability and Emerging Technologies for Smart Manufacturing*, Proceedings in Technology Transfer, https://doi.org/10.1007/978-981-97-7083-0_69

1 Introduction

Autonomous robot technology has become a popular emerging application in nowadays robotics and automation field. There are two types of mobile robots which are holonomic and non-holonomic. A holonomic mobile robot is described as a mobile robot without any kinematic constraints and capable to move at any arbitrary direction with three degrees of freedom (DOF) capability without any kinematical restriction, therefore the degree of mobility, δ_m must equal to the total DOF. If δ_m m is lesser than DOF, the mobile robot is considered as a non-holonomic system [1]. The value of δ is defined as $\delta_m = 3 - \delta_s$ where δ_s is the degree of steerability which corresponds to the number of independently steerable wheels. A holonomic kinematic constraint can be specified as a particular function of position variables only [2]. The mobile robot can move in two independent translational degrees of freedom and one rotational degree of freedom, overall, it is three degrees of motion on a planar surface.

Path planning is one of the challenges for mobile robots instead of localization and motion control, the function is to enable the selection and identification of a suitable path for the robot to traverse in the workspace [3]. The robot must know the goal position in advance to find a path for the mobile robot and to plan an appropriate addressing scheme to travel [4]. There are two branches of path planning which are local and global path planning. Global path planning is based on a mapping of a completely known environment [5]. Global path planning method normally develops a low-resolution advance path based on a recognized environmental map or its current and past sensory information of the environment [6]. In another words, it is an offline navigation, which is not suitable to react in unknown and dynamic environment. Contrarily, local path planning creates a high-resolution but low-level path over a portion of global path based on instruction from an on-board sensor. It is an online navigation and able to generate path without prior information of the environment.

However, there are several limitations of path planning. Most of the path planning algorithms assume the mobile robot is in a circular point, even the mobile robot is in a non-circular shape. The algorithm bias to select a longer distance and wider path to reach the destination. Classical approach path planning algorithms are computationally intensiveness and unable to cope with the uncertainty [7]. However, the bottleneck for robots to react fast is the classical approach requires a longer time to discover free space in a large and high-resolution map. The generated path also consists of several turning points that require a few milliseconds in motion control.

The objective in this research is to investigate the application of image processing in path planning, to introduce a modified path planning algorithm with the consideration of kinematics constraint due to the shape of holonomic mobile robot and to evaluate the performance of the proposed method through computer simulation.

The simulation done in [7] is to compare different path planning algorithm which are A* algorithm, Rapidly exploring Random Tree (RRT), bi-directional RRT (bRRT), Probabilistic Roadmap (PRM) and Genetic algorithm shows that PRM used shortest time to accomplish the shortest length of path among all the algorithm. The

A* algorithm can obtain the shortest distance among all the algorithms, but consumed most time compared to others. PRM used the shortest time to plan the shortest path. Therefore, PRM is the most efficient among these algorithms.

In this research, a modified global path planning algorithm is introduced with the consideration of kinematics constraint due to a rectangular shape holonomic mobile robot with the application of region of interest. The performance of the proposed method is evaluated through computer simulation in terms of distance, computing time and cost using MATLAB version R2018a.

2 Method

2.1 Automatic Region of Interest (AutoROI)

The image size and the real environment are being calibrated before the algorithm starts. A rectangular shaped ROI is being set up to narrow down the map into specified areas for path searching. A line, L_1, is generated using linear equations to connect the start and end point. The 4-corner point of the rectangular is perpendicular to line L_1. Then the ROI expand threshold is based on the direction of the displacement of the path. After that, a rectangular box is drawn based on the corner point. The algorithm creates a binary mask based on this shape to burn onto the original image. Therefore, the obstacle in the region is shown out and filter out the non-interest area becomes zero, as shown in Fig. 1.

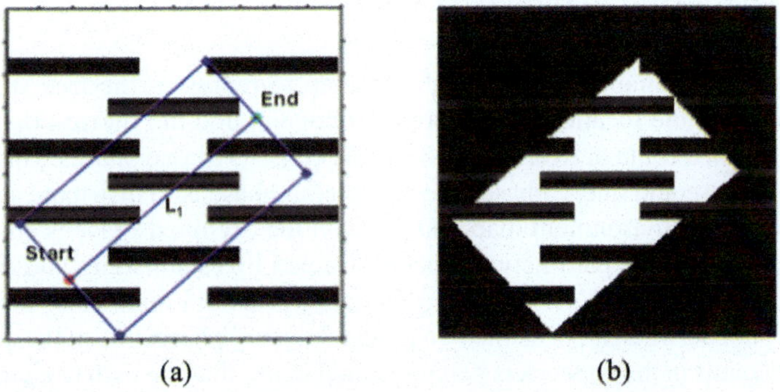

Fig. 1 Application of AutoROI: **a** Setting the ROI area, **b** Masking the ROI

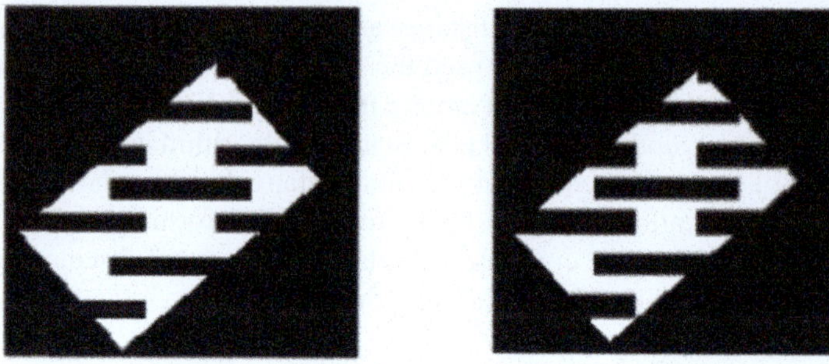

(Before) (After)

Fig. 2 Expansion of obstacles

2.2 *Expansion of Obstacles*

The obstacles are expanded based on the robot diagonal length to eliminate the path, in which the aisle width is smaller than the robot diagonal length, as shown in Fig. 2. The diagonal length of robot is the maximum safety distance for the robot to pass through the passage in a diagonal direction. This step also can avoid the algorithm plan a path close to the obstacle which can cause crashing to the wall. This technique is also able to eliminate the available free space for algorithm to allocate nodes for path planning. Therefore, it can help to shorten the elapsed time of PRM.

2.3 *Probabilistic Roadmap (PRM)*

Probabilistic Roadmap applies probability map to represent the free space of the environment for the location of the mobile robot and find the shortest distance path by construction a route network map [8]. PRM consists of two phases which are pre-processing phase and query phase. Pre-processing phase is to distribute sampling in free space of the environment map, the point on the occupied space is eliminated.

Then, the available path network is constructed by establishing undirected path network diagram $R = (N, E)$, where N denotes a random number of points and E denotes a path between all possible two points. In query phase, one of the appropriate undirected paths in the network is being selected as the shortest path from initial point to target point.

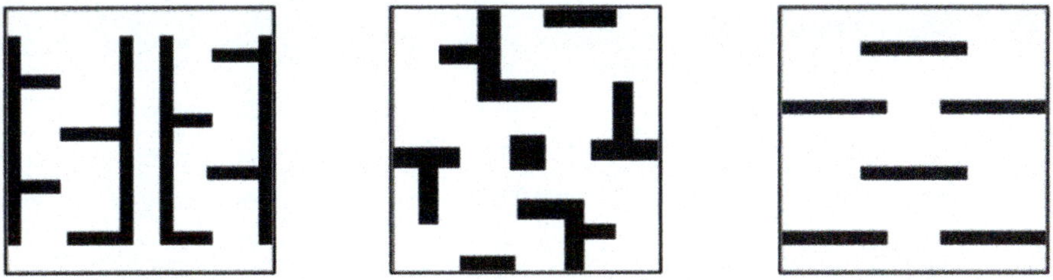

(Map 1) *(Map 2)* *(Map 3)*

Fig. 3 Environment maps

3 Simulation

To verify effectiveness and reliability of the improved PRM algorithm, there are three maps designed for simulation and the starting point and ending point is different based on the objective of testing. The overall planning efficiency simulation is mainly carried out in terms of the elapsed time of PRM, the path length, number of turning points and the planning success rate. The number of nodes setting in this study is fixed at 100 which is sufficient to generate a path. The software simulation is carried out using MATLAB version 2018a and the computer processor is Intel i5-8265U CPU.

3.1 Mapping Environment

In this experiment, there are three maps designed with different configuration of obstacles as shown in Fig. 3 to be tested by the improved PRM algorithm. The dimensions of all these maps in the real environment are 10-m long and 10-m wide. Map 1 is a model of a production line environment. The direct passage at the centre of image is 1.5-m wide, another two routes are at least 2-m wide. This map tested the algorithm whether the direct or the wide path will be chosen. Map 2 is a maze-like environment which has non-uniform size of obstacles, and the configuration of obstacles provides a large area to generate path in the middle. Map 3 contains a lot of small passages built by rectangular bar. The distance between each bar is more than 2-m. The robot is required to make more turning points to reach the target point.

3.2 Simulation Conditions

The performance of modified algorithm is evaluated based on four conditions.

Table 1 Elapsed time for PRM

Map	Elapsed time of PRM (seconds)		
	Conventional PRM	Improved PRM	Difference (%)
1	0.9183	0.0519	94
2	0.8882	0.0567	94
3	0.9124	0.0641	93

Elapsed time. The elapsed time of PRM is the time of PRM algorithm to generate shortest path on the map during simulation.

Path Distance. The shortest path distance is calculated by using Euclidean distance.

Number of Turning Points. The number of turning points is related to the wheel rotation of the holonomic mobile robot. The success rate of the path is determined by the minimum distance of the obstacles and the robot centre. The minimum distance of robot must more than the safety distance of the robot.

4 Results

4.1 Elapsed Time of PRM

AutoROI has selected the region of interest on the map which has the highest possibility. So, the available nodes are more focused in the specified area. Therefore, the elapsed time of PRM using improved PRM algorithm is shorter than the conventional PRM. The improved PRM algorithm reduced the PRM elapsed time up to 94% as given in Table 1.

4.2 Shortest Path Distance

Figure 4 shows the result of the generated PWM with obstacles being enlarged. The path distance generated by using improved PRM is longer than the conventional method and increased by more than 5% as given in Table 2. Since the obstacles in the map are being enlarged, the path generated takes longer distance to avoid the enlarged obstacles, even though it is the shortest route compared to others.

(Map 1) (Map 2) (Map 3)

Fig. 4 Shortest path distance using improved PRM

Table 2 Shortest path distance

Map	Shortest path distance (metre)		
	Conventional PRM	Improved PRM	Difference (%)
1	11.80	15.35	30
2	12.08	12.64	5
3	15.31	17.28	13

4.3 Number of Turning Points

Table 3 gives the result of the average number of turning points. The average number of turning points along the path generated by improved PRM is higher than using conventional method. The highest increment of turning point is in Map 1 which is 95%. This is because the number of nodes in the specified region of interest is near to each other, therefore the PRM algorithm takes more turns to generate a path which can avoid the obstacles.

Table 3 Average number of turning points

Map	Average number of turning points		
	Conventional PRM	Improved PRM	Difference (%)
1	4.2	8.2	95
2	5.4	6.6	22
3	6.5	9.6	48

4.4 Percentage of Success Path Planning

Table 4 gives that the percentage of success path planning of improved PRM is higher than in conventional PRM. This is due to the safety passage width that is set for improved PRM has effectively avoided the crashing between the holonomic omni-directional mobile robot and the obstacles. Figures 5, 6 and 7 show the comparison between the results obtained for the conventional and improved PRM.

Table 4 Percentage of success path planning

Map	Percentage of success path planning		
	Conventional PRM (%)	Improved PRM (%)	Difference (%)
1	0	40	40
2	0	80	80
3	20	60	40

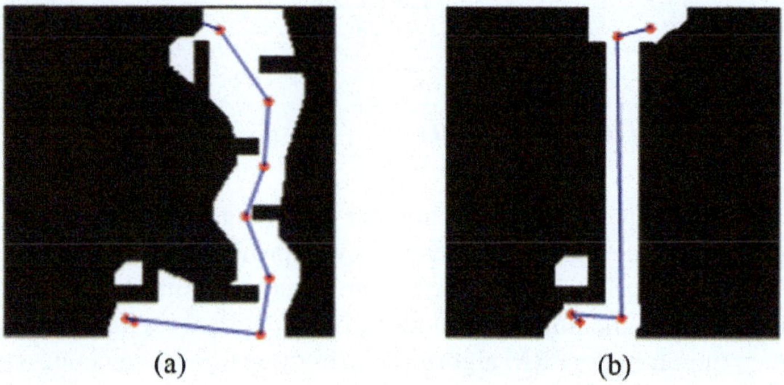

Fig. 5 Path generated in Map 1: **a** Conventional PRM, **b** Improved PRM

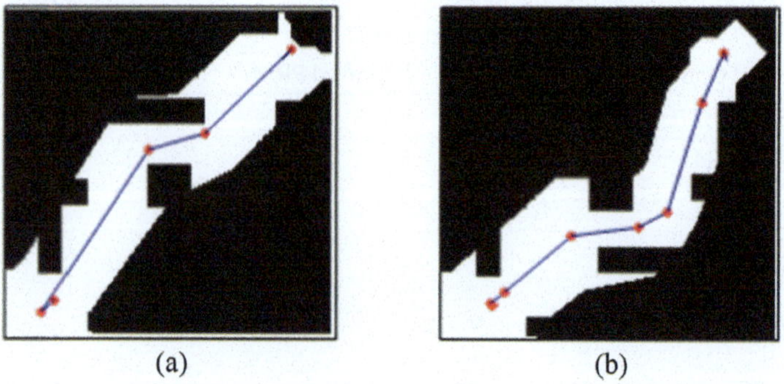

Fig. 6 Path generated in Map 2: **a** Conventional PRM, **b** Improved PRM

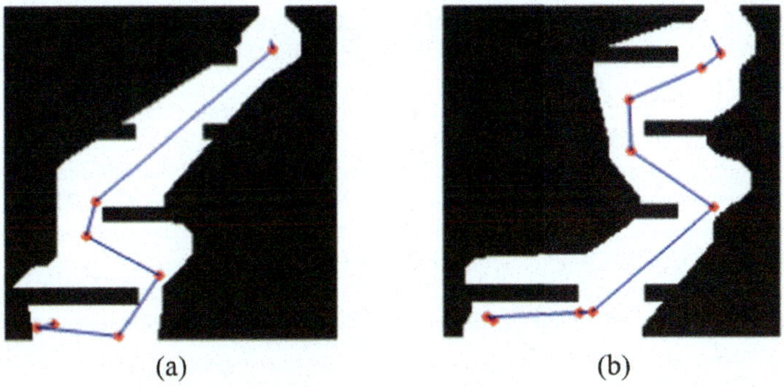

Fig. 7 Path generated in Map 3: **a** Conventional PRM, **b** Improved PRM

5 Conclusion

In conclusion, the improved Probabilistic Roadmap algorithm can reduce the elapsed time of PRM and increase the percentage of success path planning. However, the path distance generated by improved PRM is longer and the number of turning points along the path is also higher compared to the conventional PRM. Therefore, the performance of the improved path planning algorithm in terms of path distance and number of turning points did not meet the expectation.

The selection of an accurate ROI window is a very important aspect to ensure that the true shortest path can be obtained. A robust algorithm needs to be developed in the future to ensure the determination of the ROI window can be done faster and more efficiently. The current AutoROI algorithm has a limitation which the highlighted region is cut into half by the large obstacle across the ROI area. The AutoROI algorithm can be developed to be more intelligent to detect the possible path then only create the region of interest to focus the PRM nodes in the highlighted region. In addition, the performance of the proposed methods also need to be verified experimentally using actual mobile robot.

Acknowledgements This work was financed by the Ministry of Higher Education, Malaysia, through the Fundamental Research Scheme Grant (Ref: FRGS/1/2019/TK03/UNIMAP/02/6).

References

1. Safar, M.J.A.: Holonomic and omnidirectional locomotion systems for wheeled mobile robots: a review. J Teknologi **77**(28), 91–97 (2015)
2. Siegwart, R., Nourbakhsh, I.R., Scaramuzza, D.: Introduction to Autonomous Mobile Robots, 2nd ed. The MIT Press (2011)
3. Costa, M.M., Silva, M.F.: A survey on path planning algorithms for mobile robots. In: IEEE International Conference on Autonomous Robot Systems and Competitions, pp. 1–7 (2019)

4. Korkmaz, M., Durdu, A.: Comparison of optimal path planning algorithms. In 14th International Conference on Advanced Trends in Radioelectronics, Telecommunications and Computer Engineering, pp. 255–258 (2018)
5. Zeidis, I., Zimmermann, K.: Dynamics of a four-wheeled mobile robot with Mecanum wheels. J. Appl. Math. Mech. **99**(12), 1–22 (2019)
6. Jicuicui.: Path planning algorithm for mobile robot based on cellular automaton. In IEEE 4th International Conference on Control Science and Systems Engineering, pp. 71–75 (2018)
7. Atyabi, A., Powers, D.: Review of classical and heuristic-based navigation and path planning approaches. Int. J. Adv. Comput. Technol. **5**, 1–14 (2013)
8. Cao, K., Cheng, Q., Gao, S., Chen, Y., Chen, C.: Improved PRM for path planning in narrow passages. In: IEEE International Conference on Mechatronics and Automation, pp. 45–50 (2019)

Formulation of Silicone Paste Extrusion-Based for 3D Printing Technology

Khairul Salleh Basaruddin, Liyana Tajul, Nor Shafiqah Ahmat, Shah Fenner Khan Mohamad Khan, Muhammad Juhairi Aziz Safar, and Tien-Dat Hoang

Abstract The growth of 3D printer inks is in line with the growth of 3D printing technology. Finding a suitable formulation of 3D printer ink is crucial for fulfilling the demands of construction fields, bio-printing, and consumer products. Soft materials, such as silicone paste, have become one of the requested inks owing to their natural characteristics, cost effectiveness, and substantial environmental friendliness. However, because of the thixotropic behavior of silicone, the development of silicone paste as a 3D printer ink is challenging. This study aimed to develop a new soft material formulation ink made from silicone for extrusion-based 3D printing. A common silicone polymer was added with 0.2% wt of Thi-Vex and sericite mica ranging from 2 to 10% wt. The extrudability and viscosity of the silicone ink were tested by extrusion line printing. A uniaxial tensile test was conducted to evaluate the tensile strength and elongation percentage. The feasible formulation was found to be 3:1:0.2% wt:2% wt (Silicone part A, part B, Thi-Vex, sericite mica) with the tensile strength of 1.37 MPa and 148.71% of the percentage of elongation before the break.

Keywords 3D printing · Silicone ink · Paste extrusion

K. S. Basaruddin (✉) · L. Tajul · N. S. Ahmat · S. F. K. M. Khan · M. J. A. Safar
Faculty of Mechanical Engineering and Technology, Universiti Malaysia Perlis, 02600 Arau, Malaysia
e-mail: khsalleh@unimap.edu.my

T.-D. Hoang
School of Mechanical and Automotive Engineering (SMAE), Ha Noi University of Industry, Hanoi, Vietnam

D. H. Tien et al. (eds.), *Proceedings of the 1st International Conference on Sustainability and Emerging Technologies for Smart Manufacturing*, Proceedings in Technology Transfer, https://doi.org/10.1007/978-981-97-7083-0_70

1 Introduction

Three-dimensional (3D) printing or additive manufacturing (AM) is the process of creating 3D solid objects from digital files [1]. A solid object was created when the softened ink was extruded from the heated nozzle in a layer-by-layer manner. Owing to its competitive cost and ease of use, 3D printing technology is highly sought after to create objects with simple to complex geometries. To accommodate its widespread usage, the scope of its applications is changing. The aerospace, automotive, medicinal, energy, robotics, and food sectors are among the industries utilizing 3D printing. Fused deposition modeling (FDM) is frequently used for 3D printing extrusion methods, and thermoplastic filaments such as polylactic acid (PLA), acrylonitrile Butadiene styrene (ABS), polyethylene terephthalate (PET), polyethylene terephthalate glycol (PETG), and thermoplastic polyurethane (TPU) which are used as printing inks. In the recent years, FDM has shown potential for printing paste materials, including clay [2], polymers [3], and artificial living tissues [4] via paste extrusion-based 3D printing.

Paste extrusion-based 3D printing enables the use of 3D printers in the food industry [5, 6], medical industries [7], and also takes into consideration that broader materials can be printed. Currently, silicone paste is one of the soft materials being developed for 3D printer inks. Silicone has become one of the choices because it is ecologically friendly and frequently utilized in consumer goods. Silicone has great potential in various industrial sectors owing to its stable compound bonding, flexibility, and ability to withstand stress and extreme temperatures [8, 9].

Despite its widespread application and versatility, the use of silicone ink in paste extrusion-based 3D printing remains limited. It is difficult to create good printing ink owing to the viscoelasticity of silicone [10]. Furthermore, the silicone ink available on the market has a predetermined shore hardness, which is difficult to modify to fit user requirements [11]. Marketed silicone is not suitable for single-syringe dispensing and can be expensive compared with other materials. Hence, the purpose of this study was to develop a formulation that is suitable for extrusion-based 3D printing. The characteristics and texture of silicone inks were evaluated as prospective criteria for industrial applications. With this developed formulation, silicone can be fully utilized as an affordable 3D printer ink with good material properties and compatibility with single-syringe ink dispensers.

2 Methods

2.1 Formulation of Silicone and Extrusion Test

The inks used in this study were made of common silicone elastomers with a shore hardness of 10A. This silicone is suitable for all product uses, safe, and does not emit any harmful gases. There were two parts, A and B, of the silicone mixture. Part A is the polymer-based and part B is the curing agent. Since silicone is not specifically designed as a 3D printer ink, the ratio of part A to part B must be modified to obtain a suitable formulation in terms of ink flowability and viscosity.

An extrusion test was then conducted to evaluate the extrudability of the ink and its capacity to perform surface tension tests. The extrusion test was conducted under three different conditions: a single line, five lines lined horizontally, and five lines lined vertically. Figure 1 shows the three conditions used in the extrusion test. Four different silicone formulations were tested, and the best formulation was chosen as the printer ink for the paste extrusion 3D printing. The formulation used for the first-line extrusion test is presented in Table 1. The extrusion process was conducted manually using a 10 ml syringe (slip tip) with a nozzle diameter of 2 mm. Although the extrusion speed and force produced an unstable line, the purpose was to demonstrate the ability of the formulated ink to be extruded using a syringe nozzle.

The extruded line structures with different amounts of sericite mica powder are listed in Table 2. For the single line, a diameter of 0.2 mm was obtained for all formulations, which is similar to the nozzle diameter. When sericite mica was added

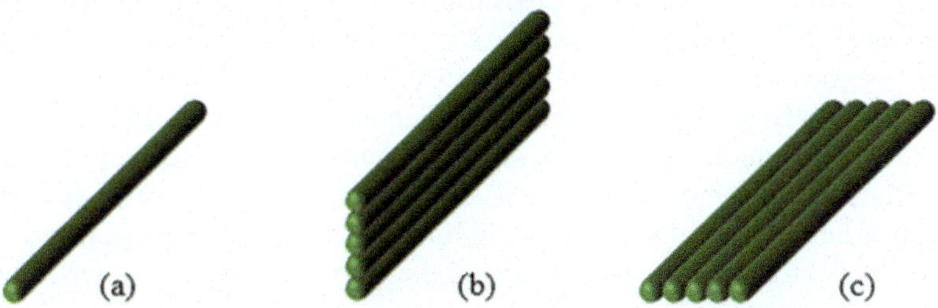

Fig. 1 **a** Single line **b** 5 lines lined together vertically **c** 5 lines lined together horizontally

Table 1 Formulation for the first-line extrusion

Formulation	Material	Method	Additive used	Ratio
1	Silicone part A + silicone part B	Let rest for 60s in refrigerator (18 °C)	None	1:1
2	Silicone part A + silicone part B + sericite mica	Let rest for 60s in room temperature (24 °C)	Sericite mica	2:1:2% wt
3	Silicone part A + silicone part B + Titanium Oxide	Let rest for 60s in room temperature (24 °C)	Titanium Oxide	2:1:2% wt
4	Silicone part A + silicone part B + sericite mica + Thi-Vex (Smooth-On.Co)	Direct extrusion through syringe	Sericite mica, Thi-Vex (Smooth-On.Co)	3:1:2% wt: 0.2% wt

to the silicone formulation, the extruded line exhibited improvement in the lines formed, resulting in correct stacking lines on one another. The initial hue of the silicone, which was clear white, changed from translucent to cloudy. These changes were influenced by the color of the sericite mica powder added to the formulation. The silicone formulation contained three parts of silicone part A, one part of silicone B, 0.2% wt Thi-Vex, and sericite mica ranging from 2 to 10% wt. These formulations were chosen as the final silicone formulation by adding sericite mica powder to improve the forming structure of the 3D printer ink.

2.2 *3D Printing*

The printer used for printing the silicone paste was a Cartesian style 3D printer (Creality3D Ender 3). The Creality3D Ender 3 head extruder is made for hot-melt extrusion, which melts the filament before it is extruded through the nozzle. Silicone paste extrusion on the other hand does not require the material to be preheated in order to be extruded. Therefore, the existing extruder head was modified to allow the 60 ml syringe to be mounted on the x-axis rail. The modified extruder was designed using the open-source website Thingiverse. A stepper motor NEMA 17 was used to rotate the rod and depress the plunger. Six components are produced and assembled. Stepper housing, guide lock, piston, slider lock, main body, and gear comprise the device. These components were created using 3D printing and were made from ABS with a layer height of 0.2 mm and 60% infill. Table 3 lists the printing parameters of the new head extruder. The printed parts of the extruded head are shown in Fig. 2.

2.3 *Mechanical Testing*

The silicone elastomer underwent significant modifications in terms of viscosity, flowability, and color owing to the presence of the additional material; therefore, tensile and hardness tests were conducted to determine its properties and characteristics. The tensile test was performed based on the ASTM D412/ ISO 37 standard using a Universal Testing Machine (Shimadzu) with a constant crosshead speed of 500 mm/min until the specimen failed. A total of 25 specimens with five different formulations were tested.

Hardness tests were then performed to determine the shore hardness of the new silicone formulations in the presence of foreign substances (sericite mica powder and Thi-Vex). The test sample was a dogbone-shaped specimen and readings were obtained at three points using a durometer.

Table 2 Line structures with different amounts of sericite mica powder added

Formulation	Single line	5 Horizontal lines	5 Vertical lines
2% wt			
4% wt		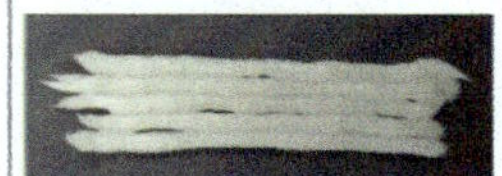	
6% wt			
8% wt		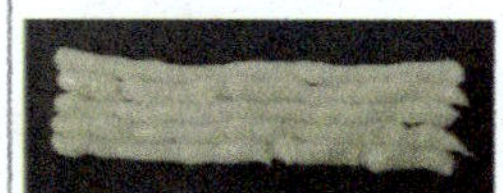	
10% wt	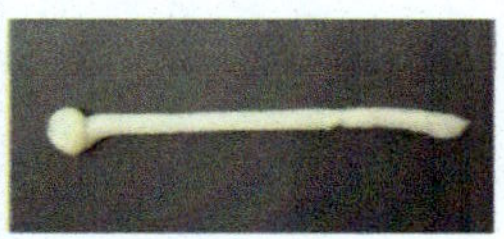	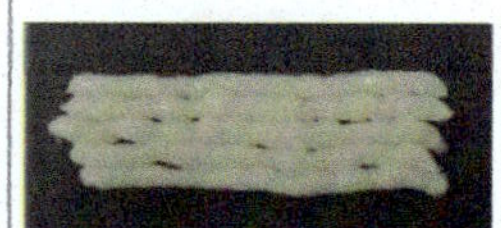	

Table 3 Printing parameter for new head extruder parts

Printing parameter	Value
Layer height	0.2 mm
Wall thickness	1.6 mm
Wall line count	3
Top/bottom thickness	1.4 mm
Infill	60%
Bed temperature (°C)	70 °C
Nozzle temperature	230 °C
Nozzle size	0.4 mm

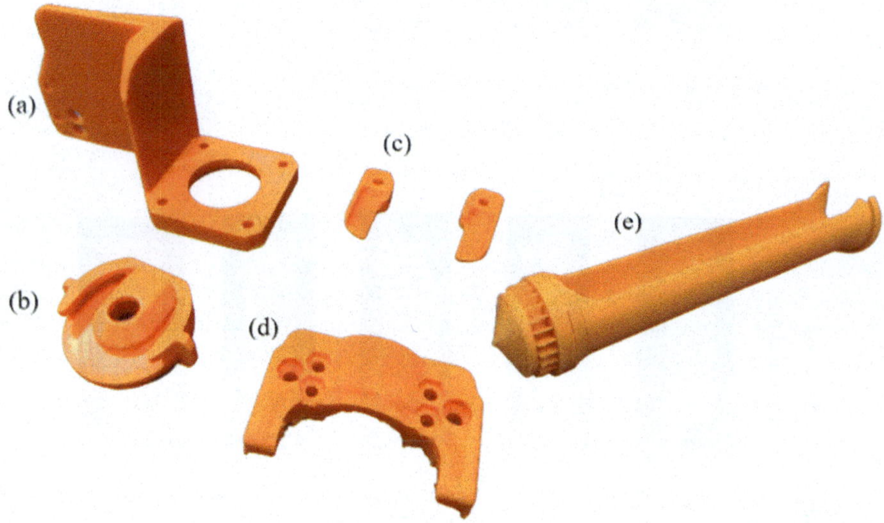

Fig. 2 Printed parts **a** Stepper housing **b** Guider lock **c** Slider lock **d** Bearing cage **e** Piston

3 Results and Discussion

The results of the tensile strength and percentage of elongation versus the percentage of sericite mica are plotted in Fig. 3. As more sericite mica was added to the formulation, the tensile strength and percentage of elongation for the silicone combination showed a decreasing trend. The highest tensile strength (1.37 MPa was recorded when sericite mica (2 wt.%) was added to pure silicone. The amount of sericite mica introduced was greater than needed, and hence, weakened the silicone tensile strength. The highest percentage of elongation at break was 148.71% at 2 wt% and the lowest percentage was 108.71% at 10 wt%. Sericite mica acts as a silicone reinforcer and reduces the elasticity of silicone. Thus, the more sericite mica is added to the silicone formulation, the lower the percentage of elongation.

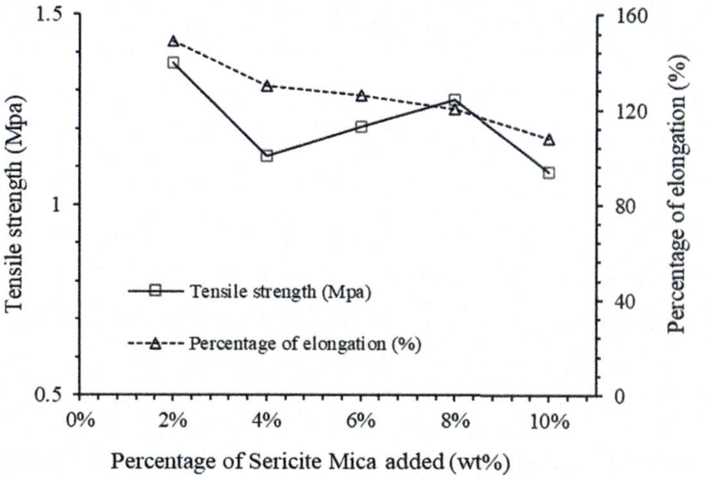

Fig. 3 Relationship of tensile strength and percentage of elongation of silicone with different percentages of sericite mica

Figure 4 shows the relationship between the silicone shore hardness and sericite mica percentage. The shore hardness of silicones increased as more sericite mica was added to the formulation. The highest measured shore hardness was 19.33A. The presence of sericite mica reinforced the crosslinking between the polymers and increased the hardness of the silicone. Table 4 summarizes the properties of the new formulation. Among these five formulations, the formulation with a ratio of 3:1:0.2 wt%: 2 wt% of silicone part A: silicone part B: Thi-Vex solution: sericite mica powder was chosen as the new 3D printer ink. This formulation exhibited the best line formation during the extrusion printing test and had the best material properties, including a high tensile strength, high percentage of elongation at break, and feasible shore hardness.

Fig. 4 Relationship between tensile strength and percentage of elongation of silicone with different percentages of sericite mica

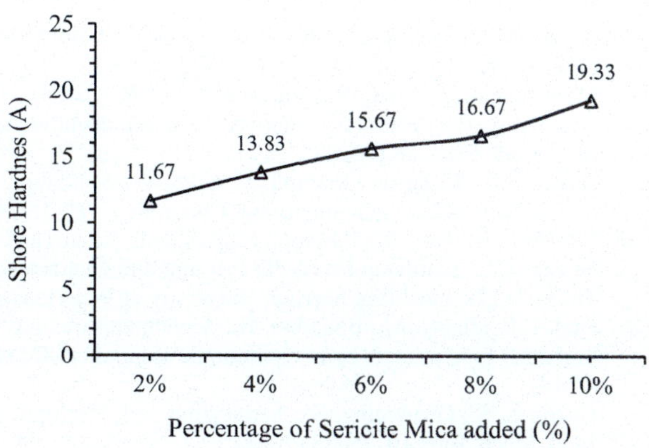

Table 4 Summary of new silicone materials properties

Formulation	2%	4%	6%	8%	10%
Tensile strength (MPa)	1.37223	1.127632	1.205336	1.277308	1.087014
Elongation at break (%)	148.7196	129.7404	125.7404	120.324	108.1782
Young's Modulus (MPa)	0.009227	0.008691	0.009586	0.010616	0.010048
Shore Hardness (A)	11.67	13.83	15.67	16.67	19.33
Curing time (min)	120	120	120	100	100

4 Conclusion

This study focused on the development of a new silicone formulation as a soft material for ink in extrusion-based 3D printing. Pure silicone elastomers were reinforced with Thi-vex and sericite mica, forming a new ink formulation with feasible material properties for general products. The new formulation contained three parts: silicone part A (polymer base), one part of silicone part B (silicone cross linker), 2% wt of sericite mica, and 0.2% wt of Thi-Vex solution. This formulation can be mixed for approximately two hours until it is completely cured, giving it a good pot life. As it is premixed and not mixed during extrusion, this formulation is suitable for users of 3D printers with a single syringe mounted on the rail. This new formulation improved the silicone properties as the tensile strength increased to 1.37 MPa and the percentage of elongation was 148.71%. The shore hardness of the silicone also increases to 11.67A, while the color of the silicone remained translucent white.

Acknowledgements The author would like to acknowledge the support from the Fundamental Research Grant Scheme (FRGS) under grant number FRGS/1/2019/TK03/UniMAP/02/3 from the Ministry of Higher Education Malaysia.

References

1. Al-Maliki, J.Q., Al-Maliki, A.J.Q.: The processes and technologies of 3D printing. Int. J. Adv. Comput. Sci. Technol. **4**(10), 161–165 (2015)
2. Manikandan, K., Jiang, X., Singh, A.A., Li, B., Qin, H.: Effects of nozzle geometries on 3D printing of clay constructs: quantifying contour deviation and mechanical properties. Procedia Manuf. **48**, 678–683 (2020)
3. Ligon, S.C., Liska, R., Stampfl, J., Gurr, M., Mülhaupt, R.: Polymers for 3D printing and customized additive manufacturing. Chem. Rev. **117**(15), 9839–10620 (2017)
4. Dey, M., Ozbolat, I.T.: 3D bioprinting of cells, tissues and organs. Sci. Rep. **10**, 14023 (2020)
5. Huang, C. .: Extrusion-based 3D Printing and Characterization of Edible Materials. Master Thesis in Chemical Engineering, University of Waterloo, Ontario, Canada (2018)
6. Guo, C.F., Zhang, M., Bhandari, B.: A comparative study between syringe-based and screw-based 3D food printers by computational simulation. Comput. Electron. Agric. **162**, 397–404 (2019)
7. Chunhua, S., Guangqing, S.: Application and development of 3D printing in medical field. Mod. Mech. Eng. **10**, 25–33 (2020)

8. Rossing, L., Scharff, R.B.N., Chömpff, B., Wang, C.C.L., Doubrovski, E.L.: Bonding between silicones and thermoplastics using 3D printed mechanical interlocking. Mater. Des. **186**, 108254 (2019)
9. Wang, Z., Li, W., Chen, L., Zhan, Z., Duan, H.: 3D printable silicone rubber for long-lasting and weather-resistant wearable devices. ACS Appl. Polym. Mater. **4**(4), 2384–2392 (2022)
10. Jindal, S.K., Sherriff, M., Waters, M.G., Smay, J.E., Coward, T.J.: Development of a 3D printable maxillofacial silicone: part II. Optimization of moderator and thixotropic agent. J. Prosthet. Dent. **119**(2), 299–304 (2018)
11. Zhao, H., Allanson, D., Ren, X.J.: Use of shore hardness tests for in-process properties estimation/monitoring of silicone rubbers. J. Mater. Sci. Chem. Eng. **3**(7), 142–147 (2015)

A Deep Learning Approach for Real-Time Detection and Evaluation of Oranges on Mobile Devices

Van Hong Son Than, Dinh Minh Khoa Tran, Thien Luong, Duc Tho Le, and Viet Hung Le

Abstract In this study, we extensively explored the most effective configurations of deep learning architectures, namely NanoDet and YOLOv8, for the real-time detection of orange fruits and assessment of their quality on mobile devices. Furthermore, we have successfully implemented a mobile application leveraging these deep learning models specifically tailored for the iOS platform. Our methodology involved the curation, annotation, and processing of a unique dataset comprising oranges. The empirical findings demonstrate that our system exhibits exceptional accuracy in both detecting orange fruits and evaluating their quality. The integration of artificial intelligence technologies presents a valuable tool for orange farms and consumers alike, facilitating rapid and precise assessments of orange fruits quality.

Keywords Object detection · Deep learning · Yolov8 · Mobile application · Core ML · Real-time

1 Introduction

The current situation of poor quality fruits in markets and stores can appear for many reasons, and this is a common problem in many countries. With the strong development of artificial intelligence technology today, the application of artificial intelligence components in manufacturing fields has become a new trend in the modernization process, including the agricultural and food processing sectors. Ensuring the quality of agricultural products, such as fruit, is important for public health and affects the sustainable development of the agricultural sector [1]. Identifying fresh fruit is a very simple task that even humans can easily recognize. However, in large-scale production, this job becomes difficult and there will be errors. For this reason, we need to apply machines in the process of classifying fresh and rotten fruit, capable of effectively recognizing and classifying them. Recently, deep learning has been expected to be a powerful tool in automatic detection and identification systems [2].

V. H. S. Than · D. M. K. Tran · T. Luong · D. T. Le · V. H. Le (✉)
The University of Da Nang—University of Science and Technology, Danang, Vietnam
e-mail: hungle.dut@gmail.com

D. H. Tien et al. (eds.), *Proceedings of the 1st International Conference on Sustainability and Emerging Technologies for Smart Manufacturing*, Proceedings in Technology Transfer, https://doi.org/10.1007/978-981-97-7083-0_71

In tandem with advancements in data and computing hardware, the utilization of deep learning has surged, becoming increasingly prevalent and applied extensively for optimization purposes [3]. In this paper, we proposed an application to detect and evaluate the quality of orange based on deep learning. We have successfully created a cutting-edge mobile application harnessing the power of artificial intelligence for real-time identification and assessment of orange quality. Our approach involved data collection, labeling, and processing of an extensive dataset, paving the way for the integration of Nanodet and YOLOv8 [4, 5] in training the system. The outcomes underscore the system's exceptional accuracy in both identification and classification processes. The core of our methodology revolves around a comprehensive analysis of peel characteristics and object shape, encompassing crucial factors such as color and form. By leveraging the deep learning model, our application has emerged as a game-changer for orange farms and consumers alike, enabling swift and precise evaluations of product quality on mobile devices. This technological innovation promises to streamline the assessment process, providing a valuable tool for enhancing efficiency and accuracy in the evaluation of oranges, benefiting both producers and consumers. The key contributions of this study include:

- Investigating the optimized versions of the YOLOv8 and NanoDet models for mobile applications.
- Successfully creating a real-time detection and evaluation mobile application on the iOS Platform.

Our code is available at https://github.com/Volcano4bit6/FreshOrangeMobile.

2 Methodology

Both YOLOv8 and NanoDet are state-of-the-art models in detection tasks. In this study, we investigated optimal versions of the YOLOv8 and NanoDet models to implement on mobile devices for real-time orange detection.

2.1 Yolov8

YOLOv8 is a modern computer vision modeling architecture built and developed by Ultralytics. The YOLOv8 model provides good support for object detection, classification, and segmentation tasks [5, 6]. YOLOv8 is available in different sizes. In this study, we use the nano version, the smallest version of YOLOv8. The advantage of this version is high processing speed, suitable for mobile applications.

Backbone:
CSPDarknet53 is a neural network architecture optimized for high-speed object detection and good accuracy [7]. CSPDarknet53 uses Cross Stage Partial Network

Fig. 1 Yolov8's backbone [7]

	Type	Filters	Size	Output
	Convolutional	32	3 × 3	256 × 256
	Convolutional	64	3 × 3 / 2	128 × 128
1x	Convolutional	32	1 × 1	
	Convolutional	64	3 × 3	
	Residual			128 × 128
	Convolutional	128	3 × 3 / 2	64 × 64
2x	Convolutional	64	1 × 1	
	Convolutional	128	3 × 3	
	Residual			64 × 64
	Convolutional	256	3 × 3 / 2	32 × 32
8x	Convolutional	128	1 × 1	
	Convolutional	256	3 × 3	
	Residual			32 × 32
	Convolutional	512	3 × 3 / 2	16 × 16
8x	Convolutional	256	1 × 1	
	Convolutional	512	3 × 3	
	Residual			16 × 16
	Convolutional	1024	3 × 3 / 2	8 × 8
4x	Convolutional	512	1 × 1	
	Convolutional	1024	3 × 3	
	Residual			8 × 8
	Avgpool		Global	
	Connected		1000	
	Softmax			

architecture to minimize parameters, improving model performance in learning features. Cross Stage Partial Network divides the input features of a layer into two groups. One group is passed through a chain of convolutional layers to learn new features, while the other group is bypassed by this chain of layers and is directly combined with the output of the convolutional layer chain [8]. Then, these two groups of features are combined together before passing to the next layer. The backbone's architecture is illustrated in Fig. 1.

Head:
YOLOv8's Head is designed to process features from the Path Aggregation Network and produce the final output for object detection. YOLOv8's Head is designed to operate at many different scales, helping the model to detect objects of many different sizes. This is often called "multi-scale prediction" or "pyramid prediction". The general architecture of Yolov8 is illustrated as Fig. 2.

2.2 *NanoDet*

NanoDet is an extremely small and lightweight object recognition model, designed to run on devices with limited resources such as mobile devices, embedded computers, or devices with limited battery capacity [4]. This model keeps its size small and

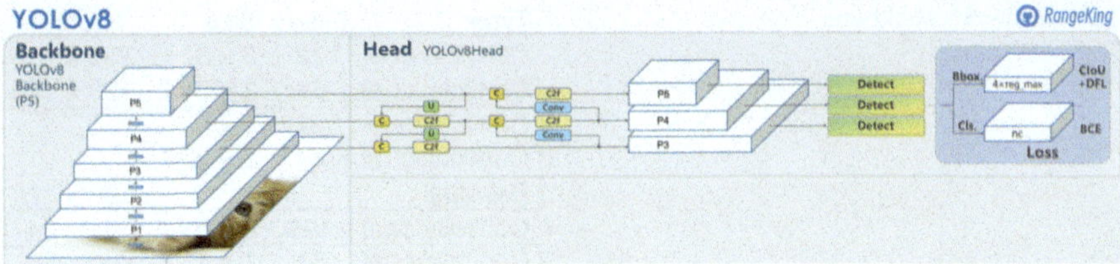

Fig. 2 Yolov8's architecture [9]

requires few computational resources while still maintaining accurate object recognition.

Backbone:
Different from most other neural network architectures, ShuffleNetV2 uses direct metrics (such as speed or memory access cost) to measure network complexity [10]. Some studies have proven that FLOPs (this is a common index to measure network performance, used as an indirect indicator) is not completely correct, although network architectures have the same FLOPs, they are different vary in speed, this may be due to memory access costs, platform in use, etc.
Channel split operation divides the channel into two groups, the first group remains the same . The other group has an equal number of input and output channels in the three convolutional layers . The 1×1 convolution layer is not groupwise. Operations on each corresponding element such as ReLU, Concat, and depth convolution layers are restricted to a single branch.
In this paper, we use the ShuffleNetV2 1.0x backbone, use stages 2, 3, and 4 to extract features and use ReLU as the activation function. The ShuffleNetV2 1.0x backbone is depicted in Fig. 3.

Feature Pyramid Networks (FPN):
FPN is a feature filter that applies the above feature pyramid principle while ensuring a balance between accuracy and processing speed. Applied to the problem of object detection. FPN allows the model to detect objects at multiple scales.

Head:
The Head of NanoDet is designed to correspond to 2 classification layers, 96 input channels, 96 output channels, 5 kernels, and 2 convolutional layers. The general model of NanoDet is depicted in Fig. 4.

2.3 Output Representation

The output of the YOLOv8 or NanoDet model is a vector that will include the following:

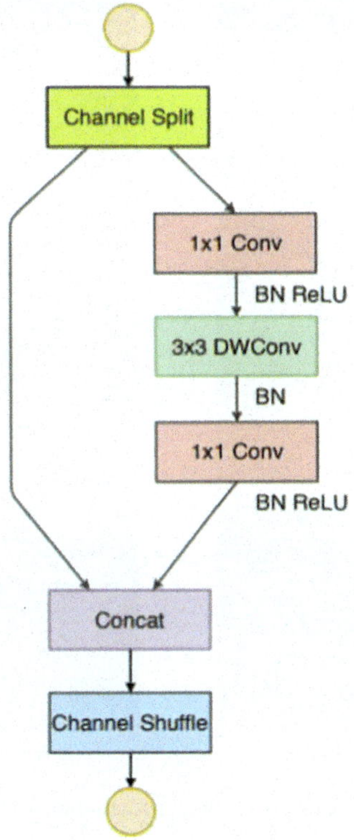

Fig. 3 ShuffleNetV2 1.0x backbone [11]

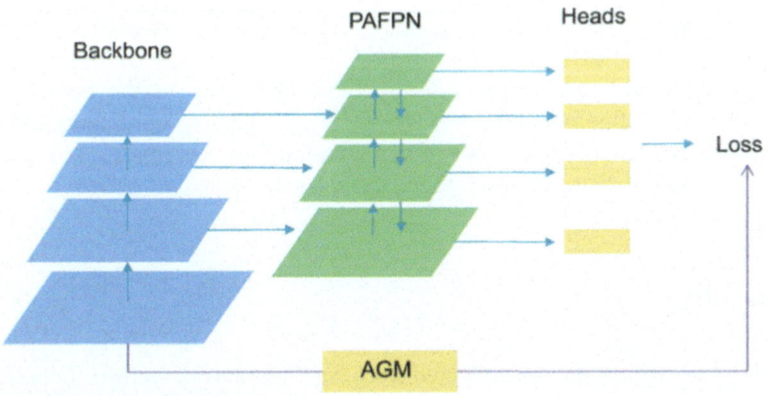

Fig. 4 NanoDet's architecture [12]

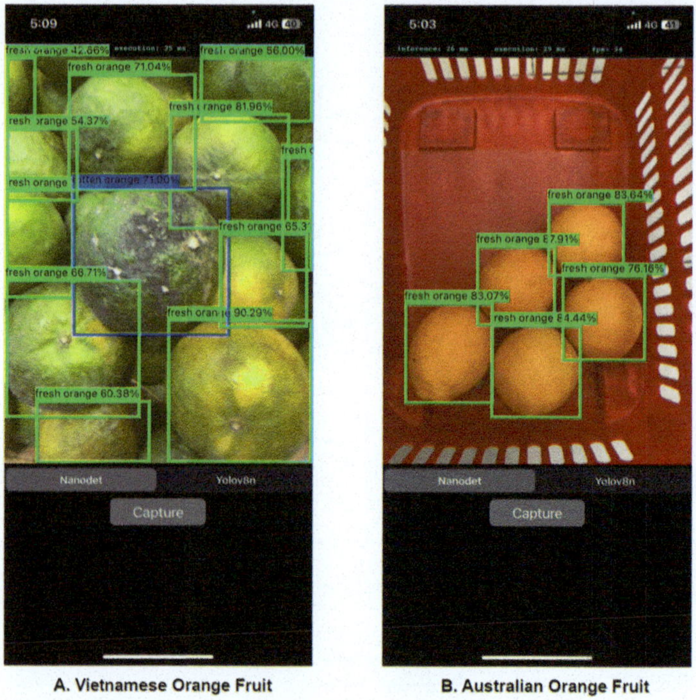

Fig. 5 Our testing on Vietnamese orange fruit and Australian orange fruit

$$\text{output}^T = [p_0, \underbrace{\langle b_x, b_y, b_w, b_h}_{\text{bounding}} \text{ box}\rangle, \underbrace{\langle p_1, p_2, \ldots, p_c\rangle}_{\text{scores of } c \text{ classes}}]$$

p_0 is the prediction probability of an object appearing in the bounding box. $\langle b_x, b_y, b_w, b_h\rangle$ define bounding box. b_x, b_y are the center coordinates and b_w, b_h is the width and length dimensions of the bounding box. $\langle p_1, p_2, \ldots, p_c\rangle$ is the prediction probability distribution vector of the classes.

3 Results

Dataset: The original dataset collected from the Shutterstock website includes 2831 images of different sizes. In the dataset, there are two classes of oranges: "fresh-orange" and "rotten-orange". We used data augmentation methods to upsample the dataset by employing rotation and flipping techniques. Finally, we obtained a new dataset with 6791 images. The data set is divided into train, valid, and test sets in a ratio of 7:2:1.

Fig. 6 Our testing in the market

Table 1 Comparison of test results between Yolov8 and Nanodet models

Model	Precision	Recall	mAP50
Yolov8	0.74	0.67	0.83
Nanodet	0.65	0.87	0.75

Mobile Application: In this paper, we use Core ML [13] known as an Apple framework introduced in 2017 for integrating artificial intelligence (AI) and machine learning into iOS, macOS, watchOS, and tvOS apps. This is a powerful tool that helps developers integrate machine learning models into their applications easily and effectively. For mobile applications, we implement IOS mobile applications using the UIKit framework. UIKit is a framework developed by Apple Inc. for building user interfaces on iOS devices such as iPhone, iPad, and iPod Touch. Our application includes functions of model selection, real-time object detection, and saving results to the user image library.We utilized the UIKit framework to successfully construct the User Interface (UI) of our mobile application, as depicted in the Figs. 5 and 6. With the high scores of YOLOv8 and NanoDet as depicted in Table 1, we can efficiently apply these models to the practical application of detecting and evaluating oranges in real-time. The Figs. 5 and 6 show that our system works efficiently and the application works smoothly, so that our mobile application can be applied to any place or situation. The model for detecting orange quality can still be improved by increasing the amount of training data and doing a better job of labeling and annotating the data. Both Yolov8 and Nanodet models have their advantages, we can create a real-time system based on these two models to simultaneously achieve those advantages. This model can be applied to many different fruits, or even developed into a general model applicable to many types of fruits. These types of machine learning models have many applications in food processing technology manufacturing.

References

1. Saha, R., Neware, S.: Orange fruit disease classification using deep learning approach. Int. J. **9**(2) (2020)
2. Van Hieu, N., et al.: PlantKViT: a combination model of vision transformer and KNN for forest plants classification. J. Univers. Comput. Sci. **29**(9) (2023)
3. Le Huy, H.N., et al.: Keyphrase extraction model: a new design and application on tourism information. Informatica **45**(4) (2021)
4. Xu, Y., Yi, J., Gao, J.: Defect detection of automotive leather based on Nanodet-Plus. In: 35th Chinese Control and Decision Conference (CCDC). IEEE, China (2023)
5. Terven, J., Córdova-Esparza, D.-M., Romero-González, J.-A.: A comprehensive review of yolo architectures in computer vision: from yolov1 to yolov8 and yolo-nas. Mach. Learn. Knowl. Extr. **5**(4), 1680–1716 (2023)
6. Shen, L., Lang, B., Song, Z.: DS-YOLOv8-based object detection method for remote sensing images. IEEE Access **11**, 125122–125137 (2023)
7. Redmon, J., Farhadi, A.: Yolov3: An Incremental Improvement. arXiv preprint arXiv:1804.02767 (2018)
8. Wang, G., et al.: UAV-YOLOv8: a small-object-detection model based on improved YOLOv8 for UAV aerial photography scenarios. Sensors **23**(16), 7190 (2023)
9. What is yolov8? The ultimate guide. blog.roboflow.com/whats-new-in-yolov8. Accessed 13 March 2024
10. Ji, W., et al.: A real-time apple targets detection method for picking robot based on ShufflenetV2-YOLOX. Agriculture **12**(6), 856 (2022)
11. Ma, N., et al.: Shufflenet v2: practical guidelines for efficient CNN architecture design. In: Proceedings of the European Conference on Computer Vision (ECCV) (2018)
12. Li, D., Zhai, J.: A real-time vehicle window positioning system based on nanodet. In: Chinese Intelligent Systems Conference. Springer Nature Singapore (2022)
13. Thakkar, M.: Introduction to Core ML Framework. Beginning Machine Learning in iOS: CoreML Framework, pp. 15–49 (2019)

Which Strategies Should Vietnam's Garment Industry Approach Leading to Sustainable Development?

Quang-Tri Tran, Nguyen Thi Le, and Tho Alang

Abstract Vietnam's garment industry ranks among the foremost industries globally, emphasizing exportation. It sits alongside China and Bangladesh as one of the top three producers and exporters of ready-made garments. Despite this status, Vietnam's garment industry is primarily recognized for its proficiency in Cut-Make-Trim (CMT) manufacturing. CMT, which constitutes around 73% of total exports, is deemed ideal for harnessing Vietnam's extensive workforce. This study seeks to identify strategies for elevating Vietnam's garment industry to higher value-added segments, thereby fostering sustainable development. The research employs a sequential mixed-method approach, comprising qualitative interviews and quantitative surveys conducted in two phases. Through this methodology, two overarching conclusions were drawn and validated. This research aims to shed light on current and future challenges within Vietnam's garment industry, facilitating the development of effective strategies while averting potential pitfalls.

Keywords Vietnam · Garment industry · Production mode · Sustainable development

Q.-T. Tran
Ho Chi Minh City University of Technology and Education, Ho Chi Minh City, Vietnam
e-mail: tritq@hcmute.edu.vn

N. T. Le (✉)
Hanoi University of Industry, Hanoi, Vietnam
e-mail: le.nguyenthi@haui.edu.vn

T. Alang
International University—Vietnam National University, Ho Chi Minh City, Vietnam
e-mail: alangtho@hcmiu.edu.vn

D. H. Tien et al. (eds.), *Proceedings of the 1st International Conference on Sustainability and Emerging Technologies for Smart Manufacturing*, Proceedings in Technology Transfer, https://doi.org/10.1007/978-981-97-7083-0_72

1 Introduction

Vietnam's textiles and garment are key industries, contributing to an average sustainable export growth rate of 8–15% per year. They crucially contribute to Vietnam's economic development strategy [1]. However, the textile and garment industries in Vietnam have currently faced many challenges due to the impact of the world and domestic economic situation. Specifically, inflation in key markets such as the United States and Europe has result in decreasing purchasing power and declining orders, rising interest rates, and differentiating exchange rate. Besides, geopolitical tensions fragmented world trade, affecting Vietnam's exports of key products including textile and garment products [2].

In the current time, to maintain economic growth and boost per capita incomes, Vietnam's economy must overcome obstacles to productivity growth and value chain advancement. FTA coordination is necessary to optimize trade creation benefits, reduce trade diversion, and guarantee development advantages from economic integration. Vietnam has been participating in and negotiating 19 free trade agreements (FTAs), 16 out of 19 FTAs have been signed and implemented. Also, Vietnam is the only country that has signed free trade agreements with all major global economic partners such as the United States, Japan, China, EU, UK, Russia. In particular, the strategy for developing Vietnam's textile, garment and footwear industries to 2030 with a vision to 2035, has been recently approved by the government to attract foreign investment in the field of yarn, weaving and dyeing. This leaded to create opportunities for the textile and garment industry to take advantage of incentives from agreements, thereby promoting sustainable growth of the industry [3].

Together with advantages original from the government, the demographic bonus and the abundant labor resources in Vietnam are also the one of comparative advantages of this sector. Obviously, this is a huge one for the development of the country if the Vietnam can take advantage of the superiority about the labor force of this period in the coming decades. However, the low-back labor quality is a big challenge for taking advantage of this opportunity. This leads to abundant labor resources in quantity but poor quality and then become an issue society. Therefore, according to the researchers, the core issue of sustainable development of Vietnam is to improve the quality of investment and proper awareness on human resources [4].

There are more than 1.6 million workers in Vietnam's textile and garment industry, accounting for more than 12% of the industrial workforce and nearly 5% of the total workforce in the country. The main production method of the garment industry is Cut-Make-Trim (CMT), which has the lowest added value. The income and capacity of workers are only average. The competitiveness of Vietnam's textile and garment industry in the international arena is low [5]. Some studies also point out contradictions and difficulties in policies to develop environmentally friendly textile and garment industries in countries with high production costs and dependence on traditional production systems [6].

However, little is known about the flow of ideas of what are underlining reasons of both government policies and human resources impacting on the sustainable development of textile and garment industry in Vietnam. Hence, this research aimed at examining the strategies that Vietnam's textile and garment industry are taking advantages to develop the sectors in the trend of sustainability.

2 Methods

The study's research methodology comprised a sequential mixed methods approach, involving two phases. The initial phase aimed to explore and categorize the opportunities, challenges, advantages, disadvantages, current status, and strategies of various textile and clothing (T&C) manufacturers and companies in Vietnam through face-to-face interviews. Subsequently, in the second phase, these categories were evaluated and refined using surveys to address any identified deficiencies.

The justification for employing both qualitative and quantitative data stemmed from the understanding that conducting a comprehensive survey within the Textile and Clothing (T&C) industry required initial exploration of the subject matter through discussions with experts, officials, professors, and entrepreneurs [7, 8]. This preliminary qualitative phase allowed for refinement and adjustment of the survey instruments to ensure their effectiveness when administered to a broader and more representative group within the T&C field.

Phase I involved qualitative investigation conducted through in-depth interviews with eight company executives chosen to represent various facets of Vietnam's garment industry. The researcher utilized interview questions crafted specifically for the study's objectives to gather relevant data. These questions were organized around a predetermined set of inquiries closely aligned with the research objectives to derive generalizations within each category. To ensure the collection of objective and robust data, the research sample encompassed a diverse group of interviewees, including directors and managers from the manufacturing sector, as well as heads of schools/colleges and principals from the educational sector (refer to Table 1).

Phase II involved quantitative research conducted via an online survey with 86 respondents from the North and South of Vietnam. Building upon the insights gained in Phase I, a survey questionnaire was crafted to delve deeper into the findings obtained during the initial phase. Similar to Phase I, the survey targeted individuals with expertise in production/manufacturing as well as training/education within the sector. The participant pool was expanded in Phase II, encompassing a broader geographical range of participants. The selection criteria for participants and informants were rooted in their extensive experience and expertise in the context of Vietnam's garment industry.

Due to a limited number of participants within the Vietnam garment industry field, in-depth interviews were initially conducted to explore key themes with a small group of participants before validating these themes with a larger sample via

Table 1 Interview participant profiles

Participants	Title of participant	Name of company
Participant 1	General Director Vice President	Saigon 3 Garment Joint Stock Company Ho Chi Minh City Vietnam Textile and Apparel Association VITAS
Participant 2	Marketing Manager	Protrade Garment Company
Participant 3	Sales Manager	Sonha Limited Company
Participant 4	Vice Dean	Faculty of Garment Industry and Fashion
Participant 5	Vice Manager	Phong Phu Corporation
Participant 6	Head of College	Viet Tien Garment Joint Stock Corporation
Participant 7	Head of College	Department of Technical Textile
Participant 8	Principal	Ho Chi Minh City Vinatex Economic Technical College

the online survey. The findings from Phase I served as the foundation for Phase II, allowing for the validation and testing of the results obtained in the Phase I.

3 Results

3.1 The Result of Phase I

Theme 1: Exploiting Effectively the Investment Policy of Government

During the today's global recession, economic stimulus policies from the government play a key role to support enterprises maintain and continuously develop.

Interviewee 1: In the fluctuation of economic situation, garment enterprises are looking forward to more practical supports from the government.

Interviewee 2: Vietnam's T&C Industry is needing to promote effective investment projects of the government as well as attract foreign investment ones in the development of textile and accessory sectors, then it is likely to be enhanced.

Specifically, strategic approaches regarding financial capital is the first critical step in those whole polices.

Interviewee 1: The government should urgently address bottlenecks on financial capital for enterprises. The high lending interest rate is not suitable for the development of both enterprises and banks at the current time. (Interviewee 4 made the same comment.)

Another strategic approach is coming from government supportive policies conquering drawbacks in connection with the textile industry. As the input of garment industry is the output of textile industry, any drawbacks from textile one will seriously impact on the garment one.

Interviewee 2: The current dilemma is that Vietnam has already built textile enterprise but cannot qualify for the dyeing process. Therefore, the color stability is not good enough to provide for the pressing needs of garment enterprise.

Theme 2: Strengthening Relationship Between Garment Enterprises and Employees

Employees are the crucial human resources in any enterprise especially Vietnam's garment enterprises which are belonging to the intensive labor forces. Mutual relationship between enterprises and employees is a key foundation for the sustainable development of this industry.

Interviewee 1: In our company, almost no workers quitted their job. It is the result of friendly relationship and trust between employers and employees in the company.

Interviewee 2: Characterized by a particular garment industry, this sector requires skills and the stability of labor workforces otherwise all production plans will not be implemented.

One of underlying reasons to be able to build this mutual relationship between them is salary.

Interviewee 1: The most important thing to make our enterprise stable is to create higher salary for employees than other enterprises.

Interviewee 7: Even though workers work effectively, enterprises do not appreciate them and evaluate their good value. So, when other garment enterprises are paying higher salary, they are ready to quit their jobs quickly.

Besides, welfare is also the key thing to convince employees choosing working and contributing for enterprises.

Interviewee 1: We make employees realize that our enterprise is really interested in them through monthly bonus policy, good social and health insurance.

Interviewee 2: To obtain a stable workforce, policies to support workers are always focused in our enterprise such as supplying accommodation for the workers, taking care of the material and spirit life of employees, implementing accumulated skills programs for them, and respecting them. These are one of the best ways to establish durable relationships of our enterprise and workforces. (Interviewee 3 made the same comment.)

Additionally, frank and honest conversation between employers and employees is a useful approach to build up this relationship stronger.

Interviewee 1: Vitas is implementing an experimental model of dialog between employers and employees to gradually stabilize labor workforces for garment industry.

Table 2 Phase II survey results

#	Question	Min value	Max value	Average value	Standard deviation
1	Exploiting effectively the investment policy of government	5.00	100.00	70.36	24.84
2	Strengthening relationship between garment enterprises and employees	14.00	100.00	74.06	22.45

Conversely, if enterprise cannot possess a stable human resource, they must spend tons of money and time for training activities which are not guarantee that whether they are skillful and loyal with garment enterprises or not.

Interviewee 2: For training of highly skilled employees, our enterprise needs more time and money, if these workforces are instable that will affect the delivery schedule, productivity, and product quality.

Interviewee 7: Typically, after each lunar new year, enterprises always have to deal with problems about the shortage of human resource. It is derived that enterprises themselves do not pay attention and do not build human resources in their enterprise culture.

3.2 *The Result of Phase II*

There were 86 respondents included in the Phase II of this survey. They were practitioners as first-line managers in textile and garment enterprises from the North and South of Vietnam. The results were presented as Table 2.

4 Discussions

With theme 1, the experts interviewed all said that textile and garment enterprises need practical support from the state with investment policies and loans at preferential interest rates to expand production and overcome limitations in the fabric production industry.

The results of interviews with experts on theme 2 show that textile and garment enterprises need to pay due attention to appropriate salary and bonus policies, training to improve professional capacity and create a friendly and respectful environment, taking care of workers to have a stable and capable workforce to meet the requirements of production and development.

The survey results in Phase II also showed that the strategy "**Strengthening relations between business with employees**" was appraised the most applicable one in current T&C Industry in Vietnam with the rate 74%. Characterized by a particular

T&C Industry, this sector requires skills and the stability of labor workforces otherwise all production plans will not be implemented. To overcome difficulties of labor shortages happening in recent years, the Vietnam Textile enterprises should respect workers and take direct cares of their material and spirit life such as: supporting accommodation, implementing wages accumulated programs, supporting travel costs … in order that these workforces have peace of mind and feel devoted to their enterprises.

The second position pertained the strategy "**Exploiting effectively the investment policy of government**" with 70%. In the fluctuation of economic situation, exporters are looking forward to more practical supports from the government. Vietnam needs to promote effectively investment projects of the government as well as attract foreign investments in the development of textile and accessory sectors. In addition, the government should urgently address bottlenecks on capital for businesses in order that they can maintain their production process.

5 Conclusions

Qualitative and quantitative studies were conducted to explore the categories of opportunities/challenges, advantages/disadvantages and current strategies of manufacturers and T&C companies in Vietnam.

In relation to sustainable development practice, the first solution Vietnamese textile and garment companies should implement is to determine appropriate salary and bonus policies, train to improve professional capacity and create a friendly, respectful and caring environment. Take care of employees to have a stable and capable workforce that meets the requirements of production and development.

Vietnamese textile and garment companies should propose specific support policies and effectively exploit the government's investment policies in key areas to improve the competitiveness of businesses.

6 Research Limitation

Both the survey and the interviews had a restricted number of participants. Only the South Vietnamese region was covered by the interviews. There were no additional parties included, such as the government or garment buyers. The study was carried out at one point of time.

Also, there is a limitation of this research regarding the generalization because the survey only conducted in the North and the South of Vietnam. It is suggested that future research should include the survey of the central region of the country. Besides, this research suggests that more research on corporate social responsibility, human resource development, green technology to gain sustainable development of Vietnam's garment industry.

References

1. Interlog Supply Chain Innovator. Accessed 10 Jan 2024. https://interlogistics.com.vn/vi/tin-tuc/blog/tu-thuong-truong-den-the-gioi-det-may-viet-nam-tren-duong-dua-xuat-khau-n-473
2. Bao dien tu Dang Cong San Viet Nam. Accessed 21 Nov 2016. https://dangcongsan.vn/kinh-te-va-hoi-nhap/det-may-viet-nam-huong-toi-muc-tieu-xuat-khau-44-ty-usd-655637.html
3. Nhan Dan. Accessed 21 Nov 2016. https://nhandan.vn/xuat-khau-hang-det-may-dat-hon-403-ty-usd-post784078.html
4. Chronicles of Vietnam. Revision 24 Mar 2019. http://vietnamtoday.net/NewsDetail.aspx?Culture=vn&catID=35&id=4503
5. Nhung, T.T.B., Thuy, T.T.P.: Vietnam's textile and garment industry: an overview. Bus. IT **VIII**(2), 45–53 (2018). https://doi.org/10.14311/bit.2018.02.05
6. Al-Fadhat, F., Nadita, S.S.: Questioning sustainability: economic expansion and contradictions behind India's eco-friendly textile policy. Asian Aff. **52**(2), 396–411 (2021). https://doi.org/10.1080/03068374.2021.1914395
7. John, W.C.: Research Design: Qualitative, Quantitative, and Mixed Methods Approaches, 2nd edn. Sage (2003)
8. John, W.C.: Research Design: Qualitative, Quantitative, and Mixed Methods Approaches, 3rd edn. Sage (2009)

Comparison of Two Methods: RAM and AROMAN

Tran Van Dua, Do Duc Trung, Nguyen Thi Phuong Giang, and Duong Van Duc

Abstract The purpose of making a decision with multiple criteria is to determine the best option among several alternatives. However, the best option determined may change if different methods are used to rank the alternatives. This study is conducted to compare two methods: *RAM* and *AROMAN*. These are two new Multi-Criteria Decision Making (*MCDM*) methods, both of which were recently discovered in 2023. Three different examples were conducted to compare these two methods, including the selection of a mushroom cultivation mixture, the choice of a metal recycling option, and the selection of an electric bike. The results show that in each example, the best option found when using the *RAM* method is consistent with the results of other *MCDM* methods. However, the best option found when using the *AROMAN* method differs from other methods. This means that *AROMAN* is deemed unsuitable for use in determining the best option in the three cases examined.

Keywords MCDM · RAM method · AROMAN method

1 Introduction

In every discipline, the selection of an option necessitates the evaluation of diverse factors, encompassing both advantageous and disadvantageous aspects. This is a *MCDM* action. There is no specific count of the number of *MCDM* methods currently available, but it is known that the number exceeds 200 and continues to grow [1]. Many studies have shown that the best option among available alternatives may vary depending on the *MCDM* method used to rank them [2–17].

The aforementioned details underscore the importance of evaluating *MCDM* techniques when utilized to prioritize alternatives within particular contexts. *RAM* [18]

T. Van Dua · D. D. Trung (✉) · D. Van Duc
Hanoi University of Industry, Hanoi, Vietnam
e-mail: doductrung@haui.edu.vn

N. T. P. Giang
Hanoi University of Science and Technology, Hanoi, Vietnam

D. H. Tien et al. (eds.), *Proceedings of the 1st International Conference on Sustainability and Emerging Technologies for Smart Manufacturing*, Proceedings in Technology Transfer, https://doi.org/10.1007/978-981-97-7083-0_73

and *AROMAN* [19] are two new methods discovered in 2023. There is currently no published literature using the *RAM* method. The number of studies applying the *AROMAN* method is also limited. This motivates the need to compare these two methods. The content in Sects. 2 and 3 of this article corresponds to the steps of applying the *RAM* and *AROMAN* methods. In Sect. 4, the comparison of these two methods has been conducted in three cases from different fields. In each case, the problem requirements were taken from published research. In each case, the alternatives were also ranked using various *MCDM* methods. The rankings of alternatives through alternative MCDM approaches provide a foundation for assessing the outcomes derived from utilizing the *RAM* and *AROMAN* methodologies to rank alternatives. The final content of this article includes the conclusion and directions for future research.

2 The MCDM Methods

For prioritizing alternatives through the *RAM* technique, the subsequent procedures ought to be implemented in a sequential manner [18].

Step 1: When confronted with m alternatives for prioritization, each alternative defined by n criteria, create a decision matrix with m rows and n columns. In this context, designate x_{ij} as the representation of criterion j for alternative i, with j varying from 1 to n and i ranging from 1 to m. The weight of criterion j is denoted as w_j. Criteria of a beneficial nature are denoted as type B criteria, while criteria of a cost nature are denoted as type C criteria.

Step 2: Apply Eq. (1) to normalize the data.

$$t_{ij} = \frac{x_{ij}}{\sum_{i=1}^{m} x_{ij}} \tag{1}$$

Step 3: The values y_{ij} are calculated according to (2).

$$y_{ij} = w_j \cdot t_{ij} \tag{2}$$

Step 4: The values S_{+i} and S_{-i} are calculated according to (3) and (4).

$$S_{+i} = \sum_{j=1}^{n} y_{+ij} \quad \text{if } j \in B \tag{3}$$

$$S_{-i} = \sum_{j=1}^{n} y_{-ij} \quad \text{if } j \in C \tag{4}$$

Step 5: The values RI_i are calculated according to (5).

$$\mathrm{RI}_i = \sqrt[2+S_{-i}]{2 + S_{+i}} \tag{5}$$

Step 6: The alternative achieving the highest score, indicated as RI_i, is considered the optimal choice.

In prioritizing alternatives through the *AROMAN* technique, the subsequent procedures ought to be implemented in a sequential manner [19].

Step 1: Similar to the primary stage of the *RAM* approach.

Step 2: Normalize the data using two different methods:

$$t_{ij} = \frac{x_{ij} - \min x_{ij}}{\max x_{ij} - \min x_{ij}} \tag{6}$$

$$t_{ij}^* = \frac{x_{ij}}{\sqrt{\sum_{i=1}^{m} x_{ij}^2}} \tag{7}$$

Step 3: Determine the mean normalized value utilizing Eq. (8), typically selecting β as 0.5 [19].

$$t_{ij}^{\mathrm{norm}} = \frac{\beta \cdot t_{ij} + (1 - \beta) \cdot t_{ij}^*}{2} \tag{8}$$

Step 4: Compute the mean normalized value taking into account the criteria weights according to Eq. (9).

$$\widehat{t_{ij}} = t_{ij}^{\mathrm{norm}} \cdot w_j \tag{9}$$

Step 5: Calculate the sum of values that have been adjusted considering the assigned weights, for criteria of type C using formula (10) and the sum of normalized values for type B criteria using formula (11).

$$L_i = \sum_{j=1}^{n} \widehat{t_{ij}}^{(\min)} \tag{10}$$

$$A_i = \sum_{j=1}^{n} \widehat{t_{ij}}^{(\max)} \tag{11}$$

Step 6: Calculate the scores of the alternatives according to (12), typically choosing λ as 0.5 [19].

$$R_i = L_i^{\lambda} + A_i^{(1-\lambda)} \quad (12)$$

Step 7: The alternative with the highest score, R_i, is deemed the most favorable, while conversely, the one with the lowest score is considered less preferable.

3 Results and Discussion

3.1 Case 1

In Table 1, data about mushroom growth when cultivated under six different mixed formulations are summarized. The mushroom mixture includes straw, maize cob, sheep dung, rice bran, and $CaCO_3$. The criteria, including cap diameter (C1), stem diameter (C2), stem length (C3), biological yield (C4), are beneficial criteria (type *B*). All four criteria have their highest values for alternative A5. In contrast, the infection rate criterion (C5) is a cost criterion (type *C*) and has its lowest value for alternative A1. The task of making a multi-criteria decision, ensuring that all four of the first criteria are considered the highest and the last criterion is the lowest, has been performed using the *MOORA* and *SAW* methods [20]. In this study, the outcomes of ranking the alternatives utilizing the *MOORA* and *SAW* methods will be juxtaposed with the rankings obtained through the *RAM* and *AROMAN* methods.

Table 2 presents the comparative ranking outcomes of the alternatives utilizing both the *RAM* and *AROMAN* methods employed in this investigation, alongside the *MOORA* and *SAW* methods utilized in the prior research [20].

Observations reveal complete consistency in the ranking of alternatives when employing the RAM method compared to the rankings obtained through the MOORA and SAW methods. When using the *AROMAN* method, only the rankings of alternatives 3, 4, 5, and 6 match those of the other methods. Alternatives 1 and 2 are reversed

Table 1 Alternative formulations for mushroom cultivation [20]

Alt	C1	C2	C3	C4	C5
A1	27.7	20.1	96.5	33.5	6.6
A2	35.2	24.3	102.6	41.7	7.1
A3	40.4	27.9	120.1	46.8	8.3
A4	46.8	30.4	132.4	51.4	9.4
A5	50.4	32.6	146.2	59.4	9.9
A6	50.3	32.5	143.4	59.1	10.8
Unit	mm	mm	mm	%	%
Weight	0.2106	0.2095	0.2095	0.1905	0.2106

Table 2 Ranking of alternatives for Case 1

Alt	MOORA	SAW	AROMAN	RAM
A1	6	6	6	6
A2	5	5	5	5
A3	4	4	4	4
A4	3	3	3	3
A5	1	1	2	1
A6	2	2	1	2

when using the *AROMAN* method compared to the other methods. Therefore, it can be concluded that the *AROMAN* method is not suitable for use in this case.

3.2 Case 2

In Table 3, values for four criteria from sixteen different non-ferrous metal machining experiments are summarized. Three criteria belong to type *C*, including surface roughness (C1), roundness error (C2), and tool wear (C3). The sole criterion classified as type B is machining rate (C4) [21].

The minimum value for criterion C1 (0.128 μm) is from experiment #8, the minimum value for criterion C2 (1500 μm) is from experiment #1, the minimum value for criterion C3 (0.663 μm) is from experiment #2, and the maximum value for criterion C4 (160 mm^3/min) is from experiment #7. The task of identifying an experiment in which all three of the first criteria are considered to be the minimum, and the last criterion is considered the maximum, has also been performed earlier using the *COCOSO*, *MABAC*, *MAIRA*, *EAMR*, and *TOPSIS* methods [21]. Figure 1 shows the ranking chart of experiments using different methods.

It is observed that when using the *RAM* method, experiments are ranked as #13, #4, #7, and #14 correspondingly, which completely matches the rankings obtained when using previous methods [21]. Nevertheless, the ranking outcomes of experiments generated by the *AROMAN* method diverge from those yielded by alternative methodologies. This further suggests that the *AROMAN* method is not appropriate for application in this instance.

3.3 Case 3

In this case, the ranking of electric bicycle models will be performed. Information about seven electric bicycle models has been summarized in Table 4. Ten criteria were used, including price (C1), range per charge (C2), charging time (C3), maximum speed (C4), weight of the bicycle (C5), load capacity (C6), seat height (C7), overall

Table 3 Non-ferrous metal machining experiments [21]

Exp	C1	C2	C3	C4
#1	0.238	1.500	2.898	21,333
#2	0.755	3.467	0.663	53,333
#3	1.271	6.233	3.941	96.000
#4	0.146	2.250	2.358	149,333
#5	0.442	5.200	2.874	53,333
#6	0.958	7.466	3.393	33,333
#7	0.389	3.000	1.812	160,000
#8	0.128	6.934	2.328	140,000
#9	0.644	8.527	2.848	96,000
#10	0.322	4.800	2.534	160.000
#11	0.838	7.654	3.048	48,000
#12	1.073	5.666	3.244	112,000
#13	0.224	2.400	0.976	149,333
#14	0.285	4.766	1.486	140,000
#15	0.520	2.833	1.678	112,000
#16	0.308	3.600	2.463	65,333
Unit	μm	μm	μm	mm^3/min
Weight	0.315	0.291	0.233	0.161

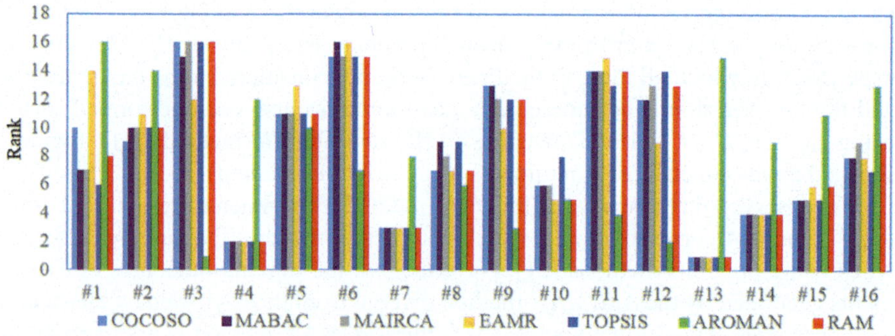

Fig. 1 Ranking of non-ferrous metal machining experiments

length of the bicycle (C8), overall width of the bicycle (C9), and overall height of the bicycle (C10). Two criteria, C1 and C3, are of type *C*, and the remaining eight criteria are of type *B* [22].

Three methods, *SAW*, *MARCOS*, and *PSI*, were used to select the best electric bicycle among the seven alternatives [22]. The graph depicted in Fig. 2 showcases the ranking results of electric bicycles achieved via different methodologies.

Table 4 Electric bicycle models [22]

Option	C1	C2	C3	C4	C5	C6	C7	C8	C9	C10
Op1	7,500,000	45	7	35	56	160	750	1593	635	1015
Op2	7,900,000	45	7	35	50	150	750	1640	640	1200
Op3	9,900,000	50	7	35	50	180	750	1640	640	1200
Op4	9,900,000	50	7	35	46	180	750	1640	640	1200
Op5	11,500,000	50	7	35	52	180	750	1640	640	1200
Op6	13,990,000	45	7	30	45	75	550	1550	650	1040
Op7	13,990,000	45	8	30	45	75	600	1530	750	1000
Unit	Viet Nam dong	km/1 charge	hour	km/h	kg	kg	mm	mm	mm	mm
Weight	0.1	0.1	0.1	0.1	0.1	0.1	0.1	0.1	0.1	0.1

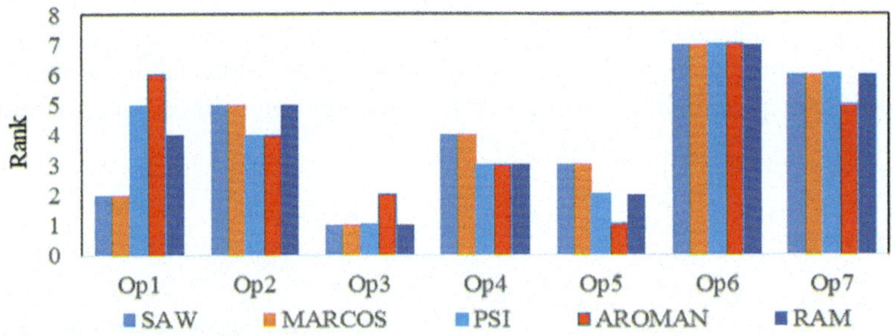

Fig. 2 Ranking of electric bicycle models

In Fig. 2, Op3 is identified as the top-ranked electric bicycle when using four methods, including *SAW*, *MARCOS*, *PSI*, and *RAM*. However, when using the *AROMAN* method, Op5 is determined to be the best electric bicycle. This also indicates that *AROMAN* is not suitable for use in this case.

The three examined cases exhibit notable diversity. Case one involved ranking six combinations of mushroom cultivation mixed with five type B criteria and one type C criterion. In case two, sixteen non-ferrous metal experiments were ranked, featuring three type C criteria and one type B criterion. Lastly, case three entailed ranking seven types of electric bikes, which included two type C criteria and eight type B criteria. Despite variations in fields, options, criteria quantity, and type across the cases, the *RAM* method consistently yielded the same optimal solution as other methods (as reported in published studies). Conversely, the optimal solution identified by the *AROMAN* method consistently diverged from other methods. This suggests that the *RAM* technique is appropriate for these scenarios, whereas the *AROMAN* approach is not suitable.

4 Conclusion

A comparative analysis between the *RAM* and *AROMAN* methods has been conducted across three distinct scenarios. Despite variations in fields, the number and type of alternatives, as well as the criteria, the optimal alternative identified through the *RAM* method consistently aligns with the optimal alternative identified through previously published methods. Conversely, the prime option pinpointed via the *AROMAN* approach consistently diverges from those acquired through alternative methods. Consequently, the *RAM* method proves suitable for all three scenarios, while the *AROMAN* method is deemed unsuitable. This conclusion is not intended as a critique of the *AROMAN* method but rather underscores its inadequacy for the specific scenarios examined. The *RAM* method is recognized to possess advantages over the *AROMAN* approach in the examined scenarios.

The study does not provide an explanation for why there are differences in the ranking of alternatives (especially the top-ranked alternative) when using the *AROMAN* method compared to other multi-criteria decision-making methods. This is a significant question that requires further investigation to be answered accurately.

References

1. Baydas, M., Eren, T., Stević, Z., Starčević, V., Parlakkaya, R.: Proposal for an objective binary benchmarking framework that validates each other for comparing MCDM methods through data analytics. Peer J. Special Issue (2023)
2. Petrović, G., Mihajlović, J., Ćojbašić, Z., Madić, M., Marinković, D.: Comparison of three fuzzy MCDM methods for solving the supplier selection problem. FACTA Universitatis, Ser. Mech. Eng. **17**(3), 455–469 (2019). https://doi.org/10.22190/FUME190420039P
3. Mulliner, E., Malys, N., Maliene, V.: Comparative analysis of MCDM methods for the assessment of sustainable housing affordability. Omega **59**(B), 146–156 (2016)
4. Aydin, F., Gümüs, B.: Comparative analysis of multi-criteria decision making methods for the assessment of optimal SVC location. Bull. Polish Acad. Sci. Tech. Sci. **70**(2) (2022)
5. Vassoney, E., Mochet, A. M., Desiderio, E., Negro, G., Pilloni, M. G., Comoglio, C.: Comparing multi-criteria decision-making methods for the assessment of flow release scenarios from small hydropower plants in the alpine area. Front. Environ. Sci. **9**(635100) (2021)
6. El-Araby, A., Sabry, I., El-Assal, A: A comparative study of using MCDM methods integrated with entropy weight method for evaluating facility location problem. Oper. Res. Eng. Sci. Theory Appl. **5**(1), 121–138 (2022)
7. Son, N.H., Hieu, T.T.: Selection of welding robot by multi-criteria decision-making method. East. Europ. J. Enterprise Technol. **1**(121), 66–72 (2023)
8. Dua, T.V.: Development of a new multicriteria decisionmaking method. East. Europ. J. Enterprise Technol. **3**(123), 33–38 (2023)
9. Tran, V.D., Hoang, X.T.: RSMVC: a new-simple method to select the cutting tool base on multi criteria. J. Appl. Eng. Sci. **21**(1), 167–175 (2023)
10. Le, D.H.: Selection of suitable data normalization method to combine with the CRADIS method for making multi-criteria decision. Appl. Eng. Lett. **8**(1), 24–35 (2023)
11. Do, D, T.: A combination method for multi-criteria decision making problem in turning process. Manufacturing review 8(26), 1–17 (2021).
12. Do, D.T.: Application of TOPSIS and PIV methods for multi-criteria decision making in hard turning process. J. Mach. Eng. **21**(4), 57–71 (2021)

13. Baydaş, M.: Comparison of the performances of MCDM Methods under uncertainty: an analysis on bist SME industry index. Opus **9**(46), 308–326 (2022)
14. Vakilipour, S., Sadeghi-Niaraki, A., Ghodousi, M., Soo-Mi, C.: Comparison between multi-criteria decision-making methods and evaluating the quality of life at different spatial levels. Sustainability **13**(4067), 1–36 (2021)
15. Srisawat, C., Payakpate, J.: Comparison of MCDM methods for intercrop selection in rubber plantations. J. ICT **15**(1), 165–182 (2016)
16. Hodgett, R.E.: Comparison of multi-criteria decision-making methods for equipment Selection. Int. J. Adv. Manuf. Technol. **85**(5–8), 1145–1157 (2016)
17. Uzun, S., Kazan, H.: Comparing MCDM methods of AHP, TOPSIS and PROMETHEE: a study on the selection of ship main engine system. J. Transp. Logistics **1**(1), 99–113 (2016)
18. Sotoudeh-Anvari, A.: Root assessment method (RAM): a novel multi-criteria decision making method and its applications in sustainability challenges. J. Cleaner Prod. **423**(138695) (2023)
19. Bošković, S., Švadlenka, L., Jovčić, S., Dobrodolac, M., Simić, V., Bačanin, N.: An alternative ranking order method accounting for two-step normalization (AROMAN)—a case study of the electric vehicle selection problem. IEEE Access **11**, 39496–39507 (2023)
20. Nguyen, T.L., Ngoc, M.C., Nguyen, X.T.: Comparison Analysis of some MCDM models and applying in decision information system. Vietnam J. Agric. Sci. **19**(4), 462–472 (2021)
21. Do, D.T., Nhu-Tung, N.: Applying COCOSO, MABAC, MAIRCA, EAMR, TOPSIS and weight determination methods for multi-criteria decision making in hole turning process. Strojnícky časopis–J. Mech. Eng. **72**(2), 15–40 (2022)
22. Trinh, V.H., Nguyen, Q.Q., Vu, H.B., Tran, M.H., Nguyen, T.T.T., Le, T.A., Dao, T.N., Nguyen, Q.D., Pham, H.T., Do, D.T.: Multi-criteria decision-making for electric bicycle selection. Adv. Eng. Lett. **1**(4), 126–135 (2022)

Experimental System Setup to Investigate the Motorcycle Performance at Various Operating Conditions

Trinh Dac Phong, Nguyen Quang Truong, Doan Cong Thanh, Le Van Anh, Nguyen Tien Han, Trinh Duy Hung, Bui Van Chinh, Nguyen Xuan Khoa, and OckTaeck Lim

Abstract The paper presents research results and design experimental system for electric motorbike. With purpose of this article is to study the input parameters that affect electric motorbikes, to better understand the important factors related to the operation and performance of electric motorbikes. To achieve this goal, an electric motorbike simulation model has been built based on MATLAB SIMULINK software. In addition, the research also focuses on designing and installing an experimental system to measure the speed, power and power consumption of motorbikes. The simulation model is used to study the influence of input factors and working conditions on the performance and power consumption of electric motorbikes. Through simulation research, the article shows that thanks to the integration of important structural parameters, such as wheel diameter and transmission ratio, this model can simulate the vehicle's operation process effectively, authentic and accurate.

Keywords Electric motorbike · Simulation model · Matlab Simulink

1 Introduction

Industrialization and modernization have led to an increasing demand for fossil fuels. The harmful emissions from internal combustion engines using fossil fuels such as gasoline and diesel are major contributors to environmental pollution [1]. This not

T. D. Phong (✉) · N. Q. Truong · D. C. Thanh · L. V. Anh · N. T. Han · T. D. Hung · B. V. Chinh · N. X. Khoa
School of Mechanical and Automotive Engineering, Ha Noi University of Industry, Hanoi, Viet Nam
e-mail: phongtd@haui.edu.vn

O. Lim
School of Mechanical Engineering, University of Ulsan, 93 Daehak-ro, Nam-gu, Ulsan 44610, South Korea

D. H. Tien et al. (eds.), *Proceedings of the 1st International Conference on Sustainability and Emerging Technologies for Smart Manufacturing*, Proceedings in Technology Transfer, https://doi.org/10.1007/978-981-97-7083-0_74

only affects the environment but also directly impacts human health due to toxic emissions like nitrogen oxides (NO_x) and carbon monoxide (CO) [2]. Therefore, there is a growing emphasis on finding, developing, and reducing dependence on fossil fuels. Electric vehicles (EVs) have emerged as a solution, with many countries, including Vietnam, pursuing this sustainable alternative [3]. With environmental friendliness, reduced noise pollution, electric motorcycles promise to be a viable replacement for conventional ones, reducing reliance on fossil fuels [4], and meeting daily transportation needs. A study by Efrén Esteba Fernandez Palomeque and colleagues [5] designed a semiconductor power module system using silicon MOSFETs to control a three-phase BLDC motor mounted on the wheels for electric motorcycles. The motor is equipped with an air-cooling system to minimize heat losses. This type of motor offers advantages such as simplicity, efficient electronic control, no need for mechanical switching, high efficiency, good dynamic characteristics, high power density and overall high performance. Depending on the voltage and current supplied by the prototype inverter module, the motor's efficiency can be increased up to 93%. Another study by Z. Zainol and collaborators [6] focused on designing and improving electric motorcycles, considering factors like motor selection and energy storage calculations based on desired speed and range. The simulation results indicated that an electric motorcycle needed an average power of 2069 W to travel within a range of approximately 65.3 km. Furthermore, intelligent energy management systems have been applied to electric motorcycles, aiming to optimize energy consumption using efficient algorithms. Real-world testing showed a 36% reduction in energy consumption compared to the initial scenario, demonstrating the potential benefits of smart energy management. In summary, previous studies in the field of electric motorcycles have predominantly focused on modeling and analyzing energy consumption. Few experimental studies have investigated the impact of input parameters on the operation of electric motorcycles currently in use in Vietnam. This research aims to provide effectiveness and high reliability in the field of electric vehicles.

2 Research and Design of Experimental System

2.1 Technical Parameters of Research Object

The current electric motorbike using lead-acid batteries is one of the models equipped with basic technology. For this vehicle used as a research subject, the power system includes five 12 V lead-acid batteries, rear wheel drive (Table 1).

Table 1 Electric vehicle specifications

Engine capacity	850 W
Battery	5 large batteries 60 V–20 Ah
Maximum speed	50 km/h
Distance travelled	90–100 km
Length × width × height	180 cm × 72 cm × 102 cm
Saddle height	77 cm
Vehicle mass	82 kg
Weight	160 kg

2.2 *Design of Experimental Jigs*

Design includes: Frame support, Data collection system. The front and rear wheel supports, where the front wheel support is fixed, and the rear wheel support lifts the wheel 20 cm away from the frame to allow smooth wheel rotation during testing for speed and wheel torque evaluation.

The minimum compressive stress < permissible compressive stress ($\sigma_{\min} < \sigma_n$) ensures that the frame support structure is designed to meet the strength requirements completely.

The fatigue limit under the load of the electric motorcycle is:

$$\sigma_b^n = \frac{P_b^n}{F_0} \tag{1}$$

In which:

P_b^n is the highest force value when under load.
F_0 is the original cross-sectional area.

Allowable stresses are:

$$\sigma_n = \frac{\sigma_b^n}{n} \tag{2}$$

where n is the safety factor.

The durability condition of the support is determined according to the following conditions:

$$|\min \sigma_z| = \min\left|\frac{N_z}{F}\right| \leq [\sigma]_n \tag{3}$$

In which:

N_z is the electric motorbike load.
F is the cross-sectional wear area.
σ_z is the bar stress.

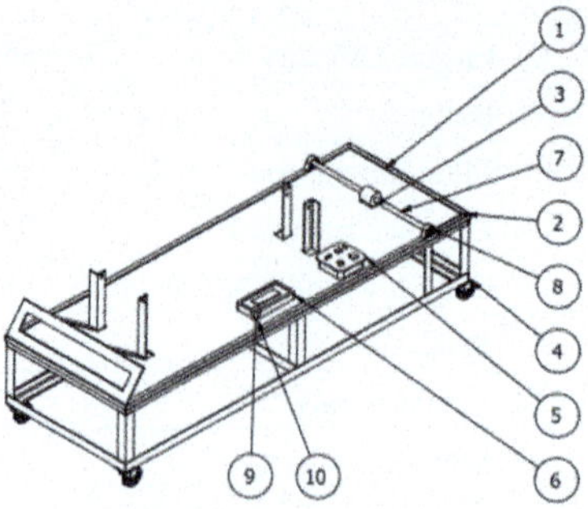

1. Main chassis; 2.Bracket; 3.Shield; 4.Wheel; 5.Electrical panel; 6.LED; 7.Axis; 8.Bearings; 9,10 Bolt

Fig. 1 Overall drawing of the price frame

Below, the loading mechanism is designed. This mechanism generates friction between the wheel and the roller, slowing down the wheel speed. Additionally, an adjustment mechanism regulates the contact between the wheel and the loading mechanism, increasing friction and resistance force for the rear wheel (Fig. 1).

2.3 Jig Details

The jig is made of a frame with front and rear brackets, convenient for placing and mounting the electric motorbike on the frame. In addition, the stand also has wheels to easily move the model.

With the electric motorbike stand designed as above, it will be suitable for the process of installing an electric motorbike model and testing system.

2.4 Design Control Circuits

The project implementation team has researched and programmed control circuits to measure the parameters of electric motorbikes in the required modes. Using Altium 21 software to design a circuit board, using multimeter to measure parameters (current, voltage, speed, etc.) and simulate a 3D circuit board. At the same time, Keil C V5 programming software is used to write code for the control circuit. The circuit design and testing process have yielded accurate and highly reliable results (Fig. 2).

In the processor, an AVR ATmega8 chip is used. The operating principle of the controller is based on the balance of the battery charging state and can control this simulation model through two control methods: rule-based control and fuzzy-based control (Fig. 3).

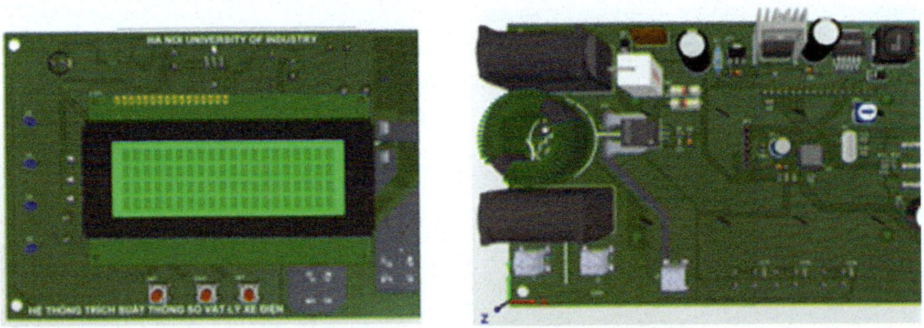

Fig. 2 3D circuit measures electric motorbike parameters

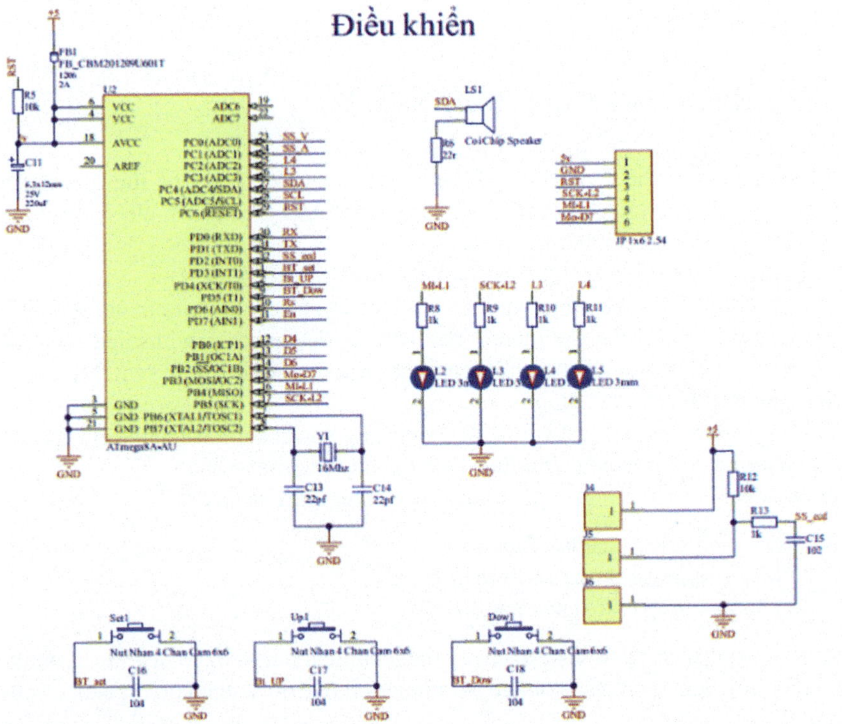

Fig. 3 Processing block

The voltage measurement circuit utilizes a voltage divider circuit to reduce the voltage from 60 V to the range of 0–5 V for the microcontroller to operate. This circuit allows for signal reading and voltage source calculations. The current measurement circuit uses the ACS758 IC, which is based on the current flowing through it and outputs a voltage level to the microcontroller.

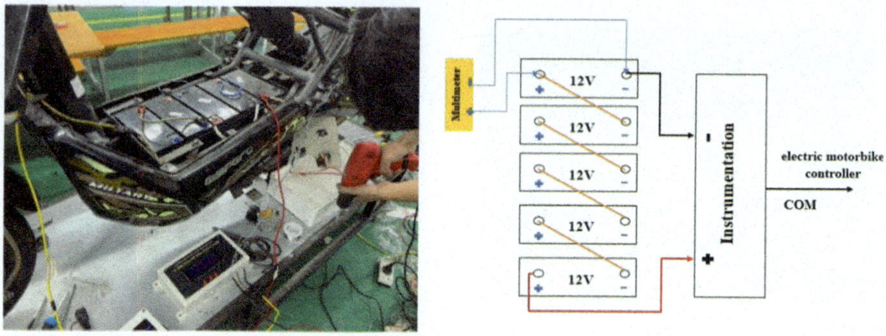

Fig. 4 The vehicle is mounted on a stand

3 Installation Process

3.1 Install the Frame and Power Switch

After testing, study the features and measurement procedures of the laboratory equipment. The research team has completed the motorbike model, electric vehicle mounting brackets and electrical circuits.

The battery connection is done as shown in the drawing (Fig. 4).

A power adapter for the meter and a 30 A fuse to protect the circuit are mounted on the control box next to the meter. The wire is designed to withstand the load in accordance with the 30 A fuse. When operating the device, this switch must be turned on and off to ensure safety.

Connect the breaker linked to the 220 V charging power source and the emergency switch, ensuring safety when charging with the electric motorbike power source.

The parameter measuring set includes 3 working buttons.

SET: setting to select display functions
DOWN: key to move down and select parameters
UP: key to move up and select parameters.

Start the experiment and install, check the technical parameters displayed on the LED screen. Including battery voltage parameters above 60 V, capacity, speed, torque (Fig. 5).

3.2 Install KEWEISI Meter

0–200 V DC Voltmeter KWS Digital DC Wattmeter can measure voltage, current, power, time, temperature, power and electricity (Fig. 6).

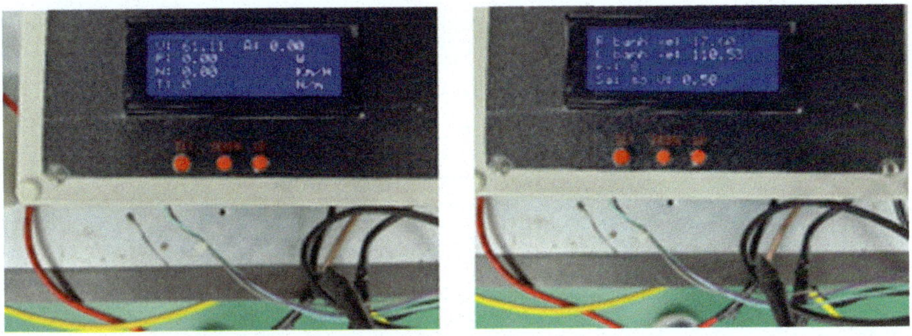

Fig. 5 Parameters displayed on the electric vehicle parameter testing kit

Fig. 6 Installed on electric motorbikes and tested

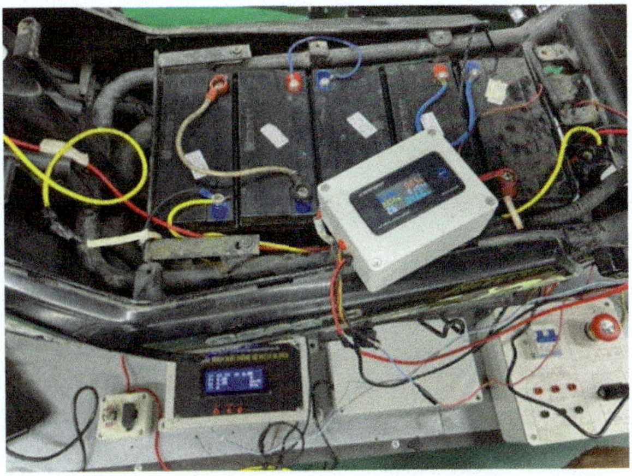

For electric vehicle control systems, it is necessary to check and install fuses to protect the circuit in case of a short circuit or electrical failure causing damage.

The potentiometer is installed, including two source pins (VCC and GRD) and one pin controlling the PWM signal. When rotating the potentiometer, the resistance changes to control the engine speed at fixed levels: 5 km/h, 10 km/h, 15 km/h, 20 km/h, 25 km/h and 30 km/h.

4 Results and Discussion

After installing the experimental system, the research team conducted experiments according to speed ranges (Table 2).

Table 2 Experimenting with electric vehicle parameters

Quantity	Measurement 1	Measurement 2	Measurement 3	Measurement 4	Measurement 5
V (km/h)	5	10	15	20	25
P (W)	5.02	18	33	47.62	64
N (N/m)	5.12	10	14.89	22	25
A (A)	0.5	0.29	0.533	0.8	1.1
t (°C)	28	28	28	28	28

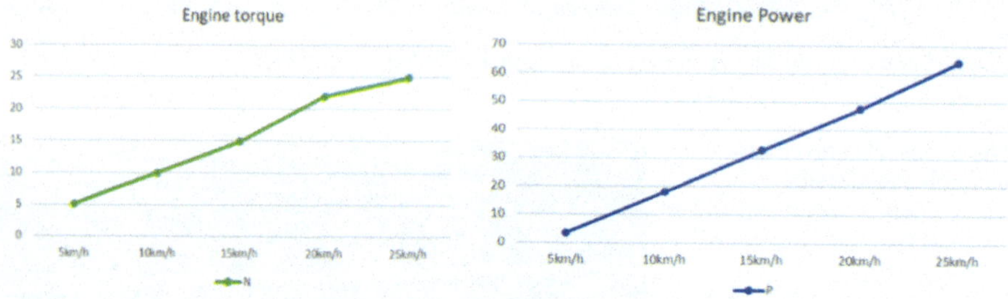

Fig. 7 Change in motor torque and engine power change

To evaluate the impact of input parameters on the work process of electric motorbikes, the research team conducted experiments on electric motorbikes with fixed wheel radius, from which it is the basis to evaluate changes in work. power, torque, current when changing wheel radius.

The results show the change of power *P*, torque *N*, and current *A* when changing the speed range (Fig. 7).

The group studied the structure and principles of electric motorbikes, the characteristics of the power source on electric cars and electric motorbikes, the structure of electric motors, and researched, designed and manufactured experimental frames for electric motorbikes. Designing mounting parts and load-generating gears helps improve the features and testing capabilities of electric motorcycles. Test the performance, reliability and durability of the device in practice.

5 Conclusion

The results of the research show that the experimental system allows measuring the vehicle's output parameters, such as vehicle speed, battery voltage, current intensity, engine capacity and the voltage drop of the electric vehicle battery. From there, experiments were conducted to evaluate the performance and influence of the initial parameters on the working process of electric motorbikes in different operating modes. The results obtained from this study will provide important information for the improvement and development of electric motorcycles.

References

1. Khoa, N.X., Lim, O.T.: The effects of bore-stroke ratio on effective release energy, residual gas, peak pressure rise and combustion duration of a V-twin engine. J. Mech. Sci. Technol. **34**(6), 2657–2666 (2020). https://doi.org/10.1007/s12206-020-0539-z
2. Le, H., Posada, F., Yang, Z.: Electric two-wheeler market growth in Viet Nam: an over view. Briefing (2022). Available from: https://theicct.org/wp-content/uploads/2022/10/asia-pacific-lvs-NDC-TIA-E2W-mkt-growth-Vietnam-nov22.pdf
3. Corbeau, A.-S.: The global energy crisis: implications of record high natural gas prices (2021). https://www.energypolicy.columbia.edu/research/commentary/global-energy-crisis-implications-record-high-natural-gas-prices
4. Liu, Y., Lai, I.K.W.: The effects of environmental policy and the perception of electric motorcycles on the acceptance of electric. In: Motorcycles: An Empirical Study in Macau, vol. 10, pp. 1–14. SAGE Open (2020)
5. Zainol, Z., Toha, S.F., Kamisan, N.A., Bukhari, W.M.S.W.: Design and development of a retrofit electric motorbike. IJRTE **3**, 2277–3878 (2019)
6. Farzaneh, A., Farjah, E., Farzaneh, A.: A novel smart energy management system in pure electric motorcycle using COA. IEEE Trans. Intell. Veh. **4**, 600–608 (2019)

The manufacturer's authorised representative in the EU is Springer Nature Customer Service Centre GmbH, Europaplatz 3, 69115 Heidelberg, Germany. If you have any concerns regarding our products, please contact ProductSafety@springernature.com

Printed and bound by CPI Group (UK) Ltd, Croydon, CR0 4YY
01/10/2026
02226532-0001